RADAR PRINCIPLES

RADAR PRINCIPLES

PEYTON Z. PEEBLES, JR., Ph.D.

A Wiley-Interscience Publication

JOHN WILEY & SONS, INC.

New York • Chichester • Weinheim • Brisbane • Singapore • Toronto

Library of Congress Cataloging-in-Publication Data:

Peebles, Peyton Z.
    Radar Principles/Peyton Z. Peebles, Jr.
       p. cm.
    "A Wiley-Interscience publication."
    Includes bibliographical references and index.
    ISBN 0-471-25205-0 (cloth: alk. paper)
    1. Radar.   I. Title.
TK6575.P36   1998
621.3848—dc21   97-39470

Printed in the United States of America.

10 9 8 7 6 5 4 3

*To the memory of
James Clerk Maxwell (1831–1879),
whose incredible insight founded all
wave-based technology
after his time*

CONTENTS

PREFACE

Many excellent books are available that cover the subject of radar in varying degrees of detail and scope. These books cover their topics in a capable and professional manner, having been authored mainly by experts working in the field. They are excellent sources of current radar technology. However, when used in a classroom environment, few existing texts provide some of the key teaching aids that students usually find valuable. Often the reader is assumed to already have some basic knowledge of radar principles such that the "why and how" are not fully developed. Most do not provide examples of the theory within the text, and few provide problems at the chapter ends to help the students exercise their use of the topics studied. Finally most of the existing books often are either highly specialized or cover a wider scope of topics with less detail than students usually need to be exposed to in an initial course on radar.

This book was written to fill some of the teaching needs listed above, and to serve as a textbook for a course giving first exposure to radar principles. Its topics are designed to meet the author's perceived need for a teaching textbook covering the fundamental principles on which radar is based. However, the book is limited in scope in order that its content be suitable for courses of not more than two semesters. The depth of topics was chosen to be typical for a graduate-level initial course sequence in these basic principles. The book should find use either in a formal university course or for an in-house course in industry for engineers having their first exposure to radar principles. With these fundamentals, students should be in a good position to address more scope and depth via on-the-job training.

It was a difficult task to choose and organize the topics to be included, and some may argue that other selections and topical order are better. There seems to be no best choice. The final choices were made based on two questions: (1) Is the topic really fundamental in the large picture and within the constraints of a book of reasonable length? (2) In what order should the chosen topics be placed to make a given subject least dependent on later topics? Because radar topics are so interdependent, the second question's answer was the most difficult to determine.

In the final analysis, those chosen fundamental principles and topics that are more-or-less external to, and not too dependent on the form of, the radar are first developed. These are elementary concepts (Chapter 1), elements of wave propagation (Chapter 2), antennas (Chapter 3), the radar equation (Chapter 4), radar cross section (Chapter 5), topics in signal theory (Chapters 6 and 7), radar resolution (Chapter 8), radar detection (Chapter 9), and the limiting accuracy of radar measurements (Chapter 10). These topics are followed by descriptions of those that are dependent on the form of the radar. These are range measurement and tracking in radar (Chapter 11), frequency (Doppler) measurement and tracking (Chapter 12), and two chapters (13, 14) on angle measurement and tracking.

A slight exception to the organization is that a chapter (15) on digital signal processing (DSP) has been placed last, even though the subject is independent of the radar's form. The placement is so that DSP topics can be studied on a stand-alone basis without having them scattered throughout other chapters. In my opinion it facilitates the learning process in earlier chapters if they are all described in "analog form" and then the conversion to "digital form" is developed separately in one place.

The book closes with coverages of some review material, such as impulse functions (Appendix A), deterministic and random signals (Appendixes B and C), various useful mathematical formulas (Appendixes D and E) to aid students in working problems at the chapter ends. The chapter-end problems that are either more advanced than typical, or require more than average time to complete, are keyed by a star (★). The book has over 950 problems.

Because little of the material of this book is new, and many sources have been drawn on for its content, I have tried to provide the reader with original sources of the material (Bibliography) whenever it is feasible. In an effort to provide a consistent notation, as much as is reasonably possible, I have adjusted the notation of some sources. A special effort has also been made to carefully proofread the text in an effort to minimize the number of in-print errors.

Finally I offer my thanks to Dr. Fred Taylor, a colleague, who reviewed part of the manuscript, students who offered comments and found errors, especially Thomas A. Corej, and to the several anonymous reviewers whose comments materially improved the book. Many persons were involved in typing the various manuscript versions. Joyce McNeill typed a large portion, while Larry Beeman, Lisa Guerino, Deanna James, Glenda Miller, Mary Turner, and Sharon Williams also contributed, and I thank the University of Florida for providing their services.

PEYTON Z. PEEBLES, JR.

Gainesville, Florida
April 1998

RADAR PRINCIPLES

1

ELEMENTARY CONCEPTS

Radar is the name of an electronic system used for the detection and location of objects. In the "language" of radar the objects are called *targets*. The word radar is an acronym for *ra*dio *d*etection *a*nd *r*anging. Early forms of radar carried various names at different places in the world. However, in modern times the name radar seems to have achieved universal acceptance.

A radar's function is intimately related to properties and characteristics of electromagnetic waves as they interface with physical objects (the targets). All early radars used radio waves, but some modern radars today are based on optical waves and the use of lasers. Thus the earliest roots of radar can be associated with the theoretical work of Maxwell[1] (in 1865) that predicted electromagnetic wave propagation and the experimental work of Hertz[2] (in 1886) that confirmed Maxwell's theory. The experimental work demonstrated that radio waves could be reflected by physical objects. This fundamental fact forms the basis by which radar performs one of its main functions; by sensing the presence of a reflected wave, the radar can determine the existence of a target (the process of *detection*).

Various early forms of radar devices were developed between about 1903 and 1925 that were also able to measure distance to a target (called the target's *range*) besides detecting the target's presence. Skolnik (1962, 1980, ch. 1) gives additional details and some historical references. In 1925 Briet and Tuve (1926) first applied pulsed-wave methods to the measurement of range; their fundamental technique is now widely used in modern systems. Radar development was accelerated during World War II. Since that time development has continued such that present-day systems are very sophisticated and advanced. In essence, radar is a maturing field, but many exciting advances are yet to be discovered.

---

[1] James Clerk Maxwell (1831–1879) was a Scottish physicist.
[2] Heinrich Rudolf Hertz (1857–1894) was a German physicist.

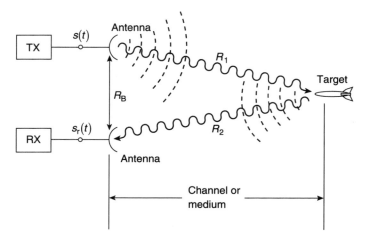

Figure 1.1-1 Basic form of radar.

In this chapter we discuss some of radar's fundamental elements, functions, and types as well as some other basic details. Following these discussions we examine the various radar principles in more detail in the remaining chapters.

1.1 FUNDAMENTAL ELEMENTS OF RADAR

There are many forms of radar that perform a variety of functions and operate with many types of targets. It is helpful first to become familiar with the most elementary structure and elements of a radar before considering details.

General Block Diagram

The most basic form of radar performs all of the typically required functions shown in the block diagram of Fig. 1.1-1. The apparatus consists of a transmitter (TX) connected to a transmitting antenna for propagating an electromagnetic wave outward from the transmitter, and a receiver (RX) connected to a receiving antenna for reception of any wave reflected from a target. In general, the target is part of the propagation *medium* (also called the *channel*) between the transmission and reception locations.

In the basic radar scheme of Fig. 1.1-1, $s(t)$ is a waveform representing the signal produced at the output terminals of the transmitter. The antenna converts the signal to a radiating electromagnetic wave having the same shape as $s(t)$. The wave radiates outwardly at the speed of light.[3] At distance R_1 it encounters the target, which reflects (scatters) some of the wave's energy (a new wave) back toward the receiving antenna. After traveling a distance R_2 at the speed of light, the reflected wave is received by the receiving antenna, which converts the wave to a received waveform $s_r(t)$ at its output terminals. The radar can detect the presence of a target by observing the presence of a signal $s_r(t)$.

---

[3] The speed of light in vacuum is said to be exactly $c = 299,792,458$ m/s (Lide, 1992, p.1-1). For many radar purposes the approximation $c \approx 3(10^8)$ m/s is adequate.

Types of Radar

In general, the transmit and receiving stations of Fig. 1.1-1 can exist at the same location (this type, where the distance R_B between stations is zero, is called a *mono-static* radar) or can have separate locations (where $R_B \neq 0$, called a *bistatic* radar). In the latter case R_1 may not be the same as R_2. In the monostatic system, which is more typical, R_1 and R_2 are equal, and a single antenna often performs both transmit and receive duty. An even more general *system* of radar involves one or more transmitting stations and more than one receiving station, all in a network; this system, called a *multistatic* radar, is the least common form.

Radars can also be typed according to their waveform $s(t)$. A *continuous-wave* (cw) type is one that transmits continuously (usually with a constant amplitude); it can contain frequency modulation (FM), the usual case, or can be constant-frequency. When the transmitted waveform is pulsed (with or without FM), we have a *pulsed radar* type. In an analogous manner, *active* and *passive* radars are types with and without transmitters, respectively.

The typing of radars involves no uniform definition. We note that almost any description can refer to type. Examples based on the main functions performed are detection type, search type, terrain avoidance type, tracking type, and so forth.

In this book we will mainly concentrate on monostatic pulsed radar, since this type of radar is most widely used and is easily understood. Applicable concepts can be applied to other types of radar.

Radar Locations

The radar components in Fig. 1.1-1 might be located on land or water (e.g., on a ship), in the earth's atmosphere (on an aircraft, missile, bomb, cannon shell, etc.), in free space (on a satellite or space vehicle), or even on other planets. Clearly there is almost no limitation on where a radar might be located. Its location does have an effect on operation because of the medium, or channel, in which the radar's waves must propagate.

Radar Medium

The most elementary and simple radar medium is free space. Here the transmitted waves simply propagate away from the radar in all directions, and there is no returned signal since there are no targets. The medium becomes more interesting if some target of interest exists in the free space (perhaps a space vehicle or satellite); this is the next most simple radar medium. The next level of medium complexity would involve addition of unwanted targets, such as returns from a nearby planet's surface when the radar is close to the surface. In this case the medium contributes unwanted target signals that can interfere with the desired target's signal. Next, the medium might contain an atmosphere with all its weather effects (rain, snow, etc.); this case might correspond to a surface-based radar that must contend with interference from a myriad of unwanted target signals, such as from land, forests, buildings, weather effects, and other propagation effects associated with the atmosphere. This last case is the most complicated, and probably the most common, form of medium.

From our discussion we see that a radar's function is to probe the medium in which it operates, to separate the desired target returns from the undesired returns due to the rest of the medium, and to make appropriate measurements on the desired signals such as to produce useful information about targets of interest. Clearly the medium has a direct effect on the performance of the radar and it is discussed in detail at various points in this book.

1.2 FUNCTIONS PERFORMED BY RADAR

The most important functions that a radar can perform are

1. Resolution (Chapter 8)
2. Detection (Chapter 9)
3. Measurement (Chapters 10–15)

Resolution corresponds to a radar's ability to separate (resolve) one desired target signal from another and to separate desired from undesired target signals (noise and clutter). Ideally we want to be able to separate target signals from each other no matter how close the targets were in physical space and no matter how close the vector velocities come to each other. In reality there is a limit to such separability that is dependent on signal design (larger bandwidths give better resolution in the range parameter, while long transmitted pulses give better resolution in frequency) and the antenna's characteristics (small spatial beamwidths give better position resolution).

The detection function consists in sensing the presence in the receiver of the reflected signal from some desired target. This sounds like a simple task, but in reality it is complicated by the presence of unwanted reflected signals and receiver noise. Noise can usually be reduced by better receiver design and by transmitting signals with larger energy per pulse. Unwanted reflected signals (usually called *clutter*) can often be reduced by proper signal design and appropriate signal-processing methods.

Measurement of target range is implicit in the name radar. However, modern radars commonly measure much more than radial range; they can measure a target's position in three-dimensional space, its velocity vector (speed in three space coordinates), angular direction, and vector angular velocity (angle rates in two angle coordinates). All these measurements can be simultaneously made on multiple desired targets in the presence of clutter and noise. Some more advanced radars may even measure target extent (size), shape, and classification (truck, tank, person, building, aircraft, etc.). In fact, as technology advances, classification may eventually become a fourth item in the list of most important radar functions.

1.3 OVERALL SYSTEM CONSIDERATIONS

When designers are called on to develop a new radar, most considerations fall into three broad classes, those related to system choices, those related to the transmitting end of the system, and those concerning the receiving end. Some of the more important considerations in making decisions are listed below. The list is, by no means, exhaustive, however.

System	Active, passive
	Monostatic, bistatic, multistatic
	Architecture
	Medium
Transmitter	Power
	Frequency
	Antenna
	Waveform
Receiver	Noise sensitivity
	Antenna
	Signal processing methods

The system topics active and passive refer to whether the system has one or more transmitters (active) or is of the receive-only type (passive). Another decision is whether the system is to be monostatic, bistatic, or multistatic. Architecture refers to the manner in which the radar must function with other systems. For example, it could be part of a missile guidance system; in this case decisions about radar structure can depend on the functions the guidance system imposes (is the radar on the missile's nose for terminal homing guidance, or is it ground-based to only provide measurements?). The medium as a topic refers to the types of unwanted targets to be encountered (rain intensity, ground reflections, etc.) as well as the desired targets to be assumed (single or multiple targets?).

In this book we will mainly discuss active monostatic radar having a "stand-alone" architecture. That is, one that operates independently of any larger system. Its purpose is therefore to provide independent resolution, detection, and measurements on targets in its vicinity.

The main considerations at the transmitter involve choosing power level (peak and average), *carrier*[4] frequency, type of antenna, and the transmitted waveform. All these choices affect performance and are not necessarily independent of either the system or receiver considerations.

The more important receiver considerations involve noise sensitivity, choice of receiving antenna, and the type of signal processing (detection logic, coherent or noncoherent processing, use of multiple pulse integration, etc.). These choices are not necessarily independent of either system or transmitter decisions.

Our discussions point out that the overall design of radar is a complicated process of trading off each design consideration with others, often in an iterative manner, until a final compromise is produced to best satisfy the overall set of system requirements. Perhaps the most important point to be made is that the system designer must not "overdesign" any one part of the system at the expense of others. Such capability is usually acquired through practical experience following a good understanding of the fundamental principles of radar. Since the purpose of this book is to develop these principles, we next briefly introduce some more specific topics to form a bridge to more detailed discussions in later chapters.

---

[4] The term "carrier" refers to the radio-frequency sinusoidal waveform that is modulated in amplitude and/or frequency to form the transmitted signal $s(t)$.

1.4 TYPES OF RADAR TARGETS

Radar targets vary greatly and require some definitions. The simplest is called a *point target*, which is one having a largest physical dimension that is small relative to the range extent of the transmitted pulse (range extent $= cT/2$ for monostatic radar) if no FM is present, or is small relative to $c/(2B)$ for waveforms with FM, where B is the 3-dB bandwidth (Hz) of $s(t)$ and T is its duration. A point target is physically small enough that no significant "smearing" or spreading in time occurs in received pulses. Many aircraft, satellites, small boats, people and animals, land vehicles, and the like, can often be considered as point targets.

Isolated targets that are too large to be point targets are often called *extended targets*. Extended targets can cause spreading in received pulses and an attendant loss in performance. Large buildings, ships, and some aircraft may behave as extended targets, depending on the radar's bandwidth.

Still larger targets are called *distributed targets*. One class of examples includes earth surfaces such as forests, farms, oceans, and mountains. These are also called *area targets* (Chapter 5). Another class of distributed target, also called a *volume target* (Chapter 5), includes rain, snow, sleet, hail, clouds, fog, smoke, and chaff.[5]

Moving targets are those having motion relative to the radar. If the radar is stationary on the earth, natural targets such as forests or grassy fields (vegetation in general) tend to have relatively low-speed motions that tend to only slightly spread the spectrum of the received signals. Rain and other forms of precipitation create a similar effect. The effect can be much more pronounced, however, for signals reflected from hurricanes, tornadoes, and other violent weather, where the severity of the effect is dependent on the radar's frequency. Moving targets such as missiles, jet aircraft, satellites, and cannon shells are often fast enough to shift the spectrum of the received signal by a significant (*Doppler*) amount in frequency relative to the transmitted signal.

When the radar is not stationary, such as when on an aircraft, all targets in the field of the radar are affected by the radar's motion as though they were moving and the radar were stationary (remember, motion is a relative quantity to the radar).

Finally we note that some targets are called *active* if they radiate energy on their own. All other targets are called *passive*. A transmitting radar site would be an active target to another radar. In another example a human body is an active target to a radar receiver operating at infrared (IR) wavelengths due to its radiation from body heat.

1.5 RADAR'S WAVEFORM, POWER, AND ENERGY

Waveform

We denote the radar's transmitted waveform by $s(t)$ and define it as the signal at the output terminals of the transmitter (see Fig. 1.1-1). In modern radar $s(t)$ may contain

---

[5] Chaff is artificial, or ribbonlike materials, usually metallic, dropped from various aircraft to create unwanted targets at a radar with the purpose of confusing the radar and masking the presence of other real targets.

modulation of both its amplitude and frequency with time. The general form of $s(t)$ can be written as

$$s(t) = a(t)\cos[\omega_0 t + \theta(t) + \phi_0] \tag{1.5-1}$$

where $a(t)$ is due to amplitude modulation, $\theta(t)$ is a phase term due to any frequency modulation,[6] and ϕ_0 is some (arbitrary) phase angle. In some radar analyses it is convenient to treat ϕ_0 as a random phase angle; in most cases, however, it is just a constant.

Theoretically $s(t)$ of (1.5-1) is to be defined for all time, that is, for $-\infty < t < \infty$. As a practical matter, however, it will sometimes be convenient to define $s(t)$ only over finite time intervals. For example, in a pulsed radar that radiates one pulse every T_R seconds (called the pulse repetition interval, or PRI)[7] we might be interested only in radar response to one pulse interval; here $s(t)$ would be defined as the transmitted waveform existing over one time interval of duration T_R. In this case system response to several intervals would be developed as the sum of responses to other intervals having transmitted waveforms that are appropriately displaced versions of $s(t)$. In still other cases it may be more convenient to define $s(t)$ itself over several intervals of time (e.g., as bursts of pulses) and develop performance for the single waveform.

Example 1.5-1   We define $s(t)$ for a pulsed radar using a single pulse with rectangular envelope of duration T, $\theta(t) = 0$ (no frequency modulation), $\phi_0 = -\pi/2$, and peak pulse amplitude A. From (1.5-1)

$$s(t) = \begin{cases} A\cos(\omega_0 t - \frac{\pi}{2}) = A\sin(\omega_0 t), & 0 < t < T \\ 0, & \text{elsewhere} \end{cases}$$

This waveform is sketched in Fig. 1.5-1a.

The above example of a rectangular pulsed waveform is so common that we introduce a special function to describe its envelope $a(t)$. Define the rectangular function rect($\cdot$) by

$$\text{rect}(t) \triangleq \begin{cases} 1, & -\frac{1}{2} < t < \frac{1}{2} \\ 0, & \text{elsewhere} \end{cases} \tag{1.5-2}$$

With this definition the signal of Example 1.5-1 is written as

$$s(t) = A\,\text{rect}\left[\frac{t - (T/2)}{T}\right]\sin(\omega_0 t) \tag{1.5-3}$$

---

[6] Recall that phase and frequency modulation are related. If $\theta_i(t)$ is the instantaneous total phase of a sinusoid $\cos[\theta_i(t)]$, then the instantaneous angular frequency of the sinusoid is $\omega_i(t) = d\theta_i(t)/dt$. In terms of (1.5-1), the instantaneous angular frequency is $\omega_i(t) = \omega_0 + d\theta(t)/dt$, and $d\theta(t)/dt$ is the instantaneous angular frequency modulation present in $s(t)$.

[7] The reciprocal of T_R is the *pulse repetition frequency* (PRF). PRF is denoted by f_R so $f_R = 1/T_R$.

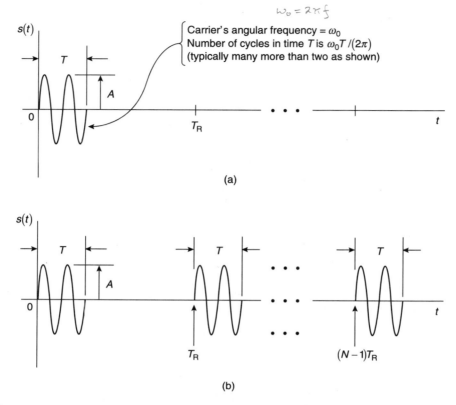

Figure 1.5-1 Two common transmitted waveforms that use rectangular pulses of duration T and peak amplitude A occurring every T_R seconds, and no frequency modulation. (*a*) $s(t)$ defined for one pulsed, and (*b*) defined for N pulses.

The rectangular function is useful in describing bursts of pulses as noted in the next example.

Example 1.5-2 We use the rectangular function to redefine $s(t)$ of the previous example to include N pulses rather than just one. Here

$$s(t) = a(t)\cos[\omega_0 t + \theta(t) + \phi_0]$$

$$= a(t)\cos(\omega_0 t - \tfrac{\pi}{2})$$

$$= a(t)\sin(\omega_0 t)$$

$$= \left\{ A \sum_{n=1}^{N} \operatorname{rect}\left[\frac{t - (n-1)T_R - (T/2)}{T}\right] \right\} \sin(\omega_0 t)$$

where the envelope is a sum of delayed rectangular functions

$$a(t) = A \sum_{n=1}^{N} \operatorname{rect}\left[\frac{t - (n-1)T_R - (T/2)}{T}\right]$$

Figure 1.5-1*b* sketches $s(t)$. Note that each pulse in the train of N pulses has the same pulse *envelope* but not the same carrier's starting phase. The angle $\phi_0 = -\pi/2$ applies to all pulses, but only the first pulse *starts* with a phase angle of zero [because $t = 0$ in $\sin(\omega_0 t)$]. Pulse n actually starts with a phase angle of $\omega_0(n-1)T_R$. For convenience in sketching the waveform of Fig. 1.5-1*b*, it has been assumed that $\omega_0(n-1)T_R$ is a multiple of 2π for all n so that all pulses are shown with initial phase angle of zero. This constraint requires ω_0 to be an integral multiple of $2\pi/T_R$. The constraint is not necessary in a real system, but it leads here to a more easily sketched signal.

Our discussions of (1.5-1) have tried to show that there is great flexibility in defining $s(t)$ for a given radar by choice of definitions for $a(t)$, $\theta(t)$, and ϕ_0.

Peak and Average Transmitted Powers

The output of the transmitter can be modeled as a Thevenin's equivalent circuit comprised of a source of voltage $s(t)$ in series with an output impedance, denoted by $R_t + jX_t$.[8] For an impedance-matched load of $R_t - jX_t$ across the output terminals, the maximum real instantaneous power that can be delivered to the load (called the *available instantaneous power*) is

$$\text{Available instantaneous power} = \frac{s^2(t)}{4R_t}$$

$$= \frac{a^2(t)}{8R_t}\{1 + \cos[2\omega_0 t + 2\theta(t) + 2\phi_0]\} \qquad (1.5\text{-}4)$$

Now the cosine term in (1.5-4) behaves almost as a pure cosine with an argument of $(2\omega_0 t + \text{constant phase})$ for any single period of the carrier frequency $\omega_0/(2\pi)$ because both $a(t)$ and $\theta(t)$ typically do not change appreciably. This fact means that the cosine term, when averaged over any carrier's period, is nearly zero. Thus the instantaneous power averaged over one period of the carrier about any value of t is approximately $a^2(t)/(8R_t)$.

We may now define *average peak transmitted power*, denoted by P_t, as the available instantaneous power at the transmitter's output terminals when averaged over one cycle of the carrier and *when $s(t)$ has its maximum amplitude*. P_t is given by[9]

$$P_t = \frac{1}{4R_t}[1 - \text{cycle-averaged } s^2(t)]_{max} = \frac{[a^2(t)]_{max}}{8R_t} \qquad (1.5\text{-}5)$$

where $[\;]_{max}$ represents the maximum value of the quantity within the brackets.

[8] In general, R_t and X_t can vary with frequency. We will presume the transmitter and its load are broadband enough relative to the bandwidth of $s(t)$ such that R_t and X_t can be approximated as constants for all important frequencies in $s(t)$. If this were not roughly true in practice, the transmitter and output load would distort $s(t)$.
[9] Frequently in radar literature, power and energy expressions are "normalized" such that impedance factors do not appear. In normalized form (1.5-5) would be written as $P_t = [1 - \text{cycle-averaged } s^2(t)]_{max} = [a^2(t)]_{max}/2$. Similarly the normalized versions of average power P_{av} (to follow) would not contain the factor $1/(4R_t)$. In much of our work we use the normalized versions of power and the "normalized" versions of energy as derived from normalized power. In most cases, where ratios of signal powers to noise powers are of most interest, the impedance factors may not be of concern and may cancel. Wherever context will not serve to keep power and energy clear, we will be careful to include the applicable impedance factors.

For a pulsed radar P_t is evaluated at the maximum of the pulse's envelope. For a constant-amplitude continuous-wave (cw) waveform, P_t will have the same value for all times.

Available average transmitted power, denoted by P_{av}, is defined as available instantaneous power averaged over a given time interval. For a time interval T_R, we have

$$P_{av} = \frac{1}{4R_t T_R} \int_0^{T_R} s^2(t)\,dt \qquad (1.5\text{-}6)$$

An alternative expression often more useful for pulsed signals is

$$P_{av} = \frac{1}{4R_t T_R} \int_{-T_R/2}^{T_R/2} s^2(t)\,dt \qquad (1.5\text{-}7)$$

For normalized versions of P_{av} the expressions (1.5-6) and (1.5-7) do not contain the factor $1/4R_t$. In a pulsed radar T_R is the interpulse period, and the average power over one period is the same as the average power over any integral number of periods if the transmitted pulses are the same in each pulse interval. For continuous-wave (cw) radar, $a(t)$ and $\theta(t)$ may both be periodic functions, typically with the same period, and T_R can be taken to be the fundamental period of the two functions.

We develop an example of average peak and available average transmitted powers[10] for a transmitter having rectangular pulses.

Example 1.5-3   We find P_t and P_{av} for a transmitted signal made up of periodic transmissions of pulses of the form shown in Fig. 1.5-1a. Here

$$P_t = \frac{[a^2(t)]_{max}}{2} = \frac{A^2}{2}$$

Next

$$P_{av} = \frac{1}{T_R} \int_0^{T_R} s^2(t)\,dt = \frac{1}{T_R} \int_0^{T_R} A^2 \, \text{rect}^2 \left[\frac{t - (T/2)}{T} \right] \sin^2(\omega_0 t)\,dt$$

$$= \frac{A^2}{T_R} \int_0^T \sin^2(\omega_0 t)\,dt = \frac{A^2}{2T_R} \int_0^T [1 - \cos(2\omega_0 t)]\,dt$$

$$= \frac{A^2 T}{2T_R} \left[1 - \frac{\sin(2\omega_0 T)}{2\omega_0 T} \right] \approx \frac{A^2 T}{2T_R} = \frac{T}{T_R} P_t$$

The approximation in the last line above uses the fact that $\omega_0 T \gg 1$ for almost all pulsed radars.

---

[10] Henceforth, except where specifically noted, we will use normalized power when referring to power.

The preceding example for rectangular transmitted pulses shows that P_{av} and P_t are related by

$$P_{av} = \frac{T}{T_R} P_t = D_t P_t, \qquad \text{rectangular pulses} \tag{1.5-8}$$

where

$$D_t = \frac{T}{T_R}, \qquad \text{rectangular pulses} \tag{1.5-9}$$

is called the *duty factor* of the transmitted waveform. Duty factor can be generalized to apply to waveforms other than those with rectangular envelopes (see Problem 1.5-9).

Energy

The total (normalized) energy, denoted by E_s transmitted in a given time interval T_R is

$$E_s = \int_0^{T_R} s^2(t) \, dt \approx \tfrac{1}{2} \int_0^{T_R} a^2(t) \, dt \tag{1.5-10}$$

As noted earlier, T_R usually represents the time between pulses in a pulsed radar, or the fundamental period of the modulation functions $a(t)$ and $\theta(t)$ in cw radar.

1.6 SOME BASIC PRINCIPLES

In this section we briefly describe several principles fundamental to radar. All of these principles are discussed more fully in later work.

Elementary Range Measurement

Measurement of the radial distance between a monostatic radar and a target is basic to nearly all radars. Although many cw radars can measure range, we describe the pulsed radar because the concepts are easier to visualize and because most radars use pulsed waveforms. The basic idea is illustrated in Fig. 1.6-1. A pulse is transmitted as shown in panel a. The pulse of electromagnetic energy radiates outward until it strikes a target. The target next reflects back some energy to the radar receiver which develops a received signal $s_r(t)$, as shown in panel b; it arrives a time (delay) t_R after the transmitted signal and is altered in amplitude by some constant factor α. The total distance from transmitter to the target and back to the receiver must equal the product of the speed of light and the delay time t_R. For a target range R the total distance is $2R$, so

$$R = \frac{c t_R}{2} \tag{1.6-1}$$

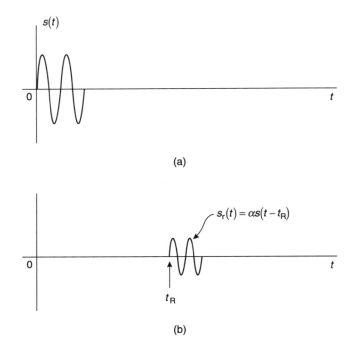

Figure 1.6-1 (*a*) A transmitted pulse, and (*b*) the received pulse in a pulsed radar.

An elementary measurement of range R can be made by forming the product of $c/2$ (a known constant) and a measurement of t_R. The result is referred to as a single-pulse range measurement. Other single-pulse measurements can be made for each pulse transmission. With a stationary target these measurements would theoretically be the same (if there were no noise or other effects) for each pulse. For a moving target the various single-pulse measurements will constantly change. Target motion also changes the radar's frequency; this change is called the *Doppler* effect.[11] In the next three subsections we first examine the Doppler effect and then return to the effect of target motion on target delay and received waveform.

Doppler Effect Due to Target Motion

We will refer to Fig. 1.6-2 to aid in understanding what happens to a transmitted wave as it is reflected by a target and returns to the receiver. We follow two pieces of the transmitted wave: one that leaves at $t = 0$, and the second radiated one carrier cycle later, in time, which is at time $T_0 = 1/f_0 = 2\pi/\omega_0$, as shown in panel *a*. The target is assumed to have a range R_0 at $t = 0$ and is moving away from the transmitter with radial speed v.

---

[11] Named for Christian Johann Doppler (1803–1853), an Austrian physicist and mathematician who made fundamental discoveries of how the frequency of sound or light waves is affected by relative motion between the wave source and the observer.

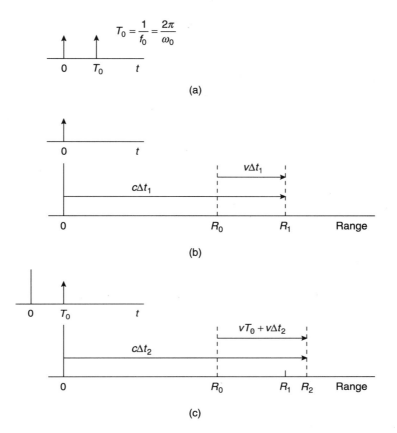

Figure 1.6-2 (*a*) Two points on the transmitted wave that are separated one carrier cycle in time. (*b*) The geometry associated with the first point, and (*c*) that of the second point.

Figure 1.6-2*b* describes what happens to the wave piece that leaves at $t = 0$. It propagates outward until it strikes the target that has moved from R_0 (its range at $t = 0$) to some range R_1 when the wave piece arrives. If Δt_1 is the time it takes the wave piece to arrive, the target has moved in range by an amount $v\Delta t_1$ to range R_1:

$$R_1 = R_0 + v\Delta t_1 \tag{1.6-2}$$

The reflected wave next travels back to the receiver over a distance R_1 at the speed of light. The time of travel must also equal Δt_1, so

$$R_1 = c\Delta t_1 \tag{1.6-3}$$

On use of (1.6-3) with (1.6-2), we have

$$\Delta t_1 = \frac{R_0}{c - v} \tag{1.6-4}$$

If t_1 is the time of arrival of the wave piece, it must equal $2\Delta t_1$, so

$$t_1 = \frac{2R_0}{c - v} \tag{1.6-5}$$

Figure 1.6-2c describes what happens to the wave piece that radiates at time T_0. Let Δt_2 be the *travel time* of the wave until it strikes the target which has now moved to a new range R_2:

$$R_2 = R_0 + vT_0 + v\Delta t_2 \tag{1.6-6}$$

Since

$$R_2 = c\Delta t_2 \tag{1.6-7}$$

then these last two expressions give

$$\Delta t_2 = \frac{R_0 + vT_0}{c - v} \tag{1.6-8}$$

The time of arrival, denoted by t_2, of the second wave piece is

$$t_2 = T_0 + 2\Delta t_2 = \frac{(c + v)T_0 + 2R_0}{c - v} \tag{1.6-9}$$

after using (1.6-8).

Since the two wave pieces were separated one period of the transmitted carrier, we find the frequency of the *received* carrier as $1/(t_2 - t_1)$:

$$f_r = \frac{1}{t_2 - t_1} = \left(\frac{c - v}{c + v}\right) f_0 \tag{1.6-10}$$

from (1.6-5) and (1.6-9). The *change* in the received frequency relative to the transmitted frequency is called the *Doppler frequency*, denoted by f_d:

$$f_d = f_r - f_0 = \frac{-2vf_0}{c + v} \approx \left(\frac{-2v}{c}\right) f_0 \qquad \text{if } |v| \ll c \tag{1.6-11}$$

From (1.6-10) we observe that outward target motion (v positive) produces a negative Doppler frequency and therefore a received carrier frequency less than that transmitted. The opposites are true for v negative.

Example 1.6-1 A radar's frequency is $\omega_0/2\pi = 4(10^9)$ Hz. We find the Doppler frequency for a target for which $v = 670$ m/s (about 1500 mi/h). From (1.6-11),

$$f_d \approx \frac{-2(670)4(10^9)}{299{,}792{,}458} = -17.879 \text{ kHz}$$

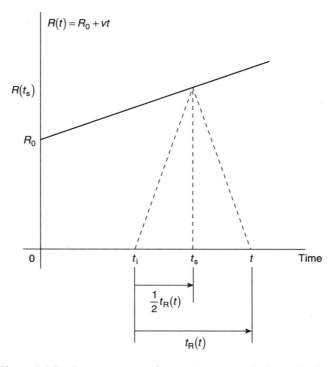

Figure 1.6-3 A constant-speed target's range variation with time.

Effect of Target Motion on Range Delay

When a target occupies a constant range its range delay t_R is constant and the received signal is

$$s_r(t) = \alpha s(t - t_R) \tag{1.6-12}$$

as shown in Fig. 1.6-1b, where α is a constant. When delay varies with time, denoted by $t_R(t)$, we write $s_r(t)$ as

$$s_r(t) = \alpha s[t - t_R(t)] \tag{1.6-13}$$

by analogy with (1.6-12), where we must solve for $t_R(t)$.

Figure 1.6-3 illustrates range versus time for a target moving at constant radial speed v.[12] If a point on the transmitted wave is radiated at some time t_i it propagates outward, strikes the target at some time t_s, at which time the target's range is $R(t_s) = R_0 + vt_s$, and then reflects back to the receiver at time t. Since $t - t_s$ must equal $t_s - t_i$ and these both must equal half the observed delay at time t (which

---

[12] We assume that target speed is constant over the time of interest to the radar, which is often relatively short. For longer time it may be necessary to model range as a second-degree polynomial function that allows for target acceleration.

corresponds to range at the reflection time t_s), we have (see DiFranco and Rubin, 1968, Appendix A)

$$t_R(t) = \frac{2}{c} R(t_s) = \frac{2}{c} \left\{ R_0 + v \left[t - \frac{t_R(t)}{2} \right] \right\}$$

(1.6-14)

so

$$t_R(t) = \frac{2(R_0 + vt)}{c + v}$$

(1.6-15)

The argument of (1.6-13) becomes

$$t - t_R(t) = \left(\frac{c - v}{c + v} \right) \left(t - \frac{2R_0}{c - v} \right)$$

(1.6-16)

The factor $(c - v)/(c + v)$ represents a stretching or compressing of the time axis, depending on the sign of v. The term $2R_0/(c - v)$ is a constant. The effects of these terms in the radar are next found by looking more carefully at the received signal itself.

Effect of Target Motion on Received Waveform

From (1.6-13) and the use of (1.5-1) and (1.6-16), we have

$$s_r(t) = \alpha a \left\{ \left(\frac{c - v}{c + v} \right) \left(t - \frac{2R_0}{c - v} \right) \right\} \cos \left\{ \omega_0 \left(\frac{c - v}{c + v} \right) \left(t - \frac{2R_0}{c - v} \right) \right.$$
$$\left. + \theta \left[\left(\frac{c - v}{c + v} \right) \left(t - \frac{2R_0}{c - v} \right) \right] + \phi_0 \right\}$$

(1.6-17)

In most radars the factor $(c - v)/(c + v)$ can be ignored and replaced by unity where it appears in the arguments of $a(\cdot)$ and $\theta(\cdot)$ because these quantities usually do not change much over the duration of $s(t)$ (see Problem 1.6-6). However, because ω_0 can be large in radar, the factor cannot be ignored in the term involving ω_0 in the argument of cosine. For this term we use (1.6-10) and (1.6-11) to write

$$\omega_0 \left(\frac{c - v}{c + v} \right) \left(t - \frac{2R_0}{c - v} \right) = (\omega_0 + \omega_d) \left(t - \frac{2R_0}{c - v} \right)$$
$$= (\omega_0 + \omega_d)(t - \tau_R)$$

(1.6-18)

where

$$\omega_d = 2\pi f_d$$

(1.6-19)

$$\tau_R \triangleq \frac{2R_0}{c - v} \approx \frac{2R_0}{c} \quad \text{if } |v| \ll c$$

(1.6-20)

If we assume the preceding approximations are all true, as they are for the most radars, the received signal can be expressed in the rather general form

$$s_r(t) = \alpha a(t - \tau_R)\cos[(\omega_0 + \omega_d)(t - \tau_R) + \theta(t - \tau_R) + \phi_0] \qquad (1.6\text{-}21)$$

where τ_R is a constant given by (1.6-20), and ω_d is the Doppler angular frequency caused by target motion.

1.7 SOME DEFINITIONS AND OTHER DETAILS

A few definitions and other details are next discussed. These topics help facilitate development of more detailed subjects in the following chapters.

Radar Coordinates

The positions of targets that are to be measured by a radar require the definition of coordinate systems. Figure 1.7-1a illustrates a rectangular coordinate system with axes x_R, y_R, z_R which define the geometry of space around a radar, assumed to be located at the origin. It is helpful to think of the x_R, z_R plane as the horizontal plane for a surface-based radar; the axis y_R then represents the local vertical direction. More generally, these directions are arbitrary, and other choices are possible (local vertical could be defined as z_R, as another choice). For the coordinates shown, we define an arbitrary point P in space by its *range r*, *azimuth angle A_T*, and *elevation angle E_T*, as shown (this point will later be that of a target).[13] We call the coordinates r, A_T, E_T the *radar coordinates* of point P. A radar typically specifies the positions of objects by specifying their radar coordinates.

Next we define a second rectangular coordinate system with axes x, y, z as shown in Fig. 1.7-1b. The x, y, z system is the result of placing the z axis in the direction defined by radar coordinates A_b, E_b while the x axis remains in the x_R, z_R plane. The coordinates x, y, z are convenient for describing most forms of radar antennas. Most radar antennas are either aperture type (radiation is from some planar area) or planar phased array type. In either case we define the radiating planar area as being in the x, y plane. The z axis is then called the *broadside* (perpendicular) *direction* for a planar phased array or the *boresight direction* (or axis) for an aperture antenna (see Chapter 3). We will collectively refer to these directions as the *reference direction* of the radar.

Our coordinate definitions are complete once we define arbitrary point P within the coordinates x, y, z by its angles ψ_x, ψ_y, and ψ_z relative to the x, y, and z axes respectively. These definitions are shown in Fig. 1.7-1c. If x_T, y_T, and z_T are the projections of point P onto axes x, y, and z, respectively, then

$$x_T = r\cos(\psi_x) \qquad (1.7\text{-}1)$$

$$y_T = r\cos(\psi_y) \qquad (1.7\text{-}2)$$

$$z_T = r\cos(\psi_z) \qquad (1.7\text{-}3)$$

---

[13] For other choices of coordinates, A_T will still be defined in whatever is called the horizontal plane; E_T will still be the angle positive upward from the horizontal plane, and r is still the radial range.

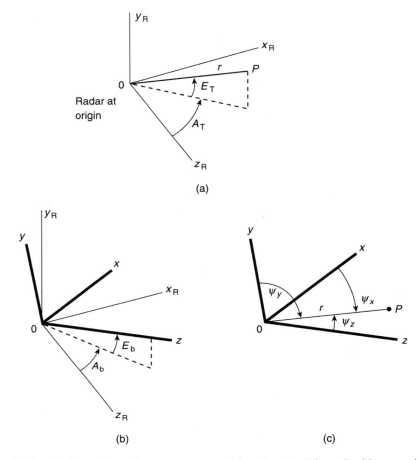

(a)

(b)   **(c)**

Figure 1.7-1 (*a*) Overall coordinates x_R, y_R, z_R with radar at origin and arbitrary point P at radar coordinates r, A_T, E_T. (*b*) Coordinates x, y, z obtained by rotating an angle A_b about axis y_R and an angle $-E_b$ about axis x; the axis z is located in direction A_b, E_b in radar coordinates. (*c*) Point P defined in coordinates x, y, z by angles ψ_x, ψ_y, and ψ_z from x, y, and z axes, respectively.

The quantities $\cos(\psi_x)$, $\cos(\psi_y)$, and $\cos(\psi_z)$ are called the *direction cosines* of point P. We may also write

$$x_T = r \sin(\theta_x) \tag{1.7-4}$$

$$y_T = r \sin(\theta_y) \tag{1.7-5}$$

$$z_T = r \sin(\theta_z) \tag{1.7-6}$$

where θ_x, θ_y, and θ_z are the complements of ψ_x, ψ_y, and ψ_z:

$$\theta_x = \frac{\pi}{2} - \psi_x \tag{1.7-7}$$

$$\theta_y = \frac{\pi}{2} - \psi_y \tag{1.7-8}$$

$$\theta_z = \frac{\pi}{2} - \psi_z \tag{1.7-9}$$

Angles θ_x, θ_y, and θ_z are the angles between the direction to point P and the (y, z), (x, z), and (x, y) planes, respectively.

From the geometry of Figs. 1.7-1a and b, it can also be shown that point P, expressed in its radar coordinates, has projections x_T, y_T, and z_T given by

$$x_T = r\cos(E_T)\sin(A_T - A_b) \tag{1.7-10}$$

$$y_T = r\sin(E_T)\cos(E_b) - r\cos(E_T)\sin(E_b)\cos(A_T - A_b) \tag{1.7-11}$$

$$z_T = r\sin(E_T)\sin(E_b) + r\cos(E_T)\cos(E_b)\cos(A_T - A_b) \tag{1.7-12}$$

By equating (1.7-10) through (1.7-12) with (1.7-4) through (1.7-6), respectively, we obtain

$$\sin(\theta_x) = \cos(E_T)\sin(A_T - A_b) \tag{1.7-13}$$

$$\sin(\theta_y) = \sin(E_T)\cos(E_b) - \cos(E_T)\sin(E_b)\cos(A_T - A_b) \tag{1.7-14}$$

$$\sin(\theta_z) = \sin(E_T)\sin(E_b) + \cos(E_T)\cos(E_b)\cos(A_T - A_b) \tag{1.7-15}$$

The first two of the last three expressions are fundamental to understanding how angles are measured by a radar.

A radar that is to measure a target's radar coordinates A_T and E_T does not typically measure these angles directly. Rather, it usually measures target angles θ_x and θ_y which can be thought of as angles *relative* to the reference direction, which the radar knows. For a small relative angle $A_T - A_b$, for example, where θ_x is small, (1.7-13) gives

$$\theta_x \approx \cos(E_T) \cdot (A_T - A_b) \tag{1.7-16}$$

so

$$A_T = A_b + \theta_x \sec(E_T) \tag{1.7-17}$$

On measurement of the relative angle θ_x, we get A_T by scaling θ_x by $\sec(E_T)$ and adding the result to the reference azimuth angle A_b.

For the elevation angle we assume that $A_T - A_b$ is small, so (1.7-14) gives

$$\sin(\theta_y) \approx \sin(E_T)\cos(E_b) - \cos(E_T)\sin(E_b) = \sin(E_T - E_b) \tag{1.7-18}$$

or

$$E_T = E_b + \theta_y \tag{1.7-19}$$

Thus E_T is equal to the known reference elevation angle E_b added to the radar's measurement of θ_y when $A_T - A_b$ is small. There can be a special case where $A_T - A_b$ is not small, and the reader is referred to Sherman (1984, p. 42), for a discussion of this case which occurs mainly as $E_T \to \pi/2$.

TABLE 1.7-1   Standard radar-frequency letter band nomenclature.

Band Designation	Nominal Frequency Range
HF	3–30 MHz
VHF	30–300 MHz
UHF	300–1000 MHz
L	1–2 GHz
S	2–4 GHz
C	4–8 GHz
X	8–12 GHz
K_u	12–18 GHz
K	18–27 GHz
K_a	27–40 GHz
V	40–75 GHz
W	75–110 GHz (see note)
mm	110–300 GHz (see note)
sub mm	300–3000 GHz

Note: Sometimes V and W band are considered part of the millimeter (mm) band when referring generally to millimeter wave radar at frequencies above 40 GHz.

To summarize, a typical radar measures (small) angles θ_x and θ_y[14] and then uses (1.7-17) and (1.7-19) to form the angle measurements of A_T and E_T for a target. For measurements of small θ_x and θ_y, the factor $\sec(E_T)$ in (1.7-17) is often replaced by $\sec(E_b)$ with only small resulting error.

Radar Frequency

An accepted practice in radar is to use letter designations as short notation for defining frequency bands of operation. Such designations have been in use during and since World War II. Standard letters for specified frequency bands are summarized in Table 1.7-1 as given in the *Supplement* to the *Record of the IEEE 1985 International Conference on Radar*.

Example 1.7-1   A radar operates at a frequency in the middle of X band. What magnitude of Doppler frequency will occur for a target with a speed of 290 m/s (about 650 mi/h)? From Table 1.7-1 the radar's frequency is $f_0 = 10.0$ GHz. From (1.6-10),

$$|f_d| = \frac{2v}{c} f_0 = \frac{2(290)10^{10}}{299,792,458} = 19,346.72 \text{ Hz}$$

---

[14] Phased array antennas that use electronic beam steering may have measurements proportional to $\sin(\theta_x) - \sin(\theta_{x0})$ and $\sin(\theta_y) - \sin(\theta_{y0})$ rather than θ_x and θ_y when the pattern is steered to a new direction defined by angles θ_{x0} and θ_{y0} from broadside.

The choice of a radar's frequency must ultimately be made for each system based on its individual requirements, and it is dangerous to generalize too much. However, some broad-based observations can be made. For example, the power capability of transmitting devices tends to decrease with frequency, which favors use of lower frequencies. On the other hand, for similar performance, antennas become smaller at higher frequencies, which favors the use of higher frequencies. To avoid cosmic noise, which increases as frequency decreases, frequencies above UHF are favored, while frequencies below K_u band are favored to avoid atmospheric absorption noise. All these observations have led to many radars operating at L, S, C, and X bands in past years. Ultimately, however, the choice of frequency is based on many other factors related to a particular radar's mission.

Radar Displays

A radar display is a device for visual presentation of target information to an operator, who may be involved with on-line operation or with maintenance of the unit. Most displays use a cathode-ray tube (CRT), although light-emitting diodes (Bryatt and Wild, 1973), liquid crystals, plasma panels (McLoughlin et al., 1977), and other solid-state devices have been used in special applications and research continues in these areas.

CRT displays are typically labeled by names such as A-scope and B-scope. A list of these devices is given in Edde (1993, pp. 502–506) and in Skolnik (1980, pp. 354–355). Display definitions are given in *IEEE Standard 686–82* (1982). Many displays are now obsolete or used mainly in special applications. We will describe only three of those extensively used in modern radar. They are the A-scope, PPI (or P-scope), and the RHI.

The *A-scope* is a display where the vertical direction is proportional to some function of the signal strength of the envelope of a target's returned waveform. The function is often linear (the envelope itself) or could be logarithmic. The horizontal direction is target range. Figure 1.7-2 illustrates the A-scope display.

A PPI display, which refers to *plan-position indicator*, can have several variations. Two of the most important PPI displays are shown in Fig. 1.7-3. In panel *a* the radar is located at the center of a polar display of the target's echo in range (in radial

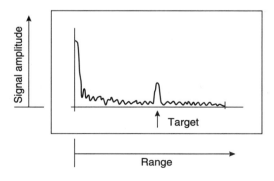

Figure 1.7-2 A radar A-scope.

distance) and azimuth angle. The result is a plan or maplike location of targets all around the radar. The reference direction can be any direction desired (e.g., north). A target's position is marked by a bright spot on the CRT which is usually dark at all other locations. The spot is achieved by intensity modulating the electron beam of the CRT. Six targets are illustrated in the figure.

Figure 1.7-3*b* shows a modification of the PPI where the radar is offset from the center. This display is commonly called a *sector PPI*. It is quite useful in surveillance (search) radar where some sector less than a full 360° is of the most interest.

The RHI (or *range-height indicator*) displays target altitude (height) on the vertical axis as a function of horizontal range along the horizontal axis, as sketched in Fig. 1.7-4. Alternatively, from the viewpoint of a polar plot, the RHI displays radial range with target elevation angle. A RHI is a useful tool for *precision approach radar* (PAR), where a particular elevation angle might correspond to a desired glide path for landing aircraft.

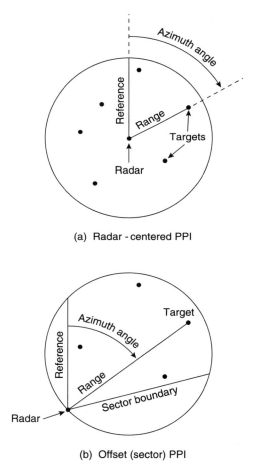

Figure 1.7-3   PPI displays, (*a*) having the radar located at display-center, and (*b*) showing only a sector of azimuth for an offset radar.

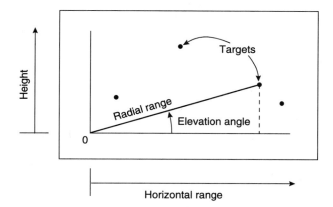

Figure 1.7-4   RHI display for target range versus height.

In the preceding discussions the vertical deflections of the A-scope, or the spots on the PPI or RHI due to the targets are called *blips*. When these blips are directly due to the received waveforms the displays are said to show *raw video*. Many modern radars process raw data through special signal-processing operations prior to generating blips; in these systems the displays show *synthetic video*. Synthetic video can also include special alphanumeric characters and other symbols to convey information to the display.

PROBLEMS

1.1-1 A transmitter and receiver are separated by 5 km as shown in Fig. P1.1-1. If a target at point T is defined by range $R_1 = 4.2$ km and elevation angle $E_1 = \pi/6$ (or 30 degrees), find R_2 and E_2.

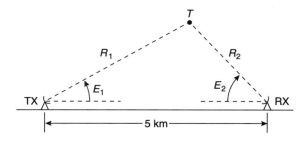

Figure P1.1-1.

1.1-2 Work Problem 1.1-1 except assume that $R_1 = 7$ km and $E_1 = \pi/9$.

1.1-3 Two receiving sites of a 2-receiver multistatic radar are separated by 1000 m. The sites report their ranges to a target are $R_2 = 1900$ m (site 1) and $R_3 = 1450$ m (site 2). What angles does the target make at each of the two sites relative to the line between sites 1 and 2?

1.4-1 The maximum physical extent of a radar target is 10 m. If one quantity is to be 0.1 or less in size compared to another quantity to be considered small, what monostatic radar pulse duration is required for the target to be called *point* if no **FM** is transmitted?

1.4-2 Work Problem 1.4-1 except assume a physical extent of 50 m.

1.4-3 Work Problem 1.4-1 except solve for pulse durations for target extents from 1 to 100 m, and plot the results.

1.4-4 A monostatic radar transmits a pulse with **FM** that has a 3-dB bandwidth of 5 MHz. What is the largest physical dimension a target can have and remain a point target?

1.5-1 Sketch the following waveform if A, T, A_1, and $\omega_0 \gg 2\pi/T$ are positive constants:

$$s(t) = A\left[A_1 + \cos^2\left(\frac{\pi t}{T}\right)\right]\text{rect}\left(\frac{t}{T}\right)\cos(\omega_0 t)$$

1.5-2 Work Problem 1.5-1 except assume the waveform

$$s(t) = A\cos\left(\frac{\pi t}{T}\right)\text{rect}\left(\frac{t}{T}\right)\cos(\omega_0 t)$$

1.5-3 Work Problem 1.5-1 except assume the waveform

$$s(t) = A\,\text{tri}\left(\frac{t}{T}\right)\cos(\omega_0 t)$$

where the function tri(t) is defined by

$$\text{tri}(t) = \begin{cases} 1 - t, & 0 \le t \le 1 \\ 1 + t, & -1 \le t \le 0 \end{cases}$$

1.5-4 Work Problem 1.5-1 except assume the waveform

$$s(t) = A[1 - A_1 t^2]\text{rect}\left(\frac{t}{T}\right)\cos(\omega_0 t)$$

where $A_1 \le 4/T^2$.

1.5-5 A transmitter uses the waveform of Problem 1.5-1 with $A_1 = 0.2$ and $T = 10\ \mu s$.

(a) Find the value of A so that the actual average peak transmitted power is 500 kW into a resistive load of 510 Ω.

(b) For the value of A found in (a) what available average transmitted power is generated if the radar's pulse rate is 800 Hz?

1.5-6 Work Problem 1.5-5 except assume the waveform of Problem 1.5-2.

1.5-7 A transmitter uses the signal of Problem 1.5-3.
 (a) Find the (normalized) average peak transmitted power in terms of A and T.
 (b) Find the (normalized) available average transmitted power in terms of A, T, and T_R.

1.5-8 Work Problem 1.5-7 except assume the waveform of Problem 1.5-4. Obtain the answer in part (b) as an expression containing A_1 as well as A, T, and T_R.

★**1.5-9** For a pulsed radar waveform $s(t)$, if ω_0 is much larger than the reciprocal of the duration of a pulse making up the envelope $a(t)$, and if an equivalent pulse duration T_{eq} is defined by

$$T_{eq} = \frac{\int_{-T_R/2}^{T_R/2} a^2(t)\, dt}{|a(t)|^2_{max}},$$

show that (1.5-9) can still be used to define D_t if T is replaced by T_{eq}.

1.5-10 Find an expression for the energy E_s for the waveform of Problem 1.5-1 that is a function of the parameters defining the waveform.

1.5-11 Work Problem 1.5-10 except assume the waveform of Problem 1.5-2.

1.5-12 Work Problem 1.5-10 except assume the waveform of Problem 1.5-3.

1.5-13 Work Problem 1.5-10 except assume the waveform of Problem 1.5-4.

1.6-1 A monostatic radar receives pulses from a target that are delayed 296 μs from the transmitted pulses. What is the target's range in meters and in statute miles?

1.6-2 A monostatic pulsed radar has a repetition rate of $f_R = 1/T_R = 1700$ pulses/s and transmits rectangular pulses of duration 15 μs. What maximum range can a target have if no part of its delayed pulse is to overlap any part of a transmitted pulse and not be delayed by more than one PRF interval?

1.6-3 A missile's speed toward a radar is 331.7 m/s (about mach 1.0 based on the speed of sound in air at one atmosphere of pressure). The radar's frequency is 12 GHz.
 (a) Find the exact Doppler frequency at the receiver.
 (b) Find the receiver's Doppler frequency assuming $v \ll c$. Is this assumption valid?

1.6-4 Plot curves of the magnitude of a receiver's Doppler frequency versus radar frequency f_0, for 0.2 GHz $\leq f_0 \leq$ 20 GHz for several speeds, including 500, 1000, and 2000 mi/h.

1.6-5 A bank of filters is to be used to measure Doppler frequency of a target about the carrier f_0 by determining which filter has the largest response. If all filters have a 3-dB bandwidth of 550 Hz and their responses cross over at the

— 3-dB points, how may filters are needed to encompass all possible targets at speeds up to $|v| = 1326.8$ m/s (about Mach 4). Assume that $f_0 = 5$ GHz.

1.6-6 What target speeds (in mi/h) are required in a monostatic radar if the stretching/compression factor is to satisfy the condition

$$0.9999 \leq \frac{c - v}{c + v} \leq 1.0001$$

1.6-7 In a radar where measurement of target speed is most important, and only radar frequency is to be considered, would a radar at 1 GHz or at 12 GHz be better? Give a reason for your answer.

1.6-8 If a radar's frequency can be measured to only an accuracy of 0.1 kHz (accuracy of measuring Doppler frequency) and this accuracy must correspond to speed measurement error of 50 mi/h, what minimum carrier frequency must be used?

1.6-9 For a monostatic pulsed radar determine a number to represent a 'rule-of-thumb' for how much target range (in meters) corresponds to each microsecond of transmitted pulse length.

1.6-10 Define the error between a radar's pulse duration T and the stretched/compressed pulse duration $T(c - v)/(c + v)$ as

$$\text{Error} = T - \frac{(c - v)T}{c + v} = \frac{2vT}{c + v} \approx \frac{2vT}{c}$$

If the transmitted signal bandwidth is B, give arguments to prove that the error is negligible if

$$BT \ll \frac{c}{2v}$$

1.7-1 A target is located at (25.2 km, 1.9 km, -0.6 km) in coordinates x_R, y_R, z_R. What is the target location in radar coordinates?

1.7-2 A target is located at $r = 56$ km, $A_T = 0.85\pi$ rad, and $E_T = 23\pi/180$ rad. What is the target's location in coordinates x_R, y_R, z_R?

1.7-3 A radar target approaches a radar over a flat earth at a constant altitude y_T and at constant speed v from a far range (approximate as $z_R = -\infty$). It flies a straight line directly over, and past, the radar to a very far range (approximate as $z_R = +\infty$).

(a) If the target passes overhead at time zero, determine an expression for the target's elevation angle E_T.

(b) Find the elevation angle rate (derivative of E_T versus time).

(c) If a radar is tracking the target, what is the largest angle rate that must be followed, and at what time does it occur?

1.7-4 The boresight axis of a tracking radar has radar coordinates $A_b = 36\pi/180$ rad, and $E_b = 47\pi/180$ rad. If the target being tracked has radar coordinates $A_T = 35.9\pi/180$ rad and $E_T = 47.2\pi/180$ rad, find the "tracking errors" θ_x and θ_y.

1.7-5 For the radar and target of Problem 1.7-4, find the target's coordinates x_T, y_T, z_T in the x, y, z coordinate system if radial range is 50 km. If the earth's atmosphere is assumed to be negligible above 20 km, is this target in free space or in the atmosphere?

1.7-6 Assume an antenna is to be designed to have a size of 30 wavelengths, regardless of frequency. Compare the antenna's size at the lowest L-band frequency to that at the highest X-band frequency.

1.7-7 Two closely spaced target blips show up on an A-scope display in a monostatic radar receiving rectangular pulses of duration $T = 1.8$ μs. How close can the two targets get in range separation before their pulse returns start to overlap (and be unresolvable)?

2

ELEMENTS OF WAVE PROPAGATION

All radars depend in some manner on the propagation of waves. A sonar (which is an acoustic radar) uses pressure waves, but all other radars use electromagnetic (EM) waves. In this chapter we introduce some elementary definitions and characteristics of electromagnetic waves as they apply to radar. No effort will be made to discuss EM wave theory in detail. Rather, plausible arguments will be given that should enable the reader to arrive at the most important results without the detailed mathematical theory usually associated with EM waves.

Our discussions will define the various types of waves that are important to radars at various frequencies. Of special interest is the effect the earth's atmosphere and various weather phenomena have on these waves. Waves near the earth are also affected by reflections from the earth's surface. We describe how reflected waves are dependent on many factors, such as wave polarization, surface irregularities, and roughness, as well as the characteristics of the radar's antennas. These various topics are developed in enough detail to allow a smooth advancement to later work. In particular, our work on wave polarization is required in later chapters on antennas and angle measurement and tracking systems.

2.1 SPHERICAL AND PLANE WAVES

Consider a physically small isolated source of radiation of energy in the form of an electromagnetic wave at an angular frequency ω. The wave radiates outwardly carrying energy in all directions at the speed of light c from the source. The locus of all points on the wave having a given phase is a sphere; the spherical surface can be considered a *wavefront* having a radius that expands at the speed of light. A line from the isolated source toward any given direction from the source is called a *ray*; it defines a direction of wave propagation. More generally, a ray is the direction normal to the surface of constant phase of a wavefront. Figure 2.1-1 illustrates these comments and

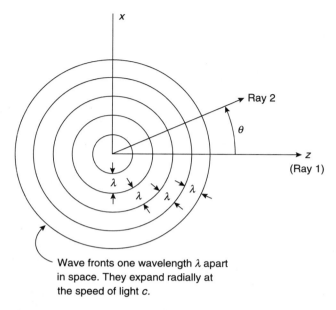

Wave fronts one wavelength λ apart in space. They expand radially at the speed of light c.

Figure 2.1-1 Waves from a small radiator of energy at the origin.

shows two rays that define the path of the wave in two directions. Wavelength λ is the distance traveled by the wave during one period T of the wave with frequency $f = 1/T$; it is given by

$$\lambda = cT = \frac{c}{f} = \frac{2\pi c}{\omega} \tag{2.1-1}$$

The intensity, or strength, of a wave depends on its electric (E) and magnetic (H) fields at any point of interest. These fields decrease with distance from the source because the radiated power is spread over a larger spherical surface with increasing distance. Thus practical waves in radar decrease in field intensity by a factor $1/R$ with range R; they decrease in power intensity by a factor $1/R^2$.

Wave intensity may also vary with direction. All sources do not radiate with equal intensity in all directions, nor do we usually want them to. A typical radar antenna (Chapter 3), for example, may concentrate most of its energy into a relatively small region in space where wave intensity is large relative to that which would have occurred with isotropic[1] radiation, and even larger relative to its own radiation in directions outside the small region. If the source of Fig. 2.1-1 is not isotropic, wave intensity along a wavefront (in the x, z plane) will vary with angle θ. It may also vary for other directions that do not lie in the x, z plane.

All the preceding discussions related to a small isolated source of radiation. Real sources, such as an antenna, are not "small." For a larger source we may view the electric and magnetic fields in the wave from the source as being the sum of

---

[1] An *isotropic* source is one that radiates energy in equal intensities in all directions over a sphere.

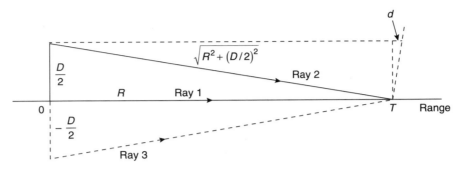

Figure 2.1-2   Geometry associated with the far field.

contributions from all the "small" radiators that make up the source. Wavefronts from the source at near distances are not spherical but become spherical at distances known as the *far field*. The far field begins at a distance R_F where the maximum dimension D of the source (normal to the ray direction of interest) is very small compared to the distance (Elliott, 1981, p. 27; Balanis, 1982, p. 23; Hansen, vol. 1, 1964, p. 29). In radar R_F is usually taken as $2D^2/\lambda$, but other values are sometimes used.[2] Thus waves are in the far field when distance R satisfies

$$R \geq R_F = \frac{2D^2}{\lambda} \tag{2.1-2}$$

Example 2.1-1   Consider two small radiators located at the edges of an antenna of maximum dimension D as shown in Fig. 2.1-2. Waves from these edges (rays 2 and 3) arrive together at a point T that is at a range R from the antenna's center. They travel an extra distance d, compared to ray 1 from the antenna's center, which is given by

$$d = \sqrt{R^2 + \left(\frac{D}{2}\right)^2} - R \approx R\left(1 + \frac{D^2}{8R^2}\right) - R$$

$$= \frac{D^2}{8R}$$

On use of R_F from (2.1-2), we have

$$d \approx \frac{D^2}{8R} \leq \frac{D^2}{8R_F} = \frac{\lambda}{16}$$

Thus the far field begins at a range R_F where the waves from the antenna are all approximately in phase ($\lambda/16$ corresponds to a phase of $\pi/8$ rad).

Even for isolated practical antennas, waves are nearly spherical at distances in the far field. Targets for most radars are in the far field and are of small enough physical

---

[2] $R_F = D^2/\lambda$ is sometimes used but it is not as conservative a value as given by (2.1-2).

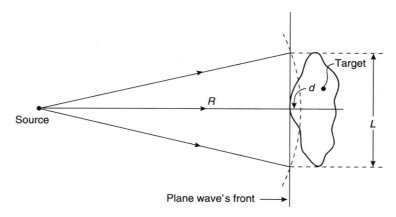

Figure 2.1-3 Geometry of spherical and plane waves at a target of physical size L.

size that the spherical wave can be approximated by a *plane wave* over the region of space occupied by the target. Figure 2.1-3 illustrates the geometry involved for such a target of physical size L and range R. If R is sufficiently large compared to L, the spherical wavefront (shown dashed) can be approximated by the plane wave front (shown solid) if d is small. For $d \leq \lambda/20$ it can be shown that R and L are related approximately by (see Problem 2.1-8)

$$R \geq \frac{2.5L^2}{\lambda} = \frac{25L^2 f_{\text{GHz}}}{3} \tag{2.1-3}$$

where R, L, and λ are in meters and f_{GHz} is f in GHz. For example, if $L = 35$ m and $f_{\text{GHz}} = 0.5$, then for $R \geq 25(35^2)0.5/3 = 5104.2$ m (or about 3.17 mi), the target can be considered to be illuminated by a plane wave.

Because most waves from radars interact with targets at distances where plane waves are good approximations, we will assume that waves are planar in the remainder of this book unless specifically indicated otherwise.

One important consequence of the plane approximation of waves is that the electric field and magnetic field vectors are spatially orthogonal to each other and both lie in the plane of the planar wavefront. There are no electric or magnetic fields in the direction of propagation. Another consequence is that the electric and magnetic field amplitudes are related to each other; the value of the electric field is η times the value of the magnetic field at any instant. Here η, called the *intrinsic impedance* of the medium in which the wave propagates, is given by

$$\eta = \sqrt{\frac{\mu}{\varepsilon}} \quad \text{(ohms)} \tag{2.1-4}$$

where μ is the *permeability* of the medium (H/m) and ε is the *permittivity* of the medium (F/m). For free (vacuum) space $\mu = \mu_0 = 4\pi(10^{-7})$ H/m and $\varepsilon = \varepsilon_0 = (10^{-9})/(36\pi)$ F/m. These and other properties (summarized by Ramo et al., 1984, pp. 274–275) are useful in various subsequent discussions.

In this section we have found that isolated sources of electromagnetic waves radiate energy in spherical wavefronts for small radiators or approximately spherical wavefronts in the far field for practical-sized radar antennas. For radar targets sufficiently into the far field, these spherical waves appear as, and can be approximated as, plane waves. However, when the source is not isolated, as would be the case where the radar exists in the earth's atmosphere and near the earth, the effect of these other objects must be considered.

2.2   WAVES NEAR THE EARTH

When a radar transmits through an antenna near the earth, the presence of the earth, as well as the atmosphere, affect the propagating waves. Figure 2.2-1 illustrates the geometry involved. As shown, the atmosphere comprises several layers. The ionosphere extends from about 50 km to about 600 km in altitude and consists of extremely thin air. The troposphere, which contains the earth's air supply and in which most weather effects occur, extends up to an altitude of about 20 km. Between the troposphere and the ionosphere from about 20 km to about 50 km, there is a region that behaves approximately as free space for radio wave propagation.

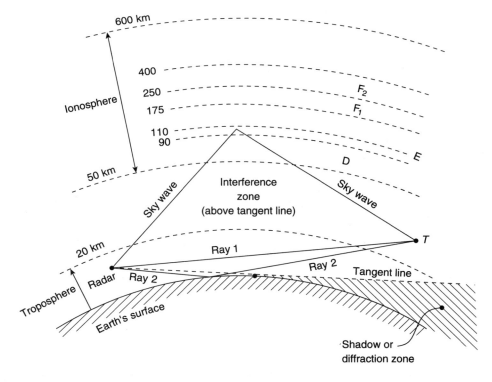

Figure 2.2-1   Geometry of earth's atmosphere.

Troposphere

In the troposphere air density, temperature, and humidity all decrease with increasing altitude. As a result the refractive index of the air decreases with altitude which causes the velocity at which waves propagate to increase. The net effect is to bend the wave's (ray) path of propagation downward or back toward the earth. The amount of bending (called *refraction*) depends on the angle of the ray path; waves propagating vertically that pass through the least amount of atmosphere suffer the least refraction. Waves directed near the horizon pass through the largest amount of atmosphere and suffer the largest amount of bending. Because of refraction waves arriving at a radar have different ray path lengths and elevation angles as compared to the direct straight-line (true) path between target and radar. The difference between the measured and straight-line quantity is an error that must be corrected in precision radars. These errors are discussed further in Section 2.3.

Ionosphere

Sunlight (mainly ultraviolet) passing through the ionosphere ionizes the rarified air and strongly affects wave propagation. The principle effects are attenuation and reflection, with both effects being very dependent on the wave's frequency. The ionization process creates free electrons. A plot of the density of free electrons with altitude shows several maximums, which lead to the definitions of several "layers" or "regions" around the maximums that affect waves differently. These layers are shown in Fig. 2.2-1 as D, E, F_1, and F_2 regions.

Waves incident on the earth side of the ionosphere at lower frequencies tend to be reflected by the lower regions. As frequency increases, waves may pass through the lowest regions, suffering mainly attenuation, and then be reflected back to earth by higher regions. When wave frequency is as high as about 30 MHz waves may penetrate the entire ionosphere while suffering mainly attenuation. In the case of severe ionospheric activity, the frequency of penetration may approach 70 MHz (David and Voge, 1969). Because of the reflective nature of the ionosphere, a practical lower frequency of about 30 to 70 MHz exists for earth-based radars that must operate with targets outside the ionosphere. Similar comments hold for communications systems between earth and spacecraft. On the other hand, *over-the-horizon* (OTH) radar is *designed* to reflect waves from the ionosphere in order to reach range points beyond the earth's horizon limit due to curvature (Fenster, 1977).

Below about 300 kHz, waves reflect from the D region which exists from about 50 to 90 km in altitude. Waves above 300 kHz are passed with attenuation. This attenuation is especially strong during daylight hours for waves at frequencies 300 kHz $< f <$ 3 MHz and is somewhat weaker for 3 MHz $< f <$ 30 MHz. At night the D region mostly disappears.

The E region is important for reflecting high frequencies, 3 MHz $< f <$ 30 MHz, in daytime and medium frequencies, 300 kHz $< f <$ 3 MHz, at night, because the layer is relatively stable at an altitude of about 110 km. Occasionally localized highly ionized areas occur (about 50% of the time). These areas are called the *sporadic E region*, and its main effect is to raise the maximum frequency at which wave reflection can occur.

Waves that penetrate the E region usually penetrate the F_1 region from about 175 to 250 km, suffering mainly some attenuation. The F_1 region exists mainly in daytime and merges with the F_2 region at night.

The main region for reflecting waves at frequencies 3 MHz $< f <$ 30 MHz is the F_2 layer from about 250 to 400 km or higher. The characteristics of the F_2 layer are more variable than those of the F_1 region, and they vary daily, seasonally, and with sunspot cycles.

Types of Waves

The far region below the radar's geometrical horizon is called the *shadow* or *diffraction zone*, as shown in Fig. 2.2-1. It is difficult for radar waves to penetrate into the shadow zone except through the mechanism of reflection from the ionosphere. The reflected waves are often called *sky waves*, and we have already seen that sky waves are most useful only at frequencies below about 30 MHz. For the usual radar frequencies a second mechanism can give *limited* coverage into the shadow zone. It is called *diffraction*, which is a wave bending caused by the presence of a physical object, the earth in our case.

The field strength of waves that penetrate the shadow zone by diffraction decreases as wave frequency increases. For the most common radar frequencies (above 300 MHz) the coverage gained is minimal. However, as a wave's frequency decreases, the refractive effect of the earth increases and the wave tends to "hug" the earth's surface for greater distances. The result is often called a *surface wave*. Surface waves are most important at frequencies below about 1 MHz where they are used both in the standard *amplitude modulation* (AM) broadcast system in the United States and in worldwide communication systems.

As shown in Fig. 2.2-1, a target (point T) is in the *interference zone* whenever it is above the radar's geometrical horizon. In this zone the wave that arrives at the target is called a *space wave*, and it is made up of two components. The two components are shown as rays 1 (the *direct ray*) and 2 (the *indirect* or *reflected ray*) in Fig. 2.2-1. Because rays 1 and 2 add together to produce the total field at the target, the sum can have any value between some maximum and minimum values that are functions of the magnitudes and phases of the two rays. This "interference" behavior gives rise to the zone's name. Our comments assume frequencies above about 30 MHz where waves penetrate the whole ionosphere. At lower frequencies the total field at some point in the interference zone might also include a sky wave reflected from the ionosphere in addition to the direct and reflected waves.

2.3   EFFECTS OF ATMOSPHERE ON WAVES

The earth's atmosphere affects waves in several ways. The most important effects are refraction, attenuation, and scattering. A discussion of scattering is reserved for Section 5.8.

Refraction

Refraction, or bending, of wave rays occurs because the atmosphere's index of refraction n decreases with height. If the atmosphere is modeled as homogeneous over

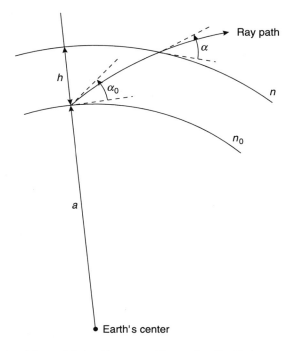

Figure 2.3-1   Geometry illustrating Snell's law.

a sphere where n varies only with height, as shown in Fig. 2.3-1, Snell's law can be used to give a simple interpretation of refraction. Snell's law states that

$$n(a + h)\cos(\alpha) = n_0 a \cos(\alpha_0)$$ (2.3-1)

Here a wave propagates from a point on the surface of the earth with true radius a. The wave's ray makes an angle α_0 from the local horizontal, and n_0 is the value of n at the surface. Due to refraction the ray bends downward. At altitude h above the surface, where the index of refraction is n, the ray makes an angle α from the local horizontal. Parameters n and α at altitude h are related to n_0 and α_0 at the surface through (2.3-1).

For a homogeneous atmosphere where a wave would travel in a straight line, n would be a constant. By Snell's law, (2.3-1), with $n = n_0$ we have

$$\left(1 + \frac{h}{a}\right)\cos(\alpha) = \cos(\alpha_0)$$ (2.3-2)

for a nonrefractive atmosphere.

Effective Earth Model for Refraction

To further develop the subject of refraction through use of (2.3-1) requires that some model be chosen for the behavior of n with height. Several models exist. Two of the more important ones are the *exponential model* (Skolnik, 1980, p. 450), which assumes $n - 1$ decreases exponentially with altitude, and the *standard model*, where n is assumed to vary linearly with altitude as

$$n = n_0 + \frac{dn}{dh} h \tag{2.3-3}$$

with dn/dh taken as constant. The standard model is in agreement with measured results mainly in the lower atmosphere where $h \leq 1$ km. In this region it allows a simple interpretation of refraction, as discussed below. The exponential model more closely represents measured results when $h > 1$ km, but it is more difficult to obtain simple results for refraction.

As a first approximation to wave refraction we will assume the standard model and substitute n from (2.3-3) into (2.3-1). We obtain

$$\left(1 + \frac{h}{a} + \frac{h}{n_0} \frac{dn}{dh} + \frac{h}{n_0} \frac{dn}{dh} \frac{h}{a}\right)\cos(\alpha) = \cos(\alpha_0) \tag{2.3-4}$$

Because n never departs greatly from unity, we may assume that $n_0 \approx 1$. Also, for the lower atmosphere $h/a \ll 1$, so the fourth left-side term is a second-order small term that can be ignored. Thus (2.3-4) becomes

$$\left[1 + \frac{h}{a}\left(1 + a\frac{dn}{dh}\right)\right]\cos(\alpha) \approx \cos(\alpha_0) \tag{2.3-5}$$

Now dn/dh is assumed constant in the standard model, so we can define a constant a_e by

$$a_e = \frac{a}{1 + a(dn/dh)} \tag{2.3-6}$$

and write (2.3-5) as

$$\left(1 + \frac{h}{a_e}\right)\cos(\alpha) \approx \cos(\alpha_0) \tag{2.3-7}$$

By comparing (2.3-7) to (2.3-2) for a nonrefracting atmosphere, we see that they both have the same form. Because the forms are the same, a simple interpretation of refraction can be made by using (2.3-7). Waves propagating in a standard atmosphere over an earth of true radius a suffer refraction (bending). However, these same waves can be treated as though they propagate through a homogeneous atmosphere (no bending) if the true earth's radius is replaced by an effective earth's radius as given by (2.3-6).

It is customary to write effective radius as $a_e = ka$, where k is a constant given by

$$k = \frac{1}{1 + a(dn/dh)} \tag{2.3-8}$$

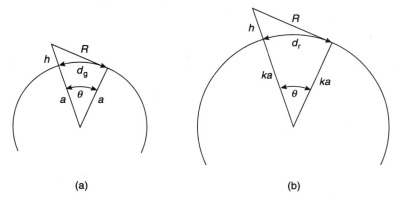

Figure 2.3-2 Geometries for horizon calculations. (*a*) For geometric horizon, and (*b*) for radio horizon.

from (2.3-6). For middle lattitudes, measurements give $dn/dh \approx -1/(4a)$ which gives $k = 4/3$ (Burrows and Attwood, 1949). This value of k has given rise to the standard model being sometimes called the *4/3-earth model*. However, k varies somewhat from 4/3 depending on climate and other factors. For colder climates k can be smaller (down to about 1.20). For warmer climates k can be larger (up to about 1.90) (Bean, 1953; Bean et al., 1966).

As an example of the effect of refraction, we consider a radar at a height h above the earth's surface and calculate both the geometrical radar horizon and the radio horizon for an effective earth radius of ka. The geometrical horizon is distance d_g defined in Fig. 2.3-2a. From the geometry, with $h \ll a$ such that θ is small,

$$d_g = a\theta \approx a\sin(\theta) = a\frac{\sqrt{(h+a)^2 - a^2}}{h+a}$$

$$= a\frac{\sqrt{h^2 + 2ah}}{h+a} \approx \sqrt{2ah} \qquad (2.3-9)$$

The radio horizon is found from the geometry of Fig. 2.3-2b. The result is given by (2.3-9) with a replaced by $ka = a_e$:

$$d_r \approx \sqrt{2kah} = \sqrt{2a_e h} \qquad (2.3-10)$$

Because of refraction the radio horizon is larger than the geometric horizon by the factor $d_r/d_g \approx k^{0.5} \approx 1.155$.

Example 2.3-1 Assume a 4/3-earth model; the radius of the earth is $a = 6371$ km,[3] and a radar over a smooth earth has an altitude of $h_1 = 0.5$ km. Let an aircraft (the

---

[3] This value of a is the radius of a sphere having the same total mass as the earth. The value is about 0.12% less than the equatorial earth's radius of 6378 km and about 0.22% larger than the polar radius of 6357 km.

target) approach the radar at a constant altitude of $h_2 = 1.2$ km. We find the surface range to the target when the target crosses the radio horizon of the radar. The surface range to the radio horizon from the radar is $[2kah_1]^{1/2}$ from (2.3-10). Similarly, for the target, its range to the radio horizon point on the surface is $[2kah_2]^{1/2}$. The total surface range d is

$$d \approx \sqrt{2kah_1} + \sqrt{2kah_2} = \sqrt{2\left(\frac{4}{3}\right)6371} \, [\sqrt{0.5} + \sqrt{1.2}] \text{ (km)}$$

$$= 234.95 \text{ km}$$

Radar Errors Due to Refraction

Wave refraction can cause errors in radar measurements, mainly of a target's range and spatial angles (elevation and azimuth). Figure 2.3-3 illustrates how errors occur in range and elevation angle measurements. A target exists at true geometric range R_0 and elevation angle E_0. Due to wave refraction the wave travels the curved path for a distance R and arrival angle E. The radar measures E, not E_0, so an error exists that equals $E - E_0$. Such angle errors vary in a complicated manner with target range and elevation angle but tend to not exceed about 10 mr (or about 0.6 degree). The largest errors occur at low-elevation angles and long ranges. Details on plots of these errors and approximations for error corrections are given in Barton (1988) and Shannon (1962).

In a similar manner, range error results because the radar measures R and not R_0 in Fig. 2.3-3. The error $R - R_0$ tends to not exceed about 100 m with the largest error for targets at low-elevation angle and large range (see Barton, 1988, for curves of errors and correction methods for errors and Shannon, 1962).

Refraction effects of the atmosphere in a horizontal direction are normally small so the associated errors in radar measurements of azimuth angle are usually negligible.

Attenuation by a Clear Atmosphere

As a radar wave passes through the atmosphere, it suffers attenuation. Part of this attenuation is due to weather effects (unclear atmosphere) such as rain, snow, and

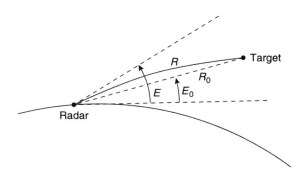

Figure 2.3-3   Geometry of a target and its associated ray path through the atmosphere.

clouds. However, even if the atmosphere is clear, there remains attenuation, mainly due to the presence of oxygen and water vapor. Straiton and Tolbert (1960) have given theoretical plots of one-way wave attenuation (in dB/km) versus frequency as shown in Fig. 2.3-4 for a reasonably typical atmosphere (one atmosphere of pressure and 1% water molecules which is 7.5 g/m^3 of water content). The plot for attenuation by oxygen is reasonably close to measured data. The two peaks are due to resonances in the oxygen molecules, and they occur at about 60 and 120 GHz. For frequencies below about 10 GHz, the attenuation rate decreases slowly.

The curve in Fig. 2.3-4 for attenuation by water vapor underestimates measured results by a factor of as much as 2 to 3 (in dB/km) for frequencies above about 30 GHz (Straiton and Tolbert, 1960, Fig. 2), and it should be used with care in this frequency range. Near the resonance at about 22.2 GHz, the curve is better supported by measurements. There is an extremely severe resonance "line" at about 184 GHz where attenuation makes it nearly impossible for a radar to operate, even in a "clear"

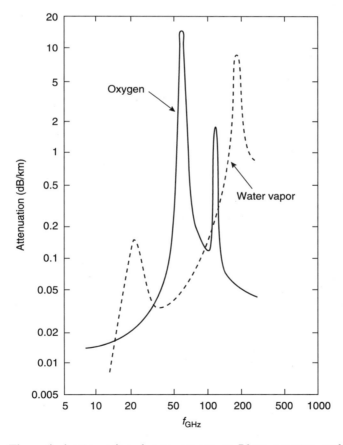

Figure 2.3-4 Theoretical attenuation due to oxygen at 76 cm pressure and water vapor at 7.5 g/m^3 content. (Adapted from A. W. Straiton and C. W. Tolbert, © 1960 IRE, with permission.)

atmosphere. Above 184 GHz as many as 588 other absorption lines have been listed with 149 between 300 GHz and 6000 GHz that have significant attenuation rates (Straiton and Tolbert, 1960, p. 898). Below 10 GHz water vapor causes negligible attenuation.

The total attenuation of the clear atmosphere at the lower altitudes can be taken as the sum of the attenuations shown in Fig. 2.3-4. These results can be used directly for targets at low altitudes and short ranges. However, the variation (decrease) of attenuation rate with altitude must be accounted for when targets are at higher altitudes. Figures 2.3-5 through 2.3-9 show the two-way attenuations through the clear atmosphere for targets at various ranges, elevation angles, and frequencies of 0.3, 1.0, 3, 10, and 30 GHz. Blake (1972) also gives curves for many other frequencies from 100 MHz to 100 GHz. For targets that are completely outside the troposphere, Fig. 2.3-10 is helpful.

Example 2.3-2   Assume the attenuations of Fig. 2.3-4 hold up to an altitude of 1 km. For a target altitude $h = 1$ km, we find the smallest radial range R such that the target's elevation angle E is not greater than 3.5 degrees. Then we find the one-way wave attenuation for range R and a frequency of 20 GHz. From geometry, $R = h/\sin(E)$. The smallest R becomes $R = 10^3/\sin(3.5\pi/180) = 16.38$ km. From Fig. 2.3-4, attenuation in dB/km (one-way) ≈ 0.019 (oxygen) $+ 0.12$ (water vapor) $= 0.139$ dB/km. Thus total one-way attenuation is $0.139(16.38) = 2.277$ dB, which is significant because frequency is so close to the water vapor resonance at 22.2 GHz.

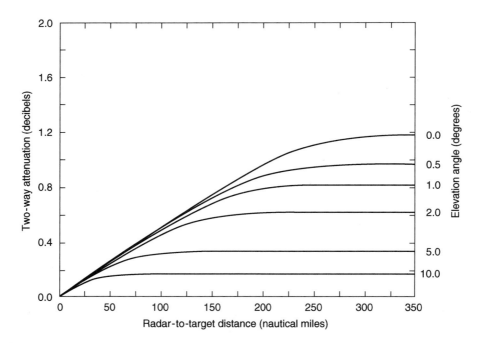

Figure 2.3-5   Wave attenuation in a standard atmosphere at 300 MHz. (From Blake, 1972.)

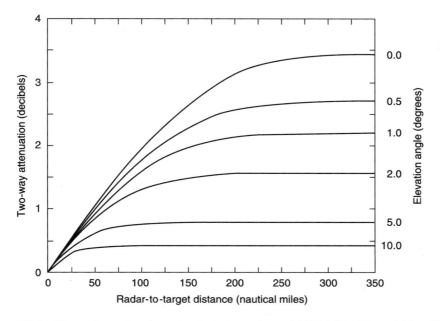

Figure 2.3-6   Wave attenuation in a standard atmosphere at 1000 MHz. (From Blake, 1972.)

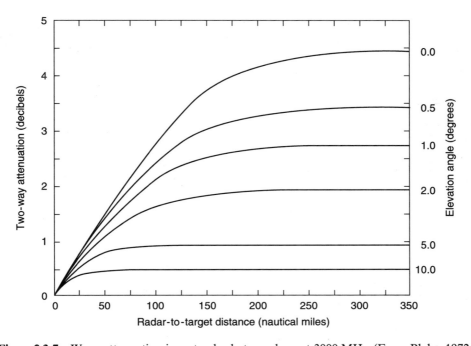

Figure 2.3-7   Wave attenuation in a standard atmosphere at 3000 MHz. (From Blake, 1972.)

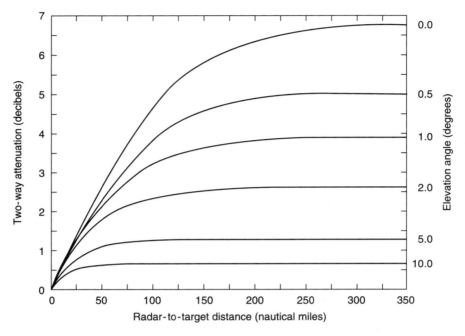

Figure 2.3-8 Wave attenuation in a standard atmosphere at 10,000 MHz. (From Blake, 1972.)

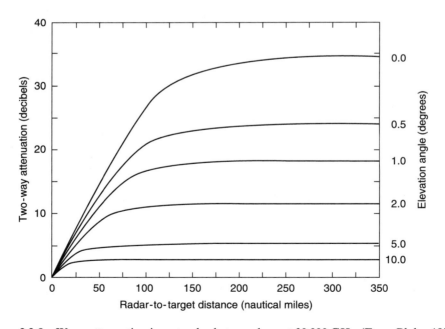

Figure 2.3-9 Wave attenuation in a standard atmosphere at 30,000 GHz. (From Blake, 1972.)

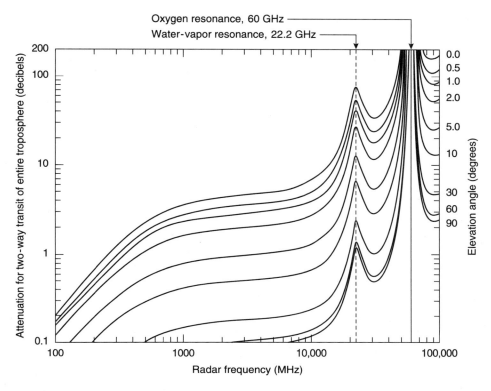

Figure 2.3-10   Wave attenuation in a standard atmosphere for a target outside the troposphere. (From Blake, 1972.)

Attenuation Due to Rainfall

When a clear atmosphere is contaminated by added water content, either in liquid or frozen form, it is called an *unclear atmosphere*. The added water content typically occurs as rain, snow (wet or dry), clouds, hail, sleet, or fog.

Attenuation of a wave due to rainfall is a complicated function of rainfall rate, frequency, temperature, size distribution of rain drops, and wave polarization. An early model for rain attenuation was given by Burrows and Attwood (1949); it used the Laws-Parsons drop size distribution and the effects of temperature were included. Since this early work many other rain models have been developed. Most of these are summarized in detail by Ippolito (1989), where an excellent discussion includes how rain is distributed over the world and within storms. Of particular modern interest is the 1986 CCIR (International Telecommunications Union) model (see CCIR, 1986) which gives rain attenuation as shown in Fig. 2.3-11 for general purpose use. The CCIR model, as well as most rain attenuation models, are based on the empirical expression

$$\text{Attenuation (in dB/km)} = ar^b \qquad (2.3\text{-}11)$$

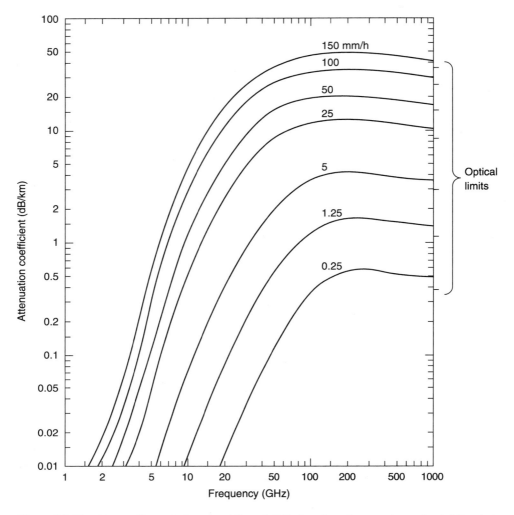

Figure 2.3-11 Attenuation rate (one-way) for rainfall at various frequencies and rainfall rates. (From Ippolito, 1989.)

where a and b are coefficients that are functions of frequency and r is rain rate in mm/h. Both a and b are mildly sensitive to wave polarization. Table 2.3-1 gives a and b for vertical (subscripts V) and horizontal (subscripts H) polarizations (Ippolito, 1989, pp. 2–15), and these values can be used to find a and b for any polarization (Flock, 1987, pp. 4–10).

The attenuations given in Fig. 2.3-11 can be applied to a region having a constant rain rate. However, it should be noted that rain rate may vary over the spatial extent of a storm. Even for generalized (constant) rainfall rate over a large area at the surface, rate decreases exponentially with the square of altitude [Atlas and Kessler, 1957; as noted in Skolnik (ed.), 1990, p. 23.8]. Hence the attenuation decreases with altitude and must be considered in some situations.

TABLE 2.3-1   Specific attenuation coefficients.

Frequency (GHz)	a_H	a_V	b_H	b_V
1	0.0000387	0.0000352	0.912	0.880
2	0.000154	0.000138	0.963	0.923
3	0.000650	0.000591	1.121	1.075
6	0.00175	0.00155	1.308	1.265
7	0.00301	0.00265	1.332	1.312
8	0.00454	0.00395	1.327	1.310
10	0.0101	0.00887	1.276	1.264
12	0.0188	0.0168	1.217	1.200
15	0.0367	0.0347	1.154	1.128
20	0.0751	0.0691	1.099	1.065
25	0.124	0.113	1.061	1.030
30	0.187	0.167	1.021	1.000
35	0.263	0.233	0.979	0.963
40	0.350	0.310	0.939	0.929
45	0.442	0.393	0.903	0.897
50	0.536	0.479	0.873	0.868
60	0.707	0.642	0.826	0.824
70	0.851	0.784	0.793	0.793
80	0.975	0.906	0.769	0.769
90	1.06	0.999	9.753	0.754
100	1.12	1.06	0.743	0.744
120	1.18	1.13	0.731	0.732
150	1.31	1.27	0.710	0.711
200	1.45	1.42	0.689	0.690
300	1.36	1.35	0.688	0.689
400	1.32	1.31	0.683	0.684

Sources: CCIR (1986) per Ippolito, 1989, pp. 2–15.

Note: Values for *a* and *b* at other frequencies can be obtained by interpolation using a logarithmic scale for *a* and frequency and a linear scale for *b*.

Example 2.3-3   Assume that rain falls at 1.25 mm/h over the entire region between the radar and target defined in Example 2.3-2, and calculate the minimum wave attenuation (one way) due to the rain if the radar operates at 20 GHz. Since the region of rain extends 16.38 km, we use Fig. 2.3-11 to obtain an attenuation of 16.38(0.09) = 1.474 dB.

Attenuation Due to Clouds and Fog

When water particles are small (diameters less than about 0.005 cm), which is usually the case for clouds and fog, the one-way attenuation imparted on a radar's wave can be written in various forms based on a result given in Kerr (1964):

$$\text{Attenuation (in dB/km)} = 0.438 \frac{m}{\lambda_{cm}^2}$$

$$= 4.867(10^{-4})mf_{GHz}^2 \qquad (2.3\text{-}12)$$

where m is the water content of the cloud or fog in grams per cubic meter, f_{GHz} is frequency in gigahertz, and

$$\lambda_{cm} = \frac{30}{f_{GHz}} \qquad (2.3\text{-}13)$$

is wavelength in centimeters. Equation (2.3-12) applies at a temperature of 18° C and is said to be accurate to about 5% for frequencies from 3 to 60 GHz. For most clouds m has a value from about 0.05 g/m³ up to about 1 to 2.5 g/m³ (Donaldson, 1955), although some isolated clouds have m up to 4 g/m³ in their upper levels (Weickmann and aufm Kampe, 1953).

Figure 2.3-12 illustrates plots of (2.3-12) for various values of m. Except possibly for some heavy sea fogs, m for fog usually does not exceed 1.0 g/m³ (Kerr, 1964, p. 677). This observation is confirmed by Burrows and Attwood (1949, p. 50), who state that m for both clouds and fog rarely exceeds 0.6 g/m³. Thus we may take the lowest three curves of Fig. 2.3-12 as being most typical of fog and clouds.

Values of attenuation given by (2.3-12) and plotted in Fig. 2.3-12 must be corrected for temperatures different than 18° C. Table 2.3-2 gives the required multiplicative correction factor (Ryde and Ryde, 1945; as referenced by Kerr, 1964, p. 677).

For frequencies above 60 GHz where use of (2.3-12) may not hold, fog and cloud attenuation can be obtained from Fig. 2.3-13 (CCIR, 1982) which is a plot of attenuation rate (in dB/km) divided by m (in g/m³). The figure's data can also be used for frequencies below 60 GHz.

Example 2.3-4   We will use Fig. 2.3-13 to find the one-way attenuation rate for a wave at 85 GHz passing through a cloud at 20° C with water density 1.4 g/m³. From the figure at 85 GHz the ordinate is 3.9 (dB/km)/(g/m³). The attenuation rate becomes 3.9(1.4) = 5.46 dB/km.

From a practical standpoint it is sometimes convenient to define visibility in fog (and clouds). If V_m is defined as visibility in meters, Kerr (1964) gives a relationship between visibility and the value of m that can be used in (2.3-12).[4] More recently, two relationships have been given, one for *advection fog* (coastal fog) and another for *radiation* (inland) fog (Ippolito, 1989, p. 6-71):

$$m = \begin{cases} \dfrac{304.1}{V_m^{1.43}}, & \text{coastal fog} \qquad (2.3\text{-}14) \\[2ex] \dfrac{131.9}{V_m^{1.54}}, & \text{inland fog} \qquad (2.3\text{-}15) \end{cases}$$

Kerr's expression is nearly identical to (2.3-14). Data plotted in Fig. 2.3-12 are derived from (2.3-14).

---

[4] Technically m given by (2.3-14) is an average value $\bar{m}$ of m "such that in 95 percent of the cases m lies between $m/2$ and $2m$," as quoted from Kerr (1964, p. 678).

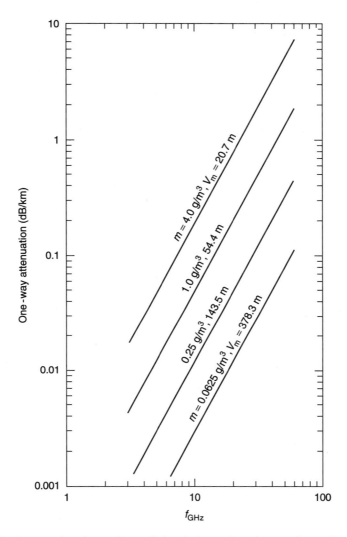

Figure 2.3-12 Attenuation due to fog and clouds for various frequencies and water content.

Example 2.3-5 Continue Example 2.3-2 by assuming that light coastal fog having a visibility of 120 m (about 394 ft) occupies the entire region between radar and target. We calculate the attenuation due to the fog. By substitution of (2.3-14) into (2.3-12), we have the one-way attenuation:

$$\text{Attenuation (in dB/km)} = \frac{4.867(10^{-4})304.1 f_{\text{GHz}}^2}{V_m^{1.43}}$$

$$= \frac{0.148 f_{\text{GHz}}^2}{V_m^{1.43}} = \frac{0.148(20^2)}{120^{1.43}} = 0.063 \text{ dB/km}$$

The total attenuation is $0.063(16.38) = 1.031$ dB, one way.

TABLE 2.3-2 Correction factors to be applied to the attenuation given by (2.3-12) or Fig. 2.3-12 when fog or cloud temperature is different from 18° C.

f_{GHz}	Multiplicative Correction Factor for Temperature Shown					
	0° C	10° C	18° C	20° C	30° C	40° C
60.0	1.59	1.20	1.0	0.95	0.73	0.59
24.0	1.93	1.29	1.0	0.95	0.73	0.57
9.4	1.98	1.30	1.0	0.95	0.70	0.56
3.0	2.00	1.25	1.0	0.95	0.67	0.59

Source: From Kerr, 1964, p. 677.

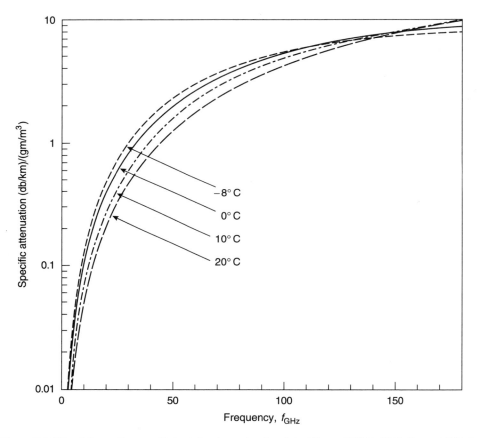

Figure 2.3-13 Attenuation coefficient due to water droplets. (From CCIR, 1982, Report 721-1; as referenced in Ippolito, 1989.)

Attenuation Due to Snow, Sleet, and Hail

As noted in Skolnik (1980), attenuation caused by dry ice particles in the atmosphere, in the form of hail, sleet, ice crystal clouds, or snow, is much less than attenuation caused by rain of the same precipitation rate, that is, for the same rate of melted water content per hour (Saxton, 1958). This conclusion is supported by an expression of Gunn and East (1954), for one-way attenuation of dry snow at 0° C and frequencies up to 20 GHz. It can be put in the form

$$\text{Attenuation (in dB/km)} = 4.568(10^{-8})r^2f_{\text{GHz}}^4 + 7.467(10^{-5})rf_{\text{GHz}} \quad (2.3\text{-}16)$$

Here r is the snowfall rate (mm/h of melted water content) and f_{GHz} is frequency in gigahertz. A plot of (2.3-16) is shown in Fig. 2.3-14 for all practical values of snowfall rate r. By comparing results of Fig. 2.3-14 for snowfall to Fig. 2.3-11 for rainfall, we find that attenuations by dry snow for $r = 1.25$, 5, and 10 mm/h are roughly the same as $r = 0.25$, 1.25, and 5 mm/h of rain, respectively, for $f_{\text{GHz}} \le 15$. Thus snowfall rates of

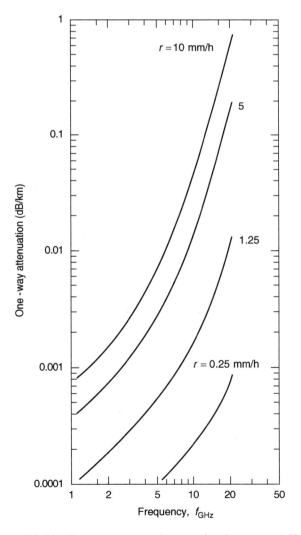

Figure 2.3-14 One-way attenuation rate for dry snow at 0° C.

about 2 to 5 times higher than rain rates are required to produce attenuations similar to rain for frequencies below 15 GHz. Above 15 GHz, snow attenuation becomes more significant compared to rain.

Example 2.3-6   Again continue Example 2.3-2 by assuming that dry snow falls at a rate of 1.25 mm/h of melted water content over the whole path between radar and target. We find the one-way attenuation caused by the snow. Since the radial path distance is 16.38 km and the attenuation rate is given by (2.3-16), we have attenuation (in dB/km) = $4.568(10^{-8})(1.25^2)(20^4) + 7.467(10^{-5})1.25(20) = 0.0114 + 0.0019 = 0.0133$, so

$$\text{Attenuation (one way)} = 0.0133(16.38) = 0.218 \text{ dB}$$

By merging results of Examples 2.3-2, 2.3-3, and 2.3-5, we see the relative significance of various precipitations over our example path:

Attenuation (one way)

Clear air	2.277 dB
Rain, 1.25 mm/h	1.474 dB
Fog, $V_m = 120$ m	1.031 dB
Snow, $r = 1.25$ mm/h	0.218 dB

In a manner similar to snow, dry hail and dry sleet cause wave attenuations that are small compared to rainfall of the same equivalent precipitation rate (mm/h of melted water content) except for frequencies in the millimeter region (Kerr, 1964, p. 687).

Attenuation through clouds of ice crystals depends on the shape of the crystals, temperature, and, of course, the density of the ice (m in grams of water content per cubic meter). For disk-shaped crystals (the worst shape) at the worst temperature ($0°$ C), the one-way attenuation is known to be $0.007 \, m/\lambda_{cm}$ (Kerr, 1964, p. 688). Since m rarely exceeds 0.5 g/m^3, the worst-case attenuation is $1.17(10^{-4})f_{GHz}$ dB/km, which is negligible even for the highest frequencies.

When ice particles (hail, sleet) fall from a freezing altitude through the $0°$ C isotherm into a melting region, they begin to melt. Larger particles can develop a water coating prior to completion of melting, and smaller particles can be completely melted. For frequencies from 10 to 30 GHz, it is stated (Skolnik, ed., 1990, p. 23.9) that attenuation increases to that which would occur for all melted particles (rain) when less than 10% of the ice particles are melted. When the melted mass was from about 10% to 20%, attenuation increased to about twice that corresponding to completely melted particles. Because of these effects attenuation due to "frozen" precipitation can be considerably larger at altitudes just below the $0°$ C isotherm than above, and perhaps even larger than in the "rain" layer below.

2.4 POLARIZATION AND REFLECTION OF WAVES

In this section we first define the polarization of an electromagnetic wave through its electric field and then use the results to help understand the reflection of waves from surfaces.

Polarization

As discussed earlier, waves in the far field of a radiator are approximately plane waves where the electric field vector lies in the plane of the wave. Since this vector can have any direction in the plane, in general, it is usualy defined by its two orthogonal components in whatever coordinate system is in use. For radar it is most convenient to use spherical coordinates located at the radiator (radar at the origin). A distant point P has radial distance R and a direction defined by the spherical angles θ and ϕ, as shown in Fig. 2.4-1. A wave passing point P will have an electric field in the plane perpendicular to the line from point 0 to point P. Its components E_θ and E_ϕ for spherical coordinates are defined in the figure. The total electric field is the vector sum of the components E_θ and E_ϕ. It has a magnitude E, and at a given instant in time, has the direction shown in Fig. 2.4-1.

As the wave passes point P the components E_θ and E_ϕ vary sinusoidally with time so that E can change both amplitude *and* direction with time. The polarization of the wave is defined according to how E varies as seen by a viewer positioned at the wave's source (radar) and who observes the behavior of E with time at point P. In general, we may write

$$E_\theta = A \cos(\omega_0 t) \tag{2.4-1}$$

$$E_\phi = B \cos(\omega_0 t + \alpha) \tag{2.4-2}$$

where ω_0 is the angular frequency of the wave, A and B are the peak amplitudes of E_θ and E_ϕ, respectively, and α is the phase of E_ϕ relative to E_θ. Figure 2.4-2 plots E_θ and E_ϕ with time. Also shown is the locus of the tip of the total field E and the actual vector E at time zero. We observe that the tip of the total electric field may trace

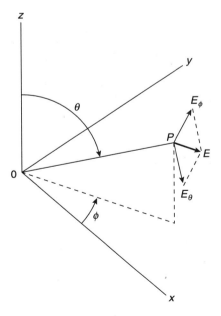

Figure 2.4-1 Typical geometry of a wave propagating from 0 to point P.

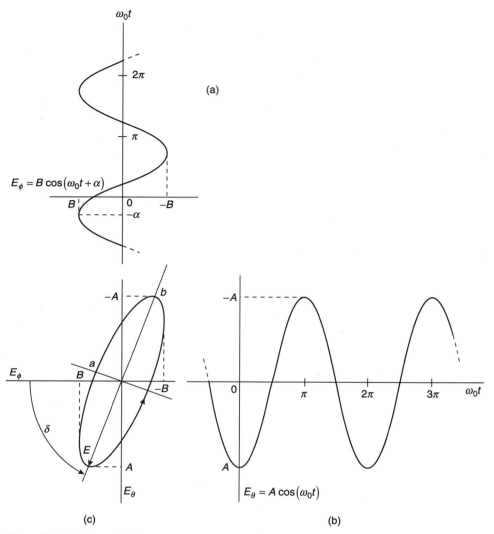

Figure 2.4-2   Plots of (*a*) field component E_ϕ, (*b*) field component E_θ, and (*c*) the total field (at $t = 0$) and its tip locus for one cycle of the wave.

out an ellipse once for each cycle of the wave's frequency. In the figure shown, the ellipse is traced in a counter clockwise direction with time.

Clearly the exact trace of the locus of E in Fig. 2.4-2 depends on A, B, and α. However, the absolute scale does not affect the *shape* of the figure or the direction of its trace. We will define the wave's polarization based only on the shape of the locus and the direction in which it is traced. For this reason only *two* parameters are needed to define polarization, the *ratio B/A* of the field component's peak amplitudes and the relative phase angle α.

A careful study of the possible trace shapes and directions of the electric field shows elliptical, circular, and linear shapes that can be traced either clockwise (CW) or counterclockwise (CCW). Table 2.4-1 is useful in seeing what combinations of B/A

TABLE 2.4-1 Values of A, B, and α that lead to various wave polarizations.

Phase Angle, α	Allowed Values of A and B	Direction of Rotation	Wave Polarization[a]
$0 < \alpha < \pi$	Any A and B with $A \neq B$	CCW	Left-hand elliptical
$0 < \alpha < \pi$ with $\alpha \neq \pi/2$	$A = B$	CCW	Left-hand elliptical
$-\pi < \alpha < 0$	Any A and B with $A \neq B$	CW	Right-hand elliptical
$-\pi < \alpha < 0$ with $\pi/2$	$A = B$	CW	Right-hand elliptical
$\pi/2$	$A = B$	CCW	Left-hand circular
$-\pi/2$	$A = B$	CW	Right-hand circular
0	Any A or B	—	Linear
$\pm\pi$	Any A or B	—	Linear
$-\pi < \alpha < 0$ or $0 < \alpha < \pi$	$A = 0$ or $B = 0$	—	Linear

[a] Defined as viewed from the source as wave propagates in the direction of point P in Fig. 2.4-1.

and α lead to the various results. From the table we see that left-hand elliptical polarization is defined by an elliptical trace in a CCW direction. Right-hand elliptical corresponds to an elliptical trace in a CW direction. Other possible polarizations are left- and right-hand circular for a circular trace, and simply linear polarization when the trace degenerates into a line.

   Another helpful tool in seeing how polarization varies is constructed in Fig. 2.4-3. Here parameters B/A and α, which define polarization, are plotted in polar form, and some examples of the polarization ellipses are sketched to show general behavior (the shapes are not to scale). Note that rotation is CCW for $0 < \alpha < \pi$ and CW for $-\pi < \alpha < 0$, regardless of trace shape. Circular polarization occurs for only two points in the parameter plane, and trace tilt angles are all large for points inside the unit circle and are all small for outside points.

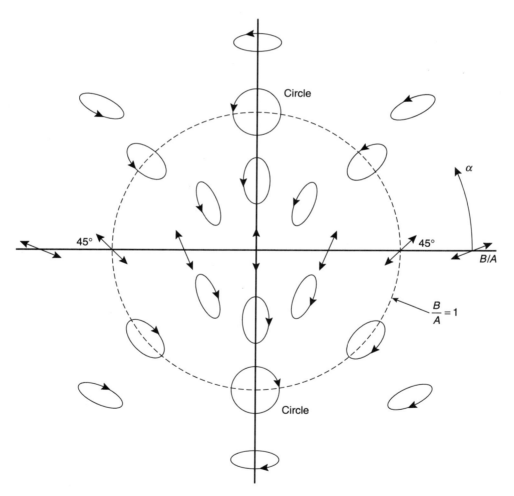

Figure 2.4-3   Polarization traces for various choices of parameters B/A and α shown in polar coordinates.

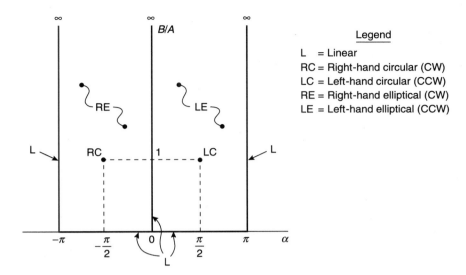

Figure 2.4-4 Wave polarizations resulting from parameter choices B/A and α shown in rectangular coordinates.

Still another representation of polarization resulting from choices of B/A and α is shown in Fig. 2.4-4.

The tilt angle of the ellipse axis as shown in Fig. 2.4-2 (relative to the axis for E_ϕ) is given by

$$\delta = \frac{1}{2} \tan^{-1}\left[\frac{2(B/A)\cos(\alpha)}{(B/A)^2 - 1}\right] \qquad (2.4\text{-}3)$$

for $B/A \geq 1$. For $B/A < 1$, δ is given by adding $\pi/2$ to the result found from direct use of (2.4-3).

Example 2.4-1   We find δ for a wave where $B/A = 0.5$ and $\alpha = 3\pi/4$ rad. From (2.4-3),

$$\delta = \frac{1}{2} \tan^{-1}\left[\frac{2(0.5)\cos(3\pi/4)}{0.25 - 1}\right] = 21.657 \text{ degrees}$$

Thus, since $B/A < 1$, $\delta = 90 + 21.657 = 111.657$ degrees, which agrees with Fig. 2.4-3.

The major and minor axis half-widths of the ellipse of Fig. 2.4-2c are, respectively, given by

$$a^2 = \frac{2(B/A)^2 \sin^2(\alpha)}{1 + (B/A)^2 + \sqrt{[1 + (B/A)^2]^2 - 4(B/A)^2 \sin^2(\alpha)}} \qquad (2.4\text{-}4)$$

$$b^2 = \frac{2(B/A)^2 \sin^2(\alpha)}{1 + (B/A)^2 - \sqrt{[1 + (B/A)^2]^2 - 4(B/A)^2 \sin^2(\alpha)}} \qquad (2.4\text{-}5)$$

The ratio of the major to minor axis length is called the *axial ratio* of the wave. Thus axial ratio is b/a using (2.4-4) and (2.4-5) if the ratio is one or more. If less than one, the axial ratio is equal to a/b.

Example 2.4-2   We find the axial ratio for the wave defined in Example 2.4-1. Here

$$a^2 = \frac{2(0.25)\sin^2(3\pi/4)}{1 + 0.25 + \sqrt{[1 + 0.25]^2 - 4(0.25)\sin^2(3\pi/4)}} = 0.1096$$

$$a = 0.3311$$

$$b^2 = \frac{2(0.25)\sin^2(3\pi/4)}{1 + 0.25 - \sqrt{[1 + 0.25]^2 - 4(0.25)\sin^2(3\pi/4)}} = 1.1404$$

$$b = 1.0679$$

so $b/a = 1.0679/0.3311 = 3.225$ is the axial ratio.

Wave Reflections from Smooth Flat Surfaces

When a radar wave strikes the surface of some object (e.g., the earth's surface), part of its energy is reflected much as a ball bounces off a road. The reflected wave is generally smaller in intensity because some of the incident wave is absorbed by the object. Objects that tend toward conductors reflect most of the incident wave, while objects that tend toward lossy dielectrics can absorb larger portions of the incident energy. The action of reflection is described by a *reflection coefficient* such that the reflected wave electric field (magnitude and phase) is the product of the reflection coefficient and the incident electric field.

When the surface is everywhere smooth and flat, each part of the incident plane wave reflects in the same direction, as shown in Fig. 2.4-5. For example, two parts marked A and B reflect, as shown, to produce parts C and D, respectively, in the reflected wave. All parts of the incident wave, after reflection, produce a plane wave. This type of reflection is called *specular*. The electric field of the reflected wave will depend on the polarization of the incident wave. It is convenient to assume the incident field has two components. One, E_{iV}, lies both in the plane normal to the surface and in the plane of the wave front (see ray 1). The other, E_{iH}, lies in the plane of the wave front and is parallel to the surface. E_{iV} and E_{iH} are commonly called the "vertical" and "horizontal" components of the incident field. An incident wave having only one of these components is said to have either *vertical* or *horizontal* *polarization*.

For the specular reflection of Fig. 2.4-5, the reflection coefficient depends on polarization. The field component relationships are (Kerr, 1964)

$$E_{rV} = \Gamma_V E_{iV} \tag{2.4-6}$$

$$E_{rH} = \Gamma_H E_{iH} \tag{2.4-7}$$

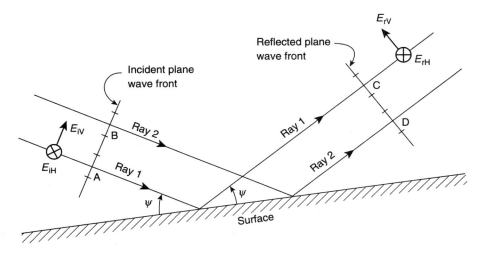

Figure 2.4-5   Plane wave reflection from a smooth plane surface.

where Γ_V and Γ_H are called *Fresnel reflection coefficients* given by

$$\Gamma_V = \frac{\varepsilon_c \sin(\psi) - \sqrt{\varepsilon_c - \cos^2(\psi)}}{\varepsilon_c \sin(\psi) + \sqrt{\varepsilon_c - \cos^2(\psi)}} = \rho_V e^{-j\phi_V} \qquad (2.4\text{-}8)$$

$$\Gamma_H = \frac{\sin(\psi) - \sqrt{\varepsilon_c - \cos^2(\psi)}}{\sin(\psi) + \sqrt{\varepsilon_c - \cos^2(\psi)}} = \rho_H e^{-j\phi_H} \qquad (2.4\text{-}9)$$

The parameters in (2.4-8) and (2.4-9) are

$$\varepsilon_c = \varepsilon_r - j\varepsilon_i = \text{complex dielectric constant of surface material} \qquad (2.4\text{-}10)$$

$$\varepsilon_r = \varepsilon/\varepsilon_0 = \text{dielectric constant of surface material} \qquad (2.4\text{-}11)$$

$$\varepsilon = \text{real part of dielectric permittivity of surface material} \qquad (2.4\text{-}12)$$

$$\varepsilon_0 = 10^{-9}/(36\pi) \text{ F/m} = \text{dielectric permittivity of free space (vacuum)} \qquad (2.4\text{-}13)$$

$$\varepsilon_i = \sigma/(\omega\varepsilon_0) \approx 60\lambda\sigma \approx 18\sigma/f_{GHz} \qquad (2.4\text{-}14)$$

$$\sigma = \text{conductivity of surface material (S/m)} \qquad (2.4\text{-}15)$$

Of course λ is wavelength in meters and ψ is called the *grazing angle* (Fig. 2.4-5) between the incident wave's ray and the surface (ψ is the same for the reflected wave). We see that the reflection coefficients depend on the wave's frequency and grazing angle, as well as the complex dielectric constant of the surface material. Since Γ_V and Γ_H are complex, in general, it is convenient to define their magnitudes by ρ_V and ρ_H and their phase angles by $-\phi_V$ and $-\phi_H$, respectively, in (2.4-8) and (2.4-9). In calculating the radicals required in Γ_V and Γ_H, principal roots must be taken because the radicals must have negative imaginary parts (Kerr, 1964, p. 399).

To illustrate the behavior of Γ_V and Γ_H, it is necessary to determine ε_c for the surface. We will use some results given by Saxton and Lane (1952) for seawater at frequencies up to 100 GHz. Their results, said to agree with experimental data to about 2%, are

$$\varepsilon_r = \frac{\varepsilon_s - \varepsilon_0}{1 + (2\pi f \tau)^2} + \varepsilon_0 \qquad (2.4\text{-}16)$$

$$\varepsilon_i = \frac{(\varepsilon_s - \varepsilon_0)2\pi f \tau}{1 + (2\pi f \tau)^2} + \frac{2\sigma_i}{f} = (\varepsilon_r - \varepsilon_0)2\pi f \tau + \frac{2\sigma_i}{f} \qquad (2.4\text{-}17)$$

Here f is frequency in hertz, ε_0 is a dielectric constant given by Saxton and Lane as 4.9 (the value for pure water without sodium chloride), ε_s is called the *static dielectric constant*, τ is a relaxation time, and σ_i is called the *ionic conductivity* (unit is the e.s.u.). The required constants ε_s, τ, and σ_i are temperature dependent. For the case of seawater, values of Table 2.4-2 were found by interpolating data of Saxton and Lane.

By using the values of ε_s, τ, and σ_i from Table 2.4-2 for a temperature of 20° C, calculations of Γ_V and Γ_H were made and are shown in Figs. 2.4-6 and 2.4-7. For comparative purposes, Γ_V and Γ_H were also found for pure water [no sodium chloride and $\sigma_i = 0$ in (2.4-17)] using the constants in Table 2.4-3. Plots are shown in Figs. 2.4-8 and 2.4-9. We note that vertical polarization produces larger variations in both magnitude and phase of the reflection coefficient with grazing angle compared to horizontal polarization (for given frequency and surface conditions, of course). In particular, these variations occur most rapidly for grazing angles near the angle of minimum magnitude (called the *Brewster angle*). The Brewster angle, denoted by ψ_b, is given by (Barton, 1988, p. 293)

$$\psi_b = \sin^{-1}\left(\frac{1}{\sqrt{\varepsilon_r + 1}}\right) \qquad (2.4\text{-}18)$$

Example 2.4-3   To demonstrate the direct solution of (2.4-8), we find Γ_V for a 10-GHz wave at a grazing angle of 3 degrees reflecting from a smooth freshwater surface

TABLE 2.4-2   Constants ε_0, τ, and σ_i for typical seawater at various temperatures.

Temperature (° C)	ε_s (unitless)	τ (s)	σ_i (e.s.u.)
0	75.08	$16.93(10^{-12})$	$2.70(10^{10})$
10	72.08	$12.10(10^{-12})$	$3.70(10^{10})$
20	69.08	$9.15(10^{-12})$	$4.70(10^{10})$
30	66.08	$7.18(10^{-12})$	$5.56(10^{10})$
40	63.32	$5.68(10^{-12})$	$6.58(10^{10})$

Note: Values shown are for seawater with a 3.6% solution of sodium chloride.
Source: Data from Saxton and Lane (1952.)

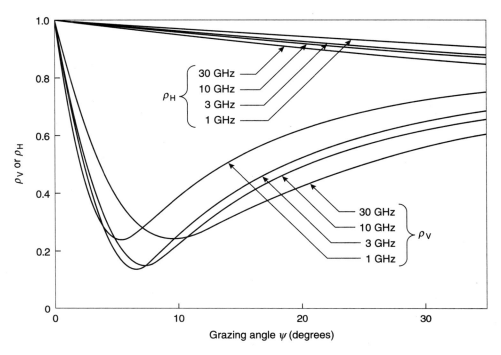

Figure 2.4-6 Magnitudes of reflection coefficients, ρ_V and ρ_H, for vertical and horizontal polarizations assuming seawater of 3.6% sodium chloride solution.

for which $\varepsilon_r = 65$ and $\sigma = 15$ S/m are assumed. We also find the Brewster angle ψ_b. Here $\varepsilon_c = 65 - j\ [18(15)/10] = 65 - j27$ from (2.4-14), $\sin(\psi) = 0.05234$, and $\cos^2(\psi) = 0.9973$, so

$$\Gamma_V = \frac{(65 - j27)0.05234 - \sqrt{65 - j27 - 0.9973}}{(65 - j27)0.05234 + \sqrt{65 - j27 - 0.9973}}$$

$$= \frac{3.4021 - j1.4132 - (8.2393 - j1.6990)}{3.4021 - j1.4132 + (8.2393 - j1.6990)}$$

$$= \frac{-4.8372 + j0.2858}{11.6414 - j3.1122} = 0.402e^{-j2.9394}$$

Hence $\rho_V = 0.402$ and $\phi_V = 2.9394$ rad (or 168.414 degrees). From (2.4-18), $\psi_b = 0.123$ rad (or 7.071 degrees).

Reflection coefficients Γ_V and Γ_H can be simplified in some special cases (see Problems 2.4-13 through 2.4-15).

Reflections from Irregular and Spherical Surfaces

When a surface is smooth but not flat, as shown in Fig. 2.4-10, each part of the incident wave corresponds to reflections in different directions, as illustrated by pieces

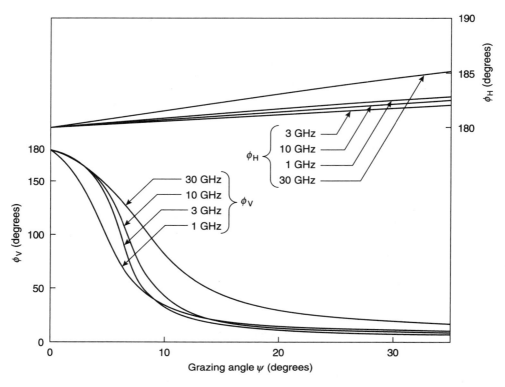

Figure 2.4-7 Phase angles of reflection coefficients, ϕ_V and ϕ_H, for vertical and horizontal polarizations assuming seawater of 3.6% sodium chloride solution.

TABLE 2.4-3   Constants ε_s and τ for pure water at various temperatures.

Temperature ($^\circ$C)	ε_s (unitless)	τ (s)
0	88	$18.7(10^{-12})$
10	84	$13.6(10^{-12})$
20	80	$10.1(10^{-12})$
30	77	$7.5(10^{-12})$
40	73	$5.9(10^{-12})$

Source: Data are from Saxton and Lane (1952).

A and B. We may refer to this type of surface as *irregular*. For irregular reflections the total electric field at some point after reflection becomes the sum of all reflections of all the portions of the incident wave.

In general, irregular reflections are difficult to analyze. However, for one case, that of a smooth spherical earth (as would apply to a smooth sea), the reflected wave

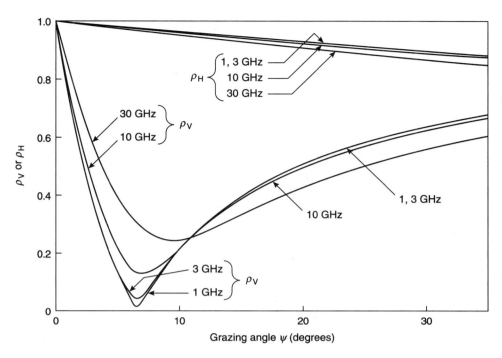

Figure 2.4-8 Magnitude of reflection coefficients, ρ_V and ρ_H, for vertical and horizontal polarizations assuming pure water.

portions diverge such that the total reflected field is less than that of the incident field. The decrease in field strength can be accounted for by a factor called the *divergence factor*, denoted by D. For the geometry and parameters defined in Fig. 2.4-11, where grazing angle ψ is assumed small such that $\sin(2\psi) \approx 2\psi$, it can be shown that

$$\psi \approx \frac{1}{r_1}\left[h_1 - \frac{r_1^2}{2a_e}\right] \tag{2.4-19}$$

This approximation for ψ can be used in an expression for D given by Kerr (1964, p. 99) to obtain

$$D \approx \left[1 + \frac{2r_1^2(r - r_1)}{a_e r(h_1 - (r_1^2/2a_e))}\right]^{-1/2} \tag{2.4-20}$$

which is valid for small grazing angles ψ.

The typical way of using (2.4-20) is to presume that h_1, h_2, and r are specified, where $h_1 \geq h_2$ is assumed. If $h_1 < h_2$, it is only necessary to relabel the target and radar in reverse manner to that of Fig. 2.4-11. With h_1, h_2, and r given, it remains to find r_1 as needed in (2.4-20).

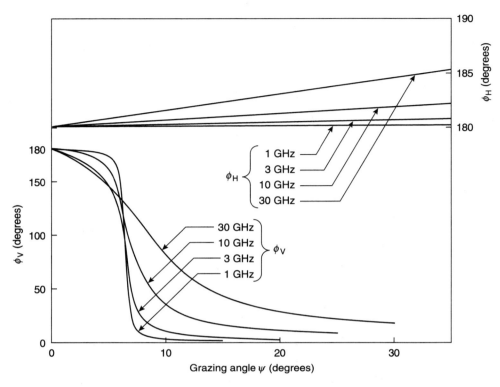

Figure 2.4-9   Phase angles of reflection coefficients, ϕ_V and ϕ_H, for vertical and horizontal polarizations assuming pure water.

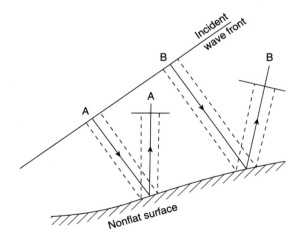

Figure 2.4-10   Irregular surface from which a plane wave reflects.

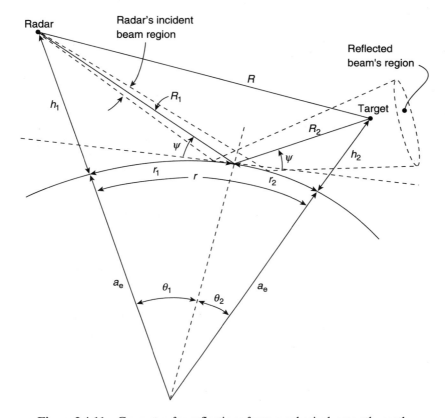

Figure 2.4-11 Geometry for reflections from a spherical, smooth, earth.

It is found by calculating, in order, the following three expressions:

$$p = \frac{2}{\sqrt{3}}\sqrt{a_e(h_1 + h_2) + \left(\frac{r}{2}\right)^2} \qquad (2.4\text{-}21)$$

$$\xi = \sin^{-1}\left[\frac{2a_e(h_2 - h_1)r}{p^3}\right] \qquad (2.4\text{-}22)$$

$$r_1 = \frac{r}{2} - p\sin\left(\frac{\xi}{3}\right) \qquad (2.4\text{-}23)$$

which are given in Skolnik (ed., 1990, p. 2.42). For all these results to be true, it is necessary that r not exceed the distance to the radio horizon of the radar, which will be true provided that

$$r \leq \sqrt{2a_e}(\sqrt{h_1} + \sqrt{h_2}) \qquad (2.4\text{-}24)$$

from use of (2.3-10). Once D is known, the reflected electric field components are given by (2.4-6) and (2.4-7) if the factor D is added to the right side of these equations. We consider an example.

Example 2.4-4   We assume a shipboard radar's height is 30 m above a smooth sea. It observes a target with a height of 1.6 km and a surface distance of 30 km from the radar. We find D. Since the target's altitude is larger than the radar's height, we define $h_1 = 1.6$ km, $h_2 = 0.03$ km, and $r = 30$ km. Initially we must show that $r = 30$ km satisfies (2.4-24) to assure the target appears to the radar above the radio horizon. We keep all distances in kilometers and find $r \le [2(4/3)6371]^{0.5}(0.03^{0.5} + 1.6^{0.5})$ $= 187.45$ km. Thus $r = 30$ km ≤ 187.45 km, so we next find r_1 from (2.4-21) through (2.4-23):

$$p = \frac{2}{\sqrt{3}}\sqrt{\frac{4}{3}(6371)(1.6 + 0.03) + \left(\frac{30}{2}\right)^2} = 136.9735 \text{ km}$$

$$\xi = \sin^{-1}\left[\frac{2(4/3)6371(0.03 - 1.6)30}{(136.9735)^3}\right] = -0.3166 \text{ rad}$$

$$r_1 = \frac{30}{2} - 136.9735 \sin\left(\frac{-0.3166}{3}\right) = 29.4285 \text{ km}$$

where a 4/3-earth model has been assumed, so $a_e = (4/3)a = (4/3)$ 6371 km. Finally

$$D \approx \cfrac{1}{\sqrt{1 + \cfrac{2(29.4285)^2(30 - 29.4285)}{\frac{4}{3}(6371)30\left[1.6 - \cfrac{(29.4285)^2}{2(4/3)6371}\right]}}} = 0.9987$$

It is of interest to also compute the grazing angle from (2.4-19):

$$\psi \approx \frac{1}{29.4285}\left[1.6 - \frac{(29.4285)^2}{2(4/3)6371}\right] = 0.05264 \text{ rad (or 3.02 degrees)}$$

Reflections from Rough Surfaces

Surfaces with irregularities that fluctuate about an "average" flat value can be called *rough*. Rough surfaces can still reflect waves in a "specular" manner. That is, there is still a significant, although reduced, field reflected at an angle ψ for an incident plane wave at a grazing angle ψ. Models for the rough surface that assume a gaussian variation of surface height about the mean, "flat" surface have yielded a *roughness* (loss) *coefficient* for wave electric field magnitude of

$$\rho_s = \exp\left[-8\left(\frac{\pi h_{rms}}{\lambda}\right)^2 \sin^2(\psi)\right] \tag{2.4-25}$$

where h_{rms} is the standard deviation of the height variations about the average flat surface, ψ is the incident wave's grazing angle, and λ is wavelength (Ament, 1953;

Beckmann and Spizzichino, 1963). This result seems to approximate measured data for various land and rough seas when

$$\frac{h_{rms}}{\lambda} \sin(\psi) < 0.1 \qquad (2.4\text{-}26)$$

For values larger than 0.1, experimental values of ρ_s are somewhat larger than predicted by (2.4-25).

In addition to the specular component of reflected wave, a rough surface produces a *diffuse* component of reflection. Here the rough surface scatters energy in all directions in a diffuse manner, such that the field strength tends to vary randomly in amplitude and phase as a function of the principal surface areas that produce reflections (these areas are those illuminated by the wave from the radar's beam). For details on diffuse scattering the reader is referred to the literature (Skolnik, 1990; Beckmann and Spizzichino, 1963).

Composite Reflections

By combining the results of the preceding four subsections, we may determine the magnitudes and phases of the reflected fields when reflections occur from a rough, spherical surface, as follows:

$$E_{rV} = \Gamma_V D \rho_s E_{iV} = \rho_V D \rho_s \exp[-j\phi_V] E_{iV} \qquad (2.4\text{-}27)$$

$$E_{rH} = \Gamma_H D \rho_s E_{iH} = \rho_H D \rho_s \exp[-j\phi_H] E_{iH} \qquad (2.4\text{-}28)$$

These two results are the extensions of (2.4-6) and (2.4-7) to account for divergence and roughness.

2.5 WAVES AND RADAR ANTENNAS

Although we will look carefully at antennas in the next chapter, it is helpful here to take a simple look at antennas to get an overall picture of how waves are generated and received by a radar.

Radar Antenna

Consider Fig. 2.5-1a where a transmitter (TX) provides a signal with peak power P_t to a transmitting antenna. This antenna transmits the power as an electromagnetic wave propagating in all directions in the space surrounding the antenna. If the wave were transmitted with the same intensity in every direction, the power density (W/m^2) at the target at distance R_1 would be $P_t/(4\pi R_1^2)$, which is the total power distributed uniformly over the surface of a sphere of radius R_1. However, antennas do not transmit uniformly; they are designed to transmit strongly in some directions and less intensely in others. The variation in power with direction is accounted for by a factor $G_D(\theta, \phi)$ called the *directive gain*, where θ and ϕ are the angles in spherical coordinates that define the direction of interest. Thus, at the target in Fig. 2.5-1a, the power density of the arriving wave is $P_t G_{Dt}/(4\pi R_1^2)$, where G_{Dt} denotes $G_D(\theta, \phi)$ in the direction of the target.

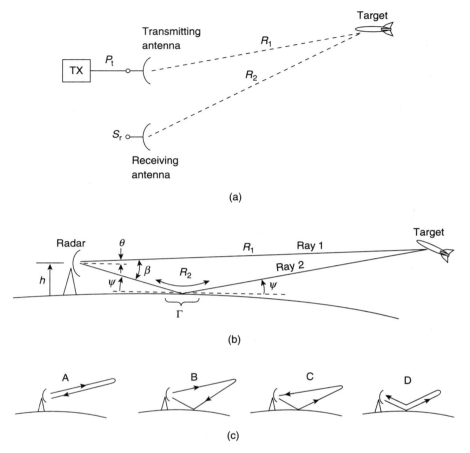

Figure 2.5-1 Geometries involving radar antennas. (*a*) Geometries helpful in determining received signal voltage. (*b*) Geometry for the multipath effect, and (*c*) paths involved in multipath.

The power, S_r, available at the output of the receiving antenna in Fig. 2.5-1*a* is equal to the product of an effective capture area, denoted by A_e, of the antenna and the power density (W/m^2) of the wave at the antenna that is reflected from the target. It is known that A_e is related to the directive gain by (see Chapter 3)

$$A_e(\theta, \phi) = \frac{\lambda^2}{4\pi} G_D(\theta, \phi) \tag{2.5-1}$$

The power density is the effective power scattered toward the receiving antenna by the target divided by the surface area of a sphere of radius R_2. Thus

$$S_r = \frac{P_t G_{Dt}}{4\pi R_1^2} \cdot \frac{\sigma}{4\pi R_2^2} \cdot \frac{\lambda^2 G_{Dr}}{4\pi} \tag{2.5-2}$$

where σ is a constant (unit is area), called the *radar cross section*, associated with the target, and G_{Dr} equals $G_D(\theta, \phi)$ of the receiving antenna in the direction of the target.

Although more precise names are given in Chapter 3 for the parameters of antennas, we can here refer to G_{Dt} or G_{Dr} simply as *power patterns* of the transmitting and receiving antennas, respectively, evaluated in the direction of the target. Furthermore, because voltage is proportional to the square root of power, we will define $\sqrt{G_{Dt}}$ and $\sqrt{G_{Dr}}$ as *voltage patterns* of the antenna evaluated in the directions of the target. From (2.5-2), the voltage, e_r, at the receiving antenna's output can be written as

$$e_r = \frac{K_1 \lambda}{4\pi} \sqrt{\frac{P_t \sigma}{4\pi}} \frac{\sqrt{G_{Dt}}}{R_1} \frac{\sqrt{G_{Dr}}}{R_2} e^{j\omega t} = K_2 \sqrt{\sigma} \frac{\sqrt{G_{Dt}}}{R_1} \frac{\sqrt{G_{Dr}}}{R_2} e^{j\omega t} \qquad (2.5\text{-}3)$$

where K_1 and K_2 are constants of proportionality. This result is very useful in understanding an important effect in radar called *multipath*.

Multipath

Consider a monostatic radar above a surface as shown in Fig. 2.5-1*b*. The transmitted wave leaves in all directions; two directions lead to the target. The first (ray 1) can lead to direct reflection back to the radar over ray 1 as well as indirect reflection back to the radar over ray 2. These two wave paths are shown as A and B, respectively, in panel *c*. The second direction over which a wave arrives at the target is ray 2. Here again there are two reflected waves arriving at the radar (paths C and D in panel *c*). Thus there are four components that produce the total received voltage e_r of the radar.

By carefully retracing the steps leading to (2.5-3), it is clear that (2.5-3) can represent each of the four voltage components of e_r if proper account is made for relative phases and reflections from the surface. Assume that σ is the same for all components and take path A as the reference for phase. Components become

$$e_A = K_2 \sqrt{\sigma} \left(\frac{\sqrt{G_{Dt}}}{R_1} \right)^2 e^{j\omega t} \qquad (2.5\text{-}4)$$

$$e_B = K_2 \sqrt{\sigma} \frac{\sqrt{G_{Dt}}}{R_1} \frac{\sqrt{G_{Dr}}}{R_2} \Gamma e^{-j2\pi \Delta R/\lambda + j\omega t} \qquad (2.5\text{-}5)$$

$$e_C = K_2 \sqrt{\sigma} \frac{\sqrt{G_{Dr}}}{R_2} \frac{\sqrt{G_{Dt}}}{R_1} \Gamma e^{-j2\pi \Delta R/\lambda + j\omega t} \qquad (2.5\text{-}6)$$

$$e_D = K_2 \sqrt{\sigma} \left(\frac{\sqrt{G_{Dr}}}{R_2} \right)^2 \Gamma^2 e^{-j4\pi \Delta R/\lambda + j\omega t} \qquad (2.5\text{-}7)$$

where

$$\Delta R = R_2 - R_1 \qquad (2.5\text{-}8)$$

and Γ is the reflection coefficient of the surface (product of the Fresnel reflection coefficient, the divergence factor and the roughness coefficient). And, as usual, λ is wavelength. The total received voltage becomes

$$e_r = e_A + e_B + e_C + e_D$$

$$= e_A \left[1 + \Gamma \frac{R_1}{R_2} \sqrt{\frac{G_{Dr}}{G_{Dt}}} \, e^{-j2\pi\Delta R/\lambda} \right]^2 \tag{2.5-9}$$

Since e_A represents the "desired" received signal that would occur if the reflecting surface were not present, the second term in the bracketed factor in (2.5-9) represents the effect of the three "undesired" signals due to reflections. The reception of waves from a target over two or more paths in radar is called *multipath*.

To see some of the principal effects of the multipath in Fig. 2.5-1b, we note that for most cases[5] $R_1 \approx R_2$. If we also assume β is not too large, then $G_{Dr} \approx G_{Dt}$. Equation (2.5-9) becomes

$$e_r \approx e_A [1 + \Gamma e^{-j2\pi\Delta R/\lambda}]^2 \tag{2.5-10}$$

Finally, to obtain some numerical results, we assume that the surface is level and flat; for this surface and a distant target it can be shown that

$$\Delta R \approx 2h \sin(\theta) \tag{2.5-11}$$

so

$$e_r = e_A [1 + \Gamma e^{-j(4\pi h/\lambda)\sin(\theta)}]^2 \tag{2.5-12}$$

By representing Γ in its polar from $\rho \exp(-j\phi)$ and reducing (2.5-12), we obtain the magnitude of e_r/e_A:

$$\left| \frac{e_r}{e_A} \right| = 1 + \rho^2 + 2\rho \cos\left[\phi + \frac{4\pi h}{\lambda} \sin(\theta) \right] \tag{2.5-13}$$

Figure 2.5-2 shows plots of (2.5-13) for vertical polarization (solid curves) and horizontal polarization (dashed curves) at 1 GHz over typical seawater. Note carefully the expanded angle scale. Since $\Gamma \approx -1$ for horizontal polarization (Figs. 2.4-6 and 2.4-7), $\rho \approx 1$ and $\phi \approx \pi$ in (2.5-13). Thus $|e_r/e_A|$ cycles between maximum and minimum values of $(1 + \rho)^2 \approx 4$ and $(1 - \rho)^2 \approx 0$, respectively (the nulls in Fig. 2.5-2 are not shown in order to make the curves less confusing). These "nulls" are all in the range of 0.02 or smaller for angles up to 60 degrees. For vertical polarization Γ varies greatly (see Figs. 2.4-6 and 2.4-7) causing the "lobes" in Fig. 2.5-2 to have smaller maximum values and larger null values (as compared to horizontal polarization), both due to the smallness of ρ. Remember that if $\rho = 0$, the plot would be 1 at all angles, the ideal case. The fact that ϕ varies significantly (as compared to ϕ for horizontal polarization) changes the lobe positions in Fig. 2.5-2.

---

[5] Multipath is most troublesome when ψ is small. In this case the distance R_2 to the target over ray path 2 is nearly the same as distance R_1 over ray path 1.

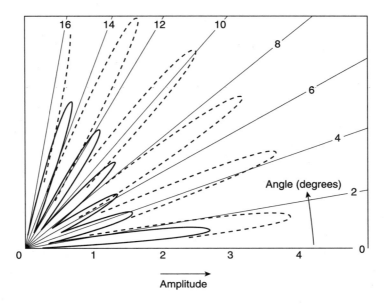

Figure 2.5-2   Plot of (2.5-13) for vertical (solid curves) and horizontal (dashed) polarizations at a frequency of 1 GHz over typical seawater when $h/\lambda = 10$.

Figure 2.5-2 applies to a flat smooth surface. If the surface roughness coefficient is added to the calculation of Γ, its principal effects are to reduce the maxima and fill in the nulls of the lobes caused by multipath. Clearly the lobing behavior of (2.5-13) means that multipath can cause the received signal to be enhanced in amplitude, as at a lobe maximum, or degraded, as at a null, depending on the target's elevation angle.

The magnitude of the bracketed factor in (2.5-12) is sometimes called the *propagation factor* (Skolnik, 1990, p. 443).

PROBLEMS

2.1-1 A wave radiates power uniformly in all directions from a small radiator isolated in free space. The power density of the wave at a distance of 5 km from the radiator is known to be $10^{-5}/\pi$ watts per square meter (W/m^2). What is the wave's power density at a distance of 20 km?

2.1-2 What total power is radiated by the source in Problem 2.1-1?

2.1-3 A small isolated source radiates 25 kW of average power uniformly into all directions in its surrounding space. What power density (in W/m^2) does the wave have at distances of 2, 8, and 32 km from the source?

2.1-4 The largest dimension of a radiating antenna is 2.6 m. If it operates at $f = 1.6$ GHz, what distance corresponds to the start of the far field?

2.1-5 For the antenna of Problem 2.1-4, what frequency will correspond to a far field distance of 50 m? What is the corresponding wavelength in free space?

2.1-6 An antenna's largest dimension, expressed in wavelengths, is $D/\lambda = 18$. Find an expression for the distance to the start of the far field as a function of frequency. Evaluate your result for $f = 2$ and 6 GHz.

2.1-7 Example 2.1-1 evaluated the departure d of a wave from planar when the wave enters the far field. Rework the example to find d if the far field is defined by D^2/λ and not $2D^2/\lambda$.

2.1-8 Show that (2.1-3) is true.

2.1-9 Evaluate (2.1-4) and find η for free space.

2.2-1 A radar is used to measure the position of a spacecraft traveling to the moon. The system includes a communication link at 30 MHz for voice contact. Explain why this frequency might be a bad choice. Would a higher or lower frequency be better?

2.3-1 An average atmosphere using the exponential model is given by the CCIR (1986) as

$$n = 1 + 315(10^{-6})e^{-0.136h}$$

where h is in kilometers. Find the maximum amount n changes as altitude increases.

2.3-2 If both the standard and average exponential model (Problem 2.3-1) give the same value of n at the surface, what must n_0 be in the standard model?

2.3-3 Suppose dn/dh for the standard model of the atmosphere is given by (2.3-3), but k is unknown. That is

$$\frac{dn}{dh} = \frac{1 - k}{ka}$$

Find k such that the standard and exponential atmospheric models (Problem 2.3-1) give the same values of n at the surface ($h = 0$) and at $h = 1$ km.

2.3-4 Equation (2.3-10) gives the distance to the radio horizon for a 4/3-earth model. If a is taken as 6371 km, show that equivalent expressions are

d_r (in statute miles) $\approx (2h)^{0.5}$, h in feet
d_r (in nautical miles) $\approx 1.23(h)^{0.5}$, h in feet
d_r (in kilometers) $= 130.3(h)^{0.5}$, h in kilometers.

2.3-5 Assume that a radar's altitude is 1.6 km, a target's altitude is 3.4 km, and the 4/3-earth model is valid. Find the ground distance between target and radar when the target just appears on the radio horizon.

2.3-6 A target flies low (altitude of 25 m) to try and avoid detection by a radar on an aircraft at a 6-km altitude. If the 4/3-earth model is valid, what is the surface range to the target when it comes over the radio horizon?

2.3-7 In the geometry of Fig. 2.3-2b for finding the radio horizon distance d_r, assume that the 4/3-earth model remains true for all θ of interest and find the largest values of d_r, h, and R such that the error in the approximation (2.3-10) for d_r is within 0.1% of the true value of d_r.

2.3-8 Use Fig. 2.3-4 to find the attenuation of clear air for a radar-to-target distance of 10 km, all in the lower atmosphere. Assume a frequency of 18 GHz.

2.3-9 Work Problem 2.3-8 for an 8-km path at 50 GHz.

2.3-10 A surface-located radar at 300 MHz must operate with targets at very long ranges and elevation angles of 2 degrees or more. What maximum clear-air attenuation should be allowed for in design?

2.3-11 Work Problem 2.3-10 except for 1000 MHz and any elevation angle (0 degrees or more).

2.3-12 An earth-based radar acquires targets at any elevation angle and operates at 10 GHz. If the system design allows a one-way clear-air attenuation of 1.0 dB, what ranges of targets will not exceed this allowance for elevation angles of 0, 0.5, 1, 2, 5, and 10 degrees?

2.3-13 A radar has a clear-air design allowance for one-way attenuation of 1.5 dB when operating at 17 GHz and targets outside the troposphere. To not exceed the allowance, what minimum elevation angle must a target have?

2.3-14 Use Fig. 2.3-11 and find the one-way rain attenuation rates (dB/km) for rainfall rates of 0.25, 1.25, and 5 mm/h when frequency is constant at 35 GHz.

2.3-15 For several frequencies up to 300 GHz use the a and b coefficients of Table 2.3-1 to calculate rain attenuation rate by use of (2.3-11) for both vertical and horizontal polarization. Compare results for $r = 0.25$, 5, and 100 mm/h to results taken from Fig. 2.3-11.

2.3-16 Rain at a rate of 25 mm/h can be considered excessive. Show that for frequencies not over 10 GHz, attenuation rate (dB/km) in a cloud with water content 4 g/m^3 (about the highest possible) is small compared to the excessive rain.

2.3-17 Both rain and cloud attenuation rates increase with frequency. At what value of m will the attenuation rate of clouds equal that of rain falling at 5 mm/h when $f_{GHz} = 60$?

2.3-18 A shipboard radar must operate at 35 GHz in fog at 18° C with visibility down to 100 m. If the fog extent is 3 km, what one-way wave attenuation occurs?

2.3-19 Suppose visibility, V_m in meters, for fog is classified as

Dense fog	$V_m < 50$ m
Thick fog	50 m $\leq V_m < 200$ m
Moderate fog	200 m $\leq V_m < 500$ m

Find the attenuation rate for the boundary values of these classifications if frequency is 35 GHz and the fog is of the coastal variety.

2.3-20 Dry snow falls over a 3.5 km extent at a rate of 10 mm/h of melted water content, which is considered to be an upper limit for snowfall rate (Gunn and East, 1954, p. 539). Find the one-way attenuation for a wave at 20 GHz. Compare the result to that of rain falling at the same rate as calculated from (2.3-11) using coefficients for vertical polarization from Table 2.3-1.

2.4-1 In (2.4-1) and (2.4-2) which determine the polarization of a transmitted wave, let $B/A = 1.5$ and $\alpha = \pi/3$ rad. Determine the polarization, tilt angle δ, and the axial ratio of the wave.

2.4-2 Work Problem 2.4-1 except assume that $B/A = 0.5$ with all else unchanged.

2.4-3 Work Problem 2.4-1 except assume that $\alpha = -2\pi/3$ rad with all else unchanged.

2.4-4 Work Problem 2.4-1 except assume that $B/A = 0.5$ and $\alpha = -3\pi/4$ rad.

2.4-5 Work Problem 2.4-1 except assume that $B/A = 1.0$ and $\alpha = \pi/2$ rad.

2.4-6 A vertically polarized wave at 3 GHz is incident on a smooth flat surface for which $\varepsilon_r = 69$ and $\sigma = 6.5$ S/m. Find Γ_V for a grazing angle of 4 degrees.

2.4-7 Work Problem 2.4-6 except assume that $f_{GHz} = 9.375$ and a dry desert soil for which $\varepsilon_r = 3.2$ and $\sigma = 0.1$ S/m.

2.4-8 Determine ε_c for seawater at $0°$ C and 6.4 GHz.

2.4-9 Work Problem 2.4-8 except assume that temperature is $20°$ C.

2.4-10 Work Problem 2.4-8 except assume that temperature is $40°$ C.

2.4-11 Plot ε_r and ε_i versus 1 GHz $\leq f \leq 100$ GHz for pure water at $0°$ C.

2.4-12 Work Problem 2.4-11 except for a temperature of $20°$ C.

2.4-13 If $|\varepsilon_c| \gg 1$ show that

$$\Gamma_V \approx \frac{-1 + \sqrt{\varepsilon_c}\,\sin(\psi)}{1 + \sqrt{\varepsilon_c}\,\sin(\psi)}$$

$$\Gamma_H \approx \frac{-\sqrt{\varepsilon_c} + \sin(\psi)}{\sqrt{\varepsilon_c} + \sin(\psi)}$$

2.4-14 If grazing angle is very small, find asymptotic forms for Γ_V and Γ_H that are valid for any value of ε_c.

2.4-15 If $|\varepsilon_c| \gg 1$, as in Problem 2.4-13, and $\psi \to \pi/2$, show that $\Gamma_V = -\Gamma_H$ and find Γ_V.

2.4-16 A radar sits on a bluff overlooking the sea. The radar is 55 m above the smooth ocean's surface. A target at altitude 1.4 km is being tracked by the radar. If the grazing angle of the reflected wave due to multipath is 2 degrees, find r_1, the distance to the reflection point, the distance r to the target over the surface, and the divergence factor D. Assume a 4/3-earth model.

2.4-17 A radar on a ship has a height of 25 m above a smooth spherical ocean. A target with altitude 580 m has a surface range of 80 km from the radar. (a) Is the target in the interference zone of the radar? If so, then calculate (b) the radar-to-multipath reflection point distance, (c) the grazing angle, and (d) the divergence factor.

2.4-18 In the geometry of Fig. 2.4-11 show that the following relationships are true.

$$R_1 = \left[h_i^2 + 4(a_e + h_1)\sin^2\left(\frac{r_1}{2a_e} \right) \right]^{1/2}$$

$$R_1 = \left[h_i^2 + 4(a_e + h_1)\sin^2\left(\frac{r_2}{2a_e} \right) \right]^{1/2}$$

$$R = \left[(h_2 - h_1)^2 + 4(a_e + h_1)(a_e + h_2)\sin^2\left(\frac{r_1 + r_2}{2a_e} \right) \right]^{1/2}$$

(*Hint*: Apply the law of cosines to appropriate triangles.)

2.4-19 Use (2.4-19), which assumes that ψ is small, and show that

$$r_i \approx [(a_e\psi)^2 + 2a_e h_i]^{1/2} - a_e\psi$$

for $i = 1, 2$.

2.4-20 A radar is located 10 m above a smooth spherical earth, and its target's altitude is 400 m. If the radar's performance becomes acceptable when the grazing angle is 3 degrees or more, what values of r_1, r_2, r, and radial range R correspond to the onset of acceptable performance?

2.4-21 The rms roughness of a surface is given to be $h_{rms} = 3$ m. If a 1.5-GHz wave is incident on the surface, what grazing angle will give a roughness coefficient of 0.4?

2.4-22 A radar system's design budget requires the roughness coefficient to not fall below 0.5. If it is known that targets will never be seen at elevation angles below 1 degree, and a flat, rough surface model applies with $h_{rms} = 0.44$ m, what frequencies are allowed in the system?

2.5-1 An isolated monostatic radar has an antenna with a directive gain that is a function of elevation angle θ only according to $G_D(\theta) = 1650\cos^2(18\theta)$. If the radar transmits a pulse with peak power 100 kW, a target has a cross section of 0.5 m^2 and is located at 30-mi range, and $\theta = 2.5$ degrees, find

(a) the power density of the wave at the target and (b) the effective power scattered toward the receiving antenna.

2.5-2 For the radar and target of Problem 2.5-1, what is the power density at the radar of the wave reflected from the target?

2.5-3 For the radar and target of Problem 2.5-1, what is the signal power available at the receiving antenna's output? Assume that the system operates at 3.6 GHz.

2.5-4 An isolated monostatic radar transmits $P_t = 10^6$ W, and operates with a target of cross section 4.6 m^2 at a range of 40 km. The target is at a direction in space such that it is at the maximum of the radar's antenna pattern where the directive gain is 2950. Find (a) the power density of the wave at the target and (b) the effective power scattered toward the radar by the target.

2.5-5 For the radar and target of Problem 2.5-4, what is the power density at the radar of the wave reflected from the target?

2.5-6 For the radar and target of Problem 2.5-4, what signal power is available at the receiving antenna's output if the system's frequency is 6.7 GHz?

2.5-7 A radar is placed on the edge of a vertical cliff at a distance of 70.2 m above a smooth, flat ocean. A target exists over the ocean at a radial range of 40.23 km and elevation angle from horizontal of 2 degrees. Determine the exact geometry of multipath reflection, and comment on whether multipath is likely to be a problem if the radar's antenna pattern gives significant directive gain for elevation angles -2.5 degrees $\leq \theta \leq 2.5$ degrees, and insignificant response for all other θ.

2.5-8 Work Problem 2.5-7, except assume that the angles over which the antenna's pattern has significant response is shifted to -1.5 degrees $\leq \theta \leq 3.5$ degrees. Note that this change is equivalent to elevating the center of the pattern's response region by 1 degree.

2.5-9 Suppose the target of Problem 2.5-7 is initially at the specified range and elevation angle relative to the radar but approaches the radar at constant altitude. Find the target's radial range and elevation angle when the multipath ray arrives at an elevation angle of -5 degrees.

2.5-10 Assume that $\Gamma = -1$ in (2.5-12), and determine the values of θ for which $|e_r/e_A| = 0$. Also find the values of θ for which $|e_r/e_A|$ is maximum. These results will be functions of h, λ, and θ, and it is permissible to assume that $\sin(\theta) \approx \theta$ for small θ. Sketch $|e_r/e_A|$ versus θ for small θ.

3

ANTENNAS

During transmission the radar antenna functions as a transducer, converting the electrical waveform from the transmitter to an electromagnetic wave for transmission. On reception, the antenna performs the inverse function; it converts the arriving electromagnetic wave into an electrical waveform. It results that an antenna is a reciprocal device, which means that its behavior during wave reception is known if its behavior as a transmitting device is known. This fact allows us to study antennas principally as a radiating device, and most of this chapter develops the antenna as a source of radiation. Where needed, we also point out the key developments necessary to use it as a receiving device. Our knowledge of antenna behavior is then used to good advantage in the next chapter where the radar equation is developed to describe the power available in the receiver of an overall radar system.

There are many forms of antennas, and the detailed design of any one is typically quite involved. Rather than become deeply involved in such detail for various antennas, this chapter will concentrate on only two forms, aperture and array antennas. Discussions of each will be sufficient to grasp the important concepts, many of which are useful in understanding other antennas, as well. Even though we develop only two forms of antennas, many of the results achieved (Sections 3.1–3.3 and 3.4) are very general and can be applied to almost any real antenna. Our work will be especially useful in later chapters that involve spatial angle measurement of target position because the antenna is intimately related to such measurements.

3.1 APERTURE ANTENNAS

Many of the important antennas for radar can be described by considering radiation from an aperture. Such antennas can broadly be defined as *aperture antennas*. A general aperture is defined in Fig. 3.1-1. The aperture consists of an area A laying in the x, y plane, and bounded by lengths L_{x1} and L_{x2} in the x coordinate, and lengths

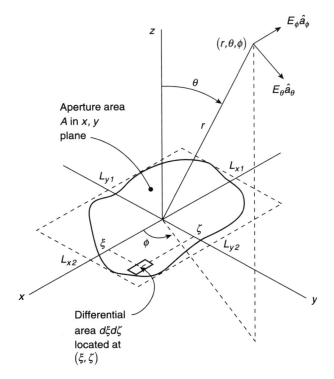

Figure 3.1-1   Geometry of radiating aperture in an aperture antenna.

L_{y1} and L_{y2} in the y coordinate. Axis z is normal to the aperture and is called the *broadside direction*. A small (differential) "patch," or area, in the aperture is located at an arbitrary point (ξ, ζ). Each patch in the aperture is radiating energy due to an electric field distribution over the aperture. Outside the aperture the only electric fields are due to radiation from the aperture. Thus at a point (r, θ, ϕ), defined in spherical coordinates, the fields are due to radiation of energy from all patches over the aperture.

Aperture Electric Field Distributions

A number of methods exists to model the radiation from an antenna. Perhaps the easiest to understand and use is that which is based on the *field equivalence principle*, due to S. A. Shelkunoff; it is a rigorous version of *Huygens's principle*, which states that "each point on a primary wavefront can be considered to be a new source of a secondary spherical wave and that a secondary wavefront can be constructed as the envelope of these secondary spherical waves," (Balanis, 1982, p. 447). In essence the actual antenna is replaced by an equivalent radiating surface (the aperture) from which distant fields can be found by proper definition of the electric and magnetic fields in the aperture.

In order to define the aperture's electric and magnetic field distributions, certain simplifying assumptions, usually true for radar, will be made. First, radiation is from the aperture only and all fields outside the aperture are due to radiation from the aperture and not due to any other sources. Second, we make the usual assumption that fields are represented in complex form. That is, a real field of the form $B \cos(\omega t + \alpha)$ is represented as the complex form $B \exp(j\omega t + j\alpha)$, where ω is the angular frequency of the field, B is its peak amplitude, α is its phase angle, and $j = \sqrt{-1}$ is the unit imaginary. By defining $Z = B \exp(j\alpha)$, we can write the complex form as $Z \exp(j\omega t)$, where Z is interpreted as the *complex envelope* (amplitude and phase) of the signal $Z \exp(j\omega t)$. Use of complex forms simplifies analyses. Clearly, since

$$B \cos(\omega t + \alpha) = \frac{B}{2} e^{j\omega t + j\alpha} + \frac{B}{2} e^{-j\omega t - j\alpha}$$

$$= \frac{Z}{2} e^{j\omega t} + \left(\frac{Z}{2} e^{j\omega t}\right)^* \tag{3.1-1}$$

where * represents the complex conjugation operation, an analysis based on the real field's form (left side) is equivalent to analysis based on half the complex form plus half the complex conjugate of the complex form (right side). Equivalently we can recognize that analysis based on the real form is the same as analysis based on *the real part* of the complex form. It results that the factor $\exp(j\omega t)$ is common to all fields encountered when studying antennas using complex forms. It is therefore common practice to suppress the common factor and represent the fields only by their complex envelopes. We will use this procedure in the following work.

The third simplifying assumption that we make is that the aperture fields are those of a plane wave. This assumption means there are no electric or magnetic fields in the z direction, and the x and y components of the magnetic fields are known if the x and y components of the electric fields are known. These assumptions mean that the vector electric and magnetic fields in the aperture, denoted by $\mathbf{E}_a(\xi, \zeta)$ and $\mathbf{H}_a(\xi, \zeta)$, respectively, can be written as

$$\mathbf{E}_a(\xi, \zeta) = E_{ax}(\xi, \zeta)\hat{a}_x + E_{ay}(\xi, \zeta)\hat{a}_y \tag{3.1-2}$$

$$\mathbf{H}_a(\xi, \zeta) = \frac{-E_{ay}(\xi, \zeta)\hat{a}_x + E_{ax}(\xi, \zeta)\hat{a}_y}{\eta} \tag{3.1-3}$$

where

$$\hat{a}_x, \hat{a}_y = \text{unit vectors in directions } x \text{ and } y, \text{ respectively} \tag{3.1-4}$$

$$\eta = \text{intrinsic impedance of the medium } (\Omega)$$

$$= \sqrt{\mu/\varepsilon} = 120\pi \text{ ohms for free space or air} \tag{3.1-5}$$

$$\mu = \text{permeability of the medium (H/m)}$$

$$= 4\pi(10^{-7}) \text{ H/m for free space or air} \tag{3.1-6}$$

$$\varepsilon = \text{permittivity of the medium (F/m)}$$

$$= 10^{-9}/(36\pi) \text{ F/m for free space or air} \qquad (3.1\text{-}7)$$

The quantities $E_{ax}(\xi, \zeta)$ and $E_{ay}(\xi, \zeta)$ are the *electric field distributions* in the aperture. $E_{ax}(\xi, \zeta)$ defines how the complex envelope (amplitude and phase) of the electric field in the x direction varies as a function of position (ξ, ζ) in the aperture. $E_{ay}(\xi, \zeta)$ is similarly defined for the electric field's component in the y direction.

$E_{ax}(\xi, \zeta)$ and $E_{ay}(\xi, \zeta)$ are not necessarily the same functions of ξ and ζ. To the extent that they differ, the polarization of the radiation will vary over the aperture. Since the ratio of the peak amplitudes of orthogonal electric fields and their relative phase are the only two quantities needed to define polarization (see Section 2.4), we can define a *polarization ratio*, $Q_a(\xi, \zeta)$, by

$$Q_a(\xi, \zeta) \triangleq \frac{E_{ay}(\xi, \zeta)}{E_{ax}(\xi, \zeta)} \qquad (3.1\text{-}8)$$

If $Q_a(\xi, \zeta) = $ constant over the aperture, then the whole aperture radiates the same polarization, and the (complex) constant defines the polarization.

Radiated Fields and Angular Spectra

The preceding three simplifying assumptions mean that the fields E_θ and E_ϕ in Fig. 3.1-1 at point (r, θ, ϕ) are entirely determined by radiation due to the aperture electric field defined by (3.1-2). We make one last assumption that is usually valid in radar. Namely the points (r, θ, ϕ) of interest all lie in the far field. This assumption means that the spherical waves at points in the far field may be approximated as plane waves. Thus the wave arriving at point (r, θ, ϕ) will have an electric field comprised only of components E_θ and E_ϕ, as shown. There is no significant component in the radial (or r) direction.

With the given assumptions, procedures given by Balanis (1982) produce the field components E_θ and E_ϕ:

$$E_\theta(r, \theta, \phi) = \frac{j2\pi k e^{-jkr}}{r} \left[\frac{1 + \cos(\theta)}{2}\right] [\cos(\phi)F_x(\theta, \phi) + \sin(\phi)F_y(\theta, \phi)] \qquad (3.1\text{-}9)$$

$$E_\phi(r, \theta, \phi) = \frac{j2\pi k e^{-jkr}}{r} \left[\frac{1 + \cos(\theta)}{2}\right] [-\sin(\phi)F_x(\theta, \phi) + \cos(\phi)F_y(\theta, \phi)] \qquad (3.1\text{-}10)$$

where

$$k = 2\pi/\lambda = \omega\sqrt{\mu\varepsilon} = \omega\mu/\eta \quad (\text{m}^{-1}) \qquad (3.1\text{-}11)$$

$$\lambda = \text{wavelength} = c/f \quad (\text{m}) \qquad (3.1\text{-}12)$$

$$c = \text{speed of light} = 299{,}792{,}458 \quad (\text{m/s}) \qquad (3.1\text{-}13)$$

$$\omega = 2\pi f \quad (\text{rad/s}) \qquad (3.1\text{-}14)$$

f is frequency in hertz (Hz), and j, μ, ε, and η were defined previously. The functions $F_x(\theta, \phi)$ and $F_y(\theta, \phi)$ are called *angular spectra* and are given by

$$F_x(\theta, \phi) = \frac{1}{(2\pi)^2} \int_A \int E_{ax}(\xi, \zeta) e^{j\xi k \sin(\theta)\cos(\phi) + j\zeta k \sin(\theta)\sin(\phi)} \, d\xi d\zeta \qquad (3.1\text{-}15)$$

$$F_y(\theta, \phi) = \frac{1}{(2\pi)^2} \int_A \int E_{ay}(\xi, \zeta) e^{j\xi k \sin(\theta)\cos(\phi) + j\zeta k \sin(\theta)\sin(\phi)} \, d\xi d\zeta \qquad (3.1\text{-}16)$$

The integrations required in (3.1-15) and (3.1-16) are taken over all points (ξ, ζ) falling inside the aperture's area A. Angular spectrum $F_x(\theta, \phi)$ is due to the component of aperture electric field that is polarized (directed) only in the x direction. Similarly the angular spectrum $F_y(\theta, \phi)$ is due to the aperture's electric field polarized only in the y direction.

The electric field vector at point (r, θ, ϕ) in the far field can be written as

$$\mathbf{E}(r, \theta, \phi) = E_\theta(r, \theta, \phi)\hat{a}_\theta + E_\phi(r, \theta, \phi)\hat{a}_\phi \qquad (3.1\text{-}17)$$

where (3.1-9) and (3.1-10) apply, $\hat{a}_\theta$ and $\hat{a}_\phi$ are unit vectors in directions θ and ϕ, respectively. Because waves in the far field are planar, the magnetic field's vector is known in terms of $\mathbf{E}(r, \theta, \phi)$:

$$\mathbf{H}(r, \theta, \phi) = \frac{1}{\eta} \left[\hat{a}_r \times \mathbf{E}(r, \theta, \phi) \right]$$

$$= \frac{1}{\eta} \left[-E_\phi(r, \theta, \phi)\hat{a}_\theta + E_\theta(r, \theta, \phi)\hat{a}_\phi \right] \qquad (3.1\text{-}18)$$

where $\hat{a}_r$ is a unit vector in direction r, and $\times$ represents the usual vector cross product.

Power Flow Due to Radiation

A well-known result from electromagnetic theory is *Poynting's vector* which describes the direction and magnitude of average power flow through a small region of space when using the complex representation of fields. At point (r, θ, ϕ) in Fig. 3.1-1 Poynting's vector, denoted by $\mathscr{P}(r, \theta, \phi)$, is

$$\mathscr{P}(r, \theta, \phi) = \frac{1}{2} \mathrm{Re} \left[\mathbf{E} \times \mathbf{H}^* \right] = \frac{1}{2\eta} \left[|E_\theta|^2 + |E_\phi|^2 \right] \hat{a}_r$$

$$= \frac{1}{2\eta} |\mathbf{E}(r, \theta, \phi)|^2 \hat{a}_r \quad (\mathrm{W/m^2}) \qquad (3.1\text{-}19)$$

where we have substituted (3.1-17) and (3.1-18), and Re[ ] represents taking the real part of the quantity in the brackets. Our result, (3.1-19), shows that the radiated average power flows only in the radial direction when the point of interest is in the far field.

3.2 RADIATION INTENSITY PATTERN

The *radiation intensity pattern* for any antenna, denoted by $\mathscr{P}(\theta, \phi)$, is defined by

$$\mathscr{P}(\theta, \phi) = \text{average power per unit solid angle in}$$
$$\text{direction } (\theta, \phi) \text{ at distance } r$$

$$= \left\{ \begin{matrix} \text{average power per unit} \\ \text{area at distance } r \end{matrix} \right\} \cdot \frac{(\text{area at distance } r)}{(\text{solid angle at distance } r)}$$

$$= |\mathscr{P}(r, \theta, \phi)| \frac{\text{area at distance } r}{\text{solid angle at distance } r} = |\mathscr{P}(r, \theta, \phi)| \frac{4\pi r^2}{4\pi}$$

$$= r^2 |\mathscr{P}(r, \theta, \phi)| \quad \text{(W/sr)} \tag{3.2-1}$$

$\mathscr{P}(\theta, \phi)$ is a function that describes how intense the power flow is in a given direction from the antenna. On substitution for the Poynting vector from (3.1-19), we have

$$\mathscr{P}(\theta, \phi) = \frac{r^2}{2\eta} |\mathbf{E}|^2 = \frac{r^2}{2\eta} \left[|E_\theta|^2 + |E_\phi|^2 \right] \quad \text{(W/sr)} \tag{3.2-2}$$

On use of the field components of (3.1-9) and (3.1-10) for an aperture antenna, we have

$$|E_\theta|^2 = \frac{(2\pi)^2 k^2}{r^2} \left[\frac{1 + \cos(\theta)}{2} \right]^2 \{ |F_x(\theta, \phi)|^2 \cos^2(\phi) + |F_y(\theta, \phi)|^2 \sin^2(\phi)$$
$$+ 2 \sin(\phi)\cos(\phi)\mathrm{Re}[F_x(\theta, \phi)F_y^*(\theta, \phi)] \} \tag{3.2-3}$$

$$|E_\phi|^2 = \frac{(2\pi)^2 k^2}{r^2} \left[\frac{1 + \cos(\theta)}{2} \right]^2 \{ |F_x(\theta, \phi)|^2 \sin^2(\phi) + |F_y(\theta, \phi)|^2 \cos^2(\phi)$$
$$- 2 \sin(\phi)\cos(\phi)\mathrm{Re}[F_x(\theta, \phi)F_y^*(\theta, \phi)] \} \tag{3.2-4}$$

$$|\mathbf{E}|^2 = \frac{(2\pi)^2 k^2}{r^2} \left[\frac{1 + \cos(\theta)}{2} \right]^2 \{ |F_x(\theta, \phi)|^2 + |F_y(\theta, \phi)|^2 \} \tag{3.2-5}$$

From (3.2-2) we may now write the radiation intensity pattern as

$$\mathscr{P}(\theta, \phi) = \frac{(2\pi)^2 k^2}{2\eta} \left[\frac{1 + \cos(\theta)}{2} \right]^2 \{ |F_x(\theta, \phi)|^2 + |F_y(\theta, \phi)|^2 \} \tag{3.2-6}$$

Since $\mathscr{P}(\theta, \phi)$ is the sum of two terms, it can be interpreted as the sum of two *component radiation intensity patterns*, denoted by $\mathscr{P}_x(\theta, \phi)$ and $\mathscr{P}_y(\theta, \phi)$, as given by

$$\mathscr{P}_x(\theta, \phi) = \frac{(2\pi)^2 k^2}{2\eta} \left[\frac{1 + \cos(\theta)}{2} \right]^2 |F_x(\theta, \phi)|^2 \tag{3.2-7}$$

$$\mathscr{P}_y(\theta, \phi) = \frac{(2\pi)^2 k^2}{2\eta} \left[\frac{1 + \cos(\theta)}{2} \right]^2 |F_y(\theta, \phi)|^2 \tag{3.2-8}$$

These are called component patterns because each derives from one component of the aperture's electric field. $\mathcal{P}_x(\theta, \phi)$ is due to the x-directed field $E_{ax}(\xi, \zeta)$ through $F_x(\theta, \phi)$ and does not depend on the y-directed field $E_{ay}(\xi, \zeta)$. Similarly $\mathcal{P}_y(\theta, \phi)$ is due to only $E_{ay}(\xi, \zeta)$ through $F_y(\theta, \phi)$. It is to be noted, however, that any aperture with linear polarization, where one component of the aperture field is zero, has a radiation intensity pattern equal to either $\mathcal{P}_x(\theta, \phi)$ or $\mathcal{P}_y(\theta, \phi)$, since one component is zero.

Vector Angular Spectrum

Perhaps a simpler and clearer representation for $\mathcal{P}(\theta, \phi)$ is obtained by defining *vector angular spectrum*, denoted by $\mathbf{F}_a(\theta, \phi)$, by

$$\mathbf{F}_a(\theta, \phi) \triangleq F_x(\theta, \phi)\hat{a}_x + F_y(\theta, \phi)\hat{a}_y$$

$$= \frac{1}{(2\pi)^2} \int_A \int \mathbf{E}_a(\xi, \zeta) e^{j\xi k \sin(\theta)\cos(\phi) + j\zeta k \sin(\theta)\sin(\phi)} \, d\xi d\zeta \qquad (3.2\text{-}9)$$

where $\mathbf{E}_a(\xi, \zeta)$ is given by (3.1-2). From (3.2-9),

$$|\mathbf{F}_a(\theta, \phi)|^2 = |F_x(\theta, \phi)|^2 + |F_y(\theta, \phi)|^2 \qquad (3.2\text{-}10)$$

so (3.2-6) becomes

$$\mathcal{P}(\theta, \phi) = \frac{(2\pi)^2 k^2}{2\eta} \left[\frac{1 + \cos(\theta)}{2} \right]^2 |\mathbf{F}_a(\theta, \phi)|^2 \quad (\text{W/sr}) \qquad (3.2\text{-}11)$$

Equation (3.2-11) indicates that $\mathcal{P}(\theta, \phi)$ is proportional to the squared magnitude of the vector angular spectrum, and the result is true for any aperture antenna. The factor $[1 + \cos(\theta)]^2$ is sometimes called an *obliquity factor* (Hansen 1964, vol. 1); it is usually not a rapidly changing function of θ, as compared to $|\mathbf{F}_a(\theta, \phi)|$, so the vector angular spectrum mainly controls the behavior of $\mathcal{P}(\theta, \phi)$. For many (large) antennas the obliquity factor can be ignored.

For radar antennas with an aperture defined as in Fig. 3.1-1, radiation intensity patterns typically behave as shown in Fig. 3.2-1. There is usually a small region of directions where the intensity is largest, as shown by the main lobes in panels a and b. A main lobe that is relatively narrow in two orthogonal planes intersecting in a line through the direction of the maximum (planes x, z and y, z), such as in panel a, is called a *pencil beam*. If the main lobe is much wider in one orthogonal plane, as in panel b, the pattern is said to have a *fan beam*. For directions outside the main beam, a typical radiation intensity pattern has sidelobes that usually (and desirably) have maximums much smaller than that of the main lobe. As an antenna is made larger (in wavelengths), its radiation intensity pattern grows in magnitude and develops more and narrower lobes. The widths of the main lobe, measured in the two orthogonal planes that intersect in the direction of the maximum, between points 3 dB down from the maximum, are called *3-dB beamwidths* of the pattern. When the planes contain coordinate axes, they are called *principal planes*, so the beamwidths are

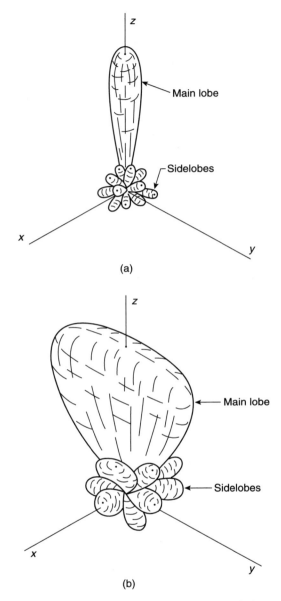

Figure 3.2-1 Radiation intensity patterns $\mathscr{P}(\theta, \phi)$ for (a) pencil beam pattern and (b) fan beam pattern.

correspondingly called *principal plane beamwidths*. Principal planes are also called *cardinal planes*. Other planes through the main beam's maximum are called *intercardinal planes*. For most aperture antennas the main lobe maximum is in the *broadside direction*, the z axis in Fig. 3.1-1.

Radiated Power

The total radiated power from an antenna can be found by integrating the radiation intensity pattern (unit is W/sr) over all solid angles in the sphere. Figure 3.2-2 illustrates the geometry for defining how a small (differential) area dA, located at a nominal point (r, θ, ϕ) on a sphere of radius r, is related to the solid angle $d\Omega$ subtended by area dA:

$$d\Omega = \sin(\theta)\,d\theta d\phi \qquad (3.2\text{-}12)$$

The radiated power for any antenna becomes

$$P_{\text{rad}} = \int_{\phi=0}^{2\pi}\int_{\theta=0}^{\pi} \mathscr{P}(\theta, \phi)\,d\Omega = \int_{\phi=0}^{2\pi}\int_{\theta=0}^{\pi} \mathscr{P}(\theta, \phi)\sin(\theta)\,d\theta d\phi \qquad (3.2\text{-}13)$$

For an aperture antenna it is also possible to find P_{rad} by a different method. Since all radiated power originates from the aperture, it can be found by integrating the aperture's Poynting vector over the aperture. Equation (3.1-19) can be used to establish Poynting's vector $\mathscr{P}(\xi, \zeta, 0)$ for the aperture's plane wave

$$\mathscr{P}(\xi, \zeta, 0) = \frac{1}{2}\text{Re}\,[\mathbf{E}_a \times \mathbf{H}_a^*] = \frac{1}{2\eta}\,[|E_{ax}(\xi, \zeta)|^2 + |E_{ay}(\xi, \zeta)|^2]\hat{a}_z$$

$$= \frac{1}{2\eta}|\mathbf{E}_a(\xi, \zeta)|^2\,\hat{a}_z \quad (\text{W/m}^2) \qquad (3.2\text{-}14)$$

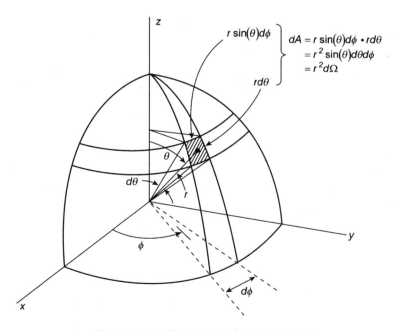

Figure 3.2-2   Geometry defining solid angle.

Here $\hat{a}_z$ is a unit vector in the z direction. Radiated power becomes

$$P_{\text{rad}} = \int_A \int |\mathscr{P}(\xi, \zeta, 0)| \, d\xi d\zeta = \frac{1}{2\eta} \int_A \int |\mathbf{E}_a(\xi, \zeta)|^2 \, d\xi d\zeta \quad \text{(W)} \qquad (3.2\text{-}15)$$

In many problems it is easier to solve (3.2-15) for P_{rad} than it is to solve (3.2-13).

3.3   PATTERN/ILLUMINATION FUNCTION RELATIONSHIP

To show the relationship between $\mathscr{P}(\theta, \phi)$ and the aperture's electric field distribution, we start with (3.2-11), which, on substitution of (3.2-9), can be written as

$$\mathscr{P}(\theta, \phi) = \frac{(2\pi)^2 k^2}{2\eta} \left[\frac{1 + \cos(\theta)}{2} \right]^2$$

$$\cdot \left| \frac{1}{(2\pi)^2} \int_A \int \mathbf{E}_a(\xi, \zeta) \, e^{j\xi k \sin(\theta)\cos(\phi) + jk\zeta\sin(\theta)\sin(\phi)} \, d\xi d\zeta \right|^2 \qquad (3.3\text{-}1)$$

Next, define new variables u_1 and u_2 by

$$u_1 = u_1(\theta, \phi) \triangleq k \sin(\theta)\cos(\phi) \qquad (3.3\text{-}2)$$

$$u_2 = u_2(\theta, \phi) \triangleq k \sin(\theta)\sin(\phi) \qquad (3.3\text{-}3)$$

so that (3.3-1) becomes

$$\mathscr{P}(\theta, \phi) = \frac{(2\pi)^2 k^2}{2\eta} \left[\frac{1 + \cos(\theta)}{2} \right]^2$$

$$\cdot \left| \frac{1}{(2\pi)^2} \int_A \int \mathbf{E}_a(\xi, \zeta) \, e^{j\xi u_1 + j\zeta u_2} \, d\xi d\zeta \right|^2 \qquad (3.3\text{-}4)$$

Fourier Transform Relationship

The integral in (3.3-4) is recognized as an inverse two-dimensional Fourier transform (ξ and ζ replace ω_1 and ω_2, respectively, and u_1 and u_2 replace t_1 and t_2, respectively, in the usual transform relationships; see Appendix B). If the inverse transform is denoted by $\mathbf{f}_a(u_1, u_2)$ then $\mathbf{f}_a(u_1, u_2)$ and $\mathbf{E}_a(\xi, \zeta)$ are a Fourier transform pair:

$$\mathbf{f}_a(u_1, u_2) = \frac{1}{(2\pi)^2} \int_A \int \mathbf{E}_a(\xi, \zeta) \, e^{j\xi u_1 + j\zeta u_2} \, d\xi d\zeta \qquad (3.3\text{-}5)$$

$$\mathbf{E}_a(\xi, \zeta) = \int_{-\infty}^{\infty} \int_{-\infty}^{\infty} \mathbf{f}_a(u_1, u_2) \, e^{-j\xi u_1 - j\zeta u_2} \, du_1 du_2 \qquad (3.3\text{-}6)$$

Hence

$$\mathscr{P}(\theta, \phi) = \frac{(2\pi)^2 k^2}{2\eta} \left[\frac{1 + \cos(\theta)}{2} \right]^2 |\mathbf{f}_a[u_1(\theta, \phi), u_2(\theta, \phi)]|^2 \qquad (3.3\text{-}7)$$

Our principal results are (3.3-7) and (3.3-5), which show that the radiation intensity pattern $\mathscr{P}(\theta, \phi)$ of any aperture antenna is proportional to the squared magnitude of the inverse transform of the aperture's vector electric field distribution, which we now call the *vector aperture illumination function.*

We will next consider two examples to demonstrate the use of (3.3-5). We will also find P_{rad} from (3.2-15).

Example of a Rectangular Aperture

Consider a rectangular aperture of width a, height b, and area $A = ab$, centered on the origin, as shown in Fig. 3.3-1. The vector aperture illumination function is assumed constant over the aperture and only in the y direction. Thus $E_{ax}(\xi, \zeta) = 0$ and

$$\mathbf{E}_a(\xi, \zeta) = E_{ay}(\xi, \zeta)\hat{a}_y$$

$$= \begin{cases} E_0 \hat{a}_y, & \text{for } \dfrac{-a}{2} < \xi < \dfrac{a}{2}, \dfrac{-b}{2} < \zeta < \dfrac{b}{2} \\ 0, & \text{elsewhere} \end{cases} \qquad (3.3\text{-}8)$$

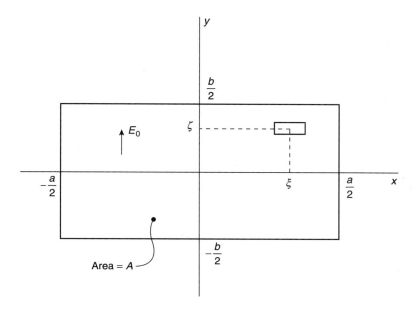

Figure 3.3-1   A radiating rectangular aperture.

From (3.3-5),

$$\mathbf{f}_a(u_1, u_2) = \frac{1}{(2\pi)^2} \int_{\zeta=-b/2}^{b/2} \int_{\xi=-a/2}^{a/2} E_0 \hat{a}_y \, e^{j\xi u_1 + j\zeta u_2} \, d\xi d\zeta$$

$$= \frac{E_0 \hat{a}_y}{(2\pi)^2} \int_{\xi=-a/2}^{a/2} e^{j\xi u_1} \, d\xi \int_{\zeta=-b/2}^{b/2} e^{j\zeta u_2} \, d\zeta$$

$$= \frac{E_0 \hat{a}_y}{(2\pi)^2} a \frac{\sin(au_1/2)}{(au_1/2)} b \frac{\sin(bu_2/2)}{(bu_2/2)} \tag{3.3-9}$$

This result is substituted into (3.3-7) to obtain

$$\mathscr{P}(\theta, \phi) = \frac{|E_0|^2 A^2}{\lambda^2 2\eta} \left[\frac{1+\cos(\theta)}{2} \right]^2 \mathrm{Sa}^2 \left[\frac{a\pi}{\lambda} \sin(\theta)\cos(\phi) \right]$$

$$\cdot \mathrm{Sa}^2 \left[\frac{b\pi}{\lambda} \sin(\theta)\sin(\phi) \right] \tag{3.3-10}$$

To put (3.3-10) in another form, we find the radiated power from (3.2-15):

$$P_{\mathrm{rad}} = \frac{1}{2\eta} \int_{\zeta=-b/2}^{b/2} \int_{\xi=-a/2}^{a/2} |E_0|^2 \, d\xi d\zeta = \frac{|E_0|^2 A}{2\eta} \tag{3.3-11}$$

$\mathscr{P}(\theta, \phi)$ now can be written as

$$\mathscr{P}(\theta, \phi) = \frac{P_{\mathrm{rad}} A}{\lambda^2} \left[\frac{1+\cos(\theta)}{2} \right]^2 \mathrm{Sa}^2 \left[\frac{a\pi}{\lambda} \sin(\theta)\cos(\phi) \right] \mathrm{Sa}^2 \left[\frac{b\pi}{\lambda} \sin(\theta)\sin(\phi) \right] \tag{3.3-12}$$

In the above, $\mathrm{Sa}(x) \triangleq \sin(x)/x$ is the usual sampling function.

Figure 3.3-2 gives a sketch of (3.3-12) versus $(a\pi/\lambda)\sin(\theta)$ for $\phi = 0$; the sketch corresponds to $\mathscr{P}(\theta, \phi)$ in the principal plane x, z. The sketch would have the same *form* in the y, z principal plane, which corresponds to a plot versus $(b\pi/\lambda)\sin(\theta)$ when $\phi = \pm \pi/2$. Note that if a/λ and b/λ are large numbers so that $\sin(\theta) \approx \theta$, the 3-dB beamwidths in the two principal planes are

$$(\text{Beamwidth}, \phi = 0) \triangleq \theta_{\mathrm{B}} = 0.886 \frac{\lambda}{a} \quad (\text{rad}) \tag{3.3-13}$$

$$\triangleq \theta_{\mathrm{B}}^\circ = 50.76 \frac{\lambda}{a} \quad (\text{degrees}) \tag{3.3-14}$$

$$\left(\text{Beamwidth}, \phi = \pm \frac{\pi}{2} \right) \triangleq \phi_{\mathrm{B}} = 0.886 \frac{\lambda}{b} \quad (\text{rad}) \tag{3.3-15}$$

$$\triangleq \phi_{\mathrm{B}}^\circ = 50.76 \frac{\lambda}{b} \quad (\text{degrees}) \tag{3.3-16}$$

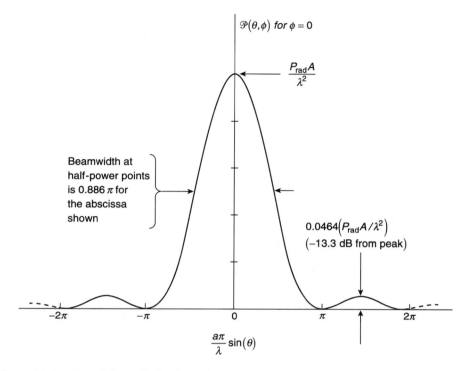

Figure 3.3-2 Plot of the radiation intensity pattern of the aperture of Fig. 3.3-1 when $\phi = 0$.

Example of a Circular Aperture

As a second example, assume a circular aperture centered on the origin having a radius R, and diameter $D = 2R$, as shown in Fig. 3.3-3. We assume a constant vector aperture illumination function with electric field only in the x direction. Thus

$$\mathbf{E}_a(\xi, \zeta) = E_{ax}(\xi, \zeta)\,\hat{a}_x = \begin{cases} E_0\,\hat{a}_x, & \xi^2 + \zeta^2 \leq R^2 \\ 0, & \text{elsewhere} \end{cases} \tag{3.3-17}$$

Again we use (3.3-5),

$$\mathbf{f}_a(u_1, u_2) = \frac{1}{(2\pi)^2} \int_{\zeta = -R}^{R} \int_{\xi = -\sqrt{R^2 - \zeta^2}}^{\sqrt{R^2 - \zeta^2}} E_0\,\hat{a}_x\, e^{j\xi u_1 + j\zeta u_2}\, d\xi d\zeta \tag{3.3-18}$$

Because of the aperture's shape, it is best to change to cylindrical coordinates ρ and ψ according to

$$\xi = \rho \cos(\psi), \quad \rho = \sqrt{\xi^2 + \zeta^2} \tag{3.3-19}$$

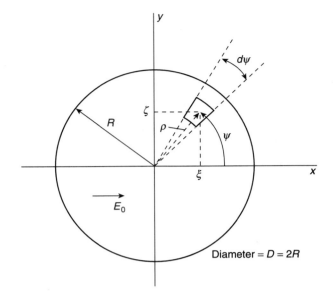

Figure 3.3-3 A radiating circular aperture.

$$\zeta = \rho \sin(\psi), \quad \psi = \tan^{-1}\left(\frac{\zeta}{\xi}\right) \tag{3.3-20}$$

$$d\xi d\zeta = \rho d\rho d\psi \tag{3.3-21}$$

Thus

$$\mathbf{f}_a(u_1, u_2) = \frac{E_0 \hat{a}_x}{(2\pi)^2} \int_{\rho=0}^{R} \rho \int_{\psi=0}^{2\pi} e^{ju_1\rho \cos(\psi) + ju_2\rho \sin(\psi)} \, d\psi d\rho \tag{3.3-22}$$

Next define parameters u and ψ_0 by

$$u = [u_1^2 + u_2^2]^{1/2} = [k^2 \sin^2(\theta) \cos^2(\phi) + k^2 \sin^2(\theta) \sin^2(\phi)]^{1/2}$$

$$= k \sin(\theta) \tag{3.3-23}$$

$$\psi_0 = \tan^{-1}\left(\frac{u_2}{u_1}\right) \tag{3.3-24}$$

so

$$\mathbf{f}_a(u_1, u_2) = \frac{E_0 \hat{a}_x}{(2\pi)^2} \int_{\rho=0}^{R} \rho \int_{\psi=0}^{2\pi} e^{j\rho u \cos(\psi - \psi_0)} \, d\psi d\rho \tag{3.3-25}$$

The integral over ψ is known to be (see Problem 3.3-9)

$$\int_0^{2\pi} e^{j\rho u \cos(\psi - \psi_0)} \, d\psi = 2\pi J_0(\rho u) \tag{3.3-26}$$

where $J_0(\cdot)$ is the Bessel function of the first kind of order zero. The second integral over ρ in (3.3-25) is also known (see Appendix D)

$$\int_0^R \rho J_0(\rho u) \, d\rho = \frac{R}{u} J_1(uR) \tag{3.3-27}$$

where $J_1(\cdot)$ is the Bessel function of order one. Hence

$$\mathbf{f}_a(u_1, u_2) = \hat{a}_x \frac{E_0 R^2}{(2\pi)} \frac{J_1(Ru)}{(Ru)} = \hat{a}_x \frac{E_0 A}{(2\pi)^2} \left[\frac{2J_1(Ru)}{Ru} \right]$$

$$= \hat{a}_x \frac{E_0 A}{(2\pi)^2} \left\{ \frac{2J_1[(2\pi R/\lambda)\sin(\theta)]}{(2\pi R/\lambda)\sin(\theta)} \right\} \tag{3.3-28}$$

On substituting (3.3-28) into (3.3-7), we get

$$\mathscr{P}(\theta, \phi) = \frac{|E_0|^2 A^2}{\lambda^2 2\eta} \left[\frac{1 + \cos(\theta)}{2} \right]^2 \left| \frac{2J_1[(2\pi R/\lambda)\sin(\theta)]}{(2\pi R/\lambda)\sin(\theta)} \right|^2 \tag{3.3-29}$$

We put this result in an alternative form after first finding P_{rad} from (3.2-15):

$$P_{\text{rad}} = \frac{1}{2\eta} \int_A \int |E_0|^2 \, d\xi d\zeta = \frac{|E_0|^2 A}{2\eta} \tag{3.3-30}$$

On using this result in (3.3-29), we have

$$\mathscr{P}(\theta, \phi) = \frac{P_{\text{rad}} A}{\lambda^2} \left[\frac{1 + \cos(\theta)}{2} \right]^2 \left| \frac{2J_1[(2\pi R/\lambda)\sin(\theta)]}{(2\pi R/\lambda)\sin(\theta)} \right|^2 \tag{3.3-31}$$

Figure 3.3-4 sketches (3.3-31). It is to be noted that $\mathscr{P}(\theta, \phi)$ does not depend on ϕ. The pattern has perfect symmetry around the z axis. For a large aperture where $\sin(\theta) \approx \theta$, the beamwidth in the variable θ becomes (see Problem 3.3-10)

$$\text{Beamwidth} \triangleq \theta_{\text{B}} = 0.514 \frac{\lambda}{R} \quad \text{(rad)} = 1.027 \frac{\lambda}{D} \quad \text{(rad)} \tag{3.3-32}$$

$$\triangleq \theta_{\text{B}}^\circ = 29.43 \frac{\lambda}{R} \quad \text{(degrees)} = 58.86 \frac{\lambda}{D} \quad \text{(degrees)} \tag{3.3-33}$$

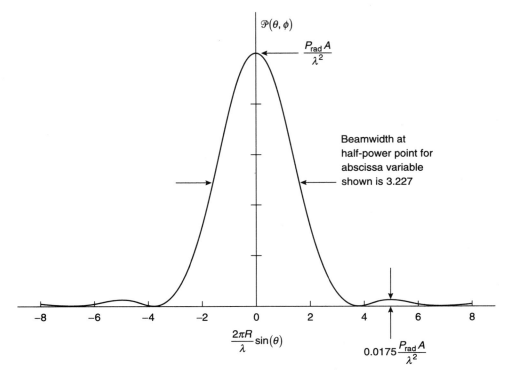

Figure 3.3-4   Plot of the radiation intensity pattern of the circular aperture of Fig. 3.3-3. The pattern is independent of angle ϕ.

3.4   FUNDAMENTAL PATTERN PARAMETERS

A number of antenna parameters are especially important. Some of these are general and apply to any antenna, not just to apertures and arrays. In this section we discuss and define these parameters and use them to develop models suitable for use in the radar equation to be developed in the next chapter.

Solid Angle

Previously Fig. 3.2-2 was used to define solid angle and to develop (3.2-13) which defines the total radiated power. Now imagine a fictitious pattern with the same value of maximum radiation intensity, $\mathscr{P}(\theta, \phi)_{\max}$, and *constant* at this value over a *beam solid angle* Ω_A; the pattern is assumed to be zero outside the solid angle Ω_A. The radiated power from the fictitious antenna is

$$P_{\text{rad}} = \mathscr{P}(\theta, \phi)_{\max} \, \Omega_A \qquad (3.4\text{-}1)$$

By requiring the fictitious and actual patterns to produce the same radiated power, we must have

$$\Omega_A = \frac{\int_{\phi=0}^{2\pi} \int_{\theta=0}^{\pi} \mathscr{P}(\theta, \phi) \sin(\theta) \, d\theta d\phi}{\mathscr{P}(\theta, \phi)_{\max}} \quad \text{(sr)} \qquad (3.4\text{-}2)$$

Beamwidths

The beam solid angle is related to a pattern's principal plane 3-dB beamwidths θ_B and ϕ_B. For a pattern with one relatively narrow main lobe and negligible sidelobes, a rough approximation is (Kraus, 1950, p. 25; see also Elliott, 1981, p. 207)

$$\Omega_A \approx \theta_B \phi_B \quad (\text{sr}) \tag{3.4-3}$$

If beamwidths are in degrees instead, denoted by θ_B° and ϕ_B°, then

$$\Omega_A \approx \theta_B^\circ \phi_B^\circ \left(\frac{\pi}{180}\right)^2 = \frac{\theta_B^\circ \phi_B^\circ}{3283} \tag{3.4-4}$$

These approximations are not to be taken as precise. They are only rough approximations that may be in error by a few percent up to a few tens of percent, as noted by Kraus (1950, p. 25), with the amount of error dependent on the form of the pattern. Accurate determination of solid angle and beamwidths must ultimately derive from the actual pattern.

Directive Gain, Directivity, and Effective Area

Antenna behavior is defined mainly through its radiation intensity pattern $\mathscr{P}(\theta, \phi)$, which has the unit of power per unit solid angle. For many radar purposes a better "pattern" is one that is unitless. Let $G_D(\theta, \phi)$ denote such a pattern. We call $G_D(\theta, \phi)$ the *directive gain* of the antenna and define it as

$$G_D(\theta, \phi) \triangleq \frac{\text{radiation intensity (W/sr) in direction } (\theta,\phi)}{\text{average radiation intensity (W/sr)}}$$

$$= \frac{\mathscr{P}(\theta, \phi)}{P_{\text{rad}}/4\pi} = \frac{4\pi \mathscr{P}(\theta, \phi)_{\text{max}}}{P_{\text{rad}}} \cdot \frac{\mathscr{P}(\theta, \phi)}{\mathscr{P}(\theta, \phi)_{\text{max}}} = G_D P(\theta, \phi) \tag{3.4-5}$$

In the last form of (3.4-5), we have defined

$$G_D \triangleq \frac{4\pi \mathscr{P}(\theta, \phi)_{\text{max}}}{P_{\text{rad}}} = \frac{4\pi \mathscr{P}(\theta, \phi)_{\text{max}}}{\int_{\phi=0}^{2\pi} \int_{\theta=0}^{\pi} \mathscr{P}(\theta, \phi) \sin(\theta)\, d\theta d\phi} = \frac{4\pi}{\Omega_A} \tag{3.4-6}$$

$$P(\theta, \phi) \triangleq \frac{\mathscr{P}(\theta, \phi)}{\mathscr{P}(\theta, \phi)_{\text{max}}} \tag{3.4-7}$$

G_D is the intensity (magnitude) of $G_D(\theta, \phi)$ at its maximum, and it is called *directivity*. $P(\theta, \phi)$ is a normalized version of $\mathscr{P}(\theta, \phi)$ such that its maximum is unity; it is usually just called the *radiation pattern*.

For relatively narrow-beam antennas (3.4-3) is used in (3.4-6) to get a rough approximation for directivity in terms of principal plane beamwidths

$$G_D \approx \frac{4\pi}{\theta_B \phi_B} = \frac{4\pi(180/\pi)^2}{\theta_B^\circ \phi_B^\circ} = \frac{41,253}{\theta_B^\circ \phi_B^\circ} \tag{3.4-8}$$

Equation (3.4-8) usually gives a value of G_D larger than the true value. The correction factor required to achieve correct results varies from about 0.6 to 1.0 depending on pattern characteristics (Kraus, 1950, p. 25). Some authors make a correction by using different constants. Elliott (1981) uses 32,400, Kraus (1988) uses 41,000, and Skolnik (1980, p. 226) uses 20,000. These variations only serve to indicate that (3.4-8) is only a rough approximation.

Effective area, denoted by $A_e(\theta, \phi)$, is a very important fundamental parameter. It is defined by

$$A_e(\theta, \phi) \triangleq \frac{\lambda^2 \mathscr{P}(\theta, \phi)}{P_{rad}} \qquad (3.4-9)$$

On using this definition in (3.4-5) and (3.4-6), we obtain most important forms for directive gain $G_D(\theta, \phi)$ and directivity G_D:

$$G_D(\theta, \phi) = \frac{4\pi A_e(\theta, \phi)}{\lambda^2} \qquad (3.4\text{-}10a)$$

$$G_D = \frac{4\pi A_e(\theta, \phi)_{max}}{\lambda^2} \qquad (3.4\text{-}10b)$$

For a general antenna, or even a general aperture antenna, it is difficult to reduce (3.4-10) and (3.4-9) further because of the difficulty of determining $\mathscr{P}(\theta, \phi)_{max}$. However, for apertures with fixed polarization or with uniform phase such that the maximum is at broadside, we may develop very practical forms. This work appears in Section 3.5 and Problem 3.4-12.

Power Gain and Antenna Efficiencies

Up to this point all our work has related to the *radiated* power of an antenna. In a real antenna not all the power applied to its input is radiated. Some power is lost as heating (I^2R loss) in the antenna's structure, and some is lost through impedance mismatch, as well as through other mechanisms. The loss due to heating is accounted for by defining the *power gain*, denoted by $G(\theta, \phi)$ as

$$G(\theta, \phi) \triangleq \frac{\text{radiation intensity in direction } (\theta, \phi)}{\begin{array}{c}\text{radiation intensity of lossless isotropic}\\\text{antenna with same power accepted}\\\text{at antenna's input}\end{array}}$$

$$= \frac{\mathscr{P}(\theta, \phi)}{P_{acc}/(4\pi)} = \frac{\mathscr{P}(\theta, \phi)}{P_{rad}/(4\pi)} \cdot \frac{P_{rad}}{P_{acc}} = G_D(\theta, \phi) \frac{P_{rad}}{P_{acc}} \qquad (3.4\text{-}11)$$

where P_{acc} is the average power[1] accepted at the antenna's input. Since $P_{rad} \leq P_{acc}$, we define *radiation efficiency*, denoted ρ_r, by

$$\rho_r \triangleq \frac{P_{rad}}{P_{acc}} \leq 1 \qquad (3.4\text{-}12)$$

---

[1] Power is averaged over one cycle of the carrier.

so

$$G(\theta, \phi) = \rho_r G_D(\theta, \phi) \tag{3.4-13}$$

Equation (3.4-13) indicates that an antenna's power gain $G(\theta, \phi)$ is never greater than its directive gain as determined by the radiation efficiency factor ρ_r.

Another efficiency factor derives from further consideration of effective area. Because the largest effective area is not larger than an aperture's true area, we can define *aperture efficiency*, denoted by ρ_a, for an aperture antenna by

$$\rho_a \triangleq \frac{A_e(\theta, \phi)_{max}}{A} = \frac{\lambda^2 \mathscr{P}(\theta, \phi)_{max}}{A P_{rad}} \tag{3.4-14}$$

With the use of (3.4-14), we can now write (3.4-10) and (3.4-13) as

$$G_D = \frac{4\pi A_e(\theta, \phi)_{max}}{\lambda^2} = \frac{4\pi \rho_a A}{\lambda^2} \tag{3.4-15}$$

$$G(\theta, \phi)_{max} = G(0, 0) = \rho_r G_D = \frac{4\pi \rho_r \rho_a A}{\lambda^2} \tag{3.4-16}$$

The factor $\rho_r \rho_a$ is often called *antenna efficiency*.

Example 3.4-1   We find ρ_a for the rectangular aperture defined in (3.3-8). From (3.3-11),

$$P_{rad} = |E_0|^2 \frac{A}{2\eta}$$

The maximum value of $\mathscr{P}(\theta, \phi)$ occurs at $\theta = 0$ in (3.3-12). It is

$$\mathscr{P}(\theta, \phi)_{max} = P_{rad} \frac{A}{\lambda^2}$$

When these two results are used in (3.4-14), we find that $\rho_a = 1$. Such a high efficiency can be explained by noting that the aperture electric field was constant. In other words, every part of the aperture was radiating power just as efficiently (at the maximum level) as every other part of the aperture. No part had a field smaller than any other part, so the antenna had maximum efficiency.

Loss Model

For purposes of modeling antennas in the next chapter, it is helpful to separate radiation loss (a practical effect) from theoretical characteristics. We define *radiation loss*, a number larger than one and denoted by L_r, by

$$L_r \triangleq \frac{1}{\rho_r} \tag{3.4-17}$$

This definition allows us to use the model of Fig. 3.4-1 where a real, lossy antenna is modeled as the loss L_r followed by a lossless antenna with input power equal to the radiated power.

Antenna as a Power-Receiving Device

Antennas are reciprocal devices, meaning that their characteristics as a transmitting device are similar to their receiving characteristics. Thus, as a radiation pattern determines how power is distributed to a particular direction in space, that pattern will also control the received power associated with a wave arriving from a particular direction in space.

Consider a wave arriving from an arbitrary direction (θ, ϕ) as defined for the (now receiving) aperture of Fig. 3.1-1. In general, the wave can have a polarization different than that to which the receiving antenna responds best. However, it can be shown (see Section 5.1) that an arbitrary wave can be decomposed into two waves; one has the polarization to which the antenna responds, and the other has an orthogonal polarization to which the antenna does *not* respond. Now assume that the power density of the wave component to which the antenna can respond is $\mathscr{S}_i(\theta, \phi)\,(W/m^2)$ as it crosses the aperture.

The maximum power the antenna can receive from the wave is the product of $\mathscr{S}_i(\theta, \phi)$ times an effective area of the antenna *for direction* (θ, ϕ). This effective area is defined in (3.4-9). A very useful form for effective area derives from (3.4-10a):

$$A_e(\theta, \phi) = \frac{\lambda^2}{4\pi} G_D(\theta, \phi) \tag{3.4-18}$$

The available power, denoted by S_r, at the receiving antenna's output becomes

$$S_r = \mathscr{S}_i(\theta, \phi) A_e(\theta, \phi) = \mathscr{S}_i(\theta, \phi) \frac{\lambda^2 G_D(\theta, \phi)}{4\pi} \tag{3.4-19}$$

This available power is realized only if the antenna has no receiving losses (some are always present) and is perfectly impedance-matched. We account for the receiving ($I^2 R$) losses by defining a receiving loss analogous to L_r in Fig. 3.4-1 for transmission. The actual notation for this loss is defined in (3.4-39) below and in Chapter 4 to be L_{rr}. The final value of S_r becomes the value in (3.4-19) divided by L_{rr}.

If $\mathscr{S}_i(\theta, \phi)$ defines the power density of the *total* arriving wave, then (3.4-19) again applies except that a new factor ρ_{pol} is to be added to the right side called a *polarization match factor* (or *polarization efficiency*). This factor is needed to account for the fact that the antenna's polarization is not fully matched to the polarization of the received wave (Mott, 1992, p. 191). We define ρ_{pol} below and in Section 5.1.

Antenna as a Voltage-Receiving Device

A received wave arriving at an antenna has an electric field strength measured in the unit of volts per meter. If the receiving antenna could multiply this field by an appropriate parameter having the unit of length (meters), the result would be voltage in volts. This is effectively what the receiving antenna does, and the parameter is called

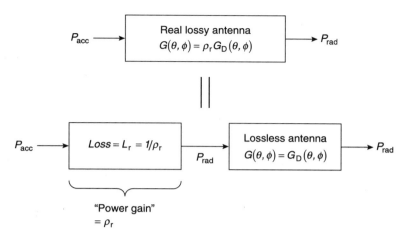

Figure 3.4-1 Model for a real, lossy antenna.

the *effective length* of the antenna. Effective length, denoted by $\mathbf{h}(\theta, \phi)$, is a vector quantity that can be derived from treating the receiving antenna as a radiating device.

Suppose that the receiving antenna is used as a radiator by driving its input terminals with a complex current I. The radiated complex vector electric field $\mathbf{E}(r, \theta, \phi)$ at a distant point (r, θ, ϕ) is given by (3.1-17) with components defined by (3.1-9) and (3.1-10). We may put $\mathbf{E}(r, \theta, \phi)$ in the following form that defines $\mathbf{h}(\theta, \phi)$:

$$\mathbf{E}(r, \theta, \phi) = \frac{j\eta I}{2\lambda r}\, e^{-jkr}\, \mathbf{h}(\theta, \phi) \tag{3.4-20}$$

where

$$\mathbf{h}(\theta, \phi) = h_\theta(\theta, \phi)\hat{a}_\theta + h_\phi(\theta, \phi)\,\hat{a}_\phi \tag{3.4-21}$$

$$h_\theta(\theta, \phi) = \frac{(2\pi)^2[1 + \cos(\theta)]}{\eta I}$$

$$\cdot[\cos(\phi)F_x(\theta, \phi) + \sin(\phi)F_y(\theta, \phi)] \tag{3.4-22}$$

$$h_\phi(\theta, \phi) = \frac{(2\pi)^2[1 + \cos(\theta)]}{\eta I}$$

$$\cdot[-\sin(\phi)F_x(\theta, \phi) + \cos(\phi)F_y(\theta, \phi)] \tag{3.4-23}$$

Here $F_x(\theta, \phi)$, $F_y(\theta, \phi)$, λ, η, k, $\hat{a}_\theta$, and $\hat{a}_\phi$ are all defined in Section 3.1. By checking the units of various terms, the reader can verify that components of $\mathbf{h}(\theta, \phi)$ both have the unit of length.

Finally it can be shown that the complex open-circuit voltage, denoted by v_r, that appears at the antenna terminals for which I is defined, is given by (Collin, 1985, p. 302; Mott, 1992, p. 193; Sinclair, 1950)

$$v_r = \mathbf{E}^s \cdot \mathbf{h} \tag{3.4-24}$$

where $\mathbf{E}^s$ represents the complex vector electric field of the scattered (arriving) wave from direction (θ, ϕ). $\mathbf{E}^s$ is to be defined *in the receiving antenna's coordinate system.* To better define $\mathbf{E}^s$, refer to Fig. 3.4-2 where a coordinate system x_2, y_2, z_2 at the target has each axis parallel with the respective radar axis. The arriving wave from the target can be written in terms of the target's spherical coordinates as

$$\mathbf{E}^s = E_{\theta_2}\hat{a}_{\theta_2} + E_{\phi_2}\hat{a}_{\phi_2} \tag{3.4-25}$$

where, as usual E_{θ_2} and E_{ϕ_2} are complex orthogonal components of the wave and $\hat{a}_{\theta_2}$ and $\hat{a}_{\phi_2}$ are unit vectors in directions θ_2 and ϕ_2, respectively. The same wave written in the antenna's spherical coordinates is

$$\mathbf{E}^s = E_\theta\hat{a}_\theta + E_\phi\hat{a}_\phi \tag{3.4-26}$$

where E_θ, E_ϕ, $\hat{a}_\theta$, and $\hat{a}_\phi$ are components and unit vectors analogous to those of the target's coordinates. From the geometry of Fig. 3.4-2,

$$E_\theta = E_{\theta_2} \tag{3.4-27}$$

$$E_\phi = -E_{\phi_2} \tag{3.4-28}$$

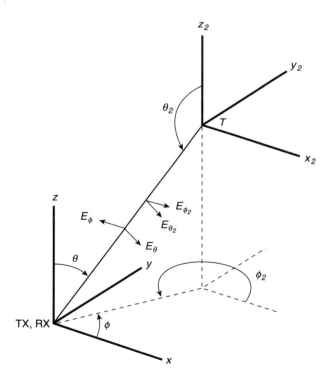

Figure 3.4-2 Geometry defining the received wave of the radar from a target at point T.

so

$$\mathbf{E}^s = E_{\theta_2}\hat{a}_\theta - E_{\phi_2}\hat{a}_\phi \tag{3.4-29}$$

represents the scattered (arriving) wave in the receiving antenna's coordinate system.
Open-circuit output voltage now derives from (3.4-24):

$$v_r = \mathbf{E}^s \cdot \mathbf{h} = (E_{\theta_2}\hat{a}_\theta - E_{\phi_2}\hat{a}_\phi) \cdot (h_\theta \hat{a}_\theta + h_\phi \hat{a}_\phi)$$

$$= E_{\theta_2} h_\theta (1 - Q_2 Q_h) \tag{3.4-30}$$

where

$$Q_2 = \frac{E_{\phi_2}}{E_{\theta_2}} \tag{3.4-31}$$

$$Q_h = \frac{h_\phi}{h_\theta} \tag{3.4-32}$$

are the polarization ratios of the incoming wave (specified from the target looking toward the antenna) and the antenna, respectively.

Consider now the available received power. Since power is proportional to the magnitude squared value of voltage, we work with voltage only. From (3.4-30),

$$|v_r|^2 = |E_{\theta_2}|^2 |h_\theta|^2 |1 - Q_2 Q_h|^2 \tag{3.4-33}$$

is proportional to the *actual* received available power. Now the absolute largest that $|v_r|^2$ can ever be is when vectors $\mathbf{E}^s$ and $\mathbf{h}$ in (3.4-24) line up exactly; in such a case

$$|v_r|^2_{\max} = |\mathbf{E}^s|^2 |\mathbf{h}|^2 = |E_{\theta_2}|^2 (1 + |Q_2|^2) |h_\theta|^2 (1 + |Q_h|^2) \tag{3.4-34}$$

We may now write (3.4-33) as

$$|v_r|^2 = |v_r|^2 \frac{|v_r|^2_{\max}}{|v_r|^2_{\max}} = |v_r|^2_{\max} \frac{|1 - Q_2 Q_h|^2}{(1 + |Q_2|^2)(1 + |Q_h|^2)}$$

$$= |v_r|^2_{\max} \, \rho_{\text{pol}} \tag{3.4-35}$$

where

$$\rho_{\text{pol}} = \frac{|1 - Q_2 Q_h|^2}{(1 + |Q_2|^2)(1 + |Q_h|^2)} \tag{3.4-36}$$

is called the *polarization match factor* or *polarization efficiency.*

The quantity ρ_{pol} represents the fraction of the total power in the received wave that is actually available to the radar. It is, in effect, a loss factor that accounts for the fact

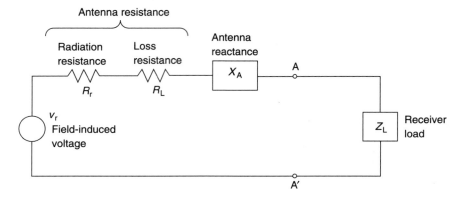

Figure 3.4-3   Equivalent circuit for antenna as a voltage-receiving device.

that the antenna's polarization may not be a complete match to the polarization of the incoming wave. By expanding the numerator of (3.4-36), it is easy to show that the antenna must have a polarization ratio of

$$Q_h = -Q_2^*$$
(3.4-37)

if $\rho_{pol} = 1$ is required (to have no polarization loss).

We complete the discussion of the antenna as a voltage-receiving device by relating received voltage v_r to available average peak received power S_r. The Thevenin's model of Fig. 3.4-3 applies, where v_r is given by (3.4-30). Resistances R_r and R_L are the antenna's radiation and loss resistances, respectively, and $R_L + R_r$ is called the antenna resistance. X_A is the antenna's reactance. *Antenna impedance* is $Z_A = (R_L + R_r) + jX_A$. The maximum power available to a receiver load impedance Z_L occurs when matched to the antenna, that is, when $Z_L = Z_A^*$. The available average peak power at point A is then

$$S_r = \frac{|v_r|^2}{8(R_L + R_r)} = \frac{|v_r|^2}{8R_r}\frac{1}{L_{rr}}$$
(3.4-38)

where L_{rr} is the antenna's loss due to ohmic effects

$$\frac{1}{L_{rr}} \triangleq \frac{R_r}{R_L + R_r}$$
(3.4-39)

The factor $|v_r|^2/8R_r$ can be interpreted as the available power that would be possible if the antenna were lossless.

3.5   APERTURES WITH CONSTANT POLARIZATION

As noted prior to (3.1-8), if the ratio of the y component of aperture electric field to the x component varies with position in the aperture, the polarization of radiation will vary over the aperture. However, if the polarization is constant over the aperture, the polarization ratio, $Q_a(\xi, \zeta)$, will be a (possibly complex) constant. Constant polarization is often present in real systems (or approximately so), and its presence considerably simplifies many antenna expressions. In this section we will assume that Q_a is a constant and summarize the important characteristics of an aperture antenna with constant polarization.

Radiated Fields and Angular Spectra

From (3.1-8) with Q_a constant, we must have

$$E_{ay}(\xi, \zeta) = Q_a E_{ax}(\xi, \zeta) \qquad \text{for all } \xi \text{ and } \zeta \text{ in } A \tag{3.5-1}$$

By substitution of (3.5-1) into (3.1-16) and (3.2-9), the various angular spectra become

$$F_y(\theta, \phi) = Q_a F_x(\theta, \phi) \tag{3.5-2}$$

$$\mathbf{F}_a(\theta, \phi) = F_x(\theta, \phi) \left[\hat{a}_x + Q_a \hat{a}_y \right] \tag{3.5-3}$$

We see that the behavior of the vector angular spectrum with direction is entirely due to the scalar angular spectrum $F_x(\theta, \phi)$. The factor $(\hat{a}_x + Q_a \hat{a}_y)$ is only a complex constant vector and has no dependance on direction (θ, ϕ).

The electric fields at the point (r, θ, ϕ) in the far field derive from (3.1-9) and (3.1-10) using (3.5-1):

$$E_\theta(r, \theta, \phi) = \frac{j2\pi k e^{-jkr}}{r} \left[\frac{1 + \cos(\theta)}{2} \right] F_x(\theta, \phi) \left[\cos(\phi) + Q_a \sin(\phi) \right] \tag{3.5-4}$$

$$E_\phi(r, \theta, \phi) = \frac{j2\pi k e^{-jkr}}{r} \left[\frac{1 + \cos(\theta)}{2} \right] F_x(\theta, \phi) \left[-\sin(\phi) + Q_a \cos(\phi) \right] \tag{3.5-5}$$

Also, from (3.2-5),

$$|\mathbf{E}|^2 = \frac{(2\pi)^2 k^2}{r^2} \left[\frac{1 + \cos(\theta)}{2} \right]^2 (1 + |Q_a|^2) |F_x(\theta, \phi)|^2 \tag{3.5-6}$$

This last expression is especially significant; it shows that except for the usually unimportant obliquity factor, the behavior of $|\mathbf{E}|$ with direction is entirely established by the angular spectrum $F_x(\theta, \phi)$.

Radiation Intensity Pattern

By using (3.5-6) in (3.2-2), we derive the radiation intensity pattern

$$\mathscr{P}(\theta, \phi) = \frac{(2\pi)^2 k^2}{2\eta} \left[\frac{1 + \cos(\theta)}{2} \right]^2 (1 + |Q_a|^2) |F_x(\theta, \phi)|^2 \tag{3.5-7}$$

A more useful form for $\mathscr{P}(\theta, \phi)$ in following work derives from writing $F_x(\theta, \phi)$ in terms of u_1 and u_2 of (3.3-2) and (3.3-3) and defining the result as $f_{ax}(u_1, u_2)$. The results are

$$f_{ax}(u_1, u_2) = \frac{1}{(2\pi)^2} \int_A \int E_{ax}(\xi, \zeta) e^{j\xi u_1 + j\zeta u_2} \, d\xi d\zeta \qquad (3.5\text{-}8)$$

$$\mathscr{P}(\theta, \phi) = \frac{(2\pi)^2 k^2}{2\eta} \left[\frac{1 + \cos(\theta)}{2}\right]^2 (1 + |Q_a|^2)|f_{ax}[u_1(\theta, \phi), u_2(\theta, \phi)]|^2 \qquad (3.5\text{-}9)$$

Now with little loss in generality, we may presume $E_{ax}(\xi, \zeta)$ to be a nonnegative function of ξ and ζ. This assumption means that $f_{ax}(u_1, u_2)$ is maximum when u_1 and u_2 are both zero (this corresponds to $\theta = 0$, so the maximum is in the antenna's broadside direction). $\mathscr{P}(\theta, \phi)_{\max}$ becomes

$$\mathscr{P}(\theta, \phi)_{\max} = \frac{(2\pi)^4}{\lambda^2 2\eta} (1 + |Q_a|^2) \left|\frac{1}{(2\pi)^2} \int_A \int E_{ax}(\xi, \zeta) \, d\xi d\zeta\right|^2 \qquad (3.5\text{-}10)$$

Radiated Power

Radiated power can now be found from (3.2-15) on using

$$\mathbf{E}_a(\xi, \zeta) = E_{ax}(\xi, \zeta) \hat{a}_x + E_{ay}(\xi, \zeta) \hat{a}_y$$
$$= E_{ax}(\xi, \zeta)(\hat{a}_x + Q_a \hat{a}_y) \qquad (3.5\text{-}11)$$

We get

$$P_{\mathrm{rad}} = \frac{1}{2\eta}(1 + |Q_a|^2) \int_A \int |E_{ax}(\xi, \zeta)|^2 \, d\xi d\zeta \qquad (3.5\text{-}12)$$

Fundamental Pattern Parameters

From (3.5-7) and (3.5-12) we may obtain other fundamental parameters for the constant-polarization aperture. From (3.4-1), (3.4-6), (3.4-7), and (3.4-9), we summarize expressions for Ω_A, G_D, $P(\theta, \phi)$, and $A_e(\theta, \phi)$:

$$\Omega_A = \frac{\lambda^2 \int_A \int |E_{ax}(\xi, \zeta)|^2 \, d\xi d\zeta}{|\int_A \int E_{ax}(\xi, \zeta) \, d\xi d\zeta|^2} \qquad (3.5\text{-}13)$$

$$G_D = \frac{4\pi |\int_A \int E_{ax}(\xi, \zeta) \, d\xi d\zeta|^2}{\lambda^2 \int_A \int |E_{ax}(\xi, \zeta)|^2 \, d\xi d\zeta} \qquad (3.5\text{-}14)$$

$$P(\theta, \phi) = \left[\frac{1 + \cos(\theta)}{2}\right]^2 \frac{|\int_A \int E_{ax}(\xi, \zeta) e^{j\xi u_1 + j\zeta u_2} \, d\xi d\zeta|^2}{|\int_A \int E_{ax}(\xi, \zeta) \, d\xi d\zeta|^2} \qquad (3.5\text{-}15)$$

$$A_e(\theta, \phi) = \left[\frac{1 + \cos(\theta)}{2}\right]^2 \frac{|\int_A \int E_{ax}(\xi, \zeta) e^{j\xi u_1 + j\zeta u_2} \, d\xi d\zeta|^2}{\int_A \int |E_{ax}(\xi, \zeta)|^2 \, d\xi d\zeta} \qquad (3.5\text{-}16)$$

Maximum Directivity and Aperture Efficiency

For a given wavelength and aperture area the largest value that G_D can have is determined by the aperture field distribution through (3.5-14). One might expect that G_D will be largest when every part of the aperture is radiating at the highest possible level, that is, when all parts radiate equally. Such an aperture is said to have *uniform illumination*, and as we will prove, it does produce the largest directivity. We begin by defining a function $R(\xi, \zeta)$ by

$$R(\xi, \zeta) = \begin{cases} 1, & (\xi, \zeta) \text{ in } A \\ 0, & \text{elsewhere} \end{cases} \tag{3.5-17}$$

so

$$E_{ax}(\xi, \zeta) = E_{ax}(\xi, \zeta) R(\xi, \zeta) \tag{3.5-18}$$

From Schwarz's inequality

$$\left| \int_A \int E_{ax}(\xi, \zeta) \, d\xi d\zeta \right|^2 = \left| \int_A \int E_{ax}(\xi, \zeta) R(\xi, \zeta) \, d\xi d\zeta \right|^2$$

$$\leq \int_A \int |E_{ax}(\xi, \zeta)|^2 \, d\xi d\zeta \int_A \int |R(\xi, \zeta)|^2 \, d\xi d\zeta$$

$$= A \int_A \int |E_{ax}(\xi, \zeta)|^2 \, d\xi d\zeta \tag{3.5-19}$$

where the equality holds only if

$$E_{ax}(\xi, \zeta) = CR(\xi, \zeta) = C \tag{3.5-20}$$

with C an arbitrary real constant. On using (3.5-19) in (3.5-14), we have

$$G_D = \frac{4\pi A_e(\theta, \phi)_{\max}}{\lambda^2} \leq \frac{4\pi A}{\lambda^2} \triangleq G_{D0} \tag{3.5-21}$$

Here we define the maximum value of G_D as G_{D0} which is

$$G_{D0} \triangleq \frac{4\pi A}{\lambda^2} \tag{3.5-22}$$

The equality in (3.5-21) holds only when (3.5-20) is true, which corresponds to uniform illumination. All other cases correspond to *nonuniform illumination*, so

$$G_D \begin{cases} = G_{D0} = \dfrac{4\pi A}{\lambda^2}, & \text{uniform illumination} \\ < G_{D0}, & \text{nonuniform illumination} \end{cases} \tag{3.5-23}$$

3.6   FACTORABLE ILLUMINATION FUNCTIONS

For rectangular apertures it is possible to have a factorable illumination; that is, the electric field distributions $E_{ax}(\xi, \zeta)$ and $E_{ay}(\xi, \zeta)$ can each be separable into the product of two factors, each of which varies as a function of only one of the two coordinates ξ and ζ. The principal advantages to be gained for factorable illuminations is that two-dimensional integrals become products of two one-dimensional integrals which are usually easier to work with. Furthermore certain optimization problems become easier to accomplish when dealing with one- rather than two-dimensional integrations.

General Rectangular Aperture

In the general case the factors of $E_{ax}(\xi, \zeta)$ and $E_{ay}(\xi, \zeta)$ can be different, which means that polarization of radiation varies over the aperture. The aperture electric field is of the form

$$\mathbf{E}_a(\xi, \zeta) = E_{1x}(\xi)E_{2x}(\zeta)\hat{a}_x + E_{1y}(\xi)E_{2y}(\zeta)\hat{a}_y \tag{3.6-1}$$

Here $E_{1x}(\xi)$ and $E_{2x}(\zeta)$ are the factors that determine how the x-directed polarization varies in the x and y directions, respectively. Similarly $E_{1y}(\xi)$ and $E_{2y}(\zeta)$ apply to the y-directed polarization of the aperture.

If (3.6-1) is used to retrace procedures outlined earlier in this chapter, it can be shown that the radiation intensity pattern has the form

$$\mathscr{P}(\theta, \phi) = \mathscr{P}_x(\theta, \phi) + \mathscr{P}_y(\theta, \phi) \tag{3.6-2}$$

where (see Problem 3.6-9)

$$\mathscr{P}_x(\theta, \phi) = \mathscr{P}_{1x}(\theta, \phi)\mathscr{P}_{2x}(\theta, \phi) \tag{3.6-3}$$

$$\mathscr{P}_y(\theta, \phi) = \mathscr{P}_{1y}(\theta, \phi)\mathscr{P}_{2y}(\theta, \phi) \tag{3.6-4}$$

are radiation intensity patterns that are factorable. $\mathscr{P}_x(\theta, \phi)$ and $\mathscr{P}_y(\theta, \phi)$ are each due only to the aperture's field in the x and y directions, respectively. Thus the component pattern due to x-directed aperture field is factorable. The component pattern due to the y-directed field is similarly factorable. However, since the four factors in (3.6-3) and (3.6-4) are not necessarily the same, the overall pattern is *not* factorable in general.

Although the detailed parameters of radiation intensity pattern, directive gain, directivity, effective area, and the like, can be worked out and given for the general factorable rectangular aperture, we leave these to the reader as exercises. In most practical cases these details are not needed because the rectangular aperture often has either only one component of electric field or the polarization is constant. These facts, and in the interest of brevity, lead us to develop details only for a rectangular aperture with constant (but arbitrary) polarization.

Constant-Polarization Rectangular Aperture

For an aperture with constant polarization, $Q_a(\xi, \zeta)$ is constant. Thus all of Section 3.5 applies. From (3.5-1) the vector aperture illumination function can be written as

$$\mathbf{E}_a(\xi, \zeta) = E_{ax}(\xi, \zeta)(\hat{a}_x + Q_a \hat{a}_y) = E_1(\xi)E_2(\zeta)(\hat{a}_x + Q_a \hat{a}_y) \qquad (3.6\text{-}5)$$

where we have also assumed the illumination is separable into two factors. $E_1(\xi)$ describes variation of $E_{ax}(\xi, \zeta)$ in the x direction only, while $E_2(\zeta)$ relates how illumination varies in the y direction. By substituting $E_{ax}(\xi, \zeta) = E_1(\xi)E_2(\zeta)$ into (3.1-15), we develop the angular spectrum

$$F_x(\theta, \phi) = F_1(\theta, \phi)F_2(\theta, \phi) \qquad (3.6\text{-}6)$$

where we define *one-dimensional angular spectra* by

$$F_1(\theta, \phi) = \frac{1}{2\pi} \int_\xi E_1(\xi)e^{j\xi \sin(\theta)\cos(\phi)} \, d\xi \qquad (3.6\text{-}7)$$

$$F_2(\theta, \phi) = \frac{1}{2\pi} \int_\zeta E_2(\zeta)e^{j\zeta k \sin(\theta)\sin(\phi)} \, d\zeta \qquad (3.6\text{-}8)$$

These are called one-dimensional because they are each due to the variation in aperture field in one coordinate (direction). Directions x and y apply to $F_1(\theta, \phi)$ and $F_2(\theta, \phi)$, respectively.

All other important characteristics of factorable-distribution, constant-polarization, rectangular apertures follow from applicable expressions in Section 3.5. From (3.5-7) the radiation intensity pattern is

$$\mathscr{P}(\theta, \phi) = \frac{(2\pi)^2 k^2}{2\eta} \left[\frac{1 + \cos(\theta)}{2} \right]^2 (1 + |Q_a|^2) |F_1(\theta, \phi)|^2 |F_2(\theta, \phi)|^2$$

$$\triangleq \mathscr{P}_1(\theta, \phi)\mathscr{P}_2(\theta, \phi) \qquad (3.6\text{-}9)$$

where we define *one-dimensional radiation intensity patterns* by

$$\mathscr{P}_1(\theta, \phi) = \frac{2\pi k}{\sqrt{2\eta}} \left| \frac{1 + \cos(\theta)}{2} \right| [1 + |Q_a|^2]^{1/2} |F_1(\theta, \phi)|^2 \qquad (3.6\text{-}10)$$

$$\mathscr{P}_2(\theta, \phi) = \frac{2\pi k}{\sqrt{2\eta}} \left| \frac{1 + \cos(\theta)}{2} \right| [1 + |Q_a|^2]^{1/2} |F_2(\theta, \phi)|^2 \qquad (3.6\text{-}11)$$

In a similar manner (3.5-12) and (3.5-14) through (3.5-16) give the radiated power, directivity, radiation pattern, and effective area, respectively:

$$P_{\text{rad}} = \frac{(1 + |Q_a|^2)}{2\eta} \int_\xi |E_1(\xi)|^2 \, d\xi \int_\zeta |E_2(\zeta)|^2 \, d\zeta \qquad (3.6\text{-}12)$$

$$G_D = \frac{4\pi |\int_\xi E_1(\xi)\,d\xi|^2 |\int_\zeta E_2(\zeta)\,d\zeta|^2}{\lambda^2 \int_\xi |E_1(\xi)|^2 \, d\xi \int_\zeta |E_2(\zeta)|^2 \, d\zeta} \qquad (3.6\text{-}13)$$

$$P(\theta, \phi) = \left[\frac{1 + \cos(\theta)}{2}\right]^2 \frac{|\int_\xi E_1(\xi) e^{j\xi u_1(\theta, \phi)} d\xi|^2 |\int_\zeta E_2(\zeta) e^{j\zeta u_2(\theta, \phi)} d\zeta|^2}{|\int_\xi E_1(\xi) d\xi|^2 |\int_\zeta E_2(\zeta) d\zeta|^2} \qquad (3.6\text{-}14)$$

$$A_e(\theta, \phi) = \left[\frac{1 + \cos(\theta)}{2}\right]^2 \frac{|\int_\xi E_1(\xi) e^{j\xi u_1(\theta, \phi)} d\xi|^2 |\int_\zeta E_2(\zeta) e^{j\zeta u_2(\theta, \phi)} d\zeta|^2}{\int_\xi |E_1(\xi)|^2 d\xi \int_\zeta |E_2(\zeta)|^2 d\zeta}$$

$$\triangleq L_{e1}(\theta, \phi) L_{e2}(\theta, \phi) \qquad (3.6\text{-}15)$$

where $u_1(\theta, \phi)$ and $u_2(\theta, \phi)$ are given by (3.3-2) and (3.3-3). We define *effective lengths* $L_{e1}(\theta, \phi)$ and $L_{e2}(\theta, \phi)$ by

$$L_{e1}(\theta, \phi) \triangleq \left[\frac{1 + \cos(\theta)}{2}\right] \frac{|\int_\xi E_1(\xi) e^{j\xi u_1(\theta, \phi)} d\xi|^2}{\int_\xi |E_1(\xi)|^2 d\xi} \qquad (3.6\text{-}16)$$

$$L_{e2}(\theta, \phi) \triangleq \left[\frac{1 + \cos(\theta)}{2}\right] \frac{|\int_\zeta E_2(\zeta) e^{j\zeta u_2(\theta, \phi)} d\zeta|^2}{\int_\zeta |E_2(\zeta)|^2 d\zeta} \qquad (3.6\text{-}17)$$

One-Dimensional Aperture

Even though they are not strictly radiation intensity patterns because their units are not W/sr, we have called $\mathscr{P}_1(\theta, \phi)$ and $\mathscr{P}_2(\theta, \phi)$ *one-dimensional* radiation intensity patterns because the overall pattern is their product and because each factor is directly due to the variation in the aperture illumination in one dimension only [variation with ξ for $\mathscr{P}_1(\theta, \phi)$ and with ζ for $\mathscr{P}_2(\theta, \phi)$]. In fact, since all the important characteristics of the factorable rectangular aperture are the products of two factors, again one for the x direction and one for the y direction, we think of the factors as each due to a *one-dimensional aperture*. Thus we may derive all important results from the behaviors of two one-dimensional apertures.

For example, the aperture's directivity from (3.6-13) can be written as

$$G_D = \frac{4\pi A_e(\theta, \phi)_{\max}}{\lambda^2} = \frac{4\pi L_{e1}(\theta, \phi)_{\max} L_{e2}(\theta, \phi)_{\max}}{\lambda^2}$$

$$= \frac{4\pi ab}{\lambda^2} \frac{L_{e1}(\theta, \phi)_{\max}}{a} \frac{L_{e2}(\theta, \phi)_{\max}}{b}$$

$$= \frac{\sqrt{4\pi}a}{\lambda} \rho_{a1} \frac{\sqrt{4\pi}b}{\lambda} \rho_{a2} = G_{D1} G_{D2} \qquad (3.6\text{-}18)$$

where for the x-directed one-dimensional aperture

$$G_{D1} \triangleq \frac{\sqrt{4\pi}a\rho_{a1}}{\lambda} \qquad (3.6\text{-}19)$$

$$\rho_{a1} \triangleq \frac{L_{e1}(\theta, \phi)_{\max}}{a} \qquad (3.6\text{-}20)$$

Similarly for the y-directed one-dimensional aperture

$$G_{D2} \triangleq \frac{\sqrt{4\pi}\, b \rho_{a2}}{\lambda} \tag{3.6-21}$$

$$\rho_{a2} \triangleq \frac{L_{e2}(\theta, \phi)_{\max}}{b} \tag{3.6-22}$$

From the preceding discussion it is clear that the pattern behavior of the aperture is determined by the pattern behaviors of the two one-dimensional apertures. In particular, control over undesired sidelobes of the overall pattern can be effectively accomplished by controlling sidelobes of the two one-dimensional apertures.

3.7 SIDELOBE CONTROL IN ONE-DIMENSIONAL APERTURES

For a rectangular aperture with constant polarization and factorable illumination, sidelobes are established through the product of the radiation intensity patterns, $\mathcal{P}_1(\theta, \phi)$ and $\mathcal{P}_2(\theta, \phi)$, of the two one-dimensional apertures. In this section we give means for controlling the sidelobes of a one-dimensional aperture. Through the product $\mathcal{P}(\theta, \phi) = \mathcal{P}_1(\theta, \phi)\mathcal{P}_2(\theta, \phi)$ we therefore effectively control the sidelobes of the aperture's pattern $\mathcal{P}(\theta, \phi)$.

We have already seen that the most efficient aperture is one with uniform illumination. Nonuniformly illuminated apertures have less efficiency and larger beamwidths but can have better (lower) sidelobe levels. The main problem in sidelobe control is then the selection of an illumination function that gives a desired sidelobe level while not producing too large an increase in beamwidth or loss in aperture efficiency.

The basic behavior of a one-dimensional aperture is obtained through its angular spectrum [either $F_1(\theta, \phi)$ or $F_2(\theta, \phi)$ from (3.6-7) or (3.6-8), respectively] and its aperture efficiency [ρ_{a1} from (3.6-20) or ρ_{a2} from (3.6-22)]. We will choose to discuss only the x-directed one-dimensional aperture. A similar set of results holds for the y-directed aperture. To make the form of our results simpler, we will show results mainly in the variable u_1, defined in (3.3-2):

$$u_1 = u_1(\theta, \phi) \triangleq k\, \sin(\theta)\cos(\phi) \tag{3.7-1}$$

Thus our main results derive from a study of the angular spectrum

$$f_{a1}(u_1) \triangleq \frac{1}{2\pi} \int_{-a/2}^{a/2} E_1(\xi)\, e^{ju_1\xi}\, d\xi \tag{3.7-2}$$

and the aperture's efficiency

$$\rho_{a1} = \frac{L_{e1}(\theta, \phi)_{\max}}{a} = \frac{|\int_{-a/2}^{a/2} E_1(\xi)\, d\xi|^2}{a \int_{-a/2}^{a/2} |E_1(\xi)|^2\, d\xi} \tag{3.7-3}$$

In writing (3.7-2) and (3.7-3), the aperture is assumed to be of length a centered on the origin.

Examples of Even Aperture Distributions

Many radar patterns of one-dimensional apertures have a single main lobe and undesired sidelobes. This behavior is characteristic of aperture distributions $E_1(\xi)$ that have even symmetry. Many even distributions have been studied to determine their aperture efficiency, peak (largest) sidelobe level, and beamwith. Of course the better patterns give high aperture efficiency, low sidelobes, and narrow beamwidth. Table 3.7-1 gives these quantities for several example distributions (see Silver, 1949, p. 187; Milligan, 1985, p. 139; Kraus, 1950, p. 350; Kirkpatrick, 1952, p. 32; Barton and Ward, 1969, pp. 251–257; Skolnik, 1980, p. 232). The forms of three of these distributions are shown in Fig. 3.7-1. Comparative performances of all are shown in Figs. 3.7-2 and 3.7-3.

Data of Figs. 3.7-1 through 3.7-3 show that distributions with heavy tapering near the aperture edges give the lowest sidelobes. Of these, distributions that have some finite illumination at the aperture edge (not zero) tend to give the best performance (lower efficiency loss, lower beamwidth spread) for a given sidelobe level. These include the Taylor, truncated gaussian, and parabolic distributions, although the latter is limited in achievable sidelobe level. As discussed below, the Taylor distribution can be taken as optimum, so our results show that the truncated gaussian distribution is near optimum in its performance.

Taylor's Distribution

The optimum distribution from the standpoint of smallest beamwidth for a given peak sidelobe level is the *Dolph-Tchebycheff distribution* (Dolph, 1946). Unfortunately, the distribution is difficult to realize in practice, and it has the undesirable property that *all* sidelobes are at the chosen level; they do not decrease with increasing angle from the main lobe. A practical compromise was given by T. T. Taylor (1955). Taylor's distribution allows all sidelobes beyond a specified number $(\bar{n} - 1)$ on each side of the main lobe to decrease from a specified peak level while producing only small increases in main lobe beamwidth compared to the Dolph-Tchebycheff distribution.

Taylor's is a two-parameter distribution. This means a designer can select the desired maximum (peak) sidelobe level relative to the main lobe's maximum, and can select $(\bar{n} - 1)$, the number of sidelobes that are nearly constant before sidelobes begin to decrease. There is a smallest value of $\bar{n}$ that should be used (Taylor, 1955, p. 23), which we denote by $\bar{n}_{\min}$. Values of $\bar{n}$ above $\bar{n}_{\min}$ can be used, and they can give some performance improvements. These improvements are small for the most part, and in many problems it is worth giving up the small benefits in order to have sidelobes begin to decrease as rapidly as possible. The data shown earlier in Figs. 3.7-2 and 3.7-3 for the Taylor distribution assumed that $\bar{n} = \bar{n}_{\min}$.

Taylor's aperture distribution is given by

$$
E_1(\xi) = \begin{cases} \dfrac{1 + 2\sum_{m=1}^{\bar{n}-1} F_m \cos(m 2\pi\xi/a)}{1 + 2\sum_{m=1}^{\bar{n}-1} F_m}, & -\dfrac{a}{2} < \xi < \dfrac{a}{2} \\ 0, & \text{elsewhere} \end{cases}
\tag{3.7-4}
$$

TABLE 3.7-1   **Aperture efficiencies, beamwidths, and sidelobe levels for various aperture illumination functions.**

$E_1(\xi)$ for $\lvert\xi\rvert < a/2$ $[E_1(\xi) = 0$ for $\lvert\xi\rvert > a/2]$	Aperture Efficiency, ρ_{a1}	Beamwidth, $\theta^\circ_B a/\lambda$ (degrees)	Peak Sidelobe Level Relative to Mainlobe Maximum in dB
Uniform: $E_1(\xi) = 1$	1.000	50.76	− 13.27
Cosine: $E_1(\xi) = \cos^N(\pi\xi/a)$			
$N = 0$ (uniform)	1.000	50.76	− 13.27
$N = 1$	0.811	68.07	− 23.00
$N = 2$	0.667	82.51	− 31.47
$N = 3$	0.576	94.84	− 39.30
$N = 4$	0.514	106.00	− 46.74
Circular: $E_1(\xi) = [1 - (2\xi/a)^2]^N$			
$N = 0.5$	0.925	58.50[††]	− 17.6[††]
$N = 1$	0.833	67.55[†]	− 21.3[†]
$N = 2$	0.700	78.21[†]	− 27.5[†]
$N = 3$	0.613	89.84[†]	− 34.7[†]
$N = 4$	0.551	99.18[†]	− 38.5[†]
Triangular: $E_1(\xi) = 1 - \lvert 2\xi/a\rvert$	0.750	72.97	− 26.52
Parabolic: $E_1(\xi) = 1 - (1 - \Delta)(2\xi/a)^2$			
$\Delta = 0.00$	0.833	66.07	− 21.29
$\Delta = 0.25$	0.918	59.63	− 19.67
$\Delta = 0.50$	0.969	55.58	− 17.08
$\Delta = 0.75$	0.993	52.76	− 14.92
$\Delta = 1.00$ (uniform)	1.000	50.76	− 13.27
Truncated gaussian: $E_1(\xi) = \exp[-1.382(n\xi/a)^2]$			
$n = 1.0$	0.990[†]	52.71[†]	− 15.5[†]
$n = 1.7$	0.930[†]	58.73[†]	− 20.8[†]
$n = 2.4$	0.808[†]	66.86[†]	− 32.1[†]
$n = 2.8$	0.727[†]	74.26[†]	− 37.0[†]
$n = 3.2$	0.650[†]	82.74[†]	− 47.5[†]
Taylor (see text)			
$\bar{n} = 2$	0.988	54.51	− 15.0
$\bar{n} = 2$	0.912	57.59	− 20.0
$\bar{n} = 3$	0.902	61.22	− 25.0
$\bar{n} = 4$	0.853	64.75	− 30.0
$\bar{n} = 5$	0.808	68.27	− 35.0
$\bar{n} = 6$	0.767	71.72	− 40.0

Sources: Data marked by † are either taken from, or calculated from, data of Barton and Ward (1969, Appendix A). Data marked †† are taken from Skolnik (1980, p. 232).

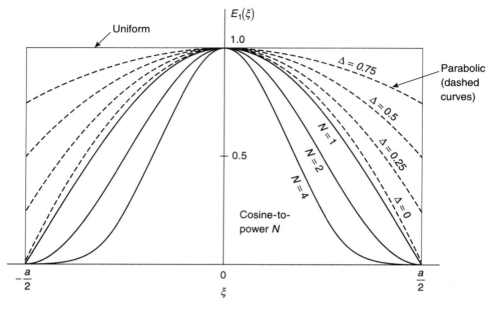

Figure 3.7-1 Uniform, parabolic, and cosine-to-power N distributions.

where the *Taylor coefficients*, F_m, are

$$F_m = \begin{cases} \dfrac{(-1)^{m+1}\prod_{n=1}^{\bar{n}-1}[1 - m^2\sigma_T^{-2}/(A^2 + (n-0.5)^2)]}{2\prod_{\substack{n=1 \\ n \neq m}}^{\bar{n}-1}(1 - (m^2/n^2))}, & m = 1, 2, \dots, (\bar{n}-1) \\ 0, & m \geq \bar{n} \end{cases}$$

(3.7-5)

The quantities σ_T and A needed in (3.7-4) and (3.7-5) are

$$\sigma_T^2 = \frac{\bar{n}^2}{A^2 + (\bar{n} - 0.5)^2} \tag{3.7-6}$$

$$A = \frac{1}{\pi}\cosh^{-1}[10^{-SLL/20}] \tag{3.7-7}$$

where SLL is the design sidelobe level in decibels.[2] Table 3.7-2 shows the solution of (3.7-7) for integer values of sidelobe level from -15 to -40 dB.

The parameter σ_T of (3.7-6) approximately equals the ratio of the beamwidth of the pattern of the Taylor distribution to the beamwidth when using a *Dolph-Tchebycheff* distribution. In effect, it is a beamwidth broadening factor. The minimum value of $\bar{n}$ also derives from (3.7-6); it is found as the largest integer value of $\bar{n}$ such that if

---

[2] For example, a design where the peak sidelobe is -30 dB compared to the main lobe's maximum corresponds to $SLL = -30$.

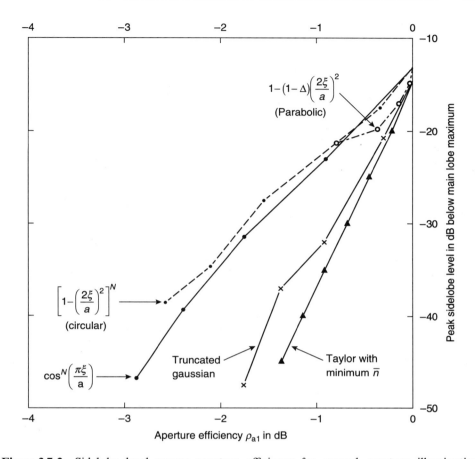

Figure 3.7-2 Sidelobe level versus aperture efficiency for several aperture illumination functions.

increased by 1, the value of σ_T does not increase (Taylor, 1955, p. 23). Some values of $\bar{n}_{\min}$ are shown in Table 3.7-2.

By using (3.7-4) in (3.7-2), we find the angular spectrum (pattern) of the Taylor distribution:

$$f_{a1}(u_1) = \frac{aK_T}{2\pi} \sum_{m=-(\bar{n}-1)}^{(\bar{n}-1)} F_m \, \mathrm{Sa}\left(\frac{au_1}{2} + m\pi\right) \tag{3.7-8}$$

where

$$F_0 = 1, \quad F_{-m} = F_m, \quad \text{for } m = 1, 2, \ldots, (\bar{n}-1) \tag{3.7-9}$$

and

$$K_T \triangleq \frac{1}{1 + 2\sum_{m=1}^{(\bar{n}-1)} F_m} \tag{3.7-10}$$

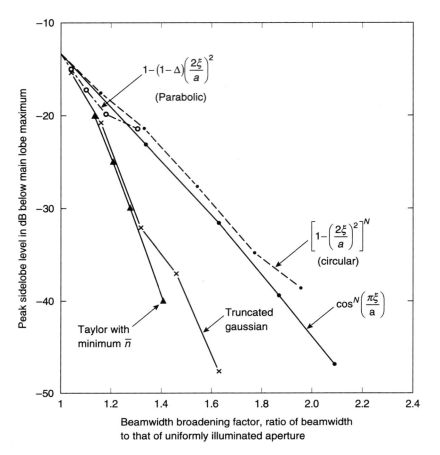

Figure 3.7-3 Sidelobe level versus main lobe beamwidth-broadening factor for several aperture illumination functions.

Next, by substituting (3.7-4) into (3.7-3), Taylor's aperture efficiency is obtained (see Problem 3.7-6)

$$\rho_{a1} = \frac{1}{1 + 2\sum_{m=1}^{(\bar{n}-1)} F_m^2} \tag{3.7-11}$$

From the preceding expressions a design procedure can be defined for Taylor's distribution:

1. Choose a desired maximum sidelobe level by choosing *SLL* (which will be a negative number in decibels).

2. Determine *A* from (3.7-7) or from Table 3.7-2 for $-40 \leq SLL$ (in 1-dB steps) ≤ -15.

3. Choose a value for $\bar{n}$. It can be the minimum value $\bar{n}_{\min}$ (usually the best choice), as derived from either σ_T or from Table 3.7-2 for integer values of *SLL*, or a value

TABLE 3.7-2 **Various parameters applicable to Taylor's one-dimensional aperture distribution.**

SLL (dB)	A	$\beta_0 a/\lambda$ (degrees)	$\bar{n}_{\min}$	SLL (dB)	A	$\beta_0 a/\lambda$ (degrees)	$\bar{n}_{\min}$
− 15	0.76779	45.93	2	− 28	1.24662	58.78	4
− 16	0.80497	47.01	2	− 29	1.28329	59.67	4
− 17	0.84203	48.07	2	− 30	1.31996	60.55	4
− 18	0.87901	49.12	2	− 31	1.35663	61.42	4
− 19	0.91592	50.15	2	− 32	1.39329	62.28	4
− 20	0.95277	51.17	2	− 33	1.42994	63.12	5
− 21	0.98958	52.17	2	− 34	1.46660	63.96	5
− 22	1.02636	53.16	3	− 35	1.50325	64.78	5
− 23	1.06311	54.13	3	− 36	1.53990	65.60	5
− 24	1.09984	55.09	3	− 37	1.57655	66.40	6
− 25	1.13656	56.04	3	− 38	1.61320	67.19	6
− 26	1.17326	56.97	3	− 39	1.64985	67.98	6
− 27	1.20994	57.88	3	− 40	1.68650	68.76	6

larger than $\bar{n}_{\min}$, if more than $(\bar{n}_{\min} - 1)$ near-in sidelobes are to be approximately constant in amplitude.

4. For the chosen value of $\bar{n}$, calculate σ_T from (3.7-6).

5. Calculate the Taylor coefficients from (3.7-5).

6. Calculate the distribution from (3.7-4).

The Taylor angular spectrum's main lobe beamwidth [in the θ direction when $\phi = 0$ in (3.7-1)] will be

$$\theta_B \approx \sigma_T \beta_0 \quad (\text{rad}) \tag{3.7-12}$$

where β_0 is the beamwidth of a similar Dolph-Tchebycheff design, as given by[3]

$$\beta_0 = \frac{2\lambda}{\pi a} \left\{ [\cosh^{-1}(\eta_T)]^2 - \left[\cosh^{-1}\left(\frac{\eta_T}{\sqrt{2}}\right)\right]^2 \right\}^{1/2} \quad (\text{rad}) \tag{3.7-13}$$

where η_T is the main lobe-to-sidelobe peak ratio

$$\eta_T \triangleq \cosh(\pi A) = 10^{-SLL/20} \tag{3.7-14}$$

We will illustrate the design procedure by an example.

Example 3.7-1 We design a Taylor's distribution to give a pattern having a maximum sidelobe level $SLL = -33$ dB. From (3.7-7) or Table 3.7-2, $A = 1.42994$. If we

---

[3] β_0 is shown in Table 3.7-2 for integer values of SLL.

elect to choose the minimum value of $\bar{n}$, then $\bar{n} = \bar{n}_{\min} = 5$. From (3.7-6).

$$\sigma_T^2 = \frac{5^2}{(1.42994)^2 + (5 - 0.5)^2} = 1.12134$$

From (3.7-5) we find that

$$F_1 = 0.324294, \quad F_2 = -0.016072$$
$$F_3 = 0.003521, \quad F_4 = -0.000227$$

From (3.7-10) and (3.7-4) we obtain Taylor's distribution

$$K_T = \{1 + 2[0.324294 - 0.016072 + 0.003521 - 0.000227]\}^{-1}$$

$$= 0.616131$$

$$E_1(\xi) = 0.616131 \ \text{rect}\left(\frac{\xi}{a}\right) \left\{1 + 0.648588 \ \cos\left(\frac{2\pi\xi}{a}\right)\right.$$

$$- 0.032144 \ \cos\left(\frac{4\pi\xi}{a}\right)$$

$$+ 0.007042 \ \cos\left(\frac{6\pi\xi}{a}\right)$$

$$\left. - 0.000454 \ \cos\left(\frac{8\pi\xi}{a}\right)\right\}$$

which is plotted in Fig. 3.7-4a. The Taylor's angular spectrum is given by (3.7-8); Fig. 3.7-4b plots $|(2\pi/aK_T)f_{a1}(u_1)|$ in decibels versus $au_1/2$ to show the sidelobe behavior. The first $(\bar{n} - 1) = 4$ sidelobes on each side of the main lobe have approximately constant peak amplitude of -33 dB but have a slight "droop" typical of Taylor patterns. Sidelobes beyond the fourth decrease as $1/u_1$ for large u_1. Finally we calculate ρ_{a1} from (3.7-11) and find that

$$\rho_{a1} = \{1 + 2[(0.324294)^2 + (0.016072)^2 + (0.003521)^2 + (0.000227)^2]\}^{-1}$$

$$= 0.82585 \quad (\text{or } -0.831 \ \text{dB})$$

Taylor's One-Parameter Distribution

Taylor has also described a distribution based on selecting only the peak sidelobe level. In this distribution the sidelobes of the angular spectrum all decrease; there is no region where their peak levels remain approximately constant (see Taylor, 1953, as discussed in Balanis, 1982, pp. 684–690). The distribution function is

$$E_1(\xi) = \text{rect}\left(\frac{\xi}{a}\right) \frac{I_0\left[\pi B \sqrt{1 - (2\xi/a)^2}\right]}{I_0(\pi B)} \tag{3.7-15}$$

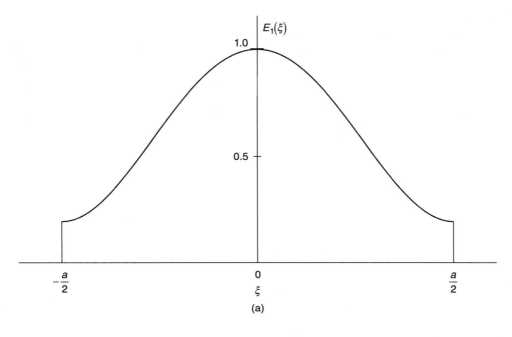

(a)

$$\frac{au_1}{2} = \left(\frac{\pi a}{\lambda}\right)\sin(\theta)\cos(\phi)$$

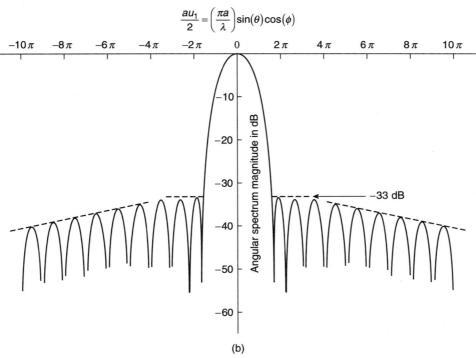

(b)

Figure 3.7-4 (*a*) Taylor distribution for $SLL = -33$ dB when $\bar{n} = \bar{n}_{\min} = 5$ and (*b*) the Taylor angular spectrum.

where $I_0(\cdot)$ is the modified Bessel function of the first kind of order zero and B is a constant established by choice of the peak sidelobe level. On inverse Fourier transformation of (3.7-15), we obtain the pattern (Balanis's results of pp. 684–685 have been put into our notation)

$$f_{a1}(u_1) = \frac{a}{2\pi I_0(\pi B)} \begin{cases} \dfrac{\sinh[\sqrt{(\pi B)^2 - (au_1/2)^2}]}{\sqrt{(\pi B)^2 - (au_1/2)^2}}, & \left(\dfrac{au_1}{2}\right)^2 < (\pi B)^2 \\[4mm] \mathrm{Sa}[\sqrt{(au_1/2)^2 - (\pi B)^2}], & \left(\dfrac{au_1}{2}\right)^2 > (\pi B)^2 \end{cases} \tag{3.7-16}$$

A relationship between B and sidelobe level can readily be established. The main lobe peak amplitude occurs when $u_1 = 0$, and its value is

$$f_{a1}(0) = \frac{a}{2\pi I_0(\pi B)} \cdot \frac{\sinh(\pi B)}{(\pi B)} \tag{3.7-17}$$

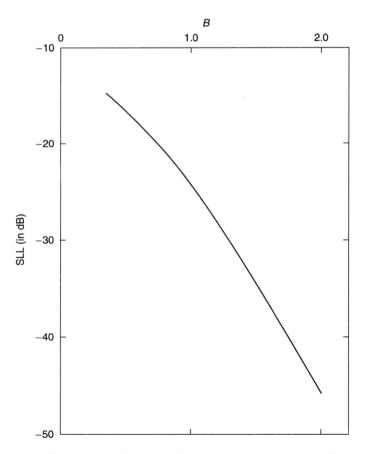

Figure 3.7-5 Variation of *SLL* versus B in Taylor's one-parameter distribution.

Sidelobes in (3.7-16) occur where $(au_1/2)^2 > (\pi B)^2$, and the first sidelobe's peak occurs at $[(au_1/2)^2 - (\pi B)^2]^{1/2} = 4.493409$ where its amplitude is $\{a/[2\pi I_0(\pi B)]\}0.217234$. The ratio of main lobe peak to first sidelobe peak amplitudes is therefore

$$\eta_T = \frac{1}{0.217234} \cdot \frac{\sinh(\pi B)}{(\pi B)} = 4.603331 \frac{\sinh(\pi B)}{(\pi B)} \qquad (3.7\text{-}18)$$

In terms of SLL (in dB), (3.7-18) becomes

$$SLL = 20 \log_{10}\left(\frac{1}{\eta_T}\right) = -13.26144 - 20 \log_{10}\left\{\frac{\sinh(\pi B)}{\pi B}\right\} \qquad (3.7\text{-}19)$$

A plot of how SLL and B are related is given in Fig. 3.7-5.

Example 3.7-2 We obtain (3.7-16) for $SLL = -33$ dB in order to compare to the similar results for the two-parameter Taylor pattern of Example 3.7-1. From (3.7-19) we find that $B = 1.420037$ corresponds to $SLL = -33$ dB. With this value of B, (3.7-16) was computed. A plot of $f_{a1}(u_1)$ normalized to have unit value at $u_1 = 0$ is shown in Fig. 3.7-6. When compared with the two-parameter case, it is found that the beamwidth is only slightly larger, but the sidelobes decrease much more rapidly in the one-parameter case.

Odd Aperture Distributions

There are radars, notably angle-tracking radars of the monopulse type, that need an antenna pattern generated from an aperture distribution that has odd symmetry. Because these distributions are so unique to angle-tracking systems, we will reserve their discussions until Chapter 14.

3.8 CIRCULARLY SYMMETRIC ILLUMINATIONS

We consider next a circular aperture of diameter a in which the illumination varies only as a function of distance ρ from the aperture's center. Since the aperture's electric field is $\hat{a}_x E_{ax}(\xi, \zeta) + \hat{a}_y E_{ay}(\xi, \zeta)$, our symmetric illumination requirement means both $E_{ax}(\xi, \zeta)$ and $E_{ay}(\xi, \zeta)$ must vary only with ρ. However, since they may, in general, be different functions of ρ, the polarization of illumination can vary over the aperture, but only as a function of ρ. Thus, although a general analysis of the symmetrically illuminated aperture involves a study of both x and y components of illumination, we will consider only one polarization, say $E_{ax}(\xi, \zeta)$. There is no real loss in generality because the radiation intensity pattern, from (3.2-6), is the sum of radiation intensity patterns due to each of the two components of illumination, and these both behave in a similar manner.

With the above assumptions in place, the radiation intensity pattern is

$$\mathscr{P}(\theta, \phi) = \mathscr{P}_x(\theta, \phi) = \frac{(2\pi)^2 k^2}{2\eta}\left[\frac{1 + \cos(\theta)}{2}\right]^2 |F_x(\theta, \phi)|^2 \qquad (3.8\text{-}1)$$

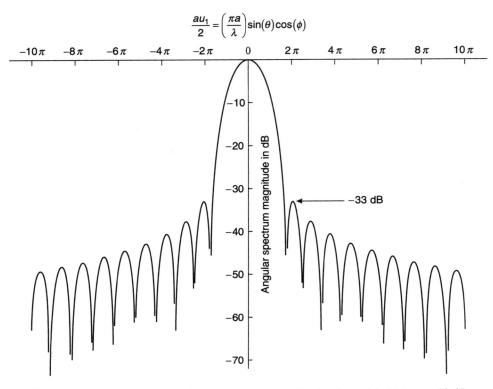

Figure 3.7-6   Taylor's pattern for a one-parameter distribution with $SLL = -33$ dB.

where

$$F_x(\theta, \phi) = \frac{1}{(2\pi)^2} \int_A \int E_{ax}(\xi, \zeta) e^{ju_1\xi + ju_2\zeta} \, d\xi d\zeta \triangleq f_{ax}(u_1, u_2) \tag{3.8-2}$$

Here u_1 and u_2 are defined by (3.3-2) and (3.3-3). With suitable changes of variables, it can be shown that (3.8-2) reduces to (see Problem 3.8-1)

$$f_{ax}(u) = f_{ax}[u_1(u), u_2(u)] = \frac{1}{2\pi} \int_0^{a/2} \rho \, E_{ax}(\rho) J_0(u\rho) \, d\rho \tag{3.8-3}$$

where

$$u = k \sin(\theta) \tag{3.8-4}$$

Equation (3.8-3) is our principal result. It is known as the *Hankel transform* of $E_{ax}(\rho)$ of order zero. The "pattern" function $f_{ax}(u)$ is the Hankel transform of the aperture illumination function $E_{ax}(\rho)$. This pattern is the principal source of behavior for $\mathscr{P}(\theta, \phi)$ through (3.8-1).

By working with (3.4-14) under the current assumptions for the circular aperture, it can be shown (see Problem 3.8-2) that the aperture efficiency is given by

$$\rho_a = \frac{8 \, | \int_0^{a/2} \rho \, E_{ax}(\rho) \, d\rho |^2}{a^2 \int_0^{a/2} \rho | E_{ax}(\rho) |^2 \, d\rho} \tag{3.8-5}$$

In a similar manner (Problem 3.8-2) the radiated power can be written as

$$P_{\text{rad}} = \frac{\pi}{\eta} \int_0^{a/2} \rho | E_{ax}(\rho) |^2 \, d\rho \tag{3.8-6}$$

Examples

Many illumination functions $E_{ax}(\rho)$ have been studied for their efficiency, sidelobe, and beamwidth behavior. We will mention just three. The circular aperture distribution is defined by

$$E_{ax}(\rho) = \begin{cases} \left[1 - \left(\dfrac{2\rho}{a} \right)^2 \right]^N, & 0 \le \rho \le \dfrac{a}{2} \\ 0, & \dfrac{a}{2} < \rho \end{cases} \tag{3.8-7}$$

It can be shown that the pattern is (see Problem 3.8-3)

$$f_{ax}(u) = \frac{a^2}{16\pi(N + 1)} \left\{ \frac{2^{N+1}(N + 1)! \, J_{N+1}(au/2)}{(au/2)^{N+1}} \right\} \tag{3.8-8}$$

where the quantity in the braces has a maximum value of unity when $u = 0$. Pertinent parameters for this distribution are shown in Table 3.8-1.

Also shown in Table 3.8-1 are data for the truncated gaussian distribution defined by

$$E_{ax}(\rho) = \begin{cases} \exp\left[-1.382 \left(\dfrac{n\rho}{a} \right)^2 \right], & 0 \le \rho \le \dfrac{a}{2} \\ 0, & \dfrac{a}{2} < \rho \end{cases} \tag{3.8-9}$$

where n is a selectable constant that determines the amount of edge illumination of the aperture.

Taylor's Distributions

Our third example distribution is due to Taylor (1960). The distribution is similar to Taylor's one-dimensional, two-parameter distribution where the smallest-beamwidth design obtained gives a prescribed peak sidelobe level where sidelobes decrease beyond sidelobe $(\bar{n} - 1)$. Here $\bar{n}$ is a parameter similar to that of the previous Taylor

TABLE 3.8-1   Aperture efficiences, beamwidths, and sidelobe levels for various circular aperture distributions.

$E_{ax}(\rho)$ for $0 \leq \rho \leq a/2$ $[E_{ax}(\rho) = 0$ for $a/2 < \rho]$	Aperture Efficiency, ρ_a	Beamwidth, $\theta_B^\circ a/\lambda$ (degrees)	Peak Sidelobe Level Relative to Maximum (dB)
Circular: $E_{ax}(\rho) = [1 - (2\rho/a)^2]^N$			
$N = 0$	1.000	57.30	-17.6
$N = 1$	0.750	70.69	-24.7
$N = 2$	0.556	82.00	-30.6
$N = 3$	0.438	92.27	-36.0
Truncated gaussian: $E_{ax}(\rho) = \exp[-1.382\,(n\rho/a)^2]$			
$n = 1.0$	0.990	60.10	-19.2
$n = 1.7$	0.924	64.17	-23.3
$n = 2.4$	0.763	70.13	-34.5
$n = 2.8$	0.646	76.66	-43.3
$n = 3.2$	0.553	84.17	-49.1
Taylor (see text)			
$\bar{n} = 2, SLL = -20$	0.966	61.65	-22.0
$\bar{n} = 3, SLL = -25$	0.914	64.17	-26.2
$\bar{n} = 4, SLL = -30$	0.846	66.69	-31.2
$\bar{n} = 5, SLL = -35$	0.774	70.70	-36.6
$\bar{n} = 6, SLL = -40$	0.706	73.74	-41.0
$\bar{n} = 8, SLL = -45$	0.651	76.43	-45.0

Sources: Beamwidths and sidelobe levels from Milligan (1985) for circular and from Barton and Ward (1969) for truncated gaussian and all data for Taylor case.

design for one-dimensional apertures. For our circular aperture of radius $a/2$, Taylor's distribution is

$$E_{ax}(\rho) = \begin{cases} \displaystyle\sum_{m=0}^{\bar{n}-1} D_m J_0\left(\frac{\mu_m 2\pi\rho}{a}\right), & 0 \leq \rho \leq \dfrac{a}{2} \\ 0, & \dfrac{a}{2} < \rho \end{cases} \tag{3.8-10}$$

Here $J_0(\cdot)$ is the Bessel function of the first kind of order zero and μ_m are constants related to zeros of $J_1(\cdot)$ as given by solutions of $J_1(\pi\mu_m) = 0$ when ordered in increasing value. The first eleven of the μ_m are shown in Table 3.8-2. The coefficients D_m are given by

$$D_m = \begin{cases} \dfrac{2}{\pi^2}, & m = 0 \\[3mm] \dfrac{2F_m}{\pi^2 [J_0(\pi\mu_m)]^2}, & m > 0 \end{cases} \tag{3.8-11}$$

TABLE 3.8-2 Values of μ_m that are the solutions of $J_1(\pi\mu) = 0$ and values of $J_0(\pi\mu_m)$.

m	μ_m	$J_0(\pi\mu_m)$
0	0.0	1.0
1	1.2196699	-0.40276
2	2.2331306	$+0.30012$
3	3.2383154	-0.24970
4	4.2410628	$+0.21836$
5	5.2427643	-0.19647
6	6.2439216	$+0.18006$
7	7.2447598	-0.16718
8	8.2453948	$+0.15672$
9	9.2458927	-0.14801
10	10.2462933	$+0.14061$

Source: $J_0(\pi\mu_m)$ values from Abramowitz and Stegun (eds., 1964, p. 409).

where

$$
F_m = \begin{cases} 1, & m = 0 \\ -J_0(\pi\mu_m)\dfrac{\prod_{n=1}^{\bar{n}-1}(1 - \{\mu_m^2/\sigma^2[A^2 + (n-0.5)^2]\})}{\prod_{\substack{n=1 \\ n \neq m}}^{\bar{n}-1}[1 - (\mu_m^2/\mu_n^2)]}, & 0 < m \leq (\bar{n}-1) \quad (3.8\text{-}12) \\ 0, & \bar{n} \leq m \end{cases}
$$

The parameters σ^2 and A in (3.8-12) are given by

$$
\sigma^2 = \frac{\mu_{\bar{n}}^2}{A^2 + (\bar{n} - 0.5)^2} \tag{3.8-13}
$$

$$
A = \frac{1}{\pi}\cosh^{-1}(10^{-SLL/20}) \tag{3.8-14}
$$

Here SLL is the design sidelobe level (a negative number of dB) and $\bar{n}$ is an integer larger than a minimum value necessary for a selected value of SLL. The choice of $\bar{n}$ will affect the shape of $E_{ax}(\rho)$, with smaller values tending to give more realistic distributions that are not peaked near the aperture edge. Hansen (1960), has given the regions of $\bar{n}$ values that lead to nonpeaked distributions; these are shown in Table 3.8-3. For a given selection of SLL in the table, the larger values of $\bar{n}$ give smaller beamwidths, and sidelobes do not decrease significantly until farther out (beyond sidelobe $\bar{n} - 1$). Often the decrease in beamwidth does not justify the larger values of $\bar{n}$.

On using (3.8-10) in (3.8-3), the Taylor pattern is found to be

$$
f_{ax}(u) = \frac{a^2}{(2\pi)^3}\sum_{m=0}^{\bar{n}-1} D_m J_0(\pi\mu_m)\frac{\pi^2(au/2)J_1(au/2)}{(au/2)^2 - (\pi\mu_m)^2} \tag{3.8-15}
$$

TABLE 3.8-3 Values of $\bar{n}$ for which the Taylor distribution is not peaked near the aperture's edge.

SLL in dB	Values of $\bar{n}$ That Give a Nonpeaked Distribution
$-20 < SLL$	None
$-25 < SLL \leq -20$	3
$-29 < SLL \leq -25$	3, 4
$-33 < SLL \leq -29$	3, 4, 5
$-36 < SLL \leq -33$	4, 5, 6
$-39 < SLL \leq -36$	4, 5, 6, 7
$-40 < SLL \leq -39$	4, 5, 6, 7, 8

Source: Data from Hansen (1960).

where (3.8-4) gives u. An alternative form for $f_{ax}(u)$ is also given by Taylor (1960):

$$f_{ax}(u) = \frac{a^2}{(2\pi)^3} \frac{2J_1(au/2)}{(au/2)} \prod_{n=1}^{\bar{n}-1} \left\{ \frac{1 - \dfrac{(au/2)^2}{\pi^2\sigma^2[A^2 + (n-0.5)^2]}}{1 - \dfrac{(au/2)^2}{(\pi\mu_n)^2}} \right\} \tag{3.8-16}$$

The 3-dB beamwidth of the pattern of (3.8-16) is

$$\theta_{\mathrm{B}} \approx \sigma\beta_0 \tag{3.8-17}$$

where σ and β_0 are given by (3.8-13) and (3.7-13), respectively. Finally, from (3.8-5) using (3.8-10), it can be shown (see Problem 3.8-5) that aperture efficiency is

$$\rho_a = \frac{1}{1 + (\pi^2/2)^2 \sum_{m=1}^{\bar{n}-1} [D_m J_0(\pi\mu_m)]^2} \tag{3.8-18}$$

Example 3.8-1 As an example we again select $SLL = -33$ dB (with $\bar{n}$ now equal to 3) and compare our distribution and pattern to those of Example 3.7-1 for the one-dimensional aperture. From Table 3.7-2, $A = 1.42994$. We use the first four μ_m from Table 3.8-2 in (3.8-13) to get $\sigma^2 = 1.26426$, and in (3.8-11) to get $D_0 = 0.20264$, $D_1 = 0.25365$, and $D_2 = -0.01683$. The distribution was calculated by computer from (3.8-10); it was almost identical to the plot of Fig. 3.7-4a for $0 \leq \xi \leq 0.4a$. For $0.4a < \xi \leq 0.5a$ the plot was progressively larger than in Fig. 3.7-4a, with a final value of 0.217 which is only 13% above the 0.192 value of Fig. 3.7-4a. The pattern was computed from (3.8-16) and compared to Fig. 3.7-4b for the one-dimensional aperture case. Beamwidth here was $\theta_{\mathrm{B}}^{\circ} = 70.06\lambda/a$ compared to $66.48\lambda/a$ for the one-dimensional aperture, an increase of 5.39%. In the current example the first sidelobe was down -33.73 dB compared to -33.25 dB for the one-dimensional aperture. Sidelobes decreased more rapidly also (since $\bar{n} = 3$ here compared to $\bar{n} = 5$ before); the ninth sidelobe was down -47.23 dB compared to -39.89 dB previously. Otherwise, the general pattern's behavior is similar to that of Fig. 3.7-4b. From (3.8-18), ρ_a was found to be 0.80.

3.9 SOME EXAMPLE ANTENNAS OF THE REFLECTOR TYPE

There are many practical radar antennas that can be modeled as an aperture antenna. To add some practicality to our basic principles, we will describe three. The second and third may be viewed as variants of the first, which is based on the characteristics of a parabola.

Parabolic Reflectors

A parabola having symmetry about the z axis is sketched in Fig. 3.9-1a. A characteristic of the parabola is that from a point called its focus, the length of a path 0A + AB to a point on the line CD is constant, regardless of where point A is taken on the parabola. This equal-length path behavior is true for any line CD that is parallel to the y axis.

If a source of radiation is placed at the focus and the parabola is made of a conducting material so that waves reflect from its surface, every ray from the source travels the same distance and all arrive at line CD with the same phase, traveling in the z direction. Together all ray paths from the surface form a wavefront along line CD. For a parabola of width b, the wavefront effectively forms the source of radiation for a one-dimensional aperture of width b. A more realistic antenna is formed if the parabola is extended in the x direction by an amount a, as shown in Fig. 3.9-1b. The resulting antenna is called a *parabolic cylinder*. The aperture for this antenna is rectangular with area ab. The source, called a *feed*, for the antenna is now a line source along the x axis. In a realistic antenna the feed is an antenna in its own right. Its purpose, however, is mainly to radiate toward the reflector, so its radiation pattern would have a beamwidth in the y, z plane that is less than the angle subtended by the parabolic cylinder. Similarly the feed's beamwidth in the x, z plane would be somewhat less than the angle subtended by the width a [which is $2 \tan^{-1}(a/2f)$, where f is the focal length of the parabola].

Feed design is important. If the feed's beamwidths are too broad, significant power is radiated toward points outside the reflector. This wasted power, called *spillover*, reduces aperture efficiency. On the other extreme, if the feed's beamwidths are too narrow, power is directed mainly toward the center area of the reflector, and the outer areas (illuminated through the feed's sidelobes) are not efficiently used. Again the result is a loss in efficiency. For most reflector-type antennas there is an optimum reached when the feed's pattern (called the *primary pattern*) illuminates the reflector's edges at a level of about 8 to 12 dB below that at the reflector's center. This level gives an overall antenna pattern (called the *secondary pattern*) with maximum aperture efficiency.

The parabolic cylinder is a useful antenna when the secondary pattern is to have a fan-type beam. An *aspect ratio*, the radio of the larger to the smaller beamwidths of the secondary pattern in two principal planes, of over 8 : 1 is achievable (Skolnik, 1980, p. 236). The smaller beamwidth is typically in the plane of the parabola. Fan beams are applicable to radars trying to detect objects at a broad range of elevation angles but in a narrow range of azimuth angle; here the parabola would lie in the horizontal plane. Alternatively, rotation by 90 degrees places the parabola and the narrow beamwidth vertically (the wide fan is horizontal). This antenna is used to detect targets in a small range of elevation angles but anywhere in a wide range of azimuth angles.

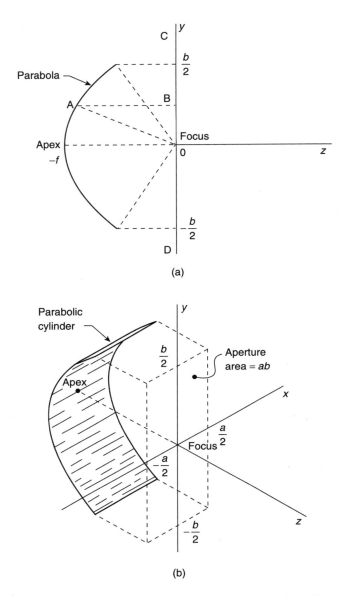

Figure 3.9-1 (*a*) A parabolic reflector and (*b*) parabolic cylinder.

Another valuable antenna derives from rotation of a parabola about its axis of symmetry (*z* axis in Fig. 3.9-1*a*). Figure 3.9-2 illustrates the resulting reflector, called a *paraboloid*. The secondary pattern of the paraboloid has a pencil-beam main lobe where beamwidths in the two principal planes are nearly equal. A paraboloid is typically fed from the focus by some form of waveguide horn or combination of horns. Because of its appearance a paraboloid is often called a *dish*. One form is often seen being used in a fixed position to receive signals from synchronous satellites. In radar, however, the paraboloid is often mechanically movable and used in angle-tracking

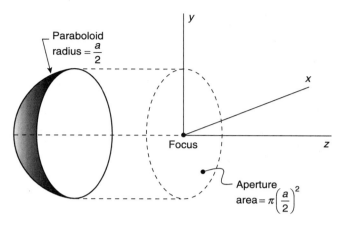

Figure 3.9-2   Geometry of a paraboloidal reflector antenna.

systems. Some extremely accurate and precise angle-tracking radars have used the paraboloid for tracking a variety of targets, even missiles. But, because of the size, mass, and inertia, the paraboloid is limited in its ability to physically follow some fast targets with high angular accelerations. Such targets are best tracked using the phased array antenna (Section 3.10 to follow) which may have electronic beam steering to eliminate mechanical motion.

The feed at the paraboloid's focus must be supported. Three or more supporting struts are often used. Each runs from the edge of the antenna to the focus to form a supporting "tripod." The arrangement presents certain practical problems, one being blockage of the transmitted wave. Even the feed itself causes blockage. Some wave energy is scattered by these structures, which reduces efficiency, while some energy reflects directly back from the reflector and attempts to return to the trans- mitter through the feed. The latter energy can be controlled somewhat by use of impedance-matching devices added to the feed. Other techniques are also possible (Skolnik, 1980, p. 238).

For a detailed analysis of the focus-fed paraboloid, the reader is referred to Balanis (1982). In particular, Balanis gives the aperture electric field needed in calculating the radiation intensity pattern from (3.2-11) using (3.2-9).

Cassegrain Reflectors

One disadvantage of the focus-fed paraboloid is the relatively long length of wave- guide needed to reach the feed. Path length can be shortened by use of the *Cassegrain reflector* illustrated in Fig. 3.9-3. In this arrangement the real feed is placed at or in front of the parabolic surface. It illuminates a second reflector, called a *subreflector*, that has the shape of a hyperbola. Rays that reflect from the hyperbola proceed to reflect normally from the parabola to form a plane wavefront along line CD, as in the parabolic reflector. In essence the subreflector has folded (reflected) rays such that the normal focal point becomes a virtual focus and the real focus is near the apex of the parabola. The actual real focal point will depend on the design of the hyperbola and is one of its foci. The most important practical antenna based on the Cassegrain

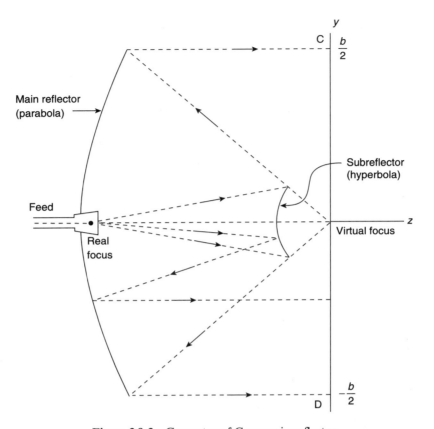

Figure 3.9-3   Geometry of Cassegrain reflector.

reflector results from rotation of the geometry of Fig. 3.9-3 about the z axis to obtain a paraboloid with a hyperboloidal subreflector.

The Cassegrain paraboloid suffers from blockage due to the subreflector and its supports, while some energy also reflects directly back to the feed from the subreflector. The former problem, blockage, can be reduced by making the subreflector smaller. Prevention of excessive spillover with a smaller subreflector requires either the feed be moved closer to the subreflector (increasing the transmission line length) or increasing the size of the feed to make its primary pattern more directive. However, increasing feed size emphasizes the latter problem of feed pickup by back-reflection. The best compromise in those two problems is said to occur when the feed and subreflector have equal areas (Hannan, 1961).

Reflectors Using Polarization Twisting

A novel technique can be used to reduce blocking due to the subreflector in a Cassegrain antenna when the antenna radiates linear polarization. The technique is based on twisting a wave's polarization by 90 degrees in space. Figure 3.9-4 illustrates how this polarization rotation occurs. We examine what happens to a wave that is linearly

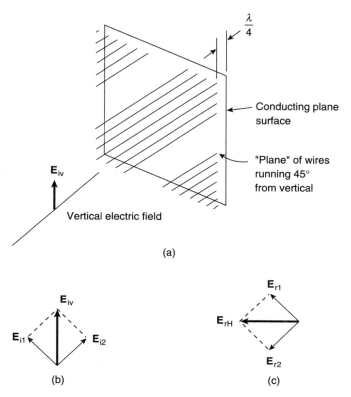

Figure 3.9-4 A polarization twisting reflector (*a*), the applicable incident (*b*), and reflected (*c*) electric fields.

polarized in the vertical direction when it reflects from a conducting surface, in front of which there are closely spaced parallel conducting wires arranged at an angle of 45 degrees from the vertical, as shown in panel *a*.

Figure 3.9-4*b* shows the incident wave's electric field E_{iV} at the instant it arrives at the plane of the wires. It can be considered as made up of components; one, E_{i1}, is perpendicular to the direction of the wires, and another, E_{i2}, is parallel with the wires. Component E_{i2} reflects directly from the wires becoming E_{r2} in panel *c* which shows fields for the wave reflecting from the wire-surface structure. Wave component E_{i1} in panel *b* passes *through* the wires unaffected; it travels a distance $\lambda/4$, reflects off the reflector, and then travels a distance $\lambda/4$ to emerge from the wires. Since the total travel is λ (the reflection is equivalent to 180 degrees or $\lambda/2$), E_{i1} emerges as E_{r1} in panel *c*. The resultant electric field reflected from the structure is E_{rH} which is now a horizontally polarized field.

The ability of the wire-surface structure to rotate a linearly polarized wave 90 degrees in space is used in the Cassegrain antenna of Fig. 3.9-5 to reduce the subreflector's blockage. The feed produces a horizontally polarized wave (electric field vector out of paper) that reflects from closely spaced parallel wires making up the subreflector. The horizontally polarized wave reflected from the subreflector then strikes the main reflector, which is equipped with parallel wires arranged to rotate

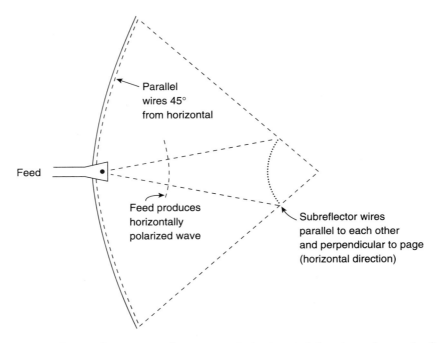

Feed

Parallel
wires 45°
from horizontal

Feed produces
horizontally
polarized wave

Subreflector wires
parallel to each other
and perpendicular to page
(horizontal direction)

Figure 3.9-5 Cassegrain antenna that uses polarization twisting to reduce subreflector blockage.

its reflected wave to vertical polarization. The vertically polarized wave now passes through the wires of the subreflector almost as though it were transparent. The net effect is greatly reduced blockage.

For other antennas and applications that use the polarization rotation of a wire-reflector structure, the reader is referred to Skolnik (1980).

3.10   ARRAY ANTENNAS

An array antenna is one that consists of a number (often large) of small radiators acting together within some overall area to produce the effect of an antenna having the overall area. In general, the overall area containing the radiators, called *elements*, does not have to be planar, but it often is in practical antennas. Because the most common phased array is planar, we will concentrate on this antenna form.

Figure 3.10-1 illustrates an area within which various small radiators act together to form a phased array. A typical radiator, the *mn*th, is located in the plane by its vector position $\mathbf{r}_{mn}$ which has a component ξ_m in the x direction and component ζ_n in the y direction. Thus

$$\mathbf{r}_{mn} = \xi_m \hat{a}_x + \zeta_n \hat{a}_y \qquad (3.10\text{-}1)$$

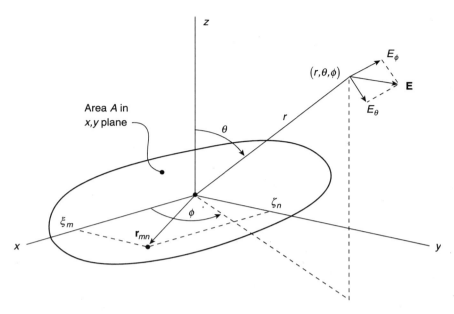

Figure 3.10-1 Geometry of planar array.

where $\hat{a}_x$ and $\hat{a}_y$ are unit vectors in the x and y directions, respectively. The projection of the radiator's location onto the line from the origin to a distant point (r, θ, ϕ) is

$$\mathbf{r}_{mn} \cdot \hat{a}_r = \xi_m \sin(\theta)\cos(\phi) + \zeta_n \sin(\theta)\sin(\phi) \tag{3.10-2}$$

where $\hat{a}_r$ is a unit vector in direction (θ, ϕ).

These results along with some reasonable assumptions will allow a simple discussion of the planar phased array.

We assume field points (r, θ, ϕ) of interest all lie in the far field. A consequence of this assumption is that the distances to the point (r, θ, ϕ) from all radiators are nearly the same and that the amplitude drop in field strength (as $1/r$) is the same for all radiators. Another consequence is that the phase shift of waves from the mnth element to point (r, θ, ϕ) is proportional to (3.10-2).

We also make three other assumptions. First, that no mutual coupling effects exist between any pairs of elements, and second, that a "reference" element radiates from the origin. The latter assumption is for ease of understanding and is not a theoretical constraint. Finally we assume that all elements (radiators) are identical with identical radiation intensity patterns that are maximum in the *broadside direction* (in the z direction).

Array Factor

Within the constraints of our assumptions, the "reference" radiator at the origin will produce a vector electric field at point (r, θ, ϕ) given by

$$\mathbf{E}_{00} = I_{00} \left[\hat{a}_\theta E_\theta + \hat{a}_\phi E_\phi \right] \tag{3.10-3}$$

where $\hat{a}_\theta$ and $\hat{a}_\phi$ are unit vectors in directions θ and ϕ, respectively (see Fig. 3.10-1), I_{00} is a (possibly complex) constant representing how "strongly" the element is being driven, and E_θ and E_ϕ are amplitudes of field components when $I_{00} = 1$. Now because all elements are assumed to have the same radiation pattern, the fields at (r, θ, ϕ) due to the mnth element are

$$\mathbf{E}_{mn} = I_{mn}[\hat{a}_\theta E_\theta + \hat{a}_\phi E_\phi] e^{jk\mathbf{r}_{mn}\cdot\hat{a}_r}$$

$$= I_{mn}[\hat{a}_\theta E_\theta + \hat{a}_\phi E_\phi] e^{jk\xi_m \sin(\theta)\cos(\phi) + jk\zeta_n \sin(\theta)\sin(\phi)} \tag{3.10-4}$$

where I_{mn} is a constant and $k = 2\pi/\lambda$, as usual. The total vector field at (r, θ, ϕ) due to all radiators is the sum of contributions from all elements

$$\mathbf{E} = [\hat{a}_\theta E_\theta + \hat{a}_\phi E_\phi] \sum_m \sum_n I_{mn} e^{jk\mathbf{r}_{mn}\cdot\hat{a}_r} \tag{3.10-5}$$

The bracketed factor in (3.10-5) is a common factor due to the (identical) elements; it is called the *element factor*. The factor involving the sums is entirely due to the array, and it is called the *array factor*,[4] denoted here by $F_{\mathrm{array}}(\theta, \phi)$:

$$F_{\mathrm{array}}(\theta, \phi) = \sum_m \sum_n I_{mn} e^{jk\mathbf{r}_{mn}\cdot\hat{a}_r} \tag{3.10-6}$$

The array factor is the principal means by which the phased array's performance and advantages are derived. Overall beamwidths and pattern shape are determined by the geometry of the area, locations of elements in the area, and the magnitudes and phases of the element illumination constants I_{mn}. The principal advantage of the phased array, that of being able to electronically move the antenna's main beam in space, results, as we will see, from controlling the phase angles of the illumination constants I_{mn}.

Radiation Intensity Pattern

We develop the Poynting vector for the wave at point (r, θ, ϕ) in Fig. 3.10-1 by using (3.10-5):

$$\mathscr{P}(r, \theta, \phi) = \frac{1}{2\eta} |\mathbf{E}|^2 \hat{a}_r$$

$$= \frac{1}{2\eta} \hat{a}_r [|E_\theta|^2 + |E_\phi|^2] |F_{\mathrm{array}}(\theta, \phi)|^2 \quad (\mathrm{W/m^2}) \tag{3.10-7}$$

---

[4] Array factor is sometimes called a *space factor*.

From (3.2-1) the radiation intensity pattern becomes

$$\mathscr{P}(\theta, \phi) = r^2 |\mathscr{P}(r, \theta, \phi)| = \mathscr{P}_e(\theta, \phi) \mathscr{P}_{array}(\theta, \phi) \tag{3.10-8}$$

where the radiation intensity pattern of the *element* is defined by

$$\mathscr{P}_e(\theta, \phi) = \frac{r^2}{2\eta} \left[|E_\theta|^2 + |E_\phi|^2 \right] \quad (\text{W/sr}) \tag{3.10-9}$$

from (3.2-2), and

$$\mathscr{P}_{array}(\theta, \phi) \triangleq |F_{array}(\theta, \phi)|^2 \quad (\text{unitless}) \tag{3.10-10}$$

will be called the *array radiation factor*.

From (3.10-8) the radiation intensity pattern of the overall array is the product of that of the element factor and the factor $\mathscr{P}_{array}(\theta, \phi)$. Usually $\mathscr{P}_e(\theta, \phi)$ is a broad pattern in θ and ϕ, while the array radiation factor typically has a much narrower main beam. As a consequence the main behavior of $\mathscr{P}(\theta, \phi)$ is determined by $\mathscr{P}_{array}(\theta, \phi)$, and we may concentrate on its characteristics, which is equivalent to concentrating on the behavior of the array factor. From use of (3.10-2) and (3.10-6) in (3.10-10), we write

$$\mathscr{P}_{array}(\theta, \phi) = |F_{array}(\theta, \phi)|^2$$
$$= \left| \sum_m \sum_n I_{mn} e^{jk\xi_m \sin(\theta)\cos(\phi) + jk\zeta_n \sin(\theta)\sin(\phi)} \right|^2 \tag{3.10-11}$$

In the most common case the constants I_{mn} are real numbers when the main beam is not being electronically controlled. For this case we use the inequality

$$\left| \sum_m \sum_n z_{mn} \right| \leq \sum_m \sum_n |z_{mn}| \tag{3.10-12}$$

(Hardy et al., 1964, p. 3) for complex numbers z_{mn} to find the maximum of $\mathscr{P}_{array}(\theta, \phi)$:

$$\mathscr{P}_{array}(\theta, \phi) \leq \left[\sum_m \sum_n |I_{mn}| \right]^2 = \mathscr{P}_{array}(\theta, \phi)_{max} \tag{3.10-13}$$

Note that the equality in (3.10-13), corresponding to the maximum of $\mathscr{P}_{array}(\theta, \phi)$, is achieved when the exponent of e in (3.10-11) is zero, and this occurs when $\theta = 0$. Thus the radiation intensity pattern of a phased array with no electronic steering is maximum in the broadside direction (z axis direction where $\theta = 0$).

Example 3.10-1   We find $\mathscr{P}_{array}(\theta, \phi)_{max}$ when all elements radiate with the same intensity and $I_{00} = 1$. In this case $I_{mn} = I_{00}$, all m and n, so (3.10-13) gives

$$\mathscr{P}_{array}(\theta, \phi)_{max} = \left[\sum_m \sum_n |I_{mn}| \right]^2 = \left[\sum_m \sum_n 1 \right]^2 = M_{array}^2$$

where M_{array} is the total number of elements in the array's area A. This example corresponds to a uniformly illuminated array.

Beam Steering

In the above discussion it was found that the maximum of the main beam in a phased array was in the direction of the z axis (broadside) when the element constants I_{mn} had no phase angles. The maximum corresponded to the exponent of e in (3.10-11) being zero. Now let us examine the case where a phase angle can be added to the elements to cause the main lobe maximum to point to a different direction, say (θ_0, ϕ_0). The new main lobe's direction will now be (θ_0, ϕ_0) provided that the exponent of e is again set to zero with the phase shifts present. If element mn has phase shift ψ_{mn}, we require that

$$k\xi_m \sin(\theta_0)\cos(\phi_0) + k\zeta_n \sin(\theta_0)\sin(\phi_0) + \psi_{mn} = 0 \qquad (3.10\text{-}14)$$

If we treat ψ_{mn} as the sum of two phases according to

$$\psi_{mn} = \psi_m + \psi_n \qquad (3.10\text{-}15)$$

then, from (3.10-14), the required values of ψ_m and ψ_n are

$$\psi_m \triangleq - k\xi_m \sin(\theta_0)\cos(\phi_0) \qquad (3.10\text{-}16)$$

$$\psi_n \triangleq - k\zeta_n \sin(\theta_0)\sin(\phi_0) \qquad (3.10\text{-}17)$$

We may view ψ_m and ψ_n as phases due to phase gradients over the array in x and y directions, respectively.

Alternatively, (3.10-16) and (3.10-17) can be solved together to give

$$\tan(\phi_0) = \frac{\xi_m \psi_n}{\zeta_n \psi_m} \qquad (3.10\text{-}18)$$

$$\sin^2(\theta_0) = \left(\frac{\psi_m}{k\xi_m}\right)^2 + \left(\frac{\psi_n}{k\zeta_n}\right)^2 \leq 1 \qquad (3.10\text{-}19)$$

which locate the direction (θ_0, ϕ_0) to which the main lobe is steered for given phases ψ_m and ψ_n and given element locations (ξ_m, ζ_n). The inequality in (3.10-19) is required, since the sine of a real angle cannot exceed unity; the condition is necessary to guarantee that a main lobe exists for larger steering angles.

With beam steering to direction (θ_0, ϕ_0), the array radiation factor of (3.10-11) becomes

$$\mathscr{P}_{array}(\theta, \phi) = \left| \sum_m \sum_n |I_{mn}| \, e^{jk\xi_m[\sin(\theta)\cos(\phi) - \sin(\theta_0)\cos(\phi_0)]} \right. $$
$$\left. \cdot e^{jk\zeta_n[\sin(\theta)\sin(\phi) - \sin(\theta_0)\sin(\phi_0)]} \right|^2 \qquad (3.10\text{-}20)$$

Directivity and Other Fundamental Parameters

The planar array can be considered to be an aperture with discrete radiators over its area. Directivity is given by (3.4-6) which can be written as

$$G_D = \frac{4\pi \mathscr{P}(\theta, \phi)_{max}}{P_{rad}} = \frac{4\pi [\mathscr{P}_e(\theta, \phi) \mathscr{P}_{array}(\theta, \phi)]_{max}}{\int_0^{2\pi} \int_0^{\pi} \mathscr{P}_e(\theta, \phi) \mathscr{P}_{array}(\theta, \phi) \sin(\theta) \, d\theta d\phi} \qquad (3.10\text{-}21)$$

where (3.10-8) has been used in the numerator and (3.2-13) applies to the denominator. Because the array factor can be steered to have a maximum different from broadside, the numerator of (3.10-21) will depend on the beam-steering direction (θ_0, ϕ_0). However, $\mathscr{P}_e(\theta, \phi)$ usually is a broad function of θ and ϕ compared to $\mathscr{P}_{array}(\theta, \phi)$ such that $[\mathscr{P}_e(\theta, \phi) \mathscr{P}_{array}(\theta, \phi)]_{max}$ will occur approximately in direction (θ_0, ϕ_0). When this condition is true, we have

$$G_D = \frac{4\pi \, \mathscr{P}_e(\theta_0, \phi_0) |F_{array}(\theta_0, \phi_0)|^2}{\int_0^{2\pi} \int_0^{\pi} \mathscr{P}_e(\theta, \phi) |F_{array}(\theta, \phi)|^2 \sin(\theta) \, d\theta d\phi} \qquad (3.10\text{-}22)$$

where

$$|F_{array}(\theta_0, \phi_0)|^2 = \left| \sum_m \sum_n |I_{mn}| \right|^2 \qquad (3.10\text{-}23)$$

from (3.10-20).

The directivity of an array is usually difficult to determine, mainly because of the integral in the denominator of (3.10-22), which is the radiated power. However, once G_D is found, either by analytical means or by computer, other fundamental parameters may be found, such as solid angle from (3.4-6), effective area from (3.4-10), and aperture efficiency from (3.4-15).

3.11 RECTANGULAR PLANAR ARRAY

We illustrate the planar array by considering a rectangular array centered on the origin as shown in Fig. 3.11-1. There are $(2N_x + 1)$ elements along any row in the x direction and $(2N_y + 1)$ rows in the y direction. The total number of elements is $(2N_x + 1)(2N_y + 1)$. Elements are placed on a *rectangular lattice*, with d_x and d_y being the separations between any adjacent pair of elements in the x and y directions, respectively. Thus element locations are defined by

$$\xi_m = md_x, \qquad -N_x \le m \le N_x \qquad (3.11\text{-}1)$$

$$\zeta_n = nd_y, \qquad -N_y \le n \le N_y \qquad (3.11\text{-}2)$$

We also assume that element illumination is factorable, that is,

$$I_{mn} = I_{mx} I_{ny}, \qquad \text{where } I_{0x} = I_{0y} = 1 \qquad (3.11\text{-}3)$$

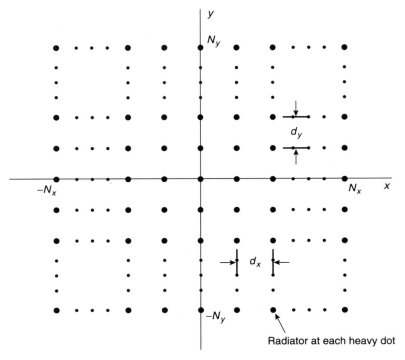

Figure 3.11-1 Rectangular array of $(2N_x + 1)(2N_y + 1)$ radiators located on a rectangular lattice.

Before discussing details, it is convenient to use Fig. 3.11-2 to define a *direction cosine* as the cosine of the angle from a coordinate axis to the line defining an arbitrary direction (θ, ϕ) in space. Two direction cosines, $\cos(\psi_x)$ and $\cos(\psi_y)$, are of most interest and are given by

$$\cos(\psi_x) = \sin(\theta)\cos(\phi) \tag{3.11-4}$$

$$\cos(\psi_y) = \sin(\theta)\sin(\phi) \tag{3.11-5}$$

The direction cosines for a specific direction, such as (θ_0, ϕ_0), the direction to which the array's main beam is steered, become

$$\cos(\psi_{x0}) = \sin(\theta_0)\cos(\phi_0) \tag{3.11-6}$$

$$\cos(\psi_{y0}) = \sin(\theta_0)\sin(\phi_0) \tag{3.11-7}$$

Array Factor

On using the preceding definitions in (3.10-20), the array factor for the rectangular array when its beam is steered to angles ψ_{x0} and ψ_{y0} from the x and y axes, respectively, becomes

$$F_{\text{array}}(\theta, \phi) = \sum_{m=-N_x}^{N_x} |I_{mx}| e^{jmkd_x[\cos(\psi_x) - \cos(\psi_{x0})]}$$

$$\cdot \sum_{n=-N_y}^{N_y} |I_{ny}| e^{jnkd_y[\cos(\psi_y) - \cos(\psi_{y0})]}$$

$$= F_{\text{array}, x}(\psi_x) \cdot F_{\text{array}, y}(\psi_y) \tag{3.11-8}$$

where

$$F_{\text{array}, x}(\psi_x) \triangleq \sum_{m=-N_x}^{N_x} |I_{mx}| e^{jmkd_x[\cos(\psi_x) - \cos(\psi_{x0})]} \tag{3.11-9}$$

$$F_{\text{array}, y}(\psi_y) \triangleq \sum_{n=-N_y}^{N_y} |I_{ny}| e^{jnkd_y[\cos(\psi_y) - \cos(\psi_{y0})]} \tag{3.11-10}$$

Our principal result is (3.11-8) which states that $F_{\text{array}}(\theta, \phi)$ is the product of two factors, each of which is equivalent to a linear array of elements. One factor, (3.11-9), is equivalent to $(2N_x + 1)$ elements linearly placed along the x axis. The other, (3.11-10), represents $(2N_y + 1)$ elements along the y axis. Each factor is a function of only one coordinate [ψ_x for (3.11-9) and ψ_y for (3.11-10)] and has its main beam steered in its coordinate. For example, (3.11-9) might appear as shown in Fig. 3.11-3 for the x, z plane cut; the full factor is a surface of revolution about the x axis, which produces a conical main beam. The factor of (3.11-10) behaves in a similar manner except that it is a surface of revolution about the y axis. The main beam of $F_{\text{array}}(\theta, \phi)$ becomes the

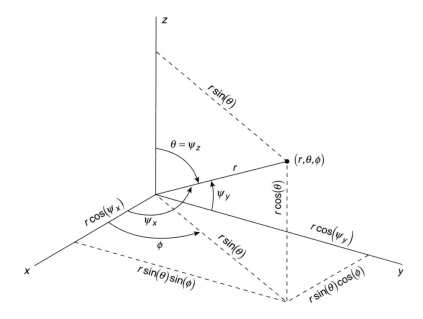

Figure 3.11-2 Geometry defining direction cosines.

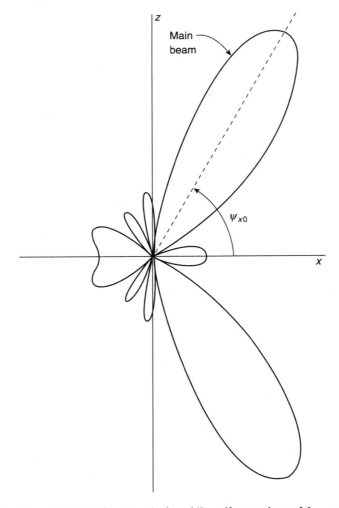

Figure 3.11-3   Plot of (3.11-9) for $N_x = 2$, $d_x = \lambda/2$, uniform values of I_{mx}, and $\psi_{x0} = \pi/3$.

intersection of the two conical beams. Two main beams result from the intersection, one in the half sphere $z \geq 0$ and the other for $z < 0$. The beam for $z < 0$ is typically suppressed by the pattern of the element or by use of a ground plane behind the radiators to prevent back-radiation.

To gain further insight into the shape of one of the factors (3.11-9) or (3.11-10), we subsequently look at the linear array in some detail (Section 3.12). First, however, we consider the beamwidths of the rectangular array.

Beamwidths

The three-decibel beamwidth of a main lobe is usually measured in two orthogonal planes that intersect in the direction of the maximum. For an array's main beam directed broadside, we denote 3-dB beamwidths by θ_B and ϕ_B when measured in the

x, z and y, z planes, respectively. For a main beam steered off broadside, the beamwidths increase. We use Fig. 3.11-4 to define 3-dB beamwidths when the beam is steered to direction (θ_0, ϕ_0). Beamwidth θ_{B0} is measured in the plane containing the z axis and the axis of the beam. Beamwidth ϕ_{B0} is measured in the orthogonal plane containing the beam's axis. Elliott (1964) has given approximate expressions for θ_{B0} and ϕ_{B0} that are applicable to large arrays where the main beam is not scanned too far off broadside:

$$\theta_{B0} = \frac{\theta_B \phi_B}{\sqrt{\cos^2(\theta_0)[\phi_B^2 \cos^2(\phi_0) + \theta_B^2 \sin^2(\phi_0)]}} \tag{3.11-11}$$

$$\phi_{B0} = \frac{\theta_B \phi_B}{\sqrt{\phi_B^2 \sin^2(\phi_0) + \theta_B^2 \cos^2(\phi_0)}} \tag{3.11-12}$$

Values of θ_B and ϕ_B are most easily found as the beamwidths of factors (3.11-9) and (3.11-10) in their respective variables ψ_x and ψ_y when the beams are broadside (i.e., for $\psi_{x0} = \pi/2$ and $\psi_{y0} = \pi/2$).

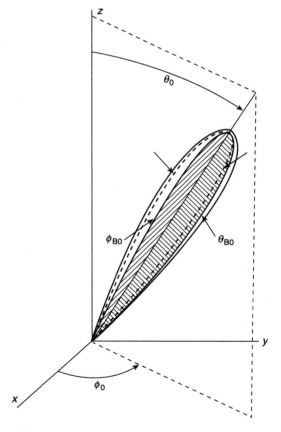

Figure 3.11-4   Beamwidths in two orthogonal planes for a main beam steered to direction (θ_0, ϕ_0).

In a square array with $N_x = N_y$ and $d_x = d_y$, the beamwidths θ_B and ϕ_B are equal, and

$$\theta_{B0} = \frac{\theta_B}{\cos(\theta_0)} \tag{3.11-13}$$

$$\phi_{B0} = \phi_B \tag{3.11-14}$$

We observe that θ_{B0} of (3.11-13) does not depend on ϕ_0, while ϕ_{B0} of (3.11-14) does not depend on either θ_0 or ϕ_0.

According to Elliott (1964, p. 78) the above beamwidth expressions are quite good for beams steered to within several beamwidths of the limiting condition of no main beam, as defined by the equality in (3.10-19).

Directivity

Equation (3.10-22) defines directivity and, as noted previously, is generally difficult to find because of the integrals involved. However, under some reasonable assumptions, Elliott (1964) has obtained an approximate solution for large arrays that have their main beam scanned to angles such that θ_0 is less than $\pi/2$ by at least several beamwidths. If the element factor of the array is approximately constant in the upper hemisphere, $\theta < \pi/2$, and is approximately zero for the lower hemisphere, G_D becomes

$$G_D \approx \frac{4\pi \left|\sum_m |I_{mx}|\right|^2 \left|\sum_n |I_{ny}|\right|^2}{\int_0^{2\pi} \int_0^{\pi/2} |F_{\text{array}}(\theta, \phi)|^2 \sin(\theta)\, d\theta d\phi} \tag{3.11-15}$$

where (3.11-3) has been used and $F_{\text{array}}(\theta, \phi)$ is given by (3.11-8). Elliott (1964) has found that (3.11-15) reduces to

$$G_D \approx \pi \cos(\theta_0)\, G_{Dx} G_{Dy} \tag{3.11-16}$$

where G_{Dx} and G_{Dy} are directivities of two linear arrays. In one, G_{Dx} corresponds to $(2N_x + 1)$ elements centered on and spaced along the x axis. In the other, corresponding to G_{Dy}, there are $(2N_y + 1)$ elements similarly positioned along the y axis. G_{Dx} and G_{Dy} are found from (3.12-14) in the following section.

3.12 LINEAR ARRAY

Since the behavior of the rectangular array of Section 3.11 is directly related to the characteristics of a linear array, we consider the linear array in more detail. We take the example of an array with elements arranged along the x axis.[5] As before, we have $(2N_x + 1)$ elements centered on the origin and uniformly separated by d_x. The array factor can be obtained from (3.10-20) when only one row is taken in the y direction.

---

[5] Discussions and results are similar for linear arrays with elements along other axes.

Element locations and illuminations are given by (3.11-1) and (3.11-3), respectively. Thus

$$F_{\text{array}}(\theta, \phi) = \sum_{m=-N_x}^{N_x} |I_{mx}| \, e^{jmkd_x[\sin(\theta)\cos(\phi) - \sin(\theta_0)\cos(\phi_0)]} \qquad (3.12\text{-}1)$$

An alternative form derives from substitution of (3.11-4) and (3.11-6):

$$F_{\text{array}}(\psi_x) = \sum_{m=-N_x}^{N_x} |I_{mx}| \, e^{jmkd_x[\cos(\psi_x) - \cos(\psi_{x0})]} \qquad (3.12\text{-}2)$$

This form is useful in interpreting the behavior of the array factor and in introducing the concept of *grating lobes*.

Beam Steering and Grating Lobes

We consider the case of uniform illumination which allows a convenient solution for the sum in (3.12-2). By using the known sum

$$\sum_{m=-N}^{N} e^{jm\beta} = \frac{\sin[(2N+1)\beta/2]}{\sin(\beta/2)} \qquad (3.12\text{-}3)$$

we can write (3.12-2) as

$$F_{\text{array}}(\psi_x) = \frac{\sin\left\{(2N_x+1)\dfrac{kd_x}{2}[\cos(\psi_x) - \cos(\psi_{x0})]\right\}}{\sin\left\{\dfrac{kd_x}{2}[\cos(\psi_x) - \cos(\psi_{x0})]\right\}} \qquad (3.12\text{-}4)$$

A plot of the magnitude of (3.12-4) is given in Fig. 3.12-1a for no beam steering (the case $\psi_{x0} = \pi/2$). The main lobe occurs at the origin for the variable $(kd_x/2)\cos(\psi_x)$, which corresponds to broadside. There are sidelobes at values of the variable out to $\pm\pi$ where replica "main lobes" occur, called *grating lobes*.

In terms of the direction cosine variable $\cos(\psi_x)$, grating lobes first appear at $\cos(\psi_x) = \pm 2\pi/kd_x = \pm \lambda/d_x$; for these lobes to be real, we must have $-1 \le \lambda/d_x \le 1$, since $|\cos(\psi_x)| \le 1$ is necessary. Thus element spacings of $d_x \ge \lambda$ can produce real grating lobes, which of course are undesirable because they represent unwanted "main lobes" in most radar cases. When $d_x = \lambda$, these grating lobes occur at $\psi_x = 0$ and π, which corresponds to directions ± 90 degrees from broadside; this case is often not a problem in practice because the element factor is often small or zero in these directions.[6] For $d_x < \lambda$ grating lobes do not occur in real space as long as there is no beam steering.

With beam steering the main lobe is shifted to a new direction. Figure 3.12-1b illustrates how the entire function $|F_{\text{array}}(\psi_x)|$ shifts without shape change when plotted

---

[6] Directions ± 90 degrees from broadside are often called *endfire* directions. Some arrays, called *endfire arrays* are purposely designed to have endfire main lobes.

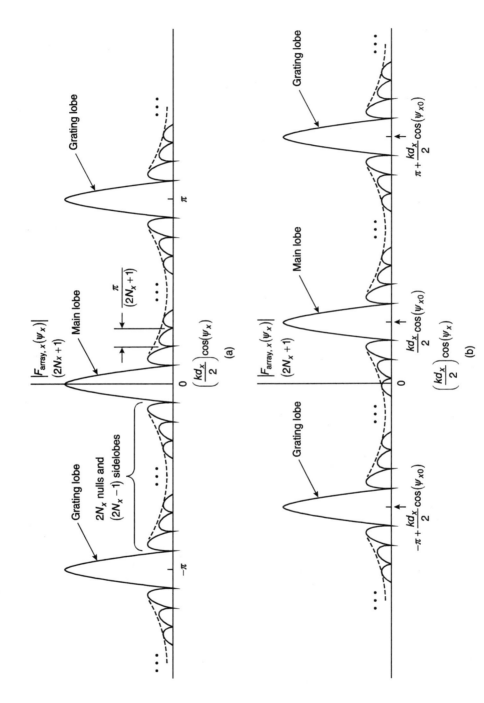

Figure 3.12-1 Magnitude of array factor of linear array (*a*) without and (*b*) with beam steering.

in the variable $(kd_x/2)\cos(\psi_x)$. The main lobe and all grating lobes shift to new locations as shown. As a consequence, grating lobes may shift into real space with beam steering, even if they existed only in imaginary space without beam steering. We demonstrate this point by an example.

Example 3.12-1   Suppose that a linear array with $d_x = \lambda$ has no problem with the grating lobes at ± 90 degrees from broadside because its element pattern has nulls in these directions. Now suppose that the main beam is steered to $\psi_{x0} = \pi/3$ (30 degrees from broadside). Since first grating lobes appear when

$$\frac{kd_x}{2}[\cos(\psi_x) - \cos(\psi_{x0})] = \frac{\pi d_x}{\lambda}\left[\cos(\psi_x) - \cos\left(\frac{\pi}{3}\right)\right] = \pm \pi$$

we have

$$\cos(\psi_x) = \pm\frac{\lambda}{d_x} + \cos\left(\frac{\pi}{3}\right) = \pm 1 + 0.5 = -0.5 \quad \text{and} \quad 1.5$$

Only one value corresponds to real ψ_x and it is $\psi_x = \cos^{-1}(-0.5) = 2\pi/3$ (or 120 degrees), which means the grating lobe is located -30 degrees from broadside.

From the plots in Fig. 3.12-1, we see that the maximum of the array factor is $(2N_x + 1)$ regardless of beam steering. There are also $(2N_x - 1)$ sidelobes between the main lobe and a nearest grating lobe. The largest sidelobe is closest to the main lobe and its peak amplitude is only about 13.2 dB below the main lobe's maximum. Other sidelobes decrease in amplitude out to a point midway between main and grating lobe maximums. Of course all this behavior applies to the assumed uniformly illuminated array. If tapered illumination is used (similar to nonuniform illumination in apertures), sidelobes may be controlled and forced to lower levels at the expense of lower aperture efficiency and larger beamwidth.

Beamwidth

For large uniform arrays ($N_x \gg 1$) the argument of sine in the denominator of (3.12-4) is much smaller than the argument of sine in the numerator. This fact means that $F_{\text{array}}(\psi_x)/(2N_x + 1)$ can be approximated by the function $\sin(x)/x$, where

$$x = (2N_x + 1)\frac{kd_x}{2}[\cos(\psi_x) - \cos(\psi_{x0})] \tag{3.12-5}$$

The approximation is close for all values of ψ_x over the 3-dB beamwidth. Thus the 3-dB beamwidth can be found as the difference between values of ψ_x, denoted as ψ_{x2} and ψ_{x1}, that correspond to respective values of x, denoted as x_2 and x_1, that correspond to 3-dB points of $\sin(x)/x$. Since $\sin(x)/x$ drops to 3 dB below its maximum at $x_1 = 0.44224\pi$ and $x_2 = -0.44224\pi$, we have

$$x_2 = (2N_x + 1)\frac{kd_x}{2}[\cos(\psi_{x2}) - \cos(\psi_{x0})] = -0.44224\pi \tag{3.12-6}$$

$$x_1 = (2N_x + 1)\frac{kd_x}{2}[\cos(\psi_{x1}) - \cos(\psi_{x0})] = 0.44224\pi \tag{3.12-7}$$

On solving these equations for ψ_{x2} and ψ_{x1}, we obtain the 3-dB beamwidth, denoted as ψ_B, in the variable ψ_x:

$$\psi_B = \psi_{x2} - \psi_{x1} = \cos^{-1}\left\{\cos(\psi_{x0}) - \frac{0.44224\lambda}{(2N_x + 1)d_x}\right\}$$

$$- \cos^{-1}\left\{\cos(\psi_{x0}) + \frac{0.44224\lambda}{(2N_x + 1)d_x}\right\}$$

(3.12-8)

Equation (3.12-8) is known to be valid for scan angles from broadside (where $\psi_{x0} = \pi/2$) down to one beamwidth from endfire (where $\psi_{x0} = 0$) called the *scan limit* (Elliott, 1963). At the scan limit the second term in (3.12-8) is zero. Equation (3.12-8) does not apply for values of ψ_{x0} from zero up to the value corresponding to the scan limit.

By using small angle approximations, (3.12-8) can be simplified considerably at a small expense in accuracy (Elliott, 1963). The result is (see Problem 3.12-1)

$$\psi_B \approx \frac{0.88448\lambda}{(2N_x + 1)d_x \sin(\psi_{x0})}$$

(3.12-9)

According to Elliott (1963), for $(2N_x + 1)d_x > 5\lambda$, (3.12-9) is accurate to 0.2% at broadside and is accurate to not more than 4% error for scan angles to within two beamwidths of endfire.

The beamwidth θ_B needed in (3.11-11) and (3.11-12) for the planar array's beamwidths is obtained from (3.12-9) when $\psi_{x0} = \pi/2$:

$$\theta_B \approx \frac{0.8845\lambda}{(2N_x + 1)d_x}$$

(3.12-10)

Elliott (1963) has also developed expressions for the beamwidth of linear arrays with cosine-on-a-pedestal and Dolph-Chebyshev[7] distributions and has shown the amount of beamwidth increase compared to (3.12-9) that results from the choice of a peak sidelobe level. In the former case beamwidth broadens by a factor as large as about 1.64 for a peak sidelobe level of about $-32\,\text{dB}$. In the latter case the beam-broadening factor, denoted here by B, is related to the peak sidelobe level in decibels, denoted here by SLL, by the empirically derived approximate expression

$$B \approx \begin{cases} 1, & -21 \le SLL \\ 1 - \dfrac{0.65}{39}(SLL + 21), & -60 \le SLL < -21 \end{cases}$$

(3-12-11)

Example 3.12-2 For Dolph-Chebyshev illumination of a linear array to produce a sidelobe level of $SLL = -40\,\text{dB}$, we find the beam-broadening factor B.

---

[7] The spelling of Chebyshev varies in the literature. We use the spelling used in the cited articles.

From (3.12-11),

$$B \approx 1 - \left(\frac{0.65}{39}\right)(-40 + 21) \approx 1.32$$

The actual beamwidth of the array will be 1.32 times the beamwidth found from (3.12-8) (Elliott, 1963, p. 56).

Directivity

The general formula for directivity can be found by substituting the array factor for the linear array into (3.10-22). The array factor can be found from that of a planar array when only one row of elements is taken in one direction. In the example case of interest here, where elements are taken along the x direction, we have only one row in the y direction. If elements phase shifts are included in the element illuminations, $F_{\text{array}}(\theta, \phi)$, can be written as

$$F_{\text{array}}(\theta, \phi) = \sum_{m = -N_x}^{N_x} I_{mx} e^{jmkd_x \sin(\theta)\cos(\phi)} \tag{3.12-12}$$

From (3.10-22),

$$G_{Dx} = \frac{4\pi \mathscr{P}_e(\theta, \phi)_{\text{max}} |\sum_{m=-N_x}^{N_x} |I_{mx}||^2}{\int_0^{2\pi} \int_0^{\pi} \mathscr{P}_e(\theta, \phi) \sum_{m=-N_x}^{N_x} \sum_{n=-N_x}^{N_x} I_{mx} I_{nx}^* e^{j(m-n)kd_x \sin(\theta)\cos(\phi)} \sin(\theta)\, d\theta d\phi} \tag{3.12-13}$$

Further reduction of (3.12-13) is difficult without taking special cases. For the important special case where $\mathscr{P}_e(\theta, \phi)$ is either isotropic (constant) or nearly constant for all directions over which the array factor is large (including steering), (3.12-13) can be reduced to (see Problem 3.12-2)

$$G_{Dx} = \frac{|\sum_{m=-N_x}^{N_x} |I_{mx}||^2}{\sum_{m=-N_x}^{N_x} \sum_{n=-N_x}^{N_x} I_{mx} I_{nx}^* \text{Sa}[(m - n)\, 2\pi d_x/\lambda]} \tag{3.12-14}$$

where

$$I_{mx} I_{nx}^* = |I_{mx} I_{nx}| e^{-j(m-n)(2\pi d_x/\lambda)\sin(\theta_0)\cos(\phi_0)} \tag{3.12-15}$$

We observe that if d_x is any integral multiple of $\lambda/2$, the sampling functions in (3.12-14) are all zero except when $m = n$, and

$$G_{Dx} = \frac{[\sum_{m=-N_x}^{N_x} |I_{mx}|]^2}{\sum_{m=-N_x}^{N_x} |I_{mx}|^2} \tag{3.12-16}$$

which is independent of scan angle (a constant). If the array has uniform illumination, (3.12-16) gives $G_D = (2N_x + 1)$, the number of elements in the array.

PROBLEMS

3.1-1 The real electric field components at a point in a radiating aperture are $E_{ax} = 100 \cos(\omega t)$ and $E_{ay} = -600 \cos(\omega t + \pi/8)$. Write an expression for the vector electric field at the aperture's point using the complex form to represent the fields.

3.1-2 Work Problem 3.1-1 except assume $E_{ax} = 100 \sin(\omega t)$ with all else unchanged.

3.1-3 The electric field over a radiating aperture has a *constant* polarization ratio of $1.2e^{j\pi/4}$. What polarization of wave is this aperture radiating? (*Hint*: Associate E_{ay} and E_{ax} with E_ϕ and E_θ, respectively, of Section 2.4, and use previous results of Chapter 2.)

3.1-4 Work Problem 3.1-3 except assume that the polarization ratio is $-j$.

3.1-5 Work Problem 3.1-3 except assume that the polarization ratio is $2 - j3$.

★3.1-6 A rectangular aperture has lengths L_x and L_y in the x and y directions, and is centered on the origin. A constant electric field $E_0 \hat{a}_x$ exists over the aperture. Find (a) the applicable angular spectra and (b) the vector electric field at a far-field point (r, θ, ϕ). (c) Does the polarization of the radiated field depend on direction? Discuss.

3.1-7 The rectangular aperture of Problem 3.1-6 has an aperture electric field

$$\mathbf{E}_a(\xi, \zeta) = \begin{cases} \hat{a}_x E_0 \cos\left(\dfrac{\pi \xi}{L_x}\right), & -\dfrac{L_x}{2} < \xi < \dfrac{L_x}{2}, \ -\dfrac{L_y}{2} < \zeta < \dfrac{L_y}{2} \\ 0, & \text{elsewhere in } \xi, \zeta \end{cases}$$

Find angular spectrums $F_x(\theta, \phi)$ and $F_y(\theta, \phi)$.

3.1-8 Use (3.1-17) and (3.1-18) and show that (3.1-19) is true.

3.1-9 At a point (r, θ, ϕ) in space, a wave has the electric field

$$\mathbf{E}(r, \theta, \phi) = 10^2 \hat{a}_\theta + 6(10^2)\hat{a}_\phi \quad (\text{V/m})$$

What is the average power in the wave, and in what direction does it flow?

3.2-1 Use (3.2-3) and (3.2-4) and show that (3.2-5) is true.

★3.2-2 Show that $\mathscr{P}(\theta, \phi)_{\max}$ is given by the expression

$$\mathscr{P}(\theta, \phi)_{\max} = \frac{P_{\text{rad}}}{\lambda^2} \frac{|\int_A \int \mathbf{E}_a(\xi, \zeta)\, d\xi d\zeta|^2}{\int_A \int |\mathbf{E}_a(\xi, \zeta)|^2\, d\xi d\zeta}$$

if $E_{ax}(\xi, \zeta)$ and $E_{ay}(\xi, \zeta)$ are real nonnegative functions.

3.2-3 What area of a rectangular aperture is needed to produce a radiated power of 5 kW if its aperture illumination is constant at

$$
\mathbf{E}_a(\xi, \zeta) = \begin{cases} 150\hat{a}_x + 200\hat{a}_y, & (\xi, \zeta) \text{ in area } A \\ 0, & \text{elsewhere} \end{cases}
$$

Assume the aperture is centered on the origin with lengths L in the x direction and $L/2$ in the y direction.

3.2-4 A circular aperture of radius R is centered on the origin and has an aperture illumination of

$$
\mathbf{E}_a(\xi, \zeta) = \begin{cases} 100\hat{a}_x + 50\left[1 - \left(\dfrac{\zeta}{R}\right)^2\right]\hat{a}_y, & (\xi, \zeta) \text{ in area } A \\ 0, & \text{elsewhere} \end{cases}
$$

Find an expression for the radiated average power and evaluate your result for $R = 2.1$ m.

3.2-5 An elliptical aperture has major and minor axis lengths of L_x and L_y in the x and y directions, respectively, and it is centered on the origin. Thus on its edges ξ and ζ are related by

$$
\left(\frac{2\xi}{L_x}\right)^2 + \left(\frac{2\zeta}{L_y}\right)^2 = 1
$$

If the aperture's electric field distribution is constant at peak amplitude E_0 and is in the y direction, find the radiated power in terms of η, L_x, L_y, and E_0.

3.2-6 If a circular aperture has an electric field whose magnitude has rotational symmetry, that is, if it is a function only of radial distance ρ from the center, show that the radiated power is given by

$$
P_{\text{rad}} = \frac{\pi}{\eta} \int_0^R \rho \, |\mathbf{E}_a(\rho)|^2 \, d\rho
$$

where R is the aperture's radius.

★3.2-7 The electric field over a circular aperture of radius R centered on the origin is

$$
\mathbf{E}_a(\xi, \zeta) = \begin{cases} E_0\left\{\Delta + (1 - \Delta)\cos^2\left[\dfrac{\pi}{2R}\sqrt{\xi^2 + \zeta^2}\right]\right\}\hat{a}_x, & \xi^2 + \zeta^2 \le R^2 \\ 0, & \text{elsewhere} \end{cases}
$$

where $0 \le \Delta \le 1$ is a constant. Find P_{rad} in terms of Δ, η, E_0, and R.

3.2-8 Work Problem 3.2-7 except assume that

$$\mathbf{E}_a(\xi, \zeta) = \begin{cases} E_0 \left[\Delta + (1 - \Delta)\sqrt{1 - \dfrac{\xi^2 + \zeta^2}{R^2}} \right] \hat{a}_x, & \xi^2 + \zeta^2 \leq R^2 \\ 0, & \text{elsewhere} \end{cases}$$

3.2-9 Work Problem 3.2-7 except assume that

$$\mathbf{E}_a(\xi, \zeta) = \begin{cases} E_0 \cos \left[\dfrac{\pi}{2R} \sqrt{\xi^2 + \zeta^2} \right] \hat{a}_x, & \xi^2 + \zeta^2 \leq R^2 \\ 0, & \text{elsewhere} \end{cases}$$

3.2-10 A circular aperture of radius R has illumination

$$\mathbf{E}_a(\xi, \zeta) = \begin{cases} E_0 \, e^{-\beta \sqrt{\xi^2 + \zeta^2}/R} \, \hat{a}_y, & \xi^2 + \zeta^2 \leq R^2 \\ 0, & \text{elsewhere} \end{cases}$$

where E_0 and β are real positive constants. Find P_{rad} as a function of E_0, β and aperture area $A = \pi R^2$. Evaluate your result to find E_0 which produces $P_{\text{rad}} = 10^5$ W when $\beta = 2.4$ and $R = 2$ m.

3.2-11 Work Problem 3.2-10 except assume

$$\mathbf{E}_a(\xi, \zeta) = \begin{cases} E_0 \, e^{-\beta (\xi^2 + \zeta^2)/R^2} \, \hat{a}_y, & \xi^2 + \zeta^2 \leq R^2 \\ 0, & \text{elsewhere} \end{cases}$$

3.2-12 A rectangular aperture has lengths L_x and L_y in the x and y directions, respectively, and is centered on the origin. It has an aperture illumination of

$$\mathbf{E}_a(\xi, \zeta) = \begin{cases} \hat{a}_x E_0 \left[1 - \left(\dfrac{2\xi}{L_x} \right)^2 \right], & -\dfrac{L_x}{2} \leq \xi \leq \dfrac{L_x}{2}, -\dfrac{L_y}{2} \leq \zeta \leq \dfrac{L_y}{2} \\ 0 & \text{elsewhere} \end{cases}$$

where E_0 is a real constant. Find the vector angular spectrum $\mathbf{F}_a(\theta, \phi)$.

★3.3-1 A circular aperture of radius R is centered on the origin and has an aperture electric field

$$\mathbf{E}_a(\xi, \zeta) = \begin{cases} \hat{a}_y E_0 \left[1 - \dfrac{\xi^2 + \zeta^2}{R^2} \right], & \xi^2 + \zeta^2 \leq R^2 \\ 0, & \text{elsewhere} \end{cases}$$

(a) Use (3.3-4) to find the radiation intensity pattern.
(b) Find $\mathscr{P}(\theta, \phi)_{\text{max}}$.
(c) Find P_{rad}.

3.3-2 A rectangular aperture of lengths $L_x = 4$ m and $L_y = 3$ m in x and y directions, respectively, is centered on the origin and is excited by a constant aperture distribution:

$$\mathbf{E}_a(\xi, \zeta) = \begin{cases} \hat{a}_y E_0, & \xi \text{ and } \zeta \text{ in aperture} \\ 0, & \text{elsewhere} \end{cases}$$

What value of E_0 is required to produce $\mathscr{P}(\theta, \phi)_{\max} = 4.8(10^7)$ W/sr when the frequency of radiation is 6 GHz?

3.3-3 What are the beamwidths θ_B and ϕ_B for the antenna of Problem 3.3-2?

3.3-4 Work Problem 3.3-2 except assume an origin-centered circular aperture of radius $R = 3.5$ m with a constant aperture electric field as given by (3.3-17).

3.3-5 Find beamwidth θ_B of the antenna of Problem 3.3-4.

3.3-6 For the rectangular aperture for which (3.3-12) applies find the 3-dB beamwidth in the variable θ when measured in the intercardinal plane $\phi = \pi/4$. Assume that $b/a = 1.5$, a and b are large enough that $\sin(\theta) \approx \theta$, and ignore the obliquity factor. (*Hint:* Use graphical or trial- and-error methods, and put the answer in terms of λ/b.)

3.3-7 Work Problem 3.3-6 except assume that $b/a = 1$.

3.3-8 Two apertures, one square with sides of length L and one circular with diameter D, are both origin centered. If both antennas are to produce the same principal plane beamwidths, what is the relationship between L and D? Sketch the two apertures using the same scale, and observe their relative shapes.

3.3-9 Show that

$$\int_0^{2\pi} e^{j\beta \cos(\alpha - \alpha_0)} \, d\alpha = 2\pi J_0(\beta)$$

where $J_0(\cdot)$ is the Bessel function of the first kind of order zero, α_0 is an arbitrary phase angle defined by $0 \leq \alpha_0 \leq 2\pi$, and β is a real constant. [*Hint:* Use the known series given below.]

$$e^{j\beta \cos(\phi)} = \sum_{n=-\infty}^{\infty} (j)^n J_n(\beta) e^{jn\phi}$$

3.3-10 With the circular aperture antenna for which (3.3-31) applies, show that beamwidth is given by (3.3-32) when R/λ is large enough so that $\sin(\theta) \approx \theta$ for all important values of θ.

3.4-1 Find the solid angle Ω_A for the rectangular antenna defined by (3.3-8) in the text.

3.4-2 For the circular antenna defined by (3.3-17) find the solid angle Ω_A.

3.4-3 An antenna has beamwidths $\theta_B^\circ = 4$ and $\phi_B^\circ = 2.5$. Find (a) the approximate solid angle Ω_A and (b) the approximate directivity.

3.4-4 The directivity of an antenna is 2600 (or 34.15 dB). If the radiated power is 3 kW, what is the radiation intensity in the direction of the main lobe's maximum?

3.4-5 An antenna has beamwidths of $\theta_B^\circ = 3$ and $\phi_B^\circ = 5$. What average radiated power is needed to produce a radiation intensity of $3.283(10^5)$ W/sr in the direction of the main lobe's maximum?

3.4-6 An antenna has an effective area 38% smaller than its true area of $3.7 \, \mathrm{m^2}$, and it operates at a frequency of 6.4 GHz. What is the antenna's directivity?

3.4-7 Work Problem 3.4-6 for an antenna with area of $2.0 \, \mathrm{m^2}$ and frequency of 12.5 GHz.

3.4-8 What effective area is needed for an antenna at 35 GHz to realize a directivity of 3600? If effective area is known to be 70% of the true area for a circular aperture, what is the required antenna diameter?

3.4-9 A rectangular aperture has sides of lengths 1 and 1.4 m. The effective area is known to be 0.52 times the area. Find an expression for the frequency that must be used to obtain any specified value of directivity. Evaluate your result for $G_D = 4000$.

3.4-10 The power gain is 3680 in an antenna excited by an average power of 2750 W from a transmitter. Radiated power is found to be 2615 W by measurements. Find this antenna's (a) radiation efficiency, (b) directivity, and (c) radiation loss in dB.

3.4-11 A transmitter feeds 5.7 kW of average power to an antenna having a radiation efficiency of 0.93, directivity of 550, and 3-dB beamwidth of $\phi_B^\circ = 15.5$ in one principal plane of the main beam. Find (a) the beamwidth θ_B° in the other principal plane, (b) the power gain, (c) the radiated power, and (d) the radiation loss.

3.4-12 For aperture illuminations that give patterns with maximums in the broadside direction (where $\theta = 0$) use (3.2-15) and (3.3-1) in (3.4-6) to show that

$$G_D = \frac{4\pi}{\lambda^2} \cdot \frac{|\int_A \int \mathbf{E}_a(\xi, \zeta) \, d\xi d\zeta|^2}{\int_A \int |\mathbf{E}_a(\xi, \zeta)|^2 \, d\xi d\zeta}$$

3.4-13 An antenna with a rectangular aperture has an aperture illumination of

$$\mathbf{E}_a(\xi, \zeta) = \begin{cases} \hat{a}_x E_0 \cos^n\left(\dfrac{\pi\xi}{L_x}\right)\cos^m\left(\dfrac{\pi\zeta}{L_y}\right), & -\dfrac{L_x}{2} \leq \xi \leq \dfrac{L_x}{2}, -\dfrac{L_y}{2} \leq \zeta \leq \dfrac{L_y}{2} \\ 0 & \text{elsewhere} \end{cases}$$

where E_0 is a real constant and L_x and L_y are the side lengths of the aperture in the x and y directions, respectively. For the special case of $n = 1$ and $m = 1$ calculate G_D. (*Hint*: Use the result of Problem 3.4-12.)

3.4-14 Work Problem 3.4-13 except for $n = 2$ and $m = 2$.

3.4-15 Work Problem 3.4-13 except for $n = 3$ and $m = 3$.

3.4-16 An antenna has principal plane beamwidths of $\theta_B^\circ = 1.6$ and $\phi_B^\circ = 1.9$. (a) Find an approximate value for G_D. (b) What fraction of the antenna's input power is radiated if the radiation efficiency is 0.938? (c) What is the approximate solid angle Ω_A of this antenna?

3.4-17 An aperture antenna with a true area of $12\,\mathrm{m}^2$ is excited by a 100-kW transmitter at 4.6 GHz. Radiated power is measured to be 96.3 kW when the antenna's power gain is 18,440 (or 42.66 dB). Find (a) radiation efficiency and loss, (b) directivity, (c) aperture efficiency, and (d) effective area.

3.4-18 A designer wishes to find the *approximate* diameter needed in a circular aperture to produce beamwidths of $\theta_B^\circ = \phi_B^\circ = 3$ when the frequency is 1.7 GHz and $\rho_a = 0.6$ is used to maintain acceptable sidelobes in the pattern. Find (a) the approximate value of G_D and (b) the required diameter D.

3.4-19 An antenna has a circular aperture and frequency of 12.5 GHz. It has $\rho_a = 0.46$ chosen to produce low pattern sidelobes. If an allowance is made for a radiation efficiency factor of 0.9, how large must the antenna's diameter be to produce a power gain of 15,560?

3.4-20 A uniformly illuminated rectangular aperture has dimensions $1.5 \times 2\,\mathrm{m}$ and operates at 6.0 GHz. The transmitter provides 500 kW to the antenna which has a radiation efficiency of 0.925. (a) What is the power gain? (b) What are the approximate beamwidths θ_B° and $\phi_B^{\circ\circ}$? (c) What is the radiated power?

3.4-21 A circular aperture of radius R has an excitation such that $\theta_B = \phi_B$. Its radiation efficiency is 0.95 and $\rho_a = 0.6$. Plot the power gain (in dB) and *approximate* beamwidth (in degrees) versus R (in meters from 0.25 to 3 m). Assume a frequency of 6 GHz.

3.4-22 What directivity is possible from an antenna with effective area of $2.6\,\mathrm{m}^2$ at a frequency of 10 GHz?

3.4-23 An antenna at 12.5 GHz is to produce a power gain of 28,510. If its radiation and aperture efficiencies are 0.9 and 0.55, respectively, find (a) the antenna's area, (b) the effective area, and (c) the directivity.

3.4-24 A circular antenna with 1-m radius operates at 8.2 GHz and is part of a target-tracking system. If $\rho_r = 0.98$ and $\rho_a = 0.70$, what power density must the wave have that arrives from the target if the antenna must produce an available power of 10^{-10} W at its output? Assume that the target is at the broadside direction.

3.5-1 If an aperture is to radiate left-circular polarization over all parts of its area what is Q_a?

3.5-2 An aperture has a constant polarization being radiated as defined by $Q_a = 1.2 \exp(j\pi/3)$. If $P_{rad} = 190$ kW, $\rho_a = 0.65$, $A_e = 3.7$ m^2, and $f = 4.3$ GHz, find (a) true area A, (b) directivity G_D, and (c) $\mathscr{P}(\theta, \phi)_{max}$.

3.6-1 A rectangular aperture with sides of lengths L_x and L_y in directions x and y, respectively, has the factorable illumination

$$
\mathbf{E}_a(\xi, \zeta)
$$
$$
= \begin{cases} \hat{a}_x E_0 \left[1 - \left(\dfrac{2|\xi|}{L_x}\right)^n\right]\left[1 - \left(\dfrac{2|\zeta|}{L_y}\right)^m\right], & -\dfrac{L_x}{2} \le \xi \le \dfrac{L_x}{2}, -\dfrac{L_y}{2} \le \zeta \le \dfrac{L_y}{2} \\ \\ 0, & \text{elsewhere} \end{cases}
$$

Find the antenna's directivity for the special case $n = 1$ and $m = 1$. Assume that E_0 is a positive constant.

3.6-2 Work Problem 3.6-1 except for $n = 2$ and $m = 2$.

3.6-3 Work Problem 3.6-1 for the general case of any nonnegative integers n and m.

3.6-4 The rectangular aperture of Problem 3.4-13 has a factorable illumination. Define the factors $E_1(\xi)$ and $E_2(\zeta)$, and use (3.6-16) and (3.6-17) to find effective lengths L_{e1} and L_{e2} for $n = 1$ and $m = 1$ when $\theta = 0$.

3.6-5 Work Problem 3.6-4 except for $n = 2$ and $m = 2$.

3.6-6 Work Problem 3.6-4 except for $n = 3$ and $m = 3$.

3.6-7 Find L_{e1} and L_{e2} for the factorable aperture defined in Problem 3.6-1. Assume that n and m are arbitrary.

3.6-8 A rectangular aperture's illumination is factorable and given by

$$
\mathbf{E}_a(\xi, \zeta) = \begin{cases} \hat{a}_x E_0 \left[1 - (1 - \Delta_1)\left(\dfrac{2\xi}{a}\right)^2\right] \\ \quad \cdot \left[1 - (1 - \Delta_2)\left(\dfrac{2\zeta}{b}\right)^2\right], & (\xi, \zeta) \text{ in aperture area} \\ \\ 0, & \text{elsewhere} \end{cases}
$$

where a and b are the aperture lengths in the x and y directions, respectively, and Δ_1 and Δ_2 are positive constants each not greater than unity. Find (a) effective lengths L_{e1} and L_{e2}, and (b) efficiencies ρ_{a1} and ρ_{a2}. Compare the values of ρ_{a1} for $\Delta_1 = 0, 0.25, 0.50, 0.75,$ and 1.0 to those of Table 3.7-1 as checks on results.

★3.6-9 Use (3.6.1) and prove that (3.6-2) is true where $\mathscr{P}_x(\theta, \phi)$ and $\mathscr{P}_y(\theta, \phi)$ are given by (3.6-3) and (3.6-4) and

$$
\mathscr{P}_{1x}(\theta, \phi) = \frac{2\pi k}{\sqrt{2\eta}}\left[\frac{1 + \cos(\theta)}{2}\right]\left|\frac{1}{2\pi}\int_{\xi} E_{1x}(\xi)\, e^{j\xi k \sin(\theta)\cos(\phi)}\, d\xi\right|^2
$$

$$\mathscr{P}_{2x}(\theta, \phi) = \frac{2\pi k}{\sqrt{2\eta}} \left[\frac{1 + \cos(\theta)}{2} \right] \left| \frac{1}{2\pi} \int_\zeta E_{2x}(\zeta) \, e^{j\zeta k \sin(\theta)\sin(\phi)} \, d\zeta \right|^2$$

$$\mathscr{P}_{1y}(\theta, \phi) = \frac{2\pi k}{\sqrt{2\eta}} \left[\frac{1 + \cos(\theta)}{2} \right] \left| \frac{1}{2\pi} \int_\xi E_{1y}(\xi) \, e^{j\xi k \sin(\theta)\cos(\phi)} \, d\xi \right|^2$$

$$\mathscr{P}_{2y}(\theta, \phi) = \frac{2\pi k}{\sqrt{2\eta}} \left[\frac{1 + \cos(\theta)}{2} \right] \left| \frac{1}{2\pi} \int_\zeta E_{2y}(\zeta) \, e^{j\zeta k \sin(\theta)\sin(\phi)} \, d\zeta \right|^2$$

3.7-1 Assume that a one-dimensional aperture of length a is defined by the triangular illumination function

$$E_1(\xi) = \begin{cases} E_0 \left[1 - \dfrac{2|\xi|}{a} \right], & -\dfrac{a}{2} < \xi < \dfrac{a}{2} \\ 0, & \text{elsewhere} \end{cases}$$

Find (a) the angular spectrum (pattern) $f_{a1}(u_1)$, (b) the 3-dB beamwidth in the variable θ, and (c) the level of the peak of the first sidelobe relative to the main lobe's peak value.

★3.7-2 A one-dimensional aperture has the illumination function

$$E_1(\xi) = \begin{cases} E_0 \left[1 - (1 - \Delta)\left(\dfrac{2\xi}{a}\right)^2 \right], & -\dfrac{a}{2} < \xi < \dfrac{a}{2} \\ 0, & \text{elsewhere} \end{cases}$$

where E_0, a and $0 \le \Delta \le 1$ are positive constants. Find the pattern function $f_{a1}(u_1)$. [*Hint*: Use the derivative property of Fourier transforms, and leave one term of $f_{a1}(u_1)$ as the second derivative of an appropriate function.]

3.7-3 Find and sketch the pattern $f_{a1}(u_1)$ for the "odd" illumination

$$E_1(\xi) = \begin{cases} E_0 \sin\left(\dfrac{2\pi\xi}{a}\right), & -\dfrac{a}{2} \le \xi \le \dfrac{a}{2} \\ 0, & \text{elsewhere} \end{cases}$$

where E_0 and a are positive constants. Note that the pattern has no maximum at $u_1 = 0$ and that it has two "main" lobes. This type of pattern is useful in angle-tracking radars about which more is said in Chapter 14.

3.7-4 Find the Taylor coefficients F_m for a Taylor's distribution using the minimum value of $\bar{n}$ and assuming a peak sidelobe level $SLL = -20$ dB. Find the pattern's 3-dB beamwidth in terms of λ/a.

3.7-5 (a) For a Taylor one-dimensional aperture designed for -25 dB sidelobes with the smallest allowable value of $\bar{n}$, find the coefficients F_m. (b) Also find the pattern's 3-dB beamwidth in terms of λ/a. (c) What is the aperture's efficiency?

★3.7-6 Carry out the steps indicated in the text to prove (3.7-11) is true.

3.7-7 Use the smallest allowable value of $\bar{n}$ and find the Taylor coefficients F_m for a Taylor design for $SLL = -30$ dB. Find the pattern's 3-dB beamwidth as a function of λ/a.

3.7-8 In a Taylor one-dimensional distribution $\bar{n} = 8$ when $SLL = -30$ dB. Find the pattern's 3-dB beamwidth in terms of λ/a.

3.7-9 A Taylor distribution for $\bar{n} = 6$ and $SLL = -40$ dB has coefficients $F_0 = 1.0$, $F_1 = 0.389117$, $F_2 = -0.009452$, $F_3 = 0.004882$, $F_4 = -0.001611$, and $F_5 = 0.000347$. Find the aperture's efficiency.

3.7-10 The constant in Taylor's one-parameter distribution is $B = 1.415$. What sidelobe level will occur for this value of B?

3.7-11 Work Problem 3.7-10 except assume that $B = 0.922$.

★3.8-1 Show that (3.8-3) derives from (3.8-2).

3.8-2 Use (3.8-3) with (3.4-14) to show that (3.8-5) is true.

★3.8-3 Use (3.8-7) and derive (3.8-8).

3.8-4 For the circular aperture where (3.8-7) applies, find the radiated power as a function of N and a.

★3.8-5 Use (3.8-10) with (3.8-5) to show that (3.8-18) is true.

3.8-6 For a circular aperture of diameter a, find (a) the coefficients D_m needed for a Taylor's distribution producing a sidelobe level $SLL = -40$ dB, (b) the 3-dB beamwidth, and (c) the aperture's efficiency. Assume that $\bar{n} = 4$.

3.10-1 An array has a total of 361 elements. Evaluate $|F_{\text{array}}(\theta, \phi)|^2$ for the broadside direction if all elements have uniform illumination, that is, for $I_{mn} = I_{00} = 1$ for all m and n.

3.10-2 An array has 49 elements located at points $(\xi_m, \zeta_n) = (md_x, nd_y)$ for $-3 \le m \le 3$ and $-3 \le n \le 3$. Elements are excited according to coefficients $I_{mn} = I_{00}[(1 + |m|)(1 + |n|)]/[(1 + m^2)(1 + n^2)]$ with I_{00} a real number. Determine $|F_{\text{array}}(0, \phi)|^2$, and compare to the case where $I_{mn} = I_{00}$ for all m and n (uniform illumination).

3.10-3 Work Problem 3.10-2 except assume that

$$\frac{I_{mn}}{I_{00}} = \left(1 - \frac{|m|}{N_x + 1}\right)\left(1 - \frac{|n|}{N_y + 1}\right), \qquad -N_x \le m \le N_x, -N_y \le n \le N_y$$

where $N_x = 3$ and $N_y = 3$.

3.10-4 Work Problem 3.10-3 for the general case of any N_x and N_y values.

3.10-5 Work Problem 3.10-2 except assume any N_x and N_y, and

$$\frac{I_{mn}}{I_{00}} = \left[1 - \frac{m^2}{(N_x + 1)^2}\right]\left[1 - \frac{n^2}{(N_y + 1)^2}\right], \qquad -N_x \leq m \leq N_x, -N_y \leq n \leq N_y$$

Evaluate results for $N_x = 3$ and $N_y = 3$.

3.10-6 An array has elements located by $\xi_m = md_x$ for $-N_x \leq m \leq N_x$ and $\zeta_n = nd_y$ for $-N_y \leq n \leq N_y$, where N_x, N_y, d_x, and d_y are positive constants. Find the phase shifts ψ_m, ψ_n, and ψ_{mn} that correspond to shifting the main beam to direction $(\theta_0, \phi_0) = (\pi/8, \pi/3)$ when $d_x = d_y = \lambda$.

3.10-7 Work Problem 3.10-6 except for a main beam shifted to $(\theta_0, \phi_0) = (\pi/3, 0)$.

3.11-1 For the array with elements and beam steering as defined in Problem 3.10-6, what are the angles ψ_{x0} and ψ_{y0} that define the main beam's direction?

3.11-2 Elements for an array are defined as shown in Fig. 3.11-1 except with coordinates labeled x', y'. The actual array is created by rotation of the x', y' coordinates by an angle α relative to coordinates x, y. (a) Find the element locations in the x, y coordinate system. (b) What phase shift must be imparted to element mn to steer the main beam to angles ψ_{x0} and ψ_{y0}?

3.11-3 For a rectangular array beamwidths at broadside are related by $\phi_B/\theta_B = 2$. Find an expression for the beamwidth ratio ϕ_{B0}/θ_{B0} corresponding to steering angles (θ_0, ϕ_0). Evaluate and plot the ratio for $\theta_0 = \pi/3$ and any ϕ_0 from zero to 2π.

★3.12-1 Define $\psi_{x2} = \psi_{x0} + \psi_{B2}$ and $\psi_{x1} = \psi_{x0} - \psi_{B1}$, where ψ_{B2} and ψ_{B1} are "half-side" beamwiths in the variable ψ_x, and assume ψ_{B1} and ψ_{B2} are small angles. Use these approximations in (3.12-8) to show that (3.12-9) is true.

★3.12-2 Assume that $\mathscr{P}_e(\theta, \phi) = 1$ in (3.12-13), and show that (3.12-14) is true. [*Hint*: Change variables θ and ϕ to new variables θ' and ϕ' according to $\theta' = \cos^{-1}[\sin(\theta)\cos(\phi)]$, $\phi' = \tan^{-1}\{1/[\tan(\theta)\sin(\phi)]\}$, and perform the integrations; inverses of these transformations are helpful and are $\theta = \cos^{-1}[\sin(\theta')\sin(\phi')]$, $\phi = \tan^{-1}[\tan(\theta')\cos(\phi')]$.]

3.12-3 A uniformly illuminated linear array has elements uniformly spaced an amount $d_x = 1.4\lambda$ and centered along the x axis. At what angles in the variable ψ_x do grating lobes appear if there is no beam steering?

3.12-4 Work Problem 3.12-3 except assume that element spacing is $d_x = 2.2\lambda$.

3.12-5 A uniformly illuminated linear array can allow first grating lobes to occur at ± 20 degrees from broadside with no beam steering. Will there be other grating lobes in real space, and if so, where are they located relative to broadside?

3.12-6 A designer wants to know the value of d_x/λ that allows exactly M pairs of grating lobes in real space with grating lobe M positioned at $\psi_x = 0$. Find the answer for the designer assuming no beam steering and uniform illumination.

3.12-7 A uniformly illuminated linear array has 21 elements centered on the origin with separation $d_x = 0.8\lambda$. (a) With no beam steering, find the 3-dB beamwidth ψ_B. (b) If the main beam is steered to an angle $\psi_{x0} = \pi/4$, find the new beamwidth by using (3.12-8). (c) Compare results of parts (a) and (b) to beamwidths found from (3.12-9), and find the percentage error in using (3.12-9).

3.12-8 Work Problem 3.12-7 except assume seven elements.

3.12-9 Find an expression for the value of ψ_{x0} that causes the main beam's 3-dB point closest to the x axis to lie on the x axis. Evaluate the result for a seven-element array with $d_x = 0.8\lambda$.

3.12-10 A linear array of 15 elements with separations $d_x = \lambda$ has its main beam scanned to $\psi_{x0} = 0.2\pi$ (or 36.0 degrees, which is 54.0 degrees from broadside). (a) If the array is uniformly illuminated, find the broadside beamwidth and sidelobe level, and the beamwidth at $\psi_{x0} = 0.2\pi$. (b) If the array has Dolph-Chebyshev illumination designed to give $SLL = -25$ dB, find beamwidths for broadside and steering to 0.2π. [*Hint*: Take 15 elements as sufficiently large so that (3.12-4) can be approximated as $\sin(x)/x$ in shape for x around the main beam and first sidelobes.]

3.12-11 A uniformly illuminated linear array has an *even number* of elements located at $\xi_m = \pm(m - 1/2)d_x$ where d_x is a positive constant and $m = 1, 2, \ldots, N_x$. Derive $F_{\text{array}}(\psi_x)$, and show that it is equivalent to (3.12-4).

3.12-12 Solve (3.12-14) for $N_x = 1$. Simplify for the special case where $|I_{1x}| = |I_{-1x}|$.

4

RADAR EQUATION

To perform its major functions a radar operates on its received signal, which is always obscured by the presence of noise and possibly other unwanted waveforms. Typically radar performance is improved when the received signal's power is increased. For this reason it is important to understand the mechanisms that lead to a particular received signal power, and show how the power can be optimized. In this chapter we study the most important of these mechanisms by development of the *radar equation*, which is fundamental to defining the amount of signal and noise power in the radar.

After some preliminary definitions we derive a basic equation to describe the signal power available to the receiver. This equation is in terms of the various system parameters (target range, antenna gain and losses, etc.). Next our work leads to models for noise in the system. Ultimately we derive measures of a system's noisiness (noise figures and noise temperatures) so that the noise power available in the receiver can be calculated. Finally overall radar system models are given so that the available received signal-to-noise power ratio can be found.

4.1 RADAR EQUATION

The target (desired) signal received by a radar can come about in several ways. Monostatic, bistatic, multistatic, and passive systems may all produce different received signals. However, by proper definitions and use, a single geometry can be used to define the received signal powers in all these systems.

Applicable Geometry

Figure 4.1-1 illustrates the geometry applicable to a bistatic system where transmitting and receiving sites are separated by a base distance R_B. Clearly the geometry applies to any up-down link in a multistatic system as well. By letting $R_B = 0$ such that

153

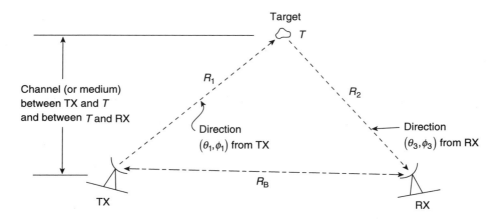

Figure 4.1-1   Geometry applicable to signal reception in radar.

$R_1 = R_2$, the geometry also can be used for the monostatic radar. Finally, by considering only the receiver's path of length R_2, and with suitable definitions of the waves originating from the target, the geometry can be used for the passive radar. The passive radar case should be relatively straightforward for the reader to develop, after having studied the other cases that are developed below. Therefore, except for a special case of a cooperative beacon-type target, we leave the passive case as a reader exercise.

Basic Equation

Consider the transmitter path in Fig. 4.1-1. Let P_t represent the *average peak power* output (see Section 1.5) of the transmitter when averaged over one cycle of the carrier located at the peak of the transmitted pulse (for a constant-amplitude cw radar P_t is the same as the power averaged over all time). This power may be reduced by mismatch and losses in the microwave elements (duplexer, circulators, isolators, etc.) and transmission line (waveguide or coaxial line) that connects the transmitter to the antenna. If

$$L_t \triangleq \text{power loss, transmitter to antenna} \qquad (4.1\text{-}1)$$

where $L_t \geq 1$, the average peak power accepted at the antenna's input, denoted by P_{acc}, is $P_{\text{acc}} = P_t/L_t$. Not all this power is radiated by the antenna; some is lost through heating effects in the structure of the antenna. This loss is denoted by L_{rt} and defined by

$$L_{rt} = \text{radiation loss of transmitting antenna}$$

$$= \frac{1}{\rho_{rt}} \geq 1 \qquad (4.1\text{-}2)$$

where

$$\rho_{rt} = \text{radiation efficiency of transmitting antenna} \qquad (4.1\text{-}3)$$

These loss definitions are consistent with the antenna model of Fig. 3.4-1 for the transmitting path. With these loss models the average peak radiated power, denoted by P_{rad}, is $P_{rad} = P_{acc}/L_{rt} = P_t/L_t L_{rt}$.

If all the average peak radiated power occurred from a nondirective (isotropic) antenna, the power density of the wave at distance R_1 would be $P_{rad}/(4\pi R_1^2)$ (W/m²) for a vacuum channel. In the real channel there is a one-way power loss, denoted by L_{ch1}:

$$L_{ch1} = \text{one-way power loss of channel on the path from}$$

$$\text{transmitting antenna to target} \qquad (4.1\text{-}4)$$

L_{ch1} is due to all clear and unclear channel effects that may be present (atmospheric attenuations, effects of rain, snow, etc.; see Chapter 2). Thus an isotropic antenna would produce a wave average peak power density at the target of $P_{rad}/(4\pi R_1^2 L_{ch1}) = P_t/(4\pi R_1^2 L_t L_{rt} L_{ch1})$.

Finally we note that a real antenna will increase the wave's power density at the target because of its directive properties. If the target is at a direction (θ_1, ϕ_1) in spherical coordinates located at the transmitting antenna (Fig. 4.1-1), the increase is given by the directive gain, denoted here by $G_{Dt}(\theta_1, \phi_1)$ [see (3.4-5)]. The wave's average peak power density at the target, denoted by $\mathscr{S}_t(R_1, \theta_1, \phi_1)$, is

$$\mathscr{S}_t(R_1, \theta_1, \phi_1) = \frac{P_t G_{Dt}(\theta_1, \phi_1)}{4\pi R_1^2 L_t L_{rt} L_{ch1}} \quad (\text{W/m}^2) \qquad (4.1\text{-}5)$$

When the transmitted wave having the average peak power density of (4.1-5) crosses the target, power is scattered by the target in various directions. To account for the power reflected back toward the receiving site, a constant, called the target's *radar cross section*[1] that is denoted by σ, is associated with the target. The constant, which has the unit of area, when multiplied by $\mathscr{S}_t(R_1, \theta_1, \phi_1)$, corresponds to an equivalent power that is reflected equally in all directions and accounts for the receiving site's actual available power.[2] Thus the equivalent average peak power reflected from the target is

$$\text{Average peak power}$$

$$\text{reflected from target} = \frac{P_t G_{Dt}(\theta_1, \phi_1)\sigma}{4\pi R_1^2 L_t L_{rt} L_{ch1}} \quad (\text{W}) \qquad (4.1\text{-}6)$$

---

[1] A more detailed discussion of cross section is given in Chapter 5.
[2] This condition implies radar cross section must include the polarization of the receiving antenna in its definition (see Chapter 5).

This power is reduced by the factor $1/(4\pi R_2^2)$ to account for intensity reduction with range as the reflected wave travels distance R_2 to the receiving site. Over the path there is also a channel loss

$$L_{\text{ch2}} = \text{one way power loss of channel on the path}$$

$$\text{from target to receiving antenna} \qquad (4.1\text{-}7)$$

so that the average peak power density of the wave at the receiving antenna becomes

$$\mathcal{S}_i = \frac{P_t G_{\text{Dt}}(\theta_1, \phi_1)\sigma}{(4\pi)^2 R_1^2 R_2^2 L_t L_{\text{rt}} L_{\text{ch1}} L_{\text{ch2}}} \quad (\text{W/m}^2) \qquad (4.1\text{-}8)$$

When the reflected wave crosses the receiving antenna, the antenna's effective area determines the available received power, as discussed in Section 3.4. In particular, the available power is given by (3.4-19) if a loss factor is included to account for the antennas radiation loss when being used as a receiving device. If this loss is defined as

$$L_{\text{rr}} = \text{radiation loss of receiving antenna}$$

$$= \frac{1}{\rho_{\text{rr}}} \geq 1 \qquad (4.1\text{-}9)$$

where

$$\rho_{\text{rr}} = \text{radiation efficiency of receiving antenna} \qquad (4.1\text{-}10)$$

then the received average peak signal power available at the output terminals of the receiving antenna is

$$S_r = \mathcal{S}_i \frac{\lambda^2 G_{\text{Dr}}(\theta_3, \phi_3)}{4\pi L_{\text{rr}}} = \frac{P_t G_{\text{Dt}}(\theta_1, \phi_1) G_{\text{Dr}}(\theta_3, \phi_3)\lambda^2\sigma}{(4\pi)^3 R_1^2 R_2^2 L_t L_{\text{rt}} L_{\text{ch1}} L_{\text{ch2}} L_{\text{rr}}} \qquad (4.1\text{-}11)$$

In writing (4.1-11), we have used (3.4-19) and observed that the directive gain of the receiving antenna, denoted here by $G_{\text{Dr}}(\theta_3, \phi_3)$, is a function of the direction of the target, (θ_3, ϕ_3), in spherical coordinates located at the receiving antenna. In some cases the maxima of the transmit and receive patterns point directly at the target. For these cases $G_{\text{Dt}}(\theta_1, \phi_1) = G_{\text{Dt}}$ and $G_{\text{Dr}}(\theta_3, \phi_3) = G_{\text{Dr}}$, where G_{Dt} and G_{Dr} are the directivities of transmit and receive antennas, respectively, and

$$S_r = \frac{P_t G_{\text{Dt}} G_{\text{Dr}} \lambda^2 \sigma}{(4\pi)^3 R_1^2 R_2^2 L_t L_{\text{rt}} L_{\text{ch1}} L_{\text{ch2}} L_{\text{rr}}} \quad (\text{W}) \qquad (4.1\text{-}12)$$

Equation (4.1-11) is the basic *radar equation*. It applies to bistatic and multistatic radars directly. With proper definitions and interpretations, we next show that it also applies to monostatic radar and to a radar that receives signals from either cooperative sources (e.g., a transponder or beacon) or undesired sources.

Monostatic Radar Equation

In a monostatic radar where separate antennas are used for transmission and reception but are close enough to be considered at the same "point," $R_1 = R_2 = R$ and $L_{ch1} = L_{ch2} = L_{ch}$ can be assumed so that (4.1-11) reduces to

$$S_r = \frac{P_t G_{Dt}(\theta_1, \phi_1) G_{Dr}(\theta_1, \phi_1) \lambda^2 \sigma}{(4\pi)^3 R^4 L_t L_{rt} L_{ch}^2 L_{rr}} \quad \text{(W)} \qquad (4.1\text{-}13)$$

If, in addition, the two antennas point directly at the target (4.1-13) reduces further to

$$S_r = \frac{P_t G_{Dt} G_{Dr} \lambda^2 \sigma}{(4\pi)^3 R^4 L_t L_{rt} L_{ch}^2 L_{rr}} \quad \text{(W)} \qquad (4.1\text{-}14)$$

As a last case, if the same antenna is used for both transmission and reception such that $G_{Dt}(\theta_1, \phi_1) = G_{Dr}(\theta_1, \phi_1) = G_D(\theta_1, \phi_1)$ and $L_{rt} = L_{rr} \triangleq L_{rad}$, then (4.1-13) and (4.1-14), respectively, become

$$S_r = \frac{P_t G_D^2(\theta_1, \phi_1) \lambda^2 \sigma}{(4\pi)^3 R^4 L_t L_{rad}^2 L_{ch}^2} \quad \text{(W)} \qquad (4.1\text{-}15)$$

$$S_r = \frac{P_t G_D^2 \lambda^2 \sigma}{(4\pi)^3 R^4 L_t L_{rad}^2 L_{ch}^2} \quad \text{(W)} \qquad (4.1\text{-}16)$$

where $G_{Dt} = G_{Dr} \triangleq G_D$ has been used in writing (4.1-16).

Example 4.1-1 We use (4.1-16) to find the maximum range R of a radar that must provide an available received average peak signal power of 10^{-12} W when frequency is 4.6 GHz, $P_t = 10^4$ W, the antenna's aperture area is 2.0 m^2, aperture efficiency is 0.64, radar cross section is 1.4 m^2, $L_t = 1.2$, $L_{rad} = 1.04$, and $L_{ch} = 1.43$. Here $\lambda = 3(10^8)/4.6(10^9) = 0.3/4.6$ m, $G_D = 4\pi A_e/\lambda^2 = 4\pi(2.0)0.64/[0.3/4.6]^2 = 3781.75$. From (4.1-16) we get

$$R = \left[\frac{10^4 (3781.75)^2 (0.3/4.6)^2 \, 1.4}{(4\pi)^3 \, 10^{-12} (1.2)(1.04)^2 (1.43)^2} \right]^{1/4} = 20.0527 \text{ km}$$

which corresponds to 12.46 mi. Target ranges larger than 20.0527 km will rapidly (as $1/R^4$) drop the available received power below the specified minimum value.

Beacon Equation

There are times when a radar operates with a cooperative (friendly) source that transmits an active signal directly to the receiver. The source may be a transponder or a beacon-type transmitter. Here there is only one path. By retracing the previous developments, it is found that S_r is given by

$$S_r = \frac{P_t G_{Dt}(\theta_2, \phi_2) G_{Dr}(\theta_3, \phi_3) \lambda^2}{(4\pi)^2 R^2 L_t L_{rt} L_{ch} L_{rr}} \quad \text{(W)} \qquad (4.1\text{-}17)$$

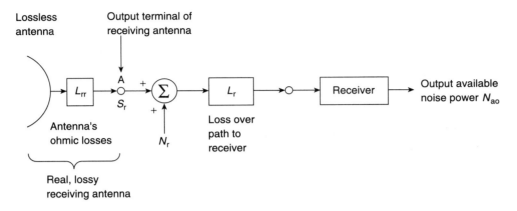

Note: $L_{rr} = L_{rad}$ for monostatic radar with one antenna

Figure 4.1-2 Model for a radar receiver.

where some terms need redefinition. The range from target to radar is now denoted by R. Definitions of losses L_t, L_{rt}, and L_{rr} are unchanged. L_{ch} now denotes the channel loss over the one-way path. Coordinates θ_2, ϕ_2 and θ_3, ϕ_3 in (4.1-17) represent spherical coordinates at the transmit (beacon) and receive antennas, respectively. However, implicit in (4.1-17) is the assumption that the polarization of the receiving antenna is perfectly matched to receive the polarization of the wave from the source. If polarizations are not matched, a factor called the *polarization efficiency* must be added to the right side of (4.1-17) to account for the loss in available power (see Section 3.4 and Mott, 1992, Chapter 4).

Equation (4.1-17) was discussed for a transponder/beacon transmitter to radar receiver path. Clearly it also applies to a radar transmitter to transponder/beacon receiver, although numerical values of parameters may be considerably different in the two cases. Of course the latter case corresponds to the interrogation of the source where a reply may be requested, while the former case corresponds to the reply.

Equation (4.1-17) may also be applied to unfriendly sources that radiate undesired signals toward the radar to confuse or jam the radar. Because S_r depends on range as $1/R^2$, rather than $1/R^4$ as with a reflective target, jammers with only small values of P_t (e.g., noise) may be especially disruptive to a radar.

Signal-to-Noise Ratio

The limiting performance of a radar is often determined by the noise power that is present.[3] A model for this noise power is shown in Fig. 4.1-2. Here the actual (lossy) receiving antenna is modeled as a lossless antenna followed by a loss to account for its radiation efficiency. The loss, denoted by L_{rr}, is defined in (4.1-9). The output of the loss is the actual output terminal of the receiving antenna and is the point where the available received average peak signal power is defined by one of the various forms (4.1-11) through (4.1-17).

---

[3] Of course noise is not always the limiting factor in performance. Cases arise where clutter, jamming signals, and interference are the limiting effects, but noise is always important, even in these cases.

A path loss L_r is shown in Fig. 4.1-2. It represents the total loss of various components that are typically used to connect the antenna to the receiver. These components include such things as waveguides, coaxial transmission lines, filters, circulators, isolators, duplexers, hybrids, attenuators, and phase shifters. Basically, L_r is used to represent the total loss of all "path" components that can be treated as "*passive elements*".

The receiver in Fig. 4.1-2 is defined as containing the *active elements* of the receiving system. Such elements often include mixers, low-noise radio-frequency (RF) amplifiers, both low-noise and high level intermediate-frequency (IF) amplifiers, phase shifters, and phase detectors. It also includes various passive elements needed to support the active circuits (filters, attenuators, isolators, etc.). The receiver includes most of the receiving path and all elements necessary to raise the signal power up to a level at which all following circuits add no appreciable noise to the system as compared to the existing noise level created from all prior circuits. The receiver's gain can easily be 100 dB or more in a practical system.

All components of a radar generate some level of noise, even the waveguide attached to the antenna's output terminal, and the antenna itself, generate noise. Thus some noise originates all through the system. Ultimately, however, all the individually generated noises combine in the output of the receiver to cause a single noise with some available average noise power level, denoted by N_{ao}.[4] It is only the final available noise power that is important because it is after the receiver that various radar functions are performed (through various additional signal processors).

Much of the remainder of this chapter is devoted to developing models to describe the output available average noise power, N_{ao}, of the receiver. For present purposes it is helpful to *model* (imagine) that *all* noise originates from one source at the output of the antenna as shown in Fig. 4.1-2. The source generates an available average noise power N_r, as shown, such that when passed through the noise-free receiving path, it emerges at the output with the correct available power level, N_{ao}. With this model the ratio of average peak signal power to average noise power at the receiver's output will be the same as the ratio S_r/N_r at the antenna's output.

The preceding comments allow us to summarize the average peak signal power to average noise power ratio available to the radar receiver at the output of the receiving antenna:

$$\left(\frac{S_r}{N_r}\right) = \begin{cases} \dfrac{P_t G_{Dt}(\theta_1,\phi_1)\, G_{Dr}(\theta_3,\phi_3)\lambda^2\sigma}{(4\pi)^3 R_1^2 R_2^2 L N_r}, & \text{bistatic or multistatic radars} \\[2ex] \dfrac{P_t G_{Dt}(\theta_1,\phi_1)\, G_{Dr}(\theta_1,\phi_1)\lambda^2\sigma}{(4\pi)^3 R^4 L N_r}, & \text{monostatic radar, two antennas} \\[2ex] \dfrac{P_t G_D^2(\theta_1,\phi_1)\lambda^2\sigma}{(4\pi)^3 R^4 L N_r}, & \text{monostatic radar, one antenna} \\[2ex] \dfrac{P_t G_{Dt}(\theta_2,\phi_2)\, G_{Dr}(\theta_3,\phi_3)\lambda^2}{(4\pi)^2 R^2 L N_r}, & \text{beacon path} \end{cases} \quad (4.1\text{-}18)$$

---

[4] The available average output noise power becomes true output average noise power to the receiver's load if the load impedance is matched to the receiver's output impedance.

where we define total loss L by

$$L \triangleq \begin{cases} L_t L_{rt} L_{ch1} L_{ch2} L_{rr}, & \text{bistatic or multistatic radars} \\ L_t L_{rt} L_{ch}^2 L_{rr}, & \text{monostatic radar, two antennas} \\ L_t L_{rad}^2 L_{ch}^2, & \text{monostatic radar, one antenna} \\ L_t L_{rt} L_{ch} L_{rr}, & \text{beacon path} \end{cases} \qquad (4.1\text{-}19)$$

Logarithmic Forms

Radar engineers often prefer to work with the radar equation when it is expressed in decibels. We will use the notation

$$(\cdot)_{dB} = 10 \log_{10}(\cdot) \qquad (4.1\text{-}20)$$

to represent in decibels (dB) the quantity within the parentheses. As long as the units in the radar equations of (4.1-18) are dimensionally correct, we can apply (4.1-20) to write

$$\left(\frac{S_r}{N_r}\right)_{dB} = \begin{cases} [(P_t)_{dB} + (G_{Dt})_{dB} + (G_{Dr})_{dB} + 2(\lambda)_{dB} \\ \quad + (\sigma)_{dB} - 32.976 - 2(R_1)_{dB} \\ \quad - 2(R_2)_{dB} - (L)_{dB} - (N_r)_{dB}], & \begin{array}{l}\text{bistatic or} \\ \text{multistatic radar}\end{array} \\[2ex] [(P_t)_{dB} + (G_{Dt})_{dB} + (G_{Dr})_{dB} + 2(\lambda)_{dB} \\ \quad + (\sigma)_{dB} - 32.976 - 4(R)_{dB} \\ \quad - (L)_{dB} - (N_r)_{dB}], & \begin{array}{l}\text{monostatic} \\ \text{radar, two} \\ \text{antennas}\end{array} \\[2ex] [(P_t)_{dB} + 2(G_D)_{dB} + 2(\lambda)_{dB} \\ \quad + (\sigma)_{dB} - 32.976 - 4(R)_{dB} \\ \quad - (L)_{dB} - (N_r)_{dB}], & \begin{array}{l}\text{monostatic} \\ \text{radar, one} \\ \text{antenna}\end{array} \\[2ex] [(P_t)_{dB} + (G_{Dt})_{dB} + (G_{Dr})_{dB} + 2(\lambda)_{dB} \\ \quad - 21.984 - 2(R)_{dB} - (L)_{dB} - (N_r)_{dB}], & \text{beacon path} \end{cases} \qquad (4.1\text{-}21)$$

Example 4.1-2 Assume that $P_t = 10^6$ W in a monostatic radar with one antenna that points at the target so that $G_{Dt}(0, 0) = G_{Dr}(0,0) = G_D^2(0, 0) = G_D^2$ with $G_D = 1000$. Also assume that $\lambda = 0.05$ m, $\sigma = 1.6$ m^2, $R = 26.5$ km, $L = 2.2$, and $N_r = 1.8(10^{-14})$ W. Then

$$\begin{aligned} (P_t)_{dB} &= 60.0, & (R)_{dB} &= 44.232 \\ (G_D)_{dB} &= 30.0, & (L)_{dB} &= 3.424 \\ (\lambda)_{dB} &= -13.010, & (N_r)_{dB} &= -137.447, \quad (\sigma)_{dB} = 2.041, \end{aligned}$$

and

$$\begin{aligned} \left(\frac{S_r}{N_r}\right)_{dB} &= 60.0 + 2(30.0) + 2(-13.010) \\ &\quad + 2.041 - 32.976 - 4(44.232) \\ &\quad - 3.424 - (-137.447) = 20.140 \text{ dB} \end{aligned}$$

4.2 IMPORTANT NETWORK DEFINITIONS AND PROPERTIES

In the following two sections we will develop models for describing noise sources and noise generated internal to a network. The work is based on describing behavior in an incremental (small) band of frequencies df centered at an arbitrary *positive* frequency f. The results obtained become the foundations for extending the models to include all frequencies, $0 \leq f$, which is of course the real-world situation.[5] These practical models are developed in Sections 4.5 and 4.6 where they are used to define the available noise power N_r needed in the signal-to-noise ratio of (4.1-18).

In this section we define and discuss several fundamental topics that are needed prior to the development of noise models.

Noise Definitions

Noise is best described through its power spectral density (see Appendix C). Let $n(t)$ be a noise voltage represented as a sample function of a noise random process $N(t)$, and let $\mathscr{S}_{NN}(\omega)$ denote the power spectral density (on $-\infty < \omega < \infty$) of the process. A possible example spectral density based on a two-sided frequency spectrum is shown in Fig. 4.2-1a. Because $\mathscr{S}_{NN}(\omega)$ is a real function and has even symmetry for a real noise, an equivalent power spectral density based on positive frequencies only ($0 \leq \omega$) is shown in panel b. The unit of $\mathscr{S}_{NN}(\omega)$ is volts squared per hertz.

Now suppose that we are interested in only a narrow band of frequencies of width $d\omega$ centered at any arbitrary angular frequency $0 \leq \omega$, as shown in Fig. 4.2-1b. The power in the small (incremental) band $d\omega$ is $2\mathscr{S}_{NN}(\omega)d\omega/2\pi = \mathscr{S}_{NN}(\omega)d\omega/\pi$. If we next let $e_s(t)$ denote the noise from a source having spectral components only in the incremental band, then the mean squared value of $e_s(t)$ is equal to the power in its spectral density:[6]

$$\overline{e_s^2(t)} = \mathscr{S}_{NN}(\omega)\frac{d\omega}{\pi} \tag{4.2-1}$$

Because $\omega = 2\pi f$, we can also write (4.2-1) as

$$\overline{e_s^2(t)} = 2\mathscr{S}_{NN}(f)\,df \tag{4.2-2}$$

if we define the exact meaning of $\mathscr{S}_{NN}(f)$. The notation $\mathscr{S}_{NN}(f)$ is used to represent the function $\mathscr{S}_{NN}(\omega)$ when ω is replaced by $2\pi f$. The two functions are the same as long as $\mathscr{S}_{NN}(f)$ is properly understood. As an example, consider the function

$$\mathscr{S}_{NN}(\omega) = \frac{100}{6\pi + \omega^2} \tag{4.2-3}$$

---

[5] To maintain some consistency with prior literature, we consider only positive frequencies. Since modern work usually uses two-sided spectral expressions (for $-\infty < f < \infty$), we will be careful to point out in each necessary instance how the adaptation to $0 \leq f$ occurs. Results are the same as though all work were based on two-sided spectra.

[6] The overbar is used to represent either the statistical expectation (average, or mean) for a stationary noise or the time average for an ergodic noise. It is also possible for $e_s(t)$ to represent a deterministic power signal of bandwidth $d\omega$; in this case the overbar represents the infinite time average.

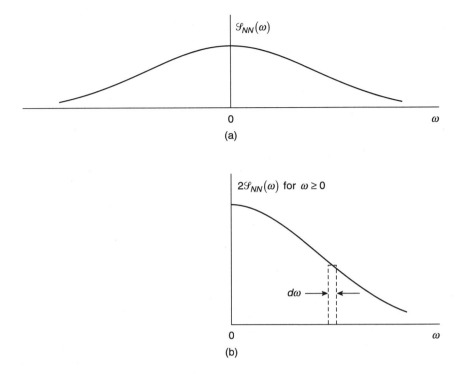

Figure 4.2-1   A two-sided noise power spectral density (*a*) and its equivalent one-sided spectral density (*b*).

By replacing ω by $2\pi f$, we have a function (the same one) of f which is *defined* as $\mathcal{S}_{NN}(f)$:

$$\mathcal{S}_{NN}(f) \triangleq \frac{100}{6\pi + (2\pi f)^2} \tag{4.2-4}$$

To be mathematically precise, $\mathcal{S}_{NN}(\omega) = \mathcal{S}_{NN}(2\pi f)$, so our *definition* is necessary so that $\mathcal{S}_{NN}(f)$ is not misconstrued as the result of setting variable ω equal to variable f.

Maximum Power Transfer Theorem

A linear source of noise (or signal, too) can be represented by its Thevenin's equivalent voltage source as shown in Fig. 4.2-2*a*. Here $e_s(t)$ is the voltage across the source's terminals a, b when the load impedance Z_L is not connected. Impedance $Z_s = R_s + jX_s$ is the complex "output" impedance of the source at the frequency f. It is well-known from steady-state circuit analysis that the maximum real average power the source can deliver to the load occurs when (see Problem 4.2-1)

$$Z_L = Z_s^* = R_s - jX_s \tag{4.2-5}$$

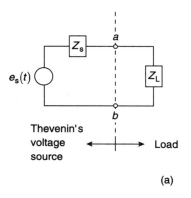

(a)

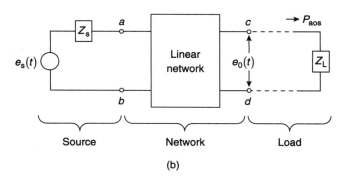

(b)

Figure 4.2-2 A source represented by its Thevenin equivalent voltage source (*a*), and (*b*) a linear network driven by the source.

This statement is known as the *maximum power transfer theorem*. The maximum load power is called the *available power* of the source, since it is always available but is achieved only when (4.2-5) is true. If P_{as} denotes the available power of the source, it can be shown (see Problem 4.2-2) that

$$P_{as} = \frac{\overline{e_s^2(t)}}{4R_s} = \frac{\mathcal{S}_{NN}(f)\,df}{2R_s} \tag{4.2-6}$$

Although (4.2-2) and (4.2-6) were developed assuming a noise source, the expression (4.2-6) can be applied to a steady-state deterministic power signal as well (see Problem 4.2-1).

Available Power Gain

Consider a linear time-invariant network (or LTI network)[7] driven by a source as shown in Fig. 4.2-2*b*. Let P_{as} be the available power from the source. When the source

---

[7] An LTI network is one in which all elements, such as capacitors, resistors, inductors, and linear amplifier gains, do not change with time.

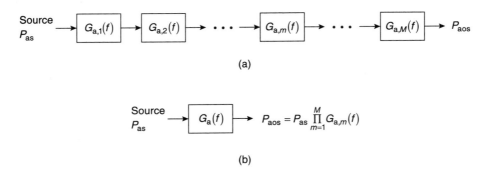

Figure 4.2-3   A cascade of M networks (a), and (b) the network equivalent to the cascade.

is connected to the network at terminals a, b but no load is connected to the output terminals c, d, the network can be considered as a new "source" for whatever load is to be used. If P_{aos} is the average power available from the *network* at its output due to excitation by the source at its input, we define *available power gain*, denoted by $G_a(f)$, of the network by

$$G_a(f) \triangleq \frac{P_{aos}}{P_{as}} \tag{4.2-7}$$

$G_a(f)$ is written as a function of frequency because P_{as} and P_{aos} may each depend on frequency. We note that available power gain does not depend on the load impedance, but does assume that the driving source *is* connected at the network's input.

The definition of power gain inherent in (4.2-7) is especially convenient to use when networks are cascaded.

Cascaded Networks

Let M linear networks be cascaded as shown in Fig. 4.2-3a. Now suppose the available power at the output of stage m *when driven by all previous stages as a source* is given by (4.2-7). Thus we let $P_{aos,m}$ correspond to output available power and $P_{as,m-1}$ correspond to prior stages as a source, and write

$$
\begin{aligned}
P_{aos} &= G_{a,M}(f) \cdot P_{as,M-1}(f) \\
&= G_{a,M}(f)\, G_{a,M-1}(f)\, P_{as,M-2}(f) \\
&= G_{a,M}(f)\, G_{a,M-1}(f)\, \cdots\, G_{a,1}(f)\, P_{as}
\end{aligned} \tag{4.2-8}
$$

so

$$G_a(f) = \frac{P_{aos}}{P_{as}} = \prod_{m=1}^{M} G_{a,m}(f) \tag{4.2-9}$$

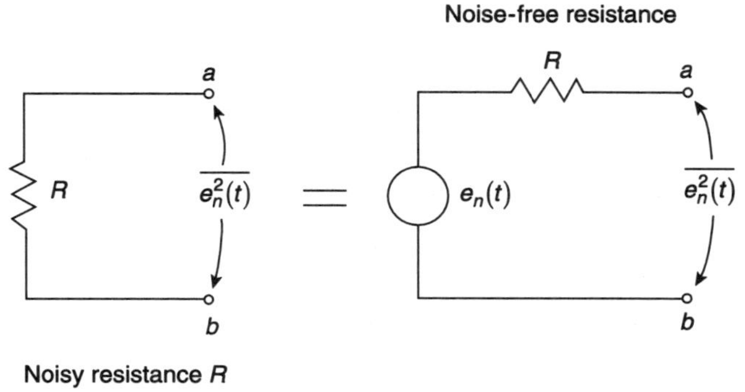

Noise-free resistance

Noisy resistance *R*

Figure 4.3-1 Equivalent circuits of a noisy resistance, *R*.

Equation (4.2-9) shows that a single network equivalent to a cascade of M networks has an available power gain equal to the product of all individual available power gains, as shown in Fig. 4.2-3*b*.

4.3 INCREMENTAL MODELING OF NOISE SOURCES

To best understand radar noise, we will first assume that noise occupies only a very narrow (incremental) band of frequencies of width df located at some (arbitrary) value of *positive* frequency.[8] In the following section we extend these results to include all (positive) frequencies, which of course is the case of a real network.

Resistor as Noise Source

A resistor of resistance R generates noise. The mere fact that the resistor has some physical temperature causes thermal agitation of electrons, resulting in random electronic motion within the resistive material. This motion (a current) causes a voltage (Ohm's law) to appear across the resistor's terminals that is random with a zero average value (because there is no net current leaving the resistor). Both theory and experiment have confirmed that the mean-squared voltage is

$$\overline{e_n^2(t)} = 2\mathscr{S}_{NN}(f)\,df = 4kTR\,df, \qquad 0 \leq f \tag{4.3-1}$$

and the Thevenin voltage model of Fig. 4.3-1 applies. Here

$$T = \text{absolute temperature in kelvin} \tag{4.3-2}$$

$$k = \text{Boltzman's constant} = 1.381(10^{-23})\ \text{J/K} \tag{4.3-3}$$

---

[8] In all our noise work we assume only positive frequencies, $0 \leq f < \infty$. We will be careful to point out any situations where double-sided spectra are used.

If the resistor is considered as a source of noise power, (4.2-6) can be used to give the available average noise power of the resistor source, denoted by dN_{as}, in an incremental band df:

$$dN_{as} = \frac{\overline{e_n^2(t)}}{4R} = \frac{4kTR\,df}{4R} = kT\,df = kT\frac{d\omega}{2\pi} \qquad (4.3\text{-}4)$$

since $d\omega = 2\pi df$. We observe that the available average noise power from the resistor source is *independent of the source's resistance*. Noise power does depend on the physical temperature of the source (as well as df, of course).

Noise from a resistor is called *thermal noise*. In our simple model, (4.3-4) is independent of where in the frequency band the band df is located. If this fact were true for $f \to \infty$, the noise would be called *white noise*. However, a somewhat better model of resistor noise (Mumford and Scheibe, 1968; Carlson, 1986) shows that $\overline{e_n^2(t)}$ varies with frequency according to

$$\overline{e_n^2(t)} = \frac{4kTR(hf/kT)\,df}{e^{hf/kT} - 1}, \qquad 0 \le f \qquad (4.3\text{-}5)$$

where $h = 6.62(10^{-34})$ J/Hz is *Planck's constant*. From (4.3-5) we find that thermal noise is approximated as white noise for frequencies up to approximately

$$f \le \frac{0.2\,kT}{h} = 4.172(10^9)\,T \qquad (4.3\text{-}6)$$

(see Problem 4.3-3). For temperatures near room temperature (where $T \approx 290$ K) thermal noise is nearly the same as white noise for all frequencies typically of interest in radar [$f \le 1.21(10^{12})$ Hz]. This result must be reevaluated for lower temperatures, such as those involved in cryogenic systems (e.g., liquid helium used to cool masers to about 4.2 K for one atmosphere of pressure).

Example 4.3-1   To gain some insight as to numbers, let $df = 10^3$ Hz, and find the noise power available from a resistor at temperature $T = 290$ K. From (4.3-4), $dN_{as} = 1.381(10^{-23})290(10^3) = 4.0049(10^{-18})$ W. To raise this level to a more useable level of 0.01 W, an available power gain of $2.497(10^{15})$ is needed (or 153.97 dB). As will be seen below, noise in a radar is typically larger than that due to a resistor alone.

Model of Arbitrary Noise Source

The noisy resistor can serve as a basis for modeling noise originating from an arbitrary source. Let an arbitrary source have an available noise power of dN_{as} in a small frequency band df. We may associate this noise power as originating in the real part of the source's impedance such that dN_{as} is still given by (4.3-4). For the arbitrary source we *assign* an appropriate temperature, denoted by T_s and called the source's *effective noise temperature*, such that (4.3-4) remains true for the actual value of dN_{as} due to the

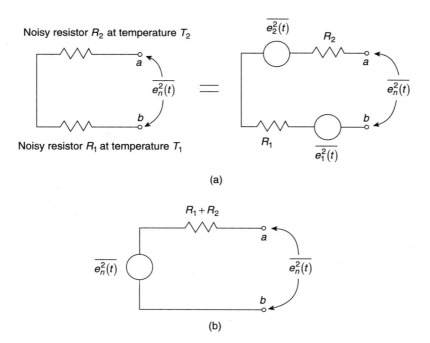

Figure 4.3-2   Two resistors in series but at different temperatures (*a*), and (*b*) their equivalent representation.

source. Thus

$$dN_{as} = kT_s\,df = kT_s\frac{d\omega}{2\pi} \qquad (4.3\text{-}7)$$

Example 4.3-2   Two noisy resistors, R_1 and R_2, are in series but have different physical temperatures, T_1 and T_2, respectively, as shown in Fig. 4.3-2*a*. Here

$$e_n(t) = e_1(t) + e_2(t)$$

and

$$\overline{e_n^2(t)} = \overline{e_1^2(t)} + \overline{e_2^2(t)} + \overline{2e_1(t)e_2(t)}$$

$$= 4kT_1R_1\,df + 4kT_2R_2\,df + \overline{2e_1(t)e_2(t)}$$

Now the two noise voltages $e_1(t)$ and $e_2(t)$ arise from different sources, and they can be considered statistically independent, so $\overline{e_1(t)e_2(t)} = \overline{e_1(t)}\cdot\overline{e_2(t)} = 0$, since $e_1(t)$ and $e_2(t)$ are both zero-mean. Thus

$$\overline{e_n^2(t)} = 4k(T_1R_1 + T_2R_2)df$$

so

$$dN_{as} = \frac{\overline{e_n^2(t)}}{4R_s} = \frac{k(T_1R_1 + T_2R_2)df}{R_1 + R_2}$$

For the equivalent noise source of Fig. 4.3-2b, $dN_{as} = kT_s df$ by (4.3-7) which must equal dN_{as} due to the actual network. Thus

$$T_s = \frac{T_1R_1 + T_2R_2}{R_1 + R_2}$$

must be assigned to the two resistors treated as a single source of noise.

We observe from this example that the effective noise temperature of a source is *not* necessarily equal to the source's physical temperature. However, for a source comprised of all passive elements all at the same physical temperature, the effective noise temperature is equal to the physical temperature (Lathi, 1968, p. 303).

An antenna is a good example of a source of noise (from ohmic, galactic, earth-radiation, and other sources) for which available power is not directly related to its physical temperature (which determines only the ohmic noise power). The effective noise temperature of an antenna treated as a source of noise is called the *antenna temperature*, denoted usually by T_a. Numerical values of T_a range typically from a few tens of kelvin to two or three hundred kelvin, but they can be as high as one to three thousand kelvin if the antenna has a narrow main beam pointing directly at the (hot) sun.

4.4 INCREMENTAL MODELING OF NOISY NETWORKS

Any real network generates noise internally. Some noise power is available at its output, even if no noise source is applied to the input. In this section we give models to easily represent the noise created by networks.

Noisy Network Model

To model a noisy network, we imagine it to be replaced with the same network but *without noise* and think of the source as having an increase in its effective noise temperature of the exact amount to account for the network's output noise. Figure 4.4-1 illustrates these notions. The temperature increase, denoted by T_e, is called the *effective input noise temperature* of the network. The available output noise power due to the network alone, denoted by ΔN_{ao}, is called *excess noise power*. For the small (incremental) band assumed in this section,

$$\Delta N_{ao} = kT_e(f)G_a(f)df \qquad (4.4\text{-}1)$$

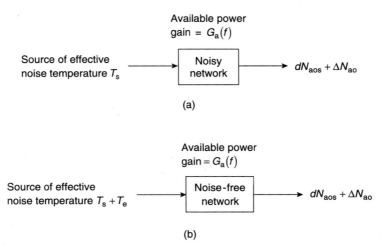

Figure 4.4-1 A noisy network driven by a noise source (*a*), and (*b*) the equivalent noise-free network driven by an equivalent noise source.

where $T_e(f)$ may, in general, depend on frequency. Similarly

$$dN_{aos} = kT_s(f)G_a(f)df \tag{4.4-2}$$

Cascaded Networks

For model purposes with cascaded networks, it is helpful to imagine the available input noise power due to the effective input noise temperature of any one stage as a separate source, as sketched in Fig. 4.4-2*a*. The equivalent model then assumes a *noise-free* cascade, as shown in panel *b*. The models of panels *a* and *b* are equivalent if the effective input noise temperature, denoted by T_e, of the overall cascade is chosen such that the two networks produce the same available output noise power $dN_{aos} + \Delta N_{ao}$. For the network of panel *a*,

$$dN_{aos} = G_{a,1} G_{a,2} \cdots G_{a,M} kT_s df = G_a kT_s df \tag{4.4-3}$$

$$\Delta N_{ao} = G_{a,1} G_{a,2} \cdots G_{a,M} kT_{e1} df$$

$$+ G_{a,2} \cdots G_{a,M} kT_{e2} df$$

$$+ \cdots + G_{a,M} kT_{eM} df$$

$$= G_a k \left[T_{e1} + \frac{T_{e2}}{G_{a,1}} + \frac{T_{e3}}{G_{a,1} G_{a,2}} + \cdots + \frac{T_{eM}}{G_{a,1} G_{a,2} \cdots G_{a,M-1}} \right] df \tag{4.4-4}$$

For the network of panel *b*,

$$dN_{aos} + \Delta N_{ao} = G_a k(T_s + T_e) df \tag{4.4-5}$$

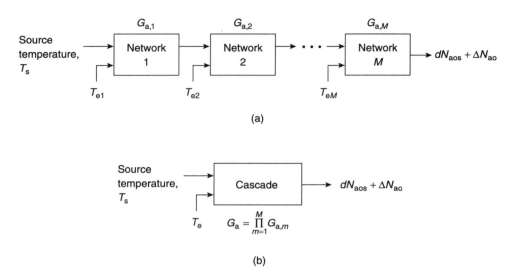

(a)

(b)

Figure 4.4-2   A cascade of M noisy networks in model form (a), and (b) the equivalent model.

On equating (4.4-5) with the sum of (4.4-3) and (4.4-4), we get

$$T_e = T_{e1} + \frac{T_{e2}}{G_{a,1}} + \frac{T_{e3}}{G_{a,1}\,G_{a,2}} + \cdots + \frac{T_{eM}}{G_{a,1}\,G_{a,2}\cdots G_{a,M-1}} \qquad (4.4\text{-}6)$$

Since T_e of (4.4-6) is a measure of the noisiness of a cascade in terms of the noisinesses of the individual networks, it is an important result. It shows that the first stage is the most important while the second is usually next most important (since T_{e2} is reduced by the first stage's available power gain). Succeeding stages are then usually progressively less important. An example helps to put these points in focus.

Example 4.4-1   Consider a four-stage cascade for which $G_{a,1} = 4.0$, $G_{a,2} = 6.0$, $G_{a,3} = 8.0$, and $G_{a,4} = 10.0$. Effective input noise temperatures are assumed to be $T_{e1} = 160$ K, $T_{e2} = 250$ K, $T_{e3} = 400$ K, and $T_{e4} = 600$ K. From (4.4-6),

$$T_e = 160 + \frac{250}{4} + \frac{400}{4(6)} + \frac{600}{4(6)8}$$

$$= 160 + 62.5 + 16.6667 + 3.125 = 242.2917 \text{ K}$$

which is only 51.43% larger than the contribution from the first stage alone.

It is clear from (4.4-6) that low-noise design of a radar receiver should ideally seek the lowest-noise, largest-gain stage as the first in the cascade.

Noise Figures

Just as effective input noise temperature is a measure of the noisiness of a network, *noise figure* is another. In general, noise figure, denoted by F, is defined by

$$F = \frac{dN_{ao}}{dN_{aos}} = \frac{dN_{aos} + \Delta N_{ao}}{dN_{aos}} = 1 + \frac{\Delta N_{ao}}{dN_{aos}}$$

$$= 1 + \frac{kT_e G_a(f)\,df}{kT_s G_a(f)\,df} = 1 + \frac{T_e}{T_s} \tag{4.4-7}$$

It is helpful to use (4.4-7) to distinguish between two noise figures. First, we define F_{op} as the value of F achieved when the network is driven by a source such as intended in practical operation. We call F_{op} the *operating noise figure* and it is equal to the right side of (4.4-7). Thus

$$F_{op} = 1 + \frac{T_e}{T_s} \tag{4.4-8}$$

The second of our two noise figures is called the *standard noise figure*; it is based on the value of F obtained in (4.4-7) when the source has the *standard temperature* 290 K, denoted symbolically by T_0. Thus standard noise figure, denoted by F_0, is

$$F_0 = 1 + \frac{T_e}{T_0} = 1 + \frac{T_e}{290} \tag{4.4-9}$$

The operating and standard noise figures are related by (see Problem 4.4-11)

$$F_0 = 1 + \frac{T_s}{T_0}(F_{op} - 1) \tag{4.4-10}$$

$$F_{op} = 1 + \frac{T_0}{T_s}(F_0 - 1) \tag{4.4-11}$$

Standard noise figure is used to help standardize noise figures associated with various practical hardware. Without some standard temperature it would be necessary for each maker of low-noise amplifiers, for example, to specify the source temperature on which their unit is based. This makes it somewhat harder for a designer to compare one unit with another. However, even if all manufacturers use the standard temperature in specifying their products, there could be a considerable difference in how a product performs in an actual system. Thus we have the need for the operating noise figure. An example will help make these points clearer.

Example 4.4-2 An engineer purchases a low-noise RF amplifier to be used in his system to amplify the output of an antenna that can be modeled as a noise source with effective noise temperature 212 K. The amplifier has a standard noise figure of

$F_0 = 4.35$ (or 6.385 dB). We find the operating noise figure from (4.4-11):

$$F_{op} = 1 + \frac{290}{212}(4.35 - 1) = 5.583 \text{ (or 7.468 dB)}$$

Clearly the system's designers should not expect the same noise figure from their system as the standard value of the amplifier. In fact a closer examination of (4.4-11) shows that F_{op} is never smaller than F_0 when $T_s \leq T_0 = 290$ K.

4.5 PRACTICAL MODELING OF NOISY SOURCES AND NETWORKS

We now extend the preceding work to the case of practical sources and networks by considering all frequencies $0 \leq f$ and not just a small incremental band.

Average Source and Effective Input Noise Temperatures

Figure 4.5-1a shows a (noisy) network modeled by the incremental methods of Sections 4.3 and 4.4. The network's noise is associated with the effective input noise temperature, which we now recognize may be a function of frequency, $T_e(f)$. Similarly the source's noise temperature may vary with frequency, $T_s(f)$. The network's available power gain is $G_a(f)$. When all frequencies are considered, the available output noise power is no longer dN_{ao} for a small band, it is now N_{ao}, the total noise

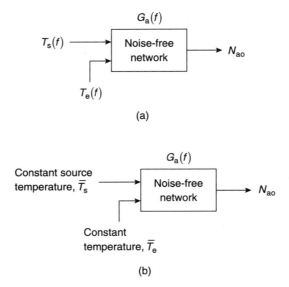

Figure 4.5-1   A model of a noisy network (a), and (b) an equivalent model.

power. For the actual sources, N_{ao} is

$$N_{ao} = \int_0^\infty dN_{ao} = k \int_0^\infty T_s(f) G_a(f) df + k \int_0^\infty T_e(f) G_a(f) df \qquad (4.5\text{-}1)$$

Next we consider the same network as redrawn in Fig. 4.5-1b, except we imagine (model) the source and effective input noise temperatures as *constant values* $\bar{T}_s$ and $\bar{T}_e$, respectively. The output noise power N_{ao} now is written as

$$N_{ao} = \int_0^\infty dN_{ao} = k\bar{T}_s \int_0^\infty G_a(f) df + k\bar{T}_e \int_0^\infty G_a(f) df \qquad (4.5\text{-}2)$$

If the model of panel b must give the same value of N_{ao} as the model of panel a, we may equate (4.5-1) and (4.5-2) term by term to determine $\bar{T}_s$ and $\bar{T}_e$:

$$\bar{T}_s = \frac{\int_0^\infty T_s(f) G_a(f) df}{\int_0^\infty G_a(f) df} \qquad (4.5\text{-}3)$$

$$\bar{T}_e = \frac{\int_0^\infty T_e(f) G_a(f) df}{\int_0^\infty G_a(f) df} \qquad (4.5\text{-}4)$$

$\bar{T}_s$ is called the *average effective noise temperature* of the source. $\bar{T}_e$ is called the *average effective input noise temperature* of the network.

In cases where $T_s(f)$ varies little (is nearly constant) over all frequencies where $G_a(f)$ is significant, the approximation $\bar{T}_s = T_s$ (a constant) can be used in (4.5-3).

Noise Bandwidth

The last step in our modeling processes is to define *noise bandwidth* with the help of Fig. 4.5-2. The preceding model of the network is shown in panel a where N_{ao} is given by (4.5-2). Now imagine (model) replacement of the network by one with an ideal rectangular available power gain, as shown in panel c, which has the same midband frequency, f_0, and same midband gain, $G_a(f_0)$, as the actual network, but bandwidth B_N centered on f_0. For the model of panel b,

$$N_{ao} = k(\bar{T}_s + \bar{T}_e) \int_0^\infty G_i(f) \, df = k(\bar{T}_s + \bar{T}_e) G_i(f_0) B_N \qquad (4.5\text{-}5)$$

By equating N_{ao} of (4.5-5) with that of (4.5-2), we find B_N.

$$B_N = \frac{\int_0^\infty G_a(f) df}{G_a(f_0)} \qquad (4.5\text{-}6)$$

The noise bandwidth of the network, B_N, is the bandwidth of an idealized network that produces the same output available noise power as the actual network when driven by the same source and having the same midband available power gain.

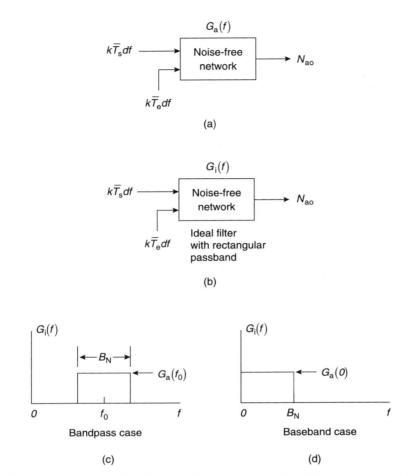

Figure 4.5-2   A network model with available power gain $G_a(f)$ of the true network (a), and (b) the equivalent network using an ideal rectangular passband for available power gain, as shown in (c) for a bandpass function and in (d) for a baseband function.

If the network happens to be baseband, a retrace of the preceding developments for the passband of Fig. 4.5-2d leads to (4.5-6) except with $f_0 = 0$.

By using the above models, we find that the total available noise power at the output of a network is

$$N_{ao} = N_{aos} + \Delta N_{ao} = k\bar{T}_s G_a(f_0) B_N + k\bar{T}_e G_a(f_0) B_N \qquad (4.5\text{-}7)$$

where N_{aos} is the output available noise power due to the source alone,

$$N_{aos} = k\bar{T}_s G_a(f_0) B_N \qquad (4.5\text{-}8)$$

and ΔN_{ao} is the total excess noise power available at the output due to the network's noise,

$$\Delta N_{ao} = k\bar{T}_e G_a(f_0) B_N \tag{4.5-9}$$

Average Noise Figures

When all frequencies are considered, it is necessary to extend the definition for noise figure as given by (4.4-7). We now define *average noise figure*, denoted by $\bar{F}$, as the ratio of *total* available output noise power, N_{ao}, to *total* available output noise power due to the source alone, which is N_{aos}. Now

$$\bar{F} = \frac{N_{ao}}{N_{aos}} = \frac{\int_0^\infty dN_{ao}}{\int_0^\infty dN_{aos}} = \frac{\int_0^\infty F dN_{aos}}{\int_0^\infty dN_{aos}} = \frac{\int_0^\infty F(f) T_s(f) G_a(f) df}{\int_0^\infty T_s(f) G_a(f) df} \tag{4.5-10}$$

The general result (4.5-10) defines two special interpretations. First, we use the notation $\bar{F}_{op}$ to imply the value of $\bar{F}$ when $F(f) = F_{op}(f)$:

$$\bar{F}_{op} = \frac{\int_0^\infty F_{op}(f) T_s(f) G_a(f) df}{\int_0^\infty T_s(f) G_a(f) df} \tag{4.5-11}$$

$\bar{F}_{op}$ is called the *average operating noise figure*. If the source has a constant noise temperature, (4.5-11) reduces to the simpler result (often at least approximately true in practice).

$$\bar{F}_{op} = \frac{\int_0^\infty F_{op}(f) G_a(f) df}{\int_0^\infty G_a(f) df}, \quad T_s(f) = \text{constant} \tag{4.5-12}$$

Equation (4.5-10) also defines *average standard noise figure*, denoted by $\bar{F}_0$, when $F(f) = F_0(f)$:

$$\bar{F}_0 = \frac{\int_0^\infty F_0(f) G_a(f) df}{\int_0^\infty G_a(f) df}, \quad \text{standard source} \tag{4.5-13}$$

Noise Figure and Noise Temperature Interrelationships

It can be shown that $\bar{T}_s$, $\bar{T}_e$, $\bar{F}_0$, and $\bar{F}_{op}$ are related by

$$\bar{F}_{op} = 1 + \frac{\bar{T}_e}{\bar{T}_s} \tag{4.5-14}$$

$$\bar{F}_0 = 1 + \frac{\bar{T}_e}{T_0} \tag{4.5-15}$$

$$\bar{F}_{op} = 1 + \frac{T_0}{\bar{T}_s}(\bar{F}_0 - 1) \tag{4.5-16}$$

$$\bar{F}_0 = 1 + \frac{\bar{T}_s}{T_0}(\bar{F}_{op} - 1) \tag{4.5-17}$$

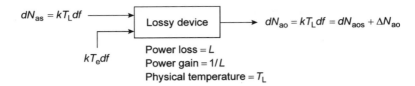

Figure 4.5-3 Model of loss L at physical temperature T_L driven by a noise source of noise temperature T_L.

Modeling of Losses

Many radar components can be modeled as a broadband, impedance-matched lossy (resistive) device. These components may be waveguides, transmission lines, isolators, hybrids, circulators, or other devices. It may even include microwave filters if they are broadband relative to the principal band-limiting networks of the radar (which are often the filters in the IF circuits). Such losses appear resistive at both input and output terminals and are modeled as shown in Fig. 4.5-3 for an incremental band df. The model is unchanged for a finite band B because of our broadband assumption, so results also apply to calculations involving the full range of frequencies.

Assume initially that the loss in Fig. 4.5-3 is driven by a source of effective noise temperature T_L which equals the physical temperature of the loss. The available power from the source is $kT_L df$. The available output noise power also equals $kT_L df$ because the output looks to any load as a resistive source at physical temperature T_L. However, this output available power is due to two sources, the noise source and the noisiness of the loss itself. These respective available powers are

$$dN_{aos} = \frac{kT_L df}{L} \tag{4.5-18}$$

$$\Delta N_{ao} = \frac{kT_e df}{L} \tag{4.5-19}$$

On equating the sum $dN_{aos} + \Delta N_{ao}$ to $kT_L df$, we find the effective input noise temperature of the loss to be

$$T_e = (L - 1)T_L \tag{4.5-20}$$

Since T_e represents only the *increase* in the source's effective noise temperature needed to represent the loss's noise, it will be the same even if the source has an effective noise temperature different than T_L. Furthermore T_e can be taken as constant due to our presumption of a broadband, matched-impedance condition. When we use it with (4.5-14) and (4.5-15), we obtain the average noise figures to be associated with a loss:

$$\bar{F}_{op} = 1 + \frac{T_L}{\bar{T}_s}(L - 1) \tag{4.5-21}$$

$$\bar{F}_0 = 1 + \frac{T_L}{T_0}(L - 1) \tag{4.5-22}$$

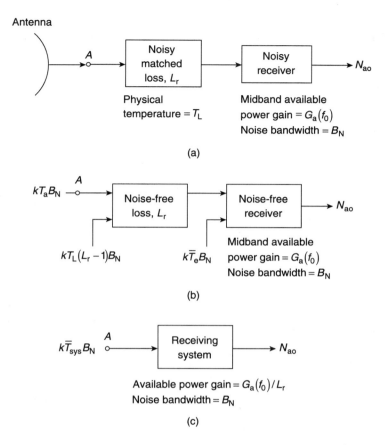

Figure 4.6-1 (*a*) Elements of receiving system showing the antenna output terminal A. (*b*) The noise-free model of the elements of the receiving system, and (*c*) the receiving system modeled as having one source of noise at point A to account for all noise in the system.

We note that if $T_L = \bar{T}_s$ or $T_L = T_0$, respectively, in these two expressions, they give $\bar{F}_{op} = L$ or $\bar{F}_0 = L$. That is, the average noise figure of a loss equals the loss when the source has the same noise temperature as the physical temperature of the loss.

4.6 OVERALL RADAR RECEIVER MODEL

By combining the various preceding concepts, we may now model noise in an overall radar receiver, as shown in Fig. 4.6-1*a*. All losses from the antenna's output (point A) to the receiver's input are represented as a single loss L_r at physical temperature T_L.[9] The receiver is modeled in panel *b* by its noise-free equivalent having an average

---

[9] In some realistic systems there may be two or more losses at different physical temperatures. For this problem L_r is separated into a cascade of losses. The extension of the results given to these more general problems should be straightforward for the reader and is left as an exercise (see Problem 4.6-10).

effective input noise temperature $\bar{T}_e$, midband available power gain $G_a(f_0)$ equal to that of the real receiver, and rectangular (idealized) passband with noise bandwidth B_N. The antenna is the primary source of noise driving the entire receiving system. Its antenna temperature is denoted by T_a, which is assumed constant over the noise bandwidth of the receiver. Figure 4.6-1a corresponds to a detailed representation of the model of Fig. 4.1-2.

Our goal is to find the output available noise power N_{ao} of the model of Fig. 4.6-1b and show that, by proper choice of the average *system noise temperature*, denoted by $\bar{T}_{sys}$ in panel c, the entire receiving system's noise can be considered to arise from one source at the antenna's output. From Fig. 4.6-1b,

$$N_{ao} = kT_a B_N \frac{G_a(f_0)}{L_r} + kT_L(L_r - 1)B_N \frac{G_a(f_0)}{L_r} + k\bar{T}_e B_N G_a(f_0) \qquad (4.6\text{-}1)$$

From panel c,

$$N_{ao} = k\bar{T}_{sys} B_N \frac{G_a(f_0)}{L_r} \qquad (4.6\text{-}2)$$

On equating N_{ao} from (4.6-1) with that of (4.6-2), we have

$$\bar{T}_{sys} = T_a + T_L(L_r - 1) + \bar{T}_e L_r \qquad (4.6\text{-}3)$$

Since the available noise power at point A in Fig. 4.6-1a is $k\bar{T}_{sys} B_N$, it is also equal to N_r of Fig. 4.1-2. Thus the available noise power, N_r, that represents the total effect of receiving system noise, and is to be used in the signal-to-noise ratios of (4.1-18), is given by

$$N_r = k\bar{T}_{sys} B_N \qquad (4.6\text{-}4)$$

Some radar engineers prefer to work with the noise figure rather than with system noise temperature. The conversion is readily made. If $\bar{T}_{e,sys}$ represents the average effective input noise temperature of the entire receiving system (at point A in Fig. 4.6-1a) and the antenna is considered the source for the system, then

$$\bar{T}_{e,sys} = T_L(L_r - 1) + \bar{T}_e L_r \qquad (4.6\text{-}5)$$

and average system operating noise figure, denoted by $\bar{F}_{op,sys}$, is given by

$$\bar{F}_{op,sys} = 1 + \frac{\bar{T}_{e,sys}}{T_a} = 1 + \frac{T_L}{T_a}(L_r - 1) + \frac{\bar{T}_e}{T_a}L_r \qquad (4.6\text{-}6)$$

from (4.5-14). System noise power now becomes

$$N_r = kT_a B_N \bar{F}_{op,sys} \qquad (4.6\text{-}7)$$

Example 4.6-1   In a radar receiver $T_a = 175$ K, the loss is $L_r = 1.45$ at physical temperature $T_L = 255$ K, $\bar{T}_e = 175$ K, and $B_N = 10^7$ Hz. We determine $\bar{T}_{e,sys}$, $\bar{T}_{sys}$, $\bar{F}_{op,sys}$, and N_r for this system and then find the necessary available power gain $G_a(f_0)$ such that the average output available power is $N_{ao} = 0.4$ W.

From (4.6-5), $\bar{T}_{e,sys} = 255(1.45 - 1) + 175(1.45) = 368.5$ K. From (4.6-3), $\bar{T}_{sys} = T_a + \bar{T}_{e,sys} = 175 + 368.5 = 543.5$ K.   From   (4.6-6), $\bar{F}_{op,sys} = 1 + (368.5/175) = 3.1057$ (or 4.92 dB). From (4.6-4), $N_r = 1.381(10^{-23})543.5(10^7) = 7.5057(10^{-14})$ W. From (4.6-2),

$$G_a(f_0) = \frac{L_r N_{ao}}{k\bar{T}_{sys} B_N} = \frac{L_r N_{ao}}{N_r} = \frac{1.45(0.4)}{7.5057(10^{-14})} = 7.727(10^{12})$$

or 128.88 dB.

PROBLEMS

4.1-1 A transmitted radar pulse is

$$s(t) = \begin{cases} A \cos\left(\dfrac{\pi t}{2\tau}\right)\cos(\omega_0 t + \theta_0), & -\tau \leq t \leq \tau \\ 0, & \text{elsewhere} \end{cases}$$

Demonstrate that P_t is given by $A^2/400$ if $s(t)$ is placed across a resistance of 50 Ω. Compare this peak power to the power found by averaging over the whole pulse of duration 2τ, assuming $\omega_0 \gg \dfrac{2\pi}{\tau}$. [*Hint*: Refer to (1.5-5).]

4.1-2 In a bistatic radar system $P_t = 500$ kW, $L_t = 1.31$, and $\lambda = 0.1$ m. The antenna has radiation efficiency 0.97, true aperture area of 2.6 m², aperture efficiency of 0.57, and its main beam points directly at the target that has a cross section of 3.9 m². The channel loss is $L_{ch1} = 1.89$. (a) If the target is 17.9 km from the transmitter what is the power density of the wave that arrives? (b) What is the equivalent power reflected from the target?

4.1-3 The two antennas in a bistatic radar are circular, have the same diameter, the same radiation efficiency of 0.97, the same aperture efficiency of 0.67, and operate at wavelength $\lambda = 0.025$ m. If $P_t = 180$ kW, $L_t = 1.24$, and distances are $R_1 = 50$ km and $R_2 = 36$ km to a target of cross section $\sigma = 5.8$ m², what antenna diameter is required to produce a received available power $S_r = 10^{-12}$ W when channel losses are $L_{ch1} = 1.8$ and $L_{ch2} = 1.3$? Assume that both antennas point toward the target.

4.1-4 A monostatic radar uses one antenna and operates with $P_t = 50$ kW, $G_D = 3(10^4)$, $\lambda = 7.5$ cm, and a total system loss $L = 1.6$. For a target range of 180 km, what minimum target radar cross section is needed to produce an available received signal power of $S_r = 2(10^{-12})$ W if the antenna points directly at the target?

4.1-5 A monostatic radar uses the same circular-aperture antenna for transmission and reception at 8 GHz; its diameter is 2.6 m, $\rho_a = 0.6$, and $L_{rad} = 1.04$. Transmit path loss is $L_t = 1.4$. The radar is to produce $S_r = 4(10^{-14})$ W when the target's radar cross section is 1 m^2 at a maximum range of 92 km. If the channel has a one-way loss $L_{ch} = 1.6$, what transmitter peak power is required if the antenna points directly at the target?

4.1-6 Rewrite (4.1-16) in the form of two right-side factors, one showing dependence on f and the other independent of f. Assume that P_t and losses do not depend on f.

4.1-7 A monostatic radar for which $P_t = 500$ W uses a single antenna at $f_0 = 35$ GHz. The antenna's diameter is 0.8 m, $\rho_r = 0.96$, and $\rho_a = 0.55$. (a) What is the antenna's directivity? (b) What is the available received power if a 10-m^2 target is at a 40-km range and losses are $L_t = 1.6$ and $L_{ch} = 1.8$? Assume that the antenna points toward the target.

4.1-8 Use (4.1-16) for a monostatic radar for which $P_t = 10^6$ W, $f_0 = 5$ GHz, total system loss is 1.95, $R = 113$ km, $\sigma = 1.0$ m^2, and the antenna has a circular aperture with aperture efficiency 0.6 in order to find the antenna's diameter that produces $S_r = 2.5(10^{-14})$ W. Is this a reasonable antenna size for the frequency used? Explain.

4.1-9 Assume that an aircraft uses a beacon at 12 GHz to transmit a power $P_t = 5$ W through an antenna for which $G_{Dt} = 10$ and has a pattern that points toward a receiver. At the receiver the antenna points toward the aircraft and has directivity $G_{Dr} = 3.5(10^4)$. Ignore losses and find the maximum aircraft range that will give an available received signal power of $5(10^{-10})$ W.

4.1-10 In a monostatic radar $P_t = 60$ kW, $G_{Dt} = G_{Dr} = G_D = 3.3(10^4)$, $\lambda = 6$ cm, total system loss is 1.8, and $\sigma = 1.6$ m^2 when $R = 80$ km. Use the logarithmic form of the radar equation to find $(S_r/N_r)_{dB}$ when $N_r = 1.25(10^{-13})$ W.

★4.2-1 Show that a load impedance Z_L must have the value given by (4.2-5) if it is to extract the largest average power from a source with a steady-state sinusoidal voltage $e_s(t) = A \cos(\omega t)$. [*Hint*: Recall that circuit analysis based on complex impedances presumes complex voltages and currents. That is, the source here is assumed to be the complex signal $e_c(t) = A \exp(j\omega t)$ which has $e_s(t)$ as its real part. Also recall that complex power in an element is equal to $v_c(t) i_c^*(t)/2$, where $v_c(t)$ and $i_c(t)$ are the complex voltage across, and the complex current through, the element, respectively. The real average power in the element is then the real part of the element's complex power. Use these results to develop the real average load power and then maximize first with respect to X_L and then with respect to R_L.]

★4.2-2 Show that the first right-side form of (4.2-6) is true by using the hints given in Problem 4.2-1.

★4.2-3 Work the maximum power transfer theorem in reverse. That is, for fixed Z_L what Z_s causes maximum real average power in Z_L? [*Hint*: Use the hints of Problem 4.2-1, the source's voltage $e_s(t)$ is not affected by changing Z_s, and the real part of Z_s cannot be negative.] For the value of Z_L found, what is the maximum load power?

★**4.2-4** A source is connected to an amplifier modeled by the network shown in Fig. P4.2-4. The constant $A_0 > 0$ is the amplifier's voltage gain. (a) Find the available power gain of the network as an expression in terms of Z_s, Z_1, A_0, and Z_2. (b) What values of Z_1 and Z_L will produce the largest real average power in Z_L? (*Hint*: Assume steady-state analysis as hinted in Problem 4.2-1.)

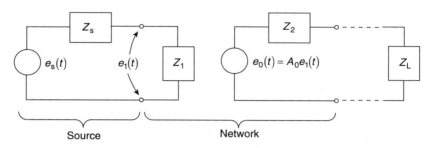

Figure P4.2-4   A source driving a network.

4.3-1 Find the maximum real average power than can be extracted from a noisy resistive source at temperature $T = 370$ K in a bandwidth of 5000 Hz.

4.3-2 Work Problem 4.3-1 except for a bandwidth of 10 MHz.

4.3-3 Find the highest frequency f in (4.3-5) that corresponds to $\overline{e_n^2(t)}$ remaining at or above 0.9 times the value at $f = 0$.

4.3-4 A resistor $R_L = 4700\ \Omega$ is connected across a noisy resistor $R = 1500\ \Omega$. Both resistors are at a temperature of $T = 300$ K. (a) What noise power is created in R_L due to the noisiness of R (in an incremental band df)? (b) What is the two-sided noise power density of the noise in part (a)?

4.3-5 A cascade of three broadband networks with available power gains of 1.8, 6.9, and 6.1 is created. A resistive noise source at temperature $T = 290$ K feeds the cascade. What is the available noise power density dN_{ao}/df at the cascade's output?

4.3-6 Two resistors, R_1 at temperature T_1 and R_2 at temperature T_2, are connected in parallel. What is the effective noise temperature of the pair as a noise source?

4.3-7 Extend Example 4.3-2 to the case of N resistors R_n at temperatures T_n all in series.

4.3-8 Work Problem 4.3-6 except assume three different resistors in parallel, all at different temperatures.

★**4.3-9** Assume a noisy resistor R is modeled as having a terminal-to-terminal capacitance C as shown in Fig. P4.3-9. If R is connected to a load resistor R_L,

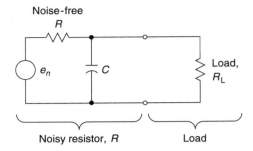

Figure P4.3-9   A noisy resistor with a load resistor across it.

what is the actual noise power (dN_{aos}) to the load due to the source resistor in a small band df?

4.4-1 A *broadband* amplifier has a midband available power gain of 10^9 (or 90 dB). It produces an available excess noise power at its output of $1.2(10^{-5})$ W in a band 1 MHz wide. What is the amplifier's effective input noise temperature for frequencies in its 1-MHz band?

4.4-2 Three amplifiers defined by $T_{e1} = 350$ K with $G_{a,1} = 3.0$, $T_{e2} = 330$ K with $G_{a,2} = 2.6$, and $T_{e3} = 300$ K with $G_{a,3} = 2.2$, can be cascaded in any order. What is the best order from a low-noise standpoint?

4.4-3 Find the operating noise figure of the cascade of Example 4.4-1 when driven by a source with an effective noise temperature of 125 K.

4.4-4 A source with effective noise temperature 180 K feeds a three-stage cascade. The first stage has an input effective noise temperature of $T_{e1} = 210$ K and available power gain of $G_{a,1} = 6.4$. For the second and third stages, respectively, $T_{e2} = 260$ K with $G_{a,2} = 5.8$, and $T_{e3} = 330$ K with $G_{a,3} = 4.5$. (a) Find T_e for the overall cascade. (b) Find the cascade's operating noise figure.

4.4-5 For a two-stage amplifier $T_{e1} = 360$ K and $T_{e2} = 510$ K. When operating with a source for which $T_s = 200$ K, the noise figure is known to be 3.82 (or 5.820 dB). What is the available power gain of the first stage?

4.4-6 A particular radar receiver has a noise figure of 2.1 dB when operating from an antenna with an effective noise temperature $T_a = 180$ K. What is the standard noise figure of the receiver?

4.4-7 Work Problem 4.4-6 except for a noise figure and noise temperature of 2.7 dB and 210 K, respectively.

4.4-8 Work Problem 4.4-6 except for a noise figure and noise temperature of 3.5 dB and 80 K, respectively.

4.4-9 An amplifier has a standard noise figure of 2.0. When used with a source, $F_{op} = 2.75$. What is the effective noise temperature of the source?

4.4-10 Work Problem 4.4-9 if $F_0 = 5.2$ and $F_{op} = 15.0$.

4.4-11 Show that (4.4-10) and (4.4-11) are true.

4.4-12 Two amplifiers are in cascade. The first (input stage) has standard noise figure $F_{01} = 2.15$ and available power gain $G_{a,1} = 5.4$. The second has standard noise figure $F_{02} = 6.36$ and available power gain $G_{a,2} = 8.9$. (a) What is the effective input noise temperature of the cascade? (b) What is the standard noise figure of the cascade? (c) What is the cascade's operating noise figure for a source for which $T_s = 127$ K? (d) What available noise power occurs at the cascade's output when driven by the source and $df = 1000$ Hz?

4.4-13 Work Problem 4.4-12 except assume that $F_{01} = 1.67$, $G_{a,1} = 3.9$, $F_{02} = 5.8$, and $G_{a,2} = 10.6$.

4.5-1 Assume that the noise temperature of a source varies with frequency as

$$T_a(f) = \frac{300}{1 + 0.4f^2}, \quad 0 \leq f$$

The source feeds an amplifier for which

$$G_a(f) = \frac{1000}{1 + 0.4f^2}, \quad 0 \leq f$$

Find the average effective source noise temperature.

4.5-2 Work Problem 4.5-1 except assume that

$$G_a(f) = \frac{1000}{1 + 0.2f^2}, \quad 0 \leq f$$

(*Hint*: Use a partial fraction expansion in one of the integrals to be evaluated.)

★4.5-3 Generalize Problems 4.5-1 and 4.5-2 by assuming that

$$G_a(f) = \frac{1000}{1 + kf^2}, \quad 0 \leq f$$

where $0 < k < \infty$ is a constant. Evaluate your result for $k = 0.6$, $k \to 0$, and $k \to \infty$.

4.5-4 A network has an available power gain given by

$$G_a(f) = \frac{G_0}{1 + (f/F)^2}, \quad 0 \leq f$$

where $0 < F$ and $0 < G_0$ are constants. Find the noise bandwidth B_N of the network in terms of G_0 and F.

4.5-5 Work Problem 4.5-4 except assume that

$$G_a(f) = \frac{G_0}{1 + (f/F)^4}, \qquad 0 \le f$$

4.5-6 Work Problem 4.5-4 except assume that

$$G_a(f) = \frac{G_0}{[1 + (f/F)^2]^2}, \qquad 0 \le f$$

4.5-7 A network has available power gain

$$G_a(f) = \frac{G_0}{\{1 + [(f-f_0)/F]^2\}^3}, \qquad 0 \le f$$

where $G_0 > 0$, $F > 0$, and $f_0 \gg F$ are constants. Find the network's noise bandwidth in terms of G_0, F, and f_0.

4.5-8 A noise source for which $\bar{T}_s = 233$ K feeds a receiving system with a noise bandwidth of 4 MHz, midband available power gain of $1.2(10^{12})$, and output available noise power of $N_{ao} = 0.05$ W. What is the average effective input noise temperature of the receiving system?

4.5-9 Work Problem 4.5-8 except assume that $\bar{T}_s = 80$ K and $B_N = 12$ MHz, with all else unchanged.

4.5-10 An engineer purchases an amplifier with an average standard noise figure of 2.8 dB to amplify the output of an antenna. When installed, the system (amplifier) measures an average operating noise figure of 6.32 dB. What is the antenna's effective noise temperature? Explain why the noise figure is not 2.8 dB in operation.

4.5-11 An amplifier has a midband available power gain of 10^{12} (at frequency f_0) and a noise bandwidth of 24 MHz. If it has an average effective input noise temperature of 260 K and is driven by a noise source for which $\bar{T}_s = 180$ K, what average noise power is available in the amplifier's output?

4.5-12 A radar's receiving path is fed from an antenna, has a noise bandwidth of 13 MHz, and available power gain $G_a(f_0) = 2.4(10^{13})$. When under operation the average noise figure is found to be 3.01 dB and the output available noise power is 0.56 W. What are the average noise temperatures $\bar{T}_s$ and $\bar{T}_e$?

4.5-13 Work Problem 4.5-12 except assume that $B_N = 4.7$ MHz.

4.5-14 An amplifier has an available power gain of $G_a(f_0) = 10^{11}$ and noise bandwidth $B_N = 12(10^6)$ Hz. When used with a source with average effective noise temperature $\bar{T}_s = 225$ K, the available output average noise power is 7.5 mW. What is the amplifier's average effective input noise temperature? What is its average operating noise figure?

4.5-15 The average standard noise figure of a network is 44/29, and when fed by its operating source, the noise figure is 23/8. What is the average effective noise temperature of the source? What is $\bar{T}_e$ for the network?

4.5-16 Work Problem 4.5-15 except assume 67/26 for the average standard noise figure.

4.5-17 An impedance-matched loss of 1.95 (or 2.90 dB) has a physical temperature of 255 K. (a) Find the effective input noise temperature of the loss. (b) What is the operating noise figure of the loss if driven by a source with effective noise temperature of 130 K? (c) What is the standard noise figure of the loss?

4.5-18 Work Problem 4.5-17 except for a loss of 1.23 at a physical temperature of 245 K.

4.5-19 An impedance-matched loss of 1.35 (or 1.303 dB) has a physical temperature of 77.4 K (liquid nitrogen at one atmosphere of pressure). The loss is driven by a thermal source of effective noise temperature of 4.3 K (liquid helium at one atmosphere of pressure). Find (a) the input effective noise temperature, (b) the operating noise figure, and (c) the standard noise figure of the loss.

4.5-20 For the loss and source of Problem 4.5-19, what is the largest loss that can be allowed if the operating noise figure of the loss is to be no larger than 2.79?

4.6-1 An antenna with noise temperature $T_a = 140$ K connects to a receiver's amplifier through a loss $L_r = 1.27$ at a physical temperature of $T_L = 275$ K. (a) If the average input effective noise temperature of the receiver is 190 K and noise bandwidth is 4 MHz, what effective noise power N_r exists at the antenna's output? (b) What is the system's noise temperature?

4.6-2 Work Problem 4.6-1 except assume that $T_a = 68$ K and $T_L = 305$ K.

4.6-3 The components connecting an antenna to a radar receiver have a physical temperature $T_L = 300$ K. For the antenna $T_a = 80$ K, and for the receiver $\bar{T}_e = 130$ K. What loss L_r exists between the antenna and the receiver if the system's average noise temperature is 412.1 K?

4.6-4 Work Problem 4.6-3 except assume that $T_L = 280$ K and $\bar{T}_{sys} = 308$ K.

4.6-5 A radar receiver has a system noise temperature of 527 K and a noise bandwidth of 3 MHz. The loss between antenna and receiver is 0.9 dB. What receiver's midband available power gain is required to give an output available noise power of 0.17 W?

4.6-6 A monostatic radar for which (4.1-16) applies has $P_t = 200$ kW, $R = 50$ km, and total loss $L_t L_{rad}^2 L_{ch}^2 = 1.75$, and operates at a frequency of 8.9 GHz with a target of 18-m² cross section. What antenna directivity is required to produce $(S_r/N_r) = 18$ when the receiving system of Problem 4.6-1 is used?

4.6-7 Suppose the antenna that produced $(S_r/N_r) = 18$ in Problem 4.6-6 has an effective area A_e. If the antenna's dimensions are fixed, and all other parameters of the system except frequency are maintained constant, what value of frequency can increase (S_r/N_r) to 22?

4.6-8 The two stations in a bistatic radar, for which (4.1-12) applies, use identical antennas at 35 GHz with directivities of 2500. In this system $P_t = 50$ kW, $\bar{T}_{sys} = 425$ K, $B_N = 12$ MHz, total loss $L = 2.8$, and $(S_r/N_r) = 20$ when $\sigma = 5$ m^2 for the target. Measurements show that $R_1 = 1.63R_2$. (a) What are the target ranges R_1 and R_2? (b) If the midband available power gain of the receiving station from the antenna's output to the receiver's output is 120 dB, what peak signal and average noise powers are available at the receiver's output?

4.6-9 What average peak transmitted power is needed in a monostatic tracking radar at 2 GHz to track a 3-m^2 target to a range of 160 km, if (S_r/N_r) of 13.7 dB is required from a system for which total loss is 2.06, $\bar{T}_{sys} = 205$ K, $B_N = 1.2(10^6)$ Hz, $G_D = 1805$, and (4.1-16) applies? If the antenna has a circular aperture with aperture efficiency of 0.5, how large is its diameter D in wavelengths?

4.6-10 A radar receiver is modeled as shown in Fig. P4.6-10, where two losses L_{r1} and L_{r2} have physical temperatures T_{L1} and T_{L2}, respectively. Find an expression for system noise temperature $\bar{T}_{sys}$ so that the equivalent model of Fig. 4.6-1c remains valid.

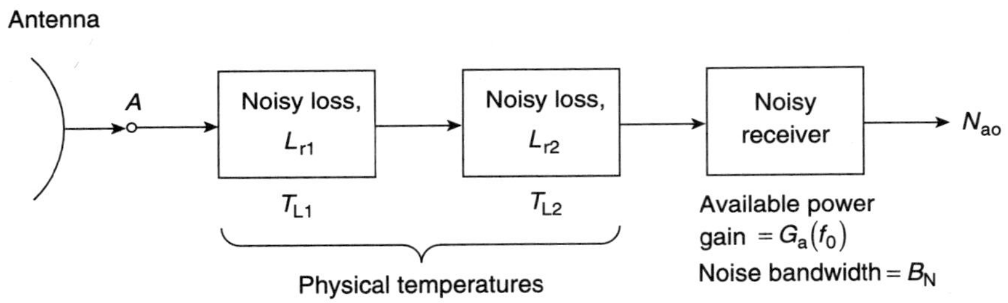

Antenna

Figure P4.6-10 A receiver model.

5

RADAR CROSS SECTION

In Chapter 1 the various types of radar targets were defined in a general way. The principal types were point, extended, and distributed (which includes area and volume) targets. In Chapter 4, in connection with the radar equation, we introduced the concept of radar cross section to account for the power scattered by a target to a receiver. In this chapter we will present a more detailed development of radar cross section. However, because the subject of cross section is deep and ultimately tied to the electromagnetic wave-scattering behavior of targets, which is beyond the intended scope of this book, our discussions will be presented at a level more consistent with reasonable applications as opposed to exact theory.

In many radar systems targets may be taken as small. Initially we discuss small (point) targets and show how scattering of waves is related to polarization and radar cross section. Some examples are then given for small and more complex extended targets. Our work next considers the most common distributed (area) targets, which are sea and land surfaces. Other distributed (volume) targets are then discussed. These include the cross sections applicable to radar scattering from weather effects, such as rain, snow, and clouds. The chapter closes with short definitions of models for fluctuations in cross section.

5.1 CROSS SECTIONS FOR SMALL TARGETS

When a radar's transmitted wave impinges on a target, some of the incident power is absorbed by the target, and part of the power is scattered (reradiated) in all directions. Scattering and absorption reduce the total power of the incident wave. If the total power scattered in all directions from the target is accounted for by a *total scattering cross section*, denoted by σ_T, and the absorbed power by an *absorption cross section*, denoted by σ_a, then the total reduction in the incident wave's power is related to an *extinction cross section*, denoted by σ_e, according to

$$\sigma_e = \sigma_T + \sigma_a \tag{5.1-1}$$

(see Bhattacharyya et al., 1991).

Although σ_e and σ_a have applications in radar to the attenuation of a wave for some targets (water or ice particles in rain, snow, clouds, etc.), the scattering problem is of main concern in this chapter.

Scattering Cross Section

Consider the scattering geometry of Fig. 5.1-1. A radar's transmitting antenna (TX), a target (T), and a radar's receiving antenna (RX), are located at the origins of right-hand coordinate systems $x_1, y_1, z_1, x_2, y_2, z_2$, and x_3, y_3, z_3, respectively. Axes z_1, z_2, and z_3 are all parallel. We also assume that x_1, x_2, and x_3 are parallel, and likewise y_1, y_2, and y_3. These last conditions are given for simplicity; they are not critical to our developments to follow. Thus the three coordinate systems differ only through displacements from each other.

The transmitted wave from TX travels a distance R_1 to T where it is scattered by the target in all directions. Our concern is for the wave that travels a distance R_2 and arrives at the receiving antenna at RX. It is convenient to define wave directions using spherical coordinates, as shown in Fig. 5.1-1. The wave incident on the target at (R_1, θ_1, ϕ_1) has a complex vector electric field, which we denote by

$$\mathbf{E}^i = E_{\theta_1}\hat{a}_{\theta_1} + E_{\phi_1}\hat{a}_{\phi_1} \tag{5.1-2}$$

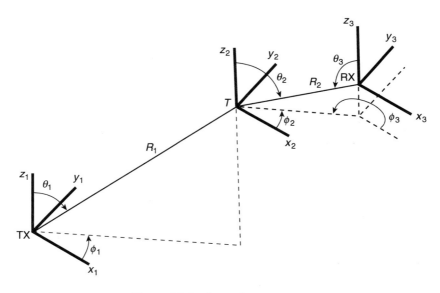

Figure 5.1-1   Scattering geometry.

where E_{θ_1} and E_{ϕ_1} are the complex amplitudes of the components of the incident field in directions θ_1 and ϕ_1, respectively, and $\hat{a}_{\theta_1}$ and $\hat{a}_{\phi_1}$ are unit vectors in these respective directions. For R_1 large so that T is in the far field, the power in $\mathbf{E}^i$ is that of a plane wave given by the magnitude of Poynting's vector:

$$\left\{\begin{array}{c}\text{Power per unit area in}\\ \text{the wave incident}\\ \text{on the target}\end{array}\right\} = \frac{|\mathbf{E}^i|^2}{2\eta} \tag{5.1-3}$$

Here η is the intrinsic impedance of the medium (usually air or vacuum).

On scattering, a portion of the scattered power travels a distance R_2 to the receiving antenna at RX in Fig. 5.1-1. The point RX is defined by (R_2, θ_2, ϕ_2) in spherical coordinates located at the target. The scattered wave at the receiving antenna has a complex vector electric field, which we denote by

$$\mathbf{E}_2^s = E_{\theta_2}\hat{a}_{\theta_2} + E_{\phi_2}\hat{a}_{\phi_2} \tag{5.1-4}$$

where E_{θ_2} and E_{ϕ_2} are the complex amplitudes of the electric field in directions θ_2 and ϕ_2, respectively, and $\hat{a}_{\theta_2}$ and $\hat{a}_{\phi_2}$ are unit vectors in these directions. For R_2 large such that the scattered wave at RX is planar, we have

$$\left\{\begin{array}{c}\text{Power per unit area in}\\ \text{total scattered wave at}\\ \text{receiving antenna}\end{array}\right\} = \frac{|\mathbf{E}_2^s|^2}{2\eta} \tag{5.1-5}$$

We are now able to define *scattering cross section* corresponding to scattering to a point (R_2, θ_2, ϕ_2) from the target as

$$\sigma_s = \sigma_s(\theta_2, \phi_2) = \lim_{R_2 \to \infty} 4\pi R_2^2 \frac{\left\{\begin{array}{c}\text{power per unit area in}\\ \text{total scattered wave at}\\ \text{receiving antenna}\end{array}\right\}}{\left\{\begin{array}{c}\text{power per unit area in}\\ \text{wave incident on}\\ \text{target}\end{array}\right\}}$$

$$= \lim_{R_2 \to \infty} 4\pi R_2^2 \frac{|\mathbf{E}_2^s|^2}{|\mathbf{E}^i|^2} \tag{5.1-6}$$

Because the wave's power per unit area at the receiving antenna decreases as $1/(4\pi R_2^2)$ with distance, the factor $4\pi R_2^2$ in (5.1-6) makes the scattering cross section a function only of θ_2 and ϕ_2 but not distance. The limit in (5.1-6) is only necessary to guarantee that RX is in the far field from T so that $\mathbf{E}_2^s$ is a plane wave.

The result (5.1-6) defines scattering in any direction (θ_2, ϕ_2). If the direction (θ_2, ϕ_2) of scattering is toward the source of the incident wave, $\sigma_s(\theta_2, \phi_2)$ is called the *back-scattering cross section*. For all other directions $\sigma_s(\theta_2, \phi_2)$ is called the *bistatic scattering cross section*.

Total cross section is found by averaging $\sigma_s(\theta_2, \phi_2)$ over all directions according to

$$\sigma_T = \frac{1}{4\pi} \int_{\phi_2=0}^{2\pi} \int_{\theta_2=0}^{\pi} \sigma_s(\theta_2, \phi_2) \sin(\theta_2)\, d\theta_2\, d\phi_2 \tag{5.1-7}$$

(Bhattacharyya et al., 1991, p. 15). Note that if $\sigma_s(\theta_2, \phi_2)$ is a constant σ_s, then $\sigma_T = \sigma_s$, which corresponds to an isotropic scatterer.

The reader should note that the word *radar* was specifically omitted in all the above cross section definitions; rather the word *scattering* was used. There is a specific reason: Scattering and radar cross sections are different, and the difference is related to the polarizations of the receiving antenna and the scattered wave.

Radar Cross Section

Referring again to Fig. 5.1-1, we now define *radar cross section* as that portion of $\sigma_s(\theta_2, \phi_2)$ that corresponds to the particular polarization of the receiving antenna. Thus, on denoting radar cross section by σ, we have

$$\sigma = \lim_{R_2 \to \infty} 4\pi R_2^2 \frac{\left\{\begin{array}{c}\text{power per unit area in scattered}\\ \text{wave at receiving antenna}\\ \text{which is in the polarization of}\\ \text{receiving antenna}\end{array}\right\}}{\left\{\begin{array}{c}\text{power per unit area in}\\ \text{wave incident on}\\ \text{target}\end{array}\right\}} \qquad (5.1\text{-}8)$$

By comparing with (4.1-8), it can be seen that (5.1-8) is consistent with our previous definition of σ used in the radar equation. Again the limit in (5.1-8) is present only to ensure that the scattered wave at the receiving antenna is planar.

Effect of Polarization on Cross Section

Equation (5.1-8) implies that only that portion of the scattered wave at the receiving antenna that has the polarization of the antenna is used in defining σ. Since the receiving antenna may have any specified polarization, the definition of σ requires that we be able to break an arbitrarily polarized scattered wave into two parts—one to which the antenna responds, which has the antenna's polarization, and another to which the antenna does not respond. The two parts are therefore orthogonal to each other.

To demonstrate the desired wave composition, we assume that $\mathbf{E}_2^s$ is the sum of two waves. One, denoted by $\mathbf{E}_A$, will have an arbitrary, but specified, polarization. The second $\mathbf{E}_B$ will be assumed orthogonal to $\mathbf{E}_A$. For $\mathbf{E}_2^s$,

$$\mathbf{E}_2^s = E_{\theta_2}\hat{a}_{\theta_2} + E_{\phi_2}\hat{a}_{\phi_2} = E_{\theta_2}(\hat{a}_{\theta_2} + Q_2\hat{a}_{\phi_2})$$

$$= E_{\theta_2}\begin{bmatrix}1\\Q_2\end{bmatrix} \qquad (5.1\text{-}9)$$

is the scattered field from (5.1-4) and[1]

$$Q_2 = \frac{E_{\phi_2}}{E_{\theta_2}} \qquad (5.1\text{-}10)$$

---

[1] We presume that $E_{\theta_2} \neq 0$.

is the polarization ratio that defines the polarization of $\mathbf{E}_2^s$ [see (3.1-8)]. The last form of (5.1-9) uses the matrix representation of a vector. The power density in $\mathbf{E}_2^s$ is

$$\text{Power density in } \mathbf{E}_2^s = \frac{|\mathbf{E}_2^s|^2}{2\eta} = E_{\theta_2} \, [1 \;\; Q_2] \begin{bmatrix} 1 \\ Q_2^* \end{bmatrix} E_{\theta_2}^* \frac{1}{2\eta}$$

$$= \frac{|E_{\theta_2}|^2}{2\eta} (1 + |Q_2|^2) \tag{5.1-11}$$

Next we represent $\mathbf{E}_A$ and $\mathbf{E}_B$ as

$$\mathbf{E}_A = A_1 \hat{a}_{\theta_2} + A_1 Q_A \hat{a}_{\phi_2} = A_1 \begin{bmatrix} 1 \\ Q_A \end{bmatrix} \tag{5.1-12}$$

$$\mathbf{E}_B = A_2 \hat{a}_{\theta_2} + A_3 \hat{a}_{\phi_2} = \begin{bmatrix} A_2 \\ A_3 \end{bmatrix} \tag{5.1-13}$$

where $A_1 \neq 0$ is assumed and

$$\mathbf{E}_2^s = \mathbf{E}_A + \mathbf{E}_B \tag{5.1-14}$$

Here Q_A is the polarization ratio that defines the polarization of the wave component $\mathbf{E}_A$ to which the antenna is to respond, A_1, A_2, and A_3 are to be determined such that (5.1-14) is true, and $\mathbf{E}_A$ and $\mathbf{E}_B$ are orthogonal. This latter condition requires that

$$\mathbf{E}_A \cdot \mathbf{E}_B^* = A_1 \, [1 \;\; Q_A] \begin{bmatrix} A_2^* \\ A_3^* \end{bmatrix} = A_1 (A_2^* + Q_A A_3^*) = 0 \tag{5.1-15}$$

from which

$$A_2 = - Q_A^* A_3 \tag{5.1-16}$$

for any A_1 and A_3. On use of (5.1-16) in (5.1-13) and then combining (5.1-12)–(5.1-14), we have

$$\mathbf{E}_2^s = E_{\theta_2} \begin{bmatrix} 1 \\ Q_2 \end{bmatrix} = \mathbf{E}_A + \mathbf{E}_B = A_1 \begin{bmatrix} 1 \\ Q_A \end{bmatrix} + A_3 \begin{bmatrix} -Q_A^* \\ 1 \end{bmatrix} \tag{5.1-17}$$

or equivalently,

$$E_{\theta_2} = A_1 - Q_A^* A_3 \tag{5.1-18}$$

$$E_{\theta_2} Q_2 = Q_A A_1 + A_3 \tag{5.1-19}$$

These last two equations are solved for A_1 and A_3, and then A_2 is obtained from (5.1-16) to get

$$A_1 = \frac{E_{\theta_2}(1 + Q_A^* Q_2)}{1 + |Q_A|^2} \tag{5.1-20}$$

$$A_2 = \frac{-E_{\theta_2} Q_A^*(Q_2 - Q_A)}{1 + |Q_A|^2} \tag{5.1-21}$$

$$A_3 = \frac{E_{\theta_2}(Q_2 - Q_A)}{1 + |Q_A|^2} \tag{5.1-22}$$

The above developments have shown that an arbitrarily polarized scattered wave $\mathbf{E}_2^s$ (with polarization ratio Q_2) can be decomposed into the sum of two waves $\mathbf{E}_A$ and $\mathbf{E}_B$ according to (5.1-14); $\mathbf{E}_A$ has any desired polarization (polarization ratio Q_A) and is given by (5.1-12) using (5.1-20). $\mathbf{E}_B$ is orthogonal to $\mathbf{E}_A$ and is given by (5.1-13) using (5.1-21) and (5.1-22).

The power in the wave component $\mathbf{E}_A$ is that which the antenna can receive; its density is

$$\text{Power density in } \mathbf{E}_A = \frac{|\mathbf{E}_A|^2}{2\eta} = \frac{|A_1|^2}{2\eta}(1 + |Q_A|^2)$$

$$= \frac{|E_{\theta_2}|^2 |1 + Q_A^* Q_2|^2}{2\eta(1 + |Q_A|^2)} \tag{5.1-23}$$

Our results can now be used to determine σ_s from (5.1-6) and σ from (5.1-8). On using (5.1-11), we have

$$\sigma_s = \lim_{R_2 \to \infty} 4\pi R_2^2 \frac{|\mathbf{E}_2^s|^2}{|\mathbf{E}^i|^2} = \lim_{R_2 \to \infty} 4\pi R_2^2 \frac{|E_{\theta_2}|^2}{|\mathbf{E}^i|^2}(1 + |Q_2|^2) \tag{5.1-24}$$

On using (5.1-12) and (5.1-20), we have

$$\sigma = \lim_{R_2 \to \infty} 4\pi R_2^2 \frac{|\mathbf{E}_A|^2}{|\mathbf{E}^i|^2} = \lim_{R_2 \to \infty} 4\pi R_2^2 \frac{|E_{\theta_2}|^2}{|\mathbf{E}^i|^2} \cdot \frac{|1 + Q_A^* Q_2|^2}{(1 + |Q_A|^2)} \tag{5.1-25}$$

By combining (5.1-24) and (5.1-25), we finally obtain

$$\sigma = \sigma_s \, \rho_{\text{pol}} \tag{5.1-26}$$

where

$$\rho_{\text{pol}} = \frac{|1 + Q_A^* Q_2|^2}{(1 + |Q_A|^2)(1 + |Q_2|^2)} \tag{5.1-27}$$

is called the *polarization efficiency* (also called the *polarization match factor*; Mott, 1992, p. 191).

The polarization efficiency is the fraction of the scattered power that the receive antenna can receive. From (5.1-26) the radar cross section is less than the scattering cross section by a factor ρ_{pol}. From Section 3.4 we know that an antenna with polarization ratio Q_3 (in coordinates x_3, y_3, z_3 of Fig. 5.1-1) must be related to the *incoming* wave's polarization ratio Q_A (in coordinates x_2, y_2, z_2 of Fig. 5.1-1) by

$$Q_3 = -Q_A^* \tag{5.1-28}$$

if it is to capture all the power in the wave component $\mathbf{E}_A$. Thus, from (5.1-27),

$$\rho_{\text{pol}} = \frac{|1 - Q_3 Q_2|^2}{(1 + |Q_3|^2)(1 + |Q_2|^2)} \tag{5.1-29}$$

which agrees with Mott (1992, p. 194).

Example 5.1-1 A target has a scattering cross section of 3 m^2 when its scattered wave is right-circularly polarized. A receiving antenna is polarized linearly in the θ_3 direction only. What cross section should be used in the radar equation?

The radar equation must use the radar cross section as given by (5.1-26). Thus $\sigma = 3\rho_{\text{pol}}$, and we must find ρ_{pol}. For a right-circular scattered wave $Q_2 = E_{\phi_2}/E_{\theta_2} = -j$ from Fig. 2.4-4. For the receiving antenna, $Q_3 = 0$ for linear polarization in the θ_3 direction, so

$$\rho_{\text{pol}} = \frac{|1 - 0|^2}{(1 + 0)(1 + |-j|^2)} = \frac{1}{2}$$

and $\sigma = 3(\frac{1}{2}) = 1.5 \text{ m}^2$ is used in the radar equation.

We take a second example where polarizations are circular.

Example 5.1-2 Assume that the receiving antenna is right-circularly polarized corresponding to $Q_3 = -j$ (see Fig. 2.4-2). Next suppose that the arriving wave is left-circularly polarized so that $Q_2 = +j$. From (5.1-29) we have $\rho_{\text{pol}} = 0$. Alternatively, if the antenna is left-circular and the wave is right-circular, we again find $\rho_{\text{pol}} = 0$. Thus a circularly polarized antenna does not respond to a scattered wave with the opposite-sense circular polarization.

On the other hand, if antenna and wave are both same-sense circular, we find that $\rho_{\text{pol}} = 1$; there is no polarization loss.

The preceding responses are valuable for a case such as a radar operating in rain. To the extent that raindrops are spherical, they backscatter opposite-sense circular polarization when illuminated circularly. Hence ideally there is no received power, and other (desired) targets embedded in the rain can be seen almost without obstruction.

The problem can also be reversed. If a radar transmits one sense of circular polarization and receives on the other, rain backscatter is enhanced. This case is important in weather radar.

All the preceding results regarding scattering and radar cross sections apply to point targets that have a largest physical dimension that is small compared to the range extent of the transmitted signal [which can be taken as $c/(2B)$, c being the speed of light and B is the 3-dB bandwidth of the transmitted waveform]. This case is known as "long pulse" illumination, which is the usual case (Knott et al., 1985, p. 49). For extended targets that do not satisfy the small-size condition, σ becomes a function of distance over the target from which scattering occurs. Larger targets may be treated as a collection of individual small scatterers that produce the overall scattered signal. Such a signal appears as a "smeared-out" version of the time waveform that would have occurred with a "point" target.

5.2   TARGET SCATTERING MATRICES

Scattering and radar cross sections of small targets do not provide all available information about a target. Full information is available only through specification of the *field components* of scattering. Since two orthogonal electric field components are needed to specify an incident wave at a target, and two field components are needed to specify a scattered wave, we see that four target "responses" are required. Each input component of field can contribute to each output field component for the total of four "responses." This scattering behavior is best described with the help of matrices, generally called *scattering matrices*. In practice, two types of scattering matrices are most useful. They differ in the coordinate systems for which they are defined.

Jones Matrix

Consider Fig. 5.1-1 again. We develop a scattering matrix, called the *Jones matrix*, which is based on defining the scattered wave in coordinates x_2, y_2, z_2. In the spherical coordinates of this coordinate system the scattered field at the receiving antenna is given by (5.1-4). The incident electric field is similarly defined by (5.1-2). It results that scattered wave components E_{θ_2} and E_{ϕ_2} can *each* have components in response (due) to both E_{θ_1} and E_{ϕ_1}. The component of E_{θ_2} (or E_{ϕ_2}) caused by E_{θ_1} (or E_{ϕ_1}) can be thought of as a *copolarized* term. The component of E_{θ_2} (or E_{ϕ_2}) caused by E_{ϕ_1} (or E_{θ_1}) can be considered as a *cross-polarized* term. On combining these responses, we write

$$E_{\theta_2} = (T_{\theta_2\theta_1}E_{\theta_1} + T_{\theta_2\phi_1}E_{\phi_1})\frac{e^{-jkR_2}}{\sqrt{4\pi}R_2} \tag{5.2-1}$$

$$E_{\phi_2} = (T_{\phi_2\theta_1}E_{\theta_1} + T_{\phi_2\phi_1}E_{\phi_1})\frac{e^{-jkR_2}}{\sqrt{4\pi}R_2} \tag{5.2-2}$$

Here the T_{pq} are complex proportionality constants where the first subscript p represents the response coordinate (θ_2 or ϕ_2) while the second subscript q represents the incident-field's coordinate (θ_1 or ϕ_1). Thus $T_{\theta_2\phi_1}$ corresponds to the *cross-polarized* component of E_{θ_2} that is due to E_{ϕ_1}. The factors $1/(\sqrt{4\pi}R_2)$ are present because field strength decreases as the scattered power is spread over the sphere of radius R_2 as it propagates to the receiving antenna. The term $\exp(-jkR_2)$, where $k = 2\pi/\lambda$, accounts for the phase shift of the scattered wave traveling distance R_2.

On using (5.2-1) and (5.2-2), the scattered vector electric field at the receiving antenna of Fig. 5.1-1 becomes

$$\mathbf{E}_2^s = E_{\theta_2}\hat{a}_{\theta_2} + E_{\phi_2}\hat{a}_{\phi_2} = [(T_{\theta_2\theta_1}E_{\theta_1} + T_{\theta_2\phi_1}E_{\phi_1})\hat{a}_{\theta_2}$$

$$+ (T_{\phi_2\theta_1}E_{\theta_1} + T_{\phi_2\phi_1}E_{\phi_1})\hat{a}_{\phi_2}]\frac{e^{-jkR_2}}{\sqrt{4\pi R_2}}$$

$$= \begin{bmatrix} T_{\theta_2\theta_1} & T_{\theta_2\phi_1} \\ T_{\phi_2\theta_1} & T_{\phi_2\phi_1} \end{bmatrix}\begin{bmatrix} E_{\theta_1} \\ E_{\phi_1} \end{bmatrix}\frac{e^{-jkR_2}}{\sqrt{4\pi R_2}} = \begin{bmatrix} E_{\theta_2} \\ E_{\phi_2} \end{bmatrix} \tag{5.2-3}$$

The 2×2 matrix of elements T_{pq} is known as the *Jones matrix* (see Mott, 1992, p. 315). In general, complete specification of this matrix requires eight quantities (amplitude and phase of each of four elements). However, the phase of one element is often factored from the matrix and absorbed in the exponential term of (5.2-3). The result is a *relative matrix* requiring specification of only four amplitudes and three relative phases. In the special case of monostatic radar (backscattering), it is known that $T_{\theta_2\phi_1} = - T_{\phi_2\theta_1}$ (Mott, 1992, p. 316), and the target is specified by three amplitudes and two relative phases. Furthermore, if the target with a monostatic radar has a surface with symmetry about a plane that contains the line from the radar to the target and the incident electric field vector, then $T_{\theta_2\phi_1} = 0$ and $T_{\phi_2\theta_1} = 0$ (Mott, 1992, p. 314) which requires only two amplitudes and one relative phase to define the target. This last case is useful in some of the examples of targets to follow in Section 5.3.

The polarization ratio of the scattered wave, denoted by Q_2, is obtained from (5.2-3),

$$Q_2 = \frac{E_{\phi_2}}{E_{\theta_2}} = \frac{T_{\phi_2\theta_1}E_{\theta_1} + T_{\phi_2\phi_1}E_{\phi_1}}{T_{\theta_2\theta_1}E_{\theta_1} + T_{\theta_2\phi_1}E_{\phi_1}} = \frac{T_{\phi_2\theta_1} + T_{\phi_2\phi_1}Q_1}{T_{\theta_2\theta_1} + T_{\theta_2\phi_1}Q_1} \tag{5.2-4}$$

where

$$Q_1 = \frac{E_{\phi_1}}{E_{\theta_1}} \quad \text{(assuming that } E_{\theta_1} \neq 0) \tag{5.2-5}$$

is the polarization ratio of the incident wave (same as that of the transmitting antenna).

Sinclair Matrix

In some radar situations it may be more convenient to describe the scattered fields at the receiving antenna in coordinates x_3, y_3, z_3 than in coordinates x_2, y_2, z_2, as done for the Jones matrix. For example, with a monostatic radar, coordinates x_3, y_3, z_3 become identical to coordinates x_1, y_1, z_1, as seen from Fig. 5.1-1. As a consequence the receiving antenna's output voltage becomes the simple dot product of the received electric field vector and the affective length of the receiving antenna (see Section 3.4).

By using procedures similar to those for the Jones matrix, we obtain scattered field components E_{θ_3} and E_{ϕ_3} in coordinates x_3, y_3, z_3:

$$E_{\theta_3} = (S_{\theta_3\theta_1} E_{\theta_1} + S_{\theta_3\phi_1} E_{\phi_1}) \frac{e^{-jkR_2}}{\sqrt{4\pi}R_2} \tag{5.2-6}$$

$$E_{\phi_3} = (S_{\phi_3\theta_1} E_{\theta_1} + S_{\phi_3\phi_1} E_{\phi_1}) \frac{e^{-jkR_2}}{\sqrt{4\pi}R_2} \tag{5.2-7}$$

Here the S_{pq} are new complex proportionality constants where p may be θ_3 or ϕ_3 and q may be θ_1 or ϕ_1. The scattered vector electric field becomes

$$
\begin{aligned}
\mathbf{E}_3^s &= E_{\theta_3} \hat{a}_{\theta_3} + E_{\phi_3} \hat{a}_{\phi_3} = [(S_{\theta_3\theta_1} E_{\theta_1} + S_{\theta_3\phi_1} E_{\phi_1}) \hat{a}_{\theta_3} \\
&\quad + (S_{\phi_3\theta_1} E_{\theta_1} + S_{\phi_3\phi_1} E_{\phi_1}) \hat{a}_{\phi_3}] \frac{e^{-jkR_2}}{\sqrt{4\pi}R_2} \\
&= \begin{bmatrix} S_{\theta_3\theta_1} & S_{\theta_3\phi_1} \\ S_{\phi_3\theta_1} & S_{\phi_3\phi_1} \end{bmatrix} \begin{bmatrix} E_{\theta_1} \\ E_{\phi_1} \end{bmatrix} \frac{e^{-jkR_2}}{\sqrt{4\pi}R_2} = \begin{bmatrix} E_{\theta_3} \\ E_{\phi_3} \end{bmatrix}
\end{aligned}
\tag{5.2-8}
$$

The matrix of (5.2-8) with elements S_{pq} is called the *Sinclair matrix*.

We may relate the Sinclair and Jones matrices by noting from Fig. 5.1-1 that $E_{\theta_3} = E_{\theta_2}$ and $E_{\phi_3} = - E_{\phi_2}$, so

$$\begin{bmatrix} E_{\theta_3} \\ E_{\phi_3} \end{bmatrix} = \begin{bmatrix} E_{\theta_2} \\ - E_{\phi_2} \end{bmatrix} \tag{5.2-9}$$

By using (5.2-9) with both (5.2-8) and (5.2-3), we find that

$$\begin{bmatrix} T_{\theta_2\theta_1} & T_{\theta_2\phi_1} \\ T_{\phi_2\theta_1} & T_{\phi_2\phi_1} \end{bmatrix} = \begin{bmatrix} S_{\theta_3\theta_1} & S_{\theta_3\phi_1} \\ - S_{\phi_3\theta_1} & - S_{\phi_3\phi_1} \end{bmatrix} \tag{5.2-10}$$

The polarization ratio, denoted by Q_3, of the field in the receiving antenna's coordinates derives from (5.2-8):

$$Q_3 = \frac{E_{\phi_3}}{E_{\theta_3}} = \frac{S_{\phi_3\theta_1} E_{\theta_1} + S_{\phi_3\phi_1} E_{\phi_1}}{S_{\theta_3\theta_1} E_{\theta_1} + S_{\theta_3\phi_1} E_{\phi_1}} = \frac{S_{\phi_3\theta_1} + S_{\phi_3\phi_1} Q_1}{S_{\theta_3\theta_1} + S_{\theta_3\phi_1} Q_1} \tag{5.2-11}$$

In terms of Q_2 defined by (5.2-4), we have

$$Q_2 = \frac{E_{\phi_2}}{E_{\theta_2}} = \frac{- E_{\phi_3}}{E_{\theta_3}} = - Q_3 \tag{5.2-12}$$

which relates the scattered wave's polarization ratio, Q_2, to the elements of the Sinclair matrix and the incident wave's polarization ratio Q_1 from (5.2-11).

Uses of Scattering Matrices

Because scattering marices define all available target information, they may be used in some modern applications to identify and classify targets. Even though radar cross section can be determined from scattering matrix elements (if known), it is usually just measured.

5.3 EXAMPLES OF TARGET CROSS SECTIONS

The calculation of radar cross section is complex, even for targets of simple shape. Few target shapes give exact solutions because of boundary value problems. Even in the few cases where exact solutions are possible, the expressions are difficult to interpret and often are evaluated by digital computer. These difficulties have lead to a variety of approximation methods described in various sources (to name just a few examples, see Fritch, ed., 1965; Harrington, 1968; Ruck, 1970; Crispin, ed., 1968; Knott et al., 1985). Neither the approximate or exact analysis methods are discussed here because they are beyond our scope. Rather, we will give some examples of backscattering cross section for some simply shaped targets in a monostatic radar.

Sphere

The perfectly conducting sphere is a good example of a target with symmetry such that the cross-polarizing elements of the scattering matrices are zero. This means that each orthogonal component of the incident wave is backscattered without cross-polarized components from the other orthogonal component. As a result the radar cross section for scattering from one component of an incident wave is the same for the other (orthogonal) component, and therefore it is adequate to give radar cross section for one component of excitation only.

The backscattering radar cross section for a perfectly conducting sphere of radius a is known exactly (see Kerr, 1964, pp. 451–452; Van Bladel, 1964, pp. 264 and 348; Currie, ed., 1989, p. 35; Ruck, ed., vol. 1, 1970, p. 141; Blake, 1980, p. 103). For linearly polarized illumination in the x direction only, as shown in Fig. 5.3-1, the backscattering radar cross section is

$$\sigma = \frac{\pi a^2}{(ka)^2} \left| \sum_{n=1}^{\infty} (-1)^n (2n+1)(a_n - b_n) \right|^2 \tag{5.3-1}$$

where

$$a_n = \frac{[\xi j_n(\xi)]'_{\xi=ka}}{[\xi h_n^{(2)}(\xi)]'_{\xi=ka}} \tag{5.3-2}$$

$$b_n = \frac{j_n(ka)}{h_n^{(2)}(ka)} \tag{5.3-3}$$

$$k = \frac{2\pi}{\lambda} \tag{5.3-4}$$

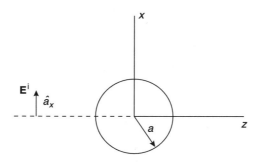

Figure 5.3-1   Perfectly conducting spherical radar target.

λ is wavelength, $j_n(\cdot)$ is the spherical Bessel function of the first kind which is related to the Bessel function of the first kind by

$$j_n(\xi) = \sqrt{\frac{\pi}{2\xi}} \, J_{n+(1/2)}(\xi) \qquad (5.3\text{-}5)$$

Also $h_n^{(2)}(\xi)$ is the spherical Hankel function of the second kind which is related to the Hankel function of the second kind by

$$h_n^{(2)}(\xi) = \sqrt{\frac{\pi}{2\xi}} \, H_{n+(1/2)}^{(2)}(\xi) \qquad (5.3\text{-}6)$$

(Kerr, 1951, p. 446). The primes in (5.3-2) denote differentiation with respect to ξ.

A plot of σ from (5.3-1) is shown in Fig. 5.3-2 as calculated by Blake (1972). There are three distinct regions of behavior. In the region where a/λ is small, called the *Rayleigh region*, cross section is

$$\frac{\sigma}{\pi a^2} = 9\left(\frac{2\pi a}{\lambda}\right)^4 = 1.4027(10^4)\left(\frac{a}{\lambda}\right)^4, \qquad 0 < \frac{a}{\lambda} \le \frac{1}{10} \qquad (5.3\text{-}7)$$

This function is shown dashed in Fig. 5.3-2. In the region where a/λ is large, called the *optical region*,

$$\frac{\sigma}{\pi a^2} = 1, \quad 1.6 \le \frac{a}{\lambda} \qquad (5.3\text{-}8)$$

Errors in using (5.3-7) and (5.3-8) to approximate cross section are not more than about 15% of the true values. The maximum errors occur at the boundaries of the middle region, called the *resonance* (or *Mie*) *region*; in this region σ oscillates about the optical cross section (πa^2) with maximum and minimum values that close together with increasing a/λ.

Because the sphere's cross section is independent of aspect (viewing) angle, it is a most valuable target for calibration and evaluation of radar systems.

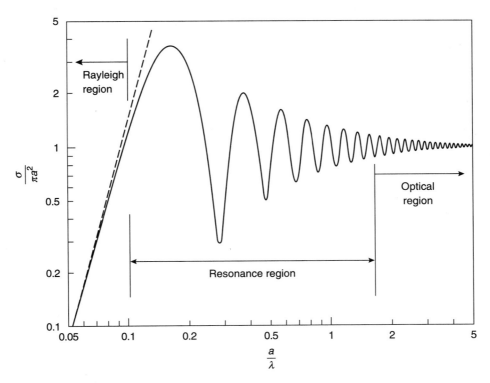

Figure 5.3-2 Radar cross section of a perfectly conducting sphere as a function of the radius-to-wavelength ratio a/λ. (Adapted from Blake, 1972.)

Flat Rectangular Plate

Ross (1966) has described the backscattering cross section of a perfectly conducting flat rectangualar plate that is assumed to be thin compared to a wavelength. The applicable geometry is shown in Fig. 5.3-3. The plate is centered on the origin and lies in the x, y plane with width $2a$ in the x direction and height $2b$ in the y direction. The radar illuminates the plate with a plane wave having a propagation direction in the x, z plane. The direction of incidence (of incoming ray) makes an angle θ from the z axis. Ross gives two radar cross sections. One applies to an incident wave with linear polarization in the y direction ($\mathbf{E_V}$ in the figure, called *vertical polarization*) and the radar measuring scattering in the same polarization. The second cross section assumes linear incident polarization with the electric field in the x, z plane ($\mathbf{E_H}$ in the figure, called *horizontal polarization*) and backscatter measured in the same polarization. The two cross sections, respectively denoted by σ_V and σ_H, are

$$\sigma_V = \frac{4b^2}{\pi}\left| Z_{1V} - \frac{e^{j[\rho - (\pi/4)]}}{\sqrt{2\pi}\rho^{3/2}}\left[\frac{1}{\cos(\theta)} + \frac{e^{j[\rho - (\pi/4)]}}{4\sqrt{2\pi}\rho^{3/2}} Z_{2V} \right] \right|^2 \tag{5.3-9}$$

$$\sigma_H = \frac{4b^2}{\pi}\left| Z_{1H} - \frac{4e^{j[\rho + (\pi/4)]}}{\sqrt{2\pi}\rho^{1/2}}\left[\frac{1}{\cos(\theta)} - \frac{e^{j[\rho + (\pi/4)]}}{2\sqrt{2\pi}\rho^{1/2}} Z_{2H} \right] \right|^2 \tag{5.3-10}$$

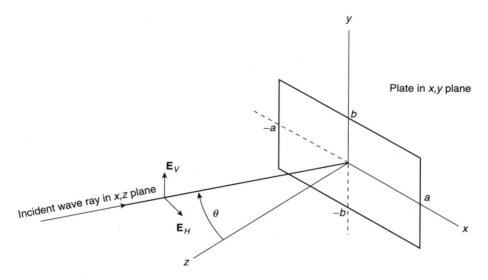

Figure 5.3-3 Geometry applicable to backscattering from a flat conducting plate.

where

$$Z_{1V} = \cos[\rho \sin(\theta)] - \frac{j \sin[\rho \sin(\theta)]}{\sin(\theta)} \tag{5.3-11}$$

$$Z_{1H} = \cos[\rho \sin(\theta)] + \frac{j \sin[\rho \sin(\theta)]}{\sin(\theta)} \tag{5.3-12}$$

$$Z_{2V} = \frac{[1 + \sin(\theta)]\,e^{-j\rho\sin(\theta)}}{[1 - \sin(\theta)]^2} + \frac{[1 - \sin(\theta)]\,e^{j\rho\sin(\theta)}}{[1 + \sin(\theta)]^2} \tag{5.3-13}$$

$$Z_{2H} = \frac{e^{-j\rho\sin(\theta)}}{[1 - \sin(\theta)]} + \frac{e^{j\rho\sin(\theta)}}{[1 + \sin(\theta)]} \tag{5.3-14}$$

$$\rho = 2ka = \frac{4\pi a}{\lambda} \tag{5.3-15}$$

The radar cross sections of (5.3-9) and (5.3-10) apply when the shortest side dimension of the plate is at least two wavelengths.[2] Although Ross (1966) *verified* his results for a plate dimensions-to-wavelength ratio from 2 to 10, there appears to be no reason higher ratios are not allowed. Figure 5.3-4 plots σ_V and σ_H (curves for geometric diffraction theory) as well as experimental results for a square plate with

[2] Ross (1966) gives more general formulas for σ_V and σ_H. We have neglected a term in the more general results that is allowed when the smallest side is at least two wavelengths.

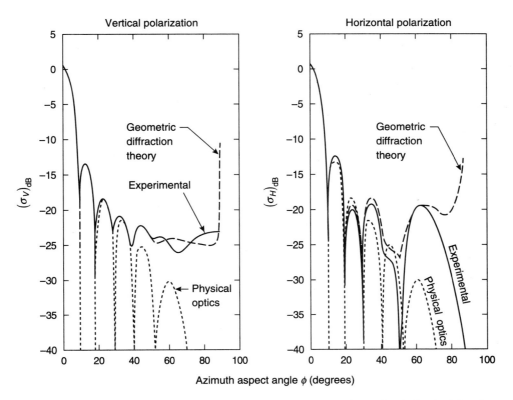

Figure 5.3-4 Cross sections σ_V (vertical polarization) and σ_H (horizontal polarization) for the flat plate of Fig. 5.3-3 when $a = b = 10.16$ cm and $f = 9.23$ GHz. (Adapted from Ross, © 1966 IEEE, with permission.)

10.16-cm sides at about 9.23 GHz ($\lambda = 3.25$ cm = 1.28 inches). Theoretical values are quite accurate for $0 \leq \theta \leq 80$ degrees. For $\theta > 80$ degrees there is a singularity (at $\theta = 90$ degrees), and the results cannot be used.

As $\theta \rightarrow \pi/2$, the true value of $\sigma_H \rightarrow 0$, but the true value of σ_V is nonzero. For $\theta = \pi/2$ Ross (1966) finds analytically the value of σ_V to be

$$\sigma_V\left(\theta = \frac{\pi}{2}\right) = \frac{ab^2}{\lambda}\left\{\left[1 + \frac{\pi}{2(2a/\lambda)^2}\right] + \left[1 - \frac{\pi}{2(2a/\lambda)^2}\right]\cos\left(2\rho - \frac{3\pi}{5}\right)\right\} \qquad (5.3\text{-}16)$$

If cross sections σ_V and σ_H are needed only for small angles θ, approximate values based on physical optics (dotted curves in Fig. 5.3-4) can be used. In this case

$$\sigma_V = \sigma_H = \frac{64\pi a^2 b^2}{\lambda^2}\left\{\frac{\sin[2ka\sin(\theta)]}{2ka\sin(\theta)}\right\}^2 \cos^2(\theta) \qquad (5.3\text{-}17)$$

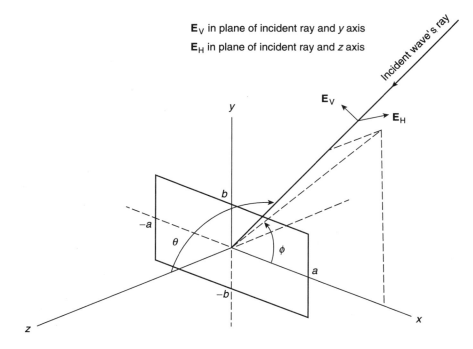

Figure 5.3-5 Geometry of scattering applicable to cross section of (5.3-18).

(Kerr, 1964, p. 457; Ross, 1966, p. 330; Blake, 1980, p. 105). Based on experimental results given in Ross, it appears that (5.3-17) is reasonably accurate for θ only up to about 30 degrees, at least for plate dimensions up to 10λ.[3]

Finally we note that for cases where physical optics analysis procedures are valid, radar backskattering cross section is known for the perfectly conducting thin rectangular plate at the arbitrary illumination direction (θ, ϕ) defined in Fig. 5.3-5; it is

$$
\sigma(\theta, \phi) = \frac{64\pi a^2 b^2}{\lambda^2} \left\{ \frac{\sin[2ka \sin(\theta) \cos(\phi)]}{2ka \sin(\theta) \cos(\phi)} \right.
$$
$$
\left. \cdot \frac{\sin[2kb \sin(\theta) \sin(\phi)]}{2kb \sin(\theta) \sin(\phi)} \right\}^2 \cos^2(\theta) \qquad (5.3\text{-}18)
$$

(see Kerr, 1964, p. 457).

---

[3] Ross also noted that the right side of (5.3-17) without the factor $\cos^2(\theta)$ is a more accurate representation for σ_V for larger angles θ than allowed for (5.3-17). A similar extension of the range of θ to σ_H is not possible. Crispin et al. (1965, p. 844) also obtained (5.3-17) without the factor $\cos^2(\theta)$.

Flat Circular Plate

Let the geometry of Fig. 5.3-3 again be used, except assume that a perfectly conducting, thin, flat circular plate of radius a is centered at the origin and replaces the rectangular plate. The radar cross section for any linearly polarized incident wave and a large (in wavelengths) plate is

$$\sigma(\theta) = \pi k^2 a^4 \left\{ \frac{2J_1 [2ka \sin(\theta)]}{2ka \sin(\theta)} \right\}^2 \cos^2(\theta) \tag{5.3-19}$$

(Kerr, 1964, p. 458) for relatively large disks (so that size corresponds to the "optical" region). When $\sigma(\theta)$ of (3.5-19) is compared with the measured data of Bechtel (1965, p. 881) for a disk for which $ka = 8.28$, there is reasonable agreement for $|\theta| \leq \pi/4$.

Circular Cylinder

As another example of backscattering radar cross section for some simple shapes, we discuss a perfectly conducting, long cylinder of circular cross section centered on the origin, as shown in Fig. 5.3-6. The cylinder has length L and radius a with its axis of symmetry forming the z axis. The cylinder is illuminated by a linearly polarized plane wave from direction (θ, ϕ_1) as shown. For $ka \sin(\theta) \gg 1$ the radar cross section is independent of the direction of the linear polarization and is given by (DiCaudo et al., 1966)

$$\sigma = kaL^2 \left\{ \frac{\sin[kL \cos(\theta)]}{kL \cos(\theta)} \right\}^2 \sin(\theta) \tag{5.3-20}$$

As a practical matter, (5.3-20) applies for ka larger than about 6 (Kerr, 1964, p. 461). However, it cannot be used as $\theta \to 0$ or π since $\sigma = 0$ according to (5.3-20), but in reality σ must be closer to the normal-incidence cross section of a circular flat plate.

Straight Wire

As a final example of backscattering from a simply shaped target, we consider the perfectly conducting wire that is long and thin relative to a wavelength. Geometry is as defined in Fig. 5.3-6 where the incident plane wave's linearly polarized electric field vector makes an angle ϕ from the plane defined by the wire's axis of symmetry (z axis) and the incident wave's ray; the axis y' is in this plane and is normal to the incident ray. The receiver's antenna also has a linear polarization in the direction ϕ. The corresponding backscattering cross section is approximately given by[4]

$$\sigma = \frac{\pi L^2 \sin^2(\theta) \left\{ \sin[kL \cos(\theta)]/kL \cos(\theta) \right\}^2 \cos^4(\phi)}{(\pi/2)^2 + \left\{ \ln[2/\gamma ka \sin(\theta)] \right\}^2} \tag{5.3-21}$$

---

[4] Equation (5.3-21) is originally attributed to Chu in unpublished work as noted by Van Vleck et al. (1947), who derived more general results from which Chu's formula is a special case.

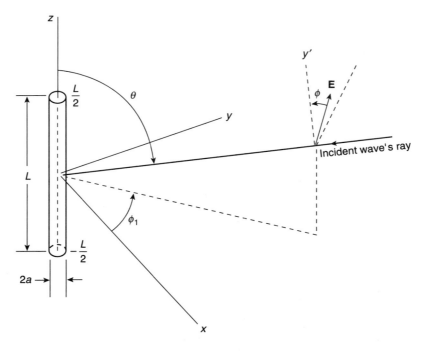

Figure 5.3-6   Geometry for backscattering from a circular cylinder.

where $\gamma = 1.781$ (Van Vleck et al., 1947, p. 277; Crispin et al., 1968, p. 107). Although (5.3-21) is simple, easy to use, and reasonably accurate, it has some limitations. More accurate results for σ were developed by Van Vleck et al. (1947), but the applicable formulas are cumbersome and difficult to interpret except through computer computations.

Other Simple Shapes

There are many other target shapes that might be called simple. Some are illustrated in Table 5.3-1. Information on these shapes are variously available in many books and in the papers of Andreasen (1965), Bechtel (1965), Blore (1964), Crispin and Maffett (1965), Senior (1965), Keller (1960), and Siegel (1958) to name only a few. Blore (1964) has summarized the nose-on backscattering cross sections for several of these shapes as reproduced in Fig. 5.3-7.

Complex Target Shapes

Simple shapes are useful in modeling more complex objects where the cross section of the complex target can be approximated as the sum of contributions from the scattering of the simple components making up the target. For example, scattering from a missile with a cylindrically shaped body, conical nose, and four rectangular fins can be modeled as the sum of the scatterings from a cone (nose), cylinder (body), flat rectangular plates (fins), and a conducting wire loop. This last shape was not discussed

Table 5.3-1 Some simple targets defined by rotation of the shaded areas about the *x* axis.

Name	Shape	Shape Definition
Oblate spheriod		Origin-centered ellipse *x*-directed minor axis
Prolate spheroid		Origin-centered ellipse *y*-directed minor axis
Ogive		Arc of circle (of radius R) above a chord in the *x* direction
Elliptic ogive		Area above *x* axis for ellipse offset-centered in $-y$ direction
Lens		Area bounded by two circles centered on *x* axis having equal chord lengths in *y* direction
Spindle		Area bounded by a parabola above the *x* axis

(*Continued*)

Table 5.3-1 (*Continued*)

Name	Shape	Shape Definition
Cone		Area of right triangle
Double-backed cone		Area of two backed right triangles; a_1 usually equal to a_2
Cone sphere		Area bounded by right triangle and circle so radius and slope of circle match triangle at junction of the two
Double-rounded cone		Area bounded by two backed right triangles and circle, so radius and slopes of circle match triangles at junctions of the pairs

above, but it represents the rocket at tail aspect where the cylindrical body looks like a hollow cylinder that presents the appearance of a wire loop. At this aspect the fins look like thin wires, incidentally. Figures 5.3-8 through 5.3-10, as taken from Crispin and Maffett (1965, p. 978), give some representative data on a missile much as described here.

Bird (1994) has made backscattering radar cross section measurements on a six-wheel Saracen armored personnel carrier (APC) using a horizontally polarized radar at 95 GHz looking down to the target 18.5 degrees below the horizontal. The APC was rotated on a turntable to obtain the data shown in Fig. 5.3-11. Data are also shown for measurements on a 0.1-scale model measured at 890 GHz (laser).

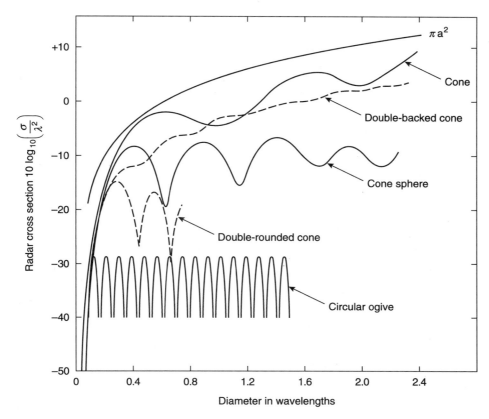

Figure 5.3-7 Nose-on backscattering radar cross sections of several simple shapes versus size in wavelengths. (Adapted from Blore, © 1964 IEEE, with permission.)

Figure 5.3-12 illustrates full scale and scale measurements of radar cross section for the MQM-107 drone aircraft. Full-scale data were taken at 10.1 GHz at a radar elevation angle of −7 degrees. The scale data are for a 0.3-scale model observed in inverted position by a radar with an elevation angle of +7 degrees and frequency of 35 GHz. Transmit and receive polarizations were vertical.

As an example of backscattering cross section for a jet fighter aircraft (a complex shape), we show Fig. 5.3-13, which is due to Wilson (1972, p. 758). Data were taken at three frequencies in L, S, and X bands on an aircraft in straight, level flight past the radar. Data shown in 10-degree intervals of aspect angle were the result of averaging a large number (40, L band to 250, X band) of independent data points and flights (5 or more) within each 10-degree interval within the "constraint that the radar views an elevation aspect angle of the aircraft between − 5 and − 15 degrees."

An example of measured backscattering radar cross section for a large jet aircraft of the middle 1950s is shown in Fig. 5.3-14. The figure is shown only for the value of the experimental data as an example of a complex target.

Follin et al. (1984) has given measured cross-sectional results for a more recent moderate-size commercial passenger jet aircraft, the Boeing 727-100C. Measurements

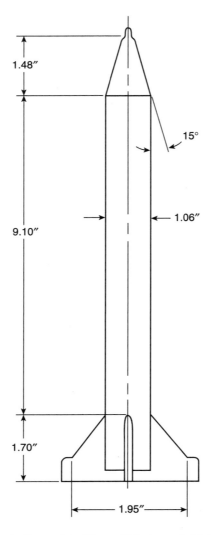

Figure 5.3-8 Sketch of missile model. (From Crispin and Maffett, © 1965 IEEE, with permission.)

were made on 1/100-scale model using horizontal polarization at 0.95 GHz, which is the scaled frequency corresponding to the model. Nose-on cross section averaged 21.8 m² over a 10° region about the nose. At side-aspect the cross section's average was about 2910 m².

As a final example of cross section of complex shapes, we note that Skolnik (1974) represented the median (50th percentile) radar cross section of naval ships by the empirical expression

$$\sigma = 52 f_{\text{MHz}}^{1/2} D^{3/2} \quad (\text{m}^2) \tag{5.3-22}$$

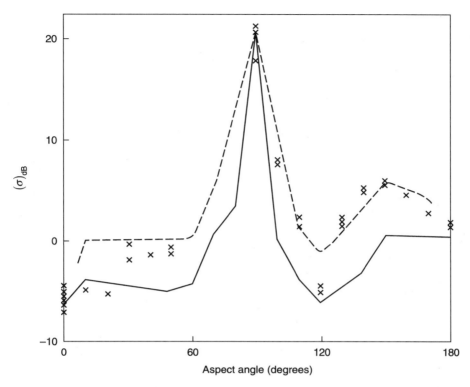

Figure 5.3-9 Missile with fins—theory and experiment for vertical polarization (E vector normal to plane in which aspect is measured, two of the fins contained in that plane). The crosses give experimental points, the dashed line is for theory (in-phase estimation); the solid line is also theory (random phase estimation). (From Crispin and Maffett, © 1965 IEEE, with permission.)

where f_{MHz} is frequency in megahertz and D is the full-load displacement of the vessel in kilotons. The expression resulted from fitting a curve to measurements on ships from 2000 to 17,000 tons at near-grazing incidence at frequencies in X ($\lambda = 3.25$ cm), S ($\lambda = 10.7$ cm), and L ($\lambda = 23$ cm) bands. Polarization was horizontal. Measured data were cross section averages about port and starboard bow and quarter aspects of the ship (omitting the peak at broadside). Figure 5.3-15 shows measured versus estimated cross sections. Since data points would fall exactly on the dashed line for perfect estimates, it is clear that (5.3-22) is a good fit to the ship data in the study.

In closing, we note that Nathanson (1991, p. 183) gives a nice table of cross sections of Soviet ships that range from 110 to $1.2(10^6)$ m² (a Kuril class carrier) for displacements of 80 to 40,000 tons, respectively. Morchin (1990) gives summaries of cross sections of many aircraft (p. 75) and other ships (p. 79). A recent book by Rihaczek and Hershkowitz (1996) gives some readable discussions of target features such as phase centers, extended size, curved surfaces, and multiple reflections among others.

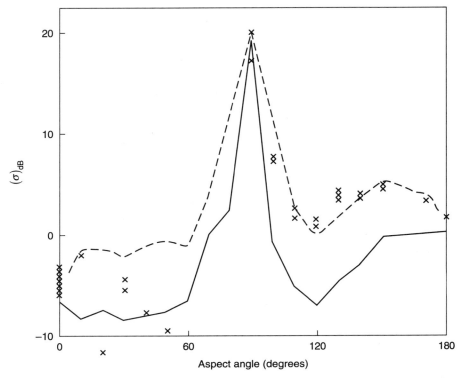

Figure 5.3-10 Missle with fins—theory and experiment for E vector slanted at a 45 degree angle to the plane in which aspect is measured (two fins located in that plane). The crosses give experimental points, the dashed line is for theory (in-phase estimation); the solid line is also theory (random phase estimation). (From Crispin and Maffett, © 1965 IEEE, with permission.)

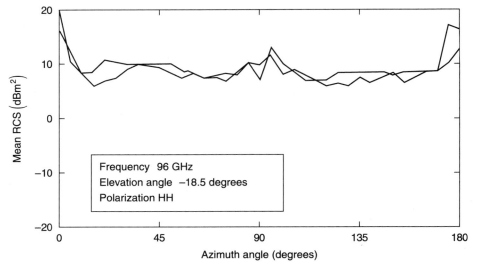

Figure 5.3-11 Comparison of scale-model and full-scale RCS of an APC. (From Bird, © 1994 IEEE, with permission.)

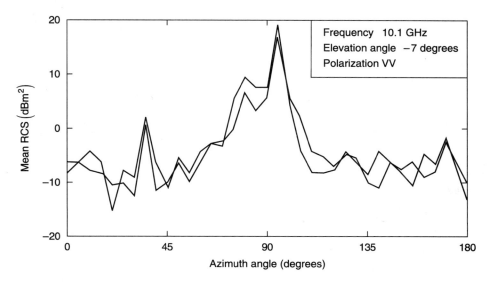

Figure 5.3-12 Comparison of scale-model and full-scale RCS of an MQM-107. (From Bird, © 1994 IEEE, with permission.)

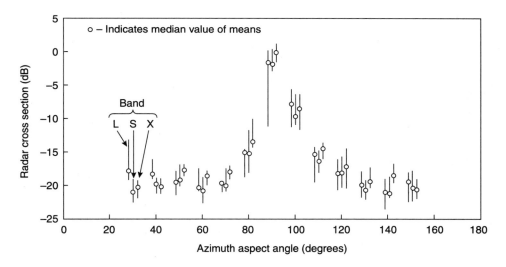

Figure 5.3.13 Radar backscattering cross section versus azimuth aspect angle for a jet fighter at L, S, and X bands. (From Wilson, © 1972 IEEE, with permission.)

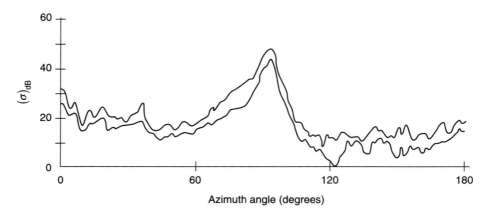

Figure 5.3-14 Radar cross section for a large manned aircraft. Measured data at 1200 MHz indicating values from both sides of aircraft in 1955. (Adapted from Crispin and Maffett, © 1965 IEEE, with permission.)

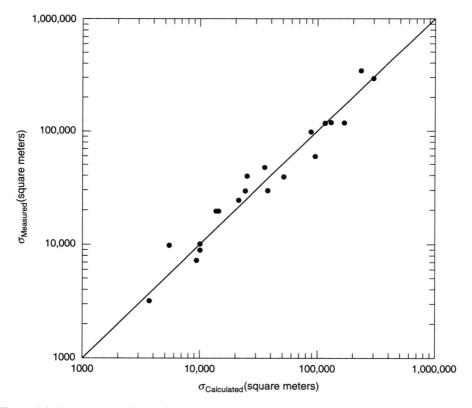

Figure 5.3-15 A comparison of the radar cross section of various naval vessels with the empirical formula of (5.3-22). (From Skolnik, © 1974 IEEE, with permission.)

5.4 CROSS SECTION OF AREA TARGETS

Radars must often work with distributed targets. Land and sea surfaces are the most common, but surfaces of cities are not uncommon targets. In many cases the backscattered signals from these surfaces are undesired, when, for example, the radar may be seeking to observe specific objects such as ships, trucks, tanks, and periscopes. These undesired signals are usually called *clutter*, specifically *land clutter* and *sea clutter* for land and sea. On the other hand, backscattering from surfaces is not always undesirable. An airborne radar doing terrain mapping would treat the surface return as a desired signal. In every case, however, we may refer to the surface as an *area target* because backscattered energy is returned by the whole surface area viewed by the radar through its antenna pattern-range resolution cell.

Surface Backscattering Coefficient

The general approach to surface backscattering is to note that total received signal power is related to the effective area of the scattering surface. This total power (in a resolution cell and in the polarization of the radar) leads to a definition of cross section, denoted by σ_c (c for clutter), given by

$$\sigma_c = \sigma^0 A_c \tag{5.4-1}$$

where A_c is the *effective clutter scattering area* of the surface being illuminated by the radar, and σ^0 is an important parameter that is characteristic of the surface. The parameter σ^0 is called the *surface backscattering coefficient* (or *backscattering cross section per unit area*; Ulaby and Dobson 1989, p. 14). The unit of σ^0 is area per unit area (m^2/m^2). It is often expressed in decibels, that is, σ^0 in dB, here denoted by $(\sigma^0)_{dB}$, is

$$(\sigma^0)_{dB} = 10 \log_{10}(\sigma^0) \tag{5.4-2}$$

For any given surface σ^0 may vary in a complicated manner with frequency, polarization, type of surface (trees, sand, water, snow, etc.), angle of incidence of the illuminating wave, wind (affects tree movement, wave height and directions for seas, etc.), and other parameters. Because of these facts, surface scattering is often characterized by use of measured data for σ^0 in tabular or graphical form. This approach is mainly used here, also. However, before giving some example data, we will first present a plausible (although not very rigorous) development to justify the definition of σ^0.

Consider Fig. 5.4-1 where a radar (point TX) transmits toward the surface through a main beam, for which the nose direction has a grazing angle ψ with the surface at a nominal distance R. The beam illuminates an area A, approximately elliptical in shape, as shown. It may be helpful to the reader to imagine A is the area confined within the intersection of the 3-dB elliptical cone of the beam and the surface. For a given radar resolution cell (in range and two space angles) the applicable scattering area, denoted by A_c, will be not greater than the illuminated area (see next subsection where the detailed cases are considered). Now let us imagine A_c as subdivided into small "patches" with scattering area ΔA_i for the ith patch located at point (R_i, θ_i, ϕ_i) in

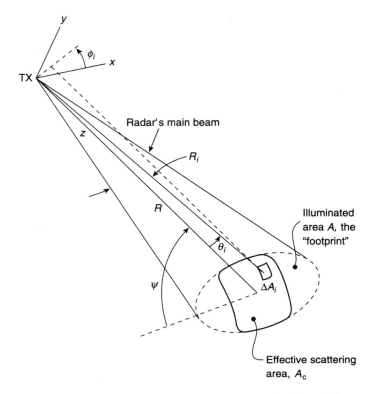

Figure 5.4-1 Geometry associated with the definition of σ^0.

spherical coordinates defined at the radar. We presume A_c is large enough that the number of independent patches is large. There will be a minimum size, of course, that will tend to make the scattering independent from patch to patch. Under this assumption the power scattered from area A_c will equal the sum of powers scattered from the various patches. Now assume that patch i has radar cross section σ_i, and use the radar equation of (4.1-15):

$$S_{ri} = \frac{P_t G_D^2(\theta_i, \phi_i) \lambda^2 \sigma_i}{(4\pi)^3 R_i^4 L_i} \tag{5.4-3}$$

where P_t is the peak transmitted power, $G_D(\theta_i, \phi_i)$ is the antenna's one-way directive gain in direction (θ_i, ϕ_i), λ is wavelength, L_i is the system loss (see Chapter 4) over the two-way path, and S_{ri} is the available received power due to patch i.

Under the reasonable assumption that $R_i \approx R$ and $L_i \approx L$ (constants at values for a patch near the center of area A_c) for all patches,[5] we sum (5.4-3) over all patches to

---

[5] This assumption means that the main beam's beamwidth is relatively small and A_c does not get too large as ψ varies, as it should not because of the radar's range resolution.

obtain the total available received clutter power, denoted by S_{rc},

$$S_{rc} = \frac{P_t \lambda^2}{(4\pi)^3 R^4 L} \sum_{i=1}^{N_c} G_D^2(\theta_i, \phi_i) \frac{\sigma_i}{\Delta A_i} \Delta A_i \tag{5.4-4}$$

where N_c is the number of independently scattering patches. We now observe that $\sigma_i/\Delta A_i$ tends to be independent of the absolute area ΔA_i (as ΔA_i increases, so does σ_i), but it can be very dependent on the surface's characteristics, and particularly on grazing angle ψ. Denote the ratio $\sigma_i/\Delta A_i$ as σ^0, and treat it as a constant for all patches. Equation (5.4-4) now becomes

$$S_{rc} \approx \frac{P_t \lambda^2}{(4\pi)^3 R^4 L} \sigma^0 \left[\sum_{i=1}^{N_c} G_D^2(\theta_i, \phi_i) \Delta A_i \right] \tag{5.4-5}$$

The bracketed quantity in (5.4-5) has the unit of area, and it depends on how directive gain varies with the directions of the patches. Whatever the value of this bracketed term, imagine it to be the same as an idealized directive gain having the *constant* nominal value of $G_D(\theta, \phi)$, which is G_D, the directivity, over some *effective clutter scattering area* A_c so that

$$\sum_{i=1}^{N_c} G_D^2(\theta_i, \phi_i) \Delta A_i = G_D^2 A_c \tag{5.4-6}$$

Total clutter available power now becomes

$$S_{rc} \approx \frac{P_t G_D^2 \lambda^2 \sigma_c}{(4\pi)^3 R^4 L} \tag{5.4-7}$$

where σ_c is given by (5.4-1) and

$$A_c = \frac{1}{G_D^2} \sum_{i=1}^{N_c} G_D^2(\theta_i, \phi_i) \Delta A_i \tag{5.4-8}$$

Equation (5.4-7) is the radar equation for area clutter.

Equation (5.4-8) for A_c needs interpretation. If the radar's pulse duration is large such that the whole surface illuminated by the main beam contributes to scattering, A_c is nearly equal to the area incribed by the 3-dB beamwidth of $G_D^2(\theta, \phi)$ (Ulaby and Dobson, 1989, p. 16). If the main beam is approximated by a gaussian function with the same beamwidths as the actual pattern, it can be shown (see Problem 5.4-6) that $A_c \approx A/[2 \ln(2)]$. The reduction factor $1/[2 \ln(2)]$ is due to the two-way action of the antenna pattern in clutter scattering.

A second case is when the radar's pulse is short such that the radar illuminates the surface in the range direction by an amount less than the range extent of the "footprint." We next consider this and the preceding case more fully.

Surface Scattering Geometry

There are two most important geometries of scattering. In one, grazing angle ψ is small so that A_c is bounded in width by the 3-dB beamwidth of $G_D^2(\theta, \phi)$, and in length by the range duration of the transmitted pulse.[6] In the second case, ψ is large enough that the effective scattering area is entirely determined by the area within the locus of the 3-dB beamwidth of the gain $G_D^2(\theta, \phi)$ [the area of intersection of the 3-dB elliptical cone of the pattern $G_D^2(\theta, \phi)$ and the surface].

We initially consider the first case. Applicable geometry is shown in Fig. 5.4-2. A radar with altitude h above its local surface (S_1) directs its directive gain $G_D(\theta, \phi)$, having beamwidths θ_B (in vertical plane) and ϕ_B (orthogonal plane), toward the clutter surface (S_c) using a *depression angle* α from radar "horizontal." The beam intersects with the surface at nominal distance R causing a "footprint" that is nearly elliptical, as shown. A single range cell distance ($c/2B$, where c = speed of light and B = pulse's 3-dB bandwidth) corresponds to an illuminated scatter area of length $c/[2B \cos(\psi)]$, where ψ is the grazing angle, and effective width $R\phi_B/\sqrt{2}$. This area is roughly a "rounded" rectangle as long as $c/[2B \cos(\psi)]$ is smaller than the effective radial extent of the footprint, $R\theta_B/\sqrt{2} \sin(\psi)$.[7] Thus

$$\sigma_c = \sigma^0 A_c \approx \frac{cR\phi_B \sigma^0}{2\sqrt{2} B \cos(\psi)}, \quad \tan(\psi) < \frac{\sqrt{2}R\theta_B B}{c} \tag{5.4-9}$$

We note that because the scattering and local surfaces are not necessarily at the same altitudes or slopes, the depression and grazing angles do not necessarily have the same magnitudes. At the scattering surface, S_c, the *angle of incidence* is defined as $(\pi/2) - \psi$.

As radar altitude increases (or, alternatively, as the depression angle is made larger), the scattering area expands to eventually fill the pattern's footprint, and we have the second case described earlier. The scattering area now becomes independent of bandwidth and is determined by the (elliptical) area of the footprint. Cross section becomes

$$\sigma_c = \sigma^0 A_c \approx \frac{\pi R^2 \theta_B \phi_B \sigma^0}{8 \ln(2) \sin(\psi)}, \quad \tan(\psi) \geq \frac{\sqrt{2}R\theta_B B}{c} \tag{5.4-10}$$

where the factor $1/[2 \ln(2)]$ is the effect of the two-way pattern on scattering (see Problem 5.4-6).

From (5.4-9), clutter cross section for low grazing angles increases proportional to R. It decreases as the radar's resolution in two coordinates (range and one space angle) improves. For large grazing angles, (5.4-10) shows that σ_c increases as R^2 and again decreases as resolution improves in two coordinates (both space angles). From (5.4-7) we see that clutter power decreases as $1/R^3$ for low grazing angles and only as $1/R^2$ for high grazing angles.

---

[6] We consider pulsed radar. The area A_c is therefore associated with the length of the radar's range resolution cell.

[7] The beamwidths of $G_D^2(\theta, \phi)$ in the θ and ϕ directions are $1/\sqrt{2}$ times the beamwidths θ_B and ϕ_B of $G_D(\theta, \phi)$ for gaussian patterns. For other patterns they should not be greatly different.

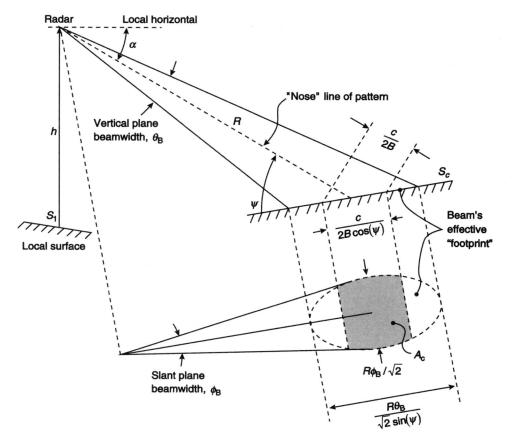

Figure 5.4-2 Geometry associated with surface clutter cross section.

General Behavior of σ^0

Although it is not possible to generalize too much about the behavior of σ^0, it does seem the following observations are reasonable:

- σ^0 varies greatly with type of surface.
- σ^0 varies greatly with frequency, polarization, grazing angle, and other parameters, even for a particular surface.
- σ^0 tends to increase with grazing angle.
- $(\sigma^0)_{dB}$ can be as large as about $+20$ dB for higher grazing angles for cities and some water surfaces.
- $(\sigma^0)_{dB}$ usually falls between -40 dB and -10 dB.
- $(\sigma^0)_{dB}$ almost always falls between -70 dB and $+20$ dB.

Some slight additional general detail can be given on behavior of σ^0 with grazing angle ψ. General behavior falls in three regions, as illustrated in Fig. 5.4-3. The low

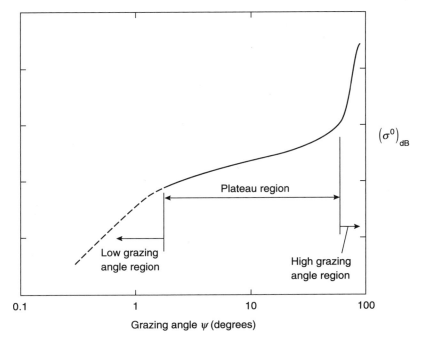

Figure 5.4-3   General behavior of σ^0 with grazing angle. (Curve not to be scaled for design.)

grazing angle region may extend up to an upper bound of about 10 degrees which depends greatly on the depth of surface variations; here shadowing and diffraction effects may be present, and multipath and ducting can also exist. Considerable variability exists in this region. In the plateau region, which extends up to about $\psi = 60$ degrees, σ^0 varies less rapidly with ψ than in other regions. In this region scattering is mainly diffuse, and σ^0 tends to increase with both roughness and frequency; vertical polarization typically gives larger σ^0 than horizontal polarization. In the high grazing angle region, scattering tends to be almost specular, and σ^0 typically increases rapidly with ψ, approaching a maximum as ψ approaches 90 degrees. In this region σ^0 does not vary as rapidly with frequency as in the plateau region, decreases as roughness increases, and tends to be independent of polarization.

Further comments on σ^0 are reserved for the examples of scattering surfaces to follow.

5.5   SEA SURFACES AS AREA TARGETS

The literature on backscattering from sea surfaces is vast, and we will give only a few examples to typify aspects of the problem. Various topics are covered in many books; some recent examples are Edde (1993), Nathanson et al. (1991), Barton et al. (1991), Morchin (1990), Skolnik, ed. (1990), Eaves and Reedy, ed. (1987), Barton (1988), Skolnik (1980), and Blake (1980). There are many others.

Effect of Sea State, Grazing Angle, and Wind

The condition of a sea, as relates to wave height (peak-to-trough), is usually defined in terms of *sea state*. Edde (1993) tabularizes data related to sea state given by Nathanson (1991). Part of these data are shown in Table 5.5-1. An empirical relationship between sea state, denoted by SS, and middle-of-the-range wind speed is

$$W_s = 3.22 \sin\left(\frac{\pi SS}{5}\right) + 21.82 \tan\left(\frac{8.77\pi SS}{180}\right) \tag{5.5-1}$$

for $0 \leq SS \leq 7$ and W_s is in knots.

Backscattering cross section is a function of sea state. Nathanson (1991) has given measured data representing averages of up-, down- and cross-wind data, where available, on about 60 experiments. Data for frequencies of 0.5, 1.25, 3.0, 5.6, 9.3, 17, and 35 GHz are given for both vertical and horizontal polarizations, and sea states from 0 to 6 in every case. Figures 5.5-1 through 5.5-3 plot the data for three frequencies and three sea states in each case. These data clearly show that σ^0 increases with sea state. The increase is larger between lower sea states and at lower frequencies. Data for sea state 0 corresponded to wave height less than 0.25 ft and winds less than 4 knots.

From Figs. 5.5-1 through 5.5-3, σ^0 increases with grazing angle ψ for all sea states, polarizations, and frequencies, although the plots do not always have the general appearance of Fig. 5.4-3. The rate of increase in σ^0 with ψ in the low grazing and plateau regions decreases with higher sea states and frequencies.

Besides contributing to sea state, wind has an affect on σ^0 because the surface is not uniform. As the radar views the sea in an upwind direction, σ^0 is largest at a value we denote by σ_u^0, for a given polarization, frequency, and grazing angle. As the aspect changes from upwind (all else constant), σ^0 decreases to a minimum at roughly a cross-wind direction; we denote this minimum value by σ_c^0. As aspect continues to $180°$ from upwind another maximum occurs, which we denote as σ_d^0 for the

TABLE 5.5-1 Sea states defined by sea wave height.

Sea State	Sea Name[a]	Wave Height (ft) (Crest to Trough)	Wind Speed (Knots)
0	Flat	–	–
1	Smooth	0–1	0–7
2	Slight	1–3	7–12
3	Moderate	3–5	12–16
4	Rough	5–8	16–19
5	Very rough	8–12	19–23
6	High	12–20	23–30
7	very high	20–40	30–45
8	Precipitous	>40	>45

Note: See Edde (1993, p. 219) for more details.
[a] Sea state 0 is not defined; it commonly describes a flat sea.

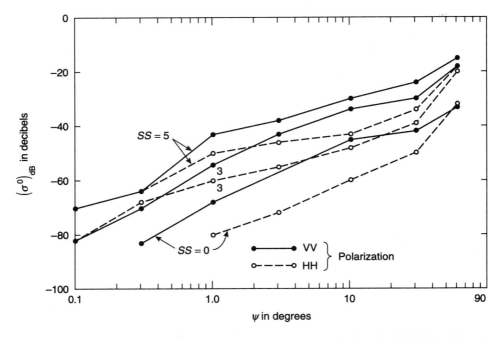

Figure 5.5-1 Mean values of σ^0 in decibels versus grazing angle for $f = 1.25$ GHz. (Plotted from data of Nathanson, F. E. et al.: *Radar Design Principles*, 1991, McGraw-Hill, Inc., with permission.)

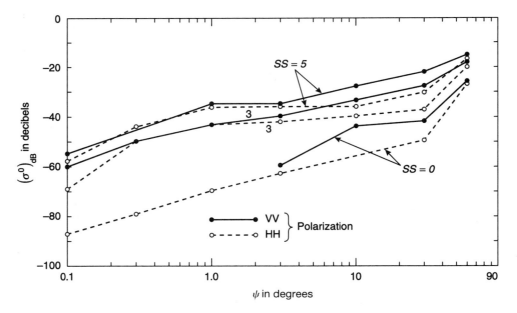

Figure 5.5-2 Mean values of σ^0 in decibels versus grazing angle for $f = 5.6$ GHz. (Plotted from data of Nathanson, F. E. et al.: *Radar Design Principles*, 1991, McGraw-Hill, Inc., with permission.)

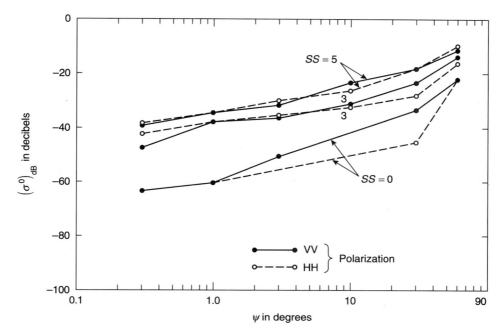

Figure 5.5-3 Mean values of σ^0 in decibels versus grazing angle for $f = 17$ GHz. (Plotted from data of Nathanson, F. E. et al.: *Radar Design Principles*, 1991, McGraw-Hill, Inc., with permission.)

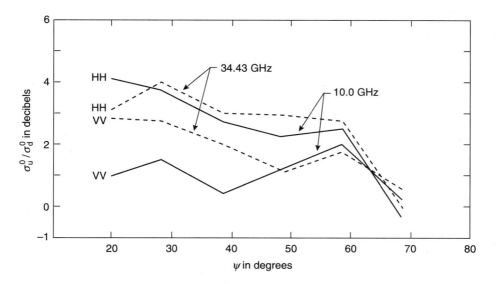

Figure 5.5-4 Averaged values (in decibels) of the ratio of the upwind peak σ_u^0 to downwind peak σ_d^0. (Plots from data of Masuko et al., 1986, *Journal of Geophysical Research*, vol. 91, No. C11, pp. 13,065–13,083, © The American Geophysical Union, with permission.)

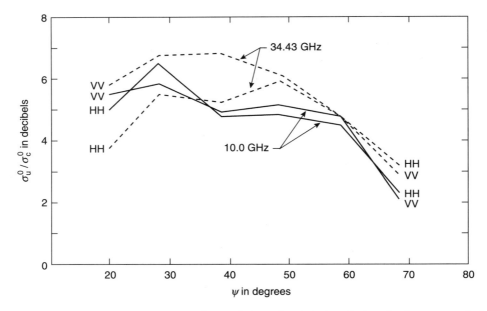

Figure 5.5-5   Averaged values (in decibels) of the ratio of the upwind peak σ_u^0 to cross-wind minimum σ_c^0. (Plots from data of Masuko et al., 1986, *Journal of Geophysical Research*, vol. 91, No. C11, pp. 13,065–13,083, © The American Geophysical Union, with permission.)

downwind direction. For aspects from 180° to 360° the behavior is the same in reverse. The downwind maximum σ_d^0 is smaller than the upwind value σ_u^0 because the radar sees the waves from the backside rather than from the front and the wave's shape is different. Masuko et al. (1986) have made extensive measurements of these effects at 10.0 and 34.43 GHz for both vertical and horizontal polarizations and various grazing angles. Figures 5.5-4 and 5.5-5 show the ratios σ_u^0/σ_d^0 and σ_u^0/σ_c^0, respectively, as plotted from data tabulated in Masuko. Note that the differences in σ^0 due to wind tend to largest values at lower grazing angles and become small for higher values of ψ. As a function of frequency the differences tend to increase with frequency and be most pronounced for vertical polarization: however, the effect does not hold for all values of ψ.

Effects of Frequency and Polarization

Data of Figs. 5.5-1 through 5.5-3 show that σ^0 tends to increase with frequency for a given sea state with the increase between states diminishing as the state increases. The effect is most pronounced for horizontal polarization. As frequency increases the effect seems to dissapear. In fact data of Guinard and Daley (1970) show that increases in σ^0 for frequencies from L to X band are relatively small (on the order of 7 dB) for most ψ and horizontal polarization, at least when sea state is around 7 (Fig. 5.5-6). For vertical polarization, similar conditions actually show σ^0 decreasing with frequency (L to X bands), as in Fig. 5.5-7. This trend has been shown by Masuko et al. (1986) to continue to K_a band (34.43 GHz) for sea state 3, as in Fig. 5.5-8.

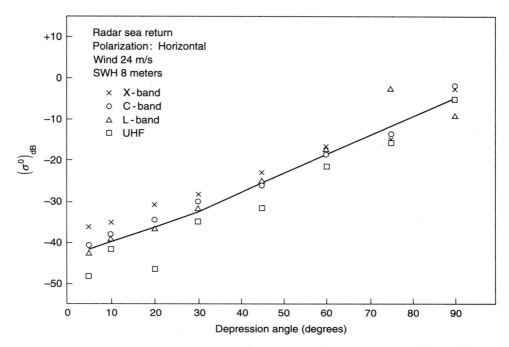

Figure 5.5-6 Sea clutter cross section, North Atlantic, February 11, 1969. (Adapted from Guinard and Daley, © 1970 IEEE, with permission.)

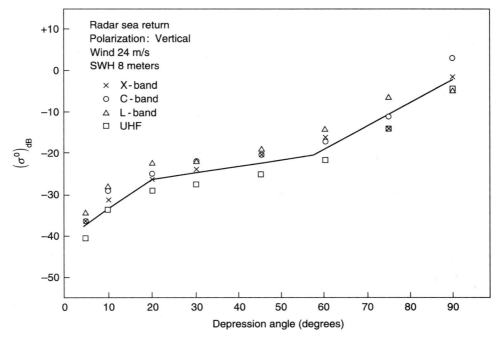

Figure 5.5-7 Sea clutter cross section, North Atlantic, February 11, 1969. (Adapted from Guinard and Daley, © 1970 IEEE, with permission.)

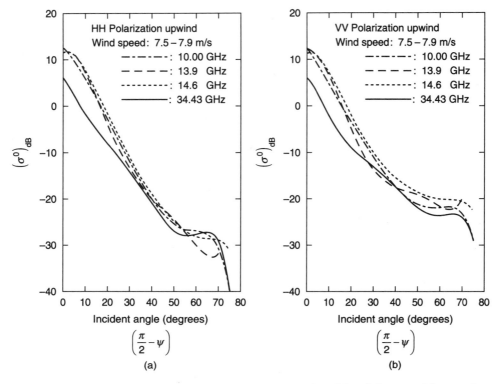

Figure 5.5-8 Comparison between incident angle dependencies of the σ^0 for several frequencies in the case of wind speed about 7.9 m/s (see state 3) with upwind aspect. (Adapted from Masuko et al., 1986, *Journal of Geophysical Reasearch,* vol. 91, No. C11, pp. 13,-65–13,083, © The American Geophysical Union, with permission.)

5.6 LAND SURFACES AS AREA TARGETS

Scattering from land surfaces is far more complicated than for sea surfaces due to the widely varying vertical structure. Surfaces can vary from relatively smooth, such as with snow over slowly undulating plains, to very rough and having high structure, such as with cities and heavily forrested and mountainous terrain. Furthermore land surfaces can change greatly with seasons (loss of leaves) and the weather (e.g., snow cover). For all these reasons we will give only limited examples of σ^0 for land surfaces. For real detail the reader is referred to the vast literature. Edde (1993), Nathanson (1991), Barton et al. (1991), Morchin (1990), and Skolnik, ed., (1990) are a few good places to start, which provide other references.

Typical Backscattering from Terrain

Nathanson (1991) has summarized data from about 50 sources for various types of terrain. The given data were averages of results for vertical and horizontal polarizations because there is not a strong dependence on polarization. Data at six frequency bands were given: L, S, C, X, K_u, and K_a bands. Figures 5.6-1 through 5.6-3 have been constructed from Nathanson's data. Only three frequencies have been used as representative examples. Also shown for extension of Nathanson's rural farmland data are results from Ulaby (1980) for larger values of ψ. The fact that the two data sets are not the same probably accounts for the value differences for $30° \le \psi \le 60°$.

Generally, the data of Figs. 5.6-1 through 5.6-3 indicate the following:

- σ^0 for most land surfaces is not a strong function of polarization (see Nathanson, 1991, p. 322; Ulaby, 1980, p. 539). Where dependence exists, σ^0 for vertical polarization is larger than for horizontal polarization (probably due to the vertical structure of some terrain, i.e., trees and grasses).

- σ^0 increases with frequency, but the frequency dependence is not strong (up to about 11 dB for desert and up to about 5 dB for urban terrain at any $1.5° \le \psi \le 60°$ for frequencies from L to X band).

- σ^0 increases with ψ from about 1° to about 60°–70°, then rapidly increases as ψ approaches 90° where the value of σ^0 is largest. The value of σ^0 at $\psi = 90°$ varies greatly; some representative values are $+15$ dB for water areas, $+11$ dB for roadways and aircraft runways, suburban and city areas about 0 dB (Nathanson, 1991, pp. 332 and 334). Ulaby (1980) shows a maximum of about 6 dB at 1.5 GHz, which falls to about 2.5 dB at 8.6 GHz, for vegitated land at $\psi = 90°$.

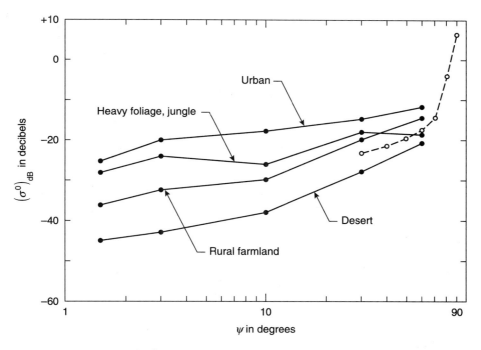

Figure 5.6-1 Behavior of σ^0 with grazing angle ψ for several types of terrain at L band (1–2 GHz). (Plots for $1.5° \le \psi \le 60°$ from data of Nathanson, F. E. et al.: *Radar Design Principles*, 1991, McGraw-Hill, Inc., with permission; plot shown dashed from Ulaby, © 1980 IEEE, with permission for vegitated farmland at 1.5 GHz.)

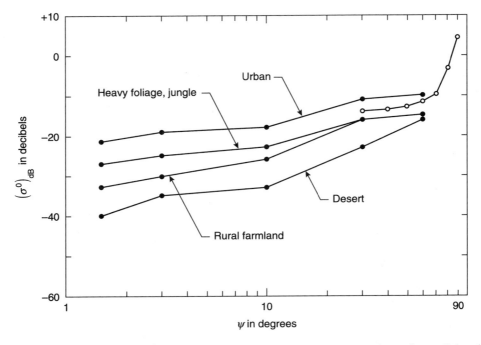

Figure 5.6-2 Behavior of σ^0 with grazing angle ψ for several types of terrain at C band (4–8 GHz). (Plots for $1.5° \leq \psi \leq 60°$ from data of Nathanson, F. E. et al.: *Radar Design Principles*, 1991, McGraw-Hll, Inc., with permission; plot shown dashed from Ulaby, © 1980 IEEE, with permission for vegitated farmland at 4.25 GHz.)

Although not indicated in the preceding data, σ^0 can also be a rather sensitive function of soil moisture content (see Ulaby et al., 1978, p. 293), where variations as large as 14 dB have been seen at 4.25 GHz with horizontal polarization at an incidence angle of 10 degrees.

A relatively simple model was empirically fit to a large body of data on various terrain by Moore et al. (1980). Although we do not show their data, we will summarize their final model which was developed using regression analysis to represent average scattering from general terrain. The resulting model for $(\sigma^0)_{dB}$ is

$$(\sigma^0)_{dB} = A + B\theta + Cf_{GHz} + Df_{GHz}\theta, \quad \theta_1 \leq \theta \leq \theta_2, f_1 \leq f_{GHz} \leq f_2 \quad (5.6\text{-}1)$$

where f_{GHz} is frequency in gigahertz, θ is angle of incidence (measured from vertical) in degrees,[8] θ_1 and θ_2 determine the range of allowed θ, f_1, and f_2 determine the range of allowed f_{GHz}. A, B, C, and D are constants that depend on polarization and the

---

[8] Angle of incidence θ is related by $\theta = 90° - \psi$ to grazing angle ψ if both θ and ψ are in degrees.

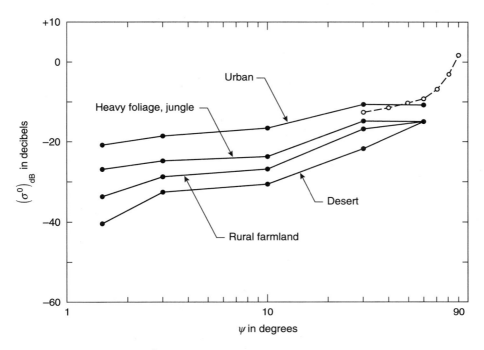

Figure 5.6-3 Behavior of σ^0 with grazing angle ψ for several types of terrain at X band (8–12 GHz). (Plots for $1.5° \leq \psi \leq 60°$ from data of Nathanson, F. E. et al.: *Radar Design Principles*, 1991, McGraw-Hill, Inc., with permission; plots shown dashed from Ulaby, © 1980 IEEE, with permission for vegitated farmland at 8.6 GHz.)

applicable ranges of θ and f_{GHz}, as listed in Table 5.6-1. Note that the model serves only two values of θ (0° and 10°) for $\theta < 20°$.

Example 5.6-1 We use (5.6-1) to compare with the data of Fig. 5.6-2. Since the plotted data average both V and H polarization, we average the coefficients in Table 5.6-1, assuming that $f_{GHz} = 4.25$ GHz, to get $A = (-14.3 - 15.0)/2 = -14.65$, $B = (-0.16 - 0.21)/2 = -0.185$, $C = 1.18$, and $D = 0.01$. From (5.6 − 1) we calculate $(\sigma^0)_{dB} = -18.17, -16.75, -15.33, -13.90$, and -12.48 dB, respectively, for $\psi = 30$, 40, 50, 60, and 70 degrees. In a similar manner $(\sigma^0)_{dB} = -6.93$ dB and $+3.22$ dB for $\psi = 80°$ and 90°, respectively. When plotted on Fig. 5.6-2, these values of $(\sigma^0)_{dB}$ compare favorably with the curves for rural farmland and with Ulaby's data for vegitated farmland.

The model of (5.6-1) is simple, easy to use, and applies as a general average for terrain. Many other models exist. Ulaby (1980) gives a more complicated model that applies for incidence angles $0° \leq \theta \leq 80°$ and $1 \leq f_{GHz} \leq 18$ with a residual mean-squared error from the true value of about 1 dB (Ulaby, 1980, p. 541). The model depends on seven coefficients and uses a four-term representation for $(\sigma^0)_{dB}$ where three of the terms are exponential (nonlinear) functions of θ. The model and

TABLE 5.6-1 Coefficients applicable to (5.6-1) when used for general terrain.

Polarization	θ_1 (degrees)	θ_2 (degrees)	f_1 (GHz)	f_2 (GHz)	A (dB)	B (dB/degree)	C (dB/GHz)	D (dB/degree GHz)
						Coefficients		
V	20	60	1	6	−14.3	−0.16	1.12	0.0051
V	20	70	6	17	−9.5	−0.13	0.32	0
H	20	60	1	6	−15.0	−0.21	1.24	0.0150
H	20	70	6	17	−9.1	−0.12	0.25	0
V and H	0	0	1	6	7.6	0	−1.03	0
V and H	0	0	6	17	0.9	0	0.10	0
V and H	10	10	1	6	−9.1	0	0.51	0
V and H	10	10	6	17	−6.5	0	0.07	0

Source: From Moore et al., © 1980 IEEE, with permission.

TABLE 5.6-2 Coefficients applicable to (5.6-1) when used for snow covered terrain.

Time of Day	Polarization	θ_1 (degrees)	θ_2 (degrees)	f_1 (GHz)	f_2 (GHz)	A (dB)	B (dB/degree)	C (dB/GHz)	D (dB/degree GHz)
								Coefficients	
Day	V	20	70	1	8	−10.00	−0.29	0.052	0.022
Day	V	20	70	13	17	0.02	−0.37	−0.50	0.021
Day	H	20	70	1	8	−11.90	−0.25	0.55	0.012
Day	H	20	70	13	17	−6.60	−0.31	0.0011	0.013
Night	V	20	70	1	8	−10.00	−0.33	−0.32	0.033
Night	V	20	70	13	17	−10.90	−0.13	0.70	0.0005
Night	H	20	70	1	8	−10.50	−0.30	0.20	0.027
Night	H	20	70	13	17	−16.90	−0.024	1.036	−0.0069
Day	V and H	0	0	1	8	3.8	0	0.092	0
Day	V and H	0	0	13	17	4.3	0	0.26	0
Day	V and H	10	10	1	8	−7.3	0	−0.26	0
Day	V and H	10	10	13	17	−14.3	0	0.66	0
Night	V and H	0	0	1	8	6.9	0	0.16	0
Night	V and H	0	0	13	17	7.1	0	0.13	0
Night	V and H	10	10	1	8	−8.0	0	−0.20	0
Night	V and H	10	10	13	17	−10.9	0	0.62	0

Source: From Moore et al., © 1980 IEEE, with permission.

coefficients are found in Ulaby's paper [model in his equation (1), p. 539, and coefficients in Table II, p. 543].

Backscattering from Snow Surfaces

Snow cover alters σ^0 for any terrain. Considerable data on this subject are available in Ulaby and Dobson (1989). Although there remains similarity to the surface being covered because of penetration of the wave to the underlying surface, there is also an effect from *volume* scattering from the snow, as well as other effects. Penetration through the snow layer is strongly dependent on frequency and other factors. For horizontal polarization at an incidence angle of 80°, for example, $(\sigma^0)_{dB}$ can vary about 27 dB for frequencies from 1.6 GHz (low σ^0) to 35.6 GHz (higher σ^0) (Ulaby and Dobson, 1989, p. 32). There can also be a great difference (as much as 5 to 15 dB) between values of σ^0 for dry versus wet snow.

According to Moore et al. (1980), the model of (5.6-1) can also be applied to snow covered terrain. Use is limited to the parameters defined in Table 5.6-2.

5.7 CROSS SECTION OF VOLUME TARGETS

When a radar views targets that are distributed over a volume in space, such as rain, snow, sleet, hail, and fog, we call these *volume targets*. The cross section seen by the radar will depend on the resolution volume of the system, polarization, frequency, the target's characteristics, and other parameters. In this section we will use some nonrigorous but plausible arguments to demonstrate that the cross section σ_c to be associated with a volume of small scatterers with individual cross sections σ_i is given by

$$\sigma_c = \sigma^1 V_c \qquad (5.7-1)$$

Here V_c is the *effective scattering volume* and σ^1 is a characteristic of the scatterers called the *volume backscattering coefficient* which has the unit of area per unit volume (m^2/m^3). Most radar literature uses the symbol η for σ^1. However, this writer believes that σ^1 is a better notation because it implies one extra spatial dimension in defining volume scattering as compared to area targets where σ^0 is used.

In measured data it is often convenient to express σ^1 in decibels. For these cases we define the notation $(\sigma^1)_{dB}$ to imply σ^1 expressed in decibels when the unit of σ^1 is $m^2/m^3 = m^{-1}$ as follows:

$$(\sigma^1)_{dB} = 10 \log_{10}(\sigma^1) \quad (dB) \qquad (5.7-2)$$

Volume-Scattering Coefficient

Consider Fig. 5.7-1. Here a radar's antenna pattern [with directive gain $G_D(\theta, \phi)$], having beamwidths θ_B and ϕ_B in orthogonal planes through the beam's maximum, has its maximum directed along the z axis. The beam illuminates many small scatterers over a volume contained within the beam over a range interval ΔR. Within the larger volume, V, we define a small volume ΔV_k having location (R_k, θ_k, ϕ_k) in the

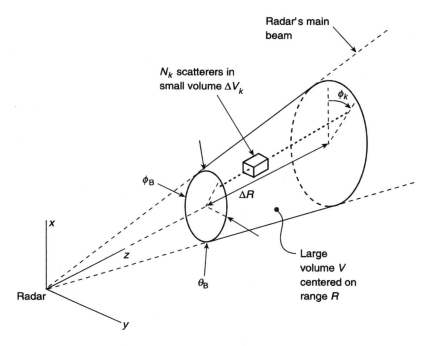

Figure 5.7-1   Geometry associated with the definition of σ^1.

radar's spherical coordinates. This volume is also assumed to contain a large number of scatterers, and we presume there to be a large number of such small volumes over the larger volume.

In the kth small volume, ΔV_k, let there be N_k scatterers. From the radar equation of (5.4-3) the power, S_{rk}, backscattered from small volume k is

$$S_{rk} = \frac{P_t G_D^2(\theta_k, \phi_k) \lambda^2}{(4\pi)^3 R_k^4 L_k} \sum_{i=1}^{N_k} \sigma_i = \frac{P_t G_D^2(\theta_k, \phi_k) \lambda^2}{(4\pi)^3 R^4 L} \sum_{i=1}^{N_k} \sigma_i \qquad (5.7\text{-}3)$$

where parameters are defined similar to those in (5.4-3). In the last form of (5.7-3), we have presumed that the large volume is small enough so that R_k, and the loss L_k, are the same for all volumes ΔV_k in the large volume.

Our main applications concern weather-related scattering. For rain it is well-known that drop diameters rarely exceed about 6 mm.[9] Because raindrops are approximately spherical (at least at smaller rainfall rates), we use (5.3-7) to find that scattering is Rayleigh as long as $f \leq 10$ GHz. Certainly scatterers in smoke, clouds, fog, and most sleet and hail satisfy the 6-mm size limit. Snow, because of its typical flake size and shape, would seem to warrant other consideration. However, studies

---

[9] Larger drops tend to break up quickly into smaller drops.

have shown (see Gunn and East, 1954, p. 535) that departures from sphericity are unimportant and cross sections of ice particles tend to not be more than twice that for spheres of the same mass. Thus we can reasonably apply Rayleigh scattering to various weather scatterers as long as $f \leq 10$ GHz. For these scatters it is known that

$$\sigma_i = \frac{\pi^5 |K|^2}{\lambda^4} D_i^6 \qquad (5.7\text{-}4)$$

(Gunn and East, 1954, p. 523), where λ is wavelength (meters), D_i is the diameter of scatterer i (meters), and $|K|^2$ is a constant that depends on the complex dielectric constant of the material. It is weakly dependent on frequency and temperature but can be taken as $|K|^2 = 0.93$ for water at temperatures from $0°$ C to $20°$ C and frequencies up to about 10 GHz. For ice particles $|K|^2 = 0.20$ can typically be assumed.

On use of (5.7-4) in the sum of (5.7-3), the cross section, σ_k, associated with small volume ΔV_k can be written as

$$\sigma_k = \sum_{i=1}^{N_k} \sigma_i = \frac{\pi^5 |K|^2}{\lambda^4} \sum_{i=1}^{N_k} D_i^6 = \frac{\pi^5 |K|^2}{\lambda^4} \frac{N_k}{\Delta V_k} \left[\frac{1}{N_k} \sum_{i=1}^{N_k} D_i^6 \right] \Delta V_k \qquad (5.7\text{-}5)$$

The bracketed term is the average of the sixth power of the diameters of the N_k particles. Since N_k is large, it is helpful to imagine the diameters are values of a continuous random variable D having a probability density function $f(D)$. On using this approach, we define and expand Z as follows:

$$Z = \frac{N_k}{\Delta V_k} \left[\frac{1}{N_k} \sum_{i=1}^{N_k} D_i^6 \right] \approx \frac{N_k}{\Delta V_k} \int_0^\infty D^6 f(D)\, dD$$

$$= \int_0^\infty D^6 \left[\frac{N_k}{\Delta V_k} f(D) \right] dD = \int_0^\infty D^6 N(D)\, dD \qquad (5.7\text{-}6)$$

where

$$N(D) = \frac{N_k}{\Delta V_k} f(D) \qquad (5.7\text{-}7)$$

is the drop number distribution; it is the distribution (with drop size D) of the number of scatterers per unit volume. Z is often called the *radar reflectivity factor*. The unit of Z is volume squared per unit volume; in radar meteorology the unit mm^6/m^3 is often used.

There are many drop number distributions. The Marshall-Palmer distribution is often used and is defined by (Marshall and Palmer, 1948)

$$N(D) = N_0 e^{-\beta D} \qquad (5.7\text{-}8)$$

where

$$N_0 = 0.08 \text{ cm}^{-4} = 8(10^6) \text{ m}^{-4} \qquad (5.7\text{-}9)$$

$$\beta = 41 r^{-0.21} \text{ cm}^{-1} = 4100 r^{-0.21} \text{ m}^{-1} \qquad (5.7\text{-}10)$$

and r is rainfall rate in mm/h.

Various experimental results and theoretical models show that Z can be put in the form (see Problems 5.7-1 through 5.7-3)

$$Z = ar^b \qquad (5.7\text{-}11)$$

where a and b are constants and r is rainfall rate. For various data and models a ranges from as low as 127 to as high as 505, while b ranges from 1.41 to 2.39. A widely accepted pair of coefficients is $a = 200 \text{ mm}^6/\text{m}^3$ and $b = 1.6$ (Skolnik, 1980, p. 500).

We now continue to develop the received clutter power S_{rc} by using (5.7-11) and (5.7-5) in the radar equation by summing (5.7-3) over the number (N_c) of small volumes:

$$S_{rc} = \sum_{k=1}^{N_c} S_{rk} = \frac{P_t \lambda^2}{(4\pi)^3 R^4 L} \cdot \frac{\pi^5 |K|^2}{\lambda^4} Z \left[\sum_{k=1}^{N_c} G_D^2(\theta_k, \phi_k) \Delta V_k \right] \qquad (5.7\text{-}12)$$

Here we have assumed that Z is independent of k. To treat the bracketed term in (5.7-12), we imagine that the antenna's power gain to be replaced by a ficticious antenna with constant gain G_D^2 (the directivity of the real antenna) that illuminates scatterers over an *effective scattering volume* V_c such that

$$\sum_{k=1}^{N_c} G_D^2(\theta_k, \phi_k) \Delta V_k = G_D^2 \sum_{k=1}^{N_c} \Delta V_k = G_D^2 V_c \qquad (5.7\text{-}13)$$

Hence (5.7-12) can finally be written as

$$S_{rc} = \frac{P_t G_D^2 \lambda^2 \sigma_c}{(4\pi)^3 R^4 L} \qquad (5.7\text{-}14)$$

where

$$\sigma_c = \sigma^1 V_c \qquad (5.7\text{-}15)$$

$$\sigma^1 = \frac{\pi^5 |K|^2 Z}{\lambda^4} = \frac{\pi^5 |K|^2 ar^b}{\lambda^4} \qquad (5.7\text{-}16)$$

$$V_c = \frac{1}{G_D^2} \sum_{k=1}^{N_c} G_D^2(\theta_k, \phi_k) \Delta V_k \qquad (5.7\text{-}17)$$

Equation (5.7-14) is the radar equation for volume clutter. The quantity σ^1 defined by (5.7-16) is a characteristic of the type of scatterer and will vary with scatterer type.

Effective Volume

The effective volume, V_c, is given by (5.7-17). From Fig. 5.7-1 we may write $\Delta V_k = \Delta R_k \Delta A_k$, where ΔR_k is in the radial direction and ΔA_k is area in the cross-beam direction. If the range resolution cell is relatively short compared to nominal range R, the two elliptical end-cap areas will be approximately equal, and

$$
\begin{aligned}
V_c &= \frac{1}{G_D^2} \sum_{k=1}^{N_c} G_D^2(\theta_k, \phi_k)\, \Delta R_k\, \Delta A_k \\[2mm]
&\approx \frac{1}{G_D^2} \int_{R-(\Delta R/2)}^{R+(\Delta R/2)} dR \iint_{\text{Area}} G_D^2(\theta, \phi)\, dA \\[2mm]
&= \frac{\Delta R}{G_D^2} \iint_{\text{Area}} G_D^2(\theta, \phi)\, d\theta\, d\phi\, R^2
\end{aligned}
\tag{5.7-18}
$$

To evaluate (5.7-18), we approximate $G_D(\theta, \phi)$ by a gaussian function (Problem 5.4-6) and assume that beamwidths θ_B and ϕ_B are small enough that limits on integrations can be approximated as infinite. Thus

$$
\begin{aligned}
V_c &\approx \Delta R \int_{-\infty}^{\infty} \int_{-\infty}^{\infty} e^{-8\ln(2)[(\theta/\theta_B)^2 + (\phi/\phi_B)^2]}\, d\theta\, d\phi\, R^2 \\[2mm]
&= \frac{\pi R^2 \theta_B \phi_B c}{16 B \ln(2)}
\end{aligned}
\tag{5.7-19}
$$

since $\Delta R = c/2B$. For other patterns (5.7-19) should not be greatly different. For a constant-frequency pulse of duration T, (5.7-19) has the form

$$
V_c \approx \frac{\pi R^2 \theta_B \phi_B c T}{16 \ln(2)}
\tag{5.7-20}
$$

5.8   METEOROLOGICAL VOLUME TARGETS

A radar's most common volume targets are due to weather. Of the various forms of precipitation, rain is usually of the most concern. Scattering from snow, hail, and the like, can also be a concern at higher frequencies and higher precipitation rates. In this section we give minimal examples of clutter due to precipitation because the subject is vast and cannot be given careful expose in the space allowed. For design details, the reader is referred to the literature.

Rain Backscatter

For linear polarization, temperate lattitudes, and lower frequencies ($f \leq 10$ GHz), a widely accepted relationship for the rain reflectivity factor is

$$
Z = 200 r^{1.6} \quad (\text{mm}^6/\text{m}^3)
\tag{5.8-1}
$$

where r is rainfall rate in mm/h (Skolnik, 1980, p. 500; Gunn and East, 1954, p. 532). However, measured data for different types of rain (from melting snow or hail, or orographic rain—as induced by presence of mountains) show variations in the coefficients of (5.8-1). The variability can be seen in data of Blanchard (1953) and

Nathanson (1991, p. 233). By using (5.8-1) in (5.7-16) with $|K|^2 = 0.93$ for rain, we have

$$\sigma^1 = 7.03(10^{-12})f_{GHz}^4 r^{1.6} \quad (m^2/m^3) \tag{5.8-2}$$

which applies for $f \leq 10$ GHz, and is plotted in Fig. 5.8-1. Also shown are curves for $f \geq 20$ GHz as replotted from Crane and Burke (1978). The data for $f \geq 20$ GHz are based on the Marshall-Palmer drop distribution of (5.7-8).

Richard et al. (1988, Fig. 11) give average measured data for frequencies of 9.375, 35, 70, and 95 GHz that compare favorably (within the quoted measurement error of 2 dB) for all frequencies but 9.375 GHz. At 9.375 GHz the errors were about 3.5 dB for higher values of r and as large as 6.5 dB for $r = 2.5$ mm/h. The discrepancies are probably due to the variations seen for different types of rain.

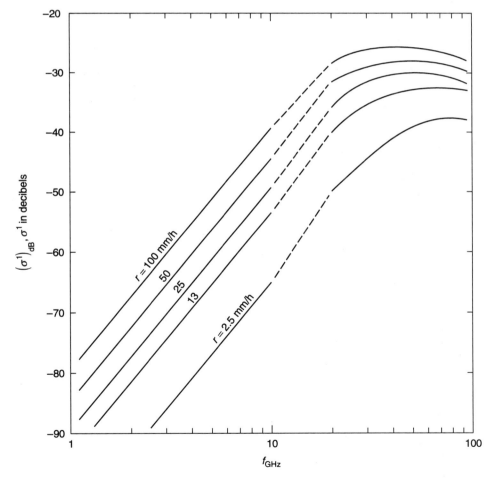

Figure 5.8-1 Volume backscattering coefficient for rain in moderate lattitudes versus frequency for various rainfall rates. (Data for $f_{GHZ} \geq 20$ from Crane and Burke, 1978.)

At higher elevations rains can result from melting of precipitation that begins as snow or hail. As the ice particles start to melt on passing the 0°C altitude, they first become water-coated ice particles and then become rain once melting is complete. In this region of altitudes radar cross section changes greatly and is referred to as the *bright band.* Skolnik (1980, p. 502) quotes values from Battan (1973) indicating that scattering in the bright band is typically 12 to 15 dB larger than that of the snow above and about 6 to 10 dB greater than the rain below it. These changes are associated with the shape changes that occur (flat, low scattering flakes to less flat water coatings that increase reflectivity, and finally spheroids of water for high reflectivity) and the density of particles (as shape changes, velocity of fall increases, which reduces the density of scatterers).

Rain Clutter Reduction

If raindrops were perfect spheres, it would be possible for a radar to transmit and receive same-sense circular polarization and have complete cancellation of rain clutter, in theory (see Example 5.1-2). However, raindrops depart from spheres as rain rate increases (they tend to first look like oblate spheroids and then approach that of a "hamburger bun"), causing a limit on the amount of cancellation. Even if rain-drops are spherical, the radar system can depart from true circular polarization, again limiting cancellation.

We will define cancellation ratio as the ratio of clutter power received by a system with linear polarization to the corresponding clutter power on using circular polarization when both systems have the same pattern, same transmitter power, and operate in the same environment. Under these constraints Peebles (1975) has examined the cancellation ratio attainable in a nonideal system operating with ideal (spherical) raindrops and found that clutter cancellation can be as poor as -20 dB for system phase errors up to ± 4 degrees (no amplitude error) or as poor as about 16.5 dB for amplitude errors of 1 dB (no phase error). For phase errors of up to ± 4 degrees *and* amplitude errors of up to 1 dB the cancellation is only about 15 dB.

For the case where the system had perfectly circular polarization, the cancellation achievable was found for rain from melting snow (worst case), melting sleet, and melting hail (best case), as shown in Fig. 5.8-2. The results were computed for the Polyakova-Shifrin (P-S) model for the probability density of raindrop diameters, as defined by

$$f(D) = \frac{\gamma^3}{2} D^2 e^{-\gamma D} \qquad (5.8\text{-}3)$$

where

$$\gamma = a_1 r^{-b_1} \qquad (5.8\text{-}4)$$

with $a_1 = 4.01$, 4.87, and 6.95, while $b_1 = 0.19$, 0.20, and 0.27 for rain from melting snow, sleet, and hail, respectively. Peebles (1975, *IEEE Region 3 Conference*) also computed cancellation ratio for the Marshall-Palmer distribution (M-P model); the result was an almost exact overlay to the curve for melting sleet.

Sekine and Lind (1983) used the basic results of Peebles (1975) to calculate clutter cancellation for both the M-P model and a Weibull model proposed by them. The M-P results overlap the curve of Fig. 5.8-2 for melting sleet almost exactly (providing

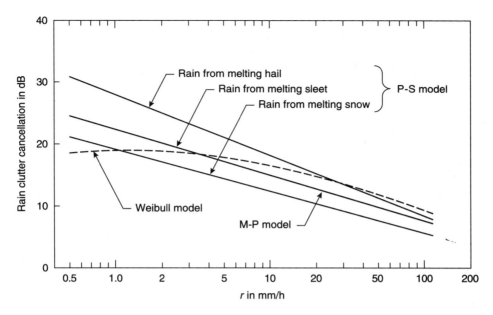

Figure 5.8-2 Attainable rain clutter cancellation ratio versus rainfall rate for various raindrop models. (Adapted from data in Peebles, © 1975 IEEE, with permission, and Sekine et al., © 1983 IEEE, with permission.)

confirmation of Peebles's results for this distribution). The results shown dashed correspond to the Weibull drop number distribution given by

$$N(D) = N_0 \frac{\eta}{\sigma} \left(\frac{D}{\sigma} \right)^{\eta-1} e^{-(D/\sigma)^\eta} \tag{5.8-5}$$

where

$$N_0 = 1000 \quad (\text{m}^{-3}) \tag{5.8-6}$$

$$\eta = 0.95 r^{0.14} \tag{5.8-7}$$

$$\sigma = 0.26 r^{0.44} \quad (\text{mm}) \tag{5.8-8}$$

were determined to fit existing data.

Cancellation data have been measured by Richard et al. (1988) at several r values from 2 to 100 mm/h at each of four frequencies (9.375, 35, 70, and 95 GHz). Except for variations that are probably due to normal variability of rain, cancellation tends to be constant with r for a given frequency, which tends to favor the Weibull model in Fig. 5.8-2 for small r. Measured cancellations were quoted as being at least 15 dB with median values of 20 dB at 9.375 GHz and 18 dB at 95 GHz; these values are consistent with all models for small r. For larger r (>20 mm/h) measured data were frequently *better* than the upper limits of all curves in Fig. 5.8-2, which may indicate better models are still possible.

Snow Backscatter

We have already seen that wet snow in the bright band (above) can give clutter cross sections larger than for rain. At lower frequencies such that particles give Rayleigh scattering the relationship of (5.7-11) is well-accepted and many values of a and b have been suggested (see Skolnik, 1980, p. 502). A corrected model due to Gunn and East (1954) is

$$Z = 2000r^2 \quad (\text{mm}^6/\text{m}^3) \tag{5.8-9}$$

For this model and using $|K|^2 = 0.2$, (5.7-16), gives

$$\sigma^1 = 1.51(10^{-11})f_{\text{GHz}}^4 r^2 \quad (\text{m}^2/\text{m}^3) \tag{5.8-10}$$

where r is the snowfall rate in melted water content (mm/h).

It is tempting to compare values of (5.8-10) for snow with Fig. 5.8-1 for rain. Indeed, if r were the same for snow and rain, snow would have higher σ^1. However, it takes a large snow rate (millimeters of *snow* accumulation) to produce a given equivalent *water* accumulation rate. Furthermore the upper limit for snow is about $r = 10$ mm/h (Gunn and East, 1954, p. 539), so most snowfalls are less troublesome in radar than most rains.

We conclude that except for the bright band and possibly wet snow at high fall rates, snow backscatter is usually less of a problem in radar than rain.

Other Types of Precipitation

As with most snow, backscattering from hail, sleet, ice and water clouds is typically small compared to rain. These comments also hold for fog. For additional detail, the reader is referred to the specialized literature.

5.9   CROSS SECTION FLUCTUATIONS AND MODELS

Suppose that a radar observes a target for a period of time, and let all target and systems parameters be constant (on the average). For a surface target this condition means that we observe the same area of surface through a constant pattern with fixed power, bandwidth, and so on. For a volume target conditions are similar, and for a point, small, or complex target the requirements mean that the target stays fixed in the same spot in a resolution cell. All these requirements would seem to imply that the backscattered signal power and cross section are constants. In reality they are not. They fluctuate with time.

Even though all parameters are fixed *on the average*, the structure of a surface can have minute changes because the wind blows through trees, grass, and bushes causing small changes to occur. With volume targets, scatterers are changing shape, size, and orientation as they fall through a resolution cell, producing changes in cross section. Even with point, small, or complex targets, such as a missile or aircraft, minute changes occur in aspect angle and positions of structural members (wings) as wind buffets the target. Even vibrations of the target's structure can cause cross section to vary with time.

Changes in cross section with time are modeled in radar as a random process. Therefore the cross section that occurs at any one time is a value of a random variable defined from the process at that time. The first-order probability density function (see Peebles, 1993, p. 169) of the process describes the amplitude fluctuations (at the one time). Changes of cross section *with* time are regulated by the autocorrelation function of the process. If changes are rapid the autocorrelation function varies quickly with time, slow changes in cross section correspond to a broad autocorrelation function. In essence, the autocorrelation function describes how "correlated" cross section is at time $t + \tau$ compared to the value at time t.

Because cross section changes with time it has spectral (bandwidth) properties. The power spectrum of the time-varying cross section is the Fourier transform of the time-averaged autocorrelation function. It gives the frequency-domain description of the cross section of targets in a manner analogous to the spectrum of an ordinary (nonrandom) signal.

In all the above discussions, all system and target parameters were constant. Now suppose that the radar moves its beam slightly in space when viewing area or volume targets (all else constant), or that the target moves a bit in space in the resolution cell for a small point or complex target. Again the received power, or cross section, changes. Now changes are over space, and correlation functions can be defined to describe these *spatial* variations.

In this section we will not give any development of temporal or spatial correlation behavior of targets. Rather, we leave these topics as extensions of the book obtained through the specialized literature. We will, however, describe several of the probability density functions that are useful in modeling amplitude fluctuations of cross section σ (or received power that is proportional to cross section). We define the following notation and terms:

$$f(\sigma) = \text{probability density function of } \sigma \tag{5.9-1}$$

$$F(\sigma) = \int_{-\infty}^{\sigma} f(\xi) \, d\xi$$

$$= \text{probability distribution function of } \sigma \tag{5.9-2}$$

$$\bar{\sigma} = \int_{-\infty}^{\infty} \sigma f(\sigma) \, d\sigma = \text{mean value of } \sigma \tag{5.9-3}$$

$$\sigma_{\text{med}} = \text{median value of } \sigma \text{ defined by (5.9-5)} \tag{5.9-4}$$

$$F(\sigma_{\text{med}}) = 0.5 \tag{5.9-5}$$

$$\Phi(j\omega) = \int_{-\infty}^{\infty} f(\sigma) e^{j\omega\sigma} \, d\sigma$$

$$= \text{characteristic function of } \sigma \tag{5.9-6}$$

$$m_n = (-j)^n \frac{d^n \Phi(j\omega)}{d\omega^n}\bigg|_{\omega=0}, \quad n = 0, 1, 2, \ldots$$

$$= \text{moment } n \text{ of the density of } \sigma \tag{5.9-7}$$

Where available, these results are summarized for the Rayleigh, Erlang, chi-square, Weibull, and log-normal distributions.

Rayleigh Model

This function describes the backscattered cross section of the sum of a large number of small scatterers, none of which dominates. Here

$$f(\sigma) = \frac{1}{\bar{\sigma}} e^{-\sigma/\bar{\sigma}} u(\sigma) \tag{5.9-8}$$

$$F(\sigma) = [1 - e^{-\sigma/\bar{\sigma}}] u(\sigma) \tag{5.9-9}$$

$$\sigma_{\text{med}} = \bar{\sigma} \ln(2) \tag{5.9-10}$$

$$\Phi(j\omega) = \frac{1}{1 - j\omega\bar{\sigma}} \tag{5.9-11}$$

$$m_n = n! \, \bar{\sigma}^n, \quad n = 0, 1, 2, \ldots \tag{5.9-12}$$

The Rayleigh density fits some land and sea data at higher grazing angles. It is a reasonable model for most volume precipitation. Rayleigh is the most "classical" of densities, and was applied early-on in radar to describe clutter fluctuations. Equation (5.9-8) is used in the Swerling detection cases I and II to be discussed in Chapter 9.

Erlang Model

The Erlang distribution is defined by

$$f(\sigma) = \left(\frac{N}{\bar{\sigma}}\right)^N \frac{\sigma^{N-1}}{(N-1)!} e^{-N\sigma/\bar{\sigma}} u(\sigma), \quad N = 1, 2, 3, \ldots \tag{5.9-13}$$

$$F(\sigma) = \left[1 - e^{-N\sigma/\bar{\sigma}} \sum_{n=0}^{N-1} \frac{(N\sigma/\bar{\sigma})^n}{n!}\right] u(\sigma) \tag{5.9-14}$$

$$\Phi(j\omega) = \left(\frac{N}{N - j\omega\bar{\sigma}}\right)^N \tag{5.9-15}$$

$$m_n = \begin{cases} 1, & n = 0 \\ \left(\frac{\bar{\sigma}}{N}\right)^n N(N+1)(N+2)\ldots(N+n-1), & n = 1, 2, 3, \ldots \end{cases} \tag{5.9-16}$$

The application of the Erlang distribution in radar has mainly been for the special cases $N = 1$ and $N = 2$. When $N = 1$, we have the Rayleigh case. For $N = 2$,

$$f(\sigma) = \frac{4\sigma}{\bar{\sigma}^2} e^{-2\sigma/\bar{\sigma}} u(\sigma) \tag{5.9-17}$$

$$F(\sigma) = \left[1 - \left(1 + \frac{2\sigma}{\bar{\sigma}} \right) e^{-2\sigma/\bar{\sigma}} \right] u(\sigma) \tag{5.9-18}$$

$$\sigma_{\text{med}} = 0.83917\bar{\sigma} \tag{5.9-19}$$

$$\Phi(j\omega) = \left(\frac{2}{2 - j\omega\bar{\sigma}} \right)^2 \tag{5.9-20}$$

$$m_n = (n + 1)!(\bar{\sigma}/2)^n, \qquad n = 0, 1, 2, \dots \tag{5.9-21}$$

The Erlang distribution seems to have no basis in nature, but it has been used as an empirical fit for some data. It is used in the Swerling detection cases III and IV to be discussed in Chapter 9.

Chi-Square Model

The chi-square distribution is defined by

$$f(\sigma) = \frac{M}{\Gamma(M)\bar{\sigma}} \left(\frac{M\sigma}{\bar{\sigma}} \right)^{M-1} e^{-M\sigma/\bar{\sigma}} u(\sigma), \qquad M > 0 \tag{5.9-22}$$

$$F(\sigma) = P\left(M, \frac{M\sigma}{\bar{\sigma}} \right) \tag{5.9-23}$$

where $2M$ is called the *degree* of the distribution,

$$P(\alpha, \beta) = \frac{1}{\Gamma(\alpha)} \int_0^\beta \xi^{\alpha-1} e^{-\xi} d\xi, \quad Re(\alpha) > 0$$

$$= \text{incomplete gamma function} \tag{5.9-24}$$

and $\Gamma(M)$ is the gamma function.

In the general density of (5.9-22), M does not need to be an integer, and it has been applied to both simple and satellite targets when M lies between 0.3 and 2 (Weinstock, 1964; as noted in Skolnik, 1980, p. 50). The chi-square density covers a broad class of targets. As M increases, the density corresponds to cross section constrained to a narrow range of values. As $M \to \infty$, the density becomes an impulse corresponding to a constant cross section. Some values of M are known to fit some data on aircraft taken over a short time period.

When M is an integer, the chi-square density reduces to other densities. For $M = 1$ the Rayleigh case results. For $M = 2$ we have (5.9-17). If M is a positive integer, the chi-square density gives the Erlang density (5.9-13) because $\Gamma(M) = (M-1)!$ for M a positive integer.

Weibull Model

The Weibull distribution is defined by

$$f(\sigma) = \frac{b\Gamma(1 + (1/b))}{\bar{\sigma}} \left[\frac{\Gamma(1 + (1/b))\sigma}{\bar{\sigma}} \right]^{b-1} e^{-[\Gamma(1 + (1/b))\,\sigma/\bar{\sigma}]^b} u(\sigma) \qquad (5.9\text{-}25)$$

$$F(\sigma) = \left\{ 1 - e^{-[\Gamma(1 + (1/b))\sigma/\bar{\sigma}]^b} \right\} u(\sigma) \qquad (5.9\text{-}26)$$

$$\sigma_{\mathrm{med}} = \bar{\sigma}\,\frac{[\ln(2)]^{1/b}}{\Gamma(1 + (1/b))} \qquad (5.9\text{-}27)$$

$$m_n = \Gamma(1 + (n/b)) \left[\frac{\bar{\sigma}}{\Gamma(1 + (1/b))} \right]^n \qquad (5.9\text{-}28)$$

The Weibull model has been useful in representing some sea clutter and some land clutter at low grazing angles.[10] It has two parameters $\bar{\sigma}$ and b (a shape parameter) that can be chosen to fit measured data. In general, two-parameter models will give better data fits than one-parameter models, such as Rayleigh and Erlang (for a specific value of N).

Log-Normal Model

The log-normal distribution of cross section σ results when a gaussian random variable x with mean a and variance b^2 is converted through the transformation $\sigma = \exp(x)$. The log-normal random cross section is defined by

$$f(\sigma) = \frac{1}{\sqrt{2\pi b^2}\,\sigma}\, e^{[\ln(\sigma) - a]^2/2b^2} u(\sigma) \qquad (5.9\text{-}29)$$

$$F(\sigma) = \left\{ 1 - Q\left[\frac{\ln(\sigma) - a}{b} \right] \right\} u(\sigma) \qquad (5.9\text{-}30)$$

where $Q(\cdot)$ is the Q-function of (E.2-1). Forms more applicable to radar result from substitution of

$$a = \ln(\sigma_{\mathrm{med}}) \qquad (5.9\text{-}31)$$

$$b = \sqrt{2\ln\left(\frac{\bar{\sigma}}{\sigma_{\mathrm{med}}} \right)} \qquad (5.9\text{-}32)$$

---

[10] In clutter models it is sometimes the voltage that is modeled as being Weibull; in such cases cross section is replaced by appropriate voltage notation.

which give

$$f(\sigma) = \frac{e^{-[\ln(\sigma/\sigma_{\text{med}})]^2/[4\ln(\bar{\sigma}/\sigma_{\text{med}})]}}{\sqrt{4\pi \ln(\bar{\sigma}/\sigma_{\text{med}})}\,\sigma} u(\sigma) \qquad (5.9\text{-}33)$$

$$F(\sigma) = \left\{1 - Q\left[\frac{\ln(\sigma/\sigma_{\text{med}})}{\sqrt{2\ln(\bar{\sigma}/\sigma_{\text{med}})}}\right]\right\} u(\sigma) \qquad (5.9\text{-}34)$$

The log-normal distribution has been applied to some targets and both terrain and sea clutter.[11] It has higher "tails" at larger σ than preceding models and is used where σ has large values a lot of the time.

Other Models

The purpose of a cross section model is to approximate real conditions that manifest themselves as a probability density function (histogram, typically) derived from measured data. Models should be as simple as possible but should lend themselves to analysis and adequately represent the real data. A model such as the Rayleigh one is simple and can usually be analyzed, but it may not adequately fit data because it only has one selectable constant. It can therefore match data at only one point ($\bar{\sigma}$ or σ_{med}).

Other models, such as the Erlang, chi-square, Weibull, log-normal, and K-distribution [see (9.10-10)], have two selectable parameters. These can match data for $\bar{\sigma}$ (or σ_{med}) and can also control the shape of the density with the second parameter. Still such models do not "fit" all data.

Clearly models based on more selectable parameters hold the promise of a better fit to data. Xu and Huang (1997) have suggested a model based on representing a real density by a series expansion using orthogonal Legendre polynomials. The coefficients of the expansion are related to the central moments of the data. A given coefficient A_n, $n = 0, 1, 2, \ldots$, is a function of all central moments up to n. An example was given where 15 to 20 series terms produced a good fit to cross section data of an airplane. Although the results were good, the new model was not evaluated for truncation to a small number of terms for comparison to simple models. Such comparative evaluation may eventually show the utility of the new model.

PROBLEMS

5.1-1 A monostatic radar is "calibrated" at a range of 8 km using a target of 1.5-m^2 known radar cross section and the available received signal power is $45(10^{-12})$ W. If this radar requires a minimum signal power of 10^{-14} W for proper performance, what radar cross section must a target have if the radar must operate properly out to 50 km?

---

[11] When (5.9-33) is used to model clutter, σ, σ_{med}, and $\bar{\sigma}$ usually apply to the voltage envelope of the clutter portion of signal plus clutter applied to an envelope detector. As such, these quantities are better represented by voltage notation, such as r, r_{med}, and $\bar{r}$, respectively.

5.1-2 If the receiver of the system in Problem 5.1-1 is improved so a minimum required signal level becomes 10^{-15} W, what minimum radar cross section is required at 50 km?

5.1-3 The backscattered wave at a monostatic radar is known to have a polarization ratio $Q_2 = 0.36 \exp(j\pi/8)$ and have a power density of 10^{-15} W/m². What is the magnitude of the scattered wave's electric field in direction θ_2; that is, what is $|E_{\theta_2}|$?

5.1-4 A scattered wave at a receiver has a polarization ratio $Q_2 = 0.2 \exp(-j\pi.12)$. The receiving antenna has a polarization defined by $Q_A = 0.4 \exp(-j\pi.12)$. Find constants A_1, A_2, and A_3, and write expressions for $\mathbf{E_A}$ and $\mathbf{E_B}$ using (5.1-12) and (5.1-13).

5.1-5 Determine what polarization ratio (Q_A) the scattered wave component $\mathbf{E_A}$ must have, in terms of the polarization of the full scattered wave, if ρ_{pol} in (5.1-27) must equal 1. Discuss the significance of your result.

5.1-6 Rework Problem 5.1-5 except assume that $\rho_{pol} = 0$.

5.1-7 A target is known to backscatter a wave that is polarized only in the θ_2 direction. If the radar is also polarized only in the θ_1 direction, justify that $\sigma_s = \sigma$ by using (5.1-24) and (5.1-25).

5.1-8 Suppose that a target scatters a wave with polarization only in the θ_2 direction. Then $Q_2 = 0$ and σ_s has some value from (5.1-24). With $Q_2 = 0$ in (5.1-25), demonstrate that $\sigma \to 0$ if the receiving antenna responds only to fields in the ϕ_2 direction.

5.1-9 The polarization ratio of a scattered wave is $Q_A = 0.8 \exp(-j\pi/3)$. The component of scattered wave corresponding to the receiving antenna's polarization has $Q_2 = 0.7 \exp(-j\pi/6)$. What fraction of the scattered wave's total power will be received by the receiving antenna?

5.1-10 Work Problem 5.1-9 except assume that $Q_2 = 0.7 \exp(j\pi/6)$.

5.1-11 A target backscatters a wave that is polarized in only the θ_2 direction such that $Q_2 = 0$. The monostatic radar's antenna is left-circularly polarized. Find ρ_{pol} for this system.

5.1-12 Work Problem 5.1-11 except assume a right-circular polarization for the receiving antenna.

★5.2-1 Show that the scattering cross section of a target can be written in terms of the elements of the Jones matrix according to

$$\sigma_s = \frac{|T_{\theta_2\theta_1} + Q_1 T_{\theta_2\phi_1}|^2 + |T_{\phi_2\theta_1} + Q_1 T_{\phi_2\phi_1}|^2}{1 + |Q_1|^2}$$

5.2-2 A monostatic radar transmits a wave with polarization only in the θ_1 direction. Find σ_s in terms of elements of the Jones matrix. What is σ_s if the radar's polarization is only in the ϕ_1 direction? (*Hint:* Use the result of Problem 5.2-1.)

5.2-3 A test facility transmits a pulse with polarization only in the θ_1 direction and independently receives polarizations in directions θ_1 and ϕ_1 (monostatic system). On the next pulse it transmits only in the ϕ_1 direction but again receives independently in polarizations θ_1 and ϕ_1. Discuss how the facility might measure the elements of the Sinclair matrix of some target.

5.3-1 A conducting sphere with a diameter of 25 cm is the target of a monostatic radar at 10 GHz. What is the target's radar cross section? Is the scattering in the Rayleigh, resonance, or optical region?

5.3-2 What is the largest radius a conducting sphere can have if its monostatic cross section must remain in the Rayleigh region? Give the answer in terms of frequency in gigahertz (f_{GHz}). What is the corresponding largest cross section?

5.3-3 A monostatic radar at 5.5 GHz is to track a conducting sphere for calibration. If $a/\lambda = 10$ for the sphere, what is the sphere's cross section?

5.3-4 The largest cross section for a conducting sphere occurs at approximately $a/\lambda = 0.165$ where σ is approximately $3.65\pi a^2$. At what radar frequency will the maximum value of σ equal 0.1 m^2?

5.3-5 What is the smallest diameter that a sphere can have and have scattering in the optical region? Find the minimum cross section in terms of frequency.

5.3-6 A flat conducting square plate 0.4 m on a side is a target for a monostatic radar. At what frequency will the plate have a maximum radar cross section of 10 m^2?

5.3-7 Work Problem 5.3-6 except for a rectangular plate for which $a = 0.08$ m and $b = 0.05$ m.

5.3-8 Assume that (5.3-17) applies to a square conducting plate. The frequency to be used is 9.5 GHz, and the plate's side length is to be chosen so that cross section has its first null at $\theta = 3°$. What side length should be used and what cross section occurs at $\theta = 0$?

5.3-9 A monostatic radar at 3.6 GHz illuminates a square conducting flat plate at an angle $\theta = 0.03\pi$ (Fig. 5.3-3). The plate has side length $2a = 5\lambda$. What is the radar cross section? What is σ for $\theta = 0$?

5.3-10 Assume that ρ is large and $\theta = 0$ in (5.3-9) and (5.3-10), and show that $\sigma_V = \sigma_H$ and are equal to the value of (5.3-17) when $\theta = 0$.

5.3-11 Find the radar cross section of a circular conducting disk when viewed normal to the disc by a radar at 12 GHz? Assume a disk radius of 15 cm.

5.3-12 Work Problem 5.3-11 except for an off-normal angle of 1.5 degrees.

5.3-13 For a circular cylinder as shown in Fig. 5.3-6, define the main scattering angle as that between first nulls on each side of $\theta = \pi/2$. Find an expression for this angle in terms of length L and frequency assuming that $ka \sin(\theta) \gg 1$ so that (5.3-20) applies. Evaluate your result for $L = 1$ m and $f = 2.3$ GHz.

5.3-14 What minimum radius must the cylinder have in Problem 5.3-13?

5.3-15 Find the ratio of the cross section of the cylinder of Fig. 5.3-6 when $\theta = \pi/2$ to the cross section when $\theta = 0$ or π. Evaluate your ratio for $a/\lambda = 6$ and $L/\lambda = 60$.

5.3-16 If $a/\lambda = 1/85$ and θ is very near $\pi/2$, show that the cross section for a long thin wire, as given by (5.3-21), can be written as

$$\sigma \approx \frac{L^2}{\pi} \cos^4(\phi)$$

5.3-17 A naval vessel has a displacement of 8.7 ktons. If another ship observes the vessel at a low grazing angle at 5.9 GHz, what cross section is to be expected?

5.3-18 A naval radar needs a cross section of at least 1900 m^2 to reliably detect another ship. If it must detect ships as small as 1000 tons of displacement, what minimum frequency should the radar use?

5.4-1 An airborne radar at 12 GHz has a peak power of 1 kW and uses an antenna having a directivity of 2100 and a total monostatic path loss of 1.8. It views a surface at nominal range 3 km. If the available received clutter power is $1.1(10^{-11})$ W, what is the effective surface scattering area if $\sigma^0 = 2.5 \, (10^{-4})$ for the surface?

5.4-2 An airborne radar has an antenna pattern with a vertical-plane beamwidth of 3.6 degrees. It views a surface at nominal range $R = 4.5$ km at a grazing angle of 5 degrees. The surface backscattering coefficient is $\sigma^0 = 7(10^{-5})$ m^2/m^2. What is the surface's radar cross section if the radar's transmitted bandwidth is 500 kHz?

5.4-3 A pulsed airborne radar flies horizontally over a flat (horizontal) surface for which $\sigma^0 = 0.01$ m^2/m^2. Altitude is 3 km, the radar's beam depression angle is $\pi/6$, and its beamwidths are $\theta_B^\circ = 12$ degrees (vertical plane) and $\phi_B^\circ = 15$ degrees (slant plane). If the radar transmits a simple rectangular, constant-frequency, pulse of duration 10 μs, what cross section of the surface does the radar see?

5.4-4 If all is constant in the system of Problem 5.4-3 except altitude, what altitude will correspond to the effective scattering area becoming exactly equal to the beam's "footprint?"

5.4-5 An aircraft's radar altimeter has its beam pointed downward at a depression angle of $\pi/2$. Beamwidths θ_B° and ϕ_B° are equal at 5 degrees. If $\sigma^0 = 2$ for the surface which is horizontal and flat, find an expression for the surface's cross section σ_c versus altitude h. Evaluate σ_c for $h = 2$ km.

★5.4-6 Assume the one-way directive gain in (5.4-8) can be approximated by the gaussian expression

$$G_D(\theta, \phi) = G_D e^{-K(\theta/\theta_B)^2 - K(\phi/\phi_B)^2}$$

where $K = 4 \ln(2)$, and θ_B and ϕ_B are the 3-dB beamwidths of $G_D(\theta, \phi)$ in directions θ and ϕ, respectively. If the radar's pulse duration is large such that it does not limit the scattering from the surface, show that $A_c \approx A/[2 \ln(2)]$ where

$$A \approx \frac{\pi R^2 \theta_B \phi_B}{4 \sin(\psi)}$$

is the elliptical area illuminated by $G_D(\theta, \phi)$.

5.5-1 Plot (5.5-1) versus sea state (SS) and show how the nominal wind is related to the wind ranges shown in Table 5.5-1 for each SS.

5.5-2 For each wave height range in Table 5.5-1 of a sea state plot the corresponding wind speed as a linear function of wave height connecting its extreme values. Show that the empirically derived expression

$$\text{Wave height(ft)} \approx 0.03 \, W_s^{1.904}$$

where W_s is in knots, is a good approximation for the wave height/wind speed relationship.

5.5-3 An airborne radar observes the sea at a grazing angle of 20 degrees searching for periscopes at a frequency of 35 GHz. It uses vertical polarization and typically flies back and forth above the surface, first in the upwind direction and then downwind. What suggestions can be made to improve the performance of the system?

5.6-1 Use the model of (5.6-1) to estimate σ^0 for general terrain at a frequency of 5.0 GHz and a 30-degree grazing angle when horizontal polarization is used.

5.6-2 Work Problem 5.6-1 except asume that $f_{GHz} = 6.0$. Note that two sets of coefficients may now be used. Compare results for both.

5.6-3 Work Problem 5.6-1 except assume snow-covered terrain observed in daytime.

5.6-4 Use the model of (5.6-1) to calculate the ratio (in dB) of σ^0 for average terrain to σ^0 for snow-covered terrain for a 30-degree grazing angle and frequencies from 1 to 6 GHz. Assume vertical polarization and daytime operation. Is there a trend present?

5.7-1 Use (5.7-8) for the Marshall-Palmer drop number distribution in (5.7-6), and show that Z has the form of (5.7-11). What are a (in mm^6/m^3) and b?

5.7-2 Show that the Polyakova-Shifrin drop number distribution of the form

$$N(D) = N_0 \frac{\gamma^3}{3} D^2 e^{-\gamma D}, \qquad D > 0$$

where $\gamma > 0$ and $N_0 > 0$ are constants, corresponds to Z having the form of (5.7-11). Find a and b in terms of N_0, a_1, and b_1 if γ has the form

$$\gamma = a_1 r^{b_1}$$

where r is a constant.

5.7-3 Show that the Weibull drop number distribution

$$N(D) = N_0 \frac{\eta}{\sigma} \left(\frac{D}{\sigma}\right)^{\eta-1} e^{-(D/\sigma)^\eta}, \quad D > 0$$

where N_0 and σ are positive constants and η is a positive integer, gives Z in the form of (5.7-11) if σ has the form

$$\sigma = a_1 r^{b_1}$$

where r, a_1, and b_1 are positive constants. Find a and b in terms of N_0, a_1, and b_1.

5.7-4 Assume that Z has the form of (5.7-11) and that a can range from 127 to 505 mm^6/m^3 and b can range from 1.41 to 2.39. Find the ratio of σ^1 at the largest extreme of Z to σ^1 at the smallest extreme of Z. Does it depend on r? Evaluate the ratio in decibels for $r = 1.25$, 5, and 50 mm/h.

5.7-5 A radar has a pencil beam with 3.2-degree beamwidth. It transmits a constant-frequency pulse of duration 2.5 μs. If precipitation is falling at a range of 30 km, what is the effective volume of the radar's resolution cell for precipitation clutter? What are the range and altitude extents of the volume?

5.8-1 Rain falls at $r = 1.9$ mm/h in the coverage volume of a radar operating at 8.5 GHz and having a pattern with beamwidths $\theta_B = \pi/90$ rad and $\phi_B = \pi/60$ rad. The radar transmits a constant-frequency rectangular pulse of 3-μs duration. (a) Find σ^1 for the rain. (b) What is the rain's radar cross section as a function of range R?

5.8-2 A radar is designed to operate with rain with values of σ^1 up to 10^{-5}. Find the maximum allowable rainfall rate r as a function of f_{GHz}. What maximum value of r is allowed at $f_{GHz} = 10$?

5.8-3 A radar transmits circular polarization in hopes of avoiding rain clutter for rain rates up to 25 mm/h. What is about the poorest cancellation ratio to be expected in practice?

5.8-4 Rain falls at 2 mm/h in the whole hemisphere around a radar operating at 8.5 GHz. If pattern beamwidths are $\theta_B^\circ = 2$ degrees and $\phi_B^\circ = 3$ degrees, and pulse duration is $T = 1.2$ μs, find σ_c for a clutter cell at range R. How large can R become before $\sigma_c = 100$ m^2?

5.8-5 In a pulsed monostatic radar $T = 6$ μs, $\theta_B = \phi_B = \pi/50$, and $f_{GHz} = 4.5$. At what range will uniform rain at a rate of 4 mm/h cause a rain cross section of 10 m^2?

5.8-6 A relatively heavy rain rate is 40 mm/h. A relatively heavy snow rate is 10 mm/h. Compare σ^1 for these two rates, and show that rain cross section is about 2.3 dB larger than for snow at any $f \leq 10$ GHz.

5.8-7 Snow falls at a rate of $r = 3$ mm/h of water content all around a radar. If snow becomes a problem when $\sigma_c = 5$ m^2 and the radar parameters are

$f_{\mathrm{GHz}} = 2$, $T = 0.6\ \mu s$, and $\theta_B = \phi_B = \pi/180$ rad, at what range will snow cross section become troublesome?

5.9-1 Show that (5.9-12) give the moments of a Rayleigh model as defined by

$$m_n = \int_0^\infty \sigma^n f(\sigma)\, d\sigma$$

5.9-2 Cross section in a radar fluctuates according to the Rayleigh model. Show that the median cross section σ_{med} is given by (5.9-10).

5.9-3 With what probability does cross section exceed its mean value for the Rayleigh model?

5.9-4 Show that (5.9-16) gives the moments of the Erlang density by using (5.9-15) in (5.9-7).

5.9-5 Show that (5.9-19) is true.

5.9-6 Work Problem 5.9-3 except for the Erlang model of (5.9-17).

5.9-7 Define σ_{90} as the 90th percentile cross section. It is the value of σ such that σ is less than or equal to σ_{90} with probability 0.90. Find σ_{90} for the Weibull distribution in terms of b and $\bar{\sigma}$.

6

RADAR SIGNALS AND NETWORKS

The study of radar signals and their passage through networks is critical to the understanding of modern radar. This chapter introduces the most fundamental aspects of these signals and networks. Our discussions will relate to only continuous-time signals and analog networks because they facilitate the learning about other radar principles introduced in later chapters. However, the reader should be aware that many network operations and signals are realized digitally in current-day systems. These digital operations require the study of discrete-time signals and digital signal-processing (DSP) methods; these subjects are developed in Chapter 15, where it is shown how to replace analog signals and networks by their digital equivalents.

We assume that the reader has already been versed in periodic waveforms (Fourier series), nonperiodic signals (impulses, Fourier transforms), and linear network theory (transfer functions, impulse responses). For those who may require some review of these subjects Appendixes A and B should prove helpful.

Our work begins with real waveforms and networks, but we will quickly extend our efforts to complex signals and networks because of the advantages they offer in radar. These results will lead naturally to the important topics of analytic signals and networks, matched filters, and ambiguity functions that are so convenient for use in radar.

6.1 REAL RADAR SIGNALS

We first consider the actual (real) waveforms used in the typical radar.

Waveform

Any radar waveform can be written in the form

$$s(t) = a(t) \cos[\omega_0 t + \theta(t) + \phi_0] \tag{6.1-1}$$

where $a(t)$ is a nonnegative function that represents any amplitude modulation that is present, $\theta(t)$ is the phase angle associated with any frequency modulation (FM), ω_0 is the nominal carrier angular frequency, and ϕ_0 is an arbitrary phase angle. The most common early radar waveform was a rectangular pulse of constant-frequency carrier. Here $a(t)$ is the rectangular function of (1.5-2) and $\theta(t) = 0$.

A convenient "quadrature" form for (6.1-1) is

$$s(t) = s_I(t) \cos(\omega_0 t + \phi_0) - s_Q(t) \sin(\omega_0 t + \phi_0) \qquad (6.1\text{-}2)$$

where

$$s_I(t) = a(t) \cos[\theta(t)] \qquad (6.1\text{-}3)$$

$$s_Q(t) = a(t) \sin[\theta(t)] \qquad (6.1\text{-}4)$$

The utility of (6.1-2) is that it brings the "slowly" varying modulation terms $a(t)$ and $\theta(t)$ together and separates them from the nominal carrier term. We will subsequently see that the grouping of $a(t)$ and $\theta(t)$ have special significance when considering the spectrum of $s(t)$.

The two representations of $s(t)$ can be brought together in another form. On using the expansion of cosine into its exponential form, we have

$$s(t) = a(t) \cos[\omega_0 t + \phi_0 + \theta(t)]$$

$$= \frac{a(t)}{2} [e^{j\omega_0 t + j\phi_0 + j\theta(t)} + e^{-j\omega_0 t - j\phi_0 - j\theta(t)}]$$

$$= \tfrac{1}{2} [g(t) e^{j\omega_0 t + j\phi_0} + g^*(t) e^{-j\omega_0 t - j\phi_0}] \qquad (6.1\text{-}5)$$

where * represents complex conjugation, and we define

$$g(t) = s_I(t) + j s_Q(t) = a(t) e^{j\theta(t)} \qquad (6.1\text{-}6)$$

The quantity $g(t)$ will be called the *complex envelope* of $s(t)$. We shortly return to (6.1-5) when the concept of a complex signal is introduced. For now, it suffices to note that even though the *terms* in (6.1-5) are complex, their *sum* is purely real. This form of $s(t)$ is useful, however, in the interpretation of the spectrum of $s(t)$.

Spectrum

Define $G(\omega)$ as the Fourier transform of $g(t)$:

$$g(t) \leftrightarrow G(\omega) \qquad (6.1\text{-}7)$$

If $S(\omega)$ denotes the Fourier transform (spectrum) of $s(t)$, then

$$S(\omega) = \tfrac{1}{2} [e^{j\phi_0} G(\omega - \omega_0) + e^{-j\phi_0} G^*(-\omega - \omega_0)] \qquad (6.1\text{-}8)$$

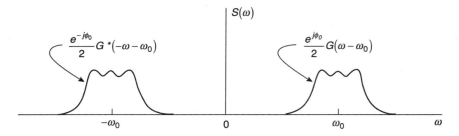

Figure 6.1-1 Spectrum of a real radar signal.

from (6.1-5) and the frequency shifting and conjugation properties of Fourier transforms (Appendix B). This result shows that $S(\omega)$ consists of two principal terms. Each term is a version of $G(\omega)$ shifted to either $+\omega_0$ or $-\omega_0$. Because $G(\omega)$ is comprised of terms [$a(t)$ and $\theta(t)$] that vary slowly relative to variations in $s(t)$ due to the carrier,[1] a plot of $S(\omega)$ shows that these two terms are nearly isolated from each other, as illustrated in Fig. 6.1-1. Thus, even though the spectrum of $S(\omega)$ exists at all $-\infty < \omega < \infty$, the dominant part at $+\omega_0$ contributes little to the dominant part at $-\omega_0$. Similarly the dominant term at $-\omega_0$ contributes little to the dominant part at $+\omega_0$. In fact little is lost by assuming that the dominant term at $+\omega_0$ is zero in the region $\omega < 0$ and the term at $-\omega_0$ is zero in the region $\omega > 0$.

Energy

By use of (6.1-1) the total energy in $s(t)$, denoted by E_s, is

$$
\begin{aligned}
E_s &= \int_{-\infty}^{\infty} s^2(t)\, dt = \tfrac{1}{2} \int_{-\infty}^{\infty} a^2(t)\{1 + \cos[2\omega_0 t + 2\theta(t) + 2\phi_0]\}\, dt \\
&\approx \tfrac{1}{2} \int_{-\infty}^{\infty} a^2(t)\, dt
\end{aligned}
\tag{6.1-9}
$$

since the term involving $2\omega_0$ is approximately zero. Note that the energy in $s(t)$ depends only on the real envelope $a(t)$ and not on any FM that may be present due to $\theta(t)$.

By using (6.1-8) in Parseval's theorem, it can be shown that

$$
\begin{aligned}
E_s &= \frac{1}{2\pi} \int_{-\infty}^{\infty} |S(\omega)|^2\, d\omega \approx \tfrac{1}{2} \frac{1}{2\pi} \int_{-\infty}^{\infty} |G(\omega)|^2\, d\omega \\
&= \tfrac{1}{2} E_g
\end{aligned}
\tag{6.1-10}
$$

where E_g is the energy in $g(t)$. In the expansion of $|S(\omega)|^2$ in (6.1-10) the cross terms were neglected as small, which is a good assumption in nearly all radar cases.

---

[1] This assumption is true in all radars where ω_0 is much greater than the bandwidth of $G(\omega)$. In very wideband, low-carrier radars, the assumption is no longer true.

Autocorrelation Functions

The autocorrelation function of $s(t)$ is defined by

$$R_{ss}(\tau) = \begin{cases} \lim\limits_{T \to \infty} \dfrac{1}{2T} \displaystyle\int_{-T}^{T} s(t)s(t+\tau)dt, & \text{power signals} \\[12pt] \displaystyle\int_{-\infty}^{\infty} s(t)\,s(t+\tau)dt, & \text{energy signals} \end{cases} \qquad (6.1\text{-}11)$$

The form for power signals is used where the waveform exists for $-\infty < t < \infty$, as would be modeled in the case of CW radar. Pulsed radar uses the definition applicable to energy signals. $R_{ss}(\tau)$ is a measure of the similarity of waveforms separated in time by an amount τ. $R_{ss}(0)$ gives average power in $s(t)$ for a power signal and total energy in $s(t)$ for an energy signal.

6.2   COMPLEX RADAR SIGNALS

We have seen above that the real signal $s(t)$ has a spectrum consisting of two terms, one clustered near ω_0 and one around $-\omega_0$, as indicated in (6.1-8). These terms originated directly from the Fourier transforms of the two terms in the last right-side form of (6.1-5). In this section we will work with a specially defined complex signal derived from the first part of (6.1-5) and find how the complex waveform is related to the real signal $s(t)$.

Waveform

Define a *complex signal*, denoted by $\psi_c(t)$, as

$$\psi_c(t) = g(t)e^{j\omega_0 t + j\phi_0} \qquad (6.2\text{-}1)$$

Its relation to $s(t)$ is

$$s(t) = \text{Re}[\psi_c(t)] \qquad (6.2\text{-}2)$$

This relationship implies that we can work with a single complex signal, when desirable to do so, with full knowledge that its real part is the real signal $s(t)$ which can be recovered at any time by just omitting the imaginary part of the complex signal. For example, we might find it more convenient to apply $\psi_c(t)$ to a real network in order to find the response to be the real part of the complex response, as compared to applying $s(t)$ and computing the response.

Two important characteristics of the complex signal are that its modulus, $|\psi_c(t)|$, and phase are equal, respectively, to the amplitude, $a(t)$, and phase of $s(t)$. Other important characteristics derive from the spectrum of $\psi_c(t)$.

Spectrum

Direct Fourier transformation of (6.2-1) gives the spectrum of $\psi_c(t)$, denoted by $\Psi_c(\omega)$:

$$\Psi_c(\omega) = G(\omega - \omega_0)e^{j\phi_0} \approx \begin{cases} 2S(\omega), & \omega > 0 \\ 0, & \omega < 0 \end{cases} = 2U(\omega)S(\omega) \qquad (6.2\text{-}3)$$

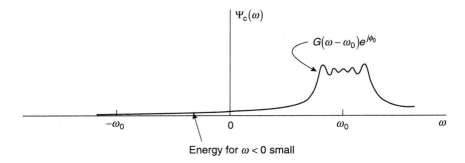

Figure 6.2-1 Spectrum of a complex signal.

The approximate equality in (6.2-3) recognizes that $G(\omega - \omega_0)$ has negligible contributions for $\omega < 0$ if ω_0 is large compared to the bandwidth of $G(\omega)$. This condition is illustrated in Fig. 6.2-1. In the special case where $G(\omega)$ is bandlimited and is exactly zero for $|\omega| > \omega_0$, the approximate equality in (6.2-3) becomes an equality. Of course no signal is perfectly bandlimited, but signals $g(t)$ do usually have negligible spectral content for $|\omega| > \omega_0$.

Even in cases where the complex signal's spectrum is not negligible for $\omega < 0$, there is a special type of complex signal that can be chosen such that its spectral content for $\omega < 0$ is exactly zero. It is called an *analytic signal*, and we consider it in the next section.

Autocorrelation Function

For complex signals the extension of (6.1-11) is

$$
R_{\psi_c \psi_c}(\tau) = \begin{cases} \lim\limits_{T \to \infty} \dfrac{1}{2T} \displaystyle\int_{-T}^{T} \psi_c^*(t)\psi_c(t + \tau)\, dt, & \text{power signals} \\[2ex] \displaystyle\int_{-\infty}^{\infty} \psi_c^*(t)\psi_c(t + \tau)\, dt, & \text{energy signals} \end{cases}
\tag{6.2-4}
$$

6.3 ANALYTIC RADAR SIGNALS

The analytic radar signal is a complex signal chosen such that its spectrum is forced to be zero for $\omega < 0$.

Spectrum and Waveform

Denote the spectrum of the analytic signal by $\Psi(\omega)$, and require it to equal the last right-side term of (6.2-3):

$$
\Psi(\omega) = 2U(\omega)S(\omega) = [1 + \text{sgn}(\omega)]S(\omega)
\tag{6.3-1}
$$

where sgn(ω) is the signum function defined by

$$\text{sgn}(\omega) = \begin{cases} +1, & \omega > 0 \\ 0, & \omega = 0 \\ -1, & \omega < 0 \end{cases} \tag{6.3-2}$$

Since waveforms and their Fourier transforms are unique, if (6.3-1) is the transform of the analytic signal, then the signal is given by the inverse transform. From the convolution property of transforms (Appendix B), we know that a spectrum given as the product of two spectra corresponds to a time function that is the convolution of the inverse transforms of the two spectra in the product. Thus, since

$$s(t) \leftrightarrow S(\omega) \tag{6.3-3}$$

$$\frac{1}{2}\delta(t) + \frac{j}{2\pi t} \leftrightarrow U(\omega) \tag{6.3-4}$$

we have

$$\psi(t) = 2 \int_{-\infty}^{\infty} s(\xi) \left[\frac{1}{2}\delta(t - \xi) + \frac{j}{2\pi(t - \xi)} \right] d\xi$$

$$= s(t) + j\frac{1}{\pi} \int_{-\infty}^{\infty} \frac{s(\xi)}{t - \xi} d\xi \tag{6.3-5}$$

where $\psi(t)$ denotes the analytic signal.

Hilbert Transforms

The integral in (6.3-5) is known to be the *Hilbert transform* of $s(t)$. Denote this transform by $\hat{s}(t)$. Then

$$\hat{s}(t) = \frac{1}{\pi} \int_{-\infty}^{\infty} \frac{s(\xi)}{t - \xi} d\xi \tag{6.3-6}$$

which gives

$$\psi(t) = s(t) + j\hat{s}(t) \tag{6.3-7}$$

Equation (6.3-7) demonstrates (1) that the real part of the analytic signal is the real radar signal $s(t)$, and (2) that the imaginary part of $\psi(t)$, which must be used to guarantee that the spectrum of $\psi(t)$ is zero for $\omega < 0$, is the Hilbert transform of the real signal.

A property of Hilbert transforms is the Hilbert transform of a Hilbert transform recovers the *negative* of the original signal. This means that we may define an inverse

Hilbert transform by

$$s(t) = -\frac{1}{\pi} \int_{-\infty}^{\infty} \frac{\hat{s}(\xi)}{t - \xi} d\xi \tag{6.3-8}$$

It is of interest to determine the Fourier transform (spectrum) of $\hat{s}(t)$, which we denote by $\hat{S}(\omega)$. On transformation of (6.3-7) we have

$$\Psi(\omega) = S(\omega) + j\hat{S}(\omega) \tag{6.3-9}$$

When (6.3-1) is used,

$$\hat{S}(\omega) = -j\{\Psi(\omega) - S(\omega)\} = -j\{[1 + \text{sgn}(\omega)]S(\omega) - S(\omega)\}$$

$$= -j\,\text{sgn}(\omega)S(\omega) = \begin{cases} -jS(\omega), & \omega > 0 \\ 0, & \omega = 0 \\ +jS(\omega), & \omega < 0 \end{cases} \tag{6.3-10}$$

This result shows that $\hat{S}(\omega)$ is equivalent to passing $s(t)$ through a constant-phase transfer function, as sketched in Fig. 6.3-1. Such a transfer function is not realizable exactly but can be approximated very accurately; it also has application to single sideband communication systems (Peebles, 1976).

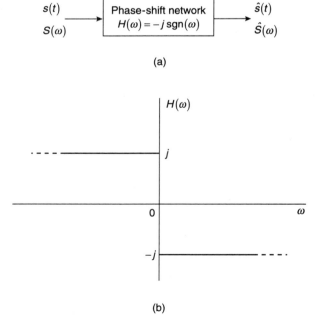

(a)

(b)

Figure 6.3-1   (a) A network that generates the Hilbert transform $\hat{s}(t)$ of the signal $s(t)$, and (b) the network's transform function.

Relationship to Complex Signal

If the complex modulation $g(t)$ of (6.1-6) is perfectly bandlimited such that $G(\omega) = 0$ for $|\omega| > \omega_0$, then the complex signal $\psi_c(t)$ of (6.2-1) is bandlimited such that $\Psi_c(\omega) = 0$ for $\omega < 0$, and the complex and analytic signals are the same. Even for $g(t)$ not strictly bandlimited, if its bandwidth is small compared to ω_0, we have seen that $\Psi_c(\omega) \approx 0$ for $\omega < 0$, a condition true in most radars. In all of our following work in this chapter, we will assume that $\psi_c(t)$ is bandlimited such that $\psi_c(t) = \psi(t)$. Thus, under either of these constraints, we may assume that the following relationships are true:

$$\psi(t) = \psi_c(t) \tag{6.3-11}$$

$$s(t) = \text{Re}[\psi_c(t)] = \text{Re}[\psi(t)] = a(t)\cos[\omega_0 t + \theta(t) + \phi_0] \tag{6.3-12}$$

$$\hat{s}(t) = \text{Im}[\psi_c(t)] = \text{Im}[\psi(t)] = a(t)\sin[\omega_0 t + \theta(t) + \phi_0] \tag{6.3-13}$$

$$|\psi_c(t)| = |\psi(t)| = a(t) \tag{6.3-14}$$

$$\arg[\psi_c(t)] = \arg[\psi(t)] = \tan^{-1}\left[\frac{\hat{s}(t)}{s(t)}\right]$$

$$= \omega_0 t + \theta(t) + \phi_0 \tag{6.3-15}$$

In summary, for most radars where the bandwidth of $s(t)$ is small compared to ω_0, the complex signal and the analytic signal can be taken as equal. Either can be used in analysis. This is especially convenient because it says that the analytic signal is found by creating the complex signal, which is a very straightforward problem of replacing the cosine with the exponential function. Even when the small-bandwidth constraint is not satisfied, the analytic signal can still be constructed by formally solving for $\hat{s}(t)$ and using (6.3-7). (For additional discussions and references, see Rihaczek, 1969; DiFranco and Rubin, 1968; and Burdic, 1968.)

Energy in Analytic Signal

The energy in $\psi(t)$, denoted by E_ψ, can most easily be found by using Parseval's theorem:

$$E_\psi = \int_{-\infty}^{\infty} |\psi(t)|^2 \, dt = \frac{1}{2\pi} \int_{-\infty}^{\infty} |\Psi(\omega)|^2 \, d\omega$$

$$= \frac{1}{2\pi} \int_0^{\infty} 4|S(\omega)|^2 \, d\omega = 2\frac{1}{2\pi} \int_{-\infty}^{\infty} |S(\omega)|^2 \, d\omega = 2E_s \tag{6.3-16}$$

In deriving (6.3-16), we used $|S(-\omega)| = |S(\omega)|$, which is true for any spectrum of a real signal $s(t)$. By noting that $E_\psi = E_{\psi_c}$ and using (6.1-10), we have

$$E_\psi = E_{\psi_c} \tag{6.3-17a}$$

$$E_{\psi_c} = E_g \tag{6.3-17b}$$

$$E_g = 2E_s \tag{6.3-17c}$$

as the relationships between the various energies.

Properties of Analytic Signals

Let $\psi_1(t)$ and $\psi_2(t)$ denote two analytic (or narrowband complex) signals. It can be shown (see Problems 6.3-4 and 6.3-5) that the following convolution and correlation integrals are zero:

$$\int_{-\infty}^{\infty} \psi_1^*(t)\psi_2(\tau - t)\,dt = 0 \qquad \text{(convolution)} \qquad (6.3\text{-}18)$$

$$\int_{-\infty}^{\infty} \psi_1(t)\psi_2(t + \tau)\,dt = 0 \qquad \text{(correlation)} \qquad (6.3\text{-}19)$$

Of course the conjugations of each of these two results are also zero. If $\psi_1(t) = \psi_2(t) = \psi(t)$, (6.3-18) and (6.3-19) remain zero.

On the other hand, the following convolution and correlation integrals are nonzero:

$$\int_{-\infty}^{\infty} \psi_1(t)\psi_2(\tau - t)\,dt = \frac{1}{2\pi}\int_{-\infty}^{\infty} \Psi_1(\omega)\Psi_2(\omega)e^{j\omega\tau}\,d\omega \neq 0 \quad \text{(convolution)} \quad (6.3\text{-}20)$$

$$\int_{-\infty}^{\infty} \psi_1^*(t)\Psi_2(t + \tau)\,dt = \frac{1}{2\pi}\int_{-\infty}^{\infty} \Psi_1^*(\omega)\Psi_2(\omega)e^{j\omega\tau}\,d\omega \neq 0 \quad \text{(correlation)} \quad (6.3\text{-}21)$$

Also the conjugations of these two results, as well as the special case where $\psi_1(t) = \psi_2(t) = \psi(t)$, are all nonzero. Here $\Psi_1(\omega)$ and $\Psi_2(\omega)$ represent the respective spectra of $\psi_1(t)$ and $\psi_2(t)$.

6.4   DURATION, FREQUENCY, AND BANDWIDTH OF SIGNALS

The origin assigned to a waveform is often taken as the center of its overall interval of nonzero values. The carrier frequency (even with FM present) is similarly taken as the "frequency" of the signal. Similarly a waveform's duration and bandwidth are typically taken as the intervals between -3-dB points in the time waveform and spectrum, respectively. However, in radar it is often desirable to define these four quantities differently, and their new definitions are the subjects of this section. Initially we establish some relationships from Parseval's theorem that we subsequently need.

Relationships from Parseval's Theorem

Let $v(t)$ be any (possibly complex) waveform having a Fourier transform $V(\omega)$. In the first of three applications from Parseval's theorem of (B.4-17), let $f_1(t) = f_2(t) = (-jt)^m v(t)$. Then we have

$$\int_{-\infty}^{\infty} t^{2m}|v(t)|^2\,dt = \frac{1}{2\pi}\int_{-\infty}^{\infty} \left|\frac{d^m V(\omega)}{d\omega^m}\right|^2 d\omega, \quad m = 0, 1, 2, \ldots \qquad (6.4\text{-}1)$$

In our second application assume that $f_1(t) = f_2(t) = d^n v(t)/dt^n$:

$$\int_{-\infty}^{\infty} \left| \frac{d^n v(t)}{dt^n} \right|^2 dt = \frac{1}{2\pi} \int_{-\infty}^{\infty} \omega^{2n} |V(\omega)|^2 \, d\omega, \quad n = 0, 1, 2, \ldots \tag{6.4-2}$$

In the final application let $f_1(t) = (-jt)^m v(t)$ and $f_2(t) = d^n v(t)/dt^n$, which give

$$(j)^m \int_{-\infty}^{\infty} t^m v^*(t) \frac{d^n v(t)}{dt^n} \, dt = \frac{(j)^n}{2\pi} \int_{-\infty}^{\infty} \omega^n V(\omega) \left[\frac{d^m V(\omega)}{d\omega^m} \right]^* d\omega \tag{6.4-3}$$

for $n = 0, 1, 2, \ldots$, and $m = 0, 1, 2, \ldots$. We use these relationships below.

Mean Time and RMS Duration

In the study of radar measurement accuracy it is often convenient to define the root-mean-squared (rms) duration of a waveform, which is a measure of the spread in a signal about a "mean time." We are variously interested in three waveforms $s(t)$, $g(t)$, and $\psi(t)$. The "mean time," $\bar{t}_\psi$, for $\psi(t)$ is defined by

$$\bar{t}_\psi = \frac{\int_{-\infty}^{\infty} t |\psi(t)|^2 \, dt}{\int_{-\infty}^{\infty} |\psi(t)|^2 \, dt} = \frac{-j \int_{-\infty}^{\infty} \Psi(\omega) [d\Psi^*(\omega)/d\omega] \, d\omega}{\int_{-\infty}^{\infty} |\Psi(\omega)|^2 \, d\omega} \tag{6.4-4}$$

which is the normalized first moment of $|\psi(t)|^2$. It is the center of gravity of $|\psi(t)|^2$. The second form of (6.4-4) derives from (6.4-1) with $m = 0$ and (6.4-3) with $m = 1$ and $n = 0$. Similar definitions can be found for $s(t)$ and $g(t)$ (see Problems 6.4-1 and 6.4-2).

Now we may *choose* the time origin for one of our waveforms, say $\psi(t)$, such that $\bar{t}_\psi = 0$. This will be done for convenience since the choice of origin will not affect the generality of other, more important, results. Once the origin is chosen so that $\bar{t}_\psi = 0$, it can be shown (Problem 6.4-15) that $\bar{t}_g = 0$ and $\bar{t}_s = 0$ also.

On having chosen the time origin so that waveforms have a first moment of zero, we now define root-mean-squared duration as the square root of the normalized second moment of the signals $\psi(t)$, $g(t)$, and $s(t)$, by $\tau_{\psi,\text{rms}}$, $\tau_{g,\text{rms}}$, and $\tau_{s,\text{rms}}$, respectively, according to

$$\tau_{\psi,\text{rms}}^2 = \frac{\int_{-\infty}^{\infty} t^2 |\psi(t)|^2 \, dt}{\int_{-\infty}^{\infty} |\psi(t)|^2 \, dt} \tag{6.4-5}$$

$$\tau_{g,\text{rms}}^2 = \frac{\int_{-\infty}^{\infty} t^2 |g(t)|^2 \, dt}{\int_{-\infty}^{\infty} |g(t)|^2 \, dt} \tag{6.4-6}$$

$$\tau_{s,\text{rms}}^2 = \frac{\int_{-\infty}^{\infty} t^2 s^2(t) \, dt}{\int_{-\infty}^{\infty} s^2(t) \, dt} \tag{6.4-7}$$

By using (6.4-1) through (6.4-3), other forms of these three results are possible (Problems 6.4-6 through 6.4-8). By using the fact that $|\psi(t)|^2 = |g(t)|^2 = a^2(t)$, it can be shown that all three rms durations are the same; that is, $\tau_{\psi,\text{rms}} = \tau_{g,\text{rms}}$ and $\tau_{g,\text{rms}} = \tau_{s,\text{rms}}$ (Problem 6.4-16).

Example 6.4-1   We show that the mean time of the rectangular pulse

$$
g(t) = \begin{cases} A, & \dfrac{-T}{2} < t < \dfrac{T}{2} \\ 0, & \text{elsewhere} \end{cases}
$$

is zero, and then we find its rms duration. Here

$$
\int_{-\infty}^{\infty} |g(t)|^2 \, dt = \int_{-T/2}^{T/2} A^2 \, dt = A^2 T
$$

$$
\int_{-\infty}^{\infty} t|g(t)|^2 \, dt = \int_{-T/2}^{T/2} A^2 t \, dt = 0
$$

$$
\int_{-\infty}^{\infty} t^2 |g(t)|^2 \, dt = \int_{-T/2}^{T/2} A^2 t^2 \, dt = \frac{A^2 T^3}{12}
$$

From Problem 6.4-2 we have $\bar{t}_g = 0$. From (6.4-6),

$$
\tau_{g,\text{rms}}^2 = \frac{T^2}{12}
$$

so $\tau_{g,\text{rms}} = T/2\sqrt{3} = T/3.4641 = 0.2887T$.

Mean Frequency and RMS Bandwidth

Analogous to mean time and rms duration in the time domain, we define mean frequency and rms bandwidth in the frequency domain, respectively, by normalized first and second *central* moments (the first moment is no longer zero as it was for the time domain). The mean frequencies of signals $\psi(t)$, $g(t)$, and $s(t)$ are defined by

$$
\bar{\omega}_\psi = \frac{\int_0^\infty \omega |\Psi(\omega)|^2 \, d\omega}{\int_0^\infty |\Psi(\omega)|^2 \, d\omega} \tag{6.4-8}
$$

$$
\bar{\omega}_g = \frac{\int_{-\infty}^\infty \omega |G(\omega)|^2 \, d\omega}{\int_{-\infty}^\infty |G(\omega)|^2 \, d\omega} \tag{6.4-9}
$$

$$
\bar{\omega}_s = \frac{\int_0^\infty \omega |S(\omega)|^2 \, d\omega}{\int_0^\infty |S(\omega)|^2 \, d\omega} \tag{6.4-10}
$$

Other forms are also possible (Problems 6.4-9 and 6.4-14). It can be shown (Problem 6.4-10) that these mean frequencies are related as

$$
\bar{\omega}_\psi = \bar{\omega}_s \tag{6.4-11}
$$

$$
\bar{\omega}_s = \omega_0 + \bar{\omega}_g \tag{6.4-12}
$$

For most waveforms $\bar{\omega}_g$ is small compared to the rms bandwidth $W_{g,\text{rms}}$ (defined below). In fact $\bar{\omega}_g = 0$ if $G(\omega)$ has even symmetry from (6.4-9). An even function $G(\omega)$ will result if $g(t)$ is even in t, which requires both $a(t)$ and $\theta(t)$ to be even. The corresponding FM must therefore be an odd function of t. The conclusion is that any complex envelope for which the amplitude modulation $a(t)$ is an even function of time and the angular *frequency* modulation $d\theta(t)/dt$ is an odd function of time will have $\bar{\omega}_g = 0$. Many radar signals satisfy these requirements. When true, the two mean frequencies $\bar{\omega}_\psi$ and $\bar{\omega}_s$ are equal to the carrier frequency ω_0.

We define rms bandwidths of the signals $\psi(t)$, $g(t)$, and $s(t)$ by

$$W^2_{\psi,\text{rms}} = \frac{\int_0^\infty (\omega - \bar{\omega}_\psi)^2 |\Psi(\omega)|^2 \, d\omega}{\int_0^\infty |\Psi(\omega)|^2 \, d\omega} = \overline{\omega_\psi^2} - (\bar{\omega}_\psi)^2 \tag{6.4-13}$$

$$W^2_{g,\text{rms}} = \frac{\int_{-\infty}^\infty (\omega - \bar{\omega}_g)^2 |G(\omega)|^2 \, d\omega}{\int_{-\infty}^\infty |G(\omega)|^2 \, d\omega} = \overline{\omega_g^2} - (\bar{\omega}_g)^2 \tag{6.4-14}$$

$$W^2_{s,\text{rms}} = \frac{\int_0^\infty (\omega - \bar{\omega}_s)^2 |S(\omega)|^2 \, d\omega}{\int_0^\infty |S(\omega)|^2 \, d\omega} = \overline{\omega_s^2} - (\bar{\omega}_s)^2 \tag{6.4-15}$$

where $\overline{\omega_\psi^2}$, $\overline{\omega_g^2}$, and $\overline{\omega_s^2}$ are normalized second moments of the appropriate functions; they are given by the right sides of (6.4-8) through (6.4-10) with the factor ω replaced by ω^2 in each equation. These second moments have several forms (Problem 6.4-11).

It can be shown (Problem 6.4-12) that all three rms bandwidths found from (6.4-13) through (6.4-15) are the same. Calculation of rms bandwidth may be facilitated by use of (6.4-14), at least for those signals for which $\overline{\omega_g} = 0$. In this case $W^2_{g,\text{rms}} = \overline{\omega_g^2}$ and any of the forms in Problem 6.4-11 are convenient.

Example 6.4-2   We find $W_{g,\text{rms}}$ for a gaussian pulse defined by

$$g(t) = A e^{-\sigma^2 t^2/2} \leftrightarrow \frac{A\sqrt{2\pi}}{\sigma} e^{-\omega^2/(2\sigma^2)} = G(\omega)$$

where $\sigma^2 = W_B^2/\ln(2)$, W_B is the -3-dB pulse bandwidth in rad/s. Since $g(t)$ is an even function of t, $\overline{\omega_g} = 0$. Two integrals are needed:

$$\int_{-\infty}^\infty |G(\omega)|^2 \, d\omega = \frac{A^2 2\pi}{\sigma^2} \int_{-\infty}^\infty e^{-\omega^2/\sigma^2} \, d\omega = \frac{A^2 (2\pi)^{3/2}}{\sqrt{2}\sigma}$$

$$\int_{-\infty}^\infty \omega^2 |G(\omega)|^2 \, d\omega = \frac{A^2 2\pi}{\sigma^2} \int_{-\infty}^\infty \omega^2 e^{-\omega^2/\sigma^2} \, d\omega = A^2 \pi^{3/2} \sigma$$

Finally from (6.4-14) we have $W_{g,\text{rms}} = \sigma/\sqrt{2} = W_B/\sqrt{2\ln(2)} = 0.849 W_B$.

6.5   TRANSMISSION OF SIGNALS THROUGH NETWORKS

In this section we consider the passage of signals through linear networks. We begin by assuming real waveforms exciting real networks, which is the real-world situation

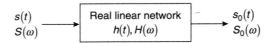

$$s(t) \qquad \boxed{\begin{array}{c}\text{Real linear network}\\ h(t), H(\omega)\end{array}} \qquad s_0(t)$$
$$S(\omega) \hspace{9.5cm} S_0(\omega)$$

Figure 6.5-1   A real linear network with real input and output time waveforms.

and the one reviewed in Appendix B. Next we consider the action of exciting the real network by an analytic signal. Finally we define an analytic *network* and determine its response to an analytic signal.

Real Signal through Real Network

A real signal $s(t)$, having a Fourier transform $S(\omega)$, is applied to a linear time-invariant network having a transfer function $H(\omega)$ and real impulse response $h(t)$. As shown in Fig. 6.5-1, the response of the network is denoted $s_0(t)$ which has a Fourier transform $S_0(\omega)$. From linear network theory the response is real and given by

$$s_0(t) = \int_{-\infty}^{\infty} s(\xi)h(t - \xi)d\xi = s(t) \star h(t) \tag{6.5-1}$$

where the star ($\star$) denotes the operation of convolution. The spectrum of the response is

$$S_0(\omega) = S(\omega)H(\omega), \qquad -\infty < \omega < \infty \tag{6.5-2}$$

Analytic Signal through Real Network

Now suppose that the input to the real network of Fig. 6.5-1 is an analytic signal

$$\psi(t) = s(t) + j\hat{s}(t) \tag{6.5-3}$$

as illustrated in Fig. 6.5-2. The complex input now evokes a complex output $\psi_0(t)$, but it is still given by the convolution operation:

$$\begin{aligned}
\psi_0(t) &= \psi(t) \star h(t) = [s(t) + j\hat{s}(t)] \star h(t) \\
&= [s(t) \star h(t)] + j[\hat{s}(t) \star h(t)] \\
&= s_0(t) + j[\hat{s}(t) \star h(t)]
\end{aligned} \tag{6.5-4}$$

Equation (6.5-4) shows that the real response is just the response to the real input, as is expected from the linearity property of linear networks.

$$\psi(t) = s(t) + j\hat{s}(t) \qquad \boxed{\begin{array}{c}\text{Real linear network}\\ h(t), H(\omega)\end{array}} \qquad \psi_0(t) = \psi(t) \star h(t)$$
$$\Psi(\omega) = 2U(\omega)S(\omega)$$

Figure 6.5-2   A real linear network with analytic input and analytic output signals.

Let us arbitrarily examine the Hilbert transform of the real response $s_0(t)$:

$$\hat{s}_0(t) = \frac{1}{\pi} \int_{-\infty}^{\infty} \frac{s_0(\alpha)}{t-\alpha} d\alpha = \frac{1}{\pi} \int_{-\infty}^{\infty} \frac{\int_{-\infty}^{\infty} h(\xi) s(\alpha - \xi) d\xi}{t - \alpha} d\alpha$$

$$= \int_{-\infty}^{\infty} h(\xi) \frac{1}{\pi} \int_{-\infty}^{\infty} \frac{s(x)}{(t-\xi) - x} dx d\xi \qquad (6.5\text{-}5)$$

$$= \int_{-\infty}^{\infty} h(\xi) \hat{s}(t - \xi) d\xi = \hat{s}(t) \star h(t)$$

The third integral in (6.5-5) was the result of a variable change $x = \alpha - \xi$, $dx = d\alpha$. When (6.5-5) is used in (6.5-4), we find that the response of a real network to an analytic input signal is an analytic output signal given by

$$\psi_0(t) = s_0(t) + j\hat{s}_0(t) = \psi(t) \star h(t) \qquad (6.5\text{-}6)$$

Because $\psi_0(t)$ is an analytic signal, its Fourier transform is readily found to be

$$\Psi_0(\omega) = \Psi(\omega) H(\omega) = 2U(\omega) S(\omega) H(\omega) = 2U(\omega) S_0(\omega) \qquad (6.5\text{-}7)$$

Analytic Signal through Analytic Network

Define an *analytic impulse response* for a real linear time-invariant network as that analytic signal for which the real part equals the network's real impulse response. When a real network is described by an analytic impulse response we refer to it as an *analytic network*.[2] Figure 6.5-3 illustrates the main points. The real impulse response $h(t)$ corresponds to the analytic impulse response

$$z(t) = h(t) + j\hat{h}(t) \qquad (6.5\text{-}8)$$

having the Fourier transform

$$Z(\omega) = 2U(\omega) H(\omega) \qquad (6.5\text{-}9)$$

where $H(\omega)$ is the Fourier transform of $h(t)$, as before.

The complex output $z_0(t)$, if taken as the convolution of $z(t)$ with the input $\psi(t)$, becomes

$$z_0(t) = \psi(t) \star z(t) = \psi(t) \star [h(t) + j\hat{h}(t)]$$

$$= [\psi(t) \star h(t)] + j[\psi(t) \star \hat{h}(t)] \qquad (6.5\text{-}10)$$

$$= \psi_0(t) + j[\psi(t) \star \hat{h}(t)]$$

---

[2] Note that there are no analytic networks in the real world. The concept is an extension of the analytic signal to the network's description to benefit from simplicities in mathematical forms and analyses.

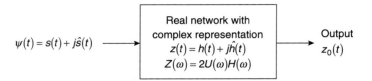

Figure 6.5-3 Some definitions for an analytic network.

where $\psi_0(t)$ is the response of the real network to $\psi(t)$ as defined by (6.5-6). To obtain the final output, we expand the last term in (6.5-10):

$$\psi(t) \star \hat{h}(t) = [s(t) + j\hat{s}(t)] \star \hat{h}(t)$$

$$= \int_{-\infty}^{\infty} [s(\alpha) + j\hat{s}(\alpha)]\hat{h}(t - \alpha)d\alpha$$

$$= \int_{-\infty}^{\infty} [s(\alpha) + j\hat{s}(\alpha)]\frac{1}{\pi}\int_{-\infty}^{\infty} \frac{h(\xi)}{t - \alpha - \xi}d\xi\,d\alpha$$

$$= \int_{-\infty}^{\infty} h(\xi)\left[\frac{1}{\pi}\int_{-\infty}^{\infty} \frac{s(\alpha) + j\hat{s}(\alpha)}{t - \xi - \alpha}d\alpha\right]d\xi \qquad (6.5\text{-}11)$$

$$= \int_{-\infty}^{\infty} h(\xi)[\hat{s}(t - \xi) + j\hat{\hat{s}}(t - \xi)]\,d\xi$$

$$= \int_{-\infty}^{\infty} h(\xi)[\hat{s}(t - \xi) - js(t - \xi)]d\xi$$

$$= \hat{s}_0(t) - js_0(t) = -j\psi_0(t)$$

Thus (6.5-10) becomes

$$z_0(t) = \psi_0(t) + j[-j\psi_0(t)] = 2\psi_0(t) \qquad (6.5\text{-}12)$$

This result indicates that the response of an analytic filter is exactly twice the response of the corresponding real filter when both are excited by the same analytic signal at their inputs.

The factor of 2 in (6.5-12) is readily explained. Remember that the input $\psi(t)$ has nonzero frequency components only for $\omega > 0$. These are processed by the real filter's transfer function $H(\omega)$ to produce the output, while the analytic filter has the transfer function $2H(\omega)$, which increases the response by a factor of two but does not change its form.

Interpretation and Summary of Responses

A powerful and practical interpretation can be made of the usefulness of the analytic filter. Suppose the real impulse response can be written either exactly or approximately in the form

$$h(t) = a_f(t)\cos[\omega_0 t + \theta_f(t) + \phi_f]$$
$$= \text{Re}[a_f(t)e^{j\omega_0 t + j\theta_f(t) + j\phi_f}] \qquad (6.5\text{-}13)$$
$$= \text{Re}[g_f(t)e^{j\omega_0 t + j\phi_f}]$$

where $a_f(t)$, $\theta_f(t)$, ϕ_f, and $g_f(t)$ are analogous to $a(t)$, $\theta(t)$, ϕ_0, and $g(t)$ defined for the real signal $s(t)$. The analytic filter's impulse response and transfer function become

$$z(t) = g_f(t)e^{j\omega_0 t + j\phi_f} \qquad (6.5\text{-}14)$$

$$Z(\omega) = G_f(\omega - \omega_0)e^{j\phi_f} \qquad (6.5\text{-}15)$$

where $G_f(\omega)$ is the Fourier transform of $g_f(t)$. Similarly the input analytic signal and its spectrum are

$$\psi(t) = g(t)e^{j\omega_0 t + j\phi_0} \qquad (6.5\text{-}16)$$

$$\Psi(\omega) = G(\omega - \omega_0)e^{j\phi_0} \qquad (6.5\text{-}17)$$

Both $g_f(t)$ and $g(t)$ are assumed band-limited here to $|\omega| < \omega_0$. Finally the spectrum of the filter's output is

$$Z_0(\omega) = Z(\omega)\Psi(\omega) = G_f(\omega - \omega_0)G(\omega - \omega_0)e^{j(\phi_f + \phi_0)} \qquad (6.5\text{-}18)$$

Our interpretation is to note that the right side of (6.5-18) is just the *baseband* function $G_f(\omega)G(\omega)\exp[j(\phi_f + \phi_0)]$ shifted to ω_0. This fact allows us to analyze the system as though it were a baseband system with no carrier. One need not consider the carrier at all, for it has no effect on the magnitude and phase of the response.

In summary, we have observed the following points in this section:

1. A real network (one having a real impulse response) responds with a real output when excited by a real input signal, as summarized in Fig. 6.5-4a.
2. A real network produces an analytic signal as its response to an analytic input signal, as summarized in Fig. 6.5-4b.
3. An analytic network (one having an impulse response which is the analytic signal with a real part that is the real impulse response of the network) produces an analytic output signal, in response to an input analytic signal, that is twice the response that would have occurred if the network were real (item 2) and not modeled as analytic. The response is summarized in Fig. 6.5-4c.

6.6 MATCHED FILTER FOR NONWHITE NOISE

Consider a radar pulse received from a target that is embedded in the ever-present receiver noise. Intuition tells us that the larger the signal amplitude relative to the noise amplitude, the better the receiver can recognize the presence of the target. This insight is perfectly correct as our later work will show. In this section we will show that a particular filter (receiving network) can be found that maximizes the ratio of peak

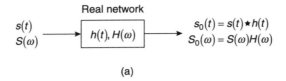

(a)

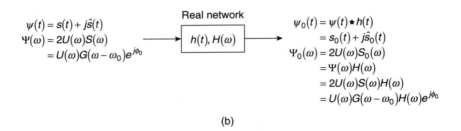

(b)

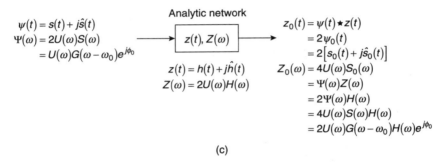

(c)

Figure 6.5-4 Summary of network responses for (a) a real filter with a real input, (b) a real filter with an analytic input, and (c) an analytic filter with an analytic input signal.

signal power to average noise power for an instant in time. For times around the optimum, the ratio decreases in relation to the shape of the filter's output signal. The optimum filter is called a *matched filter*. We first find the matched filter for arbitrary input noise and then specialize to the very important case of white noise.

We use the preceding methods based on an analytic filter with one exception. We use a factor of $\frac{1}{2}$ in calculating signal response so that the real part of the analytic output signal corresponds to the real filter's real response. Figure 6.6-1 sketches the important definitions. The real filter, having a transfer function $H(\omega)$, is to be analyzed using its analytic representation for which the transfer function is $Z(\omega)$. It is the optimum function $Z(\omega)$ that we seek. A real received signal $s_r(t)$ is represented by its analytic signal $\psi_r(t)$. It causes a real response from the real filter of $s_0(t)$ which is represented by its analytic signal $\psi_0(t)$, given by

$$\psi_0(t) = \frac{1}{2}\psi_r(t) \star z(t) = \frac{1}{2\pi}\int_{-\infty}^{\infty}\Psi_0(\omega)e^{j\omega t}d\omega$$

$$= \frac{1}{2\pi}\int_{-\infty}^{\infty}\frac{1}{2}\Psi_r(\omega)Z(\omega)e^{j\omega t}d\omega$$

(6.6-1)

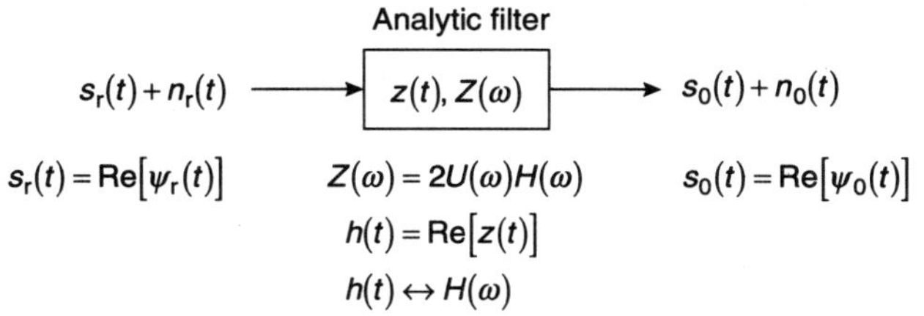

Analytic filter

$$s_r(t) + n_r(t) \longrightarrow \boxed{z(t), Z(\omega)} \longrightarrow s_0(t) + n_0(t)$$

$$s_r(t) = \mathrm{Re}[\psi_r(t)] \qquad Z(\omega) = 2U(\omega)H(\omega) \qquad s_0(t) = \mathrm{Re}[\psi_0(t)]$$
$$h(t) = \mathrm{Re}[z(t)]$$
$$h(t) \leftrightarrow H(\omega)$$

Figure 6.6-1   An analytic filter used in developing the matched filter.

for any time t. Here $\Psi_0(\omega)$ and $\Psi_r(\omega)$ are the Fourier transforms of $\psi_0(t)$ and $\psi_r(t)$, respectively. Next suppose that t_0 is defined as the time at which the optimum response occurs. At time t_0 the peak output signal power, denoted by $\hat{S}_0$, is

$$\hat{S}_0 = |\psi_0(t_0)|^2 = \frac{1}{4}\left|\frac{1}{2\pi}\int_{-\infty}^{\infty} \Psi_r(\omega)Z(\omega)e^{j\omega t_0}\,d\omega\right|^2 \tag{6.6-2}$$

Let the real input noise $n_r(t)$ be modeled as a sample function of a random process $N_r(t)$ having a two-sided $(-\infty < \omega < \infty)$ power spectral density $\mathscr{S}_{N_r N_r}(\omega)$. The power spectral density of the real output noise $n_0(t)$ is $\mathscr{S}_{N_r N_r}(\omega)|H(\omega)|^2$. The noise power formula gives the average output noise power, denoted by N_0. The formula is written in terms of the analytic filter's transfer function as follows:

$$N_0 = \frac{1}{2\pi}\int_{-\infty}^{\infty} \mathscr{S}_{N_r N_r}(\omega)|H(\omega)|^2\,d\omega = \frac{2}{2\pi}\int_0^{\infty} \mathscr{S}_{N_r N_r}(\omega)|H(\omega)|^2\,d\omega$$

$$= \frac{1}{4}\frac{2}{2\pi}\int_{-\infty}^{\infty} \mathscr{S}_{N_r N_r}(\omega)|2U(\omega)H(\omega)|^2\,d\omega \tag{6.6-3}$$

$$= \frac{1}{4}\frac{2}{2\pi}\int_{-\infty}^{\infty} \mathscr{S}_{N_r N_r}(\omega)|Z(\omega)|^2\,d\omega$$

The ratio $\hat{S}_0/N_0$ is the peak signal power to average noise power ratio that we seek to maximize

$$\frac{\hat{S}_0}{N_0} = \frac{\left|\dfrac{1}{2\pi}\displaystyle\int_{-\infty}^{\infty} \Psi_r(\omega)Z(\omega)e^{j\omega t_0}\,d\omega\right|^2}{2\dfrac{1}{2\pi}\displaystyle\int_{-\infty}^{\infty} \mathscr{S}_{N_r N_r}(\omega)|Z(\omega)|^2\,d\omega} \tag{6.6-4}$$

It would seem a difficult task to find the one function of all possible functions $Z(\omega)$ that makes (6.6-4) maximum. Fortunately it can be done almost by inspection with the help of Schwarz's inequality. The form of this inequality that we need states that

$$\left|\int_{-\infty}^{\infty} A(\omega)B(\omega)\,d\omega\right|^2 \leq \int_{-\infty}^{\infty} |A(\omega)|^2\,d\omega \int_{-\infty}^{\infty} |B(\omega)|^2\,d\omega \tag{6.6-5}$$

where $A(\omega)$ and $B(\omega)$ are, in general, complex functions of a real variable ω. The equality in (6.6-5) occurs if, and only if,

$$A(\omega) = CB^*(\omega) \tag{6.6-6}$$

where C is a nonzero real, but otherwise arbitrary, constant.

We apply (6.6-5) to the current problem by identifying

$$A(\omega) = \frac{1}{\sqrt{2\pi}} \sqrt{\mathscr{S}_{N_r N_r}(\omega)}\, Z(\omega) \tag{6.6-7}$$

$$B(\omega) = \frac{1}{\sqrt{2\pi}} \frac{\Psi_r(\omega) e^{j\omega t_0}}{\sqrt{\mathscr{S}_{N_r N_r}(\omega)}} \tag{6.6-8}$$

which give

$$\left| \frac{1}{2\pi} \int_{-\infty}^{\infty} \Psi_r(\omega) Z(\omega) e^{j\omega t_0}\, d\omega \right|^2 \leq \frac{1}{2\pi} \int_{-\infty}^{\infty} \mathscr{S}_{N_r N_r}(\omega) |Z(\omega)|^2\, d\omega \, \frac{1}{2\pi} \int_{-\infty}^{\infty} \frac{|\Psi_r(\omega)|^2}{\mathscr{S}_{N_r N_r}(\omega)}\, d\omega \tag{6.6-9}$$

When (6.6-9) is substituted into (6.6-4) and the equality is recognized as giving the largest (optimum) value of $\hat{S}_0/N_0$, denoted by $(\hat{S}_0/N_0)_{\max}$, we have

$$\left(\frac{\hat{S}_0}{N_0} \right) \leq \frac{1}{4\pi} \int_{-\infty}^{\infty} \frac{|\Psi_r(\omega)|^2}{\mathscr{S}_{N_r N_r}(\omega)}\, d\omega = \left(\frac{\hat{S}_0}{N_0} \right)_{\max} \tag{6.6-10}$$

The maximum occurs when (6.6-7) and (6.6-8) are used in (6.6-6). The value of $Z(\omega)$ obtained is the optimum analytic filter's transfer function, which we denote by $Z_{\mathrm{opt}}(\omega)$:

$$Z_{\mathrm{opt}}(\omega) = \frac{C\Psi_r{}^*(\omega) e^{-j\omega t_0}}{\mathscr{S}_{N_r N_r}(\omega)} \tag{6.6-11}$$

The transfer function of the corresponding real optimum filter can be shown (Problem 6.6-3) to be

$$H_{\mathrm{opt}}(\omega) = \frac{CS_r{}^*(\omega) e^{-j\omega t_0}}{\mathscr{S}_{N_r N_r}(\omega)} \tag{6.6-12}$$

6.7 MATCHED FILTER FOR WHITE NOISE

In this section we specialize the preceding matched filter to the case of white noise where $n_r(t)$ of Section 6.6 has a constant power spectral density defined by

$$\mathscr{S}_{N_r N_r}(\omega) = \frac{\mathscr{N}_0}{2}, \qquad -\infty < \omega < \infty \tag{6.7-1}$$

The optimum filter expressions (6.6-11) and (6.6-12) become

$$Z_{\text{opt}}(\omega) = \frac{2C}{\mathcal{N}_0} \Psi_r^*(\omega) e^{-j\omega t_0} \qquad (6.7\text{-}2)$$

$$H_{\text{opt}}(\omega) = \frac{2C}{\mathcal{N}_0} S_r^*(\omega) e^{-j\omega t_0} \qquad (6.7\text{-}3)$$

The maximum, or optimum, value of $\hat{S}_0/N_0$ derives from (6.6-10):

$$\left(\frac{\hat{S}_0}{N_0}\right)_{\text{max}} = \frac{1}{\mathcal{N}_0 2\pi} \int_{-\infty}^{\infty} |\Psi_r(\omega)|^2 \, d\omega = \frac{E_{\psi_r}}{\mathcal{N}_0} = \frac{2E_r}{\mathcal{N}_0} \qquad (6.7\text{-}4)$$

where E_{ψ_r} and E_r are the energies in $\psi_r(t)$ and $s_r(t)$, respectively.

Let us ignore Doppler (target motion) for the moment and consider a target having a range delay of τ_R. If α is a constant associated with the radar equation, we can write

$$s_r(t) = \alpha s(t - \tau_R) \leftrightarrow S_r(\omega) = \alpha S(\omega) e^{-j\omega \tau_R} \qquad (6.7\text{-}5)$$

$$\psi_r(t) = \alpha \psi(t - \tau_R) \leftrightarrow \Psi_r(\omega) = \alpha \Psi(\omega) e^{-j\omega \tau_R} \qquad (6.7\text{-}6)$$

where $s(t)$ and $\psi(t)$ are the transmitted real and analytic signals, respectively. On using these two expressions with (6.7-2) and (6.7-3), we obtain more specific filter expressions and their impulse responses:

$$z_{\text{opt}}(t) = \frac{2C\alpha}{\mathcal{N}_0} \psi^*(t_0 - \tau_R - t) \leftrightarrow Z_{\text{opt}}(\omega) = \frac{2C\alpha}{\mathcal{N}_0} \Psi^*(\omega) e^{-j\omega(t_0 - \tau_R)} \qquad (6.7\text{-}7)$$

$$h_{\text{opt}}(t) = \frac{2C\alpha}{\mathcal{N}_0} s(t_0 - \tau_R - t) \leftrightarrow H_{\text{opt}}(\omega) = \frac{2C\alpha}{\mathcal{N}_0} S^*(\omega) e^{-j\omega(t_0 - \tau_R)} \qquad (6.7\text{-}8)$$

These results allow us to find the filter's output signal.

Output Signal

The analytic output signal is

$$\psi_0(t) = \frac{1}{2} \psi_r(t) \star z_{\text{opt}}(t) = \frac{1}{2} \int_{-\infty}^{\infty} \alpha \psi(x - \tau_R) \frac{2C\alpha}{\mathcal{N}_0} \psi^*(t_0 - \tau_R - t + x) dx$$

$$= \frac{C\alpha^2}{\mathcal{N}_0} \int_{-\infty}^{\infty} \psi^*(\xi) \psi(\xi + t - t_0) d\xi \qquad (6.7\text{-}9)$$

$$= \frac{C\alpha^2}{\mathcal{N}_0} R_{\psi\psi}(t - t_0)$$

Equation (6.7-9) indicates that the optimum filter's response is proportional to the autocorrelation function of the transmitted waveform. Since autocorrelation functions reach a maximum when the argument is zero, (6.7-9) implies that $\psi_0(t)$ is maximum at $t = t_0$, the time at which $\hat{S}_0/N_0$ is largest. Such a result is not surprising, since N_0 does not depend on time so $\hat{S}_0/N_0$ must become maximum when the output signal is maximum.

Discussion of Matched Filter

The matched filter derives its name from the fact that its transfer function (6.7-8) is proportional to the transmitted signal's spectrum (conjugated). If the signal changes, the filter must change. It is matched to the signal being used. Indeed the original name applied to the white noise case but was later extended to the nonwhite noise case where it must be interpreted as being matched to both signal and noise.

Because the argument of $s(t_0 - \tau_R - t)$ in (6.7-8) has the negative of t, the impulse response is often said to be proportional to the transmitted signal "running backward." Figure 6.7-1 illustrates the fact that $h_{opt}(t)$ involves shifting $s(t)$ and then folding the shifted signal about the origin. Note that if $h_{opt}(t)$ is to have any chance of being realizable, it has to be causal, which requires that

$$(t_0 - \tau_R) - T \geq 0 \quad \text{or} \quad t_0 \geq \tau_R + T \tag{6.7-10}$$

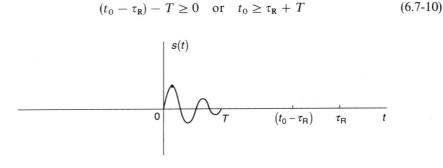

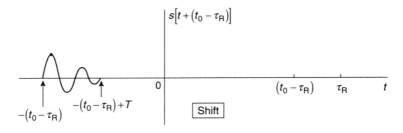

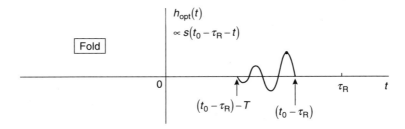

Figure 6.7-1   Waveforms applicable to the optimum matched filter's impulse response.

Since t_0 is the time at which we asked the matched filter's output to be maximum, (6.7-10) requires this time be later than the target's occurrence time by an amount equal to the time when $s(t)$ is nonzero for $t > 0$. The condition essentially means that we must wait until the entire received signal has entered the filter before we can ask that it produce a maximum response.

From (6.7-4) we observe that the maximum signal-to-noise ratio does *not* depend on the modulation type or form of transmitted signal. It depends only on the total received energy and noise level. As a consequence signal bandwidth, form, and total energy can all be independently selected when a matched filter is used. Indeed nearly every modern radar uses a matched filter or a close approximation to one in its receiver.

Finally we observe that the matched filter removes any phase variations in the spectrum of the input signal. To show this point, we find the spectrum $\Psi_0(\omega)$ of $\psi_0(t)$:

$$\Psi_0(\omega) = \tfrac{1}{2}\Psi_r(\omega)Z_{opt}(\omega) = \frac{C}{\mathcal{N}_0}|\Psi_r(\omega)|^2 e^{-j\omega t_0} \qquad (6.7\text{-}11)$$

from (6.7-2). Since the exponential factor only represents delay (a linear phase with ω), (6.7-11) shows that the output signal's spectrum depends only on the modulus of the spectrum of $\psi_r(t)$ and not on its phase.

6.8 AMBIGUITY FUNCTION

The preceding section examined the matched filter output of a white noise receiver when the input signal was a delayed replica of the transmitted signal derived from a nonmoving target. The result was (6.7-9). We are now interested in finding the white noise matched filter's response when a target moves. The response must obviously depend on the Doppler angular frequency shift ω_d caused by target motion. When $\omega_d \neq 0$, the input signal is no longer matched by the filter.

The received signal with Doppler shift ω_d for a target of nominal delay τ_R was previously defined in Chapter 1 in (1.6-21):

$$s_r(t) = \alpha a(t - \tau_R)\cos[(\omega_0 + \omega_d)(t - \tau_R) + \theta(t - \tau_R) + \phi_0] \qquad (6.8\text{-}1)$$

The corresponding analytic signal is

$$\psi_r(t) = \alpha g(t - \tau_R)e^{j(\omega_0 + \omega_d)(t - \tau_R) + j\phi_0}$$

$$= \alpha\psi(t - \tau_R)e^{j\omega_d(t - \tau_R)} \qquad (6.8\text{-}2)$$

where

$$g(t - \tau_R) = a(t - \tau_R)e^{j\theta(t - \tau_R)} \qquad (6.8\text{-}3)$$

$$\psi(t - \tau_R) = g(t - \tau_R)e^{j\omega_0(t - \tau_R) + j\phi_0} \qquad (6.8\text{-}4)$$

When $\psi_r(t)$ is applied to the optimum filter matched to $\psi(t)$, the output signal is given by the first right-side form of (6.7-9). From use of (6.8-2) and (6.7-7), we have

$$\psi_0(t) = \frac{1}{2}\int_{-\infty}^{\infty} \alpha\psi(x - \tau_R)e^{j\omega_d(x - \tau_R)}\frac{2C\alpha}{\mathcal{N}_0}\psi^*(t_0 - \tau_R - t + x)dx \qquad (6.8\text{-}5)$$

On changing variable x to $\xi = t_0 - \tau_R - t + x$, we derive

$$\psi_0(t) = \frac{C\alpha^2}{\mathcal{N}_0}e^{j\omega_d(t - t_0)}\int_{-\infty}^{\infty} \psi^*(\xi)\psi(\xi + t - t_0)e^{j\omega_d\xi}d\xi \qquad (6.8\text{-}6)$$

An alternative form, obtained by substitution for $\psi(t)$ in terms of $g(t)$, is

$$\psi_0(t) = \frac{C\alpha^2}{\mathcal{N}_0}e^{j(\omega_0 + \omega_d)(t - t_0)}\int_{-\infty}^{\infty} g^*(\xi)g(\xi + t - t_0)e^{j\omega_d\xi}d\xi \qquad (6.8\text{-}7)$$

We see that $\psi_0(t)$ depends on Doppler angular frequency as well as time. However, the time dependence is only relative to t_0, the time the filter has been designed to produce a maximum when the input has no Doppler. We will now let

$$\tau = t - t_0 \qquad (6.8\text{-}8)$$

and define a special function $\chi(\tau, \omega_d)$ by

$$\chi(\tau, \omega_d) = \int_{-\infty}^{\infty} g^*(\xi)g(\xi + \tau)e^{j\omega_d\xi}d\xi \qquad (6.8\text{-}9)$$

which also has the equivalent form

$$\chi(\tau, \omega_d) = e^{-j\omega_0\tau}\int_{-\infty}^{\infty} \psi^*(\xi)\psi(\xi + \tau)e^{j\omega_d\xi}d\xi \qquad (6.8\text{-}10)$$

The output of the filter can now be written as

$$\psi_0(\tau + t_0) = \frac{C\alpha^2}{\mathcal{N}_0}e^{j(\omega_0 + \omega_d)\tau}\chi(\tau, \omega_d) \qquad (6.8\text{-}11)$$

The function $\chi(\tau, \omega_d)$ is most important. Except for a scale constant and a "carrier" phase factor, as shown in (6.8-11), it completely determines the response to a moving target by the system (filter) when matched to a stationary target. If the system were redeveloped to be matched to a target of particular speed (value of ω_d), then $\chi(\tau, \omega_d)$ can be interpreted as describing the response over time τ of a target with Doppler ω_d *relative* to that of the matched target. We note that when $\omega_d = 0$, (6.8-6) reduces to the autocorrelation function of (6.7-9), which has lead to some sources defining the integral of (6.8-6) as a *combined* (with ω_d) *correlation function* (Berkowitz, ed., 1965, p. 205) or a similar name (Woodward, 1960, p. 120; Burdic, 1968, p. 210). Others call $\chi(\tau, \omega_d)$ an ambiguity function (Vakman, 1968, p. 36; Eaves and Reedy, eds., 1987, p. 428; Barton, 1988, p. 209; Edde, 1993, p. 428). Some authors give $\chi(\tau, \omega_d)$ no name (Cook and Bernfield, 1967; Bird, 1974) while a few call it the matched filter response (Rihaczek, 1969, p. 112; Skolnik, 1980, p. 412; Levanon, 1988, p. 117). We elect to call $\chi(\tau, \omega_d)$ the *matched filter response*.

The functions $|\chi(\tau, \omega_d)|$ and $|\chi(\tau, \omega_d)|^2$ are also important and have no generally accepted names. To prevent confusion, we call $|\chi(\tau, \omega_d)|$ the *uncertainty function*, and name $|\chi(\tau, \omega_d)|^2$ the *ambiguity function*.

Several other useful forms for $\chi(\tau, \omega_d)$ exist. These are given in Problems 6.8-1 and 6.8-2.

Properties of the Matched Filter Response

Some properties exhibited by $\chi(\tau, \omega_d)$ are listed below. For proofs, see Problems 6.8-3 through 6.8-7.

1. $\chi(0, 0) = \displaystyle\int_{-\infty}^{\infty} |\psi(t)|^2 \, dt = \int_{-\infty}^{\infty} |g(t)|^2 \, dt$

$$= E_\psi = E_g = 2E_s \tag{6.8-12}$$

2. $\chi(-\tau, -\omega_d) = e^{-j\omega_d\tau} \chi^*(\tau, \omega_d)$ \hfill (6.8-13)

3. $\chi(-\tau, \omega_d) = e^{j\omega_d\tau} \chi^*(\tau, -\omega_d)$ \hfill (6.8-14)

4. $\chi(\tau, 0) = \displaystyle\int_{-\infty}^{\infty} g^*(\xi)g(\xi + \tau)d\xi = R_{gg}(\tau)$ \hfill (6.8-15)

5. $\chi(0, \omega_d) = \displaystyle\int_{-\infty}^{\infty} |g(\xi)|^2 \, e^{j\omega_d\xi} \, d\xi$

$$\tag{6.8-16}$$

$$= \frac{1}{2\pi} \int_{-\infty}^{\infty} G^*(\omega)G(\omega - \omega_d)d\omega = \frac{1}{2\pi} R_{GG}(-\omega_d)$$

Property 1 relates the value of $\chi(\tau, \omega_d)$ at the origin to the energies of the various signals. Properties 2 and 3 relate to symmetry and show symmetry in first and third quadrants as well as second and fourth quadrants of the τ, ω_d plane. Properties 4 and 5 are "cuts" along the τ and ω_d axes, respectively. These cuts define the response with time when Doppler is zero (property 4), and the behavior of the response with Doppler for a time fixed at the time of the maximum (property 5). The cuts are equal to the autocorrelation of the complex envelope (property 4) and the correlation of the spectrum of the complex envelope (property 5). Alternatively, $\chi(0, \omega_d)$ is the Fourier transform of the squared modulus of the complex envelope.

Properties of the Ambiguity Function

Some properties of $|\chi(\tau, \omega_d)|^2$ are listed below. For proofs, see Problems 6.8-8 through 6.8-11.

1. $|\chi(\tau, \omega_d)|^2 \leq |\chi(0,0)|^2 = E_\psi^2 = E_g^2 = 4E_s^2$ \hfill (6.8-17)

2. $\displaystyle\int_{-\infty}^{\infty}\int_{-\infty}^{\infty} |\chi(\tau, \omega_d)|^2 \, d\tau d\omega_d = 2\pi E_\psi^2 = 2\pi E_g^2 = 8\pi E_s^2$ \hfill (6.8-18)

$$3. \ |\chi(-\tau, -\omega_d)|^2 = |\chi(\tau, \omega_d)|^2 \tag{6.8-19}$$

$$4. \ |\chi(-\tau, \omega_d)|^2 = |\chi(\tau, -\omega_d)|^2 \tag{6.8-20}$$

Property 1 is a boundary value showing that the maximum occurs at the origin and is determined only by the total energy in the signal. Property 2 is a volume constraint. It indicates $|\chi(\tau, \omega_d)|^2$ has a constant volume determined by total energy and is not dependent on the modulation used to generate the signal. Properties 3 and 4 are symmetry relationships; symmetry exists in the first and third, and the second and fourth, quadrants of the τ, ω_d plane.

6.9   EXAMPLES OF UNCERTAINTY FUNCTIONS

We give two examples of uncertainty functions, both for constant-frequency waveforms. The case where FM is present is so important it is reserved for discussion in its own chapter (to follow).

Rectangular Pulse

For a rectangular pulse of duration T centered on the time origin,

$$g(t) = a(t) = A \ \text{rect}\left(\frac{t}{T}\right) \tag{6.9-1}$$

$$G(\omega) = AT \ \text{Sa}\left(\frac{\omega T}{2}\right) \tag{6.9-2}$$

When (6.9-1) is substituted into (6.8-9), the matched filter response becomes

$$\chi(\tau, \omega_d) = \int_{-\infty}^{\infty} A \ \text{rect}\left(\frac{\xi}{T}\right) A \ \text{rect}\left(\frac{\xi + \tau}{T}\right) e^{j\omega_d \xi} \, d\xi$$

$$= A^2 \int_{-T/2}^{T/2} \text{rect}\left(\frac{\xi + \tau}{T}\right) e^{j\omega_d \xi} \, d\xi \tag{6.9-3}$$

This integral is zero for $\tau < -T$ and $\tau > T$. For $-T \le \tau \le T$ two cases are necessary, one where $-T \le \tau \le 0$ and the other where $0 \le \tau \le T$. On reducing the algebra, we find that

$$\chi(\tau, \omega_d) = \begin{cases} A^2 T \left(1 - \dfrac{|\tau|}{T}\right) e^{-j\omega_d \tau/2} \ \text{Sa}\left[\omega_d \dfrac{T}{2}\left(1 - \dfrac{|\tau|}{T}\right)\right], & -T \le \tau \le T \\ 0, & \text{elsewhere} \end{cases} \tag{6.9-4}$$

Several approaches are available for interpreting the behavior of the uncertainty function $|\chi(\tau, \omega_d)|$. Figure 6.9-1 illustrate one where the cuts along the τ and ω_d axes

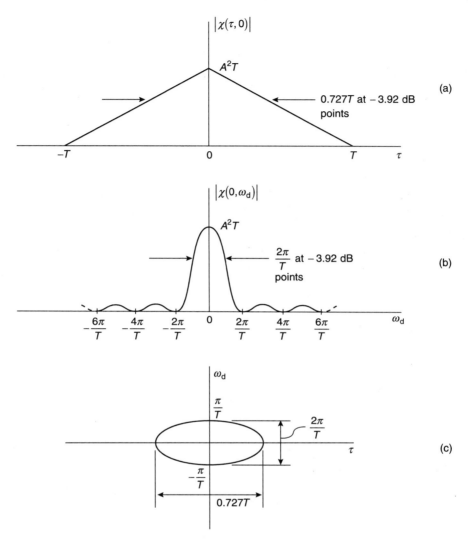

Figure 6.9-1 Plots of $|\chi(\tau, \omega_d)|$ for a constant-frequency rectangular pulse when (*a*) a cut is taken along τ for $\omega_d = 0$, (*b*) a cut is along ω_d when $\tau = 0$, and (*c*) a locus of points is defined by slicing $|\chi(\tau, \omega_d)|$ by a plane parallel to the τ, ω_d plane at a level -3.92 dB below the value at the origin.

are sketched in panels *a* and *b*. Another common method is to show a contour of $|\chi(\tau, \omega_d)|$ related to the function in both dimensions. The contour is typically produced by "slicing" $|\chi(\tau, \omega_d)|$ by a plane parallel to the τ, ω_d plane at a convenient level below the maximum at the origin. Such a slice at the -3.92-dB level is shown in panel *c*. A third approach is to show the full surface of $|\chi(\tau, \omega_d)|$ as illustrated in Fig. 6.9-2. When all approaches are taken together, a relatively good understanding of the uncertainty function is achieved.

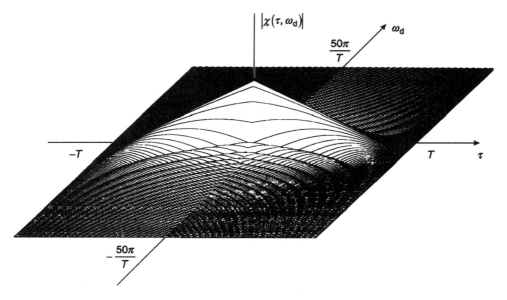

Figure 6.9-2   The uncertainty function of (6.9-4) for a constant-frequency rectangular pulse of duration T. (Adapted from Rihaczek, 1969, with permission.)

Gaussian Pulse

The gaussian pulse is defined by

$$g(t) = a(t) = Ae^{-t^2/(2\sigma^2)} \tag{6.9-5}$$

$$G(\omega) = A\sqrt{2\pi\sigma^2}\,e^{-\omega^2\sigma^2/2} \tag{6.9-6}$$

Where $A > 0$ is a constant and σ^2 is a constant related to the pulse's duration or bandwidth. If T_3 is the 3-dB pulse duration of $g(t)$, then

$$\sigma = \frac{T_3}{1.662} \tag{6.9-7}$$

If W_3 is the 3-dB bandwidth (rad/s), then σ is related by

$$\sigma = \frac{1.662}{2W_3} \tag{6.9-8}$$

By using (6.9-5) in (6.8-9), we obtain

$$\chi(\tau, \omega_d) = A^2\sqrt{\pi\sigma^2}\exp\left[-\frac{\tau^2}{4\sigma^2} - j\frac{\omega_d\tau}{2}\frac{\omega_d^2\sigma^2}{4}\right] \tag{6.9-9}$$

Figure 6.9-3 illustrates the behavior of $|\chi(\tau, \omega_d)|$ as defined by (6.9-9).

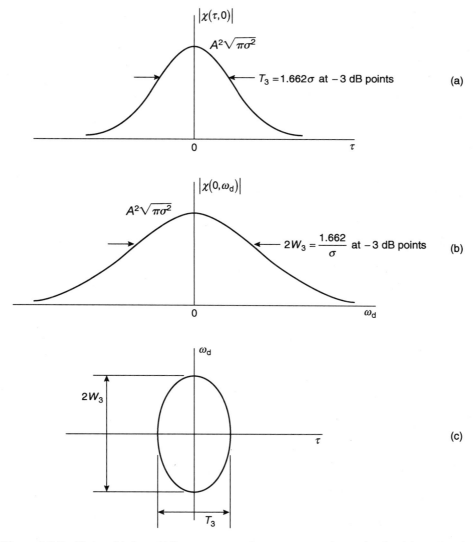

Figure 6.9-3 Plots of $|\chi(\tau, \omega_d)|$ for a constant-frequency gaussian pulse for (*a*) a cut taken along τ for $\omega_d = 0$, (*b*) a cut taken along ω_d for $\tau = 0$, and (*c*) a locus of points defined by slicing $|\chi(\tau, \omega_d)|$ by a plane parallel to the τ, ω_d plane at a level 3 dB below the value at the origin.

PROBLEMS

6.1-1 A waveform is defined by a single pulse given by

$$s(t) = A \operatorname{rect}\left(\frac{t}{T}\right) \cos\left[\omega_0 t + \frac{\Delta\omega}{\omega_m} \cos(\omega_m t)\right]$$

where A, T, ω_0, and $\omega_m \ll \omega_0$, and $\Delta\omega$ are all positive constants. Find the Fourier transform of $g(t)$ given by (6.1-6). [*Hint*: Use the identity

$$\exp[j\beta\cos(x)] = J_0(\beta) + 2 \sum_{k=1}^{\infty} j^k J_k(\beta)\cos(kx)$$

in your solution.]

6.1-2 Find and sketch the spectrum of the signal

$$s(t) = A \operatorname{rect}\left(\frac{t}{T}\right)\cos(\omega_0 t)$$

where A, T, and ω_0 are positive constants.

6.1-3 Find $g(t)$ and its spectrum for the waveform of Problem 6.1-2. Sketch the spectrum. Above what value of ω is $|G(\omega)| \leq 0.01|G(0)|$?

6.1-4 Find the time autocorrelation function of the waveform of Problem 6.1-2. Assume that ω_0 is very large relative to $2\pi/T$.

6.1-5 Find the total energy in the waveform of Problem 6.1-1. (*Hint*: Assume that ω_0 is large so that cosines involving ω_0 cycle rapidly over the time interval $-T/2 < t < T/2$.)

6.1-6 Find the total energy in the pulse of Problem 6.1-2. Assume that $\omega_0 \gg 2\pi/T$.

6.1-7 A real signal is

$$s(t) = Au(t)e^{-t/T}\cos(\omega_0 t)$$

(a) Find the spectrum $S(\omega)$ of $s(t)$ if A, T, and ω_0 are positive constants. (b) Find the amplitude at $\omega = 0$ of the term in $S(\omega)$ that dominates for $\omega > 0$ relative to the value of the term when $\omega = \omega_0$. How large must $\omega_0 T$ be if the relative amplitude must be less than $1/1000$? (c) What is $g(t)$ for the complex signal representing $s(t)$?

6.2-1 Use (6.2-1), in (6.2-4), and write expressions for the autocorrelation function in terms of $g(t)$.

6.3-1 A signal and its Fourier transform are

$$s(t) = A\cos(\omega_0 t) \leftrightarrow S(\omega) = A\pi[\delta(\omega - \omega_0) + \delta(\omega + \omega_0)]$$

Use the fact that $\hat{S}(\omega) = -j\operatorname{sgn}(\omega)S(\omega)$ to find $\hat{s}(t)$, the Hilbert transform of $s(t)$.

6.3-2 Work Problem 6.3-1 except for the signal defined by

$$s(t) = A\sin(\omega_0 t) \leftrightarrow S(\omega) = -jA\pi[\delta(\omega - \omega_0) - \delta(\omega + \omega_0)]$$

6.3-3 Let $g(t)$ be a complex function with a spectrum $G(\omega)$ bandlimited such that $G(\omega) = 0$ for $|\omega| > W_g$ where $W_g < \omega_0$. Justify, using discussion and sketches, that $g(t) \exp(j\omega_0 t)$ is an analytic signal.

6.3-4 Show that (6.3-19) is true.

6.3-5 Show that (6.3-18) is true.

6.3-6 Find the Hilbert transform of $s(t) = \pi\delta(t)$.

6.3-7 Determine the Hilbert transform of $s(t) = \operatorname{rect}(t/a)$, where $a > 0$ is a constant.

6.3-8 Solve for the Hilbert transform of the signals $s(t) = \exp[\pm j(\omega_0 t + \phi_0)]$, where ω_0 and ϕ_0 are constants.

6.4-1 For a real signal $s(t)$ the first right-side form of

$$\bar{t}_s = \frac{\int_{-\infty}^{\infty} t s^2(t)\,dt}{\int_{-\infty}^{\infty} s^2(t)\,dt} = \frac{-j\int_{-\infty}^{\infty} S(\omega)\dfrac{dS^*(\omega)}{d\omega}\,d\omega}{\int_{-\infty}^{\infty} |S(\omega)|^2\,d\omega}$$

can be taken as the definition of the mean time $\bar{t}_s$. Use (6.4-1) and (6.4-3) to prove the second right-side form is true.

6.4-2 Work Problem 6.4-1 except for the complex envelope $g(t)$ and its mean time given by

$$\bar{t}_g = \frac{\int_{-\infty}^{\infty} t|g(t)|^2\,dt}{\int_{-\infty}^{\infty} |g(t)|^2\,dt} = \frac{-j\int_{-\infty}^{\infty} G(\omega)\dfrac{dG^*(\omega)}{d\omega}\,d\omega}{\int_{-\infty}^{\infty} |G(\omega)|^2\,d\omega}$$

6.4-3 A real signal is defined by

$$s(t) = Ae^{-W|t - t_1|}\cos(\omega_0 t)$$

where A, W, t_1, and ω_0 are all positive constants with W large enough that $\psi_c(t) = \psi(t)$ can be assumed. Find the mean time $\bar{t}_\psi$ of $\psi(t)$. To make $\bar{t}_\psi = 0$, should the new time origin be right or left of the origin that defines $s(t)$ above?

6.4-4 For the signal defined in Problem 6.4-3, find the complex envelope $g(t)$ assuming that $t_1 = 0$. Use (6.4-6) to find the rms duration of $g(t)$.

6.4-5 The complex envelope of an analytic signal can be taken as

$$g(t) = A\operatorname{tri}\left(\frac{t}{T}\right) \leftrightarrow G(\omega) = ATSa^2\left(\frac{\omega T}{2}\right)$$

where T and A are positive constants. Use (6.4-6) to find the rms duration of $g(t)$. Assume that $\bar{t}_g = 0$.

6.4-6 Use (6.4-1) through (6.4-3) to show that $\tau_{\psi,\text{rms}}^2$ also has the forms

$$\tau_{\psi,\text{rms}}^2 = \frac{\int_{-\infty}^{\infty} \left|\frac{d\Psi(\omega)}{d\omega}\right|^2 d\omega}{\int_{-\infty}^{\infty} |\Psi(\omega)|^2 d\omega} = \frac{-\int_{-\infty}^{\infty} \Psi(\omega)\frac{d^2\Psi^*(\omega)}{d\omega^2} d\omega}{\int_{-\infty}^{\infty} |\Psi(\omega)|^2 d\omega}$$

6.4-7 Work Problem 6.4-6 except for $\tau_{g,\text{rms}}^2$ where

$$\tau_{g,\text{rms}}^2 = \frac{\int_{-\infty}^{\infty} \left|\frac{dG(\omega)}{d\omega}\right|^2 d\omega}{\int_{-\infty}^{\infty} |G(\omega)|^2 d\omega} = \frac{-\int_{-\infty}^{\infty} G(\omega)\frac{d^2G^*(\omega)}{d\omega^2} d\omega}{\int_{-\infty}^{\infty} |G(\omega)|^2 d\omega}$$

6.4-8 Work Problem 6.4-6 except for $\tau_{s,\text{rms}}^2$ where

$$\tau_{s,\text{rms}}^2 = \frac{\int_{-\infty}^{\infty} \left|\frac{dS(\omega)}{d\omega}\right|^2 d\omega}{\int_{-\infty}^{\infty} |S(\omega)|^2 d\omega} = \frac{-\int_{-\infty}^{\infty} S(\omega)\frac{d^2S^*(\omega)}{d\omega^2} d\omega}{\int_{-\infty}^{\infty} |S(\omega)|^2 d\omega}$$

6.4-9 Show that the mean frequencies defined in (6.4-8) and (6.4-9) are also given by

$$\bar{\omega}_\psi = \frac{-j\int_{-\infty}^{\infty} \psi^*(t)\frac{d\psi(t)}{dt} dt}{\int_{-\infty}^{\infty} |\psi(t)|^2 dt}$$

$$\bar{\omega}_g = \frac{-j\int_{-\infty}^{\infty} g^*(t)\frac{dg(t)}{dt} dt}{\int_{-\infty}^{\infty} |g(t)|^2 dt}$$

6.4-10 Show that (6.4-11) and (6.4-12) are true.

6.4-11 Define second normalized moments of $\psi(t)$, $g(t)$, and $s(t)$ by the respective first right-side expressions below, and then show that the other expressions are equivalent:

$$\overline{\omega_\psi^2} = \frac{\int_0^{\infty} \omega^2 |\Psi(\omega)|^2 d\omega}{\int_0^{\infty} |\Psi(\omega)|^2 d\omega} = \frac{\int_{-\infty}^{\infty} \left|\frac{d\psi(t)}{dt}\right|^2 dt}{\int_{-\infty}^{\infty} |\psi(t)|^2 dt}$$

$$= \frac{-\int_{-\infty}^{\infty} \psi^*(t)\frac{d^2\psi(t)}{dt^2} dt}{\int_{-\infty}^{\infty} |\psi(t)|^2 dt}$$

$$\overline{\omega_g^2} = \frac{\int_{-\infty}^{\infty} \omega^2 |G(\omega)|^2 \, d\omega}{\int_{-\infty}^{\infty} |G(\omega)|^2 \, d\omega} = \frac{\int_{-\infty}^{\infty} \left|\frac{dg(t)}{dt}\right|^2 \, dt}{\int_{-\infty}^{\infty} |g(t)|^2 \, dt}$$

$$= \frac{-\int_{-\infty}^{\infty} g^*(t) \frac{d^2 g(t)}{dt^2} \, dt}{\int_{-\infty}^{\infty} |g(t)|^2 \, dt}$$

$$\overline{\omega_s^2} = \frac{\int_{-\infty}^{\infty} \omega^2 |S(\omega)|^2 \, d\omega}{\int_{-\infty}^{\infty} |S(\omega)|^2 \, d\omega} = \frac{\int_{-\infty}^{\infty} \left[\frac{ds(t)}{dt}\right]^2 \, dt}{\int_{-\infty}^{\infty} s^2(t) \, dt}$$

$$= \frac{-\int_{-\infty}^{\infty} s(t) \frac{d^2 s(t)}{dt^2} \, dt}{\int_{-\infty}^{\infty} s^2(t) \, dt}$$

*6.4-12 Show that all the rms bandwidths of (6.4-13) through (6.4-15) are equal. Use the first right-side forms of the second moments from Problem 6.4-11, and then use (6.4-11) and (6.4-12) to show final equalities.

6.4-13 Assume that $t_1 = 0$ in the signal of Problem 6.4-3, and find its mean frequency and rms bandwidth. [*Hint:* Use the time domain versions of mean frequency and rms bandwidth after defining $g(t)$.]

*6.4-14 Show that $\bar{\omega}_s$ of (6.4-10) is also given by

$$\bar{\omega}_s = \frac{\int_{-\infty}^{\infty} s^2(t) \frac{d}{dt}\left[\frac{\hat{s}(t)}{s(t)}\right] dt}{2 \int_{-\infty}^{\infty} s^2(t) \, dt}$$

6.4-15 Let the time origin be chosen so that $\bar{t}_\psi$ of (6.4-4) is zero. With this condition true show that $\bar{t}_g$ and $\bar{t}_s$ (see Problems 6.4-1 and 6.4-2) are also zero.

6.4-16 Show that $\tau_{\psi, \text{rms}}$, $\tau_{g, \text{rms}}$, and $\tau_{s, \text{rms}}$, as given by (6.4-5) through (6.4-7) are all equal to each other.

*6.5-1 A rectangular pulse, defined by

$$s(t) = A \, \text{rect}\left(\frac{t}{T}\right) \leftrightarrow S(\omega) = AT \, \text{Sa}\left(\frac{\omega T}{2}\right)$$

where A and T are positive constants, is passed through an ideal filter defined by

$$h(t) = \frac{W_f}{\pi} \text{Sa}(W_f t) \leftrightarrow H(\omega) = \text{rect}\left(\frac{\omega}{2W_f}\right)$$

where W_f is a positive constant. Find an expression for the rms bandwidth of $s_0(t)$, the pulse emerging from the filter. Plot $W_{s_0, \text{rms}} T$ versus $W_f T$ for $0 \le W_f T \le 10$. To what does $W_{s_0, \text{rms}} T$ approach as $W_f T \to \infty$? [*Hint:* Use the known integral

$$\int_0^x \left[\frac{\sin(\xi)}{\xi} \right]^2 d\xi = -\frac{\sin^2(x)}{x} + \text{Si}(2x)$$

Where $\text{Si}(\cdot)$ is the *sine integral* defined by

$$\text{Si}(x) = \int_0^x \frac{\sin(\xi)}{\xi} d\xi$$

$\text{Si}(x)$ is a tabulated function, as it has no closed form (Abramowitz and Stegun, eds., 1964).]

6.5-2 Assume that a network has a real impulse response (ignore realizability) $h(t) = \text{rect}(t/T)$, where $T > 0$ is a constant. Find the analytic impulse response $z(t)$ and plot $|z(t)|$ for $-3 \le t/T \le 3$.

6.5-3 Work Problem 6.5-2 except assume that $h(t) = \text{Sa}(2\pi t/T)$.

6.5-4 Work Problem 6.5-2 except assume that $h(t) = (4t + 15)/(t^2 + 9)$, and plot $|z(t)|$ for $-20 < t < 20$.

6.6-1 Find the matched filter's impulse responses $z_{\text{opt}}(t)$ and $h_{\text{opt}}(t)$ for the signal defined by

$$s_r(t) = A \, \text{tri}\left(\frac{t}{T}\right) \cos(\omega_0 t)$$

$$S_r(\omega) = \frac{AT}{2}\left\{ \text{Sa}^2\left[(\omega - \omega_0)\frac{T}{2} \right] + \text{Sa}^2\left[(\omega + \omega_0)\frac{T}{2} \right] \right\}$$

when added to noise having the power spectral density

$$\mathscr{S}_{N_r N_r}(\omega) = \frac{W_2/2}{W_2^2 + (\omega - \omega_0)^2} + \frac{W_2/2}{W_2^2 + (\omega + \omega_0)^2}$$

where A, T, ω_0, and W_2 are all positive constants. Assume that $\omega_0 \gg 2\pi/T$ and $\omega_0 \gg W_2$ so that complex and analytic signals are equal. Sketch $h_{\text{opt}}(t)$.

6.6-2 Work Problem 6.6-1 except assume that the signal is

$$s_r(t) = u(t)[e^{-W_2 t} - e^{-aW_2 t}] \cos(\omega_0 t)$$

where $a > 0$ is a constant.

6.6-3 Show that (6.6-12) is true if (6.6-11) is true.

6.7-1 Find the transfer function $H_{opt}(\omega)$ and impulse response $h_{opt}(t)$ of the white noise matched filter for the signal

$$s_r(t) = A \ \text{rect}\left(\frac{t}{T}\right)\cos(\omega_0 t)$$

where A, T, and ω_0 are positive constants.

6.7-2 For the matched filter of Problem 6.7-1, find and sketch the filter's output signal. [*Hint*: Fourier transform $s_r(t)$, and use known transform pairs to obtain the output signal under the assumption $\omega_0 \gg 2\pi/T$.]

6.7-3 In the network of Fig. P6.7-3 the filter is equivalent to a narrow-band bandpass filter with transfer function

$$H(\omega) = W\left[\frac{1}{W + j2(\omega - \omega_0)} + \frac{1}{W + j2(\omega + \omega_0)}\right]$$

where W is its 3-dB bandwidth. Find and sketch the overall network's impulse response $h_0(t)$, and comment on the type of signal $s_r(t)$ to which it might be a good matched filter approximation. For convenience, assume that $\omega_0 T$ is an exact multiple of 2π.

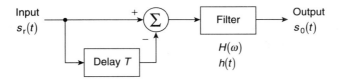

Figure P6.7-3.

6.7-4 A signal

$$s_r(t) = A \ \text{rect}\left(\frac{t}{T}\right)\cos\left(\frac{\pi t}{T}\right)\cos(\omega_0 t)$$

where A, T, and ω_0 are positive constants, is added to white noise with power spectral density $\mathcal{N}_0/2$. The sum is applied to a matched filter. Find $(\hat{S}_0/N_0)_{max}$ for this filter if $\omega_0 T \gg 1$.

6.7-5 White noise plus the signal

$$s_r(t) = Au(t)e^{-Wt}\cos(\omega_0 t)$$

are applied to a matched filter where A, W, and ω_0 are positive constants. (a) Find the transfer function $H_{opt}(\omega)$ of the matched filter. (b) Find and sketch the impulse response $h_{\text{opt}}(t)$. (c) Is there a value of t_0 for which the filter is causal? (d) Find $(\hat{S}_0/N_0)_{\text{max}}$ for this filter. Assume in every part that $\omega_0 \gg W$.

6.7-6 Work Problem 6.7-5 except for the signal

$$s_r(t) = Au(t)\,te^{-Wt}\cos(\omega_0 t)$$

6.7-7 Find the matched filter's transfer function $H_{\text{opt}}(\omega)$ for the signal

$$s_r(t) = Ae^{-W^2 t^2}\cos(\omega_0 t)$$

received with white noise, where A, W, and $\omega_0 \gg W$ are positive constants.

★6.7-8 A radar receives the signal

$$s_r(t) = \alpha A \,\text{rect}\left(\frac{t - \tau_R}{2T}\right)\left[1 - \frac{(t - \tau_R)^2}{T^2}\right]\cos[\omega_0(t - \tau_R)]$$

plus white noise, where α, τ_R, A, T, and ω_0 are positive constants. (a) Find the matched filter's output signal $s_0(t)$. (b) Plot the envelopes of $s_r(t)$ and $s_0(t)$ both normalized to have unit peak amplitudes.

6.7-9 Find $(\hat{S}_0/N_0)_{\text{max}}$ for the matched filter and signal of Problem 6.7-8.

6.7-10 Find the matched filter's transfer function $H_{\text{opt}}(\omega)$ for Problem 6.7-8. What value of arbitrary constant C will give $|H_{\text{opt}}(\omega_0)| = 1$? Assume that $\omega_0 \gg 2\pi/T$.

6.7-11 A received signal, for which $\alpha = 1$ and $\tau_R = 0$, as defined by

$$s_r(t) = \begin{cases} \dfrac{1}{5}e^{t/8}\cos(\omega_0 t), & -2 < t < 3 \\[2mm] 0, & \text{elsewhere} \end{cases}$$

is added to white noise of power spectral density $\mathcal{N}_0/2$. At the matched filter's output $(\hat{S}_0/N_0)_{\text{max}} = 20$. Find $\mathcal{N}_0/2$ if ω_0 is large compared to the bandwidth of $s_r(t)$.

6.7-12 A radar transmits a signal

$$s(t) = A\,\text{rect}\left(\frac{t}{2T}\right)\frac{1}{1 + (Wt)^2}\cos(\omega_0 t)$$

where A, T, W, and ω_0 are all positive constants. (a) What value of W is needed if the duration of the envelope of $s(t)$ between points 3-dB down from its maximum is to be $T/3$? (b) If a filter is matched to the signal in white noise for which $\mathcal{N}_0/2 = 10^{-16}$ W/Hz, what value of α^2 is associated with the radar equation if $(\hat{S}_0/N_0)_{max} = 34$ is necessary. Assume that $\tau_R = 0$ and that ω_0 is large relative to the bandwidth of $s(t)$. Evaluate your value of α for the value of W found in part (a).

6.7-13 A radar transmits the signal

$$s(t) = A \operatorname{rect}\left(\frac{t}{2T}\right)\left[1 - \left(\frac{t}{T}\right)^4\right]\cos(\omega_0 t)$$

(a) Find the 3-dB duration of the envelope of $s(t)$ in terms of T. (b) The receiver has white noise at a power level of $\mathcal{N}_0/2 = 10^{-18}$ W/Hz and $\alpha = 10^{-7}$ from the radar equation. What is $(\hat{S}_0/N_0)_{max}$ for a matched filter in terms of A and T? From your result find the value of A when $T = 1.5$ μs, such that $(\hat{S}_0)/N_0)_{max} = 220$.

6.8-1 Show that other forms for $\chi(t, \omega_d)$ are

$$\chi(\tau, \omega_d) = e^{-j(\omega_0 + \omega_d)\tau}\frac{1}{2\pi}\int_{-\infty}^{\infty}\Psi^*(\omega)\Psi(\omega - \omega_d)e^{j\omega\tau}\,d\omega$$

$$= e^{-j\omega_d\tau}\frac{1}{2\pi}\int_{-\infty}^{\infty}G^*(\omega)G(\omega - \omega_d)e^{j\omega\tau}\,d\omega$$

where $\Psi(\omega)$ and $G(\omega)$ are the Fourier transforms of $\psi(t)$ and $g(t)$, respectively.

6.8-2 Show that $\chi(\tau, \omega_d)$ can be put in the following equivalent forms:

$$\chi(\tau, \omega_d) = e^{-j\omega_d\tau}\int_{-\infty}^{\infty}g(\xi)g^*(\xi - \tau)e^{j\omega_d\xi}\,d\xi$$

$$= e^{-j(\omega_0 + \omega_d)\tau}\int_{-\infty}^{\infty}\psi(\xi)\psi^*(\xi - \tau)e^{j\omega_d\xi}\,d\xi$$

6.8-3 Show that (6.8-12) is true.

6.8-4 Show that (6.8-13) is true.

6.8-5 Show that (6.8-14) is true.

6.8-6 Show that (6.8-15) is true.

6.8-7 Show that (6.8-16) is true.

6.8-8 Show that (6.8-17) is true.

6.8-9 Show that (6.8-18) is true.

6.8-10 Show that (6.8-19) is true.

6.8-11 Show that (6.8-20) is true.

6.9-1 A radar uses a constant-frequency rectangular pulse of duration T. A moving target causes a Doppler shift in the received signal of $\omega_d = \pi/T$ rad/s. Sketch the matched filter's response to this target and compare it to the response that would have occurred if ω_d had been zero.

6.9-2 Find $\chi(\tau, \omega_d)$ for the signal

$$s(t) = Au(t)e^{-Wt}\cos(\omega_0 t)$$

where A, W, and $\omega_0 \gg W$ are all positive constants.

7

PULSE COMPRESSION WITH RADAR SIGNALS

For good detection a radar needs a large peak signal power to average noise power ratio, $\hat{S}_0/N_0$, at the time of the target's return signal. In Chapter 6 we found that a matched filter was the best of all possible filters and it produced the maximum ratio $(\hat{S}_0/N_0)_{max}$. We also found that this maximum ratio depended on the total transmitted energy and not on the presence of any frequency modulation (FM) on the transmitted signal. Thus for good detection many radars seek to transmit long-duration pulses to achieve high energy, since transmitters are typically operated near their peak power limitation.

On the other hand, for good range measurement accuracy, a radar needs short pulses. These divergent needs of long pulses for detection and short pulses for range accuracy in measurements prevented early radars from simultaneously performing both functions well. Fortunately in the late 1950s and early 1960s a new concept was developed whereby both needs could be met. The concept is called *pulse compression*. It makes use of the fact that a long-duration pulse's bandwidth can be made larger by use of FM. Large bandwidth implies narrow effective duration. With FM a waveform can be designed to have both long duration *and* small *effective* duration (large bandwidth). The waveform with small effective duration is produced when the long-duration waveform with FM is applied to its matched filter. Thus, by use of FM over long transmitted pulses and a matched filter, a system can simultaneously obtain good detection performance and highly accurate range measurements.

In this chapter we describe several types of pulse compression waveforms useful in radar. First we discuss the case where the FM is linear with time over the transmitted pulse. This is one of the most useful, and oldest, waveforms. It is found that the compressed pulse (matched filter's output) has a principal response (desired) and undesired side responses. These side responses, called sidelobes, must typically be suppressed so that they are not mistaken as small-amplitude principal responses of other targets. Methods of controlling these sidelobes are discussed, as well as ways of designing special FM laws (nonlinear) to produce compressed pulses with inherently small sidelobes.

The chapter continues by discussing waveforms based on other modulation methods (frequency hopping, binary and polyphase coding), and their special characteristics are given.

7.1   BASIC CONCEPT

If a long-duration pulse is frequency modulated, its spectrum can have a wider bandwidth than if no FM were present. Since increasing bandwidth corresponds to waveforms with decreasing effective duration, the potential exists for a long-duration, large-bandwidth, pulse to be converted to a short-duration, "effective" pulse. In effect we seek to "squeeze" the long pulse into a short pulse. If energy can be conserved, we can even expect the shorter "compressed" pulse to increase in peak amplitude compared to its amplitude as a long pulse. These effects can all be achieved by a signal-processing technique called *pulse compression*. The actual signal processor consists of a matched filter that is often followed by a second filter that minimizes certain undesired responses from the matched filter.

To visualize the process of pulse compression, imagine that a long pulse (duration T) has a linearly varying instantaneous angular frequency $\omega_i(t)$ with time, as shown in Fig. 7.1-1a. At the start of the pulse the carrier cycles at a rate $\omega_0 - (\Delta\omega/2)$. In the pulse's center the frequency is ω_0, the nominal carrier's angular frequency. At the pulse's end, frequency increases to $\omega_0 + (\Delta\omega/2)$. The total frequency deviation over time T is $\Delta\omega$ (rad/s). This pulse is applied to a pulse compression filter that has a constant-modulus transfer function but a phase that corresponds to a linearly decreasing envelope delay as shown in panel b.[1] We may visualize the low frequencies that enter the filter first as being delayed more than those that enter later. If the slope (T seconds of delay change over an angular frequency span of $\Delta\omega$) is a match to the input signal's FM, all the frequencies can be thought of as emerging at the same time and "piling up" in the output. The response can be larger in amplitude, as shown in panel c. However, because the input's bandwidth is large, these frequencies can "pile up" for only a short time and the output quickly decreases from the peak in relation to the reciprocal of bandwidth. The duration of the main response is smaller than T by the factor $1/\Delta fT$, where $\Delta f = \Delta\omega/2\pi$ and ΔfT is called the *time-bandwidth product* or the *pulse compression ratio* of $s(t)$. Similarly peak power is larger by a factor ΔfT (or $\sqrt{\Delta fT}$ in voltage; see Problem 7.2-1). Outside the region of main response, undesired responses called *sidelobes* occur for a time duration T on each side of the main response.

Sidelobes are unwanted by-products of the pulse compression process. Their form and amplitudes depend greatly on the type of FM used and whether the modulated pulse $s(t)$ contains any shaping (tapering of amplitude) across the pulse. Two principal methods exist for reduction of sidelobes. In one, the output of the pulse compression filter is passed through a compensation, or weighting filter, specifically designed to reduce the sidelobes at the expense of some loss in signal-to-noise ratio. These filters

---

[1] Recall that envelope delay (also called *group delay*) equals $-d\phi(\omega)/(d\omega)$ where $\phi(\omega)$ is the phase shift of the filter versus ω.

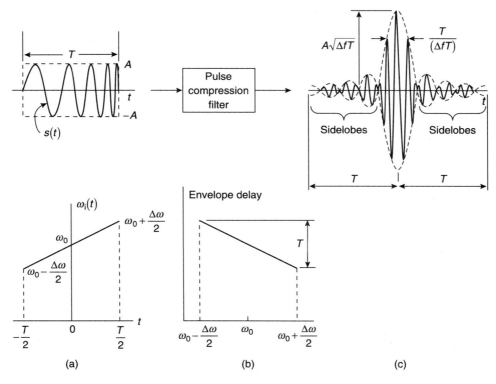

Figure 7.1-1 (*a*) A long-duration pulse with linear FM, (*b*) the compression filter and its linear delay characteristic, and (*c*) the output pulse. All are used to illustrate the concept of pulse compression.

are sometimes called *mismatch filters* because they do not preserve the performance of the matched (compression) filter. The second method relies on special choice of the FM law to provide a compressed waveform with inherently low sidelobes. Both techniques are described in this chapter.

The preceding discussions of the pulse compression concept must not be taken as precise. They are given only to impart a sort of mental picture to the process. A proper understanding of pulse compression derives only from the mathematics of a given problem. We next develop the very important case of linear FM in detail.

7.2   LINEAR FM PULSE (CHIRP)

The linear FM pulse, sometimes called a *chirp* pulse, is defined by

$$s(t) = A \, \text{rect}\left(\frac{t}{T}\right) \cos\left[\omega_0 t + \frac{\mu}{2} t^2 + \phi_0\right] \tag{7.2-1}$$

where A, T, ω_0, and μ are positive constants and ϕ_0 is an arbitrary phase angle. The constant μ is related to the frequency sweep $\Delta\omega$ over pulse duration T by

$$\mu = \frac{\Delta\omega}{T} \quad (\text{rad/s}^2) \tag{7.2-2}$$

The instantaneous angular frequency change $\omega_i(t)$ due to FM is

$$\omega_i(t) = \frac{d\theta(t)}{dt} = \frac{d}{dt}\left[\frac{\mu}{2} t^2\right] = \mu t, \qquad -\frac{T}{2} < t < \frac{T}{2} \tag{7.2-3}$$

For analysis purposes it is easiest to work with the complex envelope

$$g(t) = A \ \text{rect}\left(\frac{t}{T}\right) e^{j\mu t^2/2} \tag{7.2-4}$$

and analytic signal

$$\psi(t) = g(t) e^{j\omega_0 t + j\phi_0} \tag{7.2-5}$$

In terms of $\psi(t)$, the received analytic signal $\psi_r(t)$ corresponding to a target of nominal delay τ_R and Doppler angular frequency ω_d is

$$\psi_r(t) = \alpha\psi(t - \tau_R) e^{j\omega_d(t - \tau_R)} \tag{7.2-6}$$

from (6.8-2), where α is a constant related to the radar equation.

Matched Filter's Response

The general problem in pulse compression is sketched in Fig. 7.2-1. The real received signal $s_r(t)$ is represented by its analytic representation $\psi_r(t)$ which is applied to a white noise matched filter represented by its analytic impulse response $z(t)$. The analytic response signal $\psi_0(t)$ is given by (6.8-11) with (6.8-8) substituted:

$$\psi_0(t) = \frac{C\alpha^2}{\mathcal{N}_0} e^{j(\omega_0 + \omega_d)(t - t_0)} \chi(t - t_0, \omega_d) \tag{7.2-7}$$

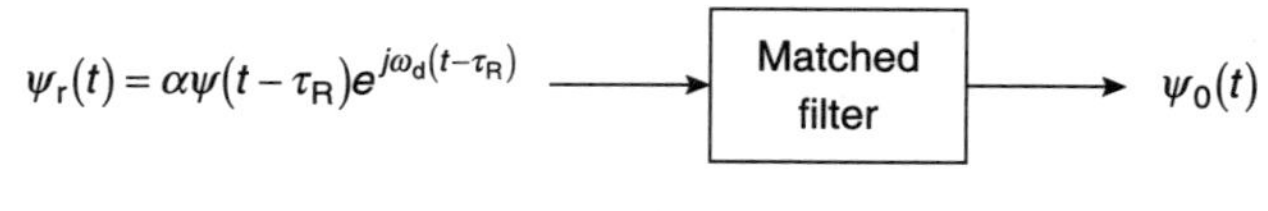

$$\psi_r(t) = \alpha\psi(t - \tau_R) e^{j\omega_d(t-\tau_R)} \longrightarrow \boxed{\begin{array}{c}\text{Matched}\\ \text{filter}\end{array}} \longrightarrow \psi_0(t)$$

$$z(t) = \frac{2C\alpha}{\mathcal{N}_0} \psi * (t_0 - \tau_R - t)$$

$$\psi_0(t) = \frac{C\alpha^2}{\mathcal{N}_0} e^{j(\omega_0 + \omega_d)(t - t_0)} \chi(t - t_0, \omega_d)$$

Figure 7.2-1 The input, filter, and output defining the pulse compression receiver.

where, with $\tau = t - t_0$,

$$\chi(\tau, \omega_d) = e^{-j\omega_0\tau} \int_{-\infty}^{\infty} \psi^*(\xi)\psi(\xi + \tau)e^{j\omega_d\xi} \, d\xi \tag{7.2-8}$$

On substitution of (7.2-5) and (7.2-4) we obtain

$$\chi(\tau, \omega_d) = \begin{cases} A^2T\left(1 - \dfrac{|\tau|}{T}\right)e^{-j\omega_d\tau/2} \, \mathrm{Sa}\left[(\mu\tau + \omega_d)\left(1 - \dfrac{|\tau|}{T}\right)\dfrac{T}{2}\right], & -T \leq \tau \leq T \\ \\ 0 & \text{elsewhere} \end{cases}$$
$$\tag{7.2-9}$$

Cuts through Response

To see the behavior of (7.2-9), we sketch cuts $\chi(\tau, 0)$ and $\chi(0, \omega_d)$ in Fig. 7.2-2, as given by

$$\chi(\tau, 0) = \begin{cases} A^2T\left(1 - \dfrac{|\tau|}{T}\right)\mathrm{Sa}\left[\dfrac{\Delta\omega\tau}{2}\left(1 - \dfrac{|\tau|}{T}\right)\right], & -T \leq \tau \leq T \\ \\ 0 & \text{elsewhere} \end{cases} \tag{7.2-10}$$

where (7.2-2) has been used, and

$$\chi(0, \omega_d) = A^2T \, \mathrm{Sa}\left(\dfrac{\omega_d T}{2}\right) \tag{7.2-11}$$

The cut of panel a shows that the filter's response is compressed to a duration $2\pi/\Delta\omega = 1/\Delta f = T/(\Delta f T)$ which can be much smaller than T because $\Delta f T \gg 1$ is easily possible. Values of $\Delta f T$ of over 100,000 are achievable, although many systems use time-bandwidth products of less than a hundred or two. The plot of panel a presumed that $\Delta f T \gg 1$ so that the factor $[1 - (|\tau|/T)]$ in (7.2-10) is approximately one for values of τ near the origin. Although sidelobes extend to $-T$ and to $+T$, the largest are near the origin and are only 13.26 dB below the main response. Such large sidelobes are easily mistaken for a small, displaced, target return, so must be reduced in practical systems that use linear FM. The mismatch filter (Section 7.3) is one commonly used method.

The cut $\chi(0, \omega_d)$ sketched in Fig. 7.2-2b shows the behavior of the matched filter's response as Doppler ω_d changes. Note that a shift of $\omega_d \geq 2\pi/T$ reduces the response by at least -13.26 dB. This observation would seem to imply that the response approximately disappears if $\omega_d \geq 2\pi/T$. However, remember the cut $\chi(0, \omega_d)$ is only for $\tau = 0$. A better view of what happens is sketched in panel c. Here the -3.92-dB contour shows coupling between τ and ω_d. If the target shifts slightly higher in ω_d, it does not disappear; rather, its point of maximum response shifts earlier (more negative) in τ where ω_d and τ are linearly coupled with slope $-\mu$.

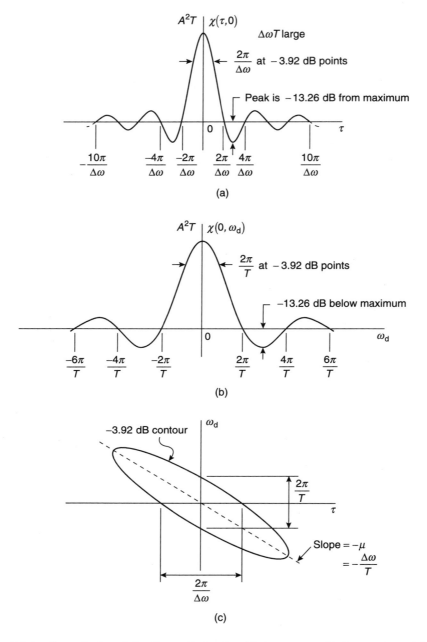

Figure 7.2-2   Cuts of $\chi(\tau, \omega_d)$ for linear FM pulse. (a) for $\omega_d = 0$, (b) for $\tau = 0$, and (c) for plane cut parallel to τ, ω_d plane and 3.92 dB below maximum at $(\tau = 0, \omega_d = 0)$.

Figure 7.2-3 shows the uncertainty function $|\chi(\tau, \omega_d)|$ for the time-bandwidth product $\Delta f T = 25$. Here the ridge of largest response is clearly tilted away from the τ axis, as indicated by the contour of Fig. 7.2-2c.

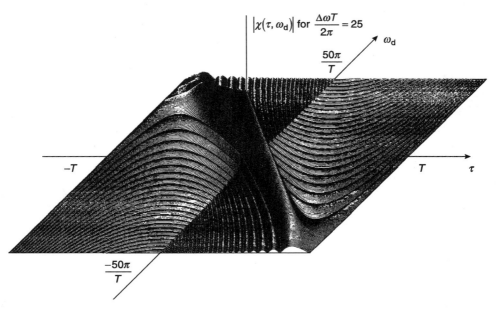

Figure 7.2-3 $|\chi(\tau, \omega_d)|$ for a linear FM rectangular pulse of duration T and frequency sweep $\Delta\omega$ (rad/s). Plot is for special case $\Delta\omega T/2\pi = 25$. (Adapted from Rihaczek, 1969, with permission.)

Spectrum of Transmitted Pulse

From (6.1-8) we see that we only need to define the spectrum $G(\omega)$ of the complex envelope of (7.2-4) in order to see the behavior of $S(\omega)$ the spectrum of the transmitted waveform $s(t)$. By Fourier transformation

$$
\begin{aligned}
G(\omega) &= \int_{-\infty}^{\infty} A \, \text{rect}\left(\frac{t}{T}\right) e^{j\mu t^2/2} e^{-j\omega t} \, dt \\
&= A \int_{-T/2}^{T/2} e^{j(\mu t^2/2) - j\omega t} \, dt
\end{aligned}
\tag{7.2-12}
$$

By use of the variable change

$$
x = \sqrt{\frac{\mu}{\pi}}\left(t - \frac{\omega}{\mu}\right), \quad dx = \sqrt{\frac{\mu}{\pi}} \, dt
\tag{7.2-13}
$$

Equation (7.2-12) becomes

$$
\begin{aligned}
G(\omega) &= A\sqrt{\frac{\pi}{\mu}} \, e^{-j\omega^2/(2\mu)} \int_{-x_1}^{x_2} e^{j\pi x^2/2} \, dx \\
&= A\sqrt{\frac{\pi}{\mu}} \, e^{-j\omega^2/(2\mu)} \left\{ \int_{0}^{x_2} e^{j\pi x^2/2} \, dx - \int_{0}^{-x_1} e^{j\pi x^2/2} \, dx \right\}
\end{aligned}
\tag{7.2-14}
$$

$$C(x) \approx \frac{1}{2} + \frac{1}{\pi x} \sin\left(\frac{\pi}{2} x^2\right) \text{ for } x \gg 1$$

$$S(x) \approx \frac{1}{2} - \frac{1}{\pi x} \cos\left(\frac{\pi}{2} x^2\right) \text{ for } x \gg 1$$

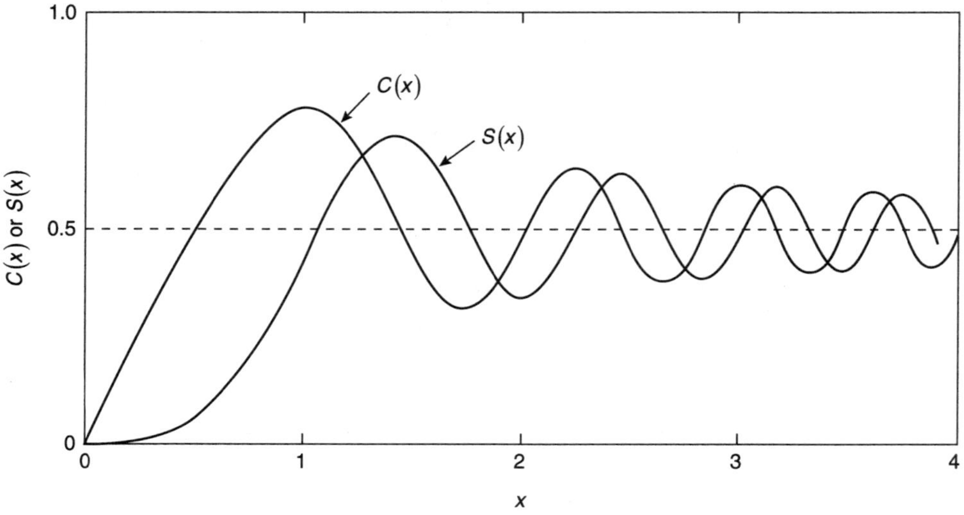

Figure 7.2-4   Fresnel integrals $C(x)$ and $S(x)$ for $0 \leq x \leq 4$.

where

$$\left.\begin{array}{c} x_1 \\ x_2 \end{array}\right\} = \sqrt{\frac{\mu}{\pi}}\left(\frac{T}{2} \pm \frac{\omega}{\mu}\right) = \sqrt{\frac{\Delta f \, T}{2}}\left[1 \pm \frac{f}{\Delta f/2}\right] \tag{7.2-15}$$

Here upper and lower signs correspond, respectively, to x_1 and x_2.

The last integrals in (7.2-14) are recognized as *Fresnel integrals*, denoted by $C(x)$ and $S(x)$, and defined by

$$C(x) = \int_0^x \cos\left(\frac{\pi}{2} \xi^2\right) d\xi \tag{7.2-16}$$

$$S(x) = \int_0^x \sin\left(\frac{\pi}{2} \xi^2\right) d\xi \tag{7.2-17}$$

$$C(-x) = -C(x) \tag{7.2-18}$$

$$S(-x) = -S(x) \tag{7.2-19}$$

Both $C(x)$ and $S(x)$ are tabulated functions (Abramowitz and Stegun, eds., 1964, pp. 321–322). Figure 7.2-4 shows $C(x)$ and $S(x)$ for $0 \leq x \leq 4$. For large x,

$$\lim_{x \to \infty} C(x) = \tfrac{1}{2} \tag{7.2-20}$$

$$\lim_{x \to \infty} S(x) = \tfrac{1}{2} \tag{7.2-21}$$

When $C(x)$ and $S(x)$ are used in (7.2-14) we finally obtain

$$G(\omega) = AT\sqrt{\frac{2\pi}{\Delta\omega T}}\,e^{-j\omega^2 T/(2\Delta\omega)}\left\{\frac{[C(x_2)+C(x_1)]+j[S(x_2)+S(x_1)]}{\sqrt{2}}\right\} \quad (7.2\text{-}22)$$

Other than the scale constant, there are two important components to $G(\omega)$. First, there is a smooth quadratic phase shift in the spectrum. Second, the factor within the braces in (7.2-22) is a rather erratic function of ω because of the cycling and phasings of the Fresnal integrals. Figures 7.2-5, 7.2-6, and 7.2-7 show the magnitude and phase of the braced factor for $\Delta f\,T = 25$, 100, and 400, respectively. Curves are shown only for positive frequencies, since $G(\omega)$ has even symmetry. In the limit as $\Delta f\,T \to \infty$, the braced factor approaches a magnitude of one and a phase of $\pi/4$ rad for $-\Delta\omega/2 < \omega < \Delta\omega/2$, and is zero for ω outside this range. In other words,

$$G(\omega) = AT\sqrt{\frac{2\pi}{\Delta\omega T}}\,\text{rect}\left(\frac{\omega}{\Delta\omega}\right)e^{-j[\omega^2 T/(2\Delta\omega)]+j(\pi/4)}, \qquad \Delta f\,T \to \infty \quad (7.2\text{-}23)$$

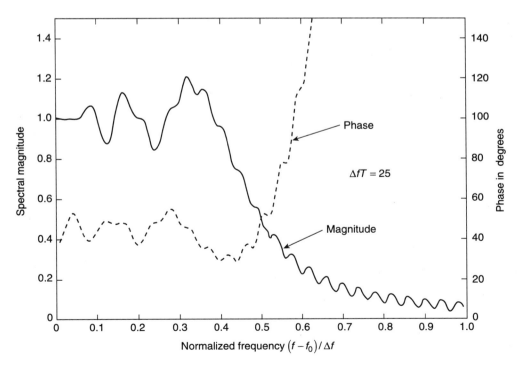

Figure 7.2-5 Magnitude and phase of factor in the linear FM signal's spectrum that is due to the fresnel integrals. Data plotted for $\Delta f\,T = 25$. (Adapted from RCA, 1965, with permission.)

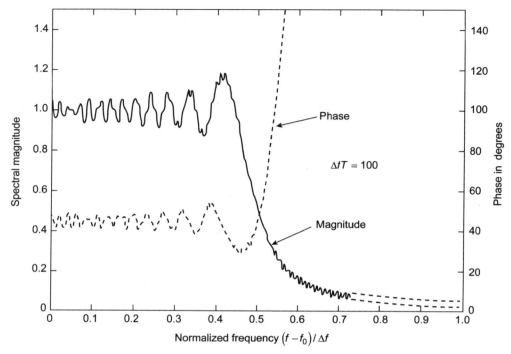

Figure 7.2-6 Magnitude and phase of factor in the linear FM signal's spectrum that is due to the Fresnel integrals. Data plotted for $\Delta f T = 100$. (Adapted from RCA, 1965, with permission.)

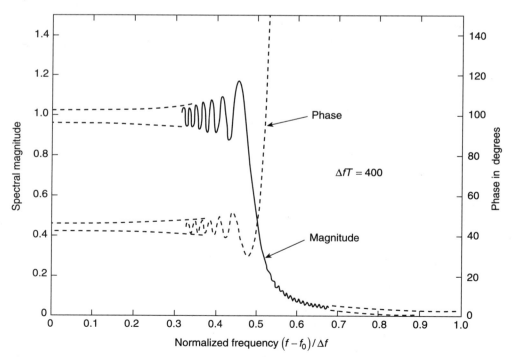

Figure 7.2-7 Magnitude and phase of factor in the linear FM signal's spectrum that is due to the Fresnel integrals. Data plotted for $\Delta f T = 400$. (Adapted from RCA, 1965, with permission.)

7.3 MISMATCH FILTERS FOR SIDELOBE CONTROL

There are several types of filters designed to be used with a pulse compression (matched) filter to reduce time sidelobes. Because the filter is in cascade with the matched filter, the overall cascade is not matched, so the added filter is often called a *mismatch filter*. It degrades the output signal-to-noise ratio slightly with a degradation that generally tends to be larger as sidelobes are more heavily suppressed.

Although mismatch filters can be designed for use with any FM, they originated from use with linear FM. We will describe these filters for use with linear FM. The applicable principles and concepts can then be easily applied in any FM problem.

The fundamental idea in sidelobe reduction is to choose a mismatch filter transfer function such that the final output signal spectrum has a taper that leads to a waveform with low sidelobes. Since the linear FM signal's spectrum is approximately rectangular (uniform), the problem is analogous to sidelobe control in antennas that have a finite aperture (Section 3.7). In fact the methods used for antennas (tapering over aperture distance to achieve low sidelobes in the angle domain) are directly applicable to waveforms (tapering over spectrum's frequencies to achieve low sidelobes in the time domain). Recall that in this type of problem the Dolph-Tchebycheff (Dolph, 1946) and Taylor (Taylor, 1955) distributions (tapers) are optimum or near-optimum methods of sidelobe reduction. We discuss these methods next.

Dolph-Tchebycheff Filter

This distribution (taper) gives the smallest pulse duration for a specified maximum sidelobe level. However, sidelobes do not decrease from the specified level for sidelobes far removed from the central response. The required distribution with frequency is also unrealizable (Cook and Bernfeld, 1967, p. 179) and difficult to even approximate. For these reasons the Doph-Tchebycheff distribution is not practical but does serve as a reference against which other distributions can be compared.

Taylor Filter

This filter is the practical approximation to the ideal Dolph-Tchebycheff distribution. Taylor's distribution gives the smallest amount of main response broadening for a specified peak sidelobe level while allowing the sidelobes to decrease below the specified level with increasing time from the central (main) response.

The normalized transfer function $H_T(\omega)$ of the Taylor filter[2] is

$$H_T(\omega) = K_T \left\{ 1 + 2 \sum_{m=1}^{\bar{n}-1} F_m \cos(m2\pi\omega/\Delta\omega) \right\} \text{rect}\left(\frac{\omega}{\Delta\omega}\right) \qquad (7.3\text{-}1)$$

---

[2] This is the lowpass equivalent obtained by suppressing the carrier ω_0. The actual bandpass version for $\omega > 0$ derives from replacement of ω by $(\omega - \omega_0)$.

where $\bar{n}$ is a selectable integer that must be larger than a minimum value for a maximum-level specified sidelobe SLL (e.g., if the largest sidelobe's peak is to be 30 dB below the main pulse's peak, then $SLL = -30$). Here

$$K_T = \frac{1}{1 + 2 \sum_{m=1}^{\bar{n}-1} F_m} \tag{7.3-2}$$

$$F_m = \begin{cases} \dfrac{(-1)^{m+1} \prod_{n=1}^{\bar{n}-1} \{1 - [m^2 \sigma_T^{-2}/(A^2 + (n-0.5)^2)]\}}{2 \prod_{\substack{n=1 \\ n \neq m}}^{\bar{n}-1} [1 - (m^2/n^2)]}, & m = 1, 2, \ldots, (\bar{n}-1) \\[2em] 0, & m \geq \bar{n} \end{cases} \tag{7.3-3}$$

$$A = \frac{1}{\pi} \cosh^{-1}(10^{-SLL/20}) \tag{7.3-4}$$

$$\sigma_T^2 = \frac{\bar{n}^2}{A^2 + (\bar{n} - 0.5)^2} > 1 \tag{7.3-5}$$

and $\Delta\omega$ is the bandwidth of the (rectangular) spectrum of the compressed pulse that feeds the mismatch filter.

The compressed, weighted (sidelobe-reduced), pulse emerging from the mismatch filter of (7.3-1) has an envelope

$$s_T(t) = K_T \frac{\Delta\omega}{2\pi} \sum_{m=-(\bar{n}-1)}^{\bar{n}-1} F_m \, \text{Sa}\left(\frac{\Delta\omega t}{2} + m\pi\right) \tag{7.3-6}$$

where $F_{-m} = F_m$ and

$$F_0 = 1 \tag{7.3-7}$$

The 3-dB pulse width τ_T of $s_T(t)$ is

$$\tau_T = \sigma_T \tau_D \tag{7.3-8}$$

where

$$\tau_D = \frac{4}{\Delta\omega} \left\{ [\cosh^{-1}(10^{-SLL/20})]^2 - \left[\cosh^{-1}\left(\frac{10^{-SLL/20}}{\sqrt{2}}\right)\right]^2 \right\}^{1/2} \tag{7.3-9}$$

We have previously given values of A and the minimum values of $\bar{n}$ (Table 3.7-2) for SLL from -15 to -40 dB. Table 7.3-1 gives some values of F_m needed in (7.3-1) for $-40 \leq SLL \leq -16$ in steps of 2 dB. Other values have been tabulated by Spelmire (1958).

Figure 7.3-1 illustrates the transfer functions of Taylor's filter for several sidelobe levels. In each case the minimum value of $\bar{n}$ was used. Clearly, as the sidelobe level is reduced, tapering of frequencies near the edges of the spectrum is more severe. There is also an increase in the duration of the compressed pulse as can be confirmed by calculation of (7.3-8).

TABLE 7.3-1 Taylor's coefficients F_m.

SLL (dB)	$\bar{n}$	F_m for Various Values of SLL and Values of $\bar{n}$ as Shown[a]				
		$m = 1$	$m = 2$	$m = 3$	$m = 4$	$m = 5$
-16	3	0.07183	0.02329	—	—	—
-18	3	0.11721	0.01115	—	—	—
-20	3	0.15615	0.00216	—	—	—
-22	3	0.18980	-0.00432	—	—	—
-24	3	0.21908	-0.00883	—	—	—
-26	3	0.24473	-0.01181	—	—	—
-28	4	0.26913	-0.01418	0.00087	—	—
-30	4	0.29266	-0.01578	0.00218	—	—
-32	4	0.31384	-0.01616	0.00305	—	—
-34	5	0.33455	-0.01576	0.00396	-0.00051	—
-36	5	0.35373	-0.01442	0.00450	-0.00092	—
-38	6	0.37189	-0.01229	0.00481	-0.00130	0.00015
-40	6	0.38912	-0.00945	0.00488	-0.00161	0.00035

[a] Original calculation of these data were by R. J. Spelmire, "Tables of Taylor Aperture Distributions," Technical Memorandum No. 581, October 1958, Hughes Aircraft Company. Tabulated values here for F_m are rounded to five digits while Spelmire's tables are given to six digits. Spelmire's tables include F_m for other values of $\bar{n}$ as well as for odd integer values of SLL. For $SLL < -20$, the values of $\bar{n}$ are the minimum values.

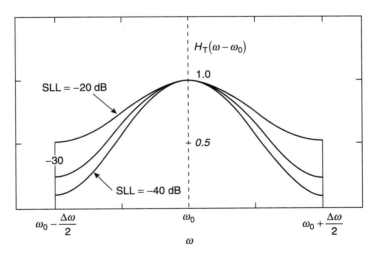

Figure 7.3-1 Transfer functions of the Taylor filter for sidelobe design levels of -20, -30, and -40 dB.

The behavior of $s_T(t)$ for $SLL = -30$ dB is shown in Fig. 7.3-2. The signal-to-noise ratio of the Taylor mismatch filter's output is related to that of the matched filter's output by (Klauder et al., 1960, p. 788)

$$\left(\frac{\hat{S}_0}{N_0}\right)_{max} = \left[1 + 2\sum_{m=1}^{\bar{n}-1} F_m^2\right]\left(\frac{\hat{S}_0}{N_0}\right)_{Taylor} \qquad (7.3\text{-}10)$$

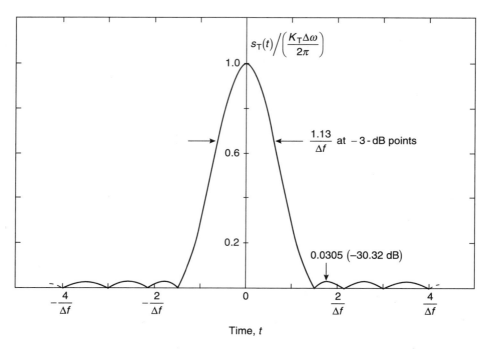

Figure 7.3-2 Envelope of pulse from Taylor mismatch filter with $SLL = -30$ dB and $\bar{n} = 4$.

where $(\hat{S}_0/N_0)_{\max}$ is given by (6.7-4). For the Taylor filter where $SLL = -30$ dB and $\bar{n} = 4$ (where the pulse of Fig. 7.3-2 applies), the bracketed factor in (7.3-10) represents a loss of 0.69 dB. For other Taylor filters with SLL as small as -47.5 dB and $\bar{n}$ up to about 20, the loss has been found to be less than 1.5 dB (Klauder et al., 1960, p. 792).

Other Filters

Many other mismatch filter transfer functions have been developed. We will discuss briefly only a few. From Table 7.3-1 the Taylor coefficients for $m \geq 2$ are small compared to that for $m = 1$, at least for the smaller sidelobe designs. One would expect performance to not greatly change if only the first term is used. In this case

$$H_T(\omega)|_{\text{truncated}} = [1 + 2F_1]^{-1} \left[1 + 2F_1 \cos\left(\frac{2\pi\omega}{\Delta\omega}\right) \right] \text{rect}\left(\frac{\omega}{\Delta\omega}\right) \quad (7.3\text{-}11)$$

It has been found (Cook and Bernfield, 1967, p. 182) that if $F_1 = 0.389$ from Table 7.3-1 is changed to $F_1 = 0.42$, the sidelobes remain below -40 dB with only slight pulse broadening.

By use of a trigonometric identity, we can put (7.3-11) in the form

$$H_T(\omega)|_{\text{truncated}} = \left[\frac{1 - 2F_1}{1 + 2F_1} + \frac{4F_1}{1 + 2F_1}\cos^2\left(\frac{\pi\omega}{\Delta\omega}\right)\right]\text{rect}\left(\frac{\omega}{\Delta\omega}\right) \qquad (7.3\text{-}12)$$

$$= \left[k + (1 - k)\cos^2\left(\frac{\pi\omega}{\Delta\omega}\right)\right]\text{rect}\left(\frac{\omega}{\Delta\omega}\right)$$

where

$$k = \frac{1 - 2F_1}{1 + 2F_1} \qquad (7.3\text{-}13)$$

In its last form (7.3-12) has the form of a *Hamming filter* where $k = 0.08$ corresponds to $SLL = -42.8$ dB and a pulse widening of about 4% compared to a Taylor filter for $SLL = -40$ dB. Hamming weighting is equivalent to using $F_1 = 0.4259$ in (7.3-11).

The *form* of (7.3-12) can be generalized as follows

$$H(\omega) = \left[k + (1 - k)\cos^n\left(\frac{\pi\omega}{\Delta\omega}\right)\right]\text{rect}\left(\frac{\omega}{\Delta\omega}\right) \qquad (7.3\text{-}14)$$

for n a parameter. Cook and Bernfeld (1967) report data that show $k = 0.08$ is optimum for $n = 2$, while $k \approx 0.2$ is best for $n = 3$, both giving SLL of -40 dB or more. They also found that pulse widening decreased as k increased but increased with increasing n. A similar behavior occurs with mismatch loss (decreases with k and increases with n).

Practical Filter Responses

If a linear FM pulse is compressed and then passed through a mismatch filter the sidelobe reduction is usually not quite the expected design value. The reason is that the spectrum of the compressed pulse is not perfectly rectangular and has "ripples" due to the Fresnel integrals. Calculated sidelobe levels of the practical response are shown in Fig. 7.3-3 for a Taylor design that ideally would produce $SLL = -40$ dB (with $\bar{n} = 6$). For small values of ΔfT the degradation from ideal (-40 dB) is large but rapidly becomes smaller as ΔfT increases. The solid curve in the figure connects calculated points where the first sidelobe on each side of the maximum is the largest (at about $\pm 2/\Delta f$, as expected). The data point for $\Delta fT = 25$ corresponded to the largest sidelobe at $\pm 7.6/\Delta f$, while the nearest (first) sidelobes were at -32.1 dB. The curve is shown dashed to this point because of insufficient number of points from $\Delta fT = 25$ to 50.

Mismatch filters are usually centered at the matched filter's design center frequency and their sidelobe reduction occurs when there is no Doppler offset frequency. Sidelobe reduction degrades with Doppler frequency. For the Taylor design for $SLL = -40$ dB with $\bar{n} = 6$ the degradation in sidelobe level from the value of Fig. 7.3-3 is about 0.72 dB per percent of shift in ω_d relative to $\Delta\omega$ when $\Delta fT = 50$. The rate increases to about 0.87 dB per percent change in $\omega_d/\Delta\omega$ for $\Delta fT = 500$. For percentages of $\omega_d/\Delta\omega$ larger than 10 the degradations continue to increase (poorer

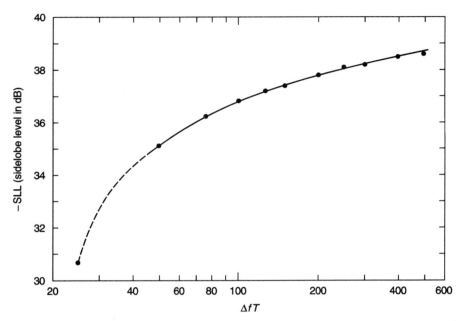

Figure 7.3-3 Practical sidelobe levels at output of Taylor mismatch filter designed for $SLL = -40$ dB with $\bar{n} = 6$. (Adapted from RCA, 1965, with permission.)

sidelobe levels) but the rates of decrease are slightly smaller (Radio Corporation of America, 1965, Fig. 2.12).

7.4 SIGNAL DESIGN FOR LOW SIDELOBES

Another method of achieving low sidelobes in the compressed pulse is to design the FM law to achieve the desired response at the matched filter's output. No mismatch filter is then necessary. Although the idea is straightforward, exact analysis is difficult because Fourier transforms of FM signals are very hard to solve for anything but the most elementary FM laws. Fortunately many radar waveforms of interest have large time-bandwidth products such that approximations may be used as found from the principle of stationary phase.

Stationary Phase Principle

Consider the complex envelope function

$$g(t) = a(t)e^{j\theta(t)} \tag{7.4-1}$$

where $a(t)$ and $\theta(t)$ represent amplitude and phase modulations (due to FM), respectively, of a radar waveform. Its Fourier transform

$$G(\omega) = \int_{-\infty}^{\infty} g(t)e^{-j\omega t}\,dt = \int_{-\infty}^{\infty} a(t)e^{j[\theta(t)-\omega t]}\,dt \tag{7.4-2}$$

is difficult to evaluate for arbitrary $a(t)$ and $\theta(t)$. However, under some conditions that we subsequently define, the *principle of stationary phase* says the integral's approximate value can be found by integrating over a small region around the time where the phase is stationary, that is, around the time when $[d\theta(t)/dt] - \omega = 0$. We next show how to apply this principle to waveform design.

Let τ and λ be values of t and ω such that $[d\theta(t)/dt] - \omega = 0$:[3]

$$\frac{d\theta(t)}{dt}\bigg|_{t=\tau} - \lambda = 0 \tag{7.4-3}$$

or, on using a dot to represent time differentiation,

$$\dot{\theta}(\tau) - \lambda = 0 \tag{7.4-4}$$

For a given λ and any t near τ, we expand the phase in (7.4-2) using a Taylor's series:

$$\theta(t) - \lambda t = [\theta(t) - \lambda t]|_{t=\tau} + \frac{d[\theta(t) - \lambda t]}{dt}\bigg|_{t=\tau}(t - \tau) + \frac{d^2[\theta(t) - \lambda t]}{dt^2}\bigg|_{t=\tau}\frac{(t - \tau)^2}{2} + \cdots$$

$$= [\theta(\tau) - \lambda\tau] + [\dot{\theta}(\tau) - \lambda](t - \tau) + [\ddot{\theta}(\tau)]\frac{(t - \tau)^2}{2} + \cdots \tag{7.4-5}$$

$$\approx \theta(\tau) - \lambda\tau + \frac{\ddot{\theta}(\tau)}{2}(t - \tau)^2$$

where (7.4-4) has been used and we neglect cubic and higher order terms.

Next we substitute (7.4-5) into (7.4-2) and integrate over a small region ε in time centered on time τ:

$$G(\lambda) \approx \int_{\tau-(\varepsilon/2)}^{\tau+(\varepsilon/2)} a(\tau)e^{j[\theta(\tau) - \lambda\tau] + j\frac{\ddot{\theta}(\tau)}{2}(t-\tau)^2}\,dt \tag{7.4-6}$$

In this integral we assume $a(t)$ changes little from its value $a(\tau)$ for $\tau - (\varepsilon/2) \le t \le \tau + (\varepsilon/2)$. We now replace $\ddot{\theta}(\tau)$ by $|\ddot{\theta}(\tau)|$ when $\ddot{\theta}(\tau) > 0$ and by $-|\ddot{\theta}(\tau)|$ when $\ddot{\theta}(\tau) < 0$. Equation (7.4-6) becomes

$$G(\lambda) \approx a(\tau)e^{j\theta(\tau) - j\lambda\tau} \int_{\tau-(\varepsilon/2)}^{\tau+(\varepsilon/2)} e^{\pm j|\ddot{\theta}(\tau)|(t - \tau)^2/2}\,dt \tag{7.4-7}$$

After making the variable change

$$x = \sqrt{\frac{|\ddot{\theta}(\tau)|}{\pi}}(t - \tau) \tag{7.4-8}$$

---

[3] Note that τ and λ are parameters here and are not related to τ and λ used elsewhere in this book.

we get

$$G(\lambda) \approx \sqrt{\frac{\pi}{|\ddot{\theta}(\tau)|}}\, a(\tau) e^{j\theta(\tau)-j\lambda\tau} \int_{-x_1}^{x_1} e^{\pm j\pi x^2/2}\, dx$$

(7.4-9)

$$= \sqrt{\frac{\pi}{|\ddot{\theta}(\tau)|}}\, a(\tau) e^{j\theta(\tau)-j\lambda\tau}\, 2[C(x_1) \pm jS(x_1)]$$

where

$$x_1 = \sqrt{\frac{|\ddot{\theta}(\tau)|}{\pi}}\left(\frac{\varepsilon}{2}\right)$$

(7.4-10)

Next let us suppose that even for small ε, $|\ddot{\theta}(\tau)|$ is large enough so that $x_1 \gg 1$. This assumption allows the Fresnel integrals of (7.4-9) to be replaced by their asymptotic values. Equation (7.4-9) becomes

$$G(\lambda) \approx \sqrt{\frac{2\pi}{|\ddot{\theta}(\tau)|}}\, a(\tau) e^{j\theta(\tau)-j\lambda\tau \pm j(\pi/4)}$$

(7.4-11)

Equation (7.4-11) is our main result. For a specific value of angular frequency λ, the spectrum $G(\lambda)$ is approximately given by the right side of (7.4-11) where τ is the solution of (7.4-4). The approximation holds for FM where the second derivative of time phase has a large magnitude.

It is also of interest to find an approximation for the inverse transform of $G(\omega)$. Represent the spectrum by the form

$$G(\omega) = A(\omega) e^{j\Theta(\omega)}$$

(7.4-12)

where $A(\omega)$ is the spectrum's modulus and $\Theta(\omega)$ is its phase angle. The inverse Fourier transform is

$$g(t) = \frac{1}{2\pi} \int_{-\infty}^{\infty} A(\omega) e^{j[\Theta(\omega) + \omega t]}\, d\omega$$

(7.4-13)

This form is identical to (7.4-2) if we note that $A(\cdot)/2\pi$, $\Theta(\cdot)$, ω, and $-t$, replace $a(\cdot)$, $\theta(\cdot)$, t, and ω, respectively, in (7.4-2). The same developments used previously lead to

$$g(\tau) \approx \sqrt{\frac{1}{2\pi|\Theta''(\lambda)|}}\, A(\lambda) e^{j\Theta(\lambda) + j\lambda\tau \pm j(\pi/4)}$$

(7.4-14)

where λ and τ are related through the solution of

$$\Theta'(\lambda) + \tau = 0$$

(7.4-15)

Here each prime denotes differentiation with respect to angular frequency λ.

Our two principal results (7.4-11) and (7.4-14) have been used to design radar signals for low sidelobes (Key et al., 1959, 1961; Fowle, 1961, 1964). By equating the two expressions with the understanding that τ and λ are related by

$$\dot{\theta}(\tau) = \frac{d\theta(\tau)}{d\tau} = \lambda \tag{7.4-16}$$

$$\Theta'(\lambda) = \frac{d\Theta(\lambda)}{d\lambda} = -\tau \tag{7.4-17}$$

we have two very important relationships

$$a(\tau) \approx \frac{A(\lambda)}{\sqrt{2\pi |d^2\Theta(\lambda)/d\lambda^2|}} \tag{7.4-18}$$

$$A(\lambda) \approx \frac{\sqrt{2\pi}\, a(\tau)}{\sqrt{|d^2\theta(\tau)/d\tau^2|}} \tag{7.4-19}$$

There is an immediate application of (7.4-16) and (7.4-17). Since instantaneous angular frequency with time is $\omega_i(\tau) = d\theta(\tau)/d\tau$, and group delay in the frequency domain is $T_d(\lambda) = -d\Theta(\lambda)/d\lambda$, we have

$$\omega_i(\tau) = \lambda = T_d^{-1}(\tau) \tag{7.4-20}$$

$$T_d(\lambda) = \tau = \omega_i^{-1}(\lambda) \tag{7.4-21}$$

These results show an inverse relationship between frequency versus time and delay versus frequency. If one has the frequency-time FM function, the inverse of the function becomes the delay versus frequency function of the spectrum. By integration the corresponding phase functions are obtained. That is,

$$\theta(\tau) = \int \frac{d\theta(\tau)}{d\tau} d\tau + C_1 = \int \omega_i(\tau)\, d\tau + C_1$$
$$= \int T_d^{-1}(\tau)\, d\tau + C_1 \tag{7.4-22}$$

$$\Theta(\lambda) = \int \frac{d\Theta(\lambda)}{d\lambda} d\lambda + C_2 = -\int T_d(\lambda)\, d\lambda + C_2$$
$$= -\int \omega_i^{-1}(\lambda)\, d\lambda + C_2 \tag{7.4-23}$$

Example 7.4-1 For linear FM, from (7.2-3), $\omega_i(\tau) = \mu\tau = \lambda$ from (7.4-20). The inverse is $\tau = \lambda/\mu = T_d(\lambda)$ from (7.4-21). Hence

$$\theta(\tau) = \int \omega_i(\tau)\, d\tau + C_1 = \frac{\mu\tau^2}{2} + C_1$$

from (7.4-22). From (7.4-23),

$$\Theta(\lambda) = -\int^{\cdot} T_{\mathrm{d}}(\lambda)\,d\lambda + C_2 = -\frac{\lambda^2}{2\mu} + C_2$$

which agrees with the expected phase from (7.2-22). Constants C_1 and C_2 can be chosen for convenience. For example, $C_1 = 0$ and $C_2 = 0$ correspond to setting the phases $\theta(\tau)$ and $\Theta(\lambda)$ to zero at midpulse and midband, respectively.

Signal Design Using Stationary Phase Approximation

Signal design typically follows independently specifying two of the quantities $a(\tau)$, $\theta(\tau)$, $A(\lambda)$, and $\Theta(\lambda)$, and then the above results are used to find the other two quantities. Technically there are six combinations. However, independent specification of $\theta(\tau)$ and $\Theta(\lambda)$ is not allowed because these functions are related through (7.4-16) and (7.4-17). Specification of $a(\tau)$ and $\theta(\tau)$ is of no interest, since the Fourier transform gives $G(\omega)$. Similarly specification of $A(\lambda)$ and $\Theta(\lambda)$ leads to $g(\tau)$ by inverse transformation.

Three combinations remain of interest. Two are relatively straightforward. In one, $a(\tau)$ and $\Theta(\lambda)$ are specified, which means that $\theta(\tau)$ is set by (7.4-22) and $A(\lambda)$ can be found by Fourier transformation or by use of (7.4-19). Similarly, if $A(\lambda)$ and $\theta(\tau)$ are given, (7.4-23) leads to $\Theta(\lambda)$, and either inverse Fourier transformation or (7.4-18) gives $a(\tau)$. Each of these combinations specifies the FM law.

The last combination is usually of the most interest. Here we specify $a(\tau)$ and $A(\lambda)$ and synthesize either $\theta(\tau)$ or $\Theta(\lambda)$, since one implies the other. Some reasons why this design is important are found from the matched filter response of (7.2-7), which is mainly given by $\chi(\tau, \omega_{\mathrm{d}})$. From (6.8-16) and Problem 6.8-1,

$$\chi(0, \omega_{\mathrm{d}}) = \int_{-\infty}^{\infty} a^2(\xi) e^{j\omega_{\mathrm{d}}\xi}\,d\xi \tag{7.4-24}$$

$$\chi(\tau, 0) = \frac{1}{2\pi}\int_{-\infty}^{\infty} A^2(\xi) e^{j\xi\tau}\,d\xi \tag{7.4-25}$$

which imply that $a(\tau)$ and $A(\lambda)$ can be chosen to establish desired cuts in $\chi(\tau, \omega_{\mathrm{d}})$ along the two axes to give low sidelobes in the two directions. The synthesis problem amounts to finding the FM law through finding either $\theta(\tau)$ or $\Theta(\lambda)$.

The procedure (see Fowle, 1961, 1964) first involves differentiation of (7.4-16) to obtain $|d\lambda/d\tau|$ and substitution into (7.4-19) to obtain

$$a^2(\tau)\,d\tau = \frac{1}{2\pi} A^2(\lambda)\,d\lambda \tag{7.4-26}$$

Alternatively, differentiation of (7.4-17) and substitution in (7.4-18) also produces (7.4-26). By integration

$$\int_{-\infty}^{\tau} a^2(\xi)\,d\xi = \frac{1}{2\pi}\int_{-\infty}^{\lambda} A^2(\eta)\,d\eta \tag{7.4-27}$$

must hold. In particular, when $\tau = \infty$ and $\lambda = \infty$, this result gives the energy in $g(t)$ from Parseval's theorem. The final procedure results from defining the indefinite integrals of $a^2(\tau)$ and $A^2(\lambda)/2\pi$ by $P(\tau)$ and $Q(\lambda)$, respectively, so

$$P(\tau) = \int a^2(\tau)\, d\tau = \frac{1}{2\pi} \int A^2(\lambda)\, d\lambda = Q(\lambda) \tag{7.4-28}$$

Thus, from (7.4-21), (7.4-17), and (7.4-23),

$$T_d(\lambda) = \tau = P^{-1}[Q(\lambda)] = -\Theta'(\lambda) \tag{7.4-29}$$

$$\Theta(\lambda) = -\int P^{-1}[Q(\lambda)]\, d\lambda + C_2 \tag{7.4-30}$$

Alternatively, from (7.4-20), (7.4-16), and (7.4-22),

$$\omega_i(\tau) = \lambda = Q^{-1}[P(\tau)] = \dot{\theta}(\tau) \tag{7.4-31}$$

$$\theta(\tau) = \int Q^{-1}[P(\tau)]\, d\tau + C_1 \tag{7.4-32}$$

We summarize the design procedure. First, $a(\tau)$ and $A(\lambda)$ are selected to be functions that give desired matched filter response cuts, as determined from (7.4-24) and (7.4-25). Next, we establish $P(\tau)$ and $Q(\lambda)$, as in (7.4-28), and solve for either τ [to obtain $\Theta(\lambda)$ from (7.4-30)] or λ [to obtain $\theta(\tau)$ from (7.4-32)]. With $a(\tau)$ and $A(\lambda)$ specified and $\theta(\tau)$ and $\Theta(\lambda)$ determined, both $g(t)$ and its spectrum are known. In the following section we illustrate the preceding design procedures by some specific examples.

7.5 EXAMPLE SIGNAL DESIGNS

We illustrate the procedures of Section 7.4 by introducing some examples.

Moduli of Same Form

If moduli $a(\tau)$ and $A(\lambda)$ are to have the same form, (7.4-19) indicates that $\ddot{\theta}(\tau)$ must be constant:

$$\ddot{\theta}(\tau) = \frac{d^2\theta(\tau)}{d\tau^2} = K_0 \qquad \text{(constant)} \tag{7.5-1}$$

Thus

$$\dot{\theta}(\tau) = \omega_i(\tau) = K_0\tau + K_1 \tag{7.5-2}$$

$$\theta(\tau) = \frac{K_0}{2}\tau^2 + K_1\tau + K_2 \tag{7.5-3}$$

where K_1 and K_2 are constants. Equation (7.5-2) indicates that linear FM is necessary to produce moduli of the same form. The same result (linear FM) is achieved if (7.4-18) were initially used, which leads to $\Theta''(\lambda) =$ constant; integration gives a delay that is a linear function of λ from (7.4-17). The two approaches are connected through (7.4-20) or (7.4-21).

Constant Envelope Pulse

When the pulse envelope is a constant A over its duration T, such as

$$a(\tau) = A \operatorname{rect}\left(\frac{\tau}{T}\right) \tag{7.5-4}$$

the procedures are conceptually straightforward. From (7.4-18),

$$\Theta''(\lambda) = \pm\frac{A^2(\lambda)}{2\pi A^2} \tag{7.5-5}$$

so

$$\Theta'(\lambda) = \pm\frac{1}{A^2 2\pi}\int A^2(\lambda)\,d\lambda + D_1 = -T_d(\lambda) \tag{7.5-6}$$

$$\Theta(\lambda) = \pm\frac{1}{A^2 2\pi}\int\int A^2(\lambda)\,d\lambda + D_1\lambda + D_2 \tag{7.5-7}$$

To make the results more specific we summarize an example developed by Key et al., 1961, where moduli have the forms

$$a(\tau) = A \operatorname{rect}\left(\frac{\tau}{T}\right) \tag{7.5-8}$$

$$A(\lambda) = \frac{2A\sqrt{T/\Delta\omega}}{\sqrt{1 + (2\lambda/\Delta\omega)^2}} \tag{7.5-9}$$

Solutions for envelope (group) delay and spectral phase follow (7.5-6) and (7.5-7), which give

$$\Theta'(\lambda) = -T_d(\lambda) = \pm\frac{T}{\pi}\tan^{-1}\left(\frac{2\lambda}{\Delta\omega}\right) \tag{7.5-10}$$

$$\Theta(\lambda) = \pm\frac{\Delta\omega T}{2}\left\{\frac{2\lambda}{\pi\Delta\omega}\tan^{-1}\left(\frac{2\lambda}{\Delta\omega}\right) - \frac{1}{2\pi}\ln\left[1 + \left(\frac{2\lambda}{\Delta\omega}\right)^2\right]\right\} \tag{7.5-11}$$

where constants of integration were chosen to be $D_1 = 0$ to make delay zero at $\lambda = 0$ and D_2 set such that $\Theta(0) = 0$. On taking the inverse of (7.5-10), we have λ, which from (7.4-20) is the FM with time $\omega_i(\tau)$:

$$\omega_i(\tau) = \lambda = \mp\frac{\Delta\omega}{2}\tan\left(\frac{\pi\tau}{T}\right) \tag{7.5-12}$$

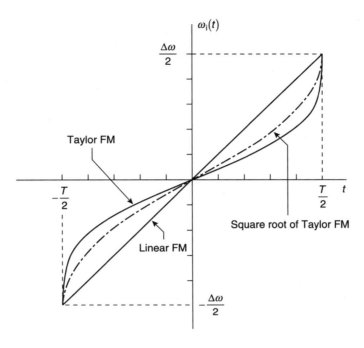

Figure 7.5-1 FM laws for linear FM, Taylor FM, and the square root of Taylor FM. (Adapted from RCA, 1965, with permission.)

This result shows that the upper sign corresponds to decreasing FM with τ (lower sign gives increasing FM with τ). The FM law is a tangent function that is not achievable because $|\omega_i(t)| \to \infty$ as $|\tau| \to T/2$. As a final item of interest, the time phase of the signal $g(t)$ is

$$\theta(\tau) = \pm \frac{\Delta\omega T}{2\pi} \ln\left[\left|\cos\left(\frac{\pi\tau}{T}\right)\right|\right] \qquad (7.5\text{-}13)$$

from (7.4-22) where C_1 has been set to zero so that $\theta(0) = 0$.

FM for Taylor Weighting

The writer (Radio Corporation of America, 1965) used the above procedures to find the FM such that a pulse with rectangular envelope of duration T would have a Taylor weighted spectral modulus for $SLL = -40$ dB when $\bar{n} = 6$. The FM law is shown in Fig. 7.5-1 along with linear FM for comparison. If a transmitted pulse of this design is passed through a receiver filter having a constant-amplitude spectral modulus and the conjugate phase of the transmitted pulse's spectrum, the response ideally would be a Taylor waveform, such as in (7.3-6), with -40 dB sidelobes. However, due to the Fresnel integrals in the exact spectrum, as sketched in Fig. 7.5-2 for $\Delta f T = 50$, the -40 dB level is not quite achieved, as shown in Fig. 7.5-3. Another

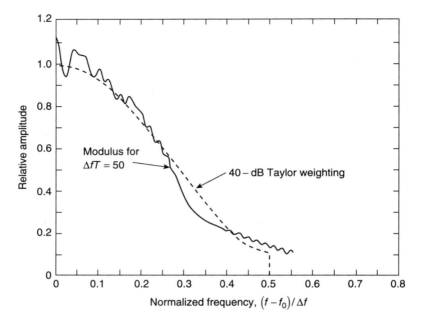

Figure 7.5-2 Spectral modulus for Taylor FM signal with $\Delta f T = 50$. (Adapted from RCA, 1965, with permission.)

problem is that the receiver filter is not a matched filter and a loss in signal-to-noise ratio occurs. This problem is removed by using the square root of Taylor FM.

Square Root of Taylor FM

In this waveform the spectral modulus is the square root of a Taylor function. If a matched filter is used in the receiver, its modulus must also be a square root. The modulus of the spectrum of the matched filter's response is therefore Taylor. Such a design was developed by the author (Radio Corporation of America, 1965) for a Taylor's response for $SLL = -40$ dB and $\bar{n} = 6$. The FM law is shown in Fig. 7.5-1. The signal modulus (solid curve) and its approximation (dashed curve) are shown in Fig. 7.5-4 for $\Delta f T = 100$. Again, because the design approach is only approximate and because of the Fresnel integral-caused ripples, the compressed pulse's envelope, shown in Fig. 7.5-5, does not achieve the desired -40 dB sidelobe level.

7.6 OTHER PULSE COMPRESSION WAVEFORMS

We have seen that linear FM is a popular and useful waveform for pulse compression. With appropriate sidelobe reduction methods added, the linear FM system has enjoyed extensive use. Other discussions (Section 7.4) have described waveforms with nonlinear FM chosen to give compressed pulses with low sidelobes. However, also important are many other FM laws that have been studied for radar and other

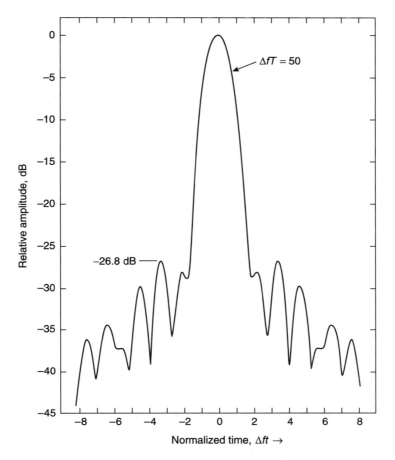

Figure 7.5-3   Compressed pulse envelope for Taylor FM signal with $\Delta fT = 50$. (Adapted from RCA, 1965, with permission.)

than single-pulse signals. In the subsequent subsections we briefly discuss these waveforms.

Other FM Laws for Single Pulses

All of the FM laws discussed above have been continuous functions of time over the pulse duration T. Although specific laws (linear, Taylor, square root of Taylor) were selected mainly because of popularity and sidelobe behavior, it should be clear that any FM law can, in principle, be used in radar. Three examples of continuous laws are shown in Fig. 7.6-1*a* through *c*. A quadratic FM having even symmetry is sketched in panel *a*, while panel *b* is for quadratic FM with odd symmetry about the carrier's frequency ω_0. "Vee" FM is shown in panel *c*; this FM law is continuous on a piecewise basis over two parts of the waveform. One can view this waveform as the sum of (two) contiguous subwaveforms. More generally, we can envision many contiguous subwaveforms, each with its own FM law, where the overall FM law is piecewise continuous with continuity at the boundaries.

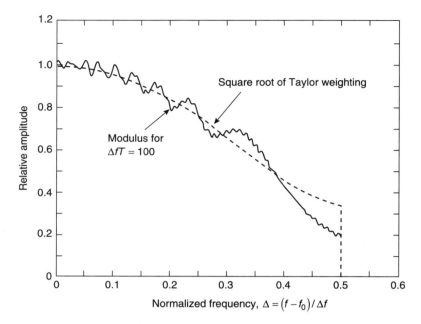

Figure 7.5-4 Spectral modulus for square root of Taylor FM signal with $\Delta f T = 100$.

If continuity at boundaries is not required, another "class" of FM laws occurs. Figure 7.6-1*d* has an FM discontinuity in the center of the pulse. Panel *e* shows the same FM law except with a constant offset in frequency added to half the waveform. More elaborate subdivided FM laws can be envisioned as in panel *f*.

There are several reasons for building an overall pulse from subpulses. Aside from the obvious fact that the matched filter response can be altered, one important reason relates to construction. With modern digital hardware, it is possible to use high-speed switching to synthesize an overall waveform from many small subpulses. Figure 7.6-1*g* attempts to illustrate the synthesis of linear FM as a digitally generated "staircase" stepped FM using short subpulses each at constant frequency. If the number of subpulses is large such that each stairstep is small, the approximation to linear can be good. Clearly, by use of nonconstant steps, any FM law can be approximated; Peebles and Stevens (1965) used this method to generate a wideband linear FM pulse.

An important class of discrete FM waveforms has recently evolved. These waveforms are similar to the staircase or stepped-FM, waveform of Fig. 7.6-1*g* in that they use N contiguous subpulses, each with one discrete frequency. However, the frequencies are not chosen to approximate linear FM. Rather, they are chosen to produce low sidelobes in the matched filter response for both delay and Doppler frequency parameters. The set of positions (values) of frequencies with time through the pulse is usually called a *Costas array*, but an equally applicable name would be *Costas FM* (Costas, 1984). We reserve further discussion of these waveforms until Section 7.7.

Rihaczek (1969) discusses many FM laws of the types shown in Fig. 7.6-1. We take only one, that sketched in panel *a*, as an example.

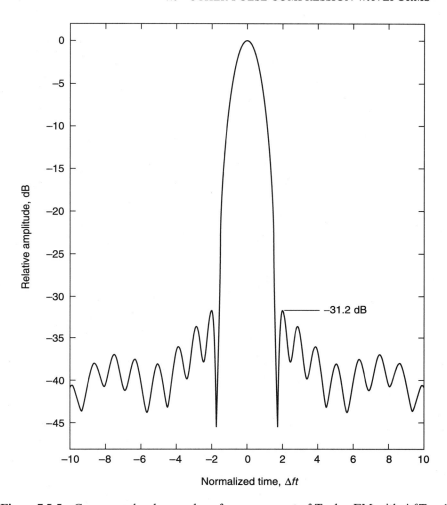

Figure 7.5-5 Compressed pulse envelope for square root of Taylor FM with $\Delta fT = 100$.

Even Quadratic FM

Here the complex envelope of the transmitted signal is assumed to have the form

$$g(t) = a(t)e^{j\theta(t)} = \frac{1}{\sqrt{T}} \text{rect}\left(\frac{t}{T}\right) e^{j\pi k t^3} \tag{7.6-1}$$

where T is the duration of the pulse,

$$k = \frac{4\Delta\omega}{3\pi T^2} \tag{7.6-2}$$

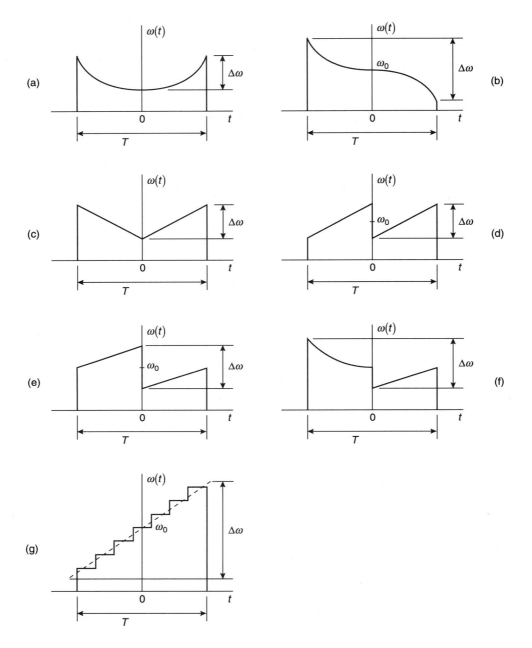

Figure 7.6-1 FM laws: (*a*) even quadratic, (*b*) odd quadratic about ω_0, (*c*) even vee, (*d*) odd vee about ω_0, (*e*) odd vee about ω_0 with offset $\Delta\omega/2$, (*f*) mixed FM laws in segments, and (*g*) stepped linear FM.

and $\Delta\omega$ is the total sweep in angular frequency that occurs over the pulse. The instantaneous angular frequency during the pulse due to modulation is

$$\omega_i(t) = \frac{d\theta(t)}{dt} = 3\pi k t^2 = \Delta\omega\left(\frac{2t}{T}\right)^2 \tag{7.6-3}$$

By substituting (7.6-1) in (6.8-9), we find the matched filter's response to be (see Problem 7.6-1)

$$\chi(\tau, \omega_d)|_{\tau \geq 0} = \frac{1/T}{\sqrt{6k|\tau|}} e^{j\pi k\tau^3 - j\frac{(3\pi k\tau^2 + \omega_d)^2}{4(3\pi k\tau)}} \tag{7.6-4}$$

$$\cdot \{[C(x_2) - C(x_1)] + j[S(x_2) - S(x_1)]\}$$

$$\chi(\tau, \omega_d)|_{\tau < 0} = \chi^*(\tau, \omega_d)|_{\tau \geq 0} \tag{7.6-5}$$

where

$$x_2 = \sqrt{\frac{3k|\tau|}{2}}\left[T - |\tau| + \frac{\omega_d}{3\pi k\tau}\right] \tag{7.6-6}$$

$$x_1 = -\sqrt{\frac{3k|\tau|}{2}}\left[T - |\tau| - \frac{\omega_d}{3\pi k\tau}\right] \tag{7.6-7}$$

and $S(\cdot)$ and $C(\cdot)$ are the Fresnel integrals of (7.2-16) and (7.2-17). For the zero-Doppler cut

$$|\chi(\tau, 0)| = \frac{1/2}{\sqrt{\Delta f|\tau|}}\{C^2(x_2) + S^2(x_2)\}^{1/2} \tag{7.6-8}$$

which is plotted in Fig. 7.6-2 for $\Delta f T = 100$. The plot clearly indicates a major reason why quadratic FM is not very practical. Sidelobes are large and decrease slowly as $|\tau|$ is increased. Other details on quadratic FM, including plots of the full surfaces $|\chi(\tau, \omega_d)|$ for both even and nonsymmetric FM, are given in Rihaczek (1969, ch. 6).

Multiple Pulses (Burst Waveforms)

In radars where nearly all functions are performed in each pulse repetition interval, it is adequate to consider the transmitted waveform as a single pulse. Other systems use several pulse transmissions to develop the desired information (e.g., measuring Doppler angular frequency). Still other systems may transmit a short burst of pulses to accomplish one task and then revert back to another mode of operation for other tasks. In either case we are interested in a sequence of pulses as shown in Fig. 7.6-3. We allow the pulses to be constant-frequency, or have FM, but will require that they have the same modulation and be coherent. That is, they all modulate the same carrier. These conditions allow the complex envelope to be stated as

$$g(t) = \sum_{n=0}^{N-1} g_1(t - nT_r) \tag{7.6-9}$$

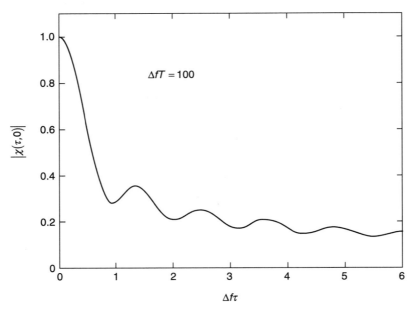

Figure 7.6-2 $|\chi(\tau, 0)|$ for a pulse of duration T having even quadratic FM with total sweep of Δf over T. The plot has even symmetry for $\Delta f \tau < 0$.

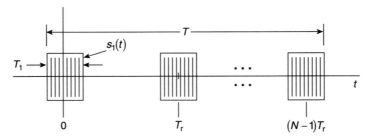

Figure 7.6-3   A transmitted waveform consisting of a train, or burst, of N pulses.

where $g_1(t)$ is the complex envelope of the pulse at the origin and T_r is the separation between adjacent pairs of N pulses.

We investigate the matched filter response of (6.8-9) when (7.6-9) applies. After direct substitution and a simple variable change, we get

$$\chi(\tau, \omega_d) = \sum_{n=0}^{N-1} \sum_{m=0}^{N-1} e^{jn\omega_d T_r} \chi_1 [\tau - (m - n) T_r, \omega_d] \qquad (7.6\text{-}10)$$

where $\chi_1(\tau, \omega_d)$ is the matched filter response of $g_1(t)$. This expression shows that $\chi(\tau, \omega_d)$ is the sum of replicas of $\chi_1(\tau, \omega_d)$ displaced in time to $(m - n) T_r$ and weighted by a phase factor.

To better see the behavior of $\chi(\tau, \omega_d)$, we let $k = m - n$ and rewrite the sums as follows:

$$\chi(\tau, \omega_d) = \sum_{k=-(N-1)}^{0} \sum_{m=0}^{(N-1)+k} e^{j(m-k)\omega_d T_r} \chi_1(\tau - kT_r, \omega_d)$$

$$+ \sum_{k=1}^{N-1} \sum_{n=0}^{(N-1)-k} e^{jn\omega_d T_r} \chi_1(\tau - kT_r, \omega_d) \tag{7.6-11}$$

Next we use the known sum

$$\sum_{n=0}^{M} e^{jn\beta} = e^{jM\beta/2} \frac{\sin[(M+1)\beta/2]}{\sin(\beta/2)} \tag{7.6-12}$$

which allows (7.6-11) to reduce to

$$\chi(\tau, \omega_d) = e^{j(N-1)\omega_d T_r/2} \sum_{k=-(N-1)}^{N-1} (N - |k|)\chi_1(\tau - kT_r, \omega_d)$$

$$\cdot e^{-jk\omega_d T_r/2} \left\{ \frac{\sin[(N - |k|)\omega_d T_r/2]}{(N - |k|)\sin(\omega_d T_r/2)} \right\} \tag{7.6-13}$$

Equation (7.6-13) shows that $\chi(\tau, \omega_d)$ is comprised of replicas of $\chi_1(\tau, \omega_d)$ centered at delays kT_r for $k = \pm 1, \pm 2, \ldots, \pm(N-1)$, with amplitudes $N - |k|$. If $T_r > 2T_1$, these replicas do not overlap in the variable τ, a common case. For this case

$$|\chi(\tau, \omega_d)| = \sum_{k=-(N-1)}^{N-1} (N - |k|)|\chi_1(\tau - kT_r, \omega_d)|$$

$$\cdot \left| \frac{\sin[(N - |k|)\omega_d T_r/2]}{(N - |k|)\sin(\omega_d T_r/2)} \right| \tag{7.6-14}$$

The function of the form $|\sin(Kx)/K\sin(x)|$ has been encountered earlier in our antenna work (see Fig. 3.12-1). It has maximum amplitude one when $\omega_d = q2\pi/T_r$, q an integer, and decreases rapidly between maxima. In (7.6-14) it acts to emphasize values of ω_d near $q2\pi/T_r$ and suppress other values.

Figure 7.6-4 illustrates a rough sketch of the -3.92 dB contour of $|\chi(\tau, \omega_d)|$ from (7.6-14) when the N pulses are constant-frequency. If a single short pulse of duration T_1 were used the small dashed contour would have applied (see Fig. 6.9-1c). However, due to the repetition of small pulses every T_r seconds the small dashed contour is repeated each $\pm kT_r$ with reduced amplitude until replicas fall below the contour of $\chi(\tau, \omega_d)$ which is the larger dashed ellipse. The small dashed contours are not realized because the $|\sin(Kx)/K\sin(x)|$-type term in (7.6-14) has responses versus ω_d that are above the contour level [which is -3.92 dB below $|\chi(0, 0)|$] only for values of ω_d around multiples of $2\pi/T_r$. Thus the small dashed contours break into "periodic" contours, as shown.

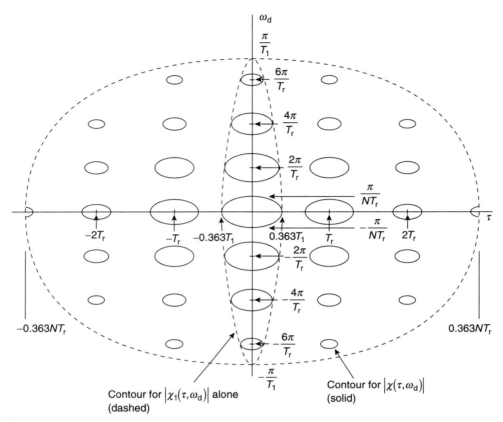

Figure 7.6-4 Uncertainty function contour for N constant-frequency rectangular pulses of duration T_1 and pulse separation T_r.

If the pulses of Fig. 7.6-3 have identical FM, the uncertainty contour is developed as in the preceding paragraph except the contour of the single small pulse is different. For linear FM, for example, the contour is tilted, as in Fig. 7.2-2c.

7.7   PULSE COMPRESSION BY COSTAS FM

In the preceding section we indicated that an important class of pulse compression waveforms involved switching frequency between discrete values during a pulse. Varying frequency discretely through the waveform is sometimes called *frequency hopping*. There are many ways of deciding which frequencies to use at the various times through the pulse. One method, here called *Costas FM*, is probably the most important and is the case we mainly consider. At the end of the section we also comment on some other methods.

Background

We consider a constant-amplitude radar pulse of duration T consisting of N contiguous sub-pulses each of duration T_1 where

$$T_1 = \frac{T}{N} \tag{7.7-1}$$

We assume that N possible frequencies are available as defined by

$$\omega_i = \omega_0 + i\delta\omega, \qquad i = 1, 2, \dots, N \tag{7.7-2}$$

where ω_0 is a constant[4] large enough so the pulse is narrowband and

$$\delta\omega = 2\pi\delta f = \frac{2\pi}{T_1} \tag{7.7-3}$$

The maximum change in frequency Δf during time T is

$$\Delta f = \frac{\Delta\omega}{2\pi} = N\delta f = \frac{N}{T_1} \tag{7.7-4}$$

The pulse has a time-bandwidth product of

$$\Delta f T = N^2 \delta f T_1 = N^2 \tag{7.7-5}$$

from (7.7-1), (7.7-3), and (7.7-4). The duration of the compressed pulse is $1/\Delta f = T/N^2$ for all Costas FM signals.

The overall pulse can be viewed as an $N \times N$ array of chosen frequencies versus time as sketched in Fig. 7.7-1 for $N = 10$. If columns are indexed by $j = 0, 1, \dots, (N-1)$ for convenience, and rows indexed by $i = 1, 2, \dots, N$, then columns refer to subpulses and rows to frequencies (frequency ω_i assigned to row i). For the example "modulation," frequency ω_1 is assigned to subpulse 0, ω_3 to subpulse 8, ω_9 to subpulse 6, and so on. Since row numbers correspond to the available frequency values, only dots have been shown in the array to imply a frequency. All frequency-time array positions (elements) with no dots are treated as zeros in the array.

The properties of the radar pulse are closely related to the logic used to assign the available frequencies to the various time slots (subpulses). This logic is often called *coding*.

Costas FM

In *Costas FM* frequencies are chosen such that the resulting waveform has an uncertainty function that rapidly decreases from its maximum in both delay and Doppler frequency coordinates, and has very low sidelobe levels over most of the

---

[4] Here ω_0 is not necessarily the carrier's angular frequency which can be designated as any convenient value, perhaps the value at the center of the frequency span.

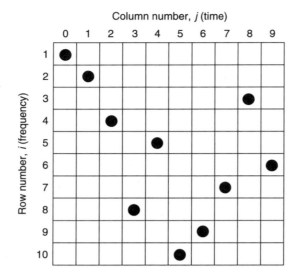

Figure 7.7-1 A frequency-time array for $N = 10$ subpulses.

delay-Doppler plane. An array such as in Fig. 7.7-1, when frequency assignment corresponds to Costas FM, is called a *Costas array*. In fact Fig. 7.7-1 defines a Costas FM array for $N = 10$. In general, Costas FM produces peak sidelobes of the uncertainty function that are down from the main response by a factor of $1/N$ for all regions in the delay-Doppler frequency plane away from the origin. For near-in regions these rise to $2/N$ at isolated points. Uncertainty functions of this type, having a sharp central spike (response) at the origin and rapidly decreasing response to a low level (plateau) in any other part of the delay-Doppler plane, are often called *thumbtack* or *ideal*.

A near-thumbtack response was obtained by Costas (1984) by using each frequency only once per pulse according to the rule: one frequency per time slot (subpulse) and one time slot per frequency. Then each frequency-time slot for which a frequency occurs (a dot as the element) corresponds to a row and column for which no other frequencies (dots) occur.

Costas's original work apparently stems from sonar research, and somewhat similar ground was covered by Sites (Costas, 1965; Sites, 1969, as referenced in Costas, 1984) and Einarsson (1980). By using computer searches and other methods, Costas found ideal frequency-time sequences for N up to 12 (Costas, 1984). Mathematical techniques of others (Golomb and Taylor, 1982; Golomb, 1984; Cooper and Yates, 1966; and others referenced in Golomb and Taylor, 1984) have extended the list of sequences. A compilation of all known sequences for N up to 360 is given by Golomb and Taylor (1984).

For a given value of N there are $N!$ possible frequency-time arrays (sequences). Although not all of these can be Costas arrays, there can be more than one Costas sequence for a value of N. Table 7.7-1 gives the number N_c of possible Costas sequences and their density $N_c/N!$ for N from 3 to 12 (Costas, 1984). Although N_c increases with N the density of sequences decreases rapidly.

TABLE 7.7-1 Number N_c of Costas sequences and their density for $3 \leq N \leq 12$.

N	Number N_c of Costas Sequences	Density, $N_c/N!$
3	4	$2/3 = 0.667$
4	12	$1/2 = 0.50$
5	40	$1/3 = 0.333$
6	116	$29/180 = 0.161$
7	200	$5/126 = 0.0397$
8	440	$11/1008 = 0.0109$
9	760	$19/9072 = 0.0021$
10	2160	$3/5040 = 5.95 \ (10^{-4})$
11	4368	$13/118{,}800 = 1.094 \ (10^{-4})$
12	7852	$1963/119{,}750{,}400 = 1.639 \ (10^{-5})$

Multiple Costas sequences raises the possibility that sequence pairs might produce low *cross-uncertainty functions.* That is, a radar using one sequence might not respond to another sequence used by a second radar. Freedman and Levanon (1985) proved that Costas signals of length $N > 3$ do not exist that have completely different sidelobe patterns. Drumheller and Titlebaum (1991), have derived bounds on the cross-uncertainty function for several Costas signals, but their application requires details on the signal's construction (more on this to follow).

Using the concept of a "cross-hit" array, which is an array of the number of sub-pulse frequency coincidences that occur for exact multiple shifts of r delay units of T_1 seconds and s units of $1/T_1$ hertz of Doppler frequency between two equal-length Costas arrays, a good idea of cross-uncertainty function behavior can be derived. The number of "hits" at position (r, s) in the cross-hit array, which has dimension $(2N - 1) \times (2N - 1)$, is manifested as a sidelobe in the cross-uncertainty function proportional to the number. For $N > 3$, Marić et al. (1994) have proved that only pairs of certain types of Costas (W-C) arrays (see Welch constructions below) can have at most two hits in their cross-uncertainty function. They have also given the maximum number of hits in the cross-hit array for families of these particular arrays that are defined by $N = p - 1$, where p is a prime[5] less than 80, as well as an approximate bound on the number of hits at any point in the cross-hit array for any odd prime p.

There are many analytical procedures for constructing a Costas FM signal. Fourteen are given by Golomb and Taylor (1984). Although Costas arrays may exist in principle for any positive integer N, these analytical constructions are typically limited to values of N related to prime numbers. We will describe only three as examples.

Welch Construction

Three methods exist, called *Welch constructions,* that lead to Costas arrays. To define the first, denoted W_1, let p be an odd prime. The number N of subpulses for this

---

[5] Recall that a prime number is an integer $p \geq 1$ that is divisible by only $+1$, -1, $+p$, and $-p$.

construction is

$$N = p - 1 \qquad (7.7\text{-}6)$$

Next let α be a primitive root of p.[6] Abramowitz and Stegun (1964) list the least positive primitive root for each prime from 3 to 9973. Next form an $N \times N$ array of all zero elements, and label columns and rows by j and i, respectively, where

$$j = 0, 1, 2, \dots, (p - 2) \qquad (7.7\text{-}7)$$

$$i = 1, 2, 3, \dots, (p - 1) \qquad (7.7\text{-}8)$$

The first Welch construction is found by placing a dot in element (i, j), corresponding to frequency ω_i of (7.7-2), if, and only if,

$$i = \alpha^j \quad (\text{modulo } p) \qquad (7.7\text{-}9)$$

We take an example construction.

Example 7.7-1 Suppose that $p = 11$ so that $N = 10$. From Abramowitz and Stegun (1964, p. 864) we use $\alpha = 2$, the smallest positive primitive root of 11. The sequence values are tabularized below.

j	$i = \alpha^j$ (Modulo 11)	Element with Dot (Frequency ω_i)
0	$2^0 = 1$	(1, 0)
1	$2^1 = 2$	(2, 1)
2	$2^2 = 4$	(4, 2)
3	$2^3 = 8$	(8, 3)
4	$2^4 = 5$	(5, 4)
5	$2^5 = 10$	(10, 5)
6	$2^6 = 9$	(9, 6)
7	$2^7 = 7$	(7, 7)
8	$2^8 = 3$	(3, 8)
9	$2^9 = 6$	(6, 9)

The sequence of frequencies in the subpulses is then $\omega_1, \omega_2, \omega_4, \omega_8, \dots, \omega_3, \omega_6$. The Costas array for this example is the same as Fig. 7.7-1.

A second Welch construction, labeled W_2, is obtained from W_1 by deleting the first row and first column from the array. The W_2 construction produces a Costas sequence of length $N = p - 2$, one less than for W_1 (Golomb and Taylor, 1984).

---

[6] For current purposes a primitive root α of odd prime p is such that powers $\alpha, \alpha^2, \alpha^3, \dots, \alpha^{p-1}$, modulo p, produce every integer from 1 to $p - 1$ (Marić et al., 1994).

A third Welch construction, labeled W_3, can be obtained only when $\alpha = 2$. When $\alpha = 2$, the array W_1 of the first construction will have dots in positions (1, 0) and (2, 1). The array W_3 of the third construction derives from deletion of first and second rows and first and second columns from the array W_1 of the first construction. The Costas sequence for W_3 has length $N = p - 3$, or two less than W_1 (Golomb and Taylor, 1984).

Sidelobes in Costas FM

Costas arrays have the property that when shifted an integer number of elements in the delay direction (along j) and an integer number of elements in the frequency direction (along i), there is never more than one coincidence of dots. An array having elements equal to the number of coincidences occurring when the shifted array is compared to the unshifted array can be called an "auto-hit" array, analogous to the cross-hit array when two equal-length costas arrays are compared. Because sidelobe levels in the uncertainty function for the same shifts is proportional to the number of coincidences, the sidelobes tend to never exceed a unit amplitude for any value N. Thus sidelobes tend to have a voltage level below the uncertainty function's maximum by the factor $1/N$. The exact sidelobe levels are a complicated function of phasings between subpulses, especially for nonintegral delay and Doppler frequency shifts. These effects become more pronounced as the origin is approached where sidelobe peaks rise to a level of about $2/N$ below the voltage maximum.

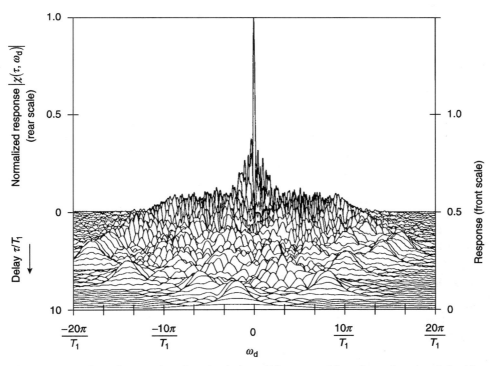

Figure 7.7-2 View of uncertainty function $|\chi(\tau, \omega_d)|$ from a position above the $+\tau$ axis looking toward the origin for a Costas FM signal using a Welch construction for 10 subpulses. (Adapted from Costas, © 1984 IEEE, with permission.)

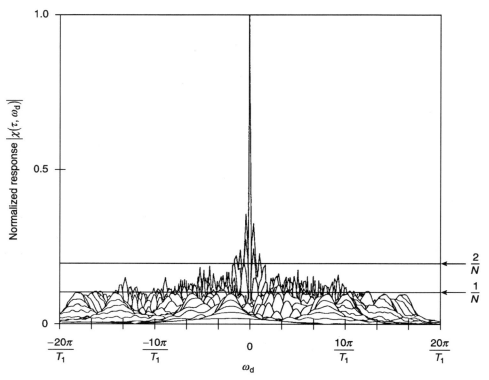

Figure 7.7-3   View of uncertainty function $|\chi(\tau, \omega_d)|$ from $+\tau$ axis looking toward origin for a Costas FM signal using a Welch construction for 10 subpulses. (Adapted from Costas, © 1984 IEEE, with permission.)

Figure 7.7-2 shows the normalized (to one at the origin) uncertainty function $|\chi(\tau, \omega_d)|$ for the 10-subpulse waveform with the array of Fig. 7.7-1. Here the viewer is looking toward the origin from a position above the τ axis in the vertical plane. Fig. 7.7-3 is another view where the viewer sits *on* the $+\tau$ axis; this view is especially helpful in visualizing sidelobe levels. Cuts of $|\chi(\tau, \omega_d)|$ along τ and ω_d axes are shown in Figs 7.7-4 and 7.7-5.

Bursts of pulses in a train have also been studied, where each pulse has Costas FM, but the sequence is different from pulse-to-pulse as much as possible (Freedman and Levanon, 1986).

Costas Design for Small Doppler Shifts

Suppose that a Costas array is shifted by r columns (left–right) and s rows (up–down). Define $C(r, s)$ as the number of dot coincidences when the array and its shifted version are compared. We call $C(r, s)$ a *coincidence function*; it is defined for $-(N - 1) \leq r \leq (N - 1)$ and $-(N - 1) \leq s \leq (N - 1)$ for an $N \times N$ Costas array. As already noted, $C(r, s) \leq 1$ for any Costas FM signal. For this subsection we will call waveforms for which $C(r, s) \leq 1$ *ideal*. In this respect any Costas signal is ideal.

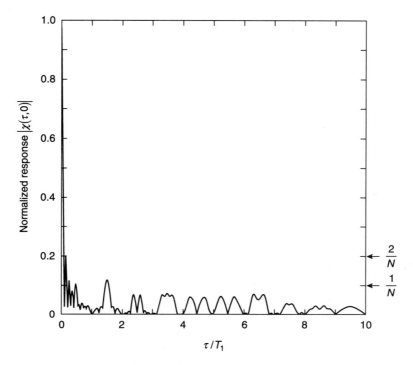

Figure 7.7-4 Delay axis cut of uncertainty function for a Costas FM signal using a Welch construction for 10 subpulses. (Adapted from Costas, © 1984 IEEE, with permission.)

Next suppose that we define a *cross-coincidence function*[7] $C_{AB}(r, s)$ exactly as above for $C(r, s)$ except that it applies to one Costas array A and another, shifted, Costas array B, both of which are $N \times N$. $C_{AB}(r, s)$ is a direct measure of the uncertainty function of a radar designed to receive Costas signal A when Costas signal B arrives. If $C_{AB}(r, s)$ is small for all r and s, then signal B will not interfere with the radar designed to operate with signal A. Ideally we would like $C_{AB}(r, s) \leq 1$ for all r and s. However, we already noted that this condition cannot be true for any $N \geq 4$ for a Costas signal (Freedman and Levanon, 1985) when r and s can have their full range of values.

On the other hand, if s (Doppler shift index) is restricted to a maximum value *less* than its full-range value $N - 1$, a number of Costas signals can be found that give ideal performance. These results are discussed in a beautiful paper by Chang and Scarbrough, 1989. We paraphrase their problem and solution.

Let q Welch-Costas signals be found that are ideal such that $C_{AB}(r, s) \leq 1$ for any pair A and B of the q signals. Signal length is to be $N = p - 1$, where p is a prime. Ideal performance is to be achieved for a restricted range of doppler frequencies f_d such that

$$|f_d| \leq |f_d|_{max} \qquad (7.7\text{-}10)$$

---

[7] Auto- and cross-coincidence functions are the same as the auto- and cross-hit arrays defined earlier.

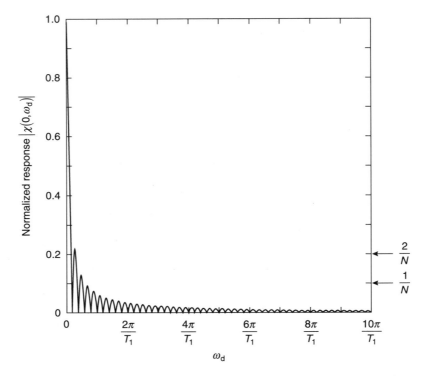

Figure 7.7-5   Doppler axis cut of uncertainty function for a Costas FM signal using a Welch construction for 10 subpulses. (Adapted from Costas, © 1984 IEEE, with permission.)

This performance is achieved if a positive integer m is chosen to satisfy

$$m \leq \frac{N - q}{q} \tag{7.7-11}$$

The chosen value of m corresponds to

$$|f_d|_{max} = \frac{m}{T_1} \tag{7.7-12}$$

The q signals will satisfy $C_{AB}(r, s) \leq 1$ for all pairs A, B, and for $|r| \leq (N - 1), |s| \leq m$.

Chang and Scarbrough (1989) also give the following procedure for finding the q Welch-Costas signals:

1. Construct an $N \times N$ Welch-Costas array using a primitive root of p, as described earlier. Call this array A_1.

2. Extend A_1 horizontally to the right by periodic extension with period N.

3. From the periodically extended array, take an $N \times N$ subarray starting at column $m + 2$ and call it A_2. Similarly A_j is defined as the $N \times N$ subarray starting at column $(m + 1)(j - 1) + 1$, for $j = 1, 2, \dots, q$.

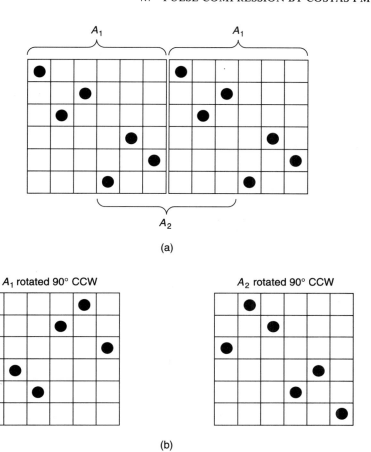

Figure 7.7-6 (*a*) A Costas array A_1 and its one-period extension. (*b*) Arrays A_1 and A_2 after rotation by 90 degrees counterclockwise.

4. Rotate all arrays $A_1, A_2, \ldots, A_q$ clockwise, or counterclockwise, by 90 degrees. The rotated arrays define the q ideal Welch-Costas FM signals.

By this design procedure, signals for q radars operating in close proximity can be chosen so that interference between any pair is minimized.

Example 7.7-2 Assume that $q = 2$ radars are to operate closely with minimal interference by using Welch-Costas FM signals with $N = 6$ ($p = 7$). From Abramowitz and Stegun (1964, p. 864) the smallest positive prime root of 7 is $\alpha = 3$. From (7.7-9) the Welch-costas array A_1 is found as shown in Fig. 7.7-6*a* with its one-period extension. Since $m = (N - q)/q = (6 - 2)/2 = 2$, we have $m + 2 = 4$, which defines A_2 as shown. The final arrays of panel *b* are the counterclockwise rotations of A_1 and A_2.

Other Frequency Hop Codes

Titlebaum (1981, pt I) introduced a method of frequency hopping based on the theory of linear congruences. Frequency hopping occurs over N pulses with N a prime. The resulting code sequences have unattractive uncertainty functions but are said to have excellent cross-uncertainty function properties (for an optimum choice of bandwidth-time product, the cross correlation function is reported to diminish as $2/N$ for large N). In a companion paper Titlebaum and Sibul (1981, pt II); and a follow-up paper Bellegarda and Titlebaum (1988; with extensive correction of errors, Bellegarda and Titlebaum, 1991), coding based on quadratic congruences are said by the authors to be a compromise between the good Costas uncertainty function properties and the good cross-uncertainty function properties of linear congruential codes. Further results on these codes are given by Titlebaum et al. (1991).

Hyperbolic Frequency Hop Codes

Finally we note that Marić et al. (1992) has introduced a set of codes called *hyperbolic frequency hop codes* that should prove valuable in radar. The reason is these codes have at most two coincidences in their coincidence functions and at most two coincidences in their cross-coincidence functions for *any* time-frequency shifts and *any* two codes (Marić et al., 1992, p. 1446). The codes are of length $N = p - 1$, where p is a prime, as in the Costas signals. Although the hyperbolic codes do not have quite as good sidelobe levels in the uncertainty function as Costas codes, they have quite good sidelobe levels in their cross-uncertainty function. For multiple radar environments these codes may prove useful.

With the hyperbolic frequency hop code, (7.7-1) defines the subpulse duration, and (7.7-3) defines the (constant) separation between adjacent available angular frequencies. The actual angular frequency to be used in time slot k, for $k = 1$, $2, \ldots, N$, is given by (7.7-2), where i is now a function $i(k)$ of k called a *placement operator*.[8] This operator derives from the theory of finite fields and is given by

$$i(k) \equiv a(k)^{-1} \quad (\text{modulo } p), \qquad k = 1, 2, \ldots, p - 1 \qquad (7.7\text{-}13)$$

where $\equiv$ is read "is congruent to," $(k)^{-1}$ is an inverse element in the field, and a is an element in the field (Marić and Titlebaum, 1992, p. 1443).

To evaluate (7.7-13) for a value of k, we use the following interpretation. Let $x = (k)^{-1}$ represent the inverse of k which we seek to find. Now every element in a field has a unique inverse, so k times its inverse $(k)^{-1}$ must equal 1 (modulo p), that is, $kx = k(k)^{-1} = 1$ (modulo p). Thus we simply multiply k by increasing positive integer values of x until we find the value that gives $kx = 1$ (modulo p); this value of x equals $(k)^{-1}$. If $a \neq 1$, we multiply this value by a and the (modulo p) result is the value of i to be used in time slot k. We illustrate with an example.

---

[8] In Costas FM we indexed time slot numbers by j for $j = 0, 1, \ldots, N - 1$, for convenience, and to be consistent with the literature's reference. Here we use k to be consistent with Marić and Titlebaum (1992) but have changed their y to i for consistency with our earlier Costas work.

Example 7.7-3 We find the hyperbolic frequency hop code for $p = 11$, $a = 1$. Here $N = p - 1 = 10$. On using the above logic we construct the following table:

k	Smallest $x = (k)^{-1}$	kx	$a(k)^{-1}$
1	1	$1\,(1) = 1 = 1 \pmod{11}$	1
2	6	$2\,(6) = 12 = 1 \pmod{11}$	6
3	4	$3\,(4) = 12 = 1 \pmod{11}$	4
4	3	$4\,(3) = 12 = 1 \pmod{11}$	3
5	9	$5\,(9) = 45 = 1 \pmod{11}$	9
6	2	$6\,(2) = 12 = 1 \pmod{11}$	2
7	8	$7\,(8) = 56 = 1 \pmod{11}$	8
8	7	$8\,(7) = 56 = 1 \pmod{11}$	7
9	5	$9\,(5) = 45 = 1 \pmod{11}$	5
10	10	$10\,(10) = 100 = 1 \pmod{11}$	10

The corresponding $N \times N = 10 \times 10$ array is shown in Fig. 7.7-7. Thus angular frequency $\omega_0 + 9\delta\omega = \omega_9$ is used in subpulse 5, for example.

We also note that earlier Marić and Titlebaum (1990) introduced frequency hop codes based on cubic congruences. These codes have at most two coincidences in their coincidence functions, as with the above hyperbolic frequency hop codes, but had at most *three* coincidences in their cross-coincidence functions (compared to only two or less for the hyperbolic codes). The cubic codes also apply only to primer integers $p = 3m + 2$, m a positive integer, for the codes to be full, and $N = p$.

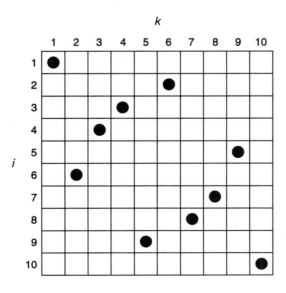

Figure 7.7-7 Hyperbolic frequency hop code for $N = 10$ subpulses ($p = 11$) and $a = 1$.

7.8   PULSE COMPRESSION BY BINARY PHASE CODING

A waveform for pulse compression can be generated by phase coding. There are mainly two approaches, binary and M-phase. In this section we emphasize binary phase coding, and discuss M-phase coding, often known as *polyphase coding*, in Section 7.9.

General Concept

The basic idea in phase coding is to use digital methods to develop a compressed pulse having low sidelobes. From (6.8-9) the matched filter's response for zero Doppler is

$$\chi(\tau, 0) = \int_{-\infty}^{\infty} g^*(\xi)g(\xi + \tau)\,d\xi \qquad (7.8\text{-}1)$$

which is the autocorrelation function of the complex envelope $g(t)$. To apply digital methods, we subdivide the transmitted pulse of duration T into N subpulses of duration $T_1 = T/N$, as sketched in Fig. 7.8-1. Phase coding amounts to maintaining the same carrier frequency in each subpulse but switching the subpulse phases from subpulse to subpulse. This modulation results in a complex modulation $g(t)$ with a constant modulus over time T, but a phase that can be different from subpulse to subpulse. Within a given subpulse phase is constant. We can associate a "code," with the sequence of phases. By proper choice of the "code," we can realize a complex envelope with low sidelobes on compression.

In *binary phase coding* the phase of any subpulse has only one of two possible values. The values 0 and π are popular choices that have the widest possible separation (since phase is modulo 2π) and generally lead to the best performance. We assume that phases are 0 and π for all following discussions of binary coding. The complex envelope can be stated as

$$g(t) = \sum_{n=1}^{N} A\,\text{rect}\left[\frac{t + \left(\dfrac{N+1}{2} - n\right)T_1}{T_1}\right]e^{j\theta_n}$$

$$= A\sum_{n=1}^{N} d_n\,\text{rect}\left[\frac{t + \left(\dfrac{N+1}{2} - n\right)T_1}{T_1}\right] \qquad (7.8\text{-}2)$$

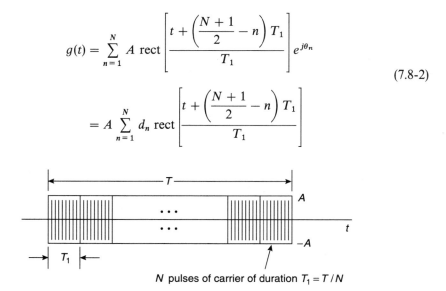

N pulses of carrier of duration $T_1 = T/N$

Figure 7.8-1   A pulse of duration T made up of N subpulses of duration T/N.

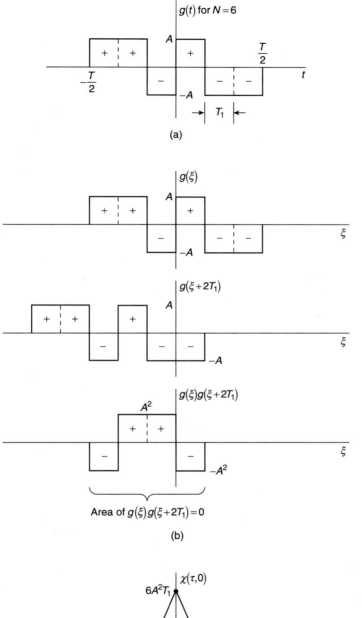

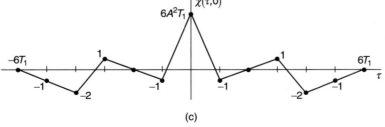

Figure 7.8-2 A complex envelope (*a*), the calculation of area for $\tau = 2T_1$ (*b*), and (*c*) the autocorrelation function.

where $A > 0$ is a constant and

$$d_n = \begin{cases} +1 & \text{for } \theta_n = 0 \\ -1 & \text{for } \theta_n = \pi \end{cases} \tag{7.8-3}$$

From (7.8-2) it is clear that $g(t)$ is just a sum of positive and negative pulses of the same magnitude, as illustrated in Fig. 7.8-2a for $N = 6$ and a sequence choice $d_n = +1$ $(n = 1)$, $+1$, -1, $+1$, -1, and -1 $(n = 6)$. For binary sequences the calculation of (7.8-1) is especially simple. For τ an exact multiple of T_1, the area of the product $g^*(\xi)g(\xi + \tau)$ is the sum of areas of rectangular pulses of areas $+A^2 T_1$ or $-A^2 T_1$. As an example, with $\tau = 2T_1$ and $g(\xi)$ given in Fig. 7.8-2a, the area is zero as shown from the product in panel b. Once the values of $\chi(\tau, 0)$ are found for multiples of T_1 in τ, the values are connected by straight lines to obtain $\chi(\tau, 0)$ because the autocorrelation functions of the subpulses are all triangular functions. The full autocorrelation function is sketched in panel c.

The main problem in binary phase coding is finding sequences $\{d_n\}$ that have good autocorrelation functions. If amplitudes $+1$ and -1 are corresponded with binary digits 1 and 0, respectively, the digits of the binary sequence are collectively called a *codeword* of *length N* for N subpulses. Thus the amplitude sequence $\{+1, +1, -1, +1, -1, -1\}$ corresponds to the binary sequence $\{1, 1, 0, 1, 0, 0\}$ which corresponds to the binary codeword 110100.

Optimal Binary Codes

We define an *optimal binary amplitude sequence* $\{d_n\}$ as one having an autocorrelation function with a peak (largest) sidelobe magnitude that is the smallest possible for a given sequence length. The corresponding binary codeword is called an *optimal binary codeword*, and if more than one such codewords exist, the collection of such codewords is called an *optimal binary code*. With this definition, the words "code" and "codeword" are synonymous if there is only a single optimal codeword. As an example, with length $N = 6$, if every possible codeword is generated,[9] it is found that the smallest possible peak sidelobe is 2 and one codeword with this sidelobe level is 110100 (see Fig. 7.8-2). It is also true that seven other codewords have the same peak sidelobe level. This code has therefore eight codewords, and it is optimal for length 6.

Optimal binary codes that are based on the smallest peak sidelobe are often called *minimum peak sidelobe*, or MPS, codes. MPS codes are summarized in Table 7.8-1 for applicable values of N (column 1). Columns 2 through 5 give, respectively, the number of optimal codewords,[10] the largest (peak) sidelobe level (PSL), the integrated sidelobe level (ISL),[11] and a sample codeword that also produces the smallest ISL of column 4. Data for N up to 40 were apparently first found by Lindner (1975) by exhaustive computer search. Cohen et al. (1989) confirmed Lindner's results for N up to 34, found

---

[9] For length N there are 2^N possible codewords.

[10] The listed number excludes three allomorphic forms, which are codewords derived from writing the codeword in reverse order, complementing the codeword, or complementing the reversed codeword.

[11] ISL is the ratio of the total sidelobe energy divided by the main response's energy. The ISL value shown is the smallest value for all optimal codewords of column 2.

TABLE 7.8-1 MPS biphase codes through length 48.

N Length	Number	PSL	ISL, dB	Sample Code[a] (Binary or Octal)
2	2	1	—	11, 10
3	1	1	—	110
4	2	1	—	1101, 1110
5	1	1	—	11101
6	8	2	—	110100
7	1	1	−9.12	047
8	16	2	−6.02	227
9	20	2	−5.28	327
10	10	2	−5.85	0547
11	1	1	−10.83	1107
12	32	2	−8.57	4657
13	1	1	−11.49	12637
14	18	2	−7.12	12203
15	26	2	−6.89	14053
16	20	2	−6.60	064167
17	8	2	−6.55	073513
18	4	2	−8.12	310365
19	2	2	−6.88	1335617
20	6	2	−7.21	1214033
21	6	2	−8.12	5535603
22	756	3	−7.93	03466537
23	1021	3	−7.50	16176511
24	1716	3	−9.03	31127743
25	2	2	−8.51	111240347
26	484	3	−8.76	216005331
27	774	3	−9.93	226735607
28	4	2	−8.94	1074210455
29	561	3	−8.31	2622500347
30	172	3	−8.82	4305222017
31	502	3	−8.56	05222306017
32	844	3	−8.52	00171325314
33	278	3	−9.30	31452454177
34	102	3	−9.49	146377415125
35	222	3	−8.79	000745525463
36	322	3	−8.38	146122404076
37	110	3	−8.44	0256411667636
38	34	3	−9.19	0007415125146
39	60	3	−8.06	1146502767474
40	114	3	−8.70	02104367035132
41	30	3	−8.75	03435224401544
42	8	3	−9.41	04210756072264
43	24	3	−8.29	000266253147034

(*Continued*)

TABLE 7.8-1   *(Continued)*

N Length	Number	PSL	ISL, dB	Sample Code[a] (Binary or Octal)
44	30	3	−7.98	017731662625327
45	8	3	−8.18	052741461555766
46	2	3	−8.12	0074031736662526
47	2	3	−8.53	0151517641214610
48	8	3	−7.87	0526554171447763

Source: Data from Skolnik (1990), for $N \le 6$ and Cohen et al. (© 1990 IEEE, with permission) for $N > 6$.

[a] Codes assign 0 for −1 and give binary representation for $2 \le N \le 6$ and octal representation for $6 < N$ where each octal digit represents three binary digits:

$$0 = 000 \qquad 4 = 100$$
$$1 = 001 \qquad 5 = 101$$
$$2 = 010 \qquad 6 = 110$$
$$3 = 011 \qquad 7 = 111$$

an optimal codeword for $N = 44$ (peak sidelobe of 3), and discovered that no optimal codes for N above 28 and less than 76 had a peak sidelobe level of 2. Data for $40 < N \le 48$ were added by Cohen et al. (1990). Some larger codewords ($N = 51, 69, 88$) have been found by Kerdock et al. (1986) but although these are "good" codes with low sidelobes, they have not been proved to be MPS codes.

Some data on code lengths equal to prime numbers from 53 to 113 are given by Boehmer, 1967. These are not MPS codes but do have relatively low sidelobes.

Barker Codes

The optimal codes having peak sidelobe levels of one are called Barker codes (Barker, 1953). It is known (Turyn and Storer, 1961) that no odd-length Barker codewords exist for $N > 13$. For a Barker code to exist with even length $N > 2$, it must have a length defined by $N = 4M^2$, where M is a natural number (Turyn, 1963). However, Turyn (1963) has proved that no values of M a power of a prime can satisfy this condition, so no Barker codewords exist for M a power of a prime. He also has shown that other values of M up to 39 produce no Barker sequence. Thus no Barker sequences exist for any values of $13 < N < 4(39^2) = 6084$ and for any $N = 4M^2$ with $M > 39$ a power of a prime. These results form nearly overwhelming support for Turyn's conjecture (1963) that no Barker sequences exist for $N > 13$.

Other Good Binary Codes

When code lengths are needed that are longer than available from an MPS code, other methods must be found. Some "good" codes are found simply by search and chance. However, one systematic study of properties of Legendre sequences (Rao and Reddy, 1986) has produced "good" codes with prime lengths from $N = 67$ to 1019.

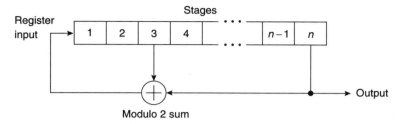

Figure 7.8-3 A shift register used to generate a maximal-length sequence. Clock input not shown.

Sometimes shorter-length codes can be combined (called *concatenation*) to form longer-length codes with good properties. Some review of these methods and references are given in Nathanson (1991).

Maximal-Length Sequences

By use of a shift register with only a few stages, some very long codewords may be generated. The method is illustrated in Fig. 7.8-3 where an *n*-stage register has feedback to the input stage generated by the sum (modulo 2) of the binary contents of various register stages. If the register is continually pulsed by a clock the output will be a sequence of binary 0s and 1s that will eventually repeat. The sequence is then periodic. If an aperiodic sequence is desired, the clock can stop at the end of one period (cycle). In either case a long, randomlike, sequence of binary digits can be generated. The generated sequence will depend on the initial values of the register (the all-zero condition being not allowed), and the number and positions of the stages used for feedback.

For cases of interest here, the proper number and position of feedback stages will generate the largest length sequence possible which will contain $N = 2^n - 1$ digits. Such sequences are variously called *maximum-length*, *m-sequences*, or *pseudonoise* (PN) *sequences*. For a given value of *n*, more than one choice of feedback setting (code) is possible. In fact *M* connections (codes) are possible (Nathanson, 1991, p. 549; Skolnik, 1990, p. 10.19):

$$M = \frac{2^n - 1}{n} \prod_{\text{all } i} \left(\frac{P_i - 1}{P_i} \right) \qquad (7.8\text{-}4)$$

Here P_i is a prime factor of N. If there is more than one prime factor of the same value it is used in (7.8-4) only once, as illustrated by the following example.

Example 7.8-1 A 12-stage shift register is used to generate a maximal-length binary code of length $N = 2^{12} - 1 = 4095$ digits. We determine how many such codes are possible. The prime factors of $2^{12} - 1$ are $2^{12} - 1 = 3^2(5)7(13)$, where the prime number 3 occurs twice. From (7.8-4) we use 3 only once to obtain the number M of possible codes:

$$M = \frac{2^{12} - 1}{12} \left(\frac{3 - 1}{3} \right) \left(\frac{5 - 1}{5} \right) \left(\frac{7 - 1}{7} \right) \left(\frac{13 - 1}{13} \right) = 144$$

TABLE 7.8-2   Values of sequence length N, number M of sequences, and example feedback stages use to generate maximal-length sequences.

Number of Stages, n	Length N of Maximal-Length Code	Number M of Maximal-Length Codes	Example of Stages Used in Feedback[a]
2	3	1	2, 1
3	7	2	3, 2
4	15	2	4, 3
5	31	6	5, 3
6	63	6	6, 5
7	127	18	7, 6
8	255	16	8, 6, 5, 4
9	511	48	9, 5
10	1,023	60	10, 7
11	2,047	176	11, 9
12	4,095	144	12, 11, 8, 6
13	8,191	630	13, 12, 10, 9
14	16,383	756	14, 13, 8, 4
15	32,767	1,800	15, 14
16	65,535	2,048	16, 15, 13, 4
17	131,071	7,710	17, 14
18	262,143	7,776	18, 11
19	524,287	27,594	19, 18, 17, 14
20	1,048,575	24,000	20, 17

[a] Values from Skolnik (ed., 1990, ch. 10, Table 10.5).

Table 7.8-2 summarizes the codeword lengths (column 2), the number of maximal-length codes (column 3) obtained from (7.8-4), and an example of register stages (column 4) that will generate a maximal-length code. Note that connections must involve the last stage and an odd number of other stages for an even total number of feedback stages.

For a given value of n and choice of feedback settings to realize a particular maximal-length code, there are still $2^n - 1$ possible codewords. These are simply shifts of each other derived by changing the chosen initial register contents (Nathanson, 1991, p. 549).

The feedback stages to be used in generating a maximal-length sequence are typically specified through an octal number.[12] The binary representation of the octal number is a sequence of 0s and 1s. The 0s and 1s are values of coefficients of terms of a primitive and irreducible polynomial of the form

$$C_n x^n + C_{n-1} x^{n-1} + \cdots + C_2 x^2 + C_1 x + 1 \tag{7.8-5}$$

[12] Octal digits $0, 1, \ldots, 7$ correspond to respective binary sequences according to $0 = 000, 1 = 001, 2 = 010, 3 = 011, 4 = 100, 5 = 101, 6 = 110,$ and $7 = 111$.

With shift register stages numbered such that n is the output stage, the coefficients C_i that equal 1 correspond to stages that are fed back. Stages where $C_i = 0$ are not fed back. We take an example.

Example 7.8-2 Suppose that feedback connections for an $n = 8$-stage sequence are specified by the octal number 543. The binary representation is 101100011. The rightmost digit (a 1) corresponds to 1 in (7.8-5). Counting $n = 8$ digits left of the rightmost digit, we have a 1 representing coefficient C_8 in (7.8-5). Other digits in between follow as $C_7 = 0$, $C_6 = 1$, $C_5 = 1$, $C_4 = C_3 = C_2 = 0$, and $C_1 = 1$. Thus stages 8, 6, 5, and 1 are all added modulo 2 and fed back to the input of stage 1.

Connections, or data for deriving connections, for generation of maximal-length sequences are available in many places. A few sources are Nathanson (1991), Skolnik (1990), Eaves and Reedy (1987), Peterson and Weldon (1972), and Golomb (1964, 1967).

If the n-stage shift register of Fig. 7.8-3 is properly connected to generage a maximal-length code and is continually pulsed by a clock the result is a periodic sequence of digits with $N = 2^n - 1$ digits per period. The autocorrelation function of the periodic sequence is periodic. Its maximum at $\tau = 0$ is N,[13] and it has $2^n - 2 = 2(2^{n-1} - 1)$ sidelobes on each side of the maximum with all sidelobes having the same level of -1. Sidelobes are down from the maximum by a factor of $1/N$ in voltage. The constant (and small) sidelobe level is very attractive for some CW radars. Even though sidelobes are relatively low for CW cases, they can be further improved in some cases by use of time weighting functions (Getz and Levanon, 1995). For pulsed radar it is not possible to take advantage of the low sidelobes of the periodic sequence.

For pulsed radar the maximal-length sequence can still be used, but in an aperiodic manner. In this case the generating shift register is clocked through the N pulses of the sequence's period and then stopped until another radar transmitted pulse is needed. This aperiodic mode alters the sidelobe structure. The autocorrelation function of the aperiodic sequence is different (generally poorer) than that for the periodic sequence. Sidelobe structure now depends on both the number and positions of stages fed back and on the initial conditions used in the register, but the general peak sidelobe level tends to be below the maximum by a factor of $1/\sqrt{N}$ in voltage for large N. Actual sidelobe levels, initial conditions, and feedback connections have been evaluated by Taylor and MacArthur (1967) for n up to 8. In some cases these sidelobe levels are too large and mismatch filtering can be used to reduce them to acceptable levels (see Hua and Oksman, 1990, and Nathanson, 1991, p. 555, who give other references).

Other Periodic Binary Codes

Many other methods exist to generate periodic binary codes. Some early methods are given in Solomon (1964, app. 2) for two-level sequences. Here we mention one recent method that is interesting because the "matched filter," or receiver sequence against which the received sequence is compared, is different from the received sequence. In

---

[13] The maximum for the complex envelope that uses this sequence is $A^2 T_1 N$ from (7.8-1). For the moment we discuss just the autocorrelation of the sequence.

essence the receiver is mismatched to the received signal, but in only one digit. The advantage of the coding procedure is that lengths unattainable by m-sequences can be achieved while all uncertainty function sidelobes are unity. The penalty paid because of mismatch is small.

The code is due to Gottesman et al. (1992). It applies to sequences of length N where

$$N = \text{prime integer} = 4m + 1 \tag{7.8-6}$$

Thus valid lengths of 5, 13, 17, 29, 37, 41, ... , may be realized. The coding procedure generates a sequence $\{a_j\}$ of digits a_j that are either $+1$ or -1 in the subpulse j where $j = 0, 1, \ldots, N - 1$. All a_j are assigned values -1 except for values of j satisfying

$$j \equiv x^2 \ (\text{mod } N), \qquad x = 0, 1, 2, \ldots, \frac{N - 1}{2} \tag{7.8-7}$$

which are assigned $+1$. Equation (7.8-7) is interpreted as follows: for each value of x from 0 to $(N - 1)/2$ the quantity x^2 is reduced modulo N to corresponding integers; these integers are the values of j to which elements a_j are assigned the value $+1$. We take an example.

Example 7.8-3   For $N = 13$ $(m = 3)$ we construct the table below:

x	x^2	$x^2 \ (\text{mod } 13)$	j for which $a_j = +1$
0	0	0	0
1	1	1	1
2	4	4	4
3	9	9	9
4	16	3	3
5	25	12	12
6	36	10	10

The final sequence is

$$\{+1, +1, -1, +1, +1, -1, -1, -1, -1, +1, +1, -1, +1\}$$

The receiver is not matched in the code of Gottesman et al. (1992). The receiver assumes that the sequence is as generated above except the sign of the initial element (for $j = 0$) is reversed. The receiver's response is now the cross-correlation of the received (code-generated) sequence and the one-digit-altered sequence. This cross-correlation will have a relative maximum of $N - 2$ and all sidelobes of -1. Compared to a relative maximum of N that would have occurred for a matched receiver there is a signal-to-noise ratio power loss, denoted SNR_{loss}, of

$$\text{SNR}_{\text{loss}} = \left(\frac{N - 2}{N}\right)^2 \tag{7.8-8}$$

which becomes negligible as N increases.

Biphase to Quadriphase Conversion

In many radars the out-of-passband spectral energy is important. It is usually desirable to have sidelobes in the power spectrum of the transmitted signal decrease rapidly out of band. Of course spectral filtering can guarantee a desired spectral response, but filtering alters the transmitted signal and can seriously degrade time sidelobes in the compressed received pulse. An alternative is to *design* for low spectral sidelobes in a way that does not degrade compressed pulse sidelobes. By using techniques similar to *minimum shift keying* (MSK) in communication systems (see Peebles, 1987, sec. 5.6), Taylor and Blinchikoff (1988) have been able to create a transmitted signal having a power spectrum that rolls off at 12 dB/octave of frequency change compared to the usual spectral decrease of 6 dB/octave. This accomplishment follows (1) the use of half-cosine subpulses of duration $2T_1$ and having pulse-to-pulse overlap instead of rectangular pulses of durations T_1 with no overlap, and (2) the conversion of a "good" biphase code to a quadriphase code of the same aperiodic length.

Let a given biphase code of length N have sequence elements b_k, $k = 1, 2, \ldots, N$, where $b_k = +1$ or -1. The quadriphase-coded sequence q_k is given by

$$q_k = j^{s(k-1)} b_k, \qquad k = 1, 2, \ldots, N \tag{7.8-9}$$

where $j = \sqrt{-1}$ and s is fixed at either $+1$ or -1 (Taylor and Blinchikoff, 1988). The coded sequence can be conceptually generated by exciting a tapped delay line, having phase-weighting on the taps, by a half-cosine pulse, as shown in Fig. 7.8-4. The quadriphase coded response has a constant magnitude over $T_1 \le t \le NT_1$ and has a cosine shape over the first and last T_1-second intervals for $0 \le t \le T_1$ and $NT_1 \le t \le (N+1)T_1$, respectively.

The quadriphase sequence has been shown to give almost the same receiver autocorrelation function as the binary code (Taylor and Blinchikoff, 1988) but does broaden the main pulse. The full uncertainty function of these codes has been investigated by Levanon and Freedman (1989) who show some plotted examples.

7.9 POLYPHASE CODING FOR PULSE COMPRESSION

Phase coding of radar waveforms has also been developed for more general cases than binary. It is similar to binary in that a waveform of duration T is subdivided into N equal-length subpulses of duration $T_1 = T/N$, but instead of assigning one of two possible phases to each subpulse (the binary case), each subpulse has one of M possible phases. This form of modulation is called *polyphase coding* or *M-phase coding*. We subsequently discuss several polyphase coding methods.

First, however, we note that an important form of modulation for some waveforms is to assign a phase to each of the N subpulses that is a value from a continuum of phase values (not discrete phases). For lack of a better name, we call this process *digital FM* because the assigned phases are chosen to be equivalent to a prescribed FM law for the waveform. We consider two such cases, digital linear FM and digital nonlinear FM, and then proceed to polyphase coding.

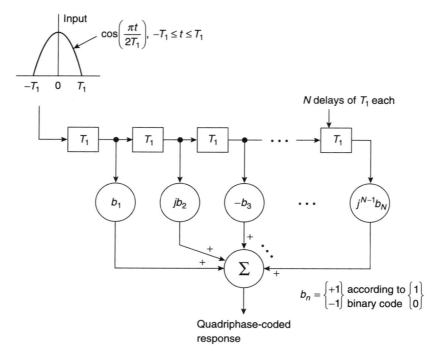

Figure 7.8-4 A concept for converting a binary sequence to a quadriphase sequence [using $s = 1$ in (7.8-9)].

Digital Linear FM

Consider a linear FM pulse of unit amplitude, duration T centered on $t = 0$, and having a downward frequency sweep of Δf (Hz) across T. Then its complex envelope can be written as

$$g(t) = \text{rect}\left(\frac{t}{T}\right) e^{-j\pi\Delta f t^2/T} \tag{7.9-1}$$

The basic idea behind digital linear FM is to transmit a sampled version of the linear FM signal rather than the continuous waveform. This is done by breaking T into N rectangular subpulses each having a constant (sample) value of phase to represent (7.9-1). The "sampled" version of $g(t)$ becomes

$$g_s(t) = \sum_{m=1}^{N} \text{rect}\left(\frac{t - t_m}{T_1}\right) e^{-j\pi\Delta f t_m^2/T} \tag{7.9-2}$$

where

$$T_1 = \frac{T}{N} \tag{7.9-3}$$

$$t_m = \left[m - \left(\frac{N+1}{2} \right) \right] T_1 \tag{7.9-4}$$

To justify (7.9-2), we treat (7.9-1) as a complex baseband waveform approximately bandlimited to $\Delta f/2$. Nyquist's theorem indicates that $g(t)$ can be conveyed by its samples as long as the sample rate is twice the spectral extent. Applying this theorem to $g(t)$ means samples at a minimum rate of Δf are needed. The total number of samples needed in T seconds is

$$N \geq \Delta f T \tag{7.9-5}$$

or

$$T_1 = \frac{T}{N} \leq \frac{1}{\Delta f} \tag{7.9-6}$$

Thus the digital representation $g_s(t)$ of (7.9-2), where N subpulses each with exact sample phase $-\pi \Delta f t_m^2/T$ (modulo 2π), should be a valid representation of $g(t)$ if N satisfies (7.9-5). The pulse constructed by $g_s(t)$ modulating a carrier is therefore a digital linear FM signal approximation derived from sampling theory.

Iglehart (1978) studied the generation of linear FM using (7.9-2). He used an analog compression filter matched to the continuous signal; that is, its impulse response was $\sqrt{j\Delta f/T} \exp(j\pi\Delta f t^2/T)$. The compression filter was followed by a gaussian mismatch filter with -3.41-dB bandwidth B_n. Data were given for loss in peak signal-to-noise ratio, peak sidelobe level, and other parameters for various values of $\Delta f T_1$. Figure 7.9-1 shows the envelope of an example compressed pulse assuming $\Delta f = 6.5$ MHz, $B_n = 0.4\Delta f$, $T = 20$ μs, and $T_1 = 0.135$ μs. Here the peak sidelobe is -35.3 dB below the peak response and most others are below -40 dB. The mismatch loss in signal-to-noise ratio was 1.67 dB.

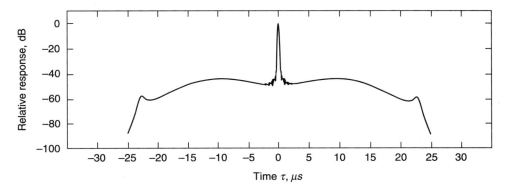

Figure 7.9-1   Example of envelope of digital linear FM waveform after compression and weighting by a gaussian filter. Parameters are $T = 20$ μs, $\Delta f = 6.5$ MHz, $T_1 = 0.135$ μs, and $B_n = 0.4\Delta f$. (Adapted from Iglehart, © 1978 IEEE, with permission.)

Digital Nonlinear FM

The above procedures for digital linear FM can be applied to any FM. If the FM law is nonlinear and corresponds to a pulse compression waveform with low side-lobes in the compressed pulse (e.g., those of Section 7.5 based on the stationary phase principle of Section 7.4), then we refer to the digital waveform as *digital nonlinear FM*.

Felhauer (1992) applied the above concepts to a nonlinear FM waveform for which the compressed pulse's complex envelope has a spectral modulus of the form

$$|G(\omega)|^2 = \text{rect}\left(\frac{\omega}{\Delta\omega}\right)\left[k + (1 - k)\cos^n\left(\frac{\pi\omega}{\Delta\omega}\right)\right] \tag{7.9-7}$$

where $0 \le k \le 1$, $n = 0, 1, \ldots$, and $\Delta\omega$ is the bandwidth (rad/s) of $g(t)$. Felhauer (1992, 1994), evaluated compressed pulse performance for $N = 16$, 64, and 100. For each N the optimum value of k (k_{opt}) was found that produced the smallest peak sidelobe level relative to the maximum response. Also found was the compressed pulse's broadening factor, Δ, at -3-dB points, relative to the pulse width when using a Frank code (described below). These data are summarized in Table 7.9-1.

Felhaurer (1994) also evaluated quantization effects using a finite number of phases to represent the FM waveform's phase function. He found that discrete results were close to continuous results when eight or more bits were used (256 phases available per subpulse). It was also found that the digital receiver had little effect on the performance when it used eight bits or more in its signal processor.

Frank Codes

In the Frank code (Frank, 1963) the length N of a codeword is the square of the number M of phases:

$$N = M^2 \tag{7.9-8}$$

TABLE 7.9-1   Optimum values of k, k_{opt}, and other parameters for digital nonlinear FM waveforms for which (7.9-7) applies.

	$N = 16$			$N = 64$			$N = 100$		
n	k_{opt}	PSR[a] (dB)	Δ	k_{opt}	PSR[a] (dB)	Δ	k_{opt}	PSR[a] (dB)	Δ
1	0.0055	-22.7	1.6	0.006	-30.7	1.5	0.01	-33.0	1.5
2	0.12	-23.5	1.7	0.039	-31.9	1.9	0.05	-34.8	1.8
3	0.085	-24.1	2.1	0.015	-33.0	2.4	0.03	-35.9	2.3
4	0.11	-24.1	2.1	0.025	-34.2	2.7	0.015	-37.1	2.8
5	0.045	-24.1	2.8	0.024	-35.1	3.1	0.015	-38.1	3.2

Source: Data from Felhauer, © 1994 IEEE, with permission.

[a] PSR is the ratio of peak sidelobe to maximum response in decibels.

"Good" codewords have been found by Frank through the following procedure: First, establish the matrix

$$
\begin{bmatrix}
0 & 0 & 0 & 0 & \cdot\ \cdot\ \cdot & 0 \\
0 & 1 & 2 & 3 & \cdot\ \cdot\ \cdot & (M-1) \\
0 & 2 & 4 & 6 & \cdot\ \cdot\ \cdot & 2(M-1) \\
0 & 3 & 6 & 9 & \cdot\ \cdot\ \cdot & 3(M-1) \\
\cdot & \cdot & \cdot & \cdot & & \cdot \\
\cdot & \cdot & \cdot & \cdot & & \cdot \\
\cdot & \cdot & \cdot & \cdot & & \cdot \\
0 & (M-1) & 2(M-1) & 3(M-1) & \cdot\ \cdot\ \cdot & (M-1)^2
\end{bmatrix} \frac{2\pi p}{M} \qquad (7.9\text{-}9)
$$

where p and M are integers; the factor $2\pi p/M$ results in matrix elements that are phases, and p is relatively prime to N,[14] the most common value of p is one. For chosen values of M and p the matrix is reduced term-by-term modulo 2π to an equivalent form. The elements of the equivalent form become the phases of the N subpulses by forming a sequence of values taken row at a time through the matrix, starting at any element. In general, there are M^2 starting points, since there are M^2 elements. However, Frank found the best codes start at the upper left element of the matrix (first element of row one). We illustrate the procedure by an example.

Example 7.9-1 Consider $M=4$ phases with $p=1$, so the phase increment is $2\pi/M = \pi/2$. We write (7.9-9) and reduce the matrix:

$$
\begin{bmatrix}
0 & 0 & 0 & 0 \\
0 & 1 & 2 & 3 \\
0 & 2 & 4 & 6 \\
0 & 3 & 6 & 9
\end{bmatrix} \frac{\pi}{2} =
\begin{bmatrix}
0 & 0 & 0 & 0 \\
0 & 1 & 2 & 3 \\
0 & 2 & 0 & 2 \\
0 & 3 & 2 & 1
\end{bmatrix} \frac{\pi}{2}
$$

The second form replaces elements where phase exceeds 2π by their equivalent values modulo 2π. A good example is the element in row 4, column 4, where $9\pi/2 = 4\pi + (\pi/2)$ has the equivalent value $\pi/2$. The codeword generated by starting at row 1, column one is 0, 0, 0, 0; 0, 1, 2, 3; 0, 2, 0, 2; 0, 3, 2, 1, and the phases form a sequence of these elements multiplied by $\pi/2$. Note that the first four codeword elements are from row 1, where the semicolon represents the end of row 1. The next four elements are from row 2. Similarly the next eight elements follow from rows 3 and 4 in order.

There are two basic modes of use for Frank codes. If the sequence of length M^2, derived from sequencing through all elements of the matrix of (7.9-9), is continually repeated, a periodic code is generated. The Frank periodic code has an autocorrelation maximum of M^2 and *zero* sidelobes. The maximum occurs periodically

[14] The largest positive integer that divides each of two integers p and M is called the greatest common divisor. Integers p and M are said to be relatively prime if their greatest common divisor is 1.

with a period of M^2 elements. This type of behavior is very attractive for CW radar. Heimiller (1961) has described multi-phase codes that are very similar to those of Frank; he shows an application to pulsed Doppler radar.

The second mode is aperiodic. The code is generated by cycling through the matrix of (7.9-9) once during the transmitted pulse, and then stopping until the next pulse is to be generated. Sidelobes of the autocorrelation function of the aperiodic code are no longer zero. By examining the possible autocorrelation functions for various starting points in the matrix, Frank found that "good" codes with the smallest peak sidelobes (for $p = 1$) occurred when the upper left corner of the matrix was the starting point. Table 7.9-2 gives these code's coefficients and their peak sidelobe ratio (peak sidelobe voltage to maximum voltage ratio) in decibels for M up to 8. The codes marked (b) and (c) are other "good" codes found by other researchers, as referenced in Frank (1963). As M increases to 8, the peak sidelobe ratio approaches about $1/(\pi M)$ [or $-20 \log_{10}(\pi M)$ in decibels]. Frank has conjectured that the sidelobe behavior holds for $M > 8$, but proof was not given.

The aperiodic autocorrelation functions of the codes defined in Table 7.9-2 have several interesting properties (Frank, 1963). It is found that for displacements from the maximum that are element multiples of M, the autocorrelation function has nulls and its magnitude is one for displacements of ± 1 about the multiples of M. Next, the magnitude of the sidelobes (on either side of the maximum) have even symmetry about the displacement $M^2/2$. Finally sidelobes defined between pairs of nulls have even

TABLE 7.9-2   Code coefficients.

Number of Phases M	Code Length $N = M^2$	Peak Sidelobe Level (dB)	Code Sequence Coefficients for a Phase Step of $2\pi/M$	
2	4	-12.04	0, 0; 0, 1	(b) 0, 0, 1, 0
3	9	-19.08	0, 0, 0; 0, 1, 2; 0, 2, 1	(b) 0, 1, 1, 2, 1, 2, 1, 1, 0 (c) 0, 0, 2, 2, 0, 0, 1, 0, 1
4	16	-21.16	0, 0, 0, 0; 0, 1, 2, 3; 0, 2, 0, 2; 0, 3, 2, 1	
5	25	-23.88	0, 0, 0, 0, 0; 0, 1, 2, 3, 4; 0, 2, 4, 1, 3; 0, 3, 1, 4, 2; 0, 4, 3, 2, 1	
6	36	-25.11	0, 0, 0, 0, 0, 0; 0, 1, 2, 3, 4, 5; 0, 2, 4, 0, 2, 4; 0, 3, 0, 3, 0, 3; 0, 4, 2, 0, 4, 2; 0, 5, 4, 3, 2, 1	
7	49	-26.76	0, 0, 0, 0, 0, 0, 0; 0, 1, 2, 3, 4, 5, 6; 0, 2, 4, 6, 1, 3, 5; 0, 3, 6, 2, 5, 1, 4; 0, 4, 1, 5, 2, 6, 3; 0, 5, 3, 1, 6, 4, 2; 0, 6, 5, 4, 3, 2, 1	
8	64	-27.82	0, 0, 0, 0, 0, 0, 0, 0; 0, 1, 2, 3, 4, 5, 6, 7; 0, 2, 4, 6, 0, 2, 4, 6; 0, 3, 6, 1, 4, 7, 2, 5; 0, 4, 0, 4, 0, 4, 0, 4; 0, 5, 2, 7, 4, 1, 6, 3; 0, 6, 4, 2, 0, 6, 4, 2; 0, 7, 6, 5, 4, 3, 2, 1	

Source: After Frank, © 1963 IEEE, with permission.

Note: The product of coefficients listed and $2\pi/M$ become the phases of subpulses in the N-pulse sequence using M-Phase coding.

symmetry about displacements half way between the nulls. Of course, since the code has M^2 elements, the aperiodic autocorrelation function is zero everywhere outside displacements of $\pm M^2$ about the maximum.

Frank codes can be considered as discrete approximations to linear FM. This fact gives the codes (for $N > 4$) better Doppler tolerance than the binary codes, and they also have relatively better sidelobe characteristics (Nathanson, 1991, p. 559; Lewis and Kretschmer, 1981, pp. 364–365). The ridge associated with the uncertainty function of linear FM shows up with the Frank code. However, for the Frank code this ridge is not a smoothly decreasing function; along the ridge there is a cyclical variation of about 4 dB between maxima and minima in the ridge response as Doppler frequency increases. For a pulse duration T the Doppler frequency shift δf_d (Hz) that occurs between adjacent points of maximum response is $\delta f_d = 1/T$.

Some additional details on Frank codes, including the pulse's spectral modulus, Doppler response, and matched filter response can be found in Cook and Bernfeld (1967).

Ideal Periodic Sequences

A sequence that repeats periodically every N digits is called *ideal* if its periodic autocorrelation function has all sidelobes equal to *zero*. Such sequences are clearly valuable to CW radar. They are also useful in pulsed radar because aperiodic sequences derived from "good" periodic sequences often have good aperiodic auto-correlation function sidelobes.

Ideal periodic sequences go back at least to Heimiller (1961), Frank and Zadoff (1962), and Frank (1963) who all used a matrix representation of the sequence similar to Frank (1963). Their sequence length N equaled the square of a positive integer. Chu (1972) gave an ideal sequence that allowed any integer length (see also Frank, 1973). Some others considering the perfect periodic sequences were Lewis and Kretschmer (1982), Kretschmer and Gerlach (1991; both for the P4 code which is discussed below), Gerlach and Kretschmer (1991), and Suehiro and Hatori (1988).

Sarwate (1979) showed that the maximum periodic cross-correlation function magnitude for any two ideal sequences each of length N cannot be smaller than $\sqrt{N}$ (lower bound). Thus the cross-correlation level cannot be smaller than a factor of $1/\sqrt{N}$ below the autocorrelation function's maximum. Codes that reach the lower bound are called *optimum*. Popović (1992) has discussed sequences that can have ideal periodic autocorrelation and optimum periodic cross-correlation functions. He calls them *generalized chirplike sequences*. These sequences are ideal for all allowed values of N, but two such sequences are optimum only for N odd and some other restrictions (to follow).

Define sequence length N by

$$N = sm^2 \tag{7.9-10}$$

where s and m are any positive integers, and let b_i, $i = 0, 1, 2, \ldots, (m - 1)$ be any sequence of complex numbers such that $|b_i| = 1$, all i. The phase angle to be assigned to subpulse k, $k = 0, 1, \ldots, (N - 1)$, for the generalized chirplike sequence is the phase angle of the complex number s_k defined as (Popović, 1992)

$$s_k = a_k b_{(k) \bmod m}, \qquad k = 0, 1, \ldots, (N - 1) \tag{7.9-11}$$

where $(k) \bmod m$ means that k is reduced modulo m,

$$a_k = \begin{cases} W^{(k^2/2) + qk}, & N \text{ even} \\ W^{[k(k+1)/2] + qk}, & N \text{ odd} \end{cases} \quad \begin{matrix} k = 0, 1, \ldots, (N - 1) \\ q = \text{any integer} \end{matrix} \tag{7.9-12}$$

and

$$W = e^{-j2\pi r/N} \tag{7.9-13}$$

is the primitive Nth root of unity and r is any integer relatively prime to N. As an example, for $N = 8$ ($s = 2$ and $m = 2$) sequence $\{s_k\}$ is

$$\{s_k\} = \{b_0, b_1 W^{0.5}, b_0 W^2, b_1 W^{4.5}, b_0, b_1 W^{4.5}, b_0 W^2, b_1 W^{0.5}\} \tag{7.9-14}$$

if we choose $q = 0$ (see, Popović, 1992, and Problem 7.9-2). The generalized P4 codes of Kretschmer and Gerlach (1991) are a special case of the generalized chirplike sequences of length $N = m^2$.

Any two generalized chirplike sequences of *odd* length N, obtained from two different primitive Nth roots of unity $\exp(j2\pi v/N)$ and $\exp(j2\pi u/N)$, have an optimum periodic cross-correlation function (absolute value is constant and equal to $\sqrt{N}$) if $(v - u)$ is relatively prime to N (Popović, 1992, p. 1408). In particular, if N is a second power of a prime number m, the set of $m - 1$ generalized chirplike sequences can be constructed using $m - 1$ different primitive Nth roots of unity. Any pair of sequences in the set will have an optimum periodic cross-correlation function.

Efficient matched filters have also been developed for generalized chirplike sequences (Popović, 1994).

Polyphase Barker Sequences

Bomer and Antweiler (1989) used constrained iteration techniques to find polyphase codes for N from 3 to 25 that have autocorrelation functions with sidelobes of magnitudes one or less. Their results are given in Table 7.9-3. These codes all have an outermost sidelobe of unit magnitude. All other sidelobes are one or less except for $N = 20$ where the largest sidelobe is 1.074. The peak sidelobe is given in column 2; the code's sequence of phases for $n = 1, 2, \ldots, N$, are listed in the next four columns opposite any value of N (column 1).

Other Codes and Some Comments

An early survey and summary of coding used for radar was given by Sarwate and Pursley (1980). In particular, they discuss m-sequence, Gold sequences, Gold-like sequences, and several others, as well as some aperiodic codes.

TABLE 7.9-3   Polyphase barker sequences for N from 3 to 25.

N	Peak Sidelobe	Phase Sequence $n = 1, 2, \dots , N$			
3	0.000	-0.0121	-1.1943	$+0.1409$	
4	0.500	-2.5016	-2.7749	-1.2319	$+2.1417$
5	0.770	$+1.3354$ $+0.2608$	-0.8992	-1.8595	-1.4368
6	1.000	$+1.3630$ $+1.6235$	$+3.0246$ $+1.1919$	-0.5804	$+3.1349$
7	0.522	$+0.8688$ $+1.7924$	$+0.9106$ $+0.0223$	-0.9048 $+2.6777$	-0.6288
8	0.662	-0.3131 -2.9788	-1.2660 -1.4428	$+2.8067$ $+2.1516$	$+2.6116$ -1.8044
9	0.112	-0.6896 -0.7098 $+2.8670$	-2.6573 -1.7158	$+7.2661$ -0.6370	-1.0342 $+1.5811$
10	0.832	-0.5887 -1.7728 $+1.6298$	-1.8322 -0.8423 $+2.8294$	$+2.1561$ -1.9530	-0.3414 -0.7544
11	0.892	$+2.0760$ -0.8944 -1.2395	-0.6452 $+1.5958$ -1.1485	$+2.3212$ $+1.8337$ -1.6531	$+1.9586$ $+1.6372$
12	0.908	$+2.2363$ -1.5832 $+1.9194$	-2.6854 $+3.0342$ -0.6520	$+3.1305$ -0.0850 -0.3623	-2.8078 $+2.2462$ -3.0102
13	0.721	$+2.7188$ $+1.6773$ $+0.6941$ $+1.2749$	$+1.3744$ -0.2793 -2.8756	$+2.0519$ $+3.0833$ -1.8306	$+0.6899$ $+2.2955$ -0.4901
14	0.968	-0.7019 $+2.0131$ $+2.5855$ $+2.9666$	-2.2825 $+2.2090$ $+2.4207$ -2.6327	$+1.2537$ $+1.7266$ -0.6031	-1.4914 $+2.9683$ $+0.1768$
15	0.803	$+0.5655$ -0.9322 $+1.8799$ -2.8457	$+0.2147$ $+2.6090$ $+2.0924$ -0.4439	-0.4469 $+1.0432$ -2.9651 $+1.5204$	-0.5837 -0.8091 -0.6062
16	0.933	-0.3146 -0.0918 $+0.0895$ -0.1218	$+0.1656$ $+1.2243$ $+2.8601$ -1.1910	$+0.1834$ -2.7713 $+0.2766$ -2.9903	$+0.4540$ -1.9528 $+2.5677$ $+1.6009$
17	0.7333	-1.6448 $+0.7623$ -2.9279 -0.3226 -1.9249	$+0.8837$ -2.4826 -2.7891 -0.0768	-2.7783 $+2.1345$ -2.5196 -1.8421	-0.0189 -0.1942 -0.3109 -1.9773

(Continued)

TABLE 7.9-3   (*Continued*)

N	Peak Sidelobe	Phase Sequence $n = 1, 2, \dots, N$			
18	0.889	+ 2.1297	− 2.4221	− 1.6995	+ 0.3822
		− 2.5280	− 0.2354	− 0.1811	+ 1.3279
		+ 0.2331	+ 2.3578	− 0.4065	+ 0.0976
		− 0.8630	+ 0.5150	− 1.6150	− 2.7582
		+ 2.4890	+ 3.7406		
19	0.980	− 1.9461	+ 2.6015	− 0.0648	− 1.3004
		+ 2.1067	+ 0.5609	+ 2.4323	− 1.6685
		− 2.7650	− 0.4415	+ 1.6424	− 1.4620
		− 0.4287	+ 1.5349	+ 2.3556	+ 1.0432
		+ 1.1451	+ 0.3446	+ 0.7259	
20	1.074	+ 0.4786	+ 0.6706	+ 1.5770	+ 1.8712
		+ 2.2135	− 2.1501	+ 0.8608	+ 1.1924
		− 2.5136	+ 0.0993	+ 1.7860	− 0.3836
		− 2.8818	+ 0.0677	− 1.3544	− 2.4389
		+ 2.4420	+ 2.5909	+ 1.1860	− 0.6338
21	0.997	+ 1.8087	+ 0.7662	− 0.5432	− 1.5944
		+ 1.3209	+ 0.0701	− 0.8463	− 1.4062
		− 0.1876	− 1.9735	+ 1.7062	− 1.4345
		+ 0.0882	+ 0.2233	+ 1.3810	+ 2.8241
		− 1.7205	+ 2.1529	− 2.4726	− 0.5451
		+ 0.9588			
22	0.995	+ 2.7208	+ 2.7943	+ 2.4522	+ 2.0044
		+ 1.5822	+ 1.7885	− 2.9352	− 1.3904
		− 2.1716	− 0.9320	+ 0.8861	+ 2.2276
		− 1.3843	+ 0.8799	− 2.1043	+ 2.4527
		− 1.6386	+ 1.3802	− 0.2948	− 1.8923
		+ 2.6576	+ 0.6222		
23	0.912	+ 0.0179	+ 1.1156	+ 2.3422	+ 1.8448
		+ 3.1247	− 2.6283	− 1.4822	+ 1.2732
		− 2.7484	− 1.6192	+ 2.3777	− 0.5946
		+ 2.3777	− 1.6192	− 2.7484	+ 1.2732
		− 1.4822	− 2.6283	+ 3.1247	+ 1.8448
		+ 2.3422	+ 1.1156	+ 0.0179	
24	0.997	+ 1.4496	− 0.3890	− 2.1404	+ 1.4554
		− 1.4167	+ 2.3177	+ 0.2333	+ 0.4296
		− 1.5464	− 2.5852	+ 0.8093	+ 2.0150
		− 1.1339	+ 1.2976	− 3.0525	− 0.8208
		− 0.8155	− 0.1505	+ 2.1155	+ 2.6104
		+ 2.8658	+ 3.0213	+ 2.3375	+ 2.6871
25	0.936	+ 2.6251	+ 2.9578	+ 1.8614	+ 2.4892
		− 1.5638	− 0.4785	− 1.0745	− 0.9767
		− 2.0911	− 0.9913	+ 1.2308	+ 2.8284
		− 1.3895	− 1.5432	− 2.6046	+ 1.4941
		− 0.1824	− 2.7280	− 2.8259	− 0.1689
		− 2.8375	+ 0.6143	− 2.2693	+ 1.4167
		− 2.0333			

Source: Adapted from Bomer and Antweiler (1989), with permission.

A series of codes, that can be operated either periodically or aperiodically, called P1, P2, P3, and P4, are due to Lewis and Kretschmer. The P1 and P2 codes (Lewis and Kretschmer, 1981) are similar to Frank polyphase codes in the behavior of the magnitude of sidelobes in the autocorrelation function, Doppler frequency tolerance, and ease of implementation. For zero Doppler shift the magnitude of the P1 code's autocorrelation function is the same as the Frank code. The P2 code has the same peak sidelobes as the Frank code for zero Doppler shift, but its mean-squared sidelobes are slightly smaller (about 1 dB for $N = 16$, decreasing with N). The main advantage of P1 and P2 codes over Frank codes is that they are said to be more tolerant of filtering (bandlimiting) prior to pulse compression.

The P3 and P4 codes (Lewis and Kretschmer, 1982) are derivable from the linear FM waveform. The P3 code is more Doppler tolerant than the Frank, P1, and P2 codes. The P4 code is a rearranged version of the P3 code; it (P4) is more tolerant to precompression filtering but has the same Doppler tolerance as the P3 code. The autocorrelation function of the P3 and P4 codes (zero Doppler frequency shift and no prefiltering) can have sidelobes 4 dB larger than for the corresponding Frank and P1 codes.

The Doppler properties of Frank, P1, P2, P3, and P4 codes have been studied (Kretschmer and Lewis, 1983). A cyclic 4-dB variation occurs in the compressed pulse's peak as Doppler frequency increases. Although the P3 and P4 codes have larger zero-Doppler peak sidelobes than other codes, they degrade less rapidly as Doppler frequency increases. Some improvements occur when compensation networks are used (Kretschmer and Lewis, 1983).

Lewis, 1993, has given an important and simple technique using a sliding window subtractor (or adder for some cases) that reduces the autocorrelation function's sidelobes of Frank, P1, P3, or P4 codes to a constant unit effective level. The word "effective" is used because the method reduces the pulse compression ratio by a factor of two (doubles the width of the compressed pulse), and the unit sidelobe is defined relative to the peak of the broadened main pulse. The sidelobe level is below the main response peak by a voltage factor of $2/N^2$ for a code of length N^2. The price paid for this low sidelobe level is very low, only a 1-dB loss in signal-to-noise ratio. For small Doppler frequency shifts (about 1% of the bandwidth) the degradation in sidelobe level is very small.

Kretschmer and Gerlach (January, 1991) discuss uniform (constant-amplitude) codes derived from orthogonal matrices. These matrices are associated with *complementary sequences* (sequences with autocorrelation functions such that when two are added the sum has zero sidelobes and a peak of twice that of one) and perfect periodic sequences (those with zero sidelobes in their autocorrelation function). These codes are said to have potential application in the elimination of ambiguous range stationary clutter. These concepts were extended (Gerlach and Kretschmer, July, 1991) to both perfect periodic and asymptotically perfect periodic codes (those with sidelobes in the autocorrelation function that approach zero as N approaches infinity). The codes are useful because good periodic codes tend to yield good aperiodic codes when only one cycle of the periodic code is used for the transmitted waveform.

Finally we note that a recent algorithm (Griep et al., 1995) helps select codes of good autocorrelation and crosscorrelation function properties. The procedure makes use of simulated annealing (Gamal et al., 1987) to find binary and polyphase sequences with minimum peak autocorrelation and crosscorrelation function sidelobe levels.

PROBLEMS

7.1-1 A radar's pulses each uses linear FM of 8.5 MHz over a duration of 10 μs. What is the system's compression ratio? What is the pulse's duration at the output of the pulse compression filter?

7.1-2 Work Problem 7.1-1 except for a frequency sweep of 24 MHz in an 18-μs pulse.

7.1-3 The pulse duration at the output of a linear FM matched filter is to be 0.5 μs. The transmitted pulse's duration is 200 μs. What frequency deviation (sweep) must the transmitted pulse have? Find the waveform's time-bandwidth product.

★7.2-1 A linear FM pulse in white noise is applied to a matched filter. Show that the ratio of the peak amplitude of the compressed pulse ($\omega_d = 0$) is larger than the peak amplitude of the uncompressed pulse by the factor $\sqrt{\Delta f T}$. [*Hint*: Assume that $\Delta f T$ is large, and choose arbitrary constant C so that $|H_{opt}(\omega_0)| = 1$.]

7.2-2 A linear FM pulse is centered on the time origin. Find the rms duration of the signal's complex envelope. [*Hint*: Use (6.4-4) and (6.4-6).]

7.2-3 A rectangular radar pulse of duration T has linear FM such that $\Delta f T = 3$. Find and plot the matched filter's response for $-T \leq t \leq T$ when $\omega_d = 0$. What is the ratio of the 3-dB width of the compressed pulse's envelope to duration T?

7.2-4 Use (7.2-10) and find a largest value of $\Delta f T$ that will give no sidelobes in $\chi(\tau, 0)$, that is, no nulls in $\chi(\tau, 0)$ for $0 \leq |\tau| < T$.

7.2-5 If delay τ and Doppler angular frequency ω_d are constrained to the ridge of maximum response in $|\chi(\tau, \omega_d)|$, that is, must correspond to points on the dashed line in Fig. 7.2-2c, find $\chi(\tau, \omega_d)$.

7.2-6 Sketch $C(x_2)$, $C(x_1)$, $S(x_2)$, and $S(x_1)$ versus $f/\Delta f$, and justify that the braced factor in (7.2-22) is unity in magnitude for $-\frac{1}{2} < f/\Delta f < \frac{1}{2}$ and zero outside this range when $\Delta f T \to \infty$.

7.2-7 Assume that $\Delta f T$ is very large, and find the rms bandwidth of the spectrum of $g(t)$ for the linear FM signal. [*Hint*: Use (7.2-23) in (6.4-14).]

7.3-1 Use (7.3-1) and determine an expression for the edge amplitude of the Taylor mismatch filter. That is, find $H_T(\Delta\omega/2)$. Plot $H_T(\Delta\omega/2)$ in decibels. Assume that $\bar{n} = \bar{n}_{min}$, and calculate data for $SLL = -24, -28, -32, -36,$ and -40.

7.3-2 Plot the factor σ_T, which is the factor by which the Taylor weighted compressed pulse's duration is larger than would be achieved by Dolph-Tchebycheff weighting, for $-40 < SLL < -20$ when $\bar{n}$ is the minimum value for each SLL.

7.3-3 The quantity τ_D is the duration of the compressed pulse's envelope when a linear FM signal is weighted by a Dolph-Tchebycheff filter. For a weighting filter to give $SLL = -20$ use (7.3-9) to find τ_D in terms of $\Delta f = \Delta\omega/2\pi$.

7.3-4 A Taylor's mismatch filter is designed for a -40 dB peak sidelobe level when $\bar{n} = 6$. What mismatch loss in signal-to-noise ratio (in dB) occurs?

7.3-5 Work Problem 7.3-4 except assume that $SLL = -24$ and $\bar{n} = 3$.

★7.3-6 Assume a compressed linear FM pulse for which $\Delta f T$ very large is applied to a mismatch filter defined by (7.3-14) for $n = 4$. Obtain an expression for the weighted compressed pulse having the form of (7.3-6). For convenience, assume a unit-amplitude spectral modulus for the unweighted signal's spectrum.

7.3-7 A radar uses linear FM with Taylor weighting for which $SLL = -40$ and $\bar{n} = 6$. Bandwidth is $\Delta f = 10$ MHz and $T = 10$ μs. What practical sidelobe level can be expected to occur?

7.3-8 If only T in the system of Problem 7.3-7 is changed to 30 μs (all else the same), what new practical sidelobe level is expected?

7.3-9 For a given value of SLL (value of A), $\bar{n}$ for a Taylor mismatch filter has its minimum value determined by the value of $\bar{n}$ such that when increased by one does not cause σ_T to increase. For $SLL = -27$ find $\bar{n}_{min}$ by this method.

★7.4-1 Apply the stationary phase principle to (7.4-13) to show that (7.4-14) is true.

7.5-1 An engineer designs a nonlinear FM signal to have a constant rectangular envelope of duration T where

$$a(t) = K \operatorname{rect}\left(\frac{t}{T}\right)$$

and spectral modulus of spectral extent $\Delta\omega_n$ defined by

$$A^2(\omega) = \operatorname{rect}\left(\frac{\omega}{\Delta\omega_n}\right) \cos^n\left(\frac{\pi\omega}{\Delta\omega_n}\right)$$

for $n = 0, 1, 2, \ldots$, where K is a constant. Use (7.4-27) to show that the envelope delay of the signal's spectrum is given by

$$T_d(\omega) = \begin{cases} -\dfrac{T}{2} + \dfrac{\Delta\omega_n}{2\pi^2 K^2} \displaystyle\int_{-\pi/2}^{\pi\omega/\Delta\omega_n} \cos^n(\xi)\,d\xi, & -\dfrac{\Delta\omega_n}{2} \leq \omega \leq \dfrac{\Delta\omega_n}{2} \\[4mm] 0, & \omega < -\dfrac{\Delta\omega_n}{2} \\[4mm] -\dfrac{T}{2} + \dfrac{\Delta\omega_n}{2\pi^2 K^2} \displaystyle\int_{-\pi/2}^{\pi/2} \cos^n(\xi)\,d\xi, & \dfrac{\Delta\omega_n}{2} < \omega \end{cases}$$

7.5-2 For $n = 0$ and 1 find (a) K necessary for (7.4-27) to satisfy Parseval's energy theorem, and (b) evaluate $T_d(\omega)$ as given in general in Problem 7.5-1.

7.5-3 Work Problem 7.5-2 except for $n = 2$ and 3.

7.5-4 Work Problem 7.5-2 except for $n = 4$.

7.5-5 A nonlinear FM signal is to have a spectral modulus such that the compressed pulse's spectrum corresponds to Hamming weighting; that is, $A^2(\omega)$ has the form

$$A^2(\omega) = \frac{1}{(1 + K_H)} \left[1 + K_H \cos \left(\frac{2\pi\omega}{\Delta\omega} \right) \right] \text{rect} \left(\frac{\omega}{\Delta\omega} \right)$$

where $K_H = 0.852$ for the Hamming case and $\Delta\omega$ is bandwidth. What envelope delay $T_d(\omega)$ must the matched filter have? Assume the signal's envelope is rectangular of duration T.

7.5-6 A pulse compression signal is to have an envelope

$$a(t) = \frac{1}{\sqrt{T}} e^{-(t/T)^2}$$

and spectral modulus

$$A(\omega) = \sqrt{\frac{2\pi}{\Delta\omega}} e^{-(\omega/\Delta\omega)^2}$$

where T and $\Delta\omega$ are signal "duration" and "bandwidth," respectively. (a) Show that Parseval's energy theorem is satisfied. (b) Find the time phase $\theta(t)$. (c) Find the spectral phase $\Theta(\omega)$.

★7.6-1 Use (7.6-1) with (6.8-9) and show that (7.6-4) and (7.6-5) are true.

7.6-2 Assume that each pulse in Fig. 7.6-3 has identical linear FM modulation. Give a rough sketch for the contour of $|\chi(\tau, \omega_d)|$ for the pulse train, and indicate how it is different from Fig. 7.6-4 for constant-frequency pulses. Assume that $T_r > 2T_1$.

7.7-1 For an $N \times N$ Costas array show that there are $N!$ possible frequency-time sequences.

7.7-2 Find the Welch construction of a Costas array for $N = 6$ ($p = 7$) when $\alpha = 3$.

7.7-3 Work Problem 7.7-2 except for $\alpha = 5$.

7.7-4 Suppose that a Welch construction for $p = 13$ is compound using the non-allowed value $\alpha = 1$. See what array results.

7.7-5 Find the Welch construction of the Costas array for $N = 4$ ($p = 5$) and $\alpha = 2$. Find the auto-hit array. Does it have at most one hit per shift position for shifts away from full overlap?

7.7-6 The Welch construction of a Costas array assumes $N = p - 1$, where p is an odd prime. N can be factored into a product of powers of primes as

$$N = p - 1 = \prod_i p_i^{m_i}$$

where p_i are primes and $m_i \geq 1$ are the powers. This result allows us to know the number of primitive roots of p according to the *Euler totient function* $\phi(p-1)$ (Drumheller and Titlebaum, 1991, p. 3)

$$\phi(p-1) = \prod_i p_i^{m_i-1}(p_i-1)$$

Use these results to find $\phi(100)$.

7.7-7 Find Welch constructions W_1, W_2, and W_3 for Costas arrays based on $p = 13$ and $\alpha = 2$.

★7.7-8 For $p = 11$ and $\alpha = 6$, find the Costas array using the W_1 construction. Find the cross-hit array for this array and that of Example 7.7-1.

7.7-9 Find the Costas array using the W_3 construction for the array of Example 7.7-1.

7.7-10 Two radars are to operate in close proximity with minimal interference by using Costas signals with $N = 10$ (see Example 7.7-1). Pulse duration is to be 120 μs and the radar's carriers all operate at 12.5 GHz. (a) Find the Costas arrays A_1 and A_2 for the two radars. (b) What is the largest target speed that can be accommodated by these radars if low interference is to be maintained?

7.7-11 Find the sequence of frequencies that corresponds to an hyperbolic frequency hop code for $N = 10$ ($p = 11$) and $a = 3$. Compare the result to the sequence of Example 7.7-3.

7.7-12 Find the auto-hit array for the code of Example 7.7-3. (*Hint*: The array is odd symmetric so only half needs to be found.)

7.7-13 In a frequency hop code based on cubic congruences (Marić and Titlebaum, 1990) the placement operator is

$$i(k) = ak^3 \quad (\text{modulo } p)$$

where $N = p$, $a = 1, 2, \ldots, (p-1)$, and $k = 0, 1, 2, \ldots, (p-1)$. Find the frequency-time array for $N = 11$ and $a = 1$.

7.7-14 Work Problem 7.7-13 except for $a = 2$.

7.8-1 A sequence for the d_n in (7.8-2) is $\{d_n\} = \{+1, -1, -1, -1, +1, +1, -1, +1, -1, -1, +1\}$. Write the corresponding binary codeword of 1s and 0s.

7.8-2 Work Problem 7.8-1 except for the sequence $\{d_n\} = \{-1, -1, +1, -1, +1, +1, -1, -1, +1, -1, +1, +1, +1, -1\}$.

7.8-3 From Table 7.8-1 find the sample code for $N = 16$ by converting the octal representation to binary.

7.8-4 Work Problem 7.8-3 except for $N = 20$.

7.8-5 Find and plot the autocorrelation for the Barker sequence of Table 7.8-1 for $N = 13$.

7.8-6 A 3-stage shift register feeds back stages 2 and 3 to the input to generate a maximal-length sequence. Assume all stages are initially zero except stage 2. (a) How long is the sequence in digits? (b) Find the binary codeword for the first full period.

7.8-7 Use (7.8-4) to confirm that a 10-stage shift register can generate as many as 60 codes.

7.8-8 Work Problem 7.8-7 except for 16 stages and 2048 codes.

7.8-9 Connections for a 6-stage shift register to generate a maximal-length binary sequence is octal 155. What stages are fed back in this register?

7.8-10 Work Problem 7.8-9 except for a 14-stage register with octal 42103 connections.

7.8-11 Find the sequence defined by (7.8-7) for $N = 17$. What signal-to-noise ratio loss occurs when this code is used?

7.8-12 Work Problem 7.8-11 except for $N = 5$.

7.9-1 For the Frank sequence of Example 7.9-1, evaluate the magnitude of the sequence's aperiodic autocorrelation function for a delay offset of 6 subpulse lengths.

7.9-2 Show that (7.9-14) results from (7.9-11) when $N = 8$, $s = 2$, $m = 2$, and $q = 0$.

8

RADAR RESOLUTION

The term resolution is used in radar to imply that two or more targets are separated, or resolved, sufficiently that measurements can be made on one or more of them of interest. Since the most important parameters that a radar measures for a target are range, Doppler frequency (proportional to range rate or speed), and two orthogonal space angles,[1] we may envision a radar *resolution cell* that contains a certain four-dimensional hypervolume that defines resolution. A target for which measurements are to be made will fall in a resolution cell. Another target, conceptually, does not interfere with measurements on the first if it occupies another resolution cell different from the first. Thus, conceptually, as long as each target occupies a resolution cell and the cells are all disjoint, the radar can make measurements on each target free of interference from others.

As we will see, it is not fruitful to try to be too precise in defining resolution. Many approaches are possible, and these may depend on the radar's design parameters (e.g., waveform used). However, for a single radar pulse, we may give a general sort of definition by considering the resolution cell to be bounded in range by the compressed pulse's duration (the time-equivalent extent of range), in Doppler by the reciprocal of the transmitted pulse's duration, and in the two angles by the antenna pattern's two orthogonal-plane beamwidths.

To exercise our general definition, consider a target inside the radar's beam, at a particular range, and moving at a particular speed. The radar's matched filter to this target will produce a maximum response at the target's delay, and Doppler frequency and measurements can be commenced. Now consider a second target; if it is outside the antenna pattern's beamwidth in either one or both angular directions, it produces essentially no receiver response, and it is resolved in angle from the first target. If both targets lay inside the antenna pattern's main beam but the second is displaced in range (time) by more than the compressed pulse's length, we can clearly separate, or resolve,

---

[1] Modern radars can often measure additional parameters, but these four are the most common and basic.

the two in range. Similarly, if both are in the main beam, both are at the same range but move at different speeds; the two are resolved by the filter's Doppler response if the difference in speeds is larger than the reciprocal of the transmitted pulse length.

If a radar seeks to simultaneously make measurements on targets resolved in Doppler frequency, it can provide a bank of matched filters operating in parallel. Each target will excite the filter matched to its Doppler frequency, and its response can be used for measurements. Targets resolved in the range (time) coordinate can be separated with range gates followed by measurements. Thus a radar can, in principle, perform simultaneous measurements on targets unresolved in angle, provided the targets are resolved in range, or Doppler frequency, or both. On the other hand, it is difficult to simultaneously measure targets resolved in angle coordinates (regardless of resolution in range or Doppler frequency). Such measurements require either a bank of "main" beams (which are possible to generate but not often implemented) or the time-sharing (switching) of one main beam among the various targets (e.g., this can be done by using a phased array antenna with electronic beam steering).

The preceding discussion implies that angle resolution can be considered independently from range and Doppler resolution. While this result is not strictly true, we will give some conditions often satisfied in radars, such that when true, the resolution properties of the radar in angle are independent of the resolution properties in range and Doppler frequency.

In the following sections we first consider range (delay) resolution alone because of its prime importance. Doppler frequency is discussed separately. The two are then developed as simultaneous range and Doppler frequency resolution, and it will be found that the ambiguity function of Chapter 6 is intimately related to resolution. Finally, we examine the radar's simultaneous resolution properties in the four coordinates of range (delay), Doppler frequency, and two space angles. The work leads to an "ambiguity function" for the two angle coordinates that describes angular resolution.

8.1   RANGE RESOLUTION

Since target range R is related to its time delay τ_R by $\tau_R = 2R/c$ (monostatic radar), resolution in range implies resolution in delay. Thus we consider resolution of delay in following work, but often refer to it as "range" resolution.

We will use the analytic signal representation of waveforms developed in Chapters 5 and 6. Let two received waveforms be represents by $\psi_{r1}(t)$ and $\psi_{r2}(t)$. In general, these signals may have delays τ_{R1} and τ_{R2} and Doppler frequencies ω_{d1} and ω_{d2}, respectively. From (6.8-2) they are given by

$$\psi_{r1}(t) = \alpha_1 \psi(t - \tau_{R1}) e^{j\omega_{d1}(t - \tau_{R1})} \tag{8.1-1}$$

$$\psi_{r2}(t) = \alpha_2 \psi(t - \tau_{R2}) e^{j\omega_{d2}(t - \tau_{R2})} \tag{8.1-2}$$

where $\psi(t)$ is the analytic transmitted signal and constants α_1 and α_2 result from the radar equation.

For present purposes we discus only range resolution where we assume no resolution is possible in the Doppler frequency parameter. In other words, we assume that both targets have the same Doppler frequency $\omega_{d1} = \omega_{d2} = \omega_d$ but different delays.

Furthermore, in order that neither target have any power advantage, we presume that $\alpha_1 = \alpha_2 = \alpha$. Thus, for discussion of range resolution only, signals are

$$\psi_{r1}(t) = \alpha\psi(t - \tau_{R1})\, e^{j\omega_d(t - \tau_{R1})} \tag{8.1-3}$$

$$\psi_{r2}(t) = \alpha\psi(t - \tau_{R2})\, e^{j\omega_d(t - \tau_{R2})} \tag{8.1-4}$$

Range resolution requires that $\psi_{r1}(t)$ and $\psi_{r2}(t)$ should be as different as possible. Their difference could then be taken as a measure of resolution. Such a measure is impractical to work with. However, it is still possible to define a reasonable resolution criterion based on signal difference.

Range Resolution Criterion

A practical, and tractable, criterion of resolution is the integrated squared difference magnitude, denoted by $|\varepsilon|^2$, and defined by

$$|\varepsilon|^2 = \int_{-\infty}^{\infty} |\psi_{r1}(t) - \psi_{r2}(t)|^2\, dt \tag{8.1-5}$$

Good range resolution requires $|\varepsilon|^2$ to be large for any target delays τ_{R1} and τ_{R2}. On substituting (8.1-3) and (8.1-4) into (8.1-5) and reducing the algebra, we have

$$|\varepsilon|^2 = 2|\alpha|^2\{E_\psi - \mathrm{Re}[e^{j(\omega_0 + \omega_d)\tau}\,\chi(\tau, 0)]\}$$
$$= 2|\alpha|^2\{E_\psi - \mathrm{Re}[e^{j(\omega_0 + \omega_d)\tau}\,R_{gg}(\tau)]\} \tag{8.1-6}$$

where $\tau = \tau_{R1} - \tau_{R2}$ and

$$\chi(\tau, 0) = \int_{-\infty}^{\infty} g^*(\xi)g(\xi + \tau)\, d\xi = R_{gg}(\tau) \tag{8.1-7}$$

from (6.8-15). The first right-side form of (8.1-6) derived from (6.8-10). E_ψ is the energy in $\psi(t)$, and ω_0, as usual, is the carrier's angular frequency. The function $R_{gg}(\tau)$ is the time autocorrelation function of $g(t)$, the complex envelope (modulation) of the carrier defined in (6.1-6).

From (8.1-6) we find that $\chi(\tau, 0)$ and $R_{gg}(\tau)$ are important quantities in maximizing $|\varepsilon|^2$ for good resolution. To interpret our result more carefully, let $\theta_{xx}(\tau)$ and $\theta_{gg}(\tau)$ be the phases of $\chi(\tau, 0)$ and $R_{gg}(\tau)$, respectively, so that we can write

$$\chi(\tau, 0) = |\chi(\tau, 0)| e^{j\theta_{xx}(\tau)} \tag{8.1-8}$$

$$R_{gg}(\tau) = |R_{gg}(\tau)| e^{j\theta_{gg}(\tau)} \tag{8.1-9}$$

These relationships allow writing (8.1-6) as

$$|\varepsilon|^2 = 2|\alpha|^2\{E_\psi - |\chi(\tau, 0)|\cos[(\omega_0 + \omega_d)\tau + \theta_{xx}(\tau)]\}$$
$$= 2|\alpha|^2\{E_\psi - |R_{gg}(\tau)|\cos[(\omega_0 + \omega_d)\tau + \theta_{gg}(\tau)]\} \tag{8.1-10}$$

Since $|\varepsilon|^2$ should be large for good range resolution, and since E_ψ is taken as constant, this result is achieved if the second term in (8.1-10) is small. Smallness is guaranteed for all angles in the cosines if

$$|\chi(\tau, 0)| = |R_{gg}(\tau)| \quad \text{is small for} \quad |\tau| \neq 0 \qquad (8.1\text{-}11)$$

Ideal Range Resolution

Ideally (8.1-10) implies that $|\chi(\tau, 0)|$, or equivalently, $|\chi(\tau, 0)|^2$, should be zero for all $\tau \neq 0$. For $\tau = 0$ $|\chi(0, 0)|^2$ must equal E_ψ^2 from (6.8-12) for any chosen waveform. Such ideal characteristics are realized by transmitting an impulse. As a practical matter, a very narrow pulse with no FM and large peak power can approach the ideal. Unfortunately, such a waveform is energy limited by the peak-power limits of transmitters. The resulting low energy degrades performance in other areas (e.g., detection). Pulse compression gives a means of achieving both high energy and *small* values of $|\chi(\tau, 0)|^2$ for $|\tau| \neq 0$. However, pulse compression always produces sidelobes so $|\chi(\tau, 0)|^2$ cannot be *zero* for $|\tau|$ near zero.

Our consideration of ideal range resolution serves to point out that some measure of the resolution ability of a practical radar waveform would be desirable. We next present one possible measure.

Range Resolution Constants

Figure 8.1-1 sketches an ambiguity function cut, $|\chi(\tau, 0)|^2$, for some possible waveform. We will define a *time resolution constant*, denoted by T_{res}, as the width of

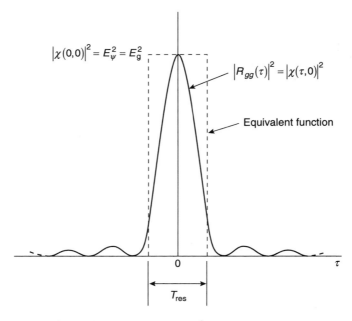

Figure 8.1-1 An ambiguity function cut $|\chi(\tau, 0)|^2$ and an equivalent rectangular function having the same peak amplitude and area.

a rectangular function equivalent to $|\chi(\tau, 0)|^2$ in the sense that it has the same maximum amplitude and same area. Then

$$T_{\text{res}} = \frac{\int_{-\infty}^{\infty} |\chi(\tau, 0)|^2 \, d\tau}{|\chi(0, 0)|^2} \tag{8.1-12}$$

can be taken as a measure of the range resolution ability of the signal $\psi(t)$ that defines $|\chi(\tau, 0)|^2$.

It is instructive to establish some other forms for T_{res}. From (8.1-7),

$$T_{\text{res}} = \frac{\int_{-\infty}^{\infty} |R_{gg}(\tau)|^2 \, d\tau}{[R_{gg}(0)]^2} = \frac{\int_{-\infty}^{\infty} |R_{gg}(\tau)|^2 \, d\tau}{E_g^2} \tag{8.1-13}$$

Problem 8.1-4 shows that

$$R_{gg}(\tau) \leftrightarrow |G(\omega)|^2 \tag{8.1-14}$$

where $G(\omega)$ is the Fourier transform of $g(t)$ which defines $\psi(t)$ through (6.5-16). From Parseval's theorem

$$\int_{-\infty}^{\infty} |R_{gg}(\tau)|^2 \, d\tau = \frac{1}{2\pi} \int_{-\infty}^{\infty} |G(\omega)|^4 \, d\omega \tag{8.1-15}$$

$$E_g = R_{gg}(0) = \frac{1}{2\pi} \int_{-\infty}^{\infty} |G(\omega)|^2 \, d\omega \tag{8.1-16}$$

so (8.1-13) can be written

$$T_{\text{res}} = \frac{2\pi \int_{-\infty}^{\infty} |G(\omega)|^4 \, d\omega}{[\int_{-\infty}^{\infty} |G(\omega)|^2 \, d\omega]^2} = \frac{\int_{-\infty}^{\infty} |G(\omega)|^4 \, d\omega}{2\pi E_g^2} \tag{8.1-17}$$

We interpret waveforms for which T_{res} is small as having good range resolution. For a given energy, (8.1-13) indicates that T_{res} can be made arbitrarily small by using a wideband signal that corresponds to a narrow autocorrelation $R_{gg}(\tau)$. Thus wide-bandwidth signals correspond to small range resolution constants. This alone does not guarantee resolution, however, because sidelobes of a large-amplitude target nearby in range can prevent resolution (separation) of the larger target from the smaller one. Because of such situations, one should not place too great an importance on a given value of T_{res}. About the best interpretation that should be taken is to say that targets separated in time by T_{res} or less will be difficult to revolve and that resolution tends to improve as the separation becomes large compared to T_{res}.

The reciprocal of T_{res} is called *effective bandwidth* (Burdic, 1968), denoted by B_{eff},

$$B_{\text{eff}} = \frac{1}{T_{\text{res}}} \quad (\text{Hz}) \tag{8.1-18}$$

B_{eff} has also been called *frequency span* (Woodward, 1960).

The values of T_{res} and B_{eff} are developed in Problems 8.1-5 through 8.1-9 for some practical waveforms.

8.2 DOPPLER FREQUENCY RESOLUTION

Suppose that we now investigate Doppler frequency resolution under the assumption that there is no range resolution. We therefore assume that the two targets of interest have the same (unresolvable) range delay so $\tau_{R1} = \tau_{R2} = \tau_R$ but different Doppler frequencies. Again we assume that no power advantage to either waveform, so $\alpha_1 = \alpha_2 = \alpha$. The analytic received signals, from (8.1-1) and (8.1-2), become

$$\psi_{r1}(t) = \alpha\psi(t - \tau_R)\,e^{j\omega_{d1}(t-\tau_R)} \tag{8.2-1}$$

$$\psi_{r2}(t) = \alpha\psi(t - \tau_R)\,e^{j\omega_{d2}(t-\tau_R)} \tag{8.2-2}$$

Doppler Frequency Resolution Criterion

Again, as for range resolution, we use the integrated squared difference magnitude of (8.1-5) as our criterion of resolution, except now the signals of (8.2-1) and (8.2-2) apply. On reduction of some algebra, we obtain

$$|\varepsilon|^2 = 2|\alpha|^2\{E_\psi - \mathrm{Re}[\chi(0, -\omega_d)]\}$$

$$= 2|\alpha|^2\left\{E_\psi - \mathrm{Re}\left[\frac{1}{2\pi}R_{GG}(\omega_d)\right]\right\} \tag{8.2-3}$$

where E_ψ is the energy in $\psi(t)$, $\omega_d = \omega_{d1} - \omega_{d2}$, and

$$\chi(0, -\omega_d) = \int_{-\infty}^{\infty} |g(\xi)|^2\, e^{-j\omega_d\xi}\,d\xi$$

$$= \frac{1}{2\pi}\int_{-\infty}^{\infty} G^*(\omega)\,G(\omega + \omega_d)\,d\omega$$

$$= \frac{1}{2\pi}R_{GG}(\omega_d) \tag{8.2-4}$$

from (6.8-16). Equation (8.2-3) indicates that good Doppler frequency resolution ($|\varepsilon|^2$ large) occurs when the second term is small for all $|\omega_d| \neq 0$, that is, when

$$|\chi(0, -\omega_d)|^2 = \left|\frac{1}{2\pi}R_{GG}(\omega_d)\right|^2 \quad \text{is small for} \quad |\omega_d| \neq 0 \tag{8.2-5}$$

Ideal Doppler Frequency Resolution

Ideally the functions in (8.2-5) would *equal* zero for all $|\omega_d| \neq 0$. Since $\chi(0, -\omega_d)$ is the Fourier transform of $|g(t)|^2$ from (8.2-4), the ideal behavior can only be approached in CW radar (transform can have an impulse at the origin). Pulsed radar, with a finite-duration pulse, always produces sidelobes in $\chi(0, -\omega_d)$ and cannot approach the ideal.

Doppler Frequency Resolution Constants

As in range resolution we develop some practical constants that are measures of the Doppler frequency resolution properties of a real radar signal. As sketched in Fig. 8.2-1, we define a *frequency resolution constant*, denoted by Ω_{res}, as the width of a rectangular function having the same peak amplitude at $\omega_d = 0$ and same area as $|\chi(0, -\omega_d)|^2$. Hence, since $|\chi(0, -\omega_d)|^2 = |\chi^*(0, \omega_d)|^2 = |\chi(0, \omega_d)|^2$,

$$\Omega_{res} = \frac{\int_{-\infty}^{\infty} |\chi(0, \omega_d)|^2 \, d\omega_d}{|\chi(0, 0)|^2} = \frac{\int_{-\infty}^{\infty} |R_{GG}(\omega_d)|^2 \, d\omega_d}{[R_{GG}(0)]^2} \qquad (8.2\text{-}6)$$

Other forms of Ω_{res} are possible. From (8.2-4),

$$|g(t)|^2 \leftrightarrow \chi(0, -\omega_d) = \frac{1}{2\pi} R_{GG}(\omega_d) \qquad (8.2\text{-}7)$$

$$\int_{-\infty}^{\infty} |g(t)|^2 \, dt = \frac{1}{2\pi} R_{GG}(0) \qquad (8.2\text{-}8)$$

On using these results in Parseval's theorem, we have

$$\int_{-\infty}^{\infty} |g(t)|^4 \, dt = \frac{1}{2\pi} \int_{-\infty}^{\infty} |\chi(0, -\omega_d)|^2 \, d\omega_d = \frac{1}{2\pi} \int_{-\infty}^{\infty} \left| \frac{1}{2\pi} R_{GG}(\omega_d) \right|^2 d\omega_d \qquad (8.2\text{-}9)$$

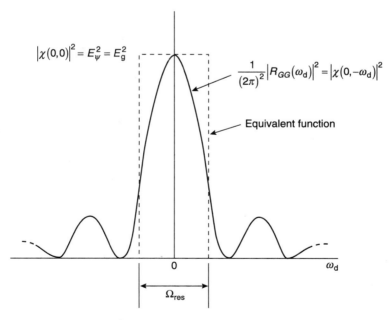

Figure 8.2-1 An ambiguity function cut $|\chi(0, -\omega_d)|^2$ and an equivalent rectangular function having the same peak amplitude and area.

These last three results allow (8.2-6) to be written as

$$\Omega_{\text{res}} = \frac{2\pi \int_{-\infty}^{\infty} |g(t)|^4 \, dt}{[\int_{-\infty}^{\infty} |g(t)|^2 \, dt]^2} = \frac{2\pi \int_{-\infty}^{\infty} |g(t)|^4 \, dt}{E_g^2} \quad \text{(rad/s)} \qquad (8.2\text{-}10)$$

From (8.2-6) we see that Ω_{res} can, in principle, be made arbitrarily small for a given energy $R_{GG}(0)/(2\pi)$ by using long-duration waveforms $g(t)$ such that $R_{GG}(\omega_d)$ is a narrow function of ω_d. The inference is that Ω_{res} can always be made small enough to have any degree of Doppler frequency resolution. Of course this is not true because there are nearly always limits on the duration of $g(t)$ in a real system, and sidelobes in $R_{GG}(\omega_d)$ cause resolution problems for targets of different amplitudes in a manner analogous to range resolution. The resolution constant Ω_{res} should be used only as a rough guide. Targets separated by Ω_{res} or less will be difficult to resolve. Resolution tends to be easier as Doppler difference becomes large with respect to Ω_{res}.

Effective duration, denoted by T_{eff}, is related to Ω_{res} by

$$T_{\text{eff}} = \frac{2\pi}{\Omega_{\text{res}}} \qquad (8.2\text{-}11)$$

Values of Ω_{res} and T_{eff} are developed for some practical waveforms in Problems 8.2-1 through 8.2-5.

8.3 SIMULTANEOUS RANGE AND DOPPLER RESOLUTION

Consider now the simultaneous resolution of targets in range (delay) and Doppler frequency.

Resolution Criterion

By repeating the procedures of the preceding two sections, this time for two signals with different delays *and* different Doppler frequencies according to (8.1-1) and (8.1-2), the integrated squared difference magnitude of (8.1-5) reduces to

$$|\varepsilon|^2 = 2|\alpha|^2 \{E_\psi - \text{Re}[e^{j(\omega_0 + \omega_{d2})\tau} \chi(\tau, - \omega_d)]\} \qquad (8.3\text{-}1)$$

where all parameters were previously defined.

On representing the phase angle of $\chi(\tau, - \omega_d)$ by $\theta_{xx}(\tau, - \omega_d)$, we write

$$\chi(\tau, - \omega_d) = |\chi(\tau, - \omega_d)| e^{j\theta_{xx}(\tau, - \omega_d)} \qquad (8.3\text{-}2)$$

Equation (8.3-1) becomes

$$|\varepsilon|^2 = 2|\alpha|^2 \{E_\psi - |X\{\tau, - \omega_d)|\cos[(\omega_0 + \omega_{d2})\tau + \theta_{xx}(\tau, - \omega_d)]\} \qquad (8.3\text{-}3)$$

As previously defined, $|\varepsilon|^2$ must be large for good resolution, which can be guaranteed for all $|\tau| \neq 0$ and all $|\omega_d| \neq 0$ by requiring $|\chi(\tau, - \omega_d)|^2$ to be small for all

$(\tau, -\omega_d) \neq (0, 0)$. This condition is equivalent to requiring a signal's ambiguity function $|\chi(\tau, \omega_d)|^2$ to be small everywhere except at the origin, where it must equal E_ψ^2 from (6.8-17).

Ideal Resolution

An ambiguity function that equals E_ψ^2 at the origin and zero for all other points away from the origin in the τ, ω_d plane can be called ideal for resolution purposes. This behavior is "impulselike." That such a function is not realizable follows from the fact that its volume is zero, in violation of the nonzero volume requirement of (6.8-18) for an ambiguity function. What then might we call "ideal" that is a step closer to reality.

One answer (among many that can be argued) is to extend the earlier "rectangular" equivalent functions that had base extents T_{res} for delay and Ω_{res} for Doppler frequency. In the present case, the height of the equivalent function would be $|\chi(0,0)|^2 = E_\psi^2$ and have a base area of 2π such that its volume is $2\pi E_\psi^2$, as required for the volume under $|\chi(\tau, \omega_d)|^2$. The shape of the base area is not defined; it could be rectangular, as shown in Fig. 8.3-1a, or elliptical with major/minor axes along τ/ω_d (or ω_d/τ) directions, as in panel b. These shapes (or others) may be interpreted as "ideal" in the sense that all the volume under the ambiguity function is pressed into the smallest region around the origin such that for (τ, ω_d) in the region there is *no resolution* possible, while outside the region there is *full resolution*. Of course practical signals do not usually approximate any of these "ideal" shapes very well.

8.4   RESOLUTION AND rms UNCERTAINTY

In earlier work we have seen that improved range and Doppler frequency resolutions follow respective increases in waveform effective bandwidth and effective duration. Since various bandwidth definitions are interrelated, as are various definitions of duration, it follows that resolution improves as bandwidth and duration increase by any definition. In particular, it is very instructive to evaluate rms bandwidth and rms duration, as discussed in Section 6.4.

Consider a complex envelope $g(t)$ for which the time origin has been chosen so that the mean-time is zero. The rms duration of $g(t)$ is given by

$$\tau_{g, rms}^2 = \frac{\int_{-\infty}^{\infty} t^2 |g(t)|^2 \, dt}{\int_{-\infty}^{\infty} |g(t)|^2 \, dt} = \frac{\int_{-\infty}^{\infty} |dG(\omega)/d\omega|^2 \, d\omega}{\int_{-\infty}^{\infty} |G(\omega)|^2 \, d\omega} \tag{8.4-1}$$

from (6.4-6) and Problem 6.4-7. Next suppose that the spectrum $G(\omega)$ of $g(t)$ has even symmetry so that its mean frequency is zero. Its rms bandwidth is given by

$$W_{g, rms}^2 = \frac{\int_{-\infty}^{\infty} \omega^2 |G(\omega)|^2 \, d\omega}{\int_{-\infty}^{\infty} |G(\omega)|^2 \, d\omega} = \frac{\int_{-\infty}^{\infty} |dg(t)/dt|^2 \, dt}{\int_{-\infty}^{\infty} |g(t)|^2 \, dt} \tag{8.4-2}$$

from (6.4-14) and Problem 6.4-11.

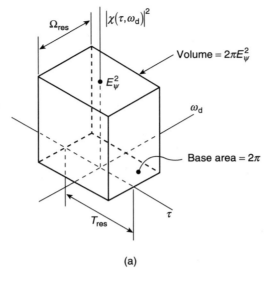

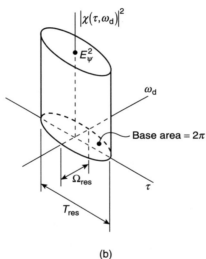

Figure 8.3-1 Some possible functions defining "ideal" resolution, (a) rectangular, and (b) elliptical.

As resolution improves as both $W_{g,\text{rms}}$ and $\tau_{g,\text{rms}}$ increase, we examine the product of the two, which is

$$W_{g,\text{rms}}^2 \, \tau_{g,\text{rms}}^2 = \frac{\int_{-\infty}^{\infty} |dG(\omega)/d\omega|^2 \, d\omega \int_{-\infty}^{\infty} \omega^2 |G(\omega)|^2 \, d\omega}{[\int_{-\infty}^{\infty} |G(\omega)|^2 \, d\omega]^2} \tag{8.4-3}$$

To obtain a useful form, we seek a substitution for the numerator of (8.4-3). We note that the square of the real part of any complex integral cannot exceed the magnitude squared of the complex integral. We apply this fact to upper-bound the left side function below:

$$\left\{ \mathrm{Re}\left[\int_{-\infty}^{\infty} \omega G^*(\omega) \frac{dG(\omega)}{d\omega} \, d\omega \right] \right\}^2 \leq \left| \int_{-\infty}^{\infty} \omega G^*(\omega) \frac{dG(\omega)}{d\omega} \, d\omega \right|^2$$

$$\leq \int_{-\infty}^{\infty} \omega^2 |G(\omega)|^2 \, d\omega \int_{-\infty}^{\infty} \left| \frac{dG(\omega)}{d\omega} \right|^2 \, d\omega$$

(8.4-4)

The last form of (8.4-4) derives from Schwarz's inequality. On substituting (8.4-4) into (8.4-3) and solving the numerator's integral by parts, under the assumption that $|G(\omega)|^2$ approaches zero faster than ω increases as $\omega \to \pm\infty$, we have (see Problem 8.4-5)

$$W_{g,\mathrm{rms}} \, \tau_{g,\mathrm{rms}} \geq \tfrac{1}{2}$$

(8.4-5)

This result shows that there is no theoretical upper limit to the range and Doppler frequency resolution ability of a radar. There is, however, a smallest resolution that is that associated with the lower bound $\frac{1}{2}$ in (8.4-5). The gaussian-shaped pulse with no FM is known to give the lower bound $\frac{1}{2}$; it gives the smallest product, thus the poorest combined resolution (see Problem 8.4-1).

8.5 OVERALL RADAR AND ANGLE RESOLUTIONS

In this section we examine the resolution properties of a radar in the four simultaneous coordinates of delay, Doppler frequency, and two space angles. From our work it will become clear how angle resolution is related to resolution in other coordinates.

Applicable Definitions

Resolution in angle is related to the antenna pattern of the radar. Because this book has mainly considered aperture antennas, we will assume this form of antenna. For convenience, we will assume a constant polarization over the aperture[2] and that the arriving waves from two targets of interest are polarization-matched to the antenna. These waves are presumed to arrive from targets in the far field such that waves are planar. For an aperture in the x, y plane (Fig. 3.1-1) the radiation pattern $P(\theta, \phi)$ for direction (θ, ϕ) is given by (3.5-15):

$$P(\theta, \phi) = \left[\frac{1 + \cos(\theta)}{2} \right]^2 \frac{|\int_A \int E_{ax}(\xi, \zeta) \, e^{j\xi k \sin(\theta)\cos(\phi) + j\zeta k \sin(\theta)\sin(\phi)} \, d\xi d\zeta|^2}{|\int_A \int E_{ax}(\xi, \zeta) d\xi d\zeta|^2}$$

(8.5-1)

---

[2] The polarization can be arbitrary, such as elliptical, but it is constant over the aperture; see Section 3.5.

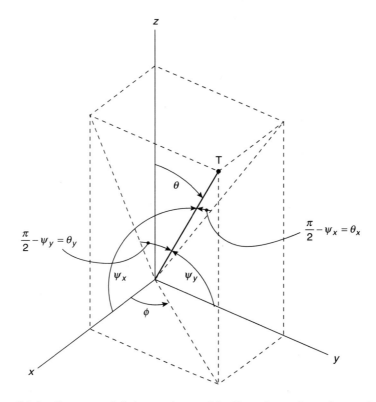

Figure 8.5-1   Geometry defining angles used in discussions of angular resolution.

where $k = 2\pi/\lambda$, λ is the wavelength of the carrier frequency, and $E_{ax}(\xi, \zeta)$ is the distribution of the x-directed component of electric field that excites the aperture when used as a radiator. As a receiving antenna it is visually helpful to define the aperture's illumination function, $I_{ax}(\xi, \zeta)$, by

$$I_{ax}(\xi, \zeta) = \frac{(2\pi)^2 E_{ax}(\xi, \zeta)}{\int_A \int E_{ax}(\xi, \zeta) \, d\xi d\zeta} \tag{8.5-2}$$

Before using (8.5-2) to simplify (8.5-1), we make some convenient definitions. From Figs. 8.5-1 and 3.11-2 we have

$$\sin(\theta_x) = \sin\left(\frac{\pi}{2} - \psi_x\right) = \cos(\psi_x) = \sin(\theta)\cos(\phi) \tag{8.5-3}$$

$$\sin(\theta_y) = \sin\left(\frac{\pi}{2} - \psi_y\right) = \cos(\psi_y) = \sin(\theta)\sin(\phi) \tag{8.5-4}$$

where $\cos(\psi_x)$ and $\cos(\psi_y)$ are the direction cosines corresponding to direction (θ, ϕ). Angles θ_x and θ_y are complementary to angles ψ_x and ψ_y, respectively. Here θ_x is the angle between the origin-to-target line and the y,z plane measured in the plane containing the x axis and origin-to-target line. Similarly θ_y is the angle between the target line and x, z plane measured in the plane containing the y axis and target line. When (8.5-2) through (8.5-4) are used in (8.5-1) the radiation pattern, denoted by $P_u(u_1, u_2)$, becomes

$$P_u(u_1, u_2) = P[u_1(\theta, \phi), u_2(\theta, \phi)]$$

$$= \left| \frac{1}{(2\pi)^2} \int_A \int I_{ax}(\xi, \zeta) e^{j\xi u_1 + j\zeta u_2} \, d\xi d\zeta \right|^2 \qquad (8.5\text{-}5)$$

where we have defined variables u_1 and u_2 by

$$u_1 = k \sin(\theta_x) \qquad (8.5\text{-}6)$$

$$u_2 = k \sin(\theta_y) \qquad (8.5\text{-}7)$$

The obliquity factor $[1 + \cos(\theta)]/2$ has also been replaced by unity because most antennas of radar interest are narrow beam enough that applicable values of θ are small.

When an antenna is used for reception, it may, or may not, have an effect on the received waveform. If the maximum dimension of the antenna projected onto the line from antenna to target, call it L_{ant},[3] is relatively small compared to the range extent of the waveshape of the received field, which is $c/(2\Delta f)$ for a monostatic radar with bandwidth Δf (Hz), the effect on the waveform emerging from the antenna is small. We will assume this is the case. Thus for a carrier frequency f_0,

$$\frac{\Delta f}{f_0} \ll \frac{c}{f_0 2L_{ant}} = \frac{\lambda}{2L_{ant}} \qquad (8.5\text{-}8)$$

is assumed where $\lambda = c/f_0$ is the wavelength. When the bandwidth-to-carrier frequency ratio $\Delta f/f_0$ does not satisfy (8.5-8) the waveform is affected. Some results for this case are given in Urkowitz et al. (1962). When (8.5-8) is satisfied, our results below agree with Urkowitz, but our procedures are slightly different.

One last definition is in order. $P_u(u_1, u_2)$ is a radiation pattern, so it is related to power. In signal reception the voltage pattern is important. We *define* $F_u(u_1, u_2)$ as a complex voltage pattern such that

$$P_u(u_1, u_2) = |F_u(u_1, u_2)|^2 \qquad (8.5\text{-}9)$$

or

$$F_u(u_1, u_2) = \frac{1}{(2\pi)^2} \int_A \int I_{ax}(\xi, \zeta) e^{j\xi u_1 + j\zeta u_2} \, d\xi d\zeta \qquad (8.5\text{-}10)$$

This voltage pattern is seen to be the inverse two-dimensional Fourier transform of the illumination function:

$$F_u(u_1, u_2) \leftrightarrow I_{ax}(\xi, \zeta) \qquad (8.5\text{-}11)$$

---

[3] For a wave received at an angle θ_T from broadside, and an aperture with largest dimension D, then $L_{ant} = D \sin(\theta_T)$.

Combined Radar Resolution Criterion

Two targets in directions (θ_1, ϕ_1) and (θ_2, ϕ_2) can be considered to be in directions $(\theta_{x1}, \theta_{y1})$ and $(\theta_{x2}, \theta_{y2})$ by use of (8.5-3) and (8.5-4). We will seek to find the resolution properties of the radar in angle coordinates θ_x and θ_y. However, it will be most convenient to first develop resolution for "angle" coordinates u_1 and u_2, and then convert the results, for some reasonable conditions, to apply to θ_x and θ_y. In coordinates u_1 and u_2 the "directions" to the two targets are

$$u_{1i} = k \sin(\theta_{xi}), \qquad i = 1 \text{ and } 2 \tag{8.5-12}$$

$$u_{2i} = k \sin(\theta_{yi}), \qquad i = 1 \text{ and } 2 \tag{8.5-13}$$

With no antenna effects, the waves arriving at the antenna can be represented by the right sides of (8.1-1) and (8.1-2). The received waveforms, under our stated antenna assumptions, are equal to these waves altered only by the voltage pattern. These waveforms are (see Problem 8.5-3)

$$\psi_{ri}(t) = \alpha F_u(u_1 - u_{1i}, u_2 - u_{2i})\psi(t - \tau_{Ri})e^{j\omega_{di}(\tau_{Ri})} \tag{8.5-14}$$

where, to give no power advantage to either signal, we have set $\alpha_1 = \alpha_2 = \alpha$.

When angles are included, the resolution criterion, which is again the integrated squared signal difference magnitude, now has the defined form

$$|\varepsilon|^2 = \int_{u_2} \int_{u_1} \int_{-\infty}^{\infty} |\psi_{r1}(t) - \psi_{r2}(t)|^2 \, dt \, du_1 \, du_2$$

$$= \int_{u_2} \int_{u_1} \int_{-\infty}^{\infty} \{|\psi_{r1}(t)|^2 + |\psi_{r2}(t)|^2 - 2 \operatorname{Re}[\psi_{r1}^*(t)\psi_{r2}(t)]\} \, dt \, du_1 \, du_2 \tag{8.5-15}$$

The first two right-side terms are constants, both closely equal to $|\alpha|^2 E_\psi (2\pi)^2/A_e$, where E_ψ is the energy in $\psi(t)$, and A_e is the antenna's maximum effective area

$$A_e = \frac{|\int_A \int E_{ax}(\xi, \zeta) \, d\xi d\zeta|^2}{\int_A \int |E_{ax}(\xi, \zeta)|^2 d\xi d\zeta} = \frac{|\int_A \int I_{ax}(\xi, \zeta) \, d\xi d\zeta|^2}{\int_A \int |I_{ax}(\xi, \zeta)|^2 \, d\xi d\zeta} \tag{8.5-16}$$

derived from use of (8.5-2) in (3.5-16) (see Problem 8.5-4). Hence,

$$|\varepsilon|^2 = 2|\alpha|^2 \left\{ \frac{E_\psi (2\pi)^2}{A_e} - \operatorname{Re}\left[\int_{u_2} \int_{u_1} \int_{-\infty}^{\infty} \frac{1}{|\alpha|^2} \psi_{r1}^*(t)\psi_{r2}(t) \, dt \, du_1 \, du_2 \right] \right\} \tag{8.5-17}$$

As before, good resolution follows making $|\varepsilon|^2$ large which is guaranteed by making the magnitude, or squared magnitude, of the integral small. We define this integral to be the *matched radar response*, denoted by $\chi(\tau, \omega_d, \Delta_{u1}, \Delta_{u2})$, where

$$\tau = \tau_{R1} - \tau_{R2} \tag{8.5-18}$$

$$\omega_d = \omega_{d1} - \omega_{d2} \tag{8.5-19}$$

$$\Delta_{u1} = u_{11} - u_{12} = k[\sin(\theta_{x1}) - \sin(\theta_{x2})] \tag{8.5-20}$$

$$\Delta_{u2} = u_{21} - u_{22} = k[\sin(\theta_{y1}) - \sin(\theta_{y2})] \tag{8.5-21}$$

Solution of (8.5-17) (see Problem 8.5-5) yields

$$\chi(\tau, \omega_d, \Delta_{u1}, \Delta_{u2}) = e^{j(\omega_0 + \omega_{d2})\tau} \, \chi(\tau, -\omega_d) \, \chi_{ant}(\Delta_{u1}, \Delta_{u2}) \tag{8.5-22}$$

where $\chi(\tau, -\omega_d)$ is defined by (6.8-10) and the *matched antenna response*, $\chi_{ant}(\Delta_{u1}, \Delta_{u2})$, is

$$\chi_{ant}(\Delta_{u1}, \Delta_{u2}) = \int_{\xi_2} \int_{\xi_1} F_u^*(\xi_1, \xi_2) \, F_u(\xi_1 + \Delta_{u1}, \xi_2 + \Delta_{u2}) \, d\xi_1 d\xi_2 \tag{8.5-23}$$

Equations (8.5-22) and (8.5-23) are our principal results. The former shows that the radar's matched response is the product of the matched filter's response (which determines delay and Doppler resolution) and the matched antenna's response. This latter response determines angle resolution properties of the radar. By the form

$$|\chi(\tau, \omega_d, \Delta_{u1}, \Delta_{u2})|^2 = |\chi(\tau, -\omega_d)|^2 \, |\chi_{ant}(\Delta_{u1}, \Delta_{u2})|^2 \tag{8.5-24}$$

we see that the *radar's ambiguity function* is the product of the signal ambiguity function and the *antenna's ambiguity function* in variables Δ_{u1} and Δ_{u2}. Thus, for good angle resolution, the antenna's ambiguity function should have the usual desirable properties of rapidly decreasing as Δ_{u1} and Δ_{u2} differ from zero. Since (8.5-23) is the complex correlation function of the aperture's voltage pattern, rapidly decreasing (narrow) patterns are desirable, which from (8.5-11) corresponds to large antennas (wide apertures) for good angle resolution. All these results serve to verify our early assumption that narrowing the beam of an antenna improves angle resolution.

Angle Resolution Constant

We define an *angle resolution constant*, denoted by $\Theta_{res,u}$, as the area around the origin of the Δ_{u1}, Δ_{u2} plane over which an equivalent constant-amplitude function has the same peak amplitude and same volume as $|\chi_{ant}(\Delta_{u1}, \Delta_{u2})|^2$:

$$\Theta_{res,u} = \frac{\int_{\Delta_{u1}} \int_{\Delta_{u2}} |\chi_{ant}(\Delta_{u1}, \Delta_{u2})|^2 \, d\Delta_{u1} d\Delta_{u2}}{|\chi_{ant}(0, 0)|^2} \tag{8.5-25}$$

To reduce (8.5-25), we use Parseval's theorem (B.7-11) with (8.5-11), and (8.5-23) with $\Delta_{u1} = 0$ and $\Delta_{u2} = 0$, to get

$$\int_{u_2} \int_{u_1} |F_u(u_1, u_2)|^2 \, du_1 \, du_2 = \chi_{\text{ant}}(0, 0)$$

$$= \frac{1}{(2\pi)^2} \int_A \int |I_{ax}(\xi, \zeta)|^2 \, d\xi d\zeta \qquad (8.5\text{-}26)$$

From the correlation property (B.7-8) of Fourier transforms, we have

$$\chi_{\text{ant}}(\Delta_{u1}, \Delta_{u2}) \leftrightarrow |I_{ax}(\xi, \zeta)|^2 \qquad (8.5\text{-}27)$$

so Parseval's theorem (B.7-11) gives

$$\int_{\Delta_{u1}} \int_{\Delta_{u2}} |\chi_{\text{ant}}(\Delta_{u1}, \Delta_{u2})|^2 \, d\Delta_{u1} \, d\Delta_{u2}$$

$$= \frac{1}{(2\pi)^2} \int_A \int |I_{ax}(\xi, \zeta)|^4 \, d\xi d\zeta \qquad (8.5\text{-}28)$$

Thus from (8.5-28) and (8.5-26), (8.5-25) becomes

$$\Theta_{\text{res},u} = \frac{(2\pi)^2 \int_A \int |I_{ax}(\xi, \zeta)|^4 \, d\xi d\zeta}{[\int_A \int |I_{ax}(\xi, \zeta)|^2 \, d\xi d\zeta]^2} = \frac{(2\pi)^2}{A_{\text{eff}}} \qquad (8.5\text{-}29)$$

where

$$A_{\text{eff}} = \frac{[\int_A \int |I_{ax}(\xi, \zeta)|^2 \, d\xi d\zeta]^2}{\int_A \int |I_{ax}(\xi, \zeta)|^4 \, d\xi d\zeta} \qquad (8.5\text{-}30)$$

is called the *effective aperture* (Vakman, 1968, p. 96); A_{eff} should not be confused with A_e, the maximum effective area of the aperture, since they are not necessarily equal (see Problem 8.5-6).

To directly relate resolution to the *angles* θ_x and θ_y, we apply (8.5-6) and (8.5-7) for small angles. The angle resolution constant, now denoted by Θ_{res}, is

$$\Theta_{\text{res}} = \frac{1}{k^2} \Theta_{\text{res},u} = \frac{\lambda^2}{A_{\text{eff}}} \qquad (8.5\text{-}31)$$

which indicates that angle resolution improves as λ is made smaller and A_{eff} is made larger; in other words, the aperture area measured in wavelengths is made larger.

PROBLEMS

8.1-1 Substitute as indicated in the text immediately prior to (8.1-6) and show that (8.1-6) is true.

8.1-2 Show that (8.1-6) can also be written as

$$|\varepsilon|^2 = 2|\alpha|^2 \{E_\psi - \text{Re}[e^{-j(\omega_0 + \omega_d)\tau} \chi*(\tau, 0)]\}$$

$$= 2|\alpha|^2 \{E_\psi - \text{Re}[e^{-j(\omega_0 + \omega_d)\tau} R_{gg}^*(\tau)]\}$$

8.1-3 Assume that a radar can transmit an impulse, that is, suppose that $g(t) = \delta(t)$. In words, describe the form of $\chi(\tau, 0)$.

8.1-4 Prove that $|G(\omega)|^2$ is the Fourier transform of $R_{gg}(\tau)$ defined by (6.8-15).

8.1-5 Determine T_{res} and B_{eff} for the constant-frequency rectangular pulse defined by

$$g(t) = A \operatorname{rect}\left(\frac{t}{T}\right) \leftrightarrow G(\omega) = AT \operatorname{Sa}\left(\frac{\omega T}{2}\right)$$

where A and T are positive constants.

8.1-6 Work Problem 8.1-5 except assume the gaussian pulse

$$g(t) = Ae^{-t^2/(2T^2)} \leftrightarrow G(\omega) = AT\sqrt{2\pi}\, e^{-\omega^2 T^2/2}$$

8.1-7 Work Problem 8.1-5 except for the sampling function pulse

$$g(t) = A \operatorname{Sa}\left(\frac{\pi t}{T}\right) \leftrightarrow G(\omega) = AT \operatorname{rect}\left(\frac{\omega T}{2\pi}\right)$$

8.1-8 Work Problem 8.1-5 except for the triangular pulse

$$g(t) = A \operatorname{tri}\left(\frac{t}{T}\right) \leftrightarrow G(\omega) = AT \operatorname{Sa}^2\left(\frac{\omega T}{2}\right)$$

8.1-9 For the linear FM pulse defined by

$$g(t) = A \operatorname{rect}\left(\frac{t}{T}\right) e^{j\mu t^2/2}$$

$$G(\omega) \approx A\sqrt{\frac{2\pi}{\mu}} \operatorname{rect}\left(\frac{\omega}{\mu T}\right) e^{-j(\omega^2/2\mu) + j(\pi/4)}$$

show that $T_{\text{res}} = 2\pi/\mu T$ and $B_{\text{eff}} = \mu T/2\pi$.

8.1-10 If T_{res} of (8.1-17) and B_{eff} of (8.1-18) are due to $G(\omega)$ constrained to a band $-\Delta\omega/2 < \omega < \Delta\omega/2$, as with bandlimiting in the system, show that $B_{eff} \leq \Delta\omega/2\pi$; that is, show that B_{eff} cannot exceed $\Delta\omega/2\pi$.

8.2-1 Find Ω_{res} and T_{eff} for the rectangular pulse of Problem 8.1-5.

8.2-2 Find Ω_{res} and T_{eff} for the pulse of Problem 8.1-6.

8.2-3 Find Ω_{res} and T_{eff} for the pulse of Problem 8.1-7.

8.2-4 Find Ω_{res} and T_{eff} for the pulse of Problem 8.1-8.

8.2-5 Show that Ω_{res} and T_{eff} for the pulse of Problem 8.1-9 are given by $\Omega_{res} = 2\pi/T$ and $T_{eff} = T$.

8.2-6 If Ω_{res} of (8.2-10) and T_{eff} of (8.2-11) are due to $g(t)$ constrained to a duration of $-T/2 < t < T/2$, as might occur with time gating in a system, show that $T_{eff} \leq T$.

8.2-7 A radar uses a linear FM pulse for which $\Delta f = 5$ MHz and $T = 60$ μs. Find T_{res}, B_{eff}, Ω_{res}, and T_{eff} for this system. How is T_{res} related to the duration (between -3.92 dB points) of the pulse at the matched filter's output if $\omega_d = 0$?

8.3-1 Suppose that an "ideal" ambiguity function has a tilted elliptically shaped base, as shown in Fig. P8.3-1. The tilt angle θ is taken relative to the τ axis and major and minor half-axis lengths are a and b as shown. Find expressions for Ω_{res} and T_{res} that are functions of θ for fixed a and b.

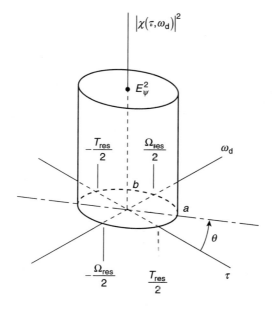

Figure P8.3-1.

8.3-2 If the cone of Fig. P8.3-2 is to have the same values of T_{res} and Ω_{res} as the shape in Fig. 8.3-1*b*, what must α be? The volumes both have elliptical bases with the same ratio of major to minor axes.

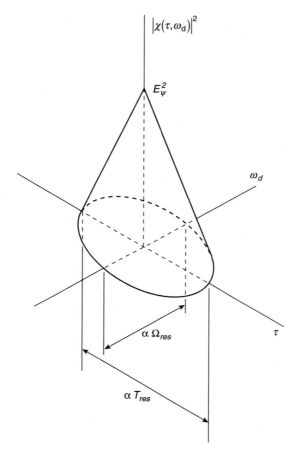

Figure P8.3-2.

8.4-1 Use (8.4-1) and (8.4-2) to find $\tau_{g,\text{rms}}$ and $W_{g,\text{rms}}$ for the gaussian pulse of Problem 8.1-6.

8.4-2 Find $W_{g,\text{rms}}$ for the sampling pulse of Problem 8.1-7. What is $\tau_{g,\text{rms}}$?

8.4-3 Find $W_{g,\text{rms}}$ and $\tau_{g,\text{rms}}$ for the triangular pulse of Problem 8.1-8.

8.4-4 For the linear FM pulse of Problem 8.1-9 show that

$$\tau_{g,\text{rms}}^2 = \frac{T^2}{12}$$

$$W_{g,\text{rms}}^2 = \frac{\mu^2 T^2}{12}$$

8.4-5 Carry out the steps indicated in the text prior to (8.4-5) and prove that (8.4-5) is true.

8.5-1 A radar operates at 35 GHz and has a signal with a bandwidth of 25 MHz. (a) What condition exists on the antenna's size if the antenna is to not be sensitive to the signal's bandwidth? (b) If the condition of (a) is satisfied, what is the ratio of $\Delta f / f_0$ for the system?

8.5-2 Work Problem 8.5-1 except assume operation at 6 GHz and bandwidth is 5 MHz.

8.5-3 For small target angles show that the voltage patterns for parameters $\theta_x - \theta_{xi}$ and $\theta_y - \theta_{yi}$, $i = 1, 2$, are equivalent to the voltage patterns for parameters $u_1 - u_{1i}$ and $u_2 - u_{2i}$, $i = 1, 2$. That is, prove that

$$k \sin(\theta_x - \theta_{xi}) \approx k \sin(\theta_x) - k \sin(\theta_{xi}) = u_1 - u_{1i}$$

$$k \sin(\theta_y - \theta_{yi}) \approx k \sin(\theta_y) - k \sin(\theta_{yi}) = u_2 - u_{2i}$$

for $i = 1, 2$.

★8.5-4 In (8.5-15) show that

$$\int_{u_2} \int_{u_1} \int_{-\infty}^{\infty} |\psi_{ri}(t)|^2 \, dt \, du_1 \, du_2 = \frac{|\alpha|^2 \, E_\psi (2\pi)^2}{A_e}, \qquad i = 1, 2$$

where E_ψ is the energy in $\psi(t)$ and A_e is given by (8.5-16).

★8.5-5 Show that the integral in (8.5-17) is equal to the right side of (8.5-22).

8.5-6 Use (8.5-16) with (8.5-30) to show how A_{eff} is related to A_e.

8.5-7 A rectangular aperture has lengths L_x and L_y centered along x and y directions and

$$E_{ax}(\xi, \zeta) = \begin{cases} E_0 \cos\left(\dfrac{\pi \xi}{L_x}\right) \cos\left(\dfrac{\pi \zeta}{L_y}\right), & -\dfrac{L_x}{2} \le \xi \le \dfrac{L_x}{2}, \ \dfrac{L_y}{2} \le \zeta \le \dfrac{L_y}{2} \\ 0, & \text{elsewhere} \end{cases}$$

where E_0 is a positive constant. (a) Find A_e. (b) Find A_{eff} and show how it differs from A_e.

★8.5-8 Use Schwarz's inequality in the numerator of (8.5-30) and show that $A_{\text{eff}} \le A$, where A is the true area of the antenna's aperture.

8.5-9 Work Problem 8.5-7 except assume $E_{ax}(\xi, \zeta) = E_0$, a constant over the aperture.

8.5-10 Work Problem 8.5-7 except assume that

$$E_{ax}(\xi, \zeta) =$$

$$\begin{cases} E_0 \left[1 - \left(\dfrac{2|\xi|}{L_x}\right)\right]^n \left[1 - \left(\dfrac{2|\zeta|}{L_y}\right)\right]^m, & -\dfrac{L_x}{2} \le \xi \le \dfrac{L_x}{2}, \dfrac{L_y}{2} \le \zeta \le \dfrac{L_y}{2} \\ 0, & \text{elsewhere} \end{cases}$$

where $n = 0, 1, 2, \ldots$, and $m = 0, 1, 2, \ldots$ Evaluate A_e/A_{eff} for $m = n = 0, 1, 2, 3$.

8.5-11 A circular aperture of radius R has a constant illumination E_0 over the aperture. (a) Find A_e. (b) Find A_{eff}. (c) Find the angle resolution constant Θ_{res} as a function of R. (d) What aperture radius corresponds to a resolution constant of one degree when the carrier frequency is 10 GHz?

9

RADAR DETECTION

Radar detection is the process whereby a radar determines if one or more targets are present within its volume of operation. The theory surrounding the subject of detection is rich and extensive, and the discussions of this chapter will make no effort to be either general or exhaustive. Rather we will consider only the most fundamental of the detection problems in an effort to lay a good foundation for the reader to explore greater detail through the literature.

In attempting to sense the presence of a target through the detection of its returned signal, the radar may have to contend with clutter, jamming, and various interference signals as well as noise. These unwanted signals may not all be present at the same time, but noise *is* always present. Because noise is always present, and ultimately establishes the bounding detection performance of the radar, we will discuss only the problem of detecting a target in the presence of noise. Many of the techniques applicable to noise analyses carry over to detection in the presence of other unwanted signals.

At the end of this chapter we will indicate how current topics have gone beyond the fundamentals of detection and give some appropriate summaries to lead the reader to the literature.

9.1 BAYES'S CONCEPTS

In this section we define the overall problem of detecting the presence of a target when the radar observes one or more pulse repetition periods. The general detection theory of Bayes (DiFranco and Rubin, 1968; Melsa and Cohn, 1978; Shanmugan and Briepohl, 1988; Helstrom, 1975) is specialized to the case of radar and leads to a criterion for defining optimum detection. In subsequent sections we apply the criterion to specific detection situations, such as detection of a steady (nonfluctuating) coherent target, and various forms of fluctuating noncoherent targets, as defined below.

Basic Definitions and Model

We presume that the radar is to observe its received waveform over a number N of pulse repetition intervals[1] each of duration T_R. At the end of the observation time, which is NT_R, the radar is called on to decide: Is there only noise being received, or is there a signal being received with the noise? If we define $y(t)$ as the received waveform, another important question is: How should the radar process the observed waveform $y(t)$ to obtain the best possible decision at time NT_R? The second question's answer defines the signal processor to be used for making the detection decision required in the first question. If the detection processor is able to make full use of the target's phase, it is called *coherent*, and the detection decision is said to be based on *coherent integration* of N pulses (N pulse repetition periods). If the processor has no knowledge of target phase, as for a target with randomly varying phase, it is called *noncoherent* and the detection decision is said to be based on *noncoherent integration* of N pulses. Most radars use noncoherent integration.

Figure 9.1-1 is helpful in establishing a model for the observations of $y(t)$ that are available to the signal processor. We assume that $y(t)$ is sampled at a very high rate with K samples in each of the N pulse repetition intervals.[2] Samples are separated by an amount Δt where

$$\Delta t = \frac{T_R}{K} \tag{9.1-1}$$

Sample times, denoted by t_{ik}, are

$$t_{ik} = (i-1)T_R + k\Delta t \tag{9.1-2}$$

where index $i = 1, 2, \dots, N$ refers to the pulse repetition interval, and index $k = 1, 2, \dots, K$ refers to samples within any given pulse repetition interval. Samples (observations) are denoted by y_{ik} and are given by

$$y_{ik} = y(t_{ik}) \tag{9.1-3}$$

The samples y_{ik} can be considered values of random variables, since $y(t)$ is random through its noise component. For convenience we group the random variables for interval i by defining a K-element row vector as

$$\underline{y}_i = (y_{i1}, y_{i2}, \dots, y_{ik}, \dots, y_{iK}), \qquad i = 1, 2, \dots, N \tag{9.1-4}$$

Samples are further grouped by an N-element row vector of these vectors according to

$$\underline{y} = (\underline{y}_1, \underline{y}_2, \dots, \underline{y}_i, \dots, \underline{y}_N) \tag{9.1-5}$$

---

[1] Here N can be as small as one interval, or for some cases, may be a large integer value.
[2] The system does not need to actually take these samples, as we will see. The sampling approach is only a modeling method that readily allows a practical analysis.

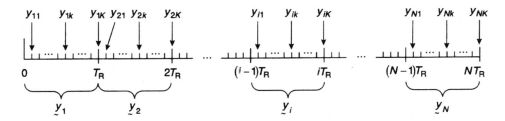

Figure 9.1-1 Sample times used for generating samples of the received waveform $y(t)$.

Thus $\underset{\sim}{y}$ is a vector random variable (see Appendix C) having all the observation random variables y_{ik} as its components.

When no target is present the joint probability density function of the observation random variables will be different than when target is present. In the former case $y(t)$ consists of noise only, while in the latter case $y(t)$ is the sum of the received signal plus noise. The joint probability density functions of the observation random variables, conditional on no signal and signal present cases, respectively, are denoted by $p(y|0)$ and $p[y|s_r(\Theta)]$. The notation $s_r(\Theta)$ is used to imply that the received signal, $s_r(t)$, may depend on some random parameters $\theta_1, \theta_2, \ldots, \theta_M$, defined by the vector

$$\Theta = (\theta_1, \theta_2, \ldots, \theta_M) \tag{9.1-6}$$

and that sample values of $s_r(t)$ are present in samples of $y(t)$ when signal is present. The corresponding signal sample vector is

$$s_r(\Theta) = (s_{r1}(\Theta), s_{r2}(\Theta), \ldots, s_{rN}(\Theta)) \tag{9.1-7}$$

where the component vectors are over intervals:

$$s_{ri}(\Theta) = (s_{ri1}(\Theta), s_{ri2}(\Theta), \ldots, s_{riK}(\Theta)) \tag{9.1-8}$$

for $i = 1, 2, \ldots, N$. The elements of (9.1-8) are just the sample values of $s_r(t)$ at times t_{ik}:

$$s_{rik}(\Theta) = s_r(t_{ik}) \tag{9.1-9}$$

where $s_r(t_{ik})$ depends implicitly on the random vector Θ. The random parameters θ_m, $m = 1, 2, \ldots, M$, could represent a variety of things. Most often they will represent randomly varying target phase angles in $s_r(t)$ and randomly fluctuating cross sections.

Bayes's Concepts for Radar

Bayes's decision theory is a systematic method of assigning cost factors to the correct and incorrect decisions that can be made in the detection process. We make use of a particular form of the general theory as usually applied to radar. Two important specializations are used. First, we will average $p[y|s_r(\Theta)]$ over all random parameters and use the result to ultimately solve the optimum detection problem. Second, we will choose special cost factors for radar (below). Let us define

$$\overline{p[y|\underline{s}_r(\Theta)]_\Theta} = \int \cdots \int_\Theta p[y|\underline{s}_r(\Theta)]p(\Theta)d\Theta \qquad (9.1\text{-}10)$$

The left side of (9.1-10) is a notational definition; the overbar represents the statistical average of the quantity with respect to all the random prameters of the vector Θ. The right side of (9.1-10) defines the actual averaging operation (Appendix C); here $p(\Theta)$ is the joint probability density function of random variables $(\theta_1, \theta_2, \ldots, \theta_M) = \Theta$. $p(\Theta)$ uses the actual density for those random variables that are known and uses the least favorable (uniform) density for those variables with unknown statistics.

Ultimately the signal processor must process the various observations to produce a single variable from which a decision is made of target plus noise present (referred to as decision $S + N$) or noise only present (referred to as decision N). If we think of any one set of observations, which is one value of y as a point in NK-dimensional hyperspace, then some points will correspond to the decision $S + N$. Call the space of such points Γ_1. All other hyperspace points, denoted by Γ_0, must correspond to the decision N. All of hyperspace denoted by Γ, is the sum of the two spaces, or $\Gamma = \Gamma_1 + \Gamma_0$. Based on these spaces we define *average detection probability* $\bar{P}_d$, *average miss probability* $\bar{P}_{\text{miss}}$,[3] *false alarm probability* P_{fa}, and *noise probability* P_{noise} by

$$\bar{P}_d = \int \cdots \int_{\Gamma_1} \overline{p[y|\underline{s}_r(\Theta)]_\Theta}\, dy = \text{average detection probability} \qquad (9.1\text{-}11)$$

$$\bar{P}_{\text{miss}} = \int \cdots \int_{\Gamma_0} \overline{p[y|\underline{s}_r(\Theta)]_\Theta}\, dy = 1 - \bar{P}_d = \text{average miss probability} \qquad (9.1\text{-}12)$$

$$P_{\text{fa}} = \int \cdots \int_{\Gamma_1} p(y|0)\, dy = \text{false alarm probability} \qquad (9.1\text{-}13)$$

$$P_{\text{noise}} = \int \cdots \int_{\Gamma_0} p(y|0)\, dy = 1 - P_{\text{fa}} = \text{noise probability} \qquad (9.1\text{-}14)$$

Clearly the final decision process can lead to only four outcomes, two are correct and two are wrong. Bayes's theory assigns costs to these decisions as follows:

1 Decide N when N is true: Cost $= C_{11}$ $\qquad (9.1\text{-}15a)$

2 Decide N when $S + N$ is true: Cost $= C_{12}$ $\qquad (9.1\text{-}15b)$

3 Decide $S + N$ when N is true: Cost $= C_{21}$ $\qquad (9.1\text{-}15c)$

4 Decide $S + N$ when $S + N$ is true: Cost $= C_{22}$ $\qquad (9.1\text{-}15d)$

---

[3] The word average is used here to imply a quantity derived from the average probability density of (9.1-10).

Now let D be a random variable representing the radar's "decision"; it can only have outcomes N and $S + N$. Similarly let T be a random variable representing the "true case;" it has outcomes N and $S + N$ with probabilities of occurrence defined as $P(T = N) = q$ and $P(T = S + N) = p = 1 - q$, respectively. The joint probabilities of the four outcomes of (9.1-15), with T independent of D, are

$$P(N, N) = P(D = N, T = N) = P_{\text{noise}} q \tag{9.1-16a}$$

$$\bar{P}(N, S + N) = \bar{P}(D = N, T = S + N) = \bar{P}_{\text{miss}} p \tag{9.1-16b}$$

$$P(S + N, N) = P(D = S + N, T = N) = P_{\text{fa}} q \tag{9.1-16c}$$

$$\bar{P}(S + N, S + N) = \bar{P}(D = S + N, T = S + N) = \bar{P}_{\text{d}} p \tag{9.1-16d}$$

The Bayes procedure defines an *average cost*, denoted by $\bar{L}$, for the four decisions of (9.1-15):

$$\bar{L} = C_{11} P(N, N) + C_{12} \bar{P}(N, S + N) + C_{21} P(S + N, N) + C_{22} \bar{P}(S + N, S + N)$$

$$= C_{11} q + C_{12} p$$

$$+ \int \cdots \int_{\Gamma_1} \{(C_{21} - C_{11}) q p(y|0) - (C_{12} - C_{22}) p \, \overline{p[y|s_r(\Theta)]_\Theta}\} \, dy \tag{9.1-17}$$

The last form of (9.1-17) has used (9.1-16) and (9.1-11) through (9.1-14).

Optimum Radar Decision Rule

The signal processor is said to be optimum if it minimizes the average cost $\bar{L}$ of decisions. Since p, q, and the cost coefficients in (9.1-17) are constants, $\bar{L}$ is minimized if the integral terms in (9.1-17) are most negative. These integrals may be made most negative by assigning to decision region Γ_1 all those points y that produce a negative result. On letting all points in the integrands of the two integrals be not greater than zero, we solve to obtain the decision rule

$$\frac{\overline{p[y|s_r(\Theta)]_\Theta}}{p(y|0)} \geq \frac{(C_{21} - C_{11}) q}{(C_{12} - C_{22}) p}, \qquad \text{decide } S + N. \tag{9.1-18}$$

The left side of (9.1-18) is called the *average likelihood ratio*. Normally $C_{21} > C_{11}$ and $C_{12} > C_{22}$ in radar because the costs of incorrect decisions (C_{21} and C_{12}) are larger than the costs of correct decisions (C_{11} and C_{22}).

In fact for the radar case we will assign *no costs* to correct decisions so $C_{22} = C_{11} = 0$, and (9.1-18) becomes

$$\overline{l(y)_\Theta} = \frac{\overline{p[y|s_r(\Theta)]_\Theta}}{p(y|0)} \begin{cases} \geq V_{\text{T}}, & \text{decide } S + N \\ < V_{\text{T}}, & \text{decide } N \end{cases} \tag{9.1-19}$$

Here $\overline{l(y)_\Theta}$ is a convenient notation for the average likelihood ratio, and V_T is a *detection threshold* defined by

$$V_T = \frac{C_{21} q}{C_{12} p} \tag{9.1-20}$$

Equation (9.1-19) is a most important result. It implies that the optimum signal processor should use the observations y to compute the average likelihood ratio and compare it to a threshold V_T, (to be established later). If the ratio equals or exceeds V_T, we decide that a target is present. If less than V_T, the decision is that no target is present. As we will ultimately show, these operations are conceptual only. The radar's signal processor does not actually compute $\overline{l(y)_\Theta}$ but rather a result derived from $\overline{l(y)_\Theta}$ to accomplish the same end result.

In all following work detection of targets in white, zero-mean, gaussian noise is presumed. The white noise, denoted by $n_w(t)$, is modeled as the limit of band-limited noise, denoted by $n(t)$. The power density spectrum and autocorrelation function of $n(t)$ are

$$\mathscr{S}_{NN}(\omega) = \frac{\mathscr{N}_0}{2} \operatorname{rect}\left(\frac{\omega}{2W_N}\right) \tag{9.1-21}$$

$$R_{NN}(\tau) = \frac{\mathscr{N}_0 W_N}{2\pi} \operatorname{Sa}(W_N \tau) \tag{9.1-22}$$

which are sketched in Fig. 9.1-2. Because of the nulls in $R_{NN}(\tau)$, samples of $n(t)$ taken Δt apart, as defined by

$$\Delta t = \frac{\pi}{W_N} \tag{9.1-23}$$

are uncorrelated and statistically independent [due to $n(t)$ assumed zero-mean and gaussian]. For W_N large enough such that the received signal is undistorted, the samples of the received waveform are defined as follows:

$$y_{ik} = y(t_{ik}) = s_r(t_{ik}) + n(t_{ik})$$
$$= s_{rik} + n_{ik}, \qquad i = 1, 2, \dots, N, \; k = 1, 2, \dots, K \tag{9.1-24}$$

Since all the noise random variables n_{ik} are statistically independent, gaussian, and of equal noise power, given by

$$\sigma_N^2 = R_{NN}(0) = \frac{\mathscr{N}_0 W_N}{2\pi} = \frac{\mathscr{N}_0}{2\Delta t} \tag{9.1-25}$$

from Fig. 9.1-2 and (9.1-23), the joint probability density functions become

$$p(y|0) = p(\underset{\sim}{n} = \underset{\sim}{y}) = \prod_{i=1}^{N} \prod_{k=1}^{K} \frac{e^{-y_{ik}^2/2\sigma_N^2}}{\sqrt{2\pi}\,\sigma_N} \tag{9.1-26}$$

$$p[\underset{\sim}{y}|\underset{\sim}{s}_r(\Theta)] = p[\underset{\sim}{y} - \underset{\sim}{s}_r(\Theta)] = \prod_{i=1}^{N} \prod_{k=1}^{K} \frac{e^{-(y_{ik} - s_{rik})^2/2\sigma_N^2}}{\sqrt{2\pi}\sigma_N} \tag{9.1-27}$$

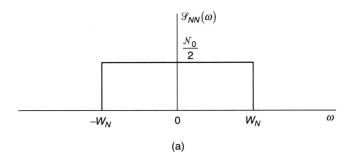

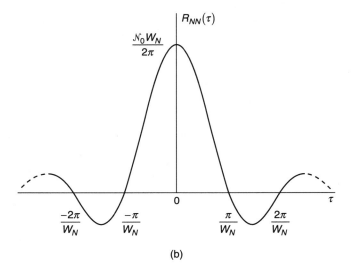

Figure 9.1-2 (*a*) Power spectrum of band-limited noise $n(t)$. As $W_N \to \infty$, $n(t) \to n_w(t)$, white noise. (*b*) Autocorrelation function of $n(t)$.

On substituting (9.1-26) and (9.1-27) into (9.1-19) and reducing some algebra, we have

$$\overline{l(y)_\Theta} = \overline{\left[\prod_{i=1}^{N} e^{-\frac{1}{\mathcal{N}_0} \sum_{k=1}^{K} s_{rik}^2 \Delta t + \frac{2}{\mathcal{N}_0} \sum_{k=1}^{K} y_{ik} s_{rik} \Delta t} \right]}_\Theta \tag{9.1-28}$$

These sums become integrals as $\Delta t \to 0$ when $W_N \to \infty$. The first is the energy E_{ri} received in interval i:

$$\sum_{k=1}^{K} s_{rik}^2 \Delta t \approx \int_{(i-1)T_R}^{iT_R} s_{ri}^2(t) \, dt = E_{ri} \tag{9.-1-29}$$

The second is the integral in interval i of the product $y(t) s_{ri}(t)$, where $s_{ri}(t)$ is the part of $s_r(t)$ received in interval i:

$$\sum_{k=1}^{k} y_{ik} s_{rik} \Delta t \approx \int_{(i-1)T_R}^{iT_R} y(t) s_{ri}(t) \, dt \tag{9.1-30}$$

The final form of the decision rule of (9.1-19) becomes

$$\overline{l(y)_{\Theta}} = E_{\Theta}\left[\prod_{i=1}^{N} \exp\left\{\frac{-E_{ri}}{\mathcal{N}_0} + \frac{2}{\mathcal{N}_0}\int_{(i-1)T_R}^{iT_R} y(t)s_{ri}(t)\,dt\right\}\right]\begin{cases}\geq V_T, & \text{decide } S+N \\ < V_T, & \text{decide } N\end{cases}$$

$$(9.1\text{-}31)$$

For our radar purposes this expression is a general form for the detection criterion. Further reduction requires that specific cases be developed where the random parameters, if any, are specified. In the following section (9.1-31) is simplified to apply to a number of specific assumptions about the target. Subsequent sections then develop each specific case to define the applicable detection signal processor and its performance.

9.2 DETECTION CRITERIA FOR SEVERAL TARGET MODELS

The assumptions made about the target's random parameters affect the detection criterion of (9.1-31). The most important parameters of interest that may be random are normally the phase angles of the target returns in the various pulse intervals, and the target's cross section. In this section we will develop the optimum decision criterion for several choices of random phase angles and cross sections. These cases will then be developed in detail in following sections to show the necessary detection processors and their performances. The cases cover both the steady-target and the Swerling models for randomly fluctuating cross section as defined below. Performance is measured through the false alarm and detection probabilities of each processor.

Completely Known Target

For the completely known type of target we assume no random parameters. The cross section is constant, and the form of the target's return waveform is totally known. Thus the presumption is that delay, Doppler frequency, and carrier phase are all known, such that the only unknown quantity is whether the target is present or absent at the specified delay time. These constraints may seem unrealistic for a typical application. However, the case is worth consideration because it forms a basis of comparison for more practical problems, and it introduces some important concepts that carry over to other cases. In fact the constraint of knowing delay can be softened by imagining that delay is divided into range cells and thinking of the problem as one of examining a specific cell to see if a target is present or not. The processor that applies to a given cell can be duplicated and put in parallel operation to apply to other cells, as we subsequently discuss.

Due to the target's cross section being constant, the pulse energy E_{ri} received in interval i is the same in all intervals at a value denoted by E_r. On defining a parameter $\mathcal{R}_p$ by[4]

$$\mathcal{R}_p = 2E_r/\mathcal{N}_0 \qquad (9.2\text{-}1)$$

---

[4] We subsequently find that $\mathcal{R}_p$ is the single-pulse peak signal-to-noise ratio that occurs at the output of a filter matched to one pulse.

the average likelihood ratio of (9.1-31) can be written as

$$\overline{l(y)_\Theta} = e^{-N\mathcal{R}_p/2} \exp\left[\sqrt{\mathcal{R}_p} \frac{2}{\mathcal{N}_0 \sqrt{\mathcal{R}_p}} \int_0^{NT_R} y(t)\, s_r(t)\, dt \right] \tag{9.2-2}$$

After using the natural logarithm, the decision criterion is

$$\left[\frac{2}{\mathcal{N}_0 \sqrt{\mathcal{R}_p}} \int_0^{NT_R} y(t)\, s_r(t)\, dt \right] - V_b \begin{cases} \geq 0, & \text{decide } S+N \\ < 0, & \text{decide } N \end{cases} \tag{9.2-3}$$

where a new threshold V_b is defined by

$$V_b = \frac{N\sqrt{\mathcal{R}_p}}{2} + \frac{1}{\sqrt{\mathcal{R}_p}} \ln(V_T) \tag{9.2-4}$$

The detection processor applicable to (9.2-3) and its performance are developed in Section 9.3.

Steady Target with Random Initial Phase

Again we will assume that the target is known except that the phase angle of its returned signal $s_r(t)$, as defined by ϕ_0 in (1.6-21), is random and uniformly distributed on $(0, 2\pi)$. Whatever the actual value of the random phase, it remains the same over all N pulse intervals. For this case (9.1-31) becomes

$$\overline{l(y)_\Theta} = e^{-N\mathcal{R}_p/2} E_{\phi_0} \left\{ \exp\left[\frac{2}{\mathcal{N}_0} \int_0^{NT_R} y(t)\, s_r(t)\, dt \right] \right\} \tag{9.2-5}$$

Here $s_r(t)$ is given by (1.6-21) if we are careful to interpret $a(t)$ and $\theta(t)$ as periodic functions over the N pulse intervals being considered. On making this presumption, we write

$$s_r(t) = s_{rI}(t)\cos(\phi_0) - s_{rQ}(t)\sin(\phi_0) \tag{9.2-6}$$

where

$$s_{rI}(t) = \alpha a(t - \tau_R)\cos\left[(\omega_0 + \omega_d)(t - \tau_R) + \theta(t - \tau_R)\right] \tag{9.2-7}$$

$$s_{rQ}(t) = \alpha a(t - \tau_R)\sin\left[(\omega_0 + \omega_d)(t - \tau_R) + \theta(t - \tau_R)\right] \tag{9.2-8}$$

are "quadrature" components of $s_r(t)$. On substitution of (9.2-6) into (9.2-5), we have

$$\overline{l(y)_\Theta} = e^{-N\mathcal{R}_p/2} E_{\phi_0} \left\{ e^{\sqrt{\mathcal{R}_p}[y_{rI}\cos(\phi_0) - y_{rQ}\sin(\phi_0)]} \right\} \tag{9.2-9}$$

where

$$y_{rI} = y_{rI}(NT_R) = \frac{2}{\sqrt{\mathscr{R}_p \mathscr{N}_0}} \int_0^{NT_R} y(t) s_{rI}(t)\, dt \qquad (9.2\text{-}10)$$

$$y_{rQ} = y_{rQ}(NT_R) = \frac{2}{\sqrt{\mathscr{R}_p \mathscr{N}_0}} \int_0^{NT_R} y(t) s_{rQ}(t)\, dt \qquad (9.2\text{-}11)$$

If we next allow the change of variables

$$r = [y_{rI}^2 + y_{rQ}^2]^{1/2} \qquad (9.2\text{-}12a)$$

$$\phi_r = \tan^{-1}\left(\frac{y_{rQ}}{y_{rI}}\right) \qquad (9.2\text{-}12b)$$

(9.2-9) reduces as follows to give the decision rule

$$\overline{l(y)_\Theta} = e^{-N\mathscr{R}_p/2} E_{\phi_0}\left\{ e^{\sqrt{\mathscr{R}_p}\, r \cos(\phi_0 + \phi_r)} \right\}$$

$$= e^{-N\mathscr{R}_p/2} \frac{1}{2\pi} \int_0^{2\pi} e^{\sqrt{\mathscr{R}_p}\, r \cos(\phi_0 + \phi_r)}\, d\phi_0$$

$$= e^{-N\mathscr{R}_p/2} I_0(\sqrt{\mathscr{R}_p}\, r) \begin{cases} \geq V_T, & \text{decide } S + N \\ < V_T, & \text{decide } N \end{cases} \qquad (9.2\text{-}13)$$

Here $I_0(\cdot)$ is the modified Bessel function of the first kind of order zero; it is given by the integral in the second right-side form for any ϕ_r (see DiFranco and Rubin, 1968, p. 301).

The detection processor and performance of the preceding target model are given in Section 9.4.

Steady Target with N Pulses Having Random Phases

In many radar problems the phase angle of the returned pulse can be random from pulse to pulse even though the target's cross section may be taken as constant. This condition can follow from small changes in the scattering phase center for a complicated target, even for relatively short times between pulses. We define $\theta_i = \phi_{0i}$ as the random phase angle for interval i, make use of statistical independence of the ϕ_{0i}, and note that each interval has the same energy E_r. Equation (9.1-31) becomes

$$\overline{l(y)_\Theta} = e^{-N\mathscr{R}_p/2} \prod_{i=1}^N E_{\phi_{0i}} \left\{ \exp\left[\frac{2}{\mathscr{N}_0} \int_{(i-1)T_R}^{iT_R} y(t) s_{ri}(t)\, dt \right] \right\} \begin{cases} \geq V_T, & \text{decide } S + N \\ < V_T, & \text{decide } N \end{cases} \qquad (9.2\text{-}14)$$

This expression is simplified in a manner similar to that used to reduce (9.2-5). By repeating the steps that lead to (9.2-13), except as they apply to interval i, we obtain

$$\overline{l(y)_\Theta} = e^{-N\mathcal{R}_p/2} \prod_{i=1}^{N} E_{\phi_{0i}}\left\{e^{\sqrt{\mathcal{R}_p}\, r_i \cos(\phi_{0i} + \phi_{ri})}\right\}$$

$$= e^{-N\mathcal{R}_p/2} \prod_{i=1}^{N} I_0(\sqrt{\mathcal{R}_p}\, r_i) \begin{cases} \geq V_T, & \text{decide } S + N \\ < V_T, & \text{decide } N \end{cases} \qquad (9.2\text{-}15)$$

After using the natural logarithm

$$\left\{\sum_{i=1}^{N} \ln[I_0(\sqrt{\mathcal{R}_p}\, r_i)]\right\} - V_{b1} \begin{cases} \geq 0, & \text{decide } S + N \\ < 0, & \text{decide } N \end{cases} \qquad (9.2\text{-}16)$$

where

$$V_{b1} = \frac{N\mathcal{R}_p}{2} + \ln(V_T) \qquad (9.2\text{-}17)$$

The detection criterion of (9.2-16) is used in Section 9.5 to find the detection processor and its performance.

Targets That Fluctuate by *N*-Pulse Groups

Consider a scanning radar where the antenna's main beam crosses a target. As the beam sweeps past, the radar receives a group of N pulses before the target passes out of the main beam. In some cases cross section is stable enough to produce N constant-amplitude pulses.[5] However, by the time the antenna returns to again search the area containing the target, cross section may have changed. This behavior is characterized by fluctuation from pulse group to pulse group but not within a group. It is sometimes referred to as *scan-to-scan fluctuation*. The problem was originally studied by Swerling (1954). Two names, *Swerling I* and *Swerling III*, are usually given to group fluctuations when cross section fluctuates according to the exponential density for case I or the chi-square density [of (5.9–22) with $M = 2$] for case III.

For group-to-group fluctuations (9.1-31) first reduces in a manner similar to a steady target with N random phases (for the expectation with respect to the phases) to give a form similar to (9.2-15). However, since $\mathcal{R}_p$ is random but does not fluctuate during the N-pulse group,

$$\overline{l(y)_\Theta} = E_{\mathcal{R}_p}\left[e^{-N\mathcal{R}_p/2} \prod_{i=1}^{N} I_0(\sqrt{\mathcal{R}_p}\, r_i)\right] \begin{cases} \geq V_T, & \text{decide } S + N \\ < V_T, & \text{decide } N \end{cases} \qquad (9.2\text{-}18)$$

This detection criterion is developed in Section 9.6 to obtain the detection processor and its performance for Swerling I and III models.

---

[5] Even with constant cross section, pulses are not exactly constant but vary with pattern gain. We presume that pulses are constant and allow a *beam shape loss* (around 1 to 2 dB) to account for pulse variations with scan.

Targets That Fluctuate Pulse to Pulse

When target cross section fluctuates rapidly enough that each pulse's cross section can be considered independent of the others in a group of N pulses, we say fluctuation is pulse to pulse. Swerling (1954) considered this type of fluctuation, which has come to be known as Swerling case II for exponential fluctuations of cross section and as Swerling case IV for chi-square fluctuations [density of (5.9-22) with $M = 2$]. With pulse-to-pulse fluctuations $\mathscr{R}_p$ is a random variable, $\mathscr{R}_{pi}$, for every interval i and the vector of random parameters in (9.1-31) is $\Theta = (\mathscr{R}_{p1}, \mathscr{R}_{p2}, \ldots, \mathscr{R}_{pN}, \phi_{01}, \phi_{02}, \ldots, \phi_{0N})$. Since all random variables are presumed statistically independent, (9.1-31) becomes

$$
\overline{l(y)_\Theta} = \prod_{i=1}^{N} E_{\mathscr{R}_{pi}} \left\{ e^{-\mathscr{R}_{pi}/2} E_{\phi_{0i}} \left[\exp \left\langle \frac{2}{\mathscr{N}_0} \int_{(i-1)T_R}^{iT_R} y(t) s_{ri}(t)\, dt \right\rangle \right] \right\}
$$

$$
= \prod_{i=1}^{N} E_{\mathscr{R}_{pi}} \left[e^{-\mathscr{R}_{pi}/2} I_0 \left(\sqrt{\mathscr{R}_{pi}}\, r_i \right) \right]
$$

$$
= \prod_{i=1}^{N} E_{\mathscr{R}_p} \left[e^{-\mathscr{R}_p/2} I_0 \left(\sqrt{\mathscr{R}_p}\, r_i \right) \right] \begin{cases} \geq V_T, & \text{decide } S+N \\ < V_T, & \text{decide } N \end{cases} \qquad (9.2\text{-}19)
$$

The second right-side form of (9.2-19) derives from the same procedure used to reduce the expectation in (9.2-14). The third right-side form is true because each random variable $\mathscr{R}_{pi}$ has the same density function.

The criterion of (9.2-19) is developed in Section 9.7 to find the detection processor and its performance for Swerling models II and IV.

9.3   DETECTION OF KNOWN TARGET

Optimum Signal Processor

The optimum detection processor for this type of target implements the optimum decision criterion of (9.2-3). It is shown in Fig. 9.3-1a where D is a decision variable. It is to be noted that the processor requires the generation of $s_r(t)$, which is a consequence of the assumption that the form of $s_r(t)$ was known. Although we presume to know the form of $s_r(t)$, it is the processor's job to determine if $s_r(t)$ is present or not, which is the detection process.

A legitimate practical question is: Suppose that we do not know the target's delay; what practical purpose does this model have? As noted earlier, one answer is to subdivide the pulse repetition interval into range resolution cells (of delay) and to implement a bank of processors in parallel; one processor for each cell. Whatever cell contains a target becomes the one that indicates a detection.

The processor of Fig. 9.3-1a uses a product device and time integrator, and is often called a *correlation detector*. Even though the integration period is shown as NT_R, the sampling can be done at the end of the target signal in the last (Nth) interval because $s_r(t)$ becomes zero after this time and the integrator ideally holds its output value after this time until discharged. The correlator (product and integrator) can be replaced by a matched filter if the filter's operation is properly interpreted. The result is the processor of panel b. To demonstrate, we assume the filter is matched to $s_r(t)$. Its

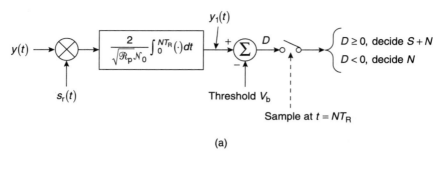

(a)

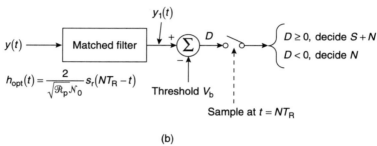

(b)

Figure 9.3-1 (a) Optimum detection signal processor for a known signal in white gaussian noise. (b) The matched filter equivalent processor.

impulse response is

$$h_{\text{opt}}(t) = \frac{2}{\sqrt{\mathcal{R}_p \mathcal{N}_0}} \, s_r(t_0 - t) \tag{9.3-1}$$

from (6.7-3) with the arbitrary constant C set to $1/\sqrt{\mathcal{R}_p}$, and from (B.4-2), and (B.4-11). The filter's response is

$$y_1(t) = \int_{-\infty}^{\infty} y(\xi) h_{\text{opt}}(t - \xi) \, d\xi = \frac{2}{\sqrt{\mathcal{R}_p \mathcal{N}_0}} \int_{-\infty}^{\infty} y(\xi) s_r(\xi + t_0 - t) \, d\xi \tag{9.3-2}$$

for any time t and any t_0 (the time at which the maximum signal-to-noise ratio and maximum *signal* amplitude occur at the filter's output). Now, if $y(t)$ is admitted to the processor only for $0 \le t \le NT_R$ (by time gating) so as to limit the system to N pulse intervals, the integration interval in (9.3-2) can be set to $0 \le \xi \le NT_R$. Finally, if the sampling time in Fig. 9.3-1b is set to t_0, that is, if $t = t_0$,

$$y_1(t_0) = \frac{2}{\sqrt{\mathcal{R}_p \mathcal{N}_0}} \int_0^{NT_R} y(\xi) s_r(\xi) \, d\xi \tag{9.3-3}$$

This result shows that the two processors of Fig. 9.3-1 are equivalent if the matched filter system samples at time $t = t_0$.

The parameter t_0 can be any value by design. However, since increasing t_0 amounts to adding delay to the matched filter, the smallest t_0 is usually desirable. If the transmitted waveform repeats in each pulse interval and is nonzero only for $0 < t < T$ in the first interval, T a constant, then $s_r(t)$ has its nonzero values only in the region $\tau_R < t < \tau_R + T + (N - 1)T_R$, where τ_R is the delay of $s_r(t)$. The smallest t_0 is then $\tau_R + T + (N - 1)T_R$. The sample times in the two processors of Fig. 9.3-1 can be equal at NT_R if t_0 in panel b is made equal to NT_R, an increase of $T_R - \tau_R - T$ over the smallest value.

System Performance

System performance is determined by the behavior of $y_1(t)$ in Fig. 9.3-1 at the sample time. By noting that

$$y(t) = s_r(t) + n_w(t) \tag{9.3-4}$$

it can be shown that the mean value of $y_1(NT_R)$ is

$$E[y_1(NT_R)] = \begin{cases} N\sqrt{\mathscr{R}_p}, & \text{target present} \\ 0, & \text{target absent} \end{cases} \tag{9.3-5}$$

where, if E_r is the energy of $s_r(t)$ in one interval,

$$\mathscr{R}_p = \frac{2E_r}{\mathscr{N}_0} \tag{9.3-6}$$

The variance, σ_{y1}^2, of $y_1(NT_R)$ is also found to be

$$\sigma_{y1}^2 = \begin{cases} E\{[y_1(NT_R) - N\sqrt{\mathscr{R}_p}]^2\} = N, & \text{target present} \\ E\{y_1^2(NT_R)\} = N, & \text{target absent} \end{cases} \tag{9.3-7}$$

These results indicate that y_1 (at $t = NT_R$) is gaussian [because $n_w(t)$ is assumed gaussian] with mean zero and variance N when there is no target. Since a false alarm occurs for $D \geq 0$, its probability is

$$P_{fa} = \int_{V_b}^{\infty} \frac{e^{-y_1^2/2N}}{\sqrt{2\pi N}} \, dy_1 = \int_{V_b/\sqrt{N}}^{\infty} \frac{e^{-\xi^2/2}}{\sqrt{2\pi}} \, d\xi = Q\left(\frac{V_b}{\sqrt{N}}\right) \tag{9.3-8}$$

where $Q(\cdot)$ is the Q function of (E.2-1). In a similar manner, y_1 is gaussian with mean $N\sqrt{\mathscr{R}_p}$ and variance N when a target is present. Detection probability is

$$P_d = \int_{V_b}^{\infty} \frac{e^{-(y_1 - N\sqrt{\mathscr{R}_p})^2/2N}}{\sqrt{2\pi N}} \, dy_1 = Q\left(\frac{V_b}{\sqrt{N}} - \sqrt{N\mathscr{R}_p}\right) \tag{9.3-9}$$

The most common mode of radar detection is to first decide on an acceptable value of P_{fa}. This choice establishes an equivalent threshold level $V_b/\sqrt{N}$ from (9.3-8) for

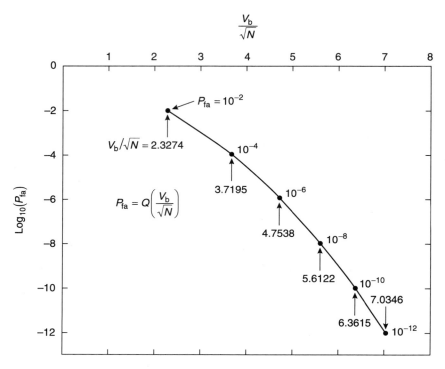

Figure 9.3-2 Relationship of P_{fa} to threshold for detection of exactly known signal in white noise.

a given choice of N. The detection probability follows from (9.3-9) for a given value of $\mathscr{R}_p$. From (9.2-1) and (6.7-4), $\mathscr{R}_p$ is seen to be the maximum peak signal-to-noise ratio at the output of the matched filter. Since $Q(\cdot)$ has no closed form, the excellent approximation of (E.2-4) is very useful for numerical calculations. Figure 9.3-2 used the approximation to determine $V_b/\sqrt{N}$ for various P_{fa}. Similarly it was used to generate Fig. 9.3-3 which describes how P_d varies with $\mathscr{R}_p$.

9.4 DETECTION OF STEADY TARGET WITH RANDOM INITIAL PHASE

The decision criterion for this type of target is (9.2-13), which can be put in the form

$$I_0(\sqrt{\mathscr{R}_p}r) - V_{b2} \begin{cases} \geq 0, & \text{decide } S + N \\ < 0, & \text{decide } N \end{cases} \tag{9.4-1}$$

where V_{b2} is a threshold defined by

$$V_{b2} = V_T e^{N\mathscr{R}_p/2} \tag{9.4-2}$$

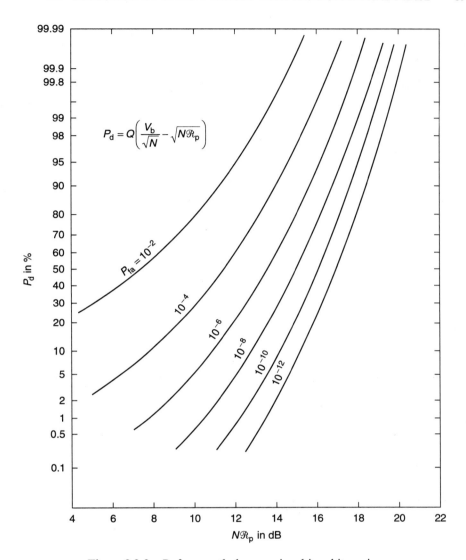

$$P_d = Q\left(\frac{V_b}{\sqrt{N}} - \sqrt{N\mathscr{R}_p}\right)$$

Figure 9.3-3 P_d for exactly known signal in white noise.

Optimum Signal Processors

The optimum detection processor implements (9.4-1) and is shown in Fig. 9.4-1a. This processor can be considerably simplified. From panel b there is an equivalent threshold, denoted by V_r, on r, for any threshold V_{b2} on $I_0(\sqrt{\mathscr{R}_p}r)$ because $I_0(\cdot)$ is a monotonic function. This fact means the nonlinear function can be dismissed and decisions based on r can be made. We next show that the processor of panel c is equivalent to the simplified version of a.

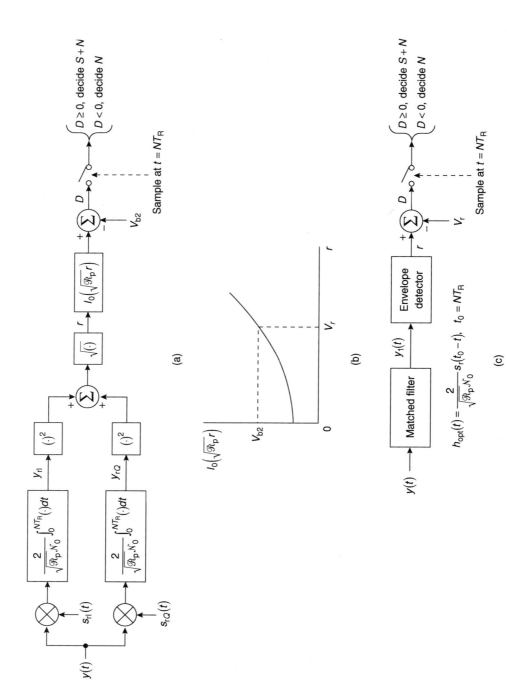

Figure 9.4-1 Optimum receivers (a) and (c) for a steady target with random initial phase angle. (b) The relationship between thresholds V_r and V_{b2}.

Assume that the filter in Fig. 9.4-1c is matched to $s_r(t)$ to produce a maximum response at $t_0 = NT_R$. It is readily shown that $y_1(NT_R)$ is

$$y_1(NT_R) = \frac{2}{\sqrt{\mathscr{R}_p \mathscr{N}_0}} \int_0^{NT_R} y(\xi)\, s_r(\xi)\, d\xi$$

$$= y_{rI} \cos(\phi_0) - y_{rQ} \sin(\phi_0) \qquad (9.4\text{-}3)$$

from use of (9.2-6), (9.2-10), and (9.2-11). From the vectors shown in Fig. 9.4-2, it is clear that

$$r = \sqrt{y_{rI}^2 + y_{rQ}^2} \qquad (9.4\text{-}4)$$

$$\phi_r = \tan^{-1}\left(\frac{y_{rQ}}{y_{rI}}\right) \qquad (9.4\text{-}5)$$

$$r\cos(\phi_r + \phi_0) = y_{rI}\cos(\phi_0) - y_{rQ}\sin(\phi_0) = y_1(NT_R) \qquad (9.4\text{-}6)$$

This last result shows that the envelope of $y_1(NT_R)$ is the same as in Fig. 9.4-1a because of (9.4-4).

Performance of Optimum System

The performances of the detection systems of Fig. 9.4-1 depend on the behavior of r which results from the behavior of y_{rI} and y_{rQ}. We summarize the results since considerable detail is needed to show all developments. The mean values, variances,

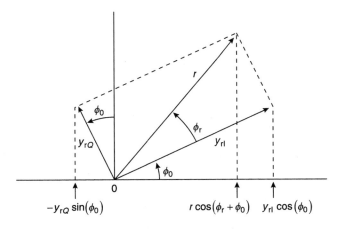

Figure 9.4-2   Geometry useful in understanding the processor of Fig. 9.4-1c.

and covariances of y_{rI} and y_{rQ} are found to be

$$E[y_{rI}] = \begin{cases} N\sqrt{\mathscr{R}_p}\cos(\phi_0), & \text{target present} \\ 0, & \text{target absent} \end{cases} \tag{9.4-7}$$

$$E[y_{rQ}] = \begin{cases} -N\sqrt{\mathscr{R}_p}\sin(\phi_0), & \text{target present} \\ 0, & \text{target absent} \end{cases} \tag{9.4-8}$$

$$\sigma_{y_{rI}}^2 = \sigma_{y_{rQ}}^2 = N, \quad \text{target present or absent} \tag{9.4-9}$$

$$C_{y_{rI}y_{rQ}} = E\{[y_{rI} - E(y_{rI})][y_{rQ} - E(y_{rQ})]\}$$
$$= 0, \quad \text{target present or absent} \tag{9.4-10}$$

By using these four results, the fact that y_{rI} and y_{rQ} are gaussian, and appropriate variable changes, it can be shown that the probability density functions $p_0(r)$ (target absent) and $p_1(r)$ (target present) are

$$p_0(r) = \frac{u(r)r}{N} e^{-r^2/(2N)}, \qquad \text{target absent} \tag{9.4-11a}$$

$$p_1(r) = \frac{u(r)r I_0(\sqrt{\mathscr{R}_p}r)}{N} e^{-(r^2 + N^2\mathscr{R}_p)/(2N)}, \qquad \text{target present} \tag{9.4-11b}$$

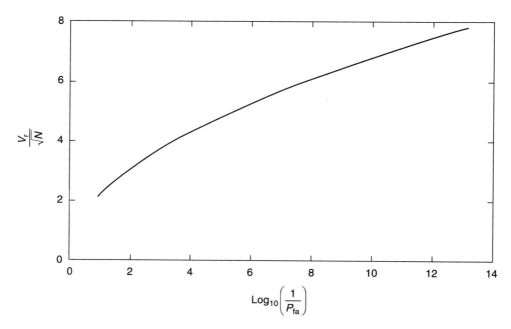

Figure 9.4-3 Normalized threshold versus false alarm probability in detection of a steady target with random initial phase angle.

Performance is established by false alarm and detection probabilities. On using (9.4-11) and the threshold V_r, these are (since $V_r > 0$)

$$P_{\mathrm{fa}} = \int_{V_r}^{\infty} \frac{r}{N} e^{-r^2/(2N)} \, dr = e^{-V_r^2/(2N)} \tag{9.4-12}$$

$$\bar{P}_{\mathrm{d}} = \int_{V_r}^{\infty} \frac{r I_0(\sqrt{\mathscr{R}_{\mathrm{p}}}\, r)}{N} e^{(r^2 + N^2 \mathscr{R}_{\mathrm{p}})/(2N)} \, dr$$

$$= Q(\sqrt{N\mathscr{R}_{\mathrm{p}}}, \sqrt{2\ln(1/P_{\mathrm{fa}})}) \tag{9.4-13}$$

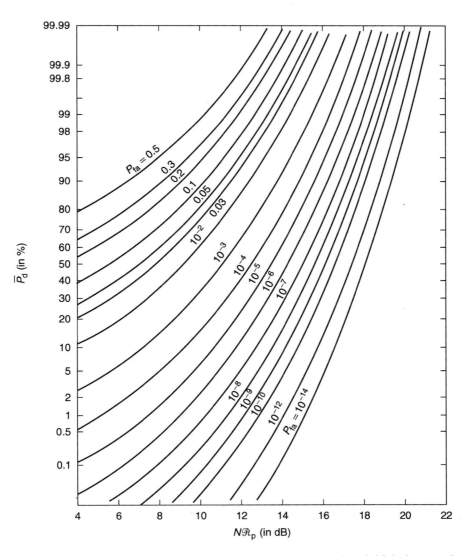

Figure 9.4-4 Detection probability for a steady target with random initial phase angle.

where the *Marcum Q-function* is defined by

$$Q(\alpha, \beta) = \int_{\beta}^{\infty} \xi I_0(\alpha\xi) e^{-(\xi^2 + \alpha^2)/2} d\xi \tag{9.4-14}$$

This Q-function is not known to have a closed form. It has been tabulated extensively (Marcum, 1950), and an easily programmed algorithm for its computation exists (Brennan and Reed, 1965).

The required threshold V_r can be found from (9.4-12):

$$V_r = \sqrt{2N \ln\left(\frac{1}{P_{fa}}\right)} \tag{9.4-15}$$

A plot of normalized threshold is shown in Fig. 9.4-3. Plots of $\bar{P}_d$ for various P_{fa} are given in Fig. 9.4-4 as calculated from (9.4-13) using the algorithm of Brennan and Reed (1965) for solving Marcum's Q-function.

9.5   DETECTION OF STEADY TARGET WITH N PULSES HAVING RANDOM PHASES

Optimum Detection Processor

When each pulse received from a nonfluctuating target has a random phase angle, (9.2-16) defines the applicable decision criterion. The optimum detection processor to implement this criterion is given in Fig. 9.5-1a. A single-pulse matched filter and linear envelope detector generate r_i for interval i in a manner analogous to Fig. 9.4-1c for the N-interval matched filter. A nonlinear function generates $\ln[I_0(\sqrt{\mathcal{R}_p} r_i)]$ which is made maximum at the end of interval i by choice of $t_0 = T_R$ in the matched filter. Samples taken at the ends of the N intervals are stored and their sum is compared to threshold V_{b1} at the end of interval N. A sum not less than V_{b1} leads to a decision that a target is present. If less than V_{b1} it is decided that no target is present. After the decision is made the process is repeated in each N-pulse interval.

If the matched filter in Fig. 9.5-1a uses the smallest allowable value of t_0 and not $t_0 = T_R$, the sample in interval i is taken at the end of the pulse corresponding to the range (delay) cell for which the processor applies. All other operations are the same. The final decision is made at the end of the target pulse being tested in interval N.

Simplified Detection Processor

Implementation of Fig. 9.5-1a is made difficult by the need for the nonlinear function $\ln[I_0(\beta)]$. The possibility of replacing the function by a simple approximation is of interest. It is well-known that

$$\ln[I_0(\beta)] \approx \begin{cases} \dfrac{\beta^2}{4}, & \beta \text{ small} \\[2mm] \beta, & \beta \text{ large} \end{cases} \tag{9.5-1}$$

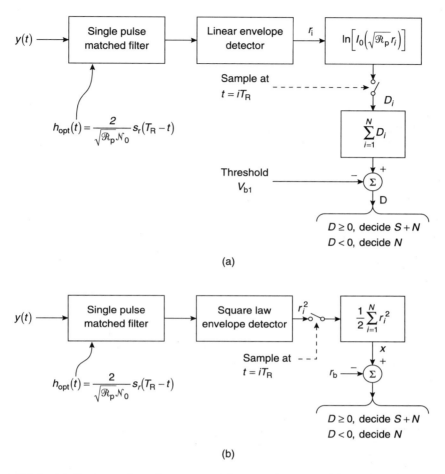

Figure 9.5-1 (*a*) Optimum detection processor for a steady target with random phase angle in each pulse interval, and (*b*) the approximately optimum processor using a square law detector.

(DiFranco and Rubin, 1968, pp. 342–343). The approximation for β small in conjunction with the linear envelope detector is equivalent to a square law detector. The second case of β large is equivalent to ignoring the nonlinearity completely. Study of the two approximations (Marcum, 1948) has shown little difference in performance. For example, with $P_d = 0.5$ and $P_{fa}/N = 10^{-6}$ the linear detector is superior by not more than 0.11 dB of required signal-to-noise ratio when N is small (N less than about 70); for larger N, the square law detector is superior by an amount not exceeding 0.19 dB (DiFranco and Rubin, 1968, pp. 370–371). Because the linear detector is more difficult to analyze and performances are similar, we discuss only the square law detector in following work.

On using the small-β approximation of (9.5-1) in (9.2-16), the decision criterion becomes

$$\left[\frac{1}{2}\sum_{i=1}^{N} r_i^2\right] - r_b \begin{cases} \geq 0, & \text{decide } S + N \\ < 0, & \text{decide } N \end{cases} \qquad (9.5\text{-}2)$$

where

$$r_b = \frac{2V_{b1}}{\mathscr{R}_p} = N + \frac{2}{\mathscr{R}_p}\ln(V_T) \tag{9.5-3}$$

The approximate optimum detection processor based on (9.5-2) is given in Fig. 9.5-1b; the developments and arguments for its validity are similar to those leading to Fig. 9.4-1c except the processor's filter is here matched to a single pulse rather than N pulses as before.

Detection Performance

The performance of the system of Fig. 9.5-1b is established by the detection and false alarm probabilities found by integrating the probability densities of x from r_b, the threshold, to ∞. The general procedure is to use $p_1(x)$, the density of x with target present to find $\bar{P}_d$, and $p_0(x)$, the density of x when noise alone is present, to find P_{fa}. Since P_{fa} can be found from $\bar{P}_d$ by letting $\mathscr{R}_p = 0$ (no signal), we only need consider finding $\bar{P}_d$.

We outline the steps needed to find $\bar{P}_d$. First, $\bar{P}_d$ is found if the density $p_1(x)$ is known, where

$$x = \tfrac{1}{2}\sum_{i=1}^{N} r_i^2 = \sum_{i=1}^{N} x_i \tag{9.5-4}$$

$$x_i = \tfrac{1}{2}r_i^2 \tag{9.5-5}$$

To proceed further, we define the *Laplace characteristic function* of a complex variable z, denoted by $C(z)$, for a random variable x having probability density $p_1(x)$, by

$$C(z) = \int_{-\infty}^{\infty} p_1(x)e^{-zx}\,dx \tag{9.5-6}$$

which is a Laplace transform. The inverse transform applies and is (Hou et al., 1987)[6]

$$p_1(x) = \frac{1}{2\pi j}\int_{c-j\infty}^{c+j\infty} C(z)e^{zx}\,dz \tag{9.5-7}$$

The contour of integration is determined by c which depends on the locations of poles in the z plane (see Hou et al., 1987). If, similar to a Fourier transform pair, we use a double-ended arrow $\overset{\mathscr{L}}{\leftrightarrow}$ to denote a Laplace transform pair, then for any $p(x)$,

$$p(x)\overset{\mathscr{L}}{\leftrightarrow} C(z) \tag{9.5-8}$$

---

[6] Hou, et al. (1987) call $C(z)$ a moment-generating function. We adopt a different name, since the moment-generating function normally assumes that z is real. $C(z)$ is also not the usual characteristic function where $z = -j\omega$ is purely imaginary.

For a sum of statistically independent random variables x_i, as in (9.5-4), it can be shown that the characteristic function of x is the product of individual characteristic functions $C_i(z)$ of the x_i. In our specific case

$$p_1(x) \overset{\mathscr{L}}{\leftrightarrow} C(z) = \prod_{i=1}^{N} C_i(z) \tag{9.5-9}$$

$$p_1(x_i) \overset{\mathscr{L}}{\leftrightarrow} C_i(z) \tag{9.5-10}$$

For the current problem it can be shown that $p_1(r_i)$ is given by (9.4-11b) with $N = 1$ (one interval here). After a change of variable to $x_i = r_i^2/2$, the new density $p_1(x_i)$ can be transformed by (9.5-6) to obtain

$$C_i(z) = (1 + z)^{-1} \exp\left[\frac{-\mathscr{R}_p z}{2(1 + z)}\right] \tag{9.5-11}$$

Because $\mathscr{R}_p$ is assumed constant (the same in every interval), $C_i(z)$ is independent of i and (9.5-9) gives

$$p_1(x) \overset{\mathscr{L}}{\leftrightarrow} C(z) = (1 + z)^{-N} \exp\left[\frac{-N\mathscr{R}_p z}{2(1 + z)}\right] \tag{9.5-12}$$

Ordinarily, the second step in our efforts to find $\bar{P}_d$ would involve inverse transformation of $C(z)$. Such an effort leads to a solution in terms of the incomplete Toronto function (DiFranco and Rubin, 1968, p. 348). However, recently Hou, Morinaga, and Namekawa (1987) used the theory of residues to solve for $\bar{P}_d$. When $C(z)$ of (9.5-12) is used in (9.5-7), $p_1(x)$ can be written as the sum of residues of $C(z)\exp(zx)$. The integral of $p_1(x)$ then is solvable in terms of these residues. Their result is

$$\bar{P}_d = \int_{r_b}^{\infty} p_1(x)\,dx$$

$$= \lim_{K \to \infty} \left\{ \sum_{l=N-1}^{K-1} \binom{K-N}{K-l-1} \frac{F^{K-l-1}}{G^{N-l-1}} \exp(-Fr_b) \sum_{m=0}^{l} \frac{(Fr_b)^m}{m!} \right\} \tag{9.5-13}$$

where

$$G = 1 - F \tag{9.5-14}$$

$$F = \left(1 + \frac{N\mathscr{R}_p}{2K}\right)^{-1} \tag{9.5-15}$$

$$\binom{K-N}{K-l-1} = \frac{(K-N)!}{(K-l-1)!(l-N+1)!} \tag{9.5-16}$$

When $\mathscr{R}_p = 0$, $\bar{P}_d$ of (9.5-13) reduces to P_{fa}. Hou et al. (1987) give the result as

$$P_{fa} = \sum_{m=0}^{N-1} \frac{r_b^m}{m!} \exp(-r_b) \qquad (9.5\text{-}17)$$

Pachares, 1958, solved (9.5-17) (another, but equivalent procedure was actually used) for threshold r_b required to give specified P_{fa} for chosen N. Figure 9.5-2 shows plots from Pachares data, which give r_b for all integer $1 \le N \le 150$, and $P_{fa} = 10^{-n}$ when $n = 1, 2, \ldots, 10$. Some useful values are shown in Table 9.5-1 (see also Problem 9.5-1).

According to Hou et al. (1987, p. 422), the calculation of $\bar{P}_d$ from (9.5-13) for the case $K = \infty$ can be approximated practically if K is larger than $5N$ to $6N$. Various data have been calculated for $P_{fa} = 10^{-2}$, 10^{-4}, 10^{-6}, 10^{-8}, and 10^{-10}, with $N = 1, 3, 10, 30,$ and 100 for each value of P_{fa}. In most cases $K = 50N$ was used, but some data for $K = 6N$ were taken when N was large to conserve computational time. To conserve space all these plots of $\bar{P}_d$ are not presented. Rather, we note that for $N = 1$ Fig. 9.4-4 applies and show how the results for $N = 1$ can be extended to $N > 1$ by use of empirical results.

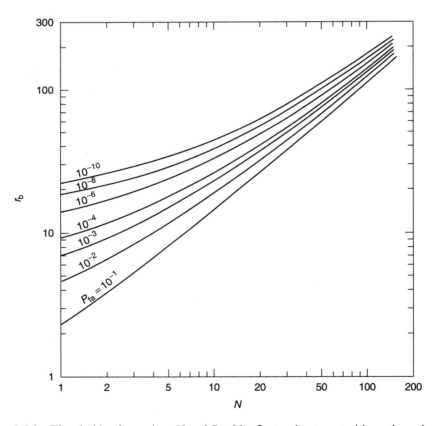

Figure 9.5-2 Threshold r_b for various N and P_{fa}. Nonfluctuating target with random phase in each pulse interval. (From data in Pachares, © 1958 IEEE, with permission.)

TABLE 9.5-1 **Threshold values that correspond to useful values of P_{fa} defined in (9.5-17).**

| P_{fa} | Threshold r_b for N Shown | | | | | |
	$N = 1$	5	10	50	100	$N = 150$
10^{-2}	4.6052	11.6046	18.7831	67.9034	124.7226	179.9532
10^{-4}	9.2103	17.7820	26.1930	80.6593	141.5301	199.8780
10^{-6}	13.8155	23.4315	32.7103	91.0634	154.9190	215.5707
10^{-8}	18.4207	28.8320	38.7990	100.3160	166.6299	229.1845
10^{-10}	23.0259	34.0838	44.6279	108.8571	177.3005	241.5082
10^{-12}	27.6310	39.2358	50.2799	116.9053	187.2480	252.9339

Source: From Data in Pachares, © 1958 IEEE, with permission.

We define the *integration improvement factor*, $I(N)$, for a given value of N, to be the ratio of $\mathscr{R}_{p1}$, the value of $\mathscr{R}_p$ needed when $N = 1$ to produce a desired value of $\bar{P}_d$ for a specified value of P_{fa}, to $\mathscr{R}_{pN}$, the value of $\mathscr{R}_p$ needed to produce the same $\bar{P}_d$ and P_{fa} except when N pulses are used. Thus

$$I(N) = \frac{\mathscr{R}_{p1}}{\mathscr{R}_{pN}} \tag{9.5-18}$$

or

$$[I(N)]_{dB} = (\mathscr{R}_{p1})_{dB} - (\mathscr{R}_{pN})_{dB} \tag{9.5-19}$$

Because $\mathscr{R}_{pN} < \mathscr{R}_{p1}$, $I(N) > 1$ and $[I(N)]_{dB}$ will be a positive number in decibels. Plots of $I(N)$ versus N are not found to vary greatly with $\bar{P}_d$ or P_{fa}. In fact, the empirically derived approximation

$$[I(N)]_{dB} = 6.79 [1 + 0.253 \bar{P}_d] \left[1 + \frac{\log_{10}(1/P_{fa})}{46.6} \right]$$

$$\cdot \log_{10}(N) [1 - 0.14 \log_{10}(N) + 0.0183 \log_{10}^2(N)] \tag{9.5-20}$$

has been found accurate to about 0.8 dB for values of $P_{fa} = 10^{-2}, 10^{-4}, 10^{-6}, 10^{-8}$, and 10^{-10} for all values of $\bar{P}_d = 0.5, 0.7, 0.9, 0.99$, and 0.999 and $N = 1, 3, 10, 30$, and 100. There is no reason to believe it is less accurate for other values of N, $\bar{P}_d$, and P_{fa} between the extremes of these parameters. Figure 9.5-3 shows results for the two extreme combinations of P_{fa} and $\bar{P}_d$ as well as for $P_{fa} = 10^{-6}$ and $\bar{P}_d = 0.9$ which is roughly a nominal combination.

We summarize the preceding results. For detection of a nonfluctuating target based on N pulses each having random phase, we use Fig. 9.5-2, or the tabulated results of Table 9.5-1 or of Pachares (1958), to establish the threshold r_b needed in Fig. 9.5-1b for a specified P_{fa}. For the chosen values of P_{fa} and $\bar{P}_d$, we find $\mathscr{R}_{p1}$ from Fig. 9.4-4 assuming $N = 1$ and use with (9.5-20) to find $\mathscr{R}_{pN}$. $\mathscr{R}_{pN}$ is the single-pulse signal-to-noise ratio required to achieve the given $\bar{P}_d$ and P_{fa} based on N-pulse detection. The procedure gives $\mathscr{R}_{pN}$ to an accuracy of about 0.8 dB or better.

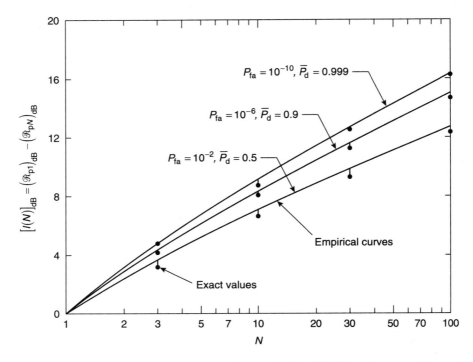

Figure 9.5-3 Empirical approximation (solid curves) for $[I(N)]_{dB}$ for a nonfluctuating target with random phase in each pulse interval, and exact results (dots) for $N = 3, 10, 30,$ and 100.

9.6 DETECTION OF TARGETS WITH GROUP FLUCTUATIONS

For a group of N pulses where the target's cross section is constant during the group such that $\mathscr{R}_p$ is constant for each pulse in the group, but fluctuates randomly from group to group, the detection criterion is (9.2-18).

Detection Criterion and Approximate Detection Processor

Again we use the small-signal approximation for the Bessel function to simplify (9.2-18). Since (with small arguments)

$$I_0(\sqrt{\mathscr{R}_p}\, r_i) \approx 1 + \frac{\mathscr{R}_p}{4} r_i^2 \tag{9.6-1}$$

we have

$$\prod_{i=1}^{N} I_0(\sqrt{\mathscr{R}_p}\, r_i) \approx 1 + \sum_{i=1}^{N} \frac{\mathscr{R}_p}{4} r_i^2 \tag{9.6-2}$$

after dropping all terms above second order. On substitution of (9.6-2) in (9.2-18), we get

$$\left[\frac{1}{2}\sum_{i=1}^{N}r_i^2\right] - r_{b1}\begin{cases}\geq 0, & \text{decide } S+N \\ < 0, & \text{decide } N\end{cases} \tag{9.6-3}$$

where

$$r_{b1} = \frac{V_T - E_{\mathscr{R}_p}[e^{-N\mathscr{R}_p/2}]}{E_{\mathscr{R}_p}\left[\frac{1}{2}\mathscr{R}_p e^{-N\mathscr{R}_p/2}\right]} \tag{9.6-4}$$

Here r_{b1} is a threshold that is independent of r_i regardless of the precise values of the expectations in (9.6-4).

Equation (9.6-3) is the decision rule of the approximately optimum detection processor. Since its *form* is exactly the same as (9.5-2), the detection processor is the same as in Fig. 9.5-1b with r_b replaced by r_{b1} of (9.6-4).

Performance with Group Fluctuations

The developments leading to $\bar{P}_d$ parallel those of Section 9.5 down to (9.5-12) except that $\mathscr{R}_p$ is now a random variable. In other words the density of x is now a conditional density, that we denote by $p_1(x|\mathscr{R}_p)$. Its Laplace transform is then a conditional Laplace characteristic function, which we denote by $C(z|\mathscr{R}_p)$. The present extension of (9.5-12) becomes

$$p_1(x|\mathscr{R}_p) \overset{\mathscr{L}}{\leftrightarrow} C(z|\mathscr{R}_p) = (1+z)^{-N}\exp\left[\frac{-N\mathscr{R}_p z}{2(1+z)}\right] \tag{9.6-5}$$

We average $p_1(x|\mathscr{R}_p)$ over $\mathscr{R}_p$ to obtain the density of x alone which is needed to compute $\bar{P}_d$, as given by

$$\bar{P}_d = \int_{r_{b1}}^{\infty} p_1(x)\,dx \tag{9.6-6}$$

Hence

$$p_1(x) = \int_{-\infty}^{\infty} p_1(x|\mathscr{R}_p)p(\mathscr{R}_p)\,d\mathscr{R}_p \tag{9.6-7}$$

Averaging $p_1(x|\mathscr{R}_p)$ over $\mathscr{R}_p$ is equivalent to replacing the left side of (9.6-5) with $p_1(x)$ if the right side is also averaged to produce an average Laplace characteristic function

$$C(z) = \int_0^{\infty} C(z|\mathscr{R}_p)p(\mathscr{R}_p)\,d\mathscr{R}_p \tag{9.6-8}$$

Again using the procedure of Hou et al. (1987), $\bar{P}_d$ is given by (9.6-6) using (9.6-7) where $p_1(x)$ is represented by the sum of the residues of its inverse Laplace characteristic function. We consider two cases.

Swerling I Fluctuation Model

In this model $\mathcal{R}_p$ is assumed to have exponential fluctuations defined by

$$p(\mathcal{R}_p) = \frac{u(\mathcal{R}_p)}{\bar{\mathcal{R}}_p} e^{-\mathcal{R}_p/\bar{\mathcal{R}}_p} \tag{9.6-9}$$

where $\bar{\mathcal{R}}_p$ is the mean (or average) value of $\mathcal{R}_p$. When used in (9.6-8) with (9.6-5), we have

$$C(z) = (1 + z)^{1-N} \left[1 + \left(1 + \frac{N\bar{\mathcal{R}}_p}{2} \right) z \right]^{-1} \tag{9.6-10}$$

This expression is a special case of a more general Laplace characteristic function solved by Hou et al. (1987) by their residue method. We give the more general results[7] and then indicate how to use them for the present problem. Hou et al. (1987) find two expressions:[8]

$$\bar{P}_d = \sum_{i=0}^{N-K-1} \binom{N-i-2}{K-1} \frac{(-F)^K \exp(-r_{b1})}{G^{N-i-1}} \sum_{l=0}^{i} \frac{r_{b1}^l}{l!}$$

$$+ \sum_{m=0}^{K-1} \binom{N-m-2}{N-K-1} \frac{(-F)^{K-m-1}\exp(-Fr_{b1})}{G^{N-m-1}} \sum_{n=0}^{m} \frac{(Fr_{b1})^n}{n!} \tag{9.6-11}$$

which applies only for $1 \le K < N$, and

$$\bar{P}_d = \sum_{l=N-1}^{K-1} \binom{K-N}{K-l-1} \frac{F^{K-l-1}\exp(-Fr_{b1})}{G^{N-l-1}} \sum_{m=0}^{l} \frac{(Fr_{b1})^m}{m!} \tag{9.6-12}$$

which applies for $1 \le N \le K$. In either case

$$F = \left(1 + \frac{N\bar{\mathcal{R}}_p}{2K} \right)^{-1} \tag{9.6-13}$$

$$G = 1 - F \tag{9.6-14}$$

---

[7] The general results are stated because they also cover all the Swerling fluctuation models to follow and not just the current case (which is Swerling I).
[8] Equivalent expressions can also be found in Shnidman (1995).

and factors of the form $\binom{m}{n}$ are binomial coefficients given by

$$\binom{m}{n} = \frac{m!}{n!(m-n)!} \tag{9.6-15}$$

For the special case considered here, $K = 1$, so (9.6-12) applies when $N = 1$ (see Problem 9.6-1) and (9.6-11) is used for $N > 1$ (see Problem 9.6-13). This case is known as the Swerling I fluctuation model.

$\bar{P}_d$ was found by solving either (9.6-12) or (9.6-11) by digital computer. Figures 9.6-1 through 9.6-5 show results for five values of P_{fa} and several values of N. P_{fa} is obtained

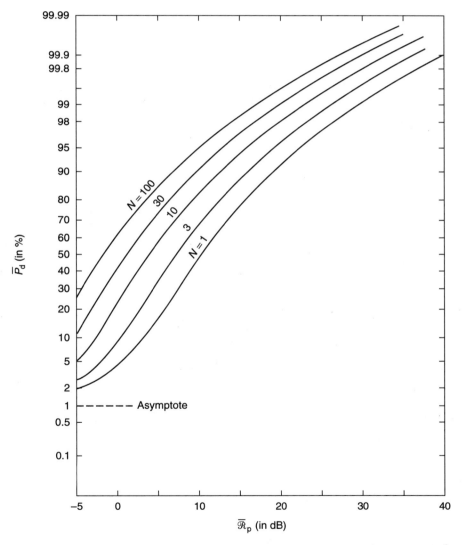

Figure 9.6-1   Average detection probability for a Swerling I target when $P_{fa} = 10^{-2}$.

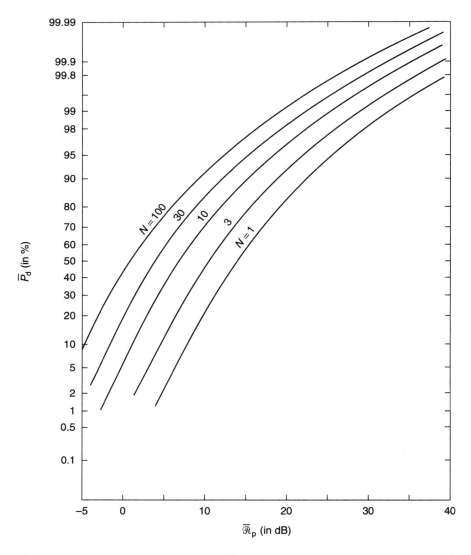

Figure 9.6-2 Average detection probability for a Swerling I target when $P_{\mathrm{fa}} = 10^{-4}$.

from (9.5-17) with r_{b} replaced by r_{b1}, since the signal processor's form is the same as in Section 9.5 for the steady target and P_{fa} is not dependent on target fluctuations.

Swerling III Fluctuation Model

In this model Swerling (1954) assumed target cross section fluctuated such that $p(\mathscr{R}_{\mathrm{p}})$ had the form

$$p(\mathscr{R}_{\mathrm{p}}) = \frac{u(\mathscr{R}_{\mathrm{p}}) 4 \mathscr{R}_{\mathrm{p}}}{\bar{\mathscr{R}}_{\mathrm{p}}^2} \, e^{-2\mathscr{R}_{\mathrm{p}}/\bar{\mathscr{R}}_{\mathrm{p}}} \qquad (9.6\text{-}16)$$

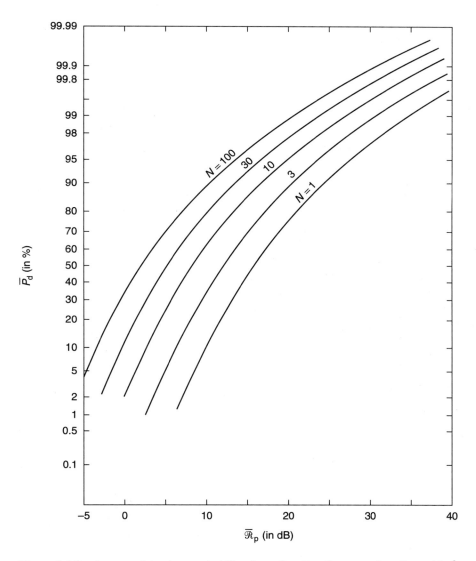

Figure 9.6-3   Average detection probability for a Swerling I target when $P_{fa} = 10^{-6}$.

This function is recognized as the chi-square density of (5.9-22) with $M = 2$. The corresponding average Laplace characteristic function, from (9.6-8) using (9.6-5), (9.6-16), and a standard table of Laplace transform pairs, is found to be

$$C(z) = (1 + z)^{2-N} \left[1 + \left(1 + \frac{N\overline{\mathcal{R}}_p}{4} \right) z \right]^{-2} \qquad (9.6\text{-}17)$$

Again, by using (9.6-17) with the residue method of Hou et al. (1987), we find $\overline{P}_d$ is given by either (9.6-11) or (9.6-12) for the special case $K = 2$. The latter is used for $N \leq 2$, the former for $N > 2$ (see Problems 9.6-14, 9.6-15, and 9.6-25). Figures 9.6-6 through 9.6-10

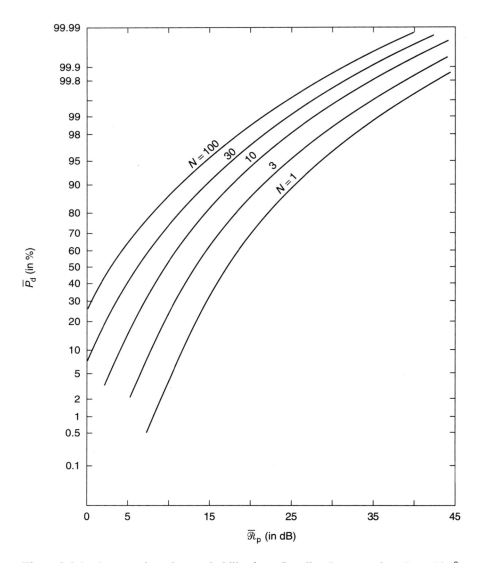

Figure 9.6-4 Average detection probability for a Swerling I target when $P_{fa} = 10^{-8}$.

show $\bar{P}_d$ for various values of N and P_{fa}. P_{fa} again derives from (9.5-17) except with r_b replaced by r_{b1}.

9.7   DETECTION OF TARGETS WITH PULSE-TO-PULSE FLUCTUATIONS

When each pulse in a group of N pulses fluctuates independently of the others, but all are governed by the same probability density $p(\mathcal{R}_p)$, the decision criterion is defined by (9.2-19).

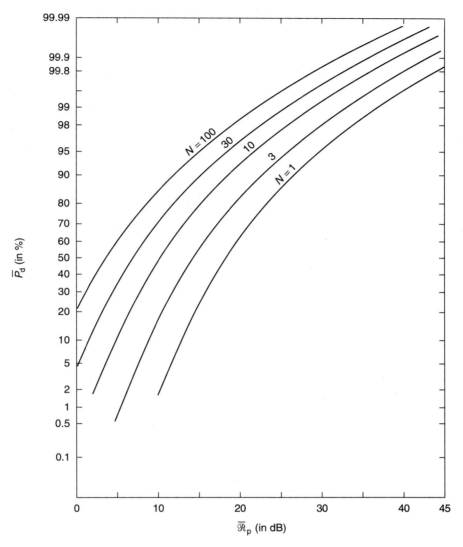

Figure 9.6-5   Average detection probability for a Swerling I target when $P_{fa} = 10^{-10}$.

Approximate Detection Processor

As in preceding work we approximate the optimum detection processor by replacing the nonlinear Bessel function in (9.2-19) by its small-signal form (9.6-1). After expanding the product and dropping terms above second order, the detection criterion reduces to

$$\left[\frac{1}{2}\sum_{i=1}^{N} r_i^2\right] - r_{b2} \begin{cases} \geq 0, & \text{decide } S + N \\ < 0, & \text{decide } N \end{cases} \qquad (9.7\text{-}1)$$

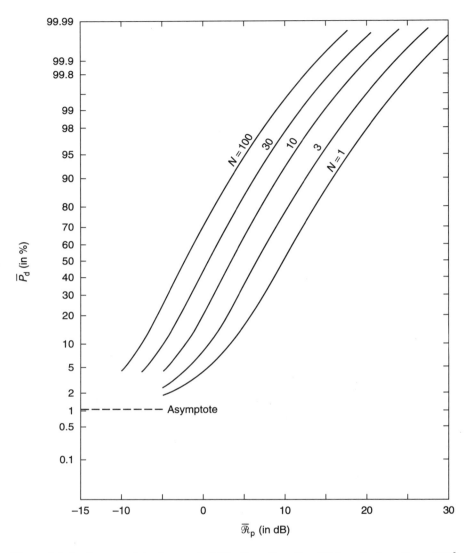

Figure 9.6-6   Average detection probability for a Swerling III target when $P_{fa} = 10^{-2}$.

where

$$r_{b2} = \frac{V_T - [E_{\mathscr{R}_p}(e^{-\mathscr{R}_p/2})]^N}{[E_{\mathscr{R}_p}(e^{-\mathscr{R}_p/2})]^{N-1} E_{\mathscr{R}_p}\left(\dfrac{\mathscr{R}_p}{2} e^{-\mathscr{R}_p/2}\right)} \tag{9.7-2}$$

is a threshold. This threshold is independent of r_i regardless of the exact values of the expectations in (9.7-2). The approximately optimum detection processor implements

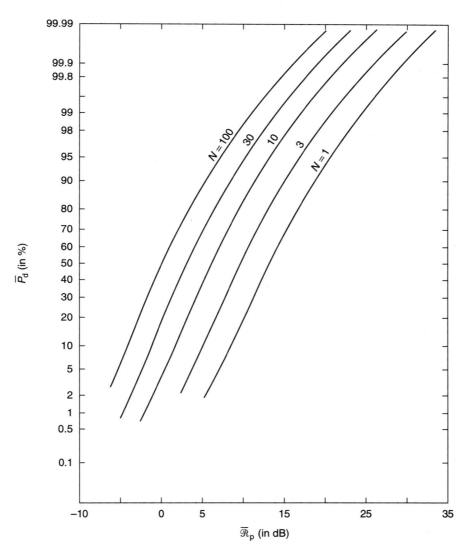

Figure 9.6-7 Average detection probability for a Swerling III target when $P_{fa} = 10^{-4}$.

(9.7-1) which is identical in *form* to (9.5-2). The processor is therefore identical to that in Fig. 9.5-1*b* except that r_b is now replaced by r_{b2} of (9.7-2).

Performance with Pulse-to-Pulse Fluctuations

Again, we apply the residue method of Hou et al. (1987), which requires finding the Laplace characteristic function $C(z)$ of x which is defined by (9.5-4). We only outline the necessary procedures, since they closely parallel those of Section 9.5. $C(z)$ is found by forming the expected value of the conditional Laplace characteristic function with respect to the random variables $\mathcal{R}_{pi}$, $i = 1, 2, \dots, N$, which are statistically

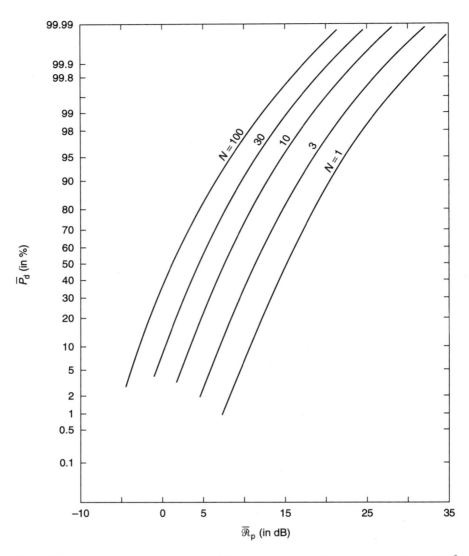

Figure 9.6-8 Average detection probability for a Swerling III target when $P_{fa} = 10^{-6}$.

independent. For independent noises in the N intervals, the conditional Laplace characteristic function is the product of all the Laplace characteristic functions for the various intervals. If all the $\mathcal{R}_{pi}$ have the same distribution, as assumed here, the average Laplace characteristic function becomes

$$C(z) = \prod_{i=1}^{N} \int_0^{\infty} C_i(z|\mathcal{R}_{pi})p(\mathcal{R}_{pi})\,d\mathcal{R}_{pi}$$
$$= \left[\int_0^{\infty} C_i(z|\mathcal{R}_p)p(\mathcal{R}_p)\,d\mathcal{R}_p\right]^N \tag{9.7-3}$$

where $C_i(z|\mathcal{R}_p)$ is given by the right-side of (9.5-11).

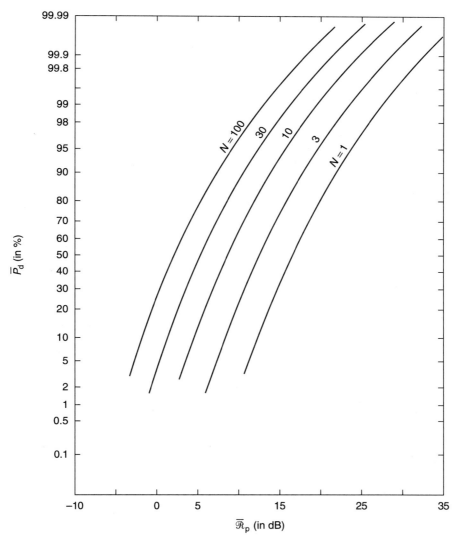

Figure 9.6-9   Average detection probability for a Swerling III target when $P_{fa} = 10^{-8}$.

We are principally concerned with two cases, namely the evaluation of (9.7-3) for $p(\mathcal{R}_p)$ given by (9.6-9) and (9.6-16). Evaluation gives

$$C(z) = \left[1 + \left(1 + \frac{\bar{\mathcal{R}}_p}{2} \right) z \right]^{-N} \tag{9.7-4}$$

for use of (9.6-9), and

$$C(z) = (1 + z)^N \left[1 + \left(1 + \frac{\bar{\mathcal{R}}_p}{4} \right) z \right]^{-2N} \tag{9.7-5}$$

for use of (9.6-16).

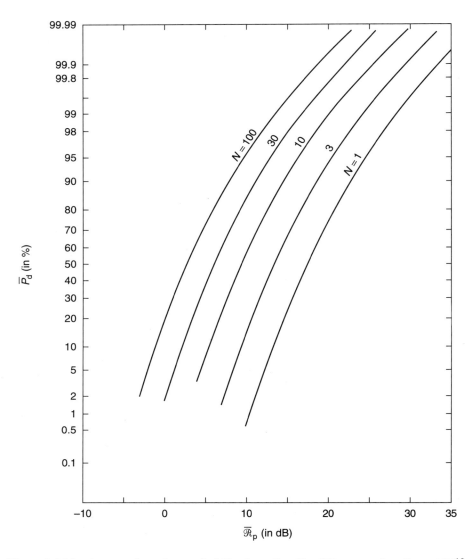

Figure 9.6-10   Average detection probability for a Swerling III target when $P_{\text{fa}} = 10^{-10}$.

Performance of the systems when detecting targets that fluctuate pulse to pulse is inherent in the behavior of $\bar{P}_{\text{d}}$. The expressions for $\bar{P}_{\text{d}}$ given previously in (9.6-11) and (9.6-12) apply to a rather general class of fluctuations (see Hou et al., 1987) for which (9.7-4) and (9.7-5) are special cases.

Swerling II Fluctuation Model

One special case for which $K = N$ applies when (9.7-4) is used. This case is called the Swerling II fluctuation model. Figures 9.7-1 through 9.7-5 illustrate the behavior of $\bar{P}_{\text{d}}$ as calculated from (9.6-12) when $K = N$.

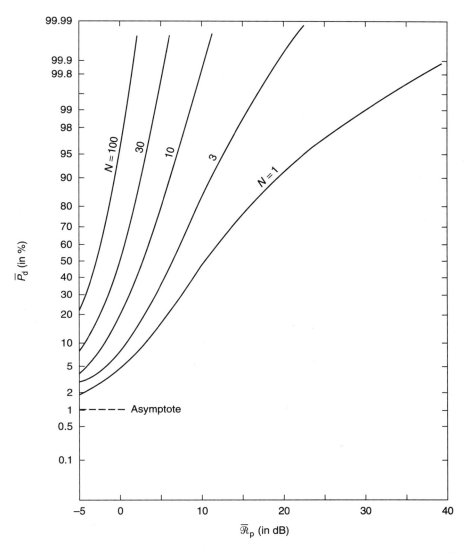

Figure 9.7-1 Average detection probability for a Swerling II target when $P_{fa} = 10^{-2}$.

Swerling IV Fluctuation Model

Swerling IV fluctuation corresponds to (9.7-5). In this case (9.6-12) with $K = 2N$ gives $\bar{P}_d$. Some calculated values are plotted in Figs. 9.7-6 through 9.7-10. For the special case of $N = 1$, Swerling (1954) has obtained an exact solution for $\bar{P}_d$ which can be put in the form

$$\bar{P}_d = \left[1 + \frac{4\bar{\mathscr{R}}_p r_{b1}}{(4 + \bar{\mathscr{R}}_p)^2} \right] \exp\left[\frac{-4r_{b1}}{4 + \bar{\mathscr{R}}_p} \right], \qquad N = 1 \text{ only} \qquad (9.7\text{-}6)$$

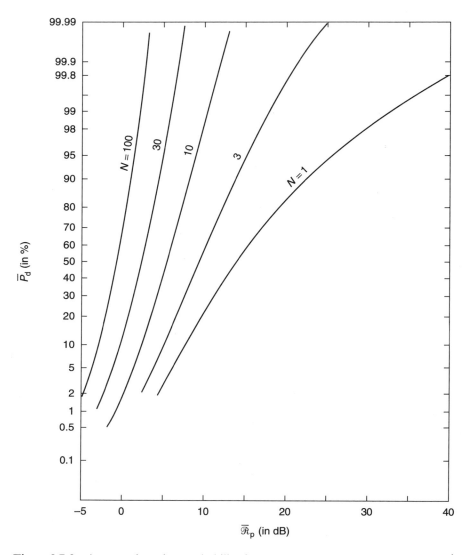

Figure 9.7-2 Average detection probability for a Swerling II target when $P_{fa} = 10^{-4}$.

This result is identical to that for the Swerling III case when $N = 1$ (see Problem 9.6-14).

P_{fa} for both Swerling models II and IV is given by (9.5-17) with appropriate relabeling of the threshold from r_b to r_{b2}.

9.8   BINARY DETECTION

Consider a detection processor as defined by Fig. 9.4-1c. It uses a filter matched to a target pulse received in a single-pulse repetition interval. Denote the single-pulse false

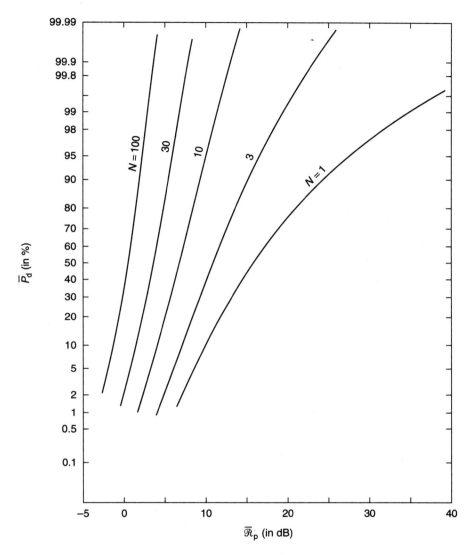

Figure 9.7-3 Average detection probability for a Swerling II target when $P_{fa} = 10^{-6}$.

alarm and detection probabilities by P_{fa1} and P_{d1}, respectively. These probabilities are given by (9.4-12) and (9.4-13) when $N = 1$:

$$P_{fa1} = \exp\left(\frac{-V_r^2}{2}\right) \tag{9.8-1}$$

$$P_{d1} = Q\left[\sqrt{\mathcal{R}_p}, \sqrt{2\ln\left(\frac{1}{P_{fa1}}\right)}\right] \tag{9.8-2}$$

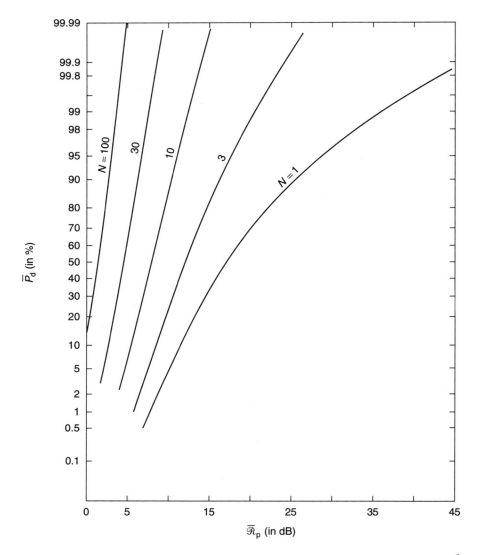

Figure 9.7-4   Average detection probability for a Swerling II target when $P_{\text{fa}} = 10^{-8}$.

Here V_{r} is the single-pulse threshold, $\mathscr{R}_{\text{p}}$ is the peak signal-to-noise power ratio at the output of the matched filter when responding to a single pulse, and $Q(.\,,.)$ is Marcum's Q-function defined in (9.4-14).

In *binary detection* the radar observes a particular range cell in each of N pulse intervals and counts the number of threshold crossings (single-pulse detections) that takes place. This number can be as small as zero, or as large as N if the range cell gives a detection in every one of the N pulse intervals. At the end of N pulse intervals the radar declares a target is present if M or more threshold crossings occur, and no target is present if less than M crossings occur. M is called a *second threshold*. Clearly the

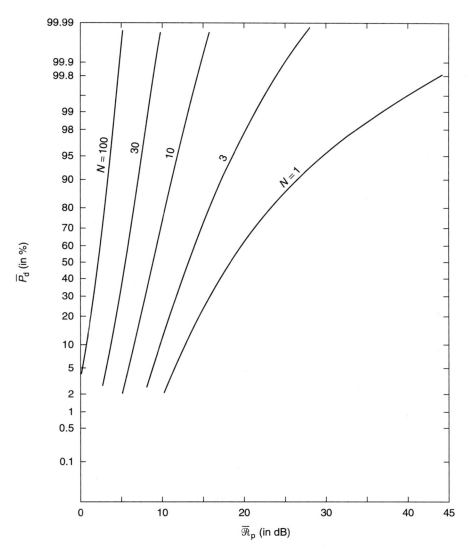

Figure 9.7-5 Average detection probability for a Swerling II target when $P_{fa} = 10^{-10}$.

important parameters in making a final detection decision are P_{fa1}, P_{d1}, M, and N. In this section we relate these parameters to the overall desired false alarm and detection probabilities, which we denote by P_{fa} and P_d, respectively. The discussions to follow center mainly on the nonfluctuating target. However, some details and literature references are also given for the fluctuating target.

Binary detection is known by various names, such as *M of N detection*, *binary integration*, *double threshold detection*, and *coincidence detection*. It appears to have been originated by Swerling (1952) and Schwartz (1956). Some early theoretical work was due to Harrington (1955) and Capon (1960). More recently Miller (1985),

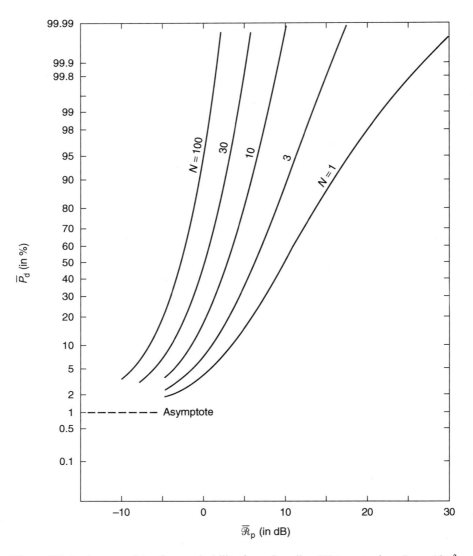

Figure 9.7-6 Average detection probability for a Swerling IV target when $P_{fa} = 10^{-2}$.

Brunner (1990), Weiner (1991), and Han et al. (1993) have extended the theory to include fluctuating targets and distributed detection systems. Others (cited below) have given useful design data.

Probabilities

False alarm probability P_{fa} corresponds to when noise only is being received. Assume that the N pulse intervals result in a sequence of exactly k threshold crossings followed by exactly $N - k$ noncrossings (for the range cell under examination). The probability

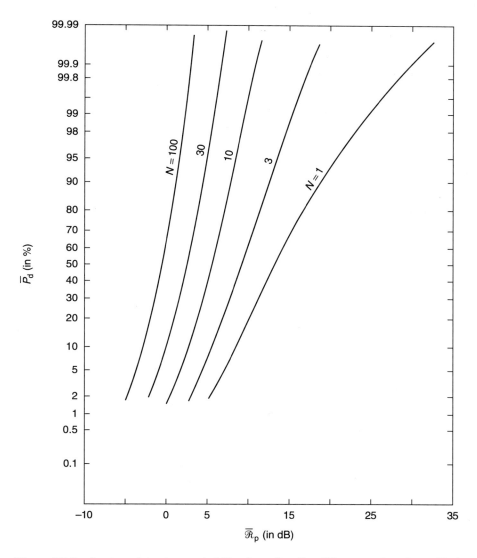

Figure 9.7-7 Average detection probability for a Swerling IV target when $P_{fa} = 10^{-4}$.

of this sequence is $P_{fa1}^k (1 - P_{fa1})^{N-k}$. The order in which threshold crossings occurs is not important, and any sequence of exactly k crossings in N intervals has the same probability. From combinatorial analysis the number of such sequences equals the binominal coefficient

$$\binom{N}{k} = \frac{N!}{k!(N-k)!} \qquad (9.8\text{-}3)$$

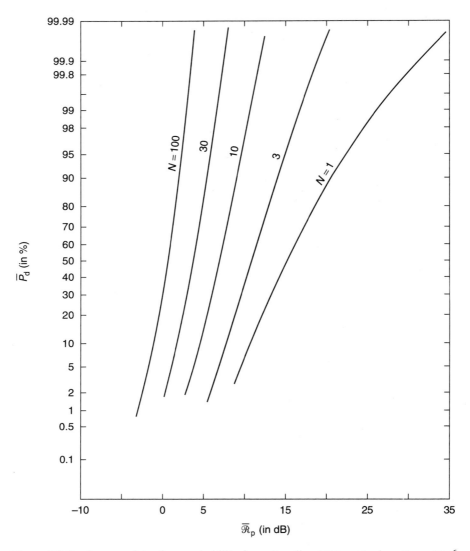

Figure 9.7-8   Average detection probability for a Swerling IV target when $P_{fa} = 10^{-6}$.

Thus the probability of a sequence of exactly k threshold crossings (of any order) in N intervals is

$$P(k \text{ of } N) = \binom{N}{k} P_{fa1}^k (1 - P_{fa1})^{N-k} \tag{9.8-4}$$

A false alarm will occur if, at the end of N intervals, M or more crossings occur. Its probability becomes

$$P_{fa} = \sum_{k=M}^{N} P(k \text{ of } N) = \sum_{k=M}^{N} \binom{N}{k} P_{fa1}^k (1 - P_{fa1})^{N-k} \tag{9.8-5}$$

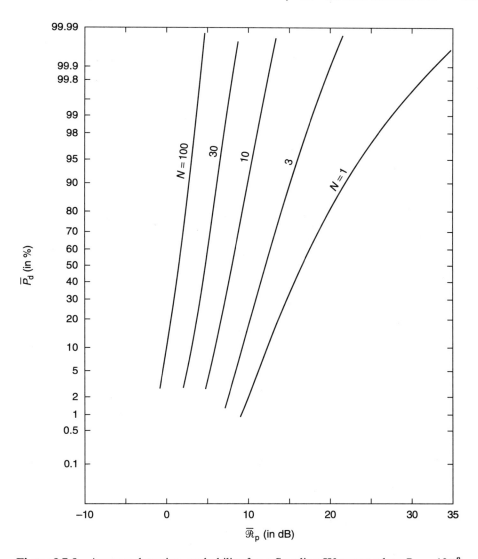

Figure 9.7-9 Average detection probability for a Swerling IV target when $P_{\text{fa}} = 10^{-8}$.

When a target with constant cross section is present, the preceding steps can be repeated to obtain the probability of detection. The steps again lead to (9.8-5) except with P_{fa1} replaced by P_{d1}:

$$P_{\text{d}} = \sum_{k=M}^{N} \binom{N}{k} P_{\text{d1}}^{k} (1 - P_{\text{d1}})^{N-k} \qquad (9.8\text{-}6)$$

This expression and (9.8-5) are the two fundamental quantities that define binary detection.

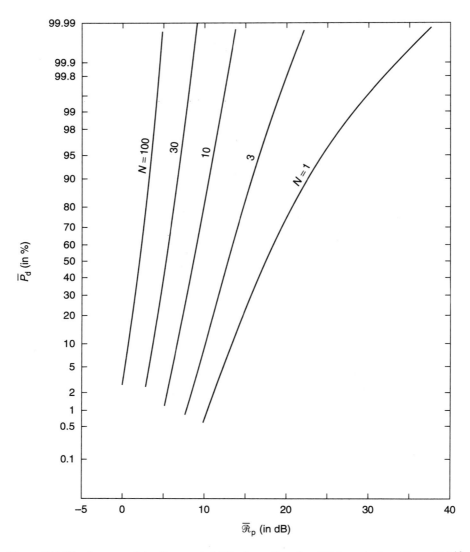

Figure 9.7-10 Average detection probability for a Swerling IV target when $P_{fa} = 10^{-10}$.

Optimum System for Nonfluctuating Target

For specified (desired) values of P_d, P_{fa}, and N, the optimum binary processor is defined as the one requiring the smallest single-pulse signal-to-noise ratio $\mathscr{R}_p$. Optimization amounts to finding optimum thresholds M and V_r. The steps involved in an optimization are:

1. Select a value of M.
2. Solve (9.8-5) for P_{fa1}.

3. Solve (9.8-6) for P_{d1} and then solve (9.8-2) for $\mathcal{R}_p$.

4. Repeat steps 1–3 for several values of M until the optimum value, denoted by M_{opt}, is found that corresponds to the smallest $\mathcal{R}_p$.

5. With $M = M_{opt}$, a solution of (9.8-5) gives P_{fa1} from which the optimum value of V_r is found from (9.8-1).

Figure 9.8-1 shows data from several sources for the optimum value of M. Although there is a dependence of M_{opt}/N to increase with decreasing P_{fa} and/or P_d, the dependence is weak, noted from the data of Worley (1968) and Walker (1971). Endresen and Hedemark (1961) also give M_{opt}/N, as taken from Swerling (1952); these results are not shown because they are almost identical to the curve due to Grasso and Guarguaglini (1967). An empirical approximation

$$\frac{M_{opt}}{N} = 0.15 + \frac{0.85}{N^{0.3}} \tag{9.8-7}$$

is also shown; it can serve as a rough approximation when more precise values are not required (accuracy is about 15% at $N = 5$, improving slowly as N increases).

Once M_{opt} (or M_{opt}/N) is found, it remains to find the applicable value of V_r. From step 5 above, V_r follows easily from (9.8-1) after P_{fa1} is found from (9.8-5). An iterative method is typically required to obtain P_{fa1}. However, some help can be obtained from Figs. 9.8-2 and 9.8-3, at least when $P_{fa} = 10^{-4}$ or 10^{-8}. These curves plot P_{fa1} versus M_{opt}/N for several values of N and the two values of P_{fa}. The heavy dots on a curve for a particular value of N correspond to integer values of M_{opt} for that value of N. For $N \geq 50$ there are many integer values so they are omitted. For parameters other than those of the two figures some interpolation is required; the interpolated value of P_{fa1} may not be accurate enough in some applications, but it can serve as an initial value to reduce the iterations needed to solve (9.8-5).

Even after the optimum binary detection processor is defined by finding M_{opt} and V_r (or P_{fa1}) it is usually necessary to determine the required value of $\mathcal{R}_p$ that produces the specified value of P_d. This value is found from (9.8-2), which typically requires iteration. The algorithm of Brennan and Reed (1965) may be helpful in the interation. Direct solution for the inverse of Marcum's Q-function is also possible (see Helstrom, 1998). Some values of $\mathcal{R}_p$ may be found from data in Worley (1968) but only for $P_{fa} = 10^{-4}$ and $P_{fa} = 10^{-8}$. However, interpolation of these results might be useful as an intial value in the iterative solution of (9.8-2). Walker (1971) has also given some data from which $\mathcal{R}_p$ can be found; the data are limited and only cover $P_{fa} = 10^{-6}$, 10^{-8}, and 10^{-10}, all for $P_d = 0.9$.

Fluctuating Targets

Some limited results are available for fluctuating targets detected by binary detection. Figures 9.8-4 and 9.8-5 show M_{opt}/N for Swerling II and IV fluctuations, respectively. By comparison with Fig. 9.8-1, we find that the optimum threshold is lowered when fluctuations are present. There is still a dependency on P_{fa} and P_d, but the dependence remains weak as with the nonfluctuating target. Weiner (1991) has given some limited data from which M_{opt} can be determined for nonfluctuating, Swerling I, and

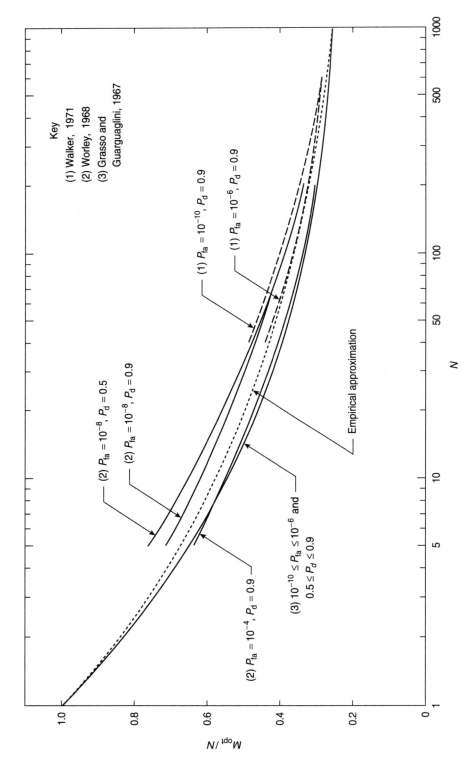

Figure 9.8-1 Optimum threshold M_{opt} versus N for binary detection of a nonfluctuating target. (Adapted from Grasso and Guarguaglini, © 1967 IEEE; Walker, © 1971 IEEE; and Worley, © 1968 IEEE, all with permission.)

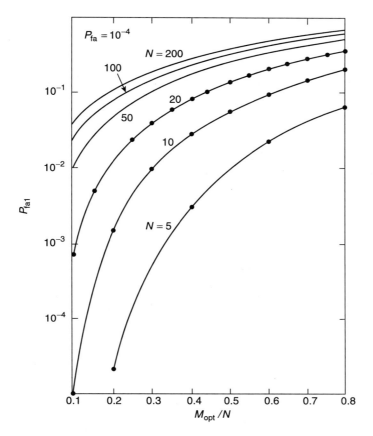

Figure 9.8-2 $P_{\text{fa}1}$ required to give $P_{\text{fa}} = 10^{-4}$ in (9.8-5) for various M_{opt} and N. (Adapted from Worley, © 1968 IEEE, with permission.)

Swerling II targets [Four cases: (1) $N = 8$, $P_{\text{fa}} = 10^{-8}$, and $P_{\text{d}} = 0.5$; (2) $N = 2$, $P_{\text{fa}} = 10^{-6}$, and $P_{\text{d}} = 0.9$; (3, 4) $N = 20$ and both $P_{\text{fa}} = 10^{-6}$ and $P_{\text{d}} = 0.9$, and $P_{\text{fa}} = 10^{-8}$ and $P_{\text{d}} = 0.5$].

Walker (1971) has given data from which the optimum values of $\mathcal{R}_{\text{p}}$ may be found for N from 3 to 600 with Swerling I and II fluctuating targets for three cases: $P_{\text{fa}} = 10^{-6}$, 10^{-8}, and 10^{-10}, all for $P_{\text{d}} = 0.9$. Similar data are available in Worley (1968), for N from 5 to 200 with Swerling II and IV targets; the only cases given are $P_{\text{fa}} = 10^{-4}$ and $P_{\text{fa}} = 10^{-8}$

9.9 OTHER CONSIDERATIONS IN CLASSICAL DETECTION

For present purposes we shall define the topics of detection, as described in the preceding portions of this chapter, as *classical detection*. The name is appropriate because the topics form the very roots of radar detection and are the most

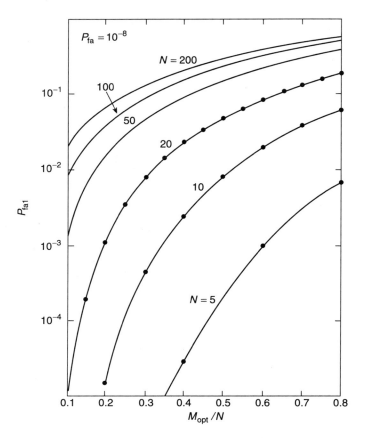

Figure 9.8-3 P_{fa1} required to give $P_{fa} = 10^{-8}$ in (9.8-5) for various M_{opt} and N. (Adapted from Worley, © 1968 IEEE, with permission.)

fundamental. However, present development of radar detection goes far beyond fundamentals. In this section we will attempt to put our prior work in current-day perspective, indicate where extensions have occurred, and lead the reader to some of the appropriate literature related to classical detection.

In the next section we attempt to lead the reader into other (current) areas of detection that may be called *nonclassical*. In particular, we show how the problem of detecting a target in *clutter* is approached, try to put the problem in perspective, and attempt to lead the reader to appropriate literature for details.

Classical detection, then, involves the detection in noise only of either a nonfluctuating (steady) target, or one with cross section fluctuating either by the exponential model of (9.6-9) or the Erlang model of (9.6-16), when using $N \geq 1$ pulses for detection. Our discussions assumed either complete correlation of the N pulses (steady target or Swerling I and III cases) or complete absence of correlation (Swerling II and IV cases). For a good alternative discussion of classical detection without resorting to a textbook, the extensive article by Rubin and DiFranco (1964) may be helpful. For some approaches to simplifications in calculating classical detection probabilities, see

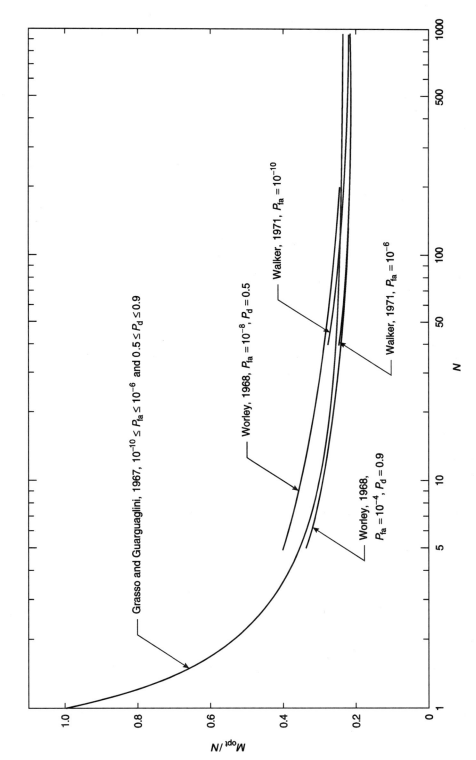

Figure 9.8-4 Optimum threshold M_{opt} versus N for binary detection of a target having Swerling II fluctuations. (Adapted from Grasso and Guarguaglini, © 1967 IEEE; Walker, © 1971 IEEE; and Worley, © 1968, all with permission.)

Grasso and Guarguaglini, 1967, $10^{-10} \leq P_{fa} \leq 10^{-6}$ and $0.5 \leq P_d \leq 0.9$

Worley, 1968, $P_{fa} = 10^{-8}$, $P_d = 0.5$

Walker, 1971, $P_{fa} = 10^{-10}$

Walker, 1971, $P_{fa} = 10^{-6}$

Worley, 1968, $P_{fa} = 10^{-4}$, $P_d = 0.9$

M_{opt}/N

N

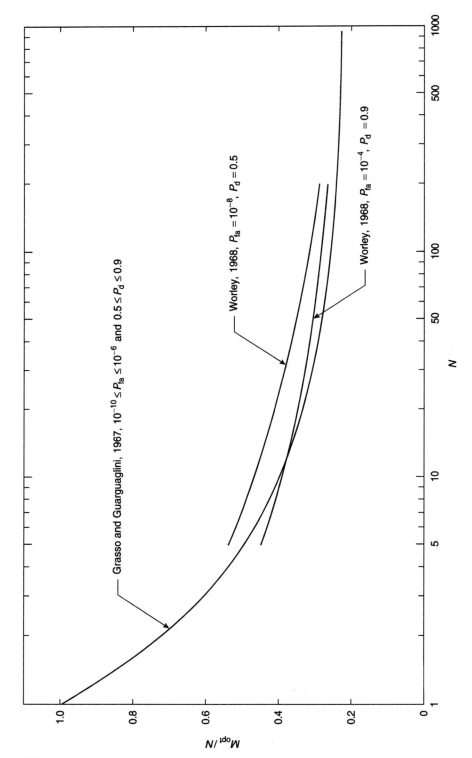

Figure 9.8-5 Optimum threshold M_{opt} versus N for binary detection of a target having Swerling IV fluctuations. (Adapted from Grasso and Guarguaglini, © 1967 IEEE; Worley, © 1968 IEEE, both with permission.)

Grasso and Guarguaglini, 1967, $10^{-10} \leq P_{fa} \leq 10^{-6}$ and $0.5 \leq P_d \leq 0.9$

Worley, 1968, $P_{fa} = 10^{-8}$, $P_d = 0.5$

Worley, 1968, $P_{fa} = 10^{-4}$, $P_d = 0.9$

M_{opt}/N

N

Barton (1969). Other sources for methods of approximating or solving the various classical expressions for thresholds and detection probabilities are Urkowitz (1985), Helstrom and Ritcey (1984), Hansen (1982), and Mitchell and Walker (1971). Some sources of tabulated and graphical detection data not previously referenced are Fehlner (1964), and the book by Meyer and Mayer (1973).

Noise and Target Models

Nearly all the literature presumes the receiving system's noise is gaussian. However, cases can arise where noise departs from gaussian, such as with some types of "impulsive noise." Apparently the only nongaussian noise with a probability density that is everywhere nonzero, and that lends itself to closed-form solutions, is Laplace noise. Detection in Laplace noise has been discussed by Dadi and Marks (1987) who also give a number of references going back to 1972. Helstrom (1989) has provided a method for conveniently computing detection probabilities.

Our earlier work all presumed the N target pulses used for the detection decision were either fully correlated or not correlated. The case of partial correlation with exponential fluctuations (the extremes give Swerling cases I and II) was solved by Kanter (1986). Helstrom, (1992) has given a saddlepoint method of integration as an alternative means to solve the detection probability integrals of Kanter. The corresponding solution for fluctuations regulated by the density of (9.6-16) was given by Weiner (1988); the extremes of this problem are the Swerling cases III and IV.

The Swerling cases I–IV are specializations of a chi-square model. In a recent paper Swerling (1997) has discussed the probability of detection for the chi-square model. Other notable solutions for this more general form of fluctuation are given by Hou, et al., (1987) and Shnidman (1995) with each giving other references. In this latter paper Shnidman also calculates detection probabilities for Weibull, Ricean, and log-normal target fluctuations. Log-normal fluctuations were also studied by Shnidman (1991) for scan-to-scan fluctuations, and by Heidbreder and Mitchell (1967) for fully correlated and uncorrelated pulses. Swerling (1965) derived the characteristic function of the N-pulse integrator's output for an *arbitrary* probability density of fluctuations; this result is important because detection probability is found by integrating the inverse transformation of the characteristic function. He did not, however, develop detection probabilities.

Other Detectors

Although the square law detector was used in most of the work of this chapter because it led to tractable analyses and because its performance was not significantly different from that of a linear detector (based on limited data of Marcum, 1948), practical detectors are often linear. However, Helstrom (1990) has shown that the difference in the two detectors is small for a wide range of N and detection probabilities. Green (1958) showed that the loss in performance of a logarithmic detector compared to a square law device is about 1.0 dB or less for integration of 100 or fewer pulses from a steady target.

Sometimes the word "detector" is used to include not only the envelope detector, be it square law, linear, logarithmic, or other law, but also the operation of summing the N pulse samples. Several summing methods are in use that are designed to follow a linear envelope detector. Good summaries of five of these methods, called *moving window, feedback integrator, two-pole filter, binary integrator*, and *batch processor*, are given by Trunk (Skolnik, ed., 1990, ch. 8). The moving window is quoted as being only about 0.5 dB (for N near 10) worse than an optimal detector; the corresponding degradations for the feedback integrator and two-pole filter are about 1 dB or less and 0.15 dB, respectively (Skolnik, ed., 1990, pp. 8.8 and 8.9). Compared to the moving window integrator, the binary integrator and batch processor are quoted as 2.0 dB and 0.5 dB poorer, respectively. More can be found on these summing methods in Mao (1981), Cantrell and Trunk (1973, 1974), and Trunk (1971, 1970).

Constant False Alarm Rate (CFAR)

In all our discussions of classical detection the threshold was considered constant. In most radar this is not a realistic assumption. For example, a look at Table 9.5-1 shows that P_{fa} is extremely sensitive to changes in threshold. At $N = 50$ and $r_b = 100.316$, we have $P_{fa} = 10^{-8}$; a change to $N = 50$ and $r_b = 80.6593$ (a drop in threshold of only 19.6%) results in $P_{fa} = 10^{-4}$, which is 10,000 times larger than initially. Of course, the radar does not purposely change its threshold; it is a variation in the *noise relative to the threshold* that occurs as the radar scans for targets. Practical radar often implements some method to adaptively cause the radar's threshold to change as the noise level changes such that constant false alarm rate (CFAR) can be approached.[9]

A basic method of approaching CFAR operation uses cell averaging (Finn and Johnson, 1968). Here several resolution cells on each side of the cell being tested for a target are sampled to develop an estimate of the noise level (the presumption is that only noise is in the adjacent cells). Threshold is then set proportional to this noise level estimate. There is a performance loss associated with all averaging that is larger for smaller numbers of resolution cells being used (poorer noise estimate accuracy) and decreases for a larger number. A compromise is needed. A suggested (Skolnik, ed., 1990, p. 8.14) compromise is to select the number of side cells large enough to keep the loss below 1 dB but not so large that the range span of the resolution cells exceeds the interval where the noise level is approximately constant.

Several key papers that develop detection probability when using cell-averaging CFAR thresholds are Shnidman (1995), Hou, et al. (1987), Helstrom and Ritcey (1984), Bird (1983), and Mitchell and Walker (1971). Most of these provide other references. When noise is not Rayleigh distributed or its power varies from pulse to pulse, other methods must be used to establish thresholds (see Trunk in Skolnik, ed., 1990, Ch. 8).

Advanced Problems

In almost all the literature on detection, it is assumed that only one target exists in the radar resolution cell being examined. The problems of how to *detect* the presence of *multiple* targets in the same range-Doppler–angle–angle cell, and how to discriminate from the single target condition so as to calculate detection probabilities, are not

---

[9] We describe CFAR as it relates to noise because classical detection considers only noise. However, CFAR can be as important, or even more important, when clutter is present (see the next section) because clutter often dominates noise and can vary more with range than noise.

treated extensively in the literature. Some work in this area is due to Bogler (1986) and Asseo (1981).

9.10 DETECTION IN CLUTTER

This chapter has concentrated on detection of targets in noise because of its fundamental limit in any radar. However, many radars operate in an environment where clutter is larger than noise and dominates the detection process. In this section we discuss the problem of detecting a target in clutter. The approach will be to put the problem in perspective, show the basic approach to analysis, and give references adequate for the reader to enter the literature for details.

Basic Approach to Analysis

When clutter is present, the radar's received waveform can be written as

$$y(t) = s_r(t) + n(t) + c(t) \tag{9.10-1}$$

where $s_r(t)$ is the target return, $n(t)$ is the "white" noise, and $c(t)$ is due to the clutter. All the analysis methods used earlier to arrive at false alarm and detection probabilities can be repeated; the procedures become more cumbersome due to the extra term in (9.10-1). We do not repeat any of these procedures here. Rather, we will indicate some fundamental ways of modeling $c(t)$ and refer to appropriate literature for detailed results.

Since $c(t)$ is a bandpass waveform centered on the carrier's frequency it can be represented in the forms

$$\begin{aligned}
c(t) &= r_c(t) \cos[\omega_0 t + \theta_c(t)] \\
&= r_c(t) \cos[\theta_c(t)] \cos(\omega_0 t) - r_c(t) \sin[\theta_c(t)] \sin(\omega_0 t) \\
&= c_I(t) \cos(\omega_0 t) - c_Q(t) \sin(\omega_0 t)
\end{aligned} \tag{9.10-2}$$

where terms are related by

$$r_c(t) = \{c_I^2(t) + c_Q^2(t)\}^{1/2} \tag{9.10-3}$$

$$\theta_c(t) = \tan^{-1}\left[\frac{c_Q(t)}{c_I(t)}\right] \tag{9.10-4}$$

$$c_I(t) = r_c(t) \cos[\theta_c(t)] \tag{9.10-5}$$

$$c_Q(t) = r_c(t) \sin[\theta_c(t)] \tag{9.10-6}$$

In these expressions $r_c(t)$ is the envelope of the clutter voltage, $\theta_c(t)$ is its phase, $c_I(t)$ and $c_Q(t)$ are the baseband in-phase and quadrature-phase components of $c(t)$, respectively, and ω_0 is the carrier's angular frequency.

Modeling of $c(t)$ consists in assigning an appropriate probability density function to describe the behaviors of $r_c(t)$ and $\theta_c(t)$ at the pulse (sample) times. If $r_c(t)$ has a log-normal or Weibull distribution, clutter is usually referred to as log-normal or Weibull. Of course any distribution might be chosen such that the clutter goes by the name of the distribution. When a definition is required, $\theta_c(t)$ is usually defined as a uniformly distributed random variable on $[0, 2\pi]$. In the following subsections we describe some of the more commonly used distributions for clutter, show their interrelationships, and cite references to appropriate literature.

Log-Normal Clutter

Experimental data on clutter of all types seems to be bracketed in the extremes by Rayleigh and log-normal amplitude (voltage) distributions (Schleher, 1975, p. 262). The latter distribution has a tail (for larger values of the random amplitude) that is larger than for most other distributions. The large tail is sometimes a better fit for sea clutter at low grazing angles when using a high-resolution radar.

There appears to be no physical mechanism of nature that is known to produce log-normal clutter. It is used in radar simply because it fits some clutter best. The probability density function of the voltage envelope (r_c) of log-normal clutter is given by (5.9-33), which can be written as

$$f(r_c) = \frac{e^{-[\ln(r_c/r_{\mathrm{med}})]^2/[4\ln(\bar{r}_c/r_{\mathrm{med}})]}}{\sqrt{4\pi \ln(\bar{r}_c/r_{\mathrm{med}})}\, r_c} \, u(r_c) \tag{9.10-7}$$

where $\bar{r}_c$ and r_{med} are the mean and median values of r_c, respectively.

Some sources for data on detection in log-normal clutter are Guida et al. (1993), Schleher (1975), Trunk (1971), and Trunk and George (1970).

Weibull Clutter

Weibull clutter has an envelope r_c described by the probability density function of (5.9-25),

$$f(r_c) = \frac{u(r_c)b}{\alpha}\left(\frac{r_c}{\alpha}\right)^{b-1} \exp\left[-\left(\frac{r_c}{\alpha}\right)^b\right] \tag{9.10-8}$$

where b is called a *shape parameter* and

$$\alpha = \frac{\bar{r}_c}{\Gamma(1 + (1/b))} \tag{9.10-9}$$

is a *scale parameter* with $\bar{r}_c$ the mean of r_c. As with the log-normal distribution, the Weibull distribution seems to have no physical basis, and it is used in radar simply because it fits some measured data. It has been suggested by Conte and Longo (1987), and proved earlier by Bochner (1937), for a limited range of shape parameter that the Weibull distribution is a special case of a Rayleigh mixture distribution (both as noted in Sangston and Gerlach, 1994), which is discussed below.

Some sources of further detail on the applicability of the Weibull distribution and the associated detection probabilities are Anastassopoulos and Lampropoulos 1995,

Rifkin (1994), Trizna (1991), Sekine and Mao (1990), Sekine et al. (1978), and Schleher (1976).

K Clutter

One of the most important recent models for clutter amplitude fluctuations is the *K-distribution* defined as

$$f(r_c) = \frac{4}{\Gamma(M)} \left(\frac{M}{2\bar{\sigma}}\right)^{(M+1)/2} r_c^M K_{M-1}\left(r_c \sqrt{\frac{2M}{\bar{\sigma}}}\right) u(r_c) \qquad (9.10\text{-}10)$$

where $K_{M-1}(\cdot)$ is the modified Bessel function of order $M - 1$, with $M > 0$, and $\bar{\sigma}$ is the average power in the bandpass clutter signal having the voltage envelope r_c.

The utility of the K-distribution is seen from an examination of its behavior with the shape parameter M. For $M \to \infty$ the K-distribution becomes the Rayleigh distribution. The case $M = 4$ has been suggested as applying to a large reflector subject to small changes in orientation (Hou and Morinaga, 1989, p. 635; Jakeman and Pusey, 1976, p. 810; Swerling, 1954). For $0.3 < M < 4$ the K-distribution conforms to certain targets of simple shape (Hou and Morinaga, 1989). Finally the small-M case produces a "long-tailed" or "spikey" distribution relative to the Rayleigh case; the worst case for measured data appears to be $M \approx 0.1$ (Watts, 1987, p. 41; Amindavar and Ritcey, 1994, p. 427).

Several important properties of the K-distribution are worth stating (Hou and Morinaga, 1989). The nth moment of r_c is

$$E[r_c^n] = \left(\frac{2\bar{\sigma}}{M}\right)^{n/2} \frac{\Gamma(M + n/2)}{\Gamma(M)} \Gamma\left(1 + \frac{1}{n}\right) \qquad (9.10\text{-}11)$$

where $E[\cdot]$ denotes the statistical expectation (average). The cumulative probability is

$$P(r_c > t) = \int_t^\infty f(r_c)\, dr_c = \frac{2}{\Gamma(M)} \left(\frac{M}{2\bar{\sigma}}\right)^{M/2} t^M K_M\left(t \sqrt{\frac{2M}{\bar{\sigma}}}\right) \qquad (9.10\text{-}12)$$

where $P(r_c > t)$ denotes the probability that r_c is greater than a value $t \geq 0$. Finally, if a_n, for $n = 1, 2, \ldots, N$, are N statistically independent random variables, each having the same K-distribution, the magnitude of the sum w of phasors

$$w = \sum_{n=1}^{N} a_n \exp(j\phi_n) \qquad (9.10\text{-}13)$$

is K-distributed, if the ϕ_n are statistically independent random phase angles uniform on $[0, 2\pi]$ and statistically independent of the a_n. If $\rho = |w|$, the density of ρ is obtained from (9.10-10) by replacing r_c and M by ρ and MN, respectively. This last property is due to Jakeman and Pusey (1976). Later they proved that ρ is K-distributed, even if the a_n have an arbitrary distribution (but with finite variance) provided N is random and has a negative binomial distribution (Jakeman and Pusey, 1978). These proofs may not appear to have much physical importance due to the arbitrary assumptions as to the distributions of the a_n or N. However, the form of (9.10-13) has practical importance because it represents a first-step in modeling real clutter as a sum of scatterers, which has some physical basis, and it shows that the K-distribution is the result, even for two widely differing conditions.

For some work on extension of the K-distribution to spatially correlated clutter, see Raghavan (1991), who also gives a good list of other references. Jao (1984) discusses the application of the K-distribution to clutter, and Watts (1985) uses it to find detection probabilities. Watts and Wicks (1990) give some simple empirical models for finding single-scan detection probabilities for K-clutter.

Rayleigh Mixture Model of Clutter

The *Rayleigh mixture model* (Sangston and Gerlach, 1994) is the most general clutter model at this time. It includes the K-distribution and Weibull distributions as special cases (the log-normal distribution is excluded, however). The fundamental concepts of the model originated from studies of sea clutter at low grazing angles with high resolution radar by Trunk (1972). The principal effect of large waves having rough surfaces composed of capillary waves is to tilt the "rough" surface when viewed at low grazing angles. By the central limit theorem, the scattering is Rayleigh distributed (sometimes caled the *speckle* component of scattering). The Rayleigh return will have a power level that slowly fluctuates more in accordance with the linear facets of waves and the larger wave patterns such as directions and height. The model that evolves is that the density of the clutter voltage envelope r_c is Rayleigh, where the power in the Rayleigh function is a random variable, denoted here by σ, that can have some (arbitrary) probability density function $f(\sigma)$. The probability density of r_c for the Rayleigh mixture model becomes

$$f(r_c) = \int_0^\infty f(r_c|\sigma) f(\sigma)\, d\sigma = \int_0^\infty f(r_c|\sigma)\, dF(\sigma) \qquad (9.10\text{-}14)$$

where

$$f(r_c|\sigma) = \frac{r_c}{\sigma} e^{-r_c^2/2\sigma} u(\sigma) \qquad (9.10\text{-}15)$$

and $F(\sigma)$ is the cumulative distribution function of σ.

To demonstrate how the K-distribution is a special case of (9.10-14), we take $f(\sigma)$ to be chi-square according to (5.9-22), substitute into (9.10-14) and make use of the known integral (Gradshteyn and Ryzhik, 1965, p. 340)

$$\int_0^\infty x^{\nu-1} e^{-\gamma x - (\beta/x)}\, dx = 2\left(\frac{\beta}{\gamma}\right)^{\nu/2} K_\nu(2\sqrt{\gamma\beta}), \qquad \mathrm{Re}(\gamma) > 0, \qquad \mathrm{Re}(\beta) > 0$$

$$(9.10\text{-}16)$$

to obtain (9.10-10) (see Problem 9.10-1), Sangston and Gerlach (1994) have computed detection probabilities for the K-distribution case for several values of M.

Final Comments

The general problem of optimal detection usually breaks down to two considerations. First, a statistical model is formulated for the noise and clutter present in the problem. Second, formal detection theory is applied to determine the structure of the detection processor modeled in the first step. However, it is possible to avoid the first step by seeking distribution-free processors. Conte et al. (1992) have done some work in this area and give other references.

Finally it often arises that algorithms are needed to generate random variables by computer for simulating radar operation, including detection in noise and clutter. Some sources for this problem are Marier (1995), who gives many other references, Peebles, Jr. (1971), and Dillard (1967). The often needed moment calculations are given in Farison (1965).

PROBLEMS

9.1-1 Show that the second right-side form of (9.1-17) results from the first when (9.1-16) and (9.1-11) through (9.1-14) are used.

9.1-2 Justify that (9.1-18) is valid by starting with (9.1-17).

9.1-3 Suppose that $s_r(t)$ has a bandwidth of 50 MHz and its carrier frequency is 5 GHz. If the sampling implied in (9.1-24) is taken literally, and W_N is required to be at least 10 times the bandwidth above the carrier frequency, what value of Δt would be necessary?

9.1-4 Use (9.1-26) and (9.1-27) and show that (9.1-28) is true.

9.2-1 Derive (9.2-3) from (9.1-31).

9.2-2 Find V_d such that (9.2-13) can be put in the form

$$\ln\left[I_0(\sqrt{\mathscr{R}_p}\,r)\right] - V_d \begin{cases} \geq 0, & \text{decide } S + N \\ < 0, & \text{decide } N \end{cases}$$

★9.2-3 Show that (9.2-15) derives from (9.1-31) if the target pulses all have uniformly distributed independent random phase angles ϕ_{0i} in intervals $i = 1, 2, \dots, N$.

9.2-4 Give arguments to properly justify that (9.2-18) derives from (9.1-31) for group-to-group cross section fluctuations.

9.2-5 Show that (9.2-19) derives from (9.1-31) for pulse-to-pulse cross section fluctuations.

9.3-1 For the detection processor of Fig. 9.3-1a, show that the mean values of $y_1(NT_R)$ are given by (9.3-5).

9.3-2 Use the results of Problem 9.3-1 and show that the variances of $y_1(NT_R)$ are defined by (9.3-7).

9.3-3 A signal's form is assumed to be known exactly in a system for which $N = 1$, $V_b = 5.3$, and $\mathscr{R}_p = 40$ (or 16.02 dB). Find P_{fa} and P_d.

9.3-4 If all else in Problem 9.3-3 is fixed except N is changed to two pulses, what do P_{fa} and P_d now become?

9.3-5 A detection system for a known signal uses $N = 8$ pulses. Its threshold is set to give $P_{fa} = 10^{-10}$. Find V_b. If a single-pulse signal-to-noise ratio of 12.5 occurs at the matched filter's output, what is P_d? Do *not* use Fig. 9.3-3 for P_d. Rather, use (E.2-4).

9.3-6 A detection receiver for a known signal has $P_{fa} = 10^{-2}$. The target return pulse has no FM, a rectangular envelope of 1 μs, peak carrier amplitude of $\sqrt{2}(10^{-6})$ volts, and $\mathcal{N}_0/2 = 4(10^{-20})$ for the noise. What is P_d if single-pulse detection is used?

9.3-7 A radar is designed to operate with a known signal and use one-pulse detection. The initial design gives $P_{fa} = 10^{-8}$ and $P_d = 0.525$. (a) What is V_b, and what value of $\mathcal{R}_p$ is needed to give $P_d = 0.525$? (b) If the radar is modified to use $N = 5$ pulses, and keep the same value of P_{fa} (by adjustment of V_b) but use a less expensive transmitter with 4-dB *less* peak power, find V_b and the new values of $\mathcal{R}_p$ and P_d.

9.3-8 Work Problem 9.3-7 except assume that $P_{fa} = 10^{-12}$ and $P_d = 0.23$ as the initial performance values.

9.3-9 A radar uses one pulse in detecting a target of known form; $P_d = 0.999$ and $P_{fa} = 0.01$ apply. What values of $\mathcal{R}_p$ are needed if P_{fa} and P_d are maintained constant but $N = 3, 10, 30, 100, 300$, and 1000 pulses are used?

★9.4-1 Show that (9.4-7) and (9.4-8) are true.

★9.4-2 Show that (9.4-9) is true.

★9.4-3 Prove that (9.4-10) is true.

9.4-4 A nonfluctuating target of unknown initial phase is detected by a radar using a single pulse. If $P_{fa} = 3(10^{-6})$ What is the system's threshold V_r?

9.4-5 Work Problem 9.4-4 except for $P_{fa} = 7(10^{-8})$.

9.4-6 In a radar designed to detect a steady target with unknown initial phase with one pulse, $P_{fa} = 10^{-10}$ and $\bar{P}_d = 0.987$. What are V_r and $\mathcal{R}_p$?

9.4-7 Work Problem 9.4-6 except for $P_{fa} = 0.01$ and $\bar{P}_d = 0.95$.

9.4-8 A radar uses N pulses in detecting a steady target with random initial phase. For $N = 1$, $\bar{P}_d = 0.999$ and $P_{fa} = 0.01$. What values of $\mathcal{R}_p$ are needed to maintain the same values of $\bar{P}_d$ and P_{fa} except when $N = 3, 10, 30, 100, 300$, and 1000?

9.4-9 A radar uses 10 pulses to detect a steady target of unknown initial phase. Threshold is set to give $P_{fa} = 10^{-8}$ and $\mathcal{R}_p$ is such as to produce $\bar{P}_d = 0.999$. (a) What are V_r and $\mathcal{R}_p$? (b) To make faster detections it is found to be necessary to use only four pulses. If threshold is adjusted to keep $P_{fa} = 10^{-8}$ and the transmitter's peak power is adjusted to maintain $\bar{P}_d = 0.999$, find the new values of V_r and $\mathcal{R}_p$.

9.5-1 For $N \gg 1$ Marcum (1948) has given the following approximation for P_{fa} for a detection processor of N pulses each with random phase:

$$P_{fa} \approx \left(\frac{N}{2\pi} \right)^{1/2} \frac{1}{(r_b - N + 1)} \exp\left\{ -r_b + N\left[1 + \ln\left(\frac{r_b}{N} \right) \right] \right\}$$

Use this expression to find P_{fa} using values of r_b taken from Table 9.5-1 for four cases: (1) $P_{fa} = 10^{-2}$, $N = 5$, (2) $P_{fa} = 10^{-2}$, $N = 150$, (3) $P_{fa} = 10^{-10}$, $N = 5$, and (4) $P_{fa} = 10^{-10}$, $N = 150$. Compare with the values of Pachares and determine roughly the accuracy of the approximation.

9.5-2 A radar has a square law detector and uses 30 pulses to detect a steady target with unknown phase in each pulse interval. The system must produce $\bar{P}_d = 0.98$. (a) Plot the required values of $\mathcal{R}_{pN}$ (in dB) versus $\log_{10}(P_{fa})$ for $P_{fa} = 10^{-2}$, 10^{-4}, 10^{-6}, and 10^{-8}. (b) If the highest available transmitter power corresponds to $\mathcal{R}_{pN} = 2.95$, what value of P_{fa} can be achieved?

9.5-3 A steady target reflects pulses with random phases to a radar that uses six pulses for detection. If $r_b = 25.9$ and $\bar{P}_d = 0.969$, use the approximation of Problem 9.5-1 to find P_{fa}. What value of $\mathcal{R}_{pN}$ is necessary?

9.5-4 Work Problem 9.5-3 except assume that $N = 100$, $r_b = 170$, and $\bar{P}_d = 0.98$.

9.5-5 What is the integration improvement factor for a detection processor for 100 pulses from a steady target reflecting pulses with random phases if $P_{fa} = 10^{-8}$ and $\bar{P}_d = 0.9$ are to be achieved?

9.5-6 Work Problem 9.5-5 except assume $N = 60$, $P_{fa} = 10^{-10}$, and $\bar{P}_d = 0.999$.

9.6-1 For detection of a target having Swerling I fluctuations, show that, for $N = 1$ pulse, (9.5-17) reduces to

$$P_{fa} = \exp(-r_{b1})$$

and (9.6-12) reduces to

$$\bar{P}_d = (P_{fa})^{2/(2+\bar{\mathcal{R}}_p)}$$

or

$$P_{fa} = (\bar{P}_d)^{1+(\bar{\mathcal{R}}_p/2)}$$

9.6-2 A target is known to fluctuate according to the exponential probability density of (9.6-9). What is the probability that $\mathcal{R}_p$ will exceed half its mean value ($\bar{\mathcal{R}}_p/2$) but not larger than three halves its mean ($3\bar{\mathcal{R}}_p/2$)?

9.6-3 Work Problem 9.6-2 except find the probability $\mathcal{R}_p$ will exceed its mean value.

9.6-4 The median is defined as the value of $\mathscr{R}_p$, denoted by $\mathscr{R}_{pm}$, for which the probability is 0.5 that $\mathscr{R}_p$ will exceed $\mathscr{R}_{pm}$. Find $\mathscr{R}_{pm}$ for $\mathscr{R}_p$ defined by (9.6-9).

9.6-5 Use the results of Problem 9.6-1 to find (for $N = 1$) an expression for the value of $\bar{\mathscr{R}}_p$ corresponding to choices of $\bar{P}_d$ and P_{fa} for Swerling I fluctuations.

9.6-6 A radar detects a Swerling I model target using one pulse. (a) Find the threshold r_{b1} that gives $P_{fa} = 10^{-8}$. (b) What value of $\bar{\mathscr{R}}_p$ is needed to give $\bar{P}_d = 0.9$? (*Hint*: Use results of Problem 9.6-1.)

9.6-7 Work Problem 9.6-6 except for $P_{fa} = 4(10^{-5})$.

9.6-8 A radar operates with group-to-group Swerling I targets. It uses $N = 10$ pulses for detection and $P_{fa} = 10^{-10}$. What value of $\bar{\mathscr{R}}_p$ is needed to achieve $\bar{P}_d = 0.99$?

9.6-9 The system of Problem 9.6-8 lowers its threshold until $P_{fa} = 10^{-2}$. If N is still 10 and the same target is being detected, what is $\bar{P}_d$?

9.6-10 A radar uses a square law detector with threshold set at $r_{b1} = 64.4$ to detect a Swerling I model target using $N = 30$ pulses. $\bar{\mathscr{R}}_p = 560$ for a single pulse. Find P_{fa} and $\bar{P}_d$. (*Hint*: Use results from Problem 9.5-1.)

9.6-11 If the threshold is raised in Problem 9.6-10 until $P_{fa} = 10^{-10}$ what value of $\bar{P}_d$ is possible?

9.6-12 A detection processor has its threshold set to give $P_{fa} = 10^{-8}$ when detecting a Swerling I model target with $N = 10$ pulses. $\bar{\mathscr{R}}_p$ for the target gives $\bar{P}_d = 0.98$. (a) What is $\bar{\mathscr{R}}_p$? (b) If the transmitter suffers a peak power loss of 2 dB over a period of time due to aging and maintenance problems, to what value does $\bar{P}_d$ drop?

★9.6-13 If $N\bar{\mathscr{R}}_p/2 \gg 1$ and P_{fa} is small, show that (9.6-11) is approximated by

$$\bar{P}_d \approx \left[1 + \frac{2}{N\bar{\mathscr{R}}_p}\right]^{N-1} \exp\left[\frac{-r_{b1}}{1 + (N\bar{\mathscr{R}}_p/2)}\right] \quad \text{(Swerling I target)}$$

This result is originally due to Swerling (1954) who indicates its applicability to $\bar{P}_d$ as low as 0.01 in some cases where P_{fa} is sufficiently small and $N\bar{\mathscr{R}}_p/2$ is large.

9.6-14 For detection of a target having Swerling III fluctuations, show that for $N = 1$ pulse, (9.5-17) reduces to

$$P_{fa} = \exp(-r_{b1})$$

and (9.6-12) reduces to

$$\bar{P}_d = \left[1 - \frac{4\bar{\mathscr{R}}_p \ln(P_{fa})}{(4 + \bar{\mathscr{R}}_p)^2}\right] \exp\left[\frac{4\ln(P_{fa})}{4 + \bar{\mathscr{R}}_p}\right]$$

9.6-15 Continue Problem 9.6-14 for $N = 2$, and show that

$$P_{fa} = (1 + r_{b1})\exp(-r_{b1})$$

$$\bar{P}_d = \left[1 + \frac{2r_{b1}}{2 + \bar{\mathscr{R}}_p}\right]\exp\left[\frac{-2r_{b1}}{2 + \bar{\mathscr{R}}_p}\right]$$

9.6-16 A target fluctuates according to the probability density function of (9.6-16). What is the probability that $\mathscr{R}_p$ exceeds half its mean $(\bar{\mathscr{R}}_p/2)$ but does not exceed three halves $\bar{\mathscr{R}}_p$?

9.6-17 Work Problem 9.6-16 except find the probability that $\mathscr{R}_p$ will exceed its mean value.

9.6-18 A radar uses one pulse to detect a target that fluctuates according to the Swerling III model. If $P_{fa} = 5(10^{-4})$ and $\bar{\mathscr{R}}_p = 160$, find the threshold r_{b1} and value of $\bar{P}_d$. (*Hint*: Use the results from Problem 9.6-14.)

9.6-19 Work Problem 9.6-18 except assume that $P_{fa} = 3(10^{-3})$ and $\bar{\mathscr{R}}_p = 40$.

9.6-20 Work Problem 9.6-18 except assume that $P_{fa} = 10^{-12}$ and $\bar{\mathscr{R}}_p = 1000$.

9.6-21 A receiver uses one pulse to detect a Swerling III model target. Threshold is set to give $P_{fa} = 10^{-6}$. The target corresponds to $\bar{\mathscr{R}}_p$ sufficient to give $\bar{P}_d = 0.85$. If threshold is raised to give $P_{fa} = 10^{-10}$, what increase in transmitter power is needed to maintain $\bar{P}_d = 0.85$?

9.6-22 Work Problem 9.6-21 except assume that the initial setting is for $P_{fa} = 10^{-2}$ and that $\bar{\mathscr{R}}_p$ is sufficient to give $\bar{P}_d = 0.95$.

9.6-23 A radar's threshold is set to give $P_{fa} = 10^{-8}$ when detecting a *steady* target with unknown phase by using one pulse. Cross section is such as to produce $\bar{P}_d = 0.95$. If the target actually fluctuates but has an average cross section the same as the steady target, find $\bar{P}_d$ when fluctuations are (a) Swerling I, and (b) Swerling III. Use graphs only in the steady-target case and make use of the results of Problems 9.6-1 and 9.6-14. Note which form of fluctuation gives the smallest value of $\bar{P}_d$.

9.6-24 As in Problem 9.6-23 a radar has $P_{fa} = 10^{-8}$ and $\bar{P}_d = 0.95$ when detecting a steady target of unknown phase using one pulse. Now, however, the system must give the same average detection probability when the target fluctuates. What change in transmitter peak power is required in the two cases? Graphical solutions are acceptable in all cases.

★9.6-25 Generalize Problems 9.6-14 and 9.6-15, which apply to Swerling III targets, so that they apply for any $N > 2$ by showing that (9.6-11) for $K = 2$ can be approximated by

$$\bar{P}_d \approx \left[1 + \frac{4}{N\bar{\mathscr{R}}_p}\right]^{N-2}\left[1 - \frac{4(N-2)}{N\bar{\mathscr{R}}_p} + \frac{4r_{b1}}{4 + N\bar{\mathscr{R}}_p}\right]\exp\left[\frac{-4r_{b1}}{4 + N\bar{\mathscr{R}}_p}\right]$$

if $N\bar{\mathscr{R}}_p/4 \gg 1$ and P_{fa} is small. This result was originally due to Swerling (1954), who also found it to be exact for $N = 1$ and $N = 2$; these facts can be seen by comparing this expression for $N = 1$ and 2, respectively with those of Problems 9.6-14 and 9.6-15.

9.6-26 A radar uses 100 pulses to detect a target that fluctuates according to the Swerling III model. If $P_{fa} = 10^{-10}$, what value of $\bar{\mathscr{R}}_p$ is needed to achieve $\bar{P}_d = 0.999$?

9.6-27 The threshold in the system of Problem 9.6-26 is lowered until $P_{fa} = 10^{-2}$. What value of $\bar{\mathscr{R}}_p$ is now required to achieve $\bar{P}_d = 0.999$?

9.6-28 A target fluctuates according to a Swerling III model. It is detected by a radar using 10 pulses when the threshold is 30. (a) What is P_{fa}? (*Hint:* Use the approximation of Problem 9.5-1.) (b) What is $\bar{P}_d$ if $\bar{\mathscr{R}}_p = 40$? (*Hint:* Use the approximation of Problem 9.6-25.)

9.6-29 Two targets, both Swerling III type, are detected by a radar using three pulses and $P_{fa} = 10^{-4}$. The first target is known to have an average cross section of 3.7 m^2 and corresponds to $\bar{P}_d = 0.9$. The second target is known to give $\bar{\mathscr{R}}_p = 63.1$. (a) What is $\bar{\mathscr{R}}_p$ for target one? (b) What is $\bar{P}_d$ for target two? (c) What is the average cross section of target two?

9.6-30 Work Problem 9.6-29 except assume that $N = 30$ pulses.

9.7-1 A radar uses 12 pulses to detect a Swerling II target when threshold is set to 29.0. (a) Find P_{fa}. (b) What is $\bar{P}_d$ if $\bar{\mathscr{R}}_p = 16$ (or 12.04 dB) for one pulse from the target?

9.7-2 Swerling II targets are detected by a radar using 100 pulses and a threshold of 166.63. Target average signal-to-noise ratio on one pulse is $\bar{\mathscr{R}}_p = 2.51$ (or 4.0 dB). Find P_{fa} and $\bar{P}_d$ for this system.

9.7-3 Find and plot the integration improvement factor $I(N)$ for $N = 3, 10, 30,$ and 100 when $P_{fa} = 10^{-4}$ and $\bar{P}_d = 0.90$ for a Swerling II target. Repeat for a Swerling IV target and plot the difference between the two plots. Which target requires a larger value of $\bar{\mathscr{R}}_p$?

9.7-4 A target fluctuates pulse-to-pulse according to the exponential probability density (Swerling II). A detection processor based on $N = 30$ pulses has its threshold set for $P_{fa} = 10^{-6}$. What value of $\bar{\mathscr{R}}_p$ is required to give $\bar{P}_d = 0.85$?

9.7-5 Find the integration improvement factor for the system of Problem 9.7-4.

9.7-6 Work Problem 9.7-4 except for $P_{fa} = 10^{-10}$.

9.7-7 A recently modified radar is designed for detecting Swerling II targets with three pulses. Threshold is set so $P_{fa} = 10^{-2}$. The earlier version gave $\bar{P}_d = 0.915$ when $\bar{\mathscr{R}}_p = 100$ (or 20 dB) for $N = 1$ and a threshold set, so $P_{fa} = 10^{-2}$. What is $\bar{P}_d$ for the modified system?

9.7-8 A radar is to detect either a Swerling III or IV target with one pulse when $P_{fa} = 10^{-10}$. (a) How much extra signal-to-noise ratio is required for the

Swerling IV target compared to the Swerling III target? (b) How does your answer in (a) change for a different value of P_{fa}?

9.7-9 A Swerling IV model target is detected by a radar using $N = 10$ pulses when $P_{fa} = 10^{-4}$. (a) If $\bar{P}_d = 0.8$ is necessary, what must $\bar{\mathscr{R}}_p$ be? (b) What would $\bar{\mathscr{R}}_p$ become if $N = 1$ and the same values of P_{fa} and $\bar{P}_d$ are assumed? (c) If, instead, the same value of $\bar{\mathscr{R}}_p$ as in (a) is used, except with $N = 1$ and $P_{fa} = 10^{-4}$, what does $\bar{P}_d$ become?

9.7-10 Work Problem 9.7-9 except for $P_{fa} = 10^{-8}$.

9.7-11 A Swerling IV target is known to give $\bar{\mathscr{R}}_p = 1.66$ (or 2.2 dB). A radar based on using 30 pulses measures $\bar{P}_d = 0.03$. It is suspected the threshold is set too high for the design value of $P_{fa} = 10^{-2}$. (a) Find P_{fa} and, if not equal to the design value, then (b) find the new value of $\bar{P}_d$ that results once the threshold is properly set. (c) With the proper threshold how much additional transmitter peak power is needed (in dB) to give $\bar{P}_d = 0.99$?

9.7-12 Work Problem 9.7-11 except where $\bar{\mathscr{R}}_p = -2.0$ dB, $N = 100$, and $\bar{P}_d = 0.1$ initially.

9.7-13 A radar engineer measures $P_{fa} = 10^{-2}$ in a search radar designed to detect Swerling IV targets by integrating 10 pulses. (a) If $\bar{\mathscr{R}}_p = 5.012$ (or 7.0 dB), what is $\bar{P}_d$? (b) If the engineer raises the threshold to get $P_{fa} = 10^{-6}$, what is $\bar{P}_d$?

9.7-14 Assume that $N = 1$ and $P_{fa} = 10^{-4}$, and plot for $0.1 \le \bar{P}_d \le 0.98$ the difference between $\mathscr{R}_p$ required by a Swerling I target and $\mathscr{R}_p$ for a steady target (with random phase) to produce the specified $\bar{P}_d$. Repeat for $P_{fa} = 10^{-6}$, 10^{-8}, and 10^{-10} all for $N = 1$. Note how much these curves depend on P_{fa}. Is there a strong dependence? Also note that for $N = 1$ the *exact* same results apply to a Swerling II target.

9.7-15 Work Problem 9.7-14 except for the Swerling III target (the curves for the Swerling IV case are identical).

9.8-1 A radar uses a binary detection processor based on three or more of five pulses from an envelope detector. For each pulse from a nonfluctuating target, $P_{fa1} = 0.05$ and $P_{d1} = 0.5$. Find P_{fa} and P_d. Is this a good system? Is three the optimum threshold?

9.8-2 Work Problem 9.8-1 except assume that a larger target cross section or more transmitter power is available so that P_{d1} increases to 0.8.

★9.8-3 In (9.8-5) assume that P_{fa1} is sufficiently small that the first two terms of the sum can be taken as the "true" value of P_{fa}. If the first term is treated as an approximation to P_{fa} and the second is considered the error in the approximation, find the fractional error of the approximation.

9.8-4 M_{opt} in a binary detection system for a nonfluctuating target is the nearest integer to the optimum value found from (9.8-7) when $N = 10$ pulses are used. If $P_{fa1} = 0.15$ and $P_{d1} = 0.7$ on each pulse, find P_{fa} and P_d.

9.8-5 A radar uses one pulse to detect a steady target; the threshold is set give $P_{fa1} = 0.094$ for which $P_{d1} = 0.88$ from Fig. 9.4-4 when $\mathcal{R}_p = 10$ dB. However, the operating engineer has enough detection time to use up to 10 pulses in a binary detector which he is considering to reduce false alarm probability. If the engineer seeks to realize $P_{fa} = 10^{-4}$ with M_{opt} equal to the nearest integer to the value found from (9.8-7), what values of M_{opt} and N should he use? With the chosen values of M_{opt} and N, find P_{fa} and P_d. How is the detection probability affected by going to a binary detector?

9.8-6 A target with Swerling II fluctuations is to be detected using a binary detector with $N = 30$ pulses. If the system is to produce $P_{fa} = 10^{-6}$ and $\bar{P}_d = 0.9$, what is the optimum threshold M_{opt}, to be used?

9.8-7 Work Problem 9.8-6 except assume a Swerling IV target.

★9.10-1 Use (9.10-16) to show that (9.10-14) gives the K-distribution when $f(\sigma)$ is defined by (5.9-22).

10

RADAR MEASUREMENTS—LIMITING ACCURACY

In the preceding chapter our attention was focused on target detection. We sought to determine if a target signal plus noise or if noise alone was present in a specified resolution cell. Once a "target-present" decision is made for a particular cell, the target's parameters, such as range delay, Doppler frequency, and angular directions are known within the dimensions of the resolution cell. These rough values of the parameters may be considered as initial estimates of the exact parameters, and the radar can proceed to refine these estimates by using the signals in one or more pulse intervals to measure the parameters more precisely. Through continuous measurements the radar's estimate of a target parameter can be made to follow the parameter value, even if it changes (a tracking radar).

In this chapter we view the problem of measuring target parameters as one of *parameter estimation*. Our principal concern will be the basic limitations on the accuracy with which the radar can estimate (measure) a parameter in the presence of noise. Our results will represent the best the radar can hope to do, since noise is always present. These results form a basis of comparison of performance for practical systems. Of course other quantities such as clutter, interference, and practical hardware and system limitations all tend to degrade the radar's accuracy. The following chapters will develop some details on the implementation of measurement systems for range delay (Chapter 11), Doppler angular frequency (Chapter 12), and spatial angles (Chapters 13, 14).

10.1 PARAMETER ESTIMATION

The study of parameter estimation hinges on the use of statistical theory to describe the radar's noise, and possibly the target's parameters if they are considered as random quantities. The theory of parameter estimation is very detailed and the literature is extensive. Its roots go all the way back to Gauss in the late 1700s and early

1800s who is credited with the conception of the method of least squares. Gauss formally derived the theory from fundamental principles and applied it to orbit determination of planets (see Deutsch, 1965, p. 5, where other references are given). Many people (Legendra, Lagrange, Bessel, Laplace, Poisson, and others) extended the least-squares concept. However, it remained for R. A. Fisher (1912, 1922, 1925) to place estimation theory on a firm foundation in several writings over a period from about 1912 to 1925.

A few of the many key early papers are: Rao (1945, 1947, 1949), Bhattacharyya (1946, 1947), Barankin (1949), Slepian (1954), Swerling (1956, 1959, 1964), and Kelley et al. (1960, pts I and II). Some early books containing discussions of parameter estimation are Cramér (1946), Wilks (1962), Mood and Graybill (1963), Berkowitz (ed., 1965), Deutsch (1965) and Van Trees (1968). Some more recent books that give various levels of detail and other references are Helstrom (1995, 1991), Shanmugan and Breipohl (1988), Stark and Woods (1986), Ibragimov and Has'minskii (1981), Sorenson (1980), and Beck and Arnold (1977). The book edited by Lainiotis (1974) gives many references to the more recent applications of estimation theory (Wiener filtering, Kalman-Bucy filtering, etc.).

Basic Definitions and Model

In estimating a target's parameters a radar can only work with a priori information and the received waveform. Generally, the more a priori knowledge the radar has, the more accurate can be its measurements (estimates). For convenience, and as a tractable mathematical model, we assume that the radar receiver is band limited to have a constant and nonzero response only for $-W_N < \omega < W_N$. As we proceed, the reader will see that this assumption is only for convenience, for we will later allow $W_N \to \infty$. For the moment W_N is assumed large enough that the received waveform reflected from a target, denoted by $s_r(t)$, passes through the band-limiting operation without distortion. For white Gaussian noise $n_w(t)$ at the receiver's input, the output noise, denoted by $n(t)$, is band-limited white Gaussian noise. We presume the radar observes the band-limited response, denoted by $y(t)$, which is the sum

$$y(t) = s_r(t) + n(t) \tag{10.1-1}$$

Parameters of interest are implicit in the received signal, since the usual parameters are target delay, Doppler angular frequency, and spatial angles. Denote these parameters by $\theta_1, \theta_2, \dots, \theta_M$. Collectively we define a parameter vector $\underline{\theta}$ by

$$\underline{\theta} = (\theta_1, \theta_2, \dots, \theta_M) \tag{10.1-2}$$

for M parameters.

Since the radar must develop its parameter estimates by observations made on $y(t)$ during some time interval of interest, we establish a model for these observations. Assume that the time interval is some multiple N of pulse repetition periods each of duration T_R. Within each period assume that the observations are in the form of K samples of $y(t)$ taken Δt apart at times

$$t_{ik} = (i-1)T_R + k\Delta t, \qquad i = 1, 2, \dots, N, \ k = 1, 2, \dots, K \tag{10.1-3}$$

Time t_{ik} is for sample k in period i. Next denote by

$$y_{ik} = y(t_{ik}) \tag{10.1-4}$$

$$s_{rik} = s_r(t_{ik}) \tag{10.1-5}$$

$$n_{ik} = n(t_{ik}) \tag{10.1-6}$$

the samples of $y(t)$, $s_r(t)$, and $n(t)$ at times t_{ik}, and define vectors of these samples by

$$\underset{\sim}{y} = (y_{11}, \dots, y_{1K}, \dots, y_{N1}, \dots, y_{NK}) \tag{10.1-7}$$

$$\underset{\sim}{s}_r = (s_{r11}, \dots, s_{r1K}, \dots, s_{rN1}, \dots, s_{rNK}) \tag{10.1-8}$$

$$\underset{\sim}{n} = (n_{11}, \dots, n_{1K}, \dots, n_{N1}, \dots, n_{NK}) \tag{10.1-9}$$

Estimators

The task of the radar is to make the observations in $\underset{\sim}{y}$ and form some function of these observations that gives an estimate of a parameter. For parameter θ_m we denote this function by $\hat{\theta}_m(\underset{\sim}{y})$, so

$$\hat{\theta}_m = \hat{\theta}_m(\underset{\sim}{y}), \qquad m = 1, 2, \dots, M \tag{10.1-10}$$

The function $\hat{\theta}_m(\underset{\sim}{y})$ is assumed to be independent of the parameters and is called an *estimator*, while its value for a particular observation $\underset{\sim}{y}$ is called the *estimate* of θ_m. There is a set of estimators when there is more than one parameter to be estimated. Because of the noise present in $y(t)$, the observation vector $\underset{\sim}{y}$ is random, and the estimate $\hat{\theta}_m$ of θ_m may not precisely *equal* θ_m for a particular observation $\underset{\sim}{y}$. However, a "good estimator" will give an estimate that does not depart greatly from the true value, and the average of a large number of estimates will converge to the true parameter value θ_m. The work of this chapter is to define the best accuracy achievable by the estimator when it is optimum in some sense.

When several ($M > 1$) parameters are to be estimated, a vector $\hat{\underset{\sim}{\theta}}$ whose components are the individual estimators is defined by

$$\hat{\underset{\sim}{\theta}} = (\hat{\theta}_1, \hat{\theta}_2, \dots, \hat{\theta}_M) \tag{10.1-11}$$

There are many ways of defining an optimum estimator. These vary greatly according to specific details and the amount of a priori information. For example, if there is no statistical description of the noise or parameters at all the least-squares approach treats the parameter estimation task as a deterministic optimization problem. This approach is also known as *least-squares curve fitting* and is related to linear regression methods (Melsa and Cohn, 1978, p. 202).

Perhaps the next step up in estimates is to presume the first- and second-order moments of the noise samples (and the parameters, if random) are known. For such a priori knowledge the *linear minimum variance* estimator is that which gives minimum estimate variances when the estimator's *form* is assumed to be a linear function of the observations y. This estimate variance is defined for parameter θ_m by

$$\sigma_{\hat{\theta}m}^2 = E[(\hat{\theta}_m - \theta_m)^2], \qquad m = 1, 2, \dots, M \tag{10.1-12}$$

where $E[\,\cdot\,]$ is the statistical expectation operation and the optimum estimator gives the smallest value of $\sigma_{\hat{\theta}m}^2$.

When full statistical descriptions of the noise and parameters are available through their respective probability density functions, the Bayes cost method of estimation gives a large class of estimation problems. The general idea is to define a cost function that assigns a unique cost to each combination of true parameter value and its estimate. This cost function can be considered somewhat analogous to that of (9.1-17) for detection. The Bayes method minimizes the cost function after averaging over all parameters and observations. Different results are obtained for different choices of the cost function.

If the cost function in the Bayes method is the magnitude squared of the difference of the vector of the estimates and the vector of the parameters, the optimum estimator is called *minimum mean-squared error*. If the mean value of the vector of estimates is equal to the true value of the vector of the parameters, they are called *unbiased estimators*. When estimates are unbiased, the minimum mean-squared error estimator is also called a *minimum variance estimator*.

If the cost function in the Bayes method is uniform, that is, if zero cost is assigned when all components of estimation error are smaller than a prescribed value and unit cost assigned when any component of estimation error exceeds the specified value, the estimator is the mode of the conditional density function $p(\theta|y)$. Since $p(\theta|y)$ is an a posteriori density, the density of θ after the observation y is obtained, the estimate is called a *maximum a posteriori* (or MAP) *estimator*. Since it is possible for $p(\theta|y)$ to be multi-modal, the MAP estimator may not be unique (see Melsa and Cohn, 1978, p. 189).

As a final example of an optimum parameter estimator, we assume a priori knowledge of $p(y|\theta)$. No knowledge of the density of the parameters is assumed. In the maximum likelihood estimation method the optimum estimators are the components of the vector $\hat{\theta}$ that equal those of the vector θ corresponding to the maximum of $p(y|\theta)$ for a given observation vector y. The *maximum likelihood* (ML) *estimator* is very useful because of its simplicity and the relatively small amount of a priori information required. For the case where the noise is gaussian (our radar case), the ML estimate is equivalent to both the least-squares estimate and the minimum variance estimate (see Deutsch, 1965, pp. 136–137). It is also known that the ML estimate is inferior to the MAP estimate if a priori information about θ is available; however, if there is no a priori knowledge about θ other than that it is in a given region, then the ML and MAP estimates are equal (see Melsa and Cohn, 1978, p. 191).

Properties of Estimators

R. A. Fisher introduced some quantities that describe estimators which we may consider as properties of estimators. The first is a very important property that defines an unbiased estimator. Let the right side of (10.1-10) define an estimator of a parameter θ_m, $m = 1, 2, \dots, M$. If the mean value of $\hat{\theta}_m$ is equal to the true value θ_m, the estimator is said to be unbiased. Thus, if

$$E[\hat{\theta}_m(y)] = \theta_m, \qquad m = 1, 2, \dots, M \tag{10.1-13}$$

for all values of θ_m, then $\hat{\theta}_m$ is unbiased. Although unbiased estimators do not always exist in theory, they are most desired for radar where we usually want our measurement system to give the true value of the quantity being measured, at least on the average. In most cases a radar's measurement systems do closely approximate the unbiased condition.

An estimator $\hat{\theta}_m(y)$ is said to be *consistent* if it converges in probability to θ_m as the total number of samples (in y) approaches infinity. A consistent estimator is always unbiased for a sufficiently large number of samples, but an unbiased estimator need not necessarily be consistent (Deutsch, 1965, p. 24).

An unbiased estimator is said to give an *efficient estimate* if no other unbiased estimator has a smaller variance as defined by (10.1-12). Since the unbiased estimator is of most interest in radar, and since the efficient estimator has the least estimate variance, we shall be concerned in the following section with the smallest possible variance to be associated with efficient estimators. This lowest value of variance is called the *Cramér-Rao bound*.

10.2 THE CRAMÉR-RAO BOUND

In this section we will demonstrate that there is a smallest value of variance for an unbiased estimator for a single parameter. This smallest variance is called the Cramér-Rao bound, and it is achieved by an efficient estimator. The case of multiple estimators of multiple parameters is also given but is not proved.

Single Parameter Cramér-Rao Bound

Consider the case of a single unknown nonrandom parameter θ_1. Denote the joint density of the observation random variables by $p(y_{11}, \ldots, y_{NK}; \theta_1) = p(y; \theta_1)$ where the notation includes θ_1 to imply that the observations depend on θ_1 through the presence of the received signal $s_r(t)$. We will require that the estimator $\hat{\theta}_1(y)$ be unbiased, which means that

$$E[\hat{\theta}_1 - \theta_1] = \int_{-\infty}^{\infty} \cdots \int_{-\infty}^{\infty} (\hat{\theta}_1 - \theta_1) p(y; \theta_1) dy = 0 \qquad (10.2\text{-}1)$$

where $dy = dy_{11} \ldots dy_{NK}$. By differentiating (10.2-1) with respect to θ_1, and noting that $\hat{\theta}_1(y)$ is not a function of θ_1 by assumption, we have (on using Leibniz's rule)

$$\frac{\partial}{\partial \theta_1} \int_{-\infty}^{\infty} \cdots \int_{-\infty}^{\infty} (\hat{\theta}_1 - \theta_1) p(y; \theta_1) dy$$
$$= \int_{-\infty}^{\infty} \cdots \int_{-\infty}^{\infty} (\hat{\theta}_1 - \theta_1) \frac{\partial p(y; \theta_1)}{\partial \theta_1} dy - \int_{-\infty}^{\infty} \cdots \int_{-\infty}^{\infty} p(y; \theta_1) dy = 0 \qquad (10.2\text{-}2)$$

The second right-side integral in (10.2-2) is unity; the first right-side integral can be rewritten by using the relationship

$$\frac{\partial p(y; \theta_1)}{\partial \theta_1} = \frac{\partial \ln[p(y; \theta_1)]}{\partial \theta_1} p(y; \theta_1) \qquad (10.2\text{-}3)$$

We have

$$\int_{-\infty}^{\infty} \cdots \int_{-\infty}^{\infty} (\hat{\theta}_1 - \theta_1) \frac{\partial \ln[p(y; \theta_1)]}{\partial \theta_1} p(y; \theta_1) dy = 1 \tag{10.2-4}$$

To develop (10.2-4) further, we use Schwarz's inequality (Deutsch, 1965, p. 139), which can be written as

$$\left[\int_{-\infty}^{\infty} \cdots \int_{-\infty}^{\infty} A(y) B(y) dy \right]^2 \le \int_{-\infty}^{\infty} \cdots \int_{-\infty}^{\infty} A^2(y) dy \int_{-\infty}^{\infty} \cdots \int_{-\infty}^{\infty} B^2(y) dy \tag{10.2-5}$$

where $A(y)$ and $B(y)$ are real functions of y. The equality in (10.2-5) holds if and only if

$$A(y) = KB(y) \tag{10.2-6}$$

with K an arbitrary real nonzero constant with respect to y. On identifying

$$A(y) = (\hat{\theta}_1 - \theta_1) \sqrt{p(y; \theta_1)} \tag{10.2-7}$$

$$B(y) = \frac{\partial \ln[p(y; \theta_1)]}{\partial \theta_1} \sqrt{p(y; \theta_1)} \tag{10.2-8}$$

(10.2-5) is used with (10.2-4) to obtain

$$\int_{-\infty}^{\infty} \cdots \int_{-\infty}^{\infty} (\hat{\theta}_1 - \theta_1)^2 p(y; \theta_1) dy \ge \frac{1}{\int_{-\infty}^{\infty} \cdots \int_{-\infty}^{\infty} \{\partial \ln[p(y; \theta_1)]/\partial \theta_1\}^2 p(y; \theta_1) dy} \tag{10.2-9}$$

The left side of (10.2-9) is recognized as the variance of the unbiased estimator $\hat{\theta}_1(y)$, which we denote by $\sigma_{\hat{\theta}_1}^2$. The right side of (10.2-9) is the reciprocal of the expected value of $\{\partial \ln[p(y; \theta_1)]/\partial \theta_1\}^2$ and is a lower bound on the variance; it is the Cramér-Rao bound which can be written as

$$\sigma_{\hat{\theta}_1}^2 \ge \frac{1}{E(\{\partial \ln[p(y; \theta_1)]/\partial \theta_1\}^2)} \tag{10.2-10}$$

We subsequently simplify (10.2-10) to a more useful form for radar. However, we first note in passing that the equality in (10.2-10), which gives the smallest estimator variance, occurs when (10.2-6) is true. On using (10.2-7) and (10.2-8), we find that the estimator must be given by

$$\hat{\theta}_1(y) = \theta_1 + K \frac{\partial \ln[p(y; \theta_1)]}{\partial \theta_1} \tag{10.2-11}$$

if it exists. Of course not all density functions and signal forms lead to a valid estimator which, as assumed, must be independent of θ_1. It can be shown that the maximum likelihood estimator, which we here denote by $\hat{\theta}_{1\mathrm{ML}} = \theta_1$, causes the derivative in (10.2-11) to be identically zero. Since K cannot be zero, we conclude that the minimum variance estimator is the same as the maximum likelihood estimator: $\hat{\theta}_1(y) = \hat{\theta}_{1\mathrm{ML}}(y)$. (See Melsa and Cohn, 1978, pp. 233–234.) Thus, if an unbiased efficient estimator exists, it can be found by the simple maximum likelihood procedure.

We return now to the simplification of (10.2-10) for our radar cases. Because the definitions of sample times, noise, and the target return signal are all the same as used in Section 9.1, we may use the right-side form of (9.1-27), which applies to the current problem and is equal to $p(y; \theta_1)$:

$$p(y; \theta_1) = \prod_{i=1}^{N} \prod_{k=1}^{K} (2\pi\sigma_N^2)^{-1/2} \exp\left\{\frac{-1}{2\sigma_N^2}(y_{ik} - s_{rik})^2\right\} \tag{10.2-12}$$

where, from (9.1-25),

$$\frac{1}{\sigma_N^2} = \frac{2\pi}{\mathcal{N}_0 W_N} = \frac{2\Delta t}{\mathcal{N}_0} \tag{10.2-13}$$

By substituting (10.2-12) into the denominator of (10.2-10) and reducing some relatively straightforward statistical algebra (Problem 10.2-2), we get

$$E\left(\left\{\frac{\partial \ln[p(y; \theta_1)]}{\partial \theta_1}\right\}^2\right) = \frac{2}{\mathcal{N}_0} \sum_{i=1}^{N} \sum_{k=1}^{K} \left(\frac{\partial s_{rik}}{\partial \theta_1}\right)^2 \Delta t \tag{10.2-14}$$

where (10.2-13) has also been used. Now, as the noise bandwidth $W_N \to \infty$, we have $\Delta t \to 0$, and the sums in (10.2-14) become integrals. The sum over index k becomes the integral of $[\partial s_r(t)/\partial \theta_1]^2$ over pulse interval i, while the sum over index i simply adds all the intervals to form the integral over N pulse intervals. Finally we write (10.2-10) as

$$\sigma_{\hat{\theta}_1}^2 \geq \frac{1}{\dfrac{2}{\mathcal{N}_0} \displaystyle\int_0^{NT_R} \left[\dfrac{\partial s_r(t)}{\partial \theta_1}\right]^2 dt} \tag{10.2-15}$$

This result will be used in Section 10.3 to compute the limiting accuracies of radar measurements caused by the presence of noise.

Multiple Parameter Cramér-Rao Bound

In the preceding subsection we considered the case of only one unknown target parameter for which only one estimator was required. More generally, a radar may need to make simultaneous measurements on several unknown nonrandom parameters by using an estimator for each parameter. Ideally each estimator would be unbiased and efficient even in the presence of the unknown parameters other than the one for which it is designed. The presence of multiple parameters can (and typically does) increase the variance of a given estimator over its value for only one parameter.

The question now becomes: What are the minimum possible estimator variances when multiple estimators are used to estimate multiple unknown parameters? In fact a deeper question is: What are the *covariances* of these multiple estimators? These questions were answered by Rao (1945, 1947) and Bhattacharyya (1946, 1947).[1]

Define an $M \times M$ covariance matrix $[V]$ having elements V_{mn} that are the covariances of the estimators defined by

$$
\begin{aligned}
V_{mn} &= E\{(\hat{\theta}_m - \theta_m)(\hat{\theta}_n - \theta_n)\} \\
&= \int_{-\infty}^{\infty} \cdots \int_{-\infty}^{\infty} (\hat{\theta}_m - \theta_m)(\hat{\theta}_n - \theta_n) p(\underset{\sim}{y}; \underset{\sim}{\theta}) d\underset{\sim}{y}
\end{aligned}
\tag{10.2-16}
$$

where $\underset{\sim}{\theta}$ is defined by (10.1-2). Next define elements I_{pq} of an $M \times M$ matrix $[I]$ according to

$$
\begin{aligned}
I_{pq} &= E\left\{ \frac{\partial \ln[p(\underset{\sim}{y}; \underset{\sim}{\theta})]}{\partial \theta_p} \frac{\partial \ln[p(\underset{\sim}{y}; \underset{\sim}{\theta})]}{\partial \theta_q} \right\} \\
&= \int_{-\infty}^{\infty} \cdots \int_{-\infty}^{\infty} \left\{ \frac{\partial \ln[p(\underset{\sim}{y}; \underset{\sim}{\theta})]}{\partial \theta_p} \frac{\partial \ln[p(\underset{\sim}{y}; \underset{\sim}{\theta})]}{\partial \theta_q} \right\} p(\underset{\sim}{y}; \underset{\sim}{\theta}) d\underset{\sim}{y}
\end{aligned}
\tag{10.2-17}
$$

Matrix $[I]$ is known as *Fisher's information matrix*. Bhattacharyya (1946, 1947), Peebles (1970), and Sorenson (1980) have shown that the covariances of unbiased estimators $\hat{\theta}_m$ of parameters θ_m satisfy

$$
V_{mn} \geq I^{mn}, \qquad m, n = 1, 2, \ldots, M
\tag{10.2-18}
$$

Here I^{mn} denotes element mn in the inverse matrix of $[I]$, denoted by $[I]^{-1}$. When $m = n$, (10.2-18) shows the minimum variance of estimator $\hat{\theta}_m$ is equal to I^{mm}. For $m \neq n$, I^{mn} is the covariance of $\hat{\theta}_m$ and $\hat{\theta}_n$.

Example 10-2-1 For $M = 2$ parameters we find the Cramér-Rao bounds for the unbiased estimators $\hat{\theta}_1$ and $\hat{\theta}_2$. We also find their covariance. The Fisher information matrix is

$$
[I] = \begin{bmatrix} I_{11} & I_{12} \\ I_{12} & I_{22} \end{bmatrix}.
$$

since $I_{pq} = I_{qp}$ from (10.2-17); that is, $[I]$ is symmetric. The inverse is readily found to be

$$
[I]^{-1} = \begin{bmatrix} I_{22} & -I_{12} \\ -I_{12} & I_{11} \end{bmatrix} \frac{1}{(I_{11}I_{22} - I_{12}^2)}
$$

---

[1] Bhattacharyya's papers consider the more general problem of multiple estimates of multiple *functions* of multiple parameters. A concise tutorial development of this problem is also given in the appendixes of Peebles (1970).

The right side of (10.2-18) is the smallest variance when $m = n$, so

$$\sigma_{\hat{\theta}_1}^2 = V_{11} \geq I^{11} = \frac{I_{22}}{I_{11}I_{22} - I_{12}^2}$$

$$\sigma_{\hat{\theta}_2}^2 = V_{22} \geq I^{22} = \frac{I_{11}}{I_{11}I_{22} - I_{12}^2}$$

The covariance of the efficient estimators, denoted by $\sigma_{\hat{\theta}_1\hat{\theta}_2}^2$, satisfies

$$\sigma_{\hat{\theta}_1\hat{\theta}_2}^2 = V_{12} \geq I^{12} = -\frac{I_{12}}{I_{11}I_{22} - I_{12}^2}$$

When the number of parameters is larger than three, it is difficult to obtain the inverse matrix $[I]^{-1}$ by analytical methods.

10.3   LIMITING ACCURACIES OF RADAR MEASUREMENTS

A radar's best measurement accuracy (smallest estimate variance for an unbiased efficient estimator) is set by the Cramér-Rao bound, as given by the right side of (10.2-15) for our white noise case. We will use this result to determine the accuracies of estimating various target parameters. In particular, we find the smallest variance for estimating signal amplitude, phase, Doppler angular frequency, time delay, and spatial angles.

The signal $s_r(t)$ to be used in (10.2-15) for all but spatial angle measurements is the response of the receiving antenna with maximum gain in the general direction of the target. It is given by (1.6-21),

$$s_r(t) = \alpha a(t - \tau_R)\cos[(\omega_0 + \omega_d)(t - \tau_R) + \theta(t - \tau_R) + \phi_0] \tag{10.3-1}$$

where the various terms are defined in Chapter 1. Of special interest here are the target parameters of α (amplitude), ϕ_0 (phase), ω_d (Doppler angular frequency), and τ_R (time delay).

For measurement of spatial angles, the current-day methods mainly utilize an antenna that produces two independent output signals. Each of these signals is proportional to the angle offset of the target from the boresight axis' direction (see Fig. 1.7-1) in one of two orthogonal directions. Let us define these offset angles as θ_x and θ_y. The two received radar signals may then be written in the form

$$s_{rx}(t) = K_x\theta_x s_r(t) \tag{10.3-2}$$

$$s_{ry}(t) = K_y\theta_y s_r(t) \tag{10.3-3}$$

where $s_r(t)$ is given by (10.3-1) and K_x and K_y are constants of proportionality. These proportionality constants will be discussed in more detail in Chapters 13 and 14 when angle-tracking systems are developed.

Amplitude Measurement Accuracy

Here the parameter θ_1 is equal to α. By differentiation of (10.3-1), we develop

$$\frac{\partial s_r(t)}{\partial \theta_1} = \frac{\partial s_r(t)}{\partial \alpha} = \frac{s_r(t)}{\alpha} \tag{10.3-4}$$

$$\int_0^{NT_R} \left[\frac{\partial s_r(t)}{\partial \theta_1}\right]^2 dt = \frac{1}{\alpha^2}\int_0^{NT_R} s_r^2(t)\,dt = \frac{NE_r}{\alpha^2} \tag{10.3-5}$$

where E_r is the energy of $s_r(t)$ in one pulse interval. From (10.2-15) the variance of estimating α, denoted by $\sigma_{\hat{\alpha}}^2$, is

$$\sigma_{\hat{\alpha}}^2 = \sigma_{\hat{\theta}_1}^2 \geq \frac{1}{2NE_r/\alpha^2\mathcal{N}_0} = \frac{\alpha^2}{N\mathcal{R}_p} = \sigma_{\hat{\alpha}(\min)}^2 \tag{10.3-6}$$

where

$$\mathcal{R}_p = \frac{2E_r}{\mathcal{N}_0} \tag{10.3-7}$$

is the single-pulse signal to noise ratio available at the output of a white noise matched filter [see (6.7-4)].

The right side of the inequality of (10.3-6) is the Cramér-Rao bound for estimating amplitude. Since accuracy improves as this bound decreases, we see that accuracy is better for larger values of $\mathcal{R}_p$. Our bound is in agreement with that given by Skolnik (1962, p. 463). This minimum value of $\sigma_{\hat{\alpha}}^2$ is denoted by $\sigma_{\hat{\alpha}(\min)}^2$.

Example 10.3-1   For a signal-to-noise ratio $\mathcal{R}_p = 12$, we find the smallest fractional (relative) variance of estimating amplitude when $N = 6$ pulse intervals are used. From (10.3-6),

$$\text{Fractional variance} = \frac{\sigma_{\hat{\alpha}}^2}{\alpha^2} \geq \frac{1}{N\mathcal{R}_p} = \frac{1}{6(12)} = 0.0139$$

The smallest fractional variance is 0.0139, which corresponds to an rms relative error in measuring α of about $\sqrt{0.0139} = 0.118$, or 11.8%.

Phase Measurement Accuracy

Here the parameter θ_1 in (10.2-15) equals ϕ_0 in (10.3-1). The derivative needed in (10.2-15) is

$$\frac{\partial s_r(t)}{\partial \theta_1} = \frac{\partial s_r(t)}{\partial \phi_0} = -\alpha a(t - \tau_R)\sin[(\omega_0 + \omega_d)(t - \tau_R) + \theta(t - \tau_R) + \phi_0] \tag{10.3-8}$$

On reducing the algebra, the integral needed is[2]

$$\int_0^{NT_R} \left[\frac{\partial s_r(t)}{\partial \theta_1} \right]^2 dt = NE_r \qquad (10.3\text{-}9)$$

Finally the variance of estimating ϕ_0, denoted by $\sigma_{\hat{\phi}_0}^2$, is given by

$$\sigma_{\hat{\phi}_0}^2 = \sigma_{\hat{\theta}_1}^2 \geq \frac{1}{2NE_r/\mathcal{N}_0} = \frac{1}{N\mathcal{R}_p} = \sigma_{\hat{\phi}_0(\min)}^2, \qquad (10.3\text{-}10)$$

As with amplitude estimation, we see that phase estimation accuracy improves as the received signal's signal-to-noise ratio increases. Our result (10.3-10) is in agreement with the bound given by Skolnik (1962, p. 463). The smallest value of $\sigma_{\hat{\phi}_0}^2$ is denoted by $\sigma_{\hat{\phi}_0(\min)}^2$.

Doppler Frequency Measurement Accuracy

In this situation θ_1 represents the Doppler angular frequency parameter ω_d in (10.3-1). On forming the derivative of $s_r(t)$ defined in (10.3-1) as required in (10.2-15) and dropping the double-carrier-frequency term (which integrates to approximately zero), we have

$$\int_0^{NT_R} \left[\frac{\partial s_r(t)}{\partial \theta_1} \right]^2 dt = \int_0^{NT_R} \left[\frac{\partial s_r(t)}{\partial \omega_d} \right]^2 dt$$

$$= \frac{N\alpha^2}{2} \int_0^{T_R} (t - \tau_R)^2 a^2(t - \tau_R) dt \qquad (10.3\text{-}11)$$

At this point it is convenient to use the complex envelope (see Section 6.1) to represent $a(t)$:

$$g(t) = a(t)e^{j\theta(t)} \qquad (10.3\text{-}12)$$

$$a^2(t) = |g(t)|^2 \qquad (10.3\text{-}13)$$

When (10.3-13) is substituted into (10.3-11), it can be shown that

$$\int_0^{NT_R} \left[\frac{\partial s_r(t)}{\partial \theta_1} \right]^2 dt = NE_r(\bar{t}_g^2 + \tau_{g,\text{rms}}^2) \qquad (10.3\text{-}14)$$

where $\bar{t}_g$ is the mean time (of occurrence) of $g(t)$ as defined in Problem 6.4-2, and $\tau_{g,\text{rms}}^2$ is the mean-squared duration of $g(t)$ (see Problem 10.3-6). The variance of estimating ω_d, denoted by $\sigma_{\hat{\omega}_d}^2$, is found from (10.2-15) using (10.3-14):

$$\sigma_{\hat{\omega}_d}^2 = \sigma_{\hat{\theta}_1}^2 \geq \frac{1}{(2/\mathcal{N}_0)NE_r(\bar{t}_g^2 + \tau_{g,\text{rms}}^2)}$$

$$= \frac{1}{N\mathcal{R}_p(\bar{t}_g^2 + \tau_{g,\text{rms}}^2)} \qquad (10.3\text{-}15)$$

---

[2] The term in the integral at twice the carrier frequency is dropped, since it integrates to approximately zero.

In most cases the time origin is chosen such that $\bar{t}_g$ is zero. For this case

$$\sigma_{\hat{\omega}_d}^2 \geq \frac{1}{N\mathscr{R}_p \tau_{g,\text{rms}}^2} = \sigma_{\hat{\omega}_d(\text{min})}^2 \tag{10.3-16}$$

Again we see that accuracy is improved as $\mathscr{R}_p$ is increased. It is important to also note that accuracy increases as the transmitted signal's rms duration, $\tau_{g,\text{rms}}$, increases. Thus good Doppler frequency measurement accuracy is achieved by use of long pulses. A similar result was derived by Manasse (1960, p. 10). The smallest value of $\sigma_{\hat{\omega}_d}^2$ is denoted by $\sigma_{\hat{\omega}_d(\text{min})}^2$.

Delay Measurement Accuracy

For delay estimation the parameter θ_1 represents the target's parameter τ_R. We use (10.3-1) and form the derivative required in (10.2-15). On squaring the derivative and dropping the double-frequency carrier terms that integrate to approximately zero, we have

$$\int_0^{NT_R} \left[\frac{\partial s_r(t)}{\partial \theta_1}\right]^2 dt = \int_0^{NT_R} \left[\frac{\partial s_r(t)}{\partial \tau_R}\right]^2 dt$$
$$= \frac{N\alpha^2}{2} \int_0^{T_R} \left| -j(\omega_0 + \omega_d)g(t - \tau_R) + \frac{\partial g(t - \tau_R)}{\partial \tau_R} \right|^2 dt \tag{10.3-17}$$

where (10.3-13) has again been used (see Problem 10.3-17).

For further reduction of (10.3-17), we note that the limits on the integral are large compared to the time span over which $g(t - \tau_R)$ is typically nonzero. On changing the variable of integration to $\xi = t - \tau_R$ and setting the limits to $-\infty$ and $+\infty$, we have

$$\int_0^{NT_R} \left[\frac{\partial s_r(t)}{\partial \theta_1}\right]^2 dt \approx \frac{N\alpha^2}{2} \int_{-\infty}^{\infty} \left| j(\omega_0 + \omega_d)g(\xi) + \frac{\partial g(\xi)}{\partial \xi} \right|^2 d\xi$$
$$= \frac{N\alpha^2}{4\pi} \int_{-\infty}^{\infty} (\omega + \omega_0 + \omega_d)^2 |G(\omega)|^2 d\omega \tag{10.3-18}$$

The second right-side form of (10.3-18) results from Parseval's theorem where $G(\omega)$ is the Fourier transform of $g(t)$. Finally (10.3-18) is expanded and reduced to

$$\int_0^{NT_R} \left[\frac{\partial s_r(t)}{\partial \theta_1}\right]^2 dt = NE_r[(\omega_0 + \omega_d + \bar{\omega}_g)^2 + W_{g,\text{rms}}^2] \tag{10.3-19}$$

where $\bar{\omega}_g$ is the mean frequency of $G(\omega)$, as defined by (6.4-9), and $W_{g,\text{rms}}$ is the rms bandwidth of $G(\omega)$, as defined by (6.4-14).

We denote the variance of estimating delay time τ_R by $\sigma_{\hat{\tau}_R}^2$ and use (10.3-19) in (10.2-15) to obtain

$$\sigma_{\hat{\tau}_R}^2 = \sigma_{\hat{\theta}_1}^2 \geq \frac{1}{N\mathscr{R}_p[(\omega_0 + \omega_d + \bar{\omega}_g)^2 + W_{g,\text{rms}}^2]} \tag{10.3-20}$$

In many (probably most) radars $(\omega_0 + \omega_d + \bar{\omega}_g) \gg W_{g,\text{rms}}$ because ω_0 is typically large. With this condition true, a direct use of (10.3-20) indicates an extremely high accuracy (small variance) that is mainly due to the carrier's angular frequency ω_0 and may be called the *fine-grain accuracy*. Because range (or delay) information is highly ambiguous (radar's response peaks in magnitude every half-cycle of ω_0 under an envelope of duration related to the rms bandwidth), the fine-grain accuracy is rarely realized in practice because the carrier's ambiguities are not resolved. Another way of viewing the problem is to note that carrier frequency and phase were presumed known in the analysis leading to (10.3-20), while a practical radar typically discards phase knowledge by using the envelope of $g(t)$ for range delay estimation. In fact other analysis procedures (see Woodward, 1953, pp. 104–112; Burdic, 1968, p. 162) that ignore the fine-grain accuracy give (10.3-20) without the term $(\omega_0 + \omega_d + \bar{\omega}_g)^2$. Thus, for all radars that use only the envelope of the received signal for delay estimation, we may take

$$\sigma_{\hat{\tau}_R}^2 \geq \frac{1}{N \mathcal{R}_p W_{g,\text{rms}}^2} = \sigma_{\hat{\tau}_R(\text{min})}^2 \qquad (10.3\text{-}21)$$

as the variance of estimating τ_R. This result is in agreement with many sources, including Manasse (1960, p. 10), Berkowitz (1965, p. 159), Skolnik (1980, p. 405), and Barton (1988, p. 426). The smallest, or lower-bounding variance (the Cramér-Rao bound), is denoted by $\sigma_{\hat{\tau}_R(\text{min})}^2$.

Example 10.3-2 To obtain some appreciation of numerical values, we determine the value of $\mathcal{R}_p$ needed to give a smallest error variance of not more than 0.04% of the compressed pulse's (unweighted) duration (between -3.92 dB points) squared when 10 linear FM pulses are used.

From Fig. 7.2-2 the compressed pulse's duration is $2\pi/\Delta\omega$, with $\Delta\omega$ the angular frequency sweep in time T. From Problem 8.4-4 $W_{g,\text{rms}} = \mu T/\sqrt{12} = \Delta\omega/\sqrt{12}$, since $\mu = \Delta\omega/T$. Thus, from (10.3-21),

$$\frac{\text{Smallest estimator variance}}{(\text{Pulse duration})^2} = \frac{12}{N \mathcal{R}_p (2\pi)^2} \leq 0.0004$$

so

$$\mathcal{R}_p \geq \frac{3(10^4)}{40\pi^2} = 75.991 \qquad (\text{or } 18.81 \text{ dB})$$

is needed. The rms relative error here is $\sqrt{0.0004} = 0.02$, or 2%.

Spatial Angle Measurement Accuracy

When θ_1 represents the spatial angle θ_x, (10.3-2) applies. When (10.3-2) is used in (10.2-15), we easily obtain the variance of estimating θ_x, which is denoted by $\sigma_{\hat{\theta}_x}^2$,

$$\sigma_{\hat{\theta}_x}^2 \geq \frac{1}{(2/\mathcal{N}_0) K_x^2 N E_r} = \frac{1}{K_x^2 N \mathcal{R}_p} = \sigma_{\hat{\theta}_x(\text{min})}^2 \qquad (10.3\text{-}22)$$

In a similar manner the variance of estimating θ_y, denoted by $\sigma_{\hat{\theta}_y}^2$, is

$$\sigma_{\hat{\theta}_y}^2 \geq \frac{1}{K_y^2 N \mathscr{R}_\mathrm{p}} = \sigma_{\hat{\theta}_y,(\min)}^2 \tag{10.3-23}$$

In (10.3-22) and (10.3-23) we denote the respective lowest (Cramér-Rao) bounds on measurement variances by $\sigma_{\hat{\theta}_x,(\min)}^2$ and $\sigma_{\hat{\theta}_y,(\min)}^2$.

Example 10.3-3   For a specific radar $K_x = 90/\pi$ rad^{-1}. For $N = 10$ and $\mathscr{R}_\mathrm{p} = 8.8$, we find the rms angle error from (10.3-22):

$$\sigma_{\hat{\theta}_x}^2 \geq \frac{\pi^2}{90^2(10)8.8} = 1.385(10^{-5}) \text{ rad}^2$$

The minimum rms angle error is then $\sqrt{1.385(10^{-5})} = 3.721(10^{-3})$ rad or 0.213 degrees.

PROBLEMS

10.1-1 Suppose that a dc voltage θ_1 is an unknown nonrandom parameter added to zero-mean gaussian noise $n(t)$ to produce $y(t) = n(t) + \theta_1$. N samples of $y(t)$ are taken. The noise samples are statistically independent and each has variance σ^2. (a) Find the joint probability density function $p(y_1, \ldots, y_N | \theta_1)$. Let $\hat{\theta}_1$ be the maximum likelihood estimator for θ_1, and (b) determine $\hat{\theta}_1$ by solving for the value of $\theta_1 = \hat{\theta}_1$ such that $p(y_1, \ldots, y_N | \theta_1)$ is maximum. (c) Show that $\hat{\theta}_1$ is unbiased.

★10.1-2 Work Problem 10.1-1, parts (a) and (b), except assume that the variance of the noise samples is a second unknown parameter. That is, assume that $\theta_2 = \sigma^2$ and find the two estimators $\hat{\theta}_1$ and $\hat{\theta}_2$ that jointly maximize $p(y_1, \ldots, y_N | \theta_1, \theta_2)$. Note that this problem is equivalent to finding joint maximum likelihood estimators for the mean and variance of a noise known only to be gaussian when using samples taken far enough apart to be statistically independent. (*Hint:* Note that $p(y_1, \ldots, y_N | \theta_1, \theta_2)$ is maximum when $\ln[p(y_1, \ldots, y_N | \theta_1, \theta_2)]$ is maximum.)

10.2-1 If the estimator of (10.2-11) must give the minimum variance equal to the right side of (10.2-10), find the necessary value of K.

★10.2-2 Show that (10.2-14) is true.

10.2-3 Show that the minimum variance of estimating θ_1, when θ_1 and θ_2 are jointly estimated as in Example 10.2-1, is larger than that when only θ_1 is estimated.

10.3-1 A radar for which $\mathscr{R}_\mathrm{p} = 15.8$ produces a single-pulse unbiased estimate of the amplitude of its received pulse. What is the minimum rms error of its estimate as a percentage of the amplitude being estimated?

10.3-2 If the minimum possible rms error in estimating a signal's amplitude by an unbiased estimator is to not exceed 5% of the true amplitude, what smallest value of $\mathcal{R}_p$ must the system have if $N = 1$ pulse is used?

10.3-3 Work Problem 10.3-2 except assume that $N = 10$ pulses are used.

10.3-4 An unbiased phase estimator must produce a minimum possible rms error of 5 degrees. What value of $\mathcal{R}_p$ is needed for (a) $N = 1$ pulse and (b) $N = 5$ pulses?

10.3-5 Work Problem 10.3-4 except for an rms error of 1 degree.

10.3-6 Substitute (10.3-13) into (10.3-11) and prove that (10.3-14) is true.

10.3-7 A radar's receiver uses an unbiased estimator of angular frequency. The transmitted signal has the form $s(t) = A \, \text{rect}(t/T)\cos(\omega_0 t)$. (a) Find an expression for $\tau_{g,\text{rms}}$. (b) What minimum variance of estimation error occurs if $\mathcal{R}_p = 21$, $N = 34$, and $T = 10 \ \mu s$?

10.3-8 Work Problem 10.3-7 except for the waveform $s(t) = A \, \text{tri}(t/T)\cos(\omega_0 t)$.

10.3-9 Work Problem 10.3-7 except assume that the waveform $s(t) = A \, \text{rect}(t/2T)$ $\cos(\pi t/2T)\cos(\omega_0 t)$.

10.3-10 Work Problem 10.3-7 except assume that the waveform $s(t) = A \, \text{rect}(t/2T)$ $\cos^2(\pi t/2T)\cos(\omega_0 t)$.

10.3-11 A monostatic radar is to be able to measure a target's radial range rate (radial speed) v_r to an rms error of P (per cent). Show that the required value of $\mathcal{R}_p$ is

$$\mathcal{R}_p \geq \frac{(900/16\pi^2)}{P^2 v_r^2 f_{\text{GHz}}^2 \tau_{g,\text{rms}}^2 N}$$

10.3-12 A monostatic radar transmits a pulse with a 20-ms rms duration. It operates at 6.5 GHz, and $\mathcal{R}_p = 14$ in the receiver. What smallest radial target speed v_r is necessary if its radial speed is to be measured to within 5% when using eight pulse intervals. (*Hint*: Use the results of Problem 10.3-11.)

10.3-13 Work Problem 10.3-12 except assume that $f_{\text{GHz}} = 1.0$ and $P = 1\%$.

10.3-14 A monostatic radar is to measure Doppler frequency to within 50 Hz of minimum rms error when $\mathcal{R}_p$ is as small as 18.6 and $N = 1$. (a) What rms waveform duration is needed? (b) For the waveform duration of part (a) if $\mathcal{R}_p$ is increased to 24, what new rms error is possible?

10.3-15 A monopulse radar transmits 1000 pulses per second and cannot devote more than 10% of each pulse interval to transmission of a rectangular pulse. If the system must measure ω_d to a minimum rms error of 100 Hz, what value of $N\mathcal{R}_p$ must be produced by the target in the receiver? (*Hint*: Use $\tau_{g,\text{rms}} = T/\sqrt{12}$ for a rectangular pulse of duration T.)

10.3-16 Work Problem 10.3-15 except assume that 500 pulses/s are transmitted.

★**10.3-17** Use (10.3-1) and show that the last form of (10.3-17) is true.

10.3-18 A monostatic radar uses the waveform of Problem 10.3-8. (a) Determine $W_{g,\text{rms}}$ for this waveform. (b) If $T = 2 \mu s$ and $N\mathcal{R}_p = 157$ in the receiver, what smallest rms error in delay measurement occurs for an unbiased estimator? (c) To what range error does this delay error correspond?

10.3-19 Work Problem 10.3-18 except assume the waveform of Problem 10.3-9.

10.3-20 Work Problem 10.3-18 except assume the waveform of Problem 10.3-10.

10.3-21 A monostatic radar has an antenna pattern with a beamwidth of 3.8 degrees in the vertical plane (radar is surface-based). It is to measure a target spatial angle in the vertical plane to a minimum rms error of 5% of its beamwidth. What minimum value of K_y is necessary if $N\mathcal{R}_p = 12.5$ in the receiver?

10.3-22 Work Problem 10.3-21 except assume a 10% rms error and that $N\mathcal{R}_p = 47.6$.

10.3-23 A particular radar has $K_x = 18.8$ in its spatial angle estimator. An rms error of 0.7 degrees of angle occurs when the received pulse has energy $E_r = 6(10^{-19})$ J. Find the input noise level $\mathcal{N}_0/2$ if $N = 1$ pulse is used for estimation.

11

RANGE MEASUREMENT AND TRACKING IN RADAR

In the preceding chapter we determined the best, or highest, accuracy that could be achieved in measuring a target parameter when noise was the limiting factor. Large, or good, accuracy was associated with small variance of measurement error. The smallest variance of error for an unbiased measurement (the type of measurements we desire in a radar system) of any parameter is the Cramér-Rao bound. In this chapter we are concerned with range as the target parameter. More precisely, we discuss delay measurement, since range is directly found from a delay measurement. Two aspects of the problem are developed.

First, we use intuitive reasoning to introduce a method of delay measurement. This practical method provides a plausible basis for defining an optimum system. In fact three forms of an optimum system are defined, and it will be shown that they realize the best accuracy (smallest, or Cramér-Rao bound on error variance). We also point out that these optimum systems have some practical problems when realization is attempted, and indicate how these may be overcome by practical systems.

Second, we use the modified optimum systems to define several forms of realistic delay measurement and tracking systems.

11.1 RANGE FROM DELAY MEASUREMENTS

Let the range of a target from a monostatic radar be R_0 at time $t = 0$. The delay of the target, denoted by τ_R, is related to R_0 by

$$\tau_R = \frac{2R_0}{c} \tag{11.1-1}$$

where c is the speed of light. The radar's problem is to measure R_0. However, it actually measures τ_R and then uses (11.1-1) to define its measurement of R_0.

461

If $\hat{\tau}_R$ denotes the radar's measurement of τ_R, the corresponding measurement of R_0, denoted by $\hat{R}_0$, is

$$\hat{R}_0 = \frac{c}{2}\hat{\tau}_R \tag{11.1-2}$$

from (11.1-1). Now suppose that we assume the radar's measurements to be unbiased, as systems typically endeavor to realize, so that the variance of measurement error and the mean-squared error are the same. Let $\sigma^2_{\hat{\tau}_R}$ denote the variance of error in the measurement $\hat{\tau}_R$. The variance of error in the measurement $\hat{R}_0$, denoted by $\sigma^2_{\hat{R}_0}$, is

$$\sigma^2_{\hat{R}_0} = \left(\frac{c}{2}\right)^2 \sigma^2_{\hat{\tau}_R} \tag{11.1-3}$$

Because the measurement of range is related directly to the measurement of delay through (11.1-2), and the accuracies (variances) of the errors made in these measurements are also directly related through (11.1-3), we will mainly be concerned in this chapter with delay measurements.

11.2   INTUITIVE DELAY MEASUREMENT USING TIME GATES

Intuition often provides powerful insight into ways of solving problems. Probably more often than not, the methods suggested by intuition will be near the optimum or, at least, will indicate the steps toward realizing the optimum system. So it is with a radar's delay measurement system. In this section we define a delay measurement method based on intuition. It is then used to define an optimum system, which is discussed in the next section.

An Intuitive Delay Measurement Method

A radar can proceed to measure a target's delay once the target has been detected. The process of detection itself provides a rough estimate of delay, which we denote by τ_0. Since the rough estimate is only accurate to approximately the delay resolution of the system, the measurement system's main purpose is to refine the rough estimate. To this end, intuition indicates that the system should produce a response proportional to the difference between the radar's estimate and the true delay τ_R. If this response is denoted by s_0 and the constant of proportionality is denoted by K_τ, the measurement system, for negligible noise, should provide the response.

$$s_0 = K_\tau(\tau_0 - \tau_R) \tag{11.2-1}$$

at least for $|\tau_0 - \tau_R|$ small. This response is shown as the solid line in Fig. 11.2-1. For larger values of $|\tau_0 - \tau_R|$, the curve may assume a different (nonlinear) behavior, as shown by the dashed lines. Our main concern is only that s_0 be linear in the region containing the largest likely values of $|\tau_0 - \tau_R|$.

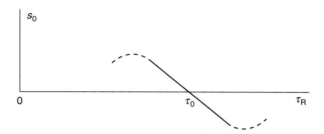

Figure 11.2-1   An intuitively developed receiver response characteristic for delay measurement.

From (11.2-1) we have

$$\tau_R = \tau_0 - \frac{s_0}{K_\tau} \tag{11.2-2}$$

which indicates that the true delay τ_R is found by adding a correction $(-s_0/K_\tau)$ to the rough estimate. Clearly, from Fig. 11.2-1, if $\tau_R = \tau_0$ the "correction" is zero because $s_0 = 0$. Thus for negligible noise the rough estimate τ_0, the real delay τ_R, and the response s_0 are related through (11.2-1).

When there is nonnegligible output noise, denoted by n_0, the response becomes $s_0 + n_0$, and (11.2-1) no longer gives τ_R but gives an estimate (denoted, as usual, by $\hat{\tau}_R$) that can be in error due to the noise. Equation (11.2-1) can now be written as

$$s_0 + n_0 = K_\tau(\tau_0 - \hat{\tau}_R) \tag{11.2-3}$$

The estimate of delay is

$$\hat{\tau}_R = \tau_0 - \frac{s_0}{K_\tau} - \frac{n_0}{K_\tau} = \tau_R - \frac{n_0}{K_\tau} \tag{11.2-4}$$

which is the true delay plus an error $(-n_0/K_\tau)$ due to the noise. The last form of (11.2-4) results from substitution of (11.2-2). If $\Delta\tau$ denotes the delay error due to noise, then

$$\Delta\tau = \hat{\tau}_R - \tau_R = -\frac{n_0}{K_\tau} \tag{11.2-5}$$

We subsequently return to the use of (11.2-4). First, however, we continue the intuitive process to show how the linear error characteristic of Fig. 11.2-1 can be realized. Consider a transmitted pulse with a symmetrical envelope centered at $t = 0$.[1] The occurrence time of the peak of the target's returned pulse is τ_R. Intuition indicates

---

[1] If the pulse is not symmetrical and the time origin is chosen to be the centroid of the transmitted pulse's envelope, the following results hold true.

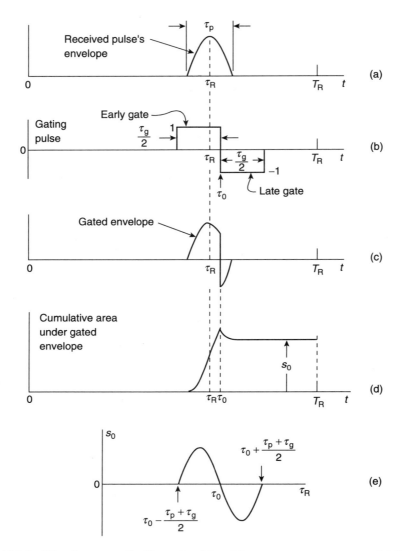

Figure 11.2-2 Waveforms applicable to intuitive delay measurement system: (*a*) Received pulse's envelope, (b) a gating pulse consisting of an early gate and a late gate, (*c*) the product of the waveforms of (*a*) and (*b*), (*d*) the cumulative (integrated) area under the waveform of (*c*), and (*e*) a plot of the area of (*d*) when $\tau_0 - \tau_R$ varies.

that τ_R can be estimated by using a gating pulse[2] and integrator with waveforms defined as in Fig. 11.2-2 for one pulse interval of duration T_R. The received pulse's envelope is shown in panel a; the time interval of the most important amplitudes is designated as τ_p (for a compressed pulse with small sidelobes, τ_p might represent the

---

[2] The name "gating" pulse derives from the practical circuit where the pulse acts to either turn on a switch (pulse present) that connects, or "gates," the input to the output, or turn the switch off (pulse absent) which disconnects the input and output.

main response lobe's duration). The gating pulse of panel b consists of an *early gate* of duration $\tau_g/2$ and a *late gate* of duration $\tau_g/2$. The gating pulse is centered at delay τ_0, and the late gate is the negative of the early gate. Gate duration τ_g is larger than τ_p.

The product of the received pulse's envelope and the gating pulse is sketched in Fig. 11.2-2c, where we assume that $\tau_g \geq 2\tau_p$ for convenience of illustration. The instantaneous response of the integrator acting on the waveform of panel c is shown in panel d. For $t \geq \tau_R + (\tau_p/2)$ this response is ideally constant at a value denoted by s_0. Clearly s_0 is equal to the net area under the gated envelope. For $\tau_0 = \tau_R$, s_0 is zero; for $\tau_0 > \tau_R$, s_0 is positive, and it is negative for $\tau_0 < \tau_R$. For small values of $|\tau_0 - \tau_R|$, the response s_0 is proportional to $\tau_0 - \tau_R$ as sketched in panel e. For larger values of $\tau_0 - \tau_R$, the response s_0 will have odd symmetry for the assumed symmetrical pulse envelope, or have roughly odd symmetry for most other radar pulse shapes. For $|\tau_0 - \tau_R| > (\tau_p + \tau_g)/2$, s_0 returns to zero for a symmetrical envelope. These behaviors are shown in panel e.

By comparing Fig. 11.2-2e with Fig. 11.2-1, it is clear that the "early–late gate" system gives a desirable measurement characteristic. This response is sometimes called a *time discriminator characteristic*, and the combination of gating pulse stage and integrator is often called a *time discriminator*.

Time Discriminator System

For a time discriminator the central response region where $|\tau_0 - \tau_R|$ and the error in $\hat{\tau}_R$ due to noise are not too large, the discriminator's characteristic can be approximated as linear, as defined by (11.2-1) when noise is negligible, or by (11.2-3) when noise is not negligible. In either case the important slope constant is defined by

$$K_\tau = \frac{ds_0}{d\tau_0}\bigg|_{\tau_0 = \tau_R} = -\frac{ds_0}{d\tau_R}\bigg|_{\tau_R = \tau_0} \tag{11.2-6}$$

Here K_τ is the measurement system's constant of proportionality. This parameter is most important in defining the accuracy of delay measurements when noise is present, as developed in the next section.

A delay measurement system that makes use of the above-described time discriminator is illustrated in Fig. 11.2-3. In this system the gating pulse is separated into its early and late gate components and the sign of the latter is accounted for by the differencing junction. Prior to the discriminator a matched filter maximizes the pulse's signal-to-noise ratio at time τ_R, and an envelope detector removes the carrier's phase. The exact shape of the time discriminator's characteristic s_0 will depend on the shape and duration of the envelope of the pulse from the matched filter, the type of detector (linear, square-law, etc.), and the shape and duration of the gating pulse. Regardless of these details, however, there will be some applicable response s_0 that will determine the slope constant K_τ defined by (11.2-6).

11.3 OPTIMUM DELAY MEASUREMENT SYSTEM

We next use the intuitive concepts introduced in the foregoing sections as a starting point for the definition of an optimum delay measurement system for any one-pulse

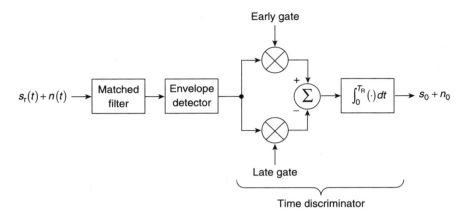

Figure 11.2-3 Delay measurement system based on an intuitively developed early–late gating method for centroid estimation.

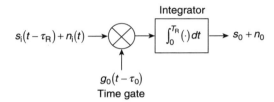

Figure 11.3-1 Time gate and integrator that define the optimum time discriminator portion of a delay measurement system.

interval. The optimum system is defined as that which contains an optimum time gate and integrator, as depicted in Fig. 11.3-1. The gate-integrator, which is just a generalized time discriminator, is optimum when the time gate $g_0(t - \tau_0)$ is chosen to produce the smallest possible variance of error in the measurement of target delay τ_R, when driven by the sum of a signal $s_i(t - \tau_R)$ and noise $n_i(t)$. The input $s_i(t - \tau_R) + n_i(t)$ can originate directly from the radar's antenna, or it could be from another point after some signal processing, such as the output of a matched filter. Whatever the source of the input the system is optimized *for that input* by choice of the time gate $g_0(t - \tau_0)$. In subsequent work the optimum system is developed following mainly the procedures of Mallinckrodt and Sollenberger (1954), although some steps have been modified slightly.

Optimum Accuracy

The time gate in Fig. 11.3-1 is some function $g_0(t)$ that is shifted to the time τ_0 of the rough estimate of delay. Delay τ_0 differs from the true target's delay τ_R by a small "error" defined as $\tau_0 - \tau_R$. The function $g_0(t)$ is not restricted to baseband, and in general, it can be bandpass.

At any time after the target's pulse has arrived, the integrator's *signal* response in Fig. 11.3-1 is ideally constant at a value s_0. For any given value of $\tau_0 - \tau_R$, s_0 will be accompanied by a noise n_0. As previously described, s_0 is related to τ_0 and τ_R by (11.2-1), while the noise contributes an error $\Delta\tau$ to the estimate $\hat{\tau}_R$ of true delay τ_R according to (11.2-5). On squaring both sides of (11.2-5) and taking the statistical average, we get

$$E[(\Delta\tau)^2] = \frac{E[n_0^2]}{K_\tau^2} \tag{11.3-1}$$

For the small errors assumed here, the estimator $\hat{\tau}_R$ is unbiased from (11.2-5), since the noise is assumed to have zero mean. Hence the variance of $\hat{\tau}_R$ and the mean-squared error are the same, and we write (11.3-1) as

$$\sigma_{\hat{\tau}_R}^2 = E[(\hat{\tau}_R - \tau_R)^2] = E[(\Delta\tau)^2] = \frac{E[n_0^2]}{K_\tau^2} \tag{11.3-2}$$

To reduce (11.3-2), we first determine K_τ. For any one-pulse interval the limits of the integrator in Fig. 11.3-1 completely span the input signal's waveform for any τ_R. For convenience these limits can be replaced by $-\infty$ to ∞. We use this limit substitution and define $S_i(\omega)$ and $G_0(\omega)$ as the Fourier transforms of $s_i(t)$ and $g_0(t)$, respectively, to write the signal response s_0 as

$$
\begin{aligned}
s_0 &= \int_{-\infty}^{\infty} s_i(t - \tau_R) g_0(t - \tau_0)\, dt \\
&= \int_{-\infty}^{\infty} \frac{1}{2\pi} \int_{-\infty}^{\infty} S_i(\omega) e^{j\omega(t-\tau_R)}\, d\omega \frac{1}{2\pi} \int_{-\infty}^{\infty} G_0(\eta) e^{j\eta(t-\tau_0)}\, d\eta\, dt \\
&= \frac{1}{2\pi} \int_{-\infty}^{\infty} S_i(\omega) e^{-j\omega\tau_R} \frac{1}{2\pi} \int_{-\infty}^{\infty} G_0(\eta) e^{-j\eta\tau_0} \int_{-\infty}^{\infty} e^{j(\omega+\eta)t}\, dt\, d\eta\, d\omega
\end{aligned}
\tag{11.3-3}
$$

The rightmost integral in (11.3-3) is recognized as $2\pi\delta(\omega + \eta)$ from (A.3-11). The integral over η is then easily evaluated using the definition (A.3-1) of the impulse function. Finally, on observing that $G_0(-\omega) = G_0^*(\omega)$ when $g_0(t)$ is real, as assumed here, s_0 becomes (Problem 11.3-1)

$$s_0 = \frac{1}{2\pi} \int_{-\infty}^{\infty} S_i(\omega) G_0^*(\omega) e^{j\omega(\tau_0 - \tau_R)}\, d\omega \tag{11.3-4}$$

By differentiating (11.3-4) according to (11.2-6), we have

$$K_\tau = \frac{ds_0}{d\tau_0}\bigg|_{\tau_0 = \tau_R} = \frac{1}{2\pi} \int_{-\infty}^{\infty} j\omega S_i(\omega) G_0^*(\omega)\, d\omega \tag{11.3-5}$$

Next, consider the response noise n_0 in Fig. 11.3-1:

$$n_0 = \int_0^{T_R} n_i(t)g_0(t - \tau_0)dt \tag{11.3-6}$$

The mean-squared value of n_0 is the noise power:

$$E[n_0^2] = E\left\{ \int_0^{T_R} n_i(t)g_0(t - \tau_0)dt \int_0^{T_R} n_i(\xi)g_0(\xi - \tau_0)d\xi \right\}$$

$$= \int_0^{T_R} g_0(t - \tau_0) \int_0^{T_R} g_0(\xi - \tau_0)E[n_i(t)n_i(\xi)]d\xi\, dt \tag{11.3-7}$$

The expectation is recognized as the autocorrelation function of $n_i(t)$, denoted here by $R_{N_i N_i}(t - \xi)$, which is the inverse Fourier transform of the noise's power spectrum, denoted by $\mathscr{S}_{N_i N_i}(\omega)$, if we assume that $n_i(t)$ is at least wide-sense stationary. Thus we can write (11.3-7) in the form

$$E[n_0^2] = \frac{1}{2\pi} \int_{-\infty}^{\infty} \mathscr{S}_{N_i N_i}(\omega) \int_0^{T_R} g_0(t - \tau_0)e^{j\omega t}dt \int_0^{T_R} g_0(\xi - \tau_0)e^{-j\omega\xi}d\xi\, d\omega \tag{11.3-8}$$

Since the limits on the second and third integrals completely span their integrands for any possible value of τ_0, they can be replaced by $-\infty$ to ∞. These integrals are then recognized as $G_0(-\omega)\exp(j\omega\tau_0)$ and $G_0(\omega)\exp(-j\omega\tau_0)$, respectively. Again, using $G_0(-\omega) = G_0^*(\omega)$ for real $g_0(t)$, we have

$$E[n_0^2] = \frac{1}{2\pi} \int_{-\infty}^{\infty} \mathscr{S}_{N_i N_i}(\omega)|G_0(\omega)|^2 d\omega \tag{11.3-9}$$

Equations (11.3-9) and (11.3-5) are now substituted into (11.3-2) to obtain $\sigma_{\hat{\tau}_R}^2$. Our goal is to minimize $\sigma_{\hat{\tau}_R}^2$, which is equivalent to *maximizing* the quantity

$$\frac{1}{\sigma_{\hat{\tau}_R}^2} = \frac{K_\tau^2}{E[n_0^2]} = \frac{\left| \dfrac{1}{2\pi} \displaystyle\int_{-\infty}^{\infty} j\omega S_i(\omega)G_0^*(\omega)d\omega \right|^2}{\dfrac{1}{2\pi} \displaystyle\int_{-\infty}^{\infty} \mathscr{S}_{N_i N_i}(\omega)|G_0(\omega)|^2 d\omega} \tag{11.3-10}$$

To maximize (11.3-10), we use Schwarz's inequality with

$$A(\omega) = \frac{1}{\sqrt{2\pi}} \sqrt{\mathscr{S}_{N_i N_i}(\omega)}\, G_0^*(\omega) \tag{11.3-11}$$

$$B(\omega) = \frac{j\omega S_i(\omega)}{\sqrt{2\pi}\sqrt{\mathscr{S}_{N_i N_i}(\omega)}} \tag{11.3-12}$$

We have

$$\left| \frac{1}{2\pi} \int_{-\infty}^{\infty} j\omega S_i(\omega) G_0^*(\omega) d\omega \right|^2 = \left| \int_{-\infty}^{\infty} A(\omega) B(\omega) d\omega \right|^2$$

$$\leq \int_{-\infty}^{\infty} |A(\omega)|^2 d\omega \int_{-\infty}^{\infty} |B(\omega)|^2 d\omega \qquad (11.3\text{-}13)$$

$$= \frac{1}{2\pi} \int_{-\infty}^{\infty} \mathscr{S}_{N_i N_i}(\omega) |G_0(\omega)|^2 d\omega \frac{1}{2\pi} \int_{-\infty}^{\infty} \frac{\omega^2 |S_i(\omega)|^2}{\mathscr{S}_{N_i N_i}(\omega)} d\omega$$

On using (11.3-13) with (11.3-10), the error variance becomes

$$\sigma_{\hat{\tau}_R}^2 \geq \sigma_{\hat{\tau}_R(min)}^2 = \left\{ \frac{1}{2\pi} \int_{-\infty}^{\infty} \frac{\omega^2 |S_i(\omega)|^2}{\mathscr{S}_{N_i N_i}(\omega)} d\omega \right\}^{-1} \qquad (11.3\text{-}14)$$

Optimum Time Gate

From Schwarz's inequality we know that the equality in (11.3-14), which defines the minimum achievable variance, occurs only if $A(\omega) = C_g B^*(\omega)$, where C_g is any real nonzero, but otherwise arbitrary, constant. From (11.3-11) and (11.3-12) we solve for $G_0(\omega)$:

$$G_0(\omega) = C_g \frac{j\omega S_i(\omega)}{\mathscr{S}_{N_i N_i}(\omega)} \qquad (11.3\text{-}15)$$

which is the Fourier transform of the optimum time gate $g_0(t)$.

We summarize our developments. For some specified input signal $s_i(t - \tau_R)$ and noise $n_i(t)$ in Fig. 11.3-1, the time gate $g_0(t)$ that yields the minimum variance of estimating target delay τ_R, as given by (11.3-14), has the Fourier transform defined by (11.3-15). Further definition of the optimum system requires specification of $s_i(t - \tau_R)$ and $n_i(t)$. We consider several cases in the following sections.

11.4   OPTIMUM WIDEBAND RECEIVER

As a first example of an optimum system, we let $s_i(t - \tau_R)$ and $n_i(t)$ of Fig. 11.3-1 equal, respectively, the received signal and noise directly out of the receiving antenna. We refer to this case as a wideband receiver, since there is no bandwidth restriction on the receiver as it feeds the optimum gate-integrator. Then

$$s_i(t - \tau_R) = s_r(t) \qquad (11.4\text{-}1)$$

$$n_i(t) = n(t) \qquad (11.4\text{-}2)$$

where $s_r(t)$ is given by (10.3-1) and $n(t)$ is white noise. If $S_i(\omega)$ and $S_r(\omega)$ are the Fourier transforms of $s_i(t)$ and $s_r(t)$, respectively, then from (11.4-1),

$$S_i(\omega) = S_r(\omega)e^{j\omega\tau_R} \tag{11.4-3}$$

The power spectra of $n_i(t)$ and $n(t)$, as respectively denoted by $\mathscr{S}_{N_iN_i}(\omega)$ and $\mathscr{S}_{NN}(\omega)$, are given by

$$\mathscr{S}_{N_iN_i}(\omega) = \mathscr{S}_{NN}(\omega) = \frac{\mathcal{N}_0}{2}, \qquad -\infty < \omega < \infty \tag{11.4-4}$$

where $\mathcal{N}_0$ is a constant.

Optimum Gate

On substitution of (11.4-3) and (11.4-4) into (11.3-15), the spectrum of the optimum time gate is found to be

$$G_0(\omega) = \frac{2C_g}{\mathcal{N}_0} j\omega S_r(\omega)e^{j\omega\tau_R} \tag{11.4-5}$$

This expression is inverse transformed by use of the time differentiation and time-shifting properties of Fourier transforms (Appendix B):

$$g_0(t) = \frac{2C_g}{\mathcal{N}_0} \frac{ds_r(t + \tau_R)}{dt} \tag{11.4-6}$$

This result indicates that the optimum time gate is proportional to the derivative of the received waveform from the target. Since $s_r(t)$ is a bandpass waveform, so is the gate $g_0(t)$.

For the special case where the received signal has no Doppler frequency, or the Doppler frequency has been removed by a mixer,[3] $s_r(t) = \alpha s(t - \tau_R)$, and it can be shown that (11.4-6) becomes

$$g_0(t) = \frac{2C_g\alpha}{\mathcal{N}_0} \frac{ds(t)}{dt} \tag{11.4-7}$$

where $s(t)$ is the transmitted waveform.

Signal and Noise Responses

For later use, it is of interest to find the signal response s_0 for the optimum system. From the operations of Fig. 11.3-1, (11.4-1), and the optimum gate of (11.4-6), we can

---

[3] Doppler removal is possible if there is a Doppler measurement system such that Doppler frequency is known. Doppler frequency measurement is discussed in Chapter 12.

derive (Problem 11.4-3)

$$s_0 = \frac{2C_g}{\mathcal{N}_0} \int_0^{T_R} s_r(t) \frac{ds_r(t - \tau_0 + \tau_R)}{dt} dt \tag{11.4-8}$$

Similarly, for later use, the output noise power of the optimum system is found to be (Problem 11.4-4)

$$E[n_0^2] = \frac{2C_g^2}{\mathcal{N}_0} \int_0^{T_R} \left[\frac{ds_r(t - \tau_0 + \tau_R)}{dt} \right]^2 dt \tag{11.4-9}$$

An alternative form of (11.4-9) derives from Parseval's theorem:

$$E[n_0^2] = \frac{C_g^2}{\pi \mathcal{N}_0} \int_{-\infty}^{\infty} \omega^2 |S_r(\omega)|^2 d\omega \tag{11.4-10}$$

(Problem 11.4-5). Based on the power formula for noise (Appendix C), it is clear from (11.4-10) that the power spectrum of the output noise is

$$\mathcal{S}_{N_0 N_0}(\omega) = \frac{2C_g^2}{\mathcal{N}_0} \omega^2 |S_r(\omega)|^2 \tag{11.4-11}$$

Noise Performance

The optimum gate of (11.4-6) must realize the minimum variance of measurement error defined by (11.3-14). We next evaluate this minimum variance and show that it equals the Cramér-Rao bound, so that no other unbiased estimator exists with better performance.

We substitute (11.4-3) and (11.4-4) into (11.3-14) to obtain

$$\sigma_{\hat{\tau}_R (\min)}^2 = \frac{1}{\frac{1}{\pi \mathcal{N}_0} \int_{-\infty}^{\infty} \omega^2 |S_r(\omega)|^2 d\omega} \tag{11.4-12}$$

This expression is reduced in Problem 11.4-7 where it is found, for $\omega_0 - |\omega_d| \gg$ [bandwidth of $g(t)$], that

$$\sigma_{\hat{\tau}_R (\min)}^2 = \frac{1}{\mathcal{R}_p [W_{g, \text{rms}}^2 + (\omega_0 + \omega_d + \bar{\omega}_g)^2]} \tag{11.4-13}$$

where

$$\mathcal{R}_p = \frac{2E_r}{\mathcal{N}_0} \tag{11.4-14}$$

with E_r being the energy in one pulse of $s_r(t)$. Other parameters are $W_{g,\text{rms}}$ is the rms bandwidth of $g(t)$ as defined by (6.4-14); ω_d is the Doppler angular frequency present in $s_r(t)$; $\bar{\omega}_g$ is the mean frequency of $g(t)$ defined by (6.4-9); and $g(t)$ is the complex envelope of the transmitted waveform $s(t)$, as defined by (6.1-6). We note that the variance of delay estimation error given by (11.4-13) for the optimum delay measurement system is equal to the right side of (10.3-20) which is the Cramér-Rao (lowest) bound for all unbiased estimators for $N = 1$ pulse interval, as discussed here. Thus the optimum wideband delay estimator achieves the lowest variance of error (the Cramér-Rao bound).

Some qualifying comments regarding the optimum system of this section are appropriate. Since the received signal $s_r(t)$ is bandpass with a "carrier" angular frequency $\omega_0 + \omega_d$, the optimum time gate of (11.4-6) is bandpass. Its use can, theoretically, lead to the error variance defined by (11.4-13). However, because the carrier frequency is typically large in radar, the ambiguities in delay due to the carrier can rarely be resolved. The consequence is that the accuracy offered by the term $(\omega_0 + \omega_d + \bar{\omega}_g)^2$ is not realized. The reason is that practical systems (as we later discuss) typically use an envelope detector, which essentially discards phase information, such that the only term in (11.4-13) that is effective is that due to the rms bandwidth ($W_{g,\text{rms}}$) of the complex envelope $g(t)$. A system with an envelope detector is analogous (but not identical) to a system with zero carrier ($\omega_0 + \omega_d = 0$). More is said on this subject in Section 11.7.

11.5   OPTIMUM RECEIVER WITH MATCHED FILTER

As a second example of an optimum delay measurement system, let the optimum gate-integrator be designed to work with its input equal to the response of a matched filter for the received pulse in any one pulse interval. The system's block diagram is given in Fig. 11.5-1.

Time Discriminator's Inputs

The signal input $s_r(t)$ in Fig. 11.5-1 is defined by (10.3-1); we define its Fourier transform as $S_r(\omega)$. The input noise $n(t)$ is assumed white, with a power spectrum given by (11.4-4). Since the filter is assumed matched to $s_r(t)$, its transfer function is

$$H_{\text{opt}}(\omega) = \frac{2C}{\mathcal{N}_0} S_r^*(\omega) e^{-j\omega t_0} \tag{11.5-1}$$

from (6.7-3), where C is an arbitrary nonzero real constant and t_0 is the (selectable) time at which the peak signal-to-noise ratio occurs at the filter's output.

If $S_i(\omega)$ denotes the Fourier transform of $s_i(t)$, then the input signal to the time discriminator is defined by

$$s_i(t - \tau_R) \leftrightarrow S_i(\omega) e^{-j\omega\tau_R} \tag{11.5-2}$$

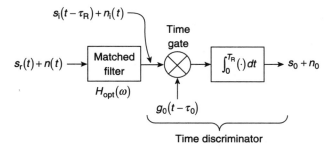

Figure 11.5-1 An optimum delay measurement system that uses a matched filter.

Since the product of the input signal's spectrum $S_r(\omega)$ and $H_{opt}(\omega)$ must equal the spectrum of $s_i(t - \tau_R)$, as given by the right side of (11.5-2), we find that

$$S_i(\omega) = \frac{2C}{\mathcal{N}_0}|S_r(\omega)|^2 e^{-j\omega(t_0 - \tau_R)} \qquad (11.5\text{-}3)$$

Finally we choose $t_0 = \tau_R$ so that the matched filter's peak response occurs at the delay time τ_R of the target's signal:

$$S_i(\omega) = \frac{2C}{\mathcal{N}_0}|S_r(\omega)|^2 \qquad (11.5\text{-}4)$$

The power spectrum of $n_i(t)$ is readily found to be $(\mathcal{N}_0/2)|H_{opt}(\omega)|^2$ from the theory of random processes (Appendix C). On use of (11.5-1), we have

$$\mathcal{S}_{N_iN_i}(\omega) = \frac{2C^2}{\mathcal{N}_0}|S_r(\omega)|^2 \qquad (11.5\text{-}5)$$

Having found spectra to describe signal and noise inputs to the time discriminator, we next determine the optimum gate function.

Optimum Time Gate

The optimum time gate for any input signal $s_i(t - \tau_R)$ and wide sense stationary noise $n_i(t)$ was found to have the spectrum of (11.3-15). For the present input definitions, we use (11.5-4) and (11.5-5) to obtain

$$G_0(\omega) = \frac{C_g}{C}j\omega \qquad (11.5\text{-}6)$$

where, of course, C_g/C is just an arbitrary real constant because both C_g and C were arbitrary and real.

Equation (11.5-6) is recognized as the Fourier transform of the derivative of an impulse function. Thus

$$g_0(t) = \frac{C_g}{C} \frac{d\delta(t)}{dt} = \frac{C_g}{C} \delta^{(1)}(t) \tag{11.5-7}$$

where $\delta^{(1)}(t)$ represents the first derivation of $\delta(t)$; it is sometimes called a *unit doublet*.

It may seem strange that the optimum gate is the derivative of an impulse. Perhaps some comments can help the reader derive some physical justification from (11.3-15), which applies for any inputs to the time discriminator. The presence of the factor $j\omega$ implies that $g_0(t)$ is the time derivative of the waveform having the Fourier transform $[S_i(\omega)/\mathcal{S}_{N_iN_i}(\omega)]$. This waveform is simply proportional to the derivative of $s_i(t)$ when the noise is white. However, as the power spectrum $\mathcal{S}_{N_iN_i}(\omega)$ becomes more shaped; its reciprocal tends to "crispen," or enhance, higher frequencies in $S_i(\omega)$,[4] which tends to narrow the waveform having the spectrum $S_i(\omega)/\mathcal{S}_{N_iN_i}(\omega)$. Thus $g_0(t)$ remains a derivative, but of a function that tends to become narrower in time. In the present special case the shapes of $\mathcal{S}_{N_iN_i}(\omega)$ and $S_i(\omega)$ become the same, so their ratio is a constant, corresponding to the spectrum of an impulse.

System Responses

It can be shown (Problem 11.5-1) that s_0, $E[n_0^2]$, and the power spectrum of n_0 for the system of Fig. 11.5-1, having the optimum gate of (11.5-7), are, respectively, given by (11.4-8), (11.4-10), and (11.4-11). Since these results are identical to those of the wideband system, which previously was shown to realize the Cramér-Rao lowest error bound on estimation error, they serve as confirmations of the fact that both the wideband and matched filter systems are optimum.

11.6 OPTIMUM RECEIVER WITH DIFFERENTIATOR

As a third example of an optimum delay measurement system, we consider that of Fig. 11.6-1. The input to this system is assumed to be the radar antenna's output. That is, $s_r(t)$ is defined by (10.3-1). The noise $n(t)$ is assumed to be white with a constant power spectrum as defined by (11.4-4). The matched filter has the transfer function of (11.5-1) so we can easily obtain its impulse response by inverse Fourier transformation:

$$h_{\text{opt}}(t) = \frac{2C}{\mathcal{N}_0} s_r(t_0 - t) \leftrightarrow H_{\text{opt}}(\omega) = \frac{2C}{\mathcal{N}_0} S_r^*(\omega) e^{-j\omega t_0} \tag{11.6-1}$$

---

[4] For bandpass $S_i(\omega)$ this crispening is at frequencies displaced by larger amounts to each side of the "carrier" frequency.

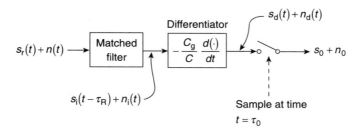

Figure 11.6-1 An optimum delay measurement system that uses a matched filter and differentiator.

Both the matched filter's signal response and noise response are given by appropriate convolutions:

$$s_i(t - \tau_R) = \frac{2C}{\mathcal{N}_0} \int_{-\infty}^{\infty} s_r(\xi) s_r(\xi + t_0 - t) d\xi \qquad (11.6\text{-}2)$$

$$n_i(t) = \frac{2C}{\mathcal{N}_0} \int_{-\infty}^{\infty} n(\xi) s_r(\xi + t_0 - t) d\xi \qquad (11.6\text{-}3)$$

Signal Output

On using (11.6-2), the differentiator's output signal becomes

$$s_d(t) = \frac{-2C_g}{\mathcal{N}_0} \int_{-\infty}^{\infty} s_r(\xi) \frac{ds_r(t_0 - t + \xi)}{dt} d\xi \qquad (11.6\text{-}4)$$

On choosing t_0 to produce a matched filter maximum output at $t_0 = \tau_R$ and using proper variable changes, we can rewrite (11.6-4) as

$$s_d(t) = \frac{2C_g}{\mathcal{N}_0} \int_{-\infty}^{\infty} s_r(\xi) \frac{ds_r(\xi + \tau_R - t)}{d\xi} d\xi \qquad (11.6\text{-}5)$$

This result is valid for any time t. The final signal output in Fig. 11.6-1 is the value of (11.6-5) at $t = \tau_0$, the time the receiver believes to be the target's delay time:

$$s_0 = \frac{2C_g}{\mathcal{N}_0} \int_{-\infty}^{\infty} s_r(\xi) \frac{ds_r(\xi - \tau_0 + \tau_R)}{d\xi} d\xi \qquad (11.6\text{-}6)$$

Noise Response

In a manner similar to the signal output's calculation, we have

$$n_d(t) = \frac{2C_g}{\mathcal{N}_0} \int_{-\infty}^{\infty} n(\xi) \frac{ds_r(\xi + \tau_R - t)}{d\xi} d\xi \qquad (11.6\text{-}7)$$

where $t_0 = \tau_R$ has again been assumed. The overall noise in the output at $t = \tau_0$ is

$$n_0 = n_d(\tau_0) = \frac{2C_g}{\mathcal{N}_0} \int_{-\infty}^{\infty} n(\xi) \frac{ds_r(\xi - \tau_0 + \tau_R)}{d\xi} d\xi \qquad (11.6\text{-}8)$$

Bound on Measurement Error

We demonstrate that the measurement system of Fig. 11.6-1 is optimum by showing that the mean-squared measurement error, which is given by (11.3-2) in general, is equal to the Cramér-Rao bound.

From (11.6-6) the time discriminator's slope is

$$K_\tau = \frac{ds_0}{d\tau_0}\bigg|_{\tau_0 = \tau_R} = \frac{2C_g}{\mathcal{N}_0} \int_{-\infty}^{\infty} s_r(\xi) \frac{d^2 s_r(\xi - \tau_0 + \tau_R)}{d\xi d\tau_0}\bigg|_{\tau_0 = \tau_R} d\xi$$

$$= -\frac{2C_g}{\mathcal{N}_0} \int_{-\infty}^{\infty} s_r(\xi) \frac{d^2 s_r(\xi)}{d\xi^2} d\xi \qquad (11.6\text{-}9)$$

The overall output noise power is, from (11.6-8),

$$E[n_0^2] = \left(\frac{2C_g}{\mathcal{N}_0}\right)^2 \int_{-\infty}^{\infty}\int_{-\infty}^{\infty} E[n(\xi)n(\beta)] \frac{ds_r(\xi - \tau_0 + \tau_R)}{d\xi} \frac{ds_r(\beta - \tau_0 + \tau_R)}{d\beta} d\beta d\xi$$

$$= \left(\frac{2C_g}{\mathcal{N}_0}\right)^2 \int_{-\infty}^{\infty} \frac{ds_r(\xi - \tau_0 + \tau_R)}{d\xi} \frac{\mathcal{N}_0}{2} \int_{-\infty}^{\infty} \delta(\beta - \xi) \frac{ds_r(\beta - \tau_0 + \tau_R)}{d\beta} d\beta d\xi$$

$$= \frac{2C_g^2}{\mathcal{N}_0} \int_{-\infty}^{\infty} \left[\frac{ds_r(\xi - \tau_0 + \tau_R)}{d\xi}\right]^2 d\xi \qquad (11.6\text{-}10)$$

By using Parseval's theorem, (11.6-10) can be written as

$$E[n_0^2] = \frac{C_g^2}{\pi \mathcal{N}_0} \int_{-\infty}^{\infty} \omega^2 |S_r(\omega)|^2 d\omega \qquad (11.6\text{-}11)$$

where $S_r(\omega)$ is the Fourier transform of $s_r(t)$.

Similarly Parseval's theorem allows (11.6-9) to be written as

$$K_\tau = \frac{C_g}{\pi \mathcal{N}_0} \int_{-\infty}^{\infty} \omega^2 |S_r(\omega)|^2 d\omega \qquad (11.6\text{-}12)$$

We substitute this expression and (11.6-11) into (11.3-2) to obtain

$$\sigma_{\hat{\tau}_R}^2 = \frac{E[n_0^2]}{K_\tau^2} = \frac{1}{\left(\dfrac{2}{\mathcal{N}_0}\right) \dfrac{1}{2\pi} \displaystyle\int_{-\infty}^{\infty} \omega^2 |S_r(\omega)|^2 d\omega} \qquad (11.6\text{-}13)$$

This expression is identical to the right side of (11.4-12) which was previously shown to equal the Cramér-Rao bound on error variance of (11.4-13). Thus the system of Fig. 11.6-1 is optimum, and it achieves the smallest possible value of variance of error in delay measurement.

11.7 PRACTICAL DELAY MEASUREMENT AND TRACKING

Any of the preceding systems, the intuitive as well as the optimum, can form a basis for constructing a practical delay measurement method. In this section we define and discuss some realistic systems.

The main difference between the optimum systems discussed earlier and practical systems is in the use of the carrier's phase. Although the optimum (wideband, matched filter, and differentiator) systems all suggest extremely low delay measurement error variance because of the large carrier term in (11.4-13), this result is not achieved in practice because of carrier-caused delay ambiguities that are difficult to resolve. What the practical systems usually achieve is (11.4-13) with the carrier term absent, that is, with $\omega_0 + \omega_d + \bar{\omega}_g$ set equal to zero. In fact one approach to describing accuracy *assumes* a "zero-frequency IF" (see Barton and Ward, 1969, p. 82).

Practical systems can achieve the "zero-carrier" effect in mainly two ways. In one, an envelope detector is used to remove carrier phase. Systems that use an envelope detector can only approach the accuracy defined by the Cramér-Rao bound for large signal-to-noise ratios; these systems are subsequently described more carefully. In the following subsection we briefly describe the second of the two main ways of achieving the "zero-carrier" effect. Here a phase-coherent detector is used at some point in the receiving path to remove the effects of the carrier. The detector's output is baseband (video). The error variance when using a coherent detector can equal the Cramér-Rao bound of (11.4-13) with $\omega_0 + \omega_d = 0$, but proper performance requires the radar, target, and medium combine to produce target pulse returns that are phase coherent.

Systems with a Coherent Detector

A *coherent detector* is basically a product device followed by a wideband low-pass real filter (WBLPF) as shown in Fig. 11.7-1. A signal $s_1(t)$ and noise[5] $n_1(t)$ are multiplied by a local oscillator signal having a phase $\phi_{LO}(t)$ that is the same as (coherent with) the phase term present in $s_1(t)$ due to its carrier. The WBLPF serves only to remove any signal or noise components in the output of the product device that are not baseband. In general, the bandwidth of the WBLPF can be as large as half the carrier's frequency, although from a practical standpoint it only needs to be large enough to pass the baseband component of $s_p(t)$ with negligible distortion.

The general analysis of the coherent detector of Fig. 11.7-1 is done in Problems 11.7-1 and 11.7-2. There the input signal has the general form

$$s_1(t) = a_1(t)\cos[\omega_1 t + \theta_1(t) + \phi_1] \qquad (11.7\text{-}1)$$

---

[5] Noise $n_1(t)$ is assumed at least wide-sense stationary.

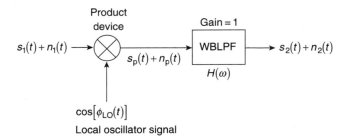

Figure 11.7-1 Components that constitute a coherent detector.

where $a_1(t)$ and $\theta_1(t)$ are terms that account for amplitude and phase modulations, respectively, while $\omega_1 t + \phi_1$ is the carrier's phase. For a local oscillator signal defined by

$$\phi_{\text{LO}}(t) = \omega_1 t + \phi_1 \tag{11.7-2}$$

where ω_1 is large compared to the bandwidth of $s_1(t)$, it is found in Problem 11.7-1 that

$$s_2(t) = \tfrac{1}{2} a_1(t) \cos[\theta_1(t)] = \tfrac{1}{2} s_1(t)|_{\omega_1 t + \phi_1 = 0} \tag{11.7-3}$$

In other words, the response $s_2(t)$ is half the input signal with its carrier's phase set to zero.

Next let $\mathcal{S}_{N_1 N_1}(\omega)$ represent the power spectrum of the bandpass noise $n_1(t)$; it has components centered at $\pm \omega_1$. If the spectral extent of $\mathcal{S}_{N_1 N_1}(\omega)$ is not greater than ω_1, then the power spectrum of $\mathcal{S}_{N_2 N_2}(\omega)$ of $n_2(t)$ is found in Problem 11.7-2 to be

$$\mathcal{S}_{N_2 N_2}(\omega) = \tfrac{1}{4}[\mathcal{S}_{N_1 N_1}(\omega - \omega_1) + \mathcal{S}_{N_1 N_1}(\omega + \omega_1)]|H(\omega)|^2 \tag{11.7-4}$$

The presence of the squared magnitude of the transfer function $H(\omega)$ of the WBLPF serves to preserve only the baseband portion of the power spectrum of the noise from the product device in Fig. 11.7-1.

Equation (11.7-3) shows that the coherent detector can remove the carrier from a signal anywhere in the receiving path that is desired [since $s_1(t)$ was arbitrary]. Thus, in principle, a coherent detector can be used at RF or IF prior to any significant signal processing (at wideband points). However, depending on the form of the transmitted signal, such use may affect realization of the Cramér-Rao bound on accuracy. The problem is avoided if matched filtering is used prior to coherent detection. Since nearly all radars use some form of matched filter, this condition normally constitutes no problem. Thus the systems illustrated in Fig. 11.7-2 can be shown to be optimum and realize the Cramér-Rao bound on variance of error in delay measurement given by (see Problem 11.7-3)

$$\sigma_{\hat{\tau}_{\text{R}}(\text{min})}^2 = \frac{1}{\mathcal{R}_{\text{p}}(W_{g,\text{rms}}^2 + \bar{\omega}_g^2)} \tag{11.7-5}$$

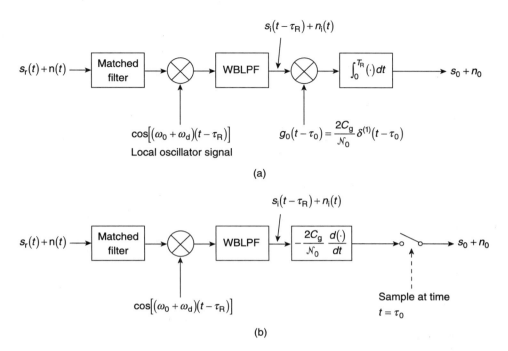

Figure 11.7-2 (*a*) An optimum delay measurement system that uses a coherent detector and matched filter, and (*b*) its equivalent structure using a differentiator and sampler.

Here terms are defined as in (11.4-13). In many systems the mean frequency $\bar{\omega}_g$ of the transmitted signal's complex envelope $g(t)$ is zero.

Systems, such as those of Fig. 11.7-2, that use a coherent detector rely on knowledge of the received carrier's phase. Best performance then presumes Doppler to be known. In many cases Doppler is unknown until range tracking is established. For these cases the systems based on a coherent detector may not be as desirable as those based on an envelope detector.

Systems with an Envelope Detector

Many forms of practical delay measurement systems exist that use an envelope detector to remove the carrier's phase. In these systems if a matched filter is used and $\mathscr{R}_p$ is large, the smallest variance of the delay measurement error is

$$\sigma^2_{\hat{\tau}_R(\text{min})} = \frac{1}{\mathscr{R}_p W^2_{g,\text{rms}}} \tag{11.7-6}$$

However, if $\mathscr{R}_p$ decreases below about 8, a loss factor L_x that accounts for the action of the envelope detector must be included. In this case $\mathscr{R}_p$ in (11.7-6) is replaced by $\mathscr{R}_p/L_x$. Barton and Ward (1969, pp. 82, 88) have given L_x as

$$L_x = \frac{1 + (S/N)}{(S/N)} \tag{11.7-7}$$

where for the matched filter case

$$\left(\frac{S}{N}\right) = \frac{1}{2}\mathcal{R}_p \tag{11.7-8}$$

Equation (11.7-6) becomes

$$\sigma_{\hat{\tau}_R}^2 = \frac{1 + (S/N)}{W_{g,\text{rms}}^2 2(S/N)^2} = \frac{2 + \mathcal{R}_p}{W_{g,\text{rms}}^2 \mathcal{R}_p^2} \tag{11.7-9}$$

Other practical effects may alter the use of (11.7-9), which presumes the system is optimum except for the allowance of the detector's loss. For example, if the receiving filter is not matched perfectly, a power loss L_m (Barton and Ward, 1969) is used to account for the loss in signal-to-noise ratio. This mismatch not only lowers the effective value of $\mathcal{R}_p$, it changes the value of $W_{g,\text{rms}}$ for the system. For more detail on these and other practical effects the reader is referred to Barton and Ward (1969) and Barton (1988).

One very practical form of delay measurement system based on an envelope detector and differentiator is sketched in Fig. 11.7-3. The system develops a delay error measurement in each pulse interval. Initially (on detection) the counter C_2 is designated to the currently estimated value (τ_0) of τ_R. The delay word in counter C_2 which represents τ_0 is available at the comparator. At the transmission time a transmitter trigger sets counter C_1 to zero. C_1 counts a high-frequency clock during the delay time until its delay "word" is identical to that in C_2, whereupon a trigger is initiated to form a sample of the differentiator's output. In accordance with earlier descriptions, this sample is proportional to the delay error $\tau_0 - \tau_R$. The delay error voltage (at point A) is converted to a pulse stream with frequency increasing as the error voltage increases in magnitude. When the voltage at A is positive, counter C_2 counts down to lower the delay word in C_2. When the error voltage is negative, it counts up to increase the delay word. If τ_0 (the delay in C_2) equals the true delay τ_R, the error voltage at A is zero and the delays in both counters are equal.

The precision with which delay can be measured in the system of Fig. 11.7-3 is set by the number of bits (stages) used in counters C_1 and C_2. Since range is proportional to delay the "word" in C_2 can be scaled to represent range. As a consequence a *delay tracker* is synonymously known as a *range tracker*.

To this point nothing has been said about the loop filter in Fig. 11.7-3. If this filter is broadband (relative to the pulse rate), the delay corrections are essentially pulse to pulse, and the delay measurement variance is that achieved with one-pulse measurement. If, however, the loop filter has a small bandwidth, it can provide integration (averaging) of the error voltage at point A over several pulse periods. If N pulses are effectively integrated, the variance of delay error is the variance when using one pulse divided by N (see Barton, 1964, p. 43). For a continuously tracking loop (point A through point G back to A) of closed-loop bandwidth B_n (Hz), the effective number of pulses integrated is

$$N = \frac{f_r}{(2B_n)} \tag{11.7-10}$$

where f_r is the pulse rate (Barton and Ward, 1969, p. 88).

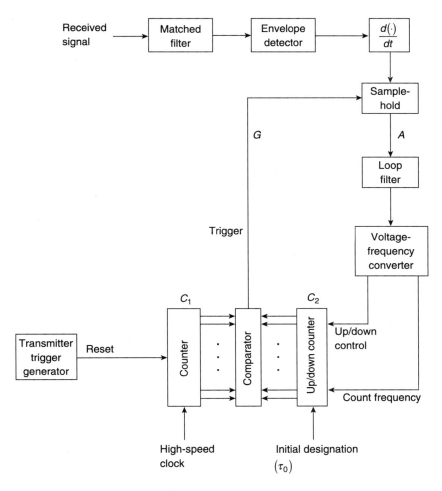

Figure 11.7-3 Delay measurement system.

Alternative practical differentiator systems are possible (see Barton, 1988, p. 426) based on detection of the zero crossing associated with the derivative of the pulse's envelope. One possible system is shown in Fig. 11.7-4, which illustrates a few key waveforms without noise. This system measures delay directly and does not measure a delay *error* $(\tau_0 - \tau_R)$ as in Fig. 11.7-3. Here the target pulse's envelope reaches a peak at the true delay time. Its derivative passes through a null at this time. A zero-crossing detector senses when the null occurs and issues a trigger pulse that passes through a gate to become a stop-count trigger for a counter. The counter begins to count a high-frequency clock at the time of the transmitted pulse. When the counter is stopped by the stop-count trigger, the counter's contents (word) represents the target's delay, or range if desired. The threshold shown is set at a level high enough to block out most noise while low enough to be crossed by the target's pulse. The response of the threshold is a rectangular pulse approximately centered at delay τ_R and of

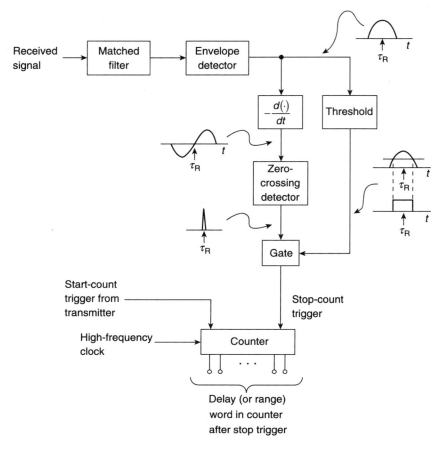

Figure 11.7-4   A single delay measurement method that uses a differentiator and counter.

duration equal to the time the detected envelope exceeds the threshold.[6] The pulse acts as an enabling source for the gate such that spurious outputs from the zero-crossing detector, as generated by noise, do not stop the delay counter.

The single-pulse method of Fig. 11.7-4 can be modified to work over N pulse periods, or to give continuous delay tracking. For N pulses it is only necessary to add a delay averaging function that accepts N single-pulse delay measurements from the counter and provides the average in the stages of another "averaging" counter. For continuous operation the averaging function might be designed to drop the oldest of the N single-pulse measurements and average in the newest measurement in sort of a "sliding average" method. Many other averaging schemes are also possible.

In the continuous (delay tracking) systems of the preceding paragraph, a gate generator might be added to provide a "range gate" centered on the average delay time. This gate, once continuous operation is established, could replace the threshold

---

[6] This threshold might actually be that used in the detection operation, and its level is set by the desired single-pulse false alarm probability.

response in Fig. 11.7-4 as a gating pulse. The principle advantage to be gained is that the gate's duration can be reduced, thereby admitting fewer spurious noise responses from the zero-crossing detector. The reduction is possible because the gate's "jitter" due to noise is smaller when many pulses are averaged. The target's dynamics (speed, acceleration) will also have an effect on the gate's minimum allowed width and may constitute a more important effect than noise.

Early–Late Gate System

One of the most commonly used practical delay measurement systems is shown in Fig. 11.7-5. It uses the early–late gate concept that was earlier developed using intuition. The system develops a delay error measurement in each pulse interval; the integrator is shown with limits for the first period, but they are reset for each interval

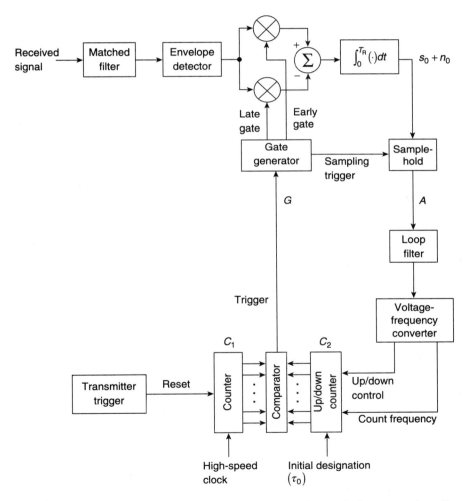

Figure 11.7-5 A practical delay measurement system using early–late gates (a split-gate system).

in which a measurement is to be made. From the received signal input to the integrator's output the system works as in Fig. 11.2-2. At a time shortly after the received pulse has arrived (a little after the late gate) a sampling trigger from the gate generator causes a sample-hold circuit to generate a *signal* voltage at point A equal to s_0, and its value is held until the next sample time. The gate generator uses an input trigger (point G) that is timed to generate gates centered at time τ_0, the "current" receiver's estimate of τ_R. The remaining portions of Fig. 11.7-5 from points A to G are the same as in Fig. 11.7-3.

The early–late gate system is a popular method of delay measurement described in many places (Barton and Ward, 1969, pp. 72–78; Cross, 1975; Skolnik, 1980, pp. 176–177; Eaves and Reedy, eds., 1987, pp. 545–556; Barton, 1988, pp. 435–436; Skolnik, ed., 1990, pp. 18.27–18.30; Biernson, 1990, pp. 391–394; Edde, 1993, pp. 352–355). Various versions from pure analog to all digital are possible. Our version in Fig. 11.7-5 has an analog time discriminator but uses a digital system to develop gate positions. An all-digital tracking system is described by Eaves and Reedy (eds., 1987, pp. 556–558).

For the early/late gate system of Fig. 11.7-5 to approach optimum performance signal-to-noise ratio must be large to avoid the loss (L_x) of the envelope detector, and the gates must be very narrow such that together they approximate a unit doublet. To avoid the use of narrow gates, other approaches are possible in which the matched filter's transfer function is separated into two filtering operations with the gating operation inserted between the two. In these approaches the gates may have wider durations, and their shapes are related to the signal and noise that emerge from the earliest filter. For further details on these approaches, including the effect of using incorrect gate durations, the reader is referred to Barton (1988, pp. 426–429).

PROBLEMS

11.1-1 A target's range from a monostatic radar is 17.5 km. What delay will the system measure?

11.1-2 Work Problem 11.1-1 except assume a range of 24 km.

11.1-3 A monostatic radar measures a target's delay with an rms error due to noise of 0.05 μs. What is the corresponding rms error in measuring range?

11.1-4 Work Problem 11.1-3 except assume an rms delay error of 0.01 μs.

11.1-5 A monostatic radar can measure the angular position of a target at 65-km range to an rms error of 0.5 mrad. What distance does this error represent in an "off-axis" direction from the radar's boresight axis? If the delay measurement system is allowed to have an rms range error due to noise of as much as half the "off-axis" error, how large can the delay's rms error be?

11.1-6 Work Problem 11.1-5 except assume that the range is 125 km and the rms angle error is 0.8 mrad.

11.1-7 Discuss whether the system of Problem 11.1-5 is able to maintain the delay performance limit specified at 65 km for ranges larger than 65 km. [*Hint*: Recall the behavior of the radar equation with range and make use of (10.3-21) and (10.3-22).]

11.2-1 Suppose that a radar uses the gating system of Fig. 11.2-2 when the target pulse's envelope defined in panel *a* is

$$a_0(t) = \begin{cases} A \cos\left[\dfrac{\pi(t - \tau_R)}{\tau_p}\right], & \tau_R - \dfrac{\tau_p}{2} \le t \le \tau_R + \dfrac{\tau_p}{2} \\ 0, & \text{elsewhere} \end{cases}$$

Find s_0 when $\tau_g = 2\tau_p$. What is K_τ for this waveform and gate?

★11.2-2 A radar uses the intuitive gating method of Fig. 11.2-2 when the target pulse's envelope of panel *a* is defined by

$$a_0(t) = \begin{cases} A\left(1 - \dfrac{|t - \tau_R|}{\tau_p/2}\right), & \dfrac{-\tau_p}{2} \le t - \tau_R \le \dfrac{\tau_p}{2} \\ 0, & \text{elsewhere} \end{cases}$$

and $\tau_g = \tau_p$. Find s_0 and the slope K_τ.

★11.2-3 Work Problem 11.2-2 except assume $\tau_g = 2\tau_p$.

11.3-1 Carry out the steps indicated in the text to show that (11.3-4) results from (11.3-3).

11.3-2 At the input to the optimum time discriminator of Fig. 11.3-1 assume that the signal is

$$s_i(t - \tau_R) = A \cos[\omega_0(t - \tau_R)] \operatorname{rect}\left(\frac{t - \tau_R}{2\tau_p}\right) \cos\left[\frac{\pi(t - \tau_R)}{2\tau_p}\right]$$

where $A > 0$ is a constant, ω_0 is the carrier's angular frequency, and the pulse's extent is $2\tau_p$. Assume that the noise $n_i(t)$ is white with power density $\mathcal{N}_0/2$. Find (a) the spectrum $G_0(\omega)$ of the optimum gate $g_0(t)$ and (b) $g_0(t)$. Assume that $\omega_0 \gg \pi/(2\tau_p)$ in writing $g_0(t)$.

11.3-3 Work Problem 11.3-2 except assume that $\omega_0 = 0$ as if $s_i(t - \tau_R)$ were at an envelope detector's output.

11.4-1 An optimum wideband time discriminator is to be found for white input noise $n_i(t)$ having a constant power density $\mathcal{N}_0/2$ and an input signal

$$s_r(t) = s_i(t - \tau_R) = \alpha \cos^2\left[\frac{\pi(t - \tau_R)}{2\tau_p}\right] \operatorname{rect}\left[\frac{t - \tau_R}{2\tau_p}\right]$$

where τ_p and α are positive constants. Find (a) $g_0(t)$, and (b) s_0. (c) Plot s_0 versus $(\tau_0 - \tau_R)/\tau_p$.

11.4-2 Find the power density spectrum of the output noise in the system of Problem 11.4-1.

11.4-3 Show that s_0 in Fig. 11.3-1 is given by (11.4-8) when the input signal and noise are defined by (11.4-1) and (11.4-2), respectively.

11.4-4 Show that (11.4-9) is true.

11.4-5 Show that (11.4-10) derives from (11.4-9).

11.4-6 By using (11.4-9) and (11.4-8), show that (11.3-2), when applied to the optimum wideband delay measurement system, gives

$$\sigma_{\hat{\tau}_R}^2 = \frac{1}{\dfrac{2}{\mathcal{N}_0} \displaystyle\int_{-\infty}^{\infty} \left[\frac{ds_r(t)}{dt}\right]^2 dt} = \sigma_{\hat{\tau}_R(\min)}^2$$

where $\sigma_{\hat{\tau}_R(\min)}^2$ is given by either (11.4-12) or (11.4-13).

11.4-7 Show that (11.4-12) will reduce to (11.4-13).

★11.5-1 For the optimum delay measurement system with a matched filter, show that s_0, $E[n_0^2]$, and $\mathcal{S}_{N_0 N_0}(\omega)$ are given by (11.4-8), (11.4-10), and (11.4-11) which apply to the wideband system.

★11.5-2 A receiver uses a matched filter for white noise to process the signal from the antenna which is given by

$$s_r(t) = \alpha\left[1 - \left(\frac{t - \tau_R}{\tau_p}\right)^2\right] \text{rect}\left(\frac{t - \tau_R}{2\tau_p}\right)\cos[(\omega_0 + \omega_d)(t - \tau_R) + \phi_0]$$

where $\alpha > 0$ is a constant, ω_0 is the radar's angular frequency, ϕ_0 is an arbitrary constant phase angle, ω_d and τ_R are the target's Doppler angular frequency and delay at time zero, respectively, and $2\tau_p$ is the time duration of the transmitted pulse. Show that the filter's output signal is

$$s_i(t - \tau_R) = \frac{C\alpha^2 \tau_p}{30\mathcal{N}_0}[32 - 40\Delta^2 + 20\Delta^3 - \Delta^5]\left\{\text{rect}\left(\frac{t - \tau_R}{4\tau_p}\right)\right.$$
$$\left. \cdot \cos[(\omega_0 + \omega_d)(t - \tau_R)]\right\}$$

where

$$\Delta = \frac{t - \tau_R}{\tau_p}$$

$\mathcal{N}_0/2$ is the power density of the white input noise, and C is the arbitrary constant associated with the matched filter [see (6.6-6)].

11.5-3 Work Problem 11.5-2 except show that

$$s_i(t - \tau_R) = \frac{C\alpha^2 \tau_p}{\mathcal{N}_0}\text{tri}\left(\frac{t - \tau_R}{\tau_p}\right)\cos[(\omega_0 + \omega_d)(t - \tau_R)]$$

when

$$s_r(t) = \alpha \, \text{rect}\left(\frac{t - \tau_R}{\tau_p}\right) \cos[(\omega_0 + \omega_d)(t - \tau_R)]$$

11.6-1 In Fig. P11.6-1 let $s_i(t - \tau_R)$ be any real pulse having a nonzero voltage extent falling entirely in one pulse repetition interval from 0 to T_R. Show that s_0 (at $t = T_R$) in (a) is equal to s_0 in (b). The two systems are therefore signal-equivalent.

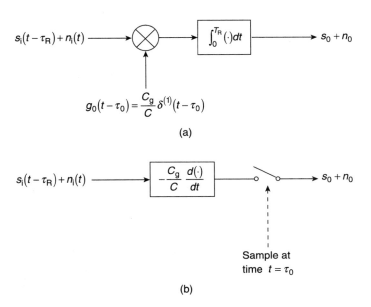

$$g_0(t - \tau_0) = \frac{C_g}{C} \delta^{(1)}(t - \tau_0)$$

(a)

(b)

Sample at
time $t = \tau_0$

Figure   P11.6-1.

11.6-2 In Fig. P11.6-1 show that noise n_0 in (a) is the same as the noise n_0 in (b). The two systems are therefore noise-equivalent. With the result of Problem 11.6-1, we see that the two systems are performance-equivalent.

11.7-1 For the input signal of (11.7-1), show that the output signal $s_2(t)$ of Fig. 11.7-1 is given by (11.7-3).

★11.7-2 For an arbitrary power density spectrum $\mathcal{S}_{N_1 N_1}(\omega)$ applicable to input noise $n_1(t)$ in Fig. 11.7-1, show that the output noise has the power density spectrum of (11.7-4).

★11.7-3 From Problems 11.6-1 and 11.6-2 we conclude that the two systems of Fig. 11.7-2 must give the same variance of delay measurement error. Find the Fourier transform $S_i(\omega)$ of $s_i(t)$ and the power density spectrum $\mathcal{S}_{N_i N_i}(\omega)$ of $n_i(t)$. Show that the differentiator of (b) is optimum to process $s_i(t - \tau_R)$ and $n_i(t)$, and then prove that $\sigma^2_{\tau_R(\text{min})}$ for the optimum system is given by (11.7-5). Assume that $s_r(t)$ is defined by (10.3-1) and that $n(t)$ is white noise with power density $\mathcal{N}_0/2$. Since (11.7-5) is the Cramér-Rao bound, we conclude that the systems realize this bound, in principle.

11.7-4 A radar uses the delay measurement system of Fig. 11.7-2a. If

$$s_r(t) = A \cos\left[\frac{\pi(t - \tau_R)}{\tau_p}\right] \text{rect}\left[\frac{t - \tau_R}{\tau_p}\right] \cos[(\omega_0 + \omega_d)(t - \tau_R) + \phi_0]$$

where A, τ_R, τ_p, ω_0, and ϕ_0 are positive constants and ω_d is the target's Doppler angular frequency. Show that

$$s_i(t - \tau_R) = \frac{CA^2\tau_p}{4\pi\mathcal{N}_0}[\pi(1 - \Delta x)\cos(\pi\Delta x) + \sin(\pi\Delta x)]\text{rect}\left(\frac{\Delta x}{2}\right)$$

where

$$\Delta x = \frac{t - \tau_R}{\tau_p}$$

and C is the arbitrary constant to be associated with the matched filter [see (6.6-6)].

11.7-5 A radar's transmitted pulse has the complex envelope defined in Fig. P11.7-5 (there is no FM used). Find an expression for the rms bandwidth of $g(t)$.

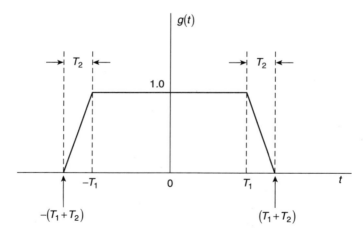

Figure P11.7-5.

11.7-6 (a) Work Problem 11.7-5 except assume that

$$g(t) = \begin{cases} A\left[1 - \left(\frac{t}{\tau}\right)^2\right], & -\tau \leq t \leq \tau \\ 0, & \text{elsewhere} \end{cases}$$

where A and τ are real positive constants. (b) What delay and range measurement rms errors are possible if $\mathscr{R}_p = 36$ and $\tau = 1\ \mu s$?

11.7-7 Find $W_{g,\text{rms}}$ for a radar's transmitted pulse defined by the complex envelope

$$g(t) = \begin{cases} A\left[1 - \left(\dfrac{t}{\tau}\right)^N\right], & -\tau \le t \le \tau \\ 0, & \text{elsewhere} \end{cases}$$

where A and τ are real positive constants and $N = 2, 4, 6, \ldots$. Plot $W_{g,\text{rms}}\,\tau$ versus N. To what does $W_{g,\text{rms}}\,\tau$ approach in the limit as $N \to \infty$?

11.7-8 The complex envelope of a radar's transmitted pulse is

$$g(t) = \begin{cases} A\left[1 - \left(\dfrac{t}{\tau}\right)^2\right]^2, & -\tau \le t \le \tau \\ 0, & \text{elsewhere} \end{cases}$$

where A and τ are real positive constants. (a) Find $W_{g,\text{rms}}$. (b) What rms range and delay errors are possible if $\mathscr{R}_p = 24$ and $\tau = 5\ \mu s$?

11.7-9 The transmitted pulse in a radar has the complex envelope

$$g(t) = \begin{cases} A\left[1 - \left(\dfrac{t}{\tau}\right)^N\right]^2, & -\tau \le t \le \tau \\ 0, & \text{elsewhere} \end{cases}$$

where A and τ are real positive constants and $N = 2, 4, 6, \ldots$. (a) Find $W_{g,\text{rms}}$. (b) Plot $W_{g,\text{rms}}\,\tau$ versus N, and determine the limiting value as $N \to \infty$. (c) For $\tau = 4\ \mu s$, $N = 4$, and $\mathscr{R}_p = 10$, what rms delay and range measurement errors are possible?

12

FREQUENCY (DOPPLER) MEASUREMENT AND TRACKING

Whenever a target has motion relative to a radar, its returned signal has a frequency that differs from that of the transmitted signal by the Doppler frequency offset. Previously from (1.6-11) we have seen that a target's radial speed and Doppler frequency are proportional to each other. Thus a measurement of Doppler frequency affords a convenient means of measuring radial speed. Such measurements can be more accurate than when derived from other methods such as differentiation of range data to produce the speed measurement.

Besides providing an alternative means of speed measurement, the use of Doppler measurement methods allows a new dimension in radar design. In essence the Doppler effect allows the discrimination between moving and stationary targets through their difference in frequency. The frequency discriminant can facilitate both detection and tracking of moving targets.

In this chapter we discuss the most basic principles in the optimum measurement of Doppler frequency. After carefully defining the problem and system, we determine the best measurement accuracy possible and show, for the white noise case, that this accuracy equals the Cramér-Rao bound of Chapter 10. The optimum system is then derived and some practical effects are discussed. These include mismatch effects and practical ways of approximating the optimum system by a real system. Finally some noncoherent and coherent system realizations are outlined that show how a receiver can make Doppler measurements based either on a finite number of pulses or on continuous measurement (tracking).

As we will find, the accuracy of Doppler measurement increases as the signal-processing time (number of pulse intervals) increases. There is a limit, however, that is typically related to the dynamics (acceleration, and higher range rates) of the target. To realize high Doppler accuracy, it is found that complexity of the measurement system increases as the initial knowledge of the true Doppler frequency is poorer. For the most accurate systems (fine-line measurements) a means for reducing some of the complexity is given. Both coherent and noncoherent practical systems are discussed.

Finally we note that our work is mathematical, since it deals with the basics of Doppler measurement and the mathematics naturally leads to an analog description of systems. The mathematical (analog) developments are also most easily understood on first exposure to the material. However, it should be pointed out that many of the systems ("filters") to be described are actually *implemented* using digital signal-processing (DSP) techniques done in a digital computer. A summary of the more important aspects of DSP is contained in Chapter 15.

12.1   DEFINITION OF OPTIMUM FREQUENCY MEASUREMENT

In this section we define the approach to be used in finding a system (filter) for optimally measuring Doppler angular frequency. After preliminary definitions, some criteria are developed to define an optimum system. The best possible accuracy in Doppler measurement by the optimum system is then found and is shown to equal the Cramér-Rao bound of Chapter 10. The determination of the optimum filter itself is developed in Section 12.2.

Problem Definition

The general problem is defined with the help of Fig. 12.1-1. The filter represents the key part of the signal processor needed to measure ω_d, the Doppler angular frequency of the target's returned signal to a monostatic radar.[1] This signal, $s_r(t)$, is defined by (1.6-21):

$$s_r(t) = \alpha a(t - \tau_R)\cos\left[(\omega_0 + \omega_d)(t - \tau_R) + \theta(t - \tau_R) + \phi_0\right] \qquad (12.1\text{-}1)$$

Here α is a constant determined by the radar equation, ω_0 is the transmitted carrier's angular frequency, τ_R is the target's range delay (defined at time $t = 0$), $a(t)$ and $\theta(t)$ are the envelope and phase modulations of the transmitted waveform, and ϕ_0 is a constant phase angle. The target's signal $s_r(t)$ arrives with input noise $n(t)$. We mainly assume that $n(t)$ is white, gaussian, zero-mean, stationary noise in following work. However, initially we allow $n(t)$ to be nonwhite with a power density spectrum $\mathcal{S}_{NN}(\omega)$ and then specialize to the white noise case.

For purposes of anaysis we represent $s_r(t)$ by its analytic signal $\psi_r(t)$, where

$$s_r(t) = \text{Re}\left[\psi_r(t)\right] \qquad (12.1\text{-}2)$$

$$\psi_r(t) = \alpha\psi(t - \tau_R)e^{j\omega_d(t - \tau_R)} \qquad (12.1\text{-}3)$$

$$\psi(t) = g(t)e^{j\omega_0 t + j\phi_0} \qquad (12.1\text{-}4)$$

$$g(t) = a(t)e^{j\theta(t)} \qquad (12.1\text{-}5)$$

The filter in Fig. 12.1-1 is presumed to be linear and real; its transfer function is $H(\omega)$, and its real impulse response is $h(t)$, in general. The analytic representations of

---

[1] There are other components necessary for a complete measurement system, as described in later parts of this chapter.

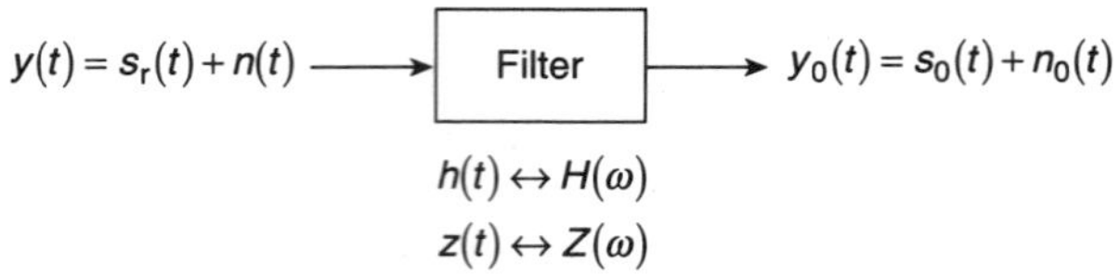

$$y(t) = s_r(t) + n(t) \longrightarrow \boxed{\text{Filter}} \longrightarrow y_0(t) = s_0(t) + n_0(t)$$

$$h(t) \leftrightarrow H(\omega)$$
$$z(t) \leftrightarrow Z(\omega)$$

Figure 12.1-1 Signals and filter applicable to optimum Doppler frequency measurement.

$H(\omega)$ and $h(t)$ are denoted by $Z(\omega)$ and $z(t)$, respectively. In our analyses it is convenient and instructive to use the analytic representations. Optimum values of these quantities are indicated by using the subscript "opt". Thus optimum functions to be found are $H_{\text{opt}}(\omega)$, $h_{\text{opt}}(t)$, $Z_{\text{opt}}(\omega)$, and $z_{\text{opt}}(t)$. In general, the real filter and analytic filter are related by

$$h(t) = \operatorname{Re}[z(t)] = \tfrac{1}{2}[z(t) + z^*(t)] \tag{12.1-6}$$

$$H(\omega) = \tfrac{1}{2}[Z(\omega) + Z^*(-\omega)] \tag{12.1-7}$$

$$Z(\omega) = 2U(\omega)H(\omega) \tag{12.1-8}$$

In general, the real responses of the real filter in Fig. 12.1-1 are denoted by $s_0(t)$ for the signal and $n_0(t)$ for the noise. From Fig. 6.5-4c we know that $s_0(t)$ is the real part of half the convolution of $\psi_r(t)$ and $z(t)$:

$$s_0(t) = \tfrac{1}{2}\operatorname{Re}[\psi_r(t) \star z(t)] \tag{12.1-9}$$

Overall, our problem is to determine the one filter of all possible filters [by finding either $z(t)$ or $Z(\omega)$] that best estimates ω_d. This determination is done in the next section. Here we first define some reasonable criteria of optimality and then determine the best accuracy achievable by the optimum filter.

Optimum Doppler Measurement Filter Criteria

It is assumed that the filter has available some initial estimate of ω_d. This initial "guess" can be derived by differentiation of range measurements made by a range measurement system or by observing which filter in a parallel bank of filters (at different center frequencies) responds best to the received signal. In some cases where $|\omega_d|$ is expected to be small in relation to the signal's half-bandwidth, the initial estimate of ω_d might be taken as zero. Whatever the value of the initial estimate, we denote it by ω_{d0}.

First, let us assume that output noise is negligible. The system can only work with its output signal to estimate Doppler frequency. If ω_{d0} is exactly equal to ω_d, the true Doppler frequency, it is reasonable to ask that the signal's response be zero, implying that there is no error between the initial estimate and the true value ω_d. For $\omega_{d0} \neq \omega_d$ we can also reasonably ask that the output signal increase proportionately with the difference $\omega_{d0} - \omega_d$, at least for small difference magnitudes $|\omega_{d0} - \omega_d|$, as sketched in Fig. 12.1-2. Now because the output signal $s_0(t)$ varies with time, it is necessary to

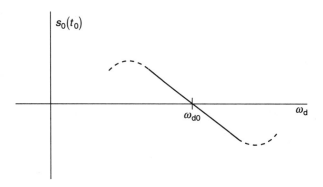

Figure 12.1-2   System response for various target Doppler angular frequencies.

specify a time at which our criteria are to be attained. For the present we label this time as t_0.

If K_ω is the proportionality constant for the linear region in Fig. 12.1-2 (the curve's slope at the crossover point at time t_0) and if $\omega_{d0} \neq \omega_d$, the response $s_0(t_0)$ can be used to correct ω_{d0} to equal ω_d. In the linear region $s_0(t_0) = K_\omega(\omega_{d0} - \omega_d)$, so

$$\omega_d = \omega_{d0} - \frac{s_0(t_0)}{K_\omega} \tag{12.1-10}$$

where the correction is $[-s_0(t_0)/K_\omega]$. Ideally for a linear characteristic, as assumed when $|\omega_{d0} - \omega_d|$ is small, the correction is exact (no bias).

Next suppose that conditions are as described in the preceding paragraphs except that there is now output noise, denoted by $n_0(t_0)$, at the time of observation t_0. The receiver still presumes a linear relationship between the filter's response and ω_d. However, the response is now $s_0(t_0) + n_0(t_0)$ and instead of solving for ω_d, as in (12.1-10), the solution produces $\hat{\omega}_d$ an estimate of ω_d. The linear relationship becomes $s_0(t_0) + n_0(t_0) = K_\omega(\omega_{d0} - \hat{\omega}_d)$, so

$$\hat{\omega}_d = \omega_{d0} - \frac{s_0(t_0) + n_0(t_0)}{K_\omega} = \omega_d - \frac{n_0(t_0)}{K_\omega} \tag{12.1-11}$$

where (12.1-10) has been used. This expression shows that the estimate $\hat{\omega}_d$ of ω_d equals the true value ω_d plus an "error" due to noise. Denote this error by $\Delta\omega_d$; it is given by

$$\Delta\omega_d = \hat{\omega}_d - \omega_d = -\frac{n_0(t_0)}{K_\omega} \tag{12.1-12}$$

The relationships (12.1-11) and (12.1-12) hold as long as $s_0(t_0) + n_0(t_0)$ corresponds to the linear region of the response characteristic of Fig. 12.1-2. For nonsmall values of $|\omega_{d0} - \omega_d|$, the response typically behaves as shown dashed. The overall response is often called a *frequency discriminator characteristic*. The filter is therefore known as a *frequency discriminator*. All useful response functions have either odd or nearly odd symmetry about the point ω_{d0}. Generally, the slope (K_ω) of the frequency discriminator characteristic for small $|\omega_{d0} - \omega_d|$ is given by

$$K_\omega = \frac{ds_0(t_0)}{d\omega_{d0}}\bigg|_{\omega_{d0} = \omega_d} = -\frac{ds_0(t_0)}{d\omega_d}\bigg|_{\omega_d = \omega_{d0}} \tag{12.1-13}$$

Our last criterion of optimality is to ask that the optimum filter produce the smallest possible mean-squared error in estimating Doppler frequency. This mean-squared estimation error, from (12.1-12), is

$$E[(\Delta\omega_d)^2] = E[(\hat{\omega}_d - \omega_d)^2] = \frac{E[n_0^2(t_0)]}{K_\omega^2} \tag{12.1-14}$$

If the input noise is zero-mean, as assumed, and the filter produces no constant response, $n_0(t_0)$ will be zero-mean and $\hat{\omega}_d$ will be unbiased from (12.1-11). For the unbiased estimate, (12.1-14) is equal to the estimator's variance, denoted by $\sigma_{\hat{\omega}_d}^2$:

$$\sigma_{\hat{\omega}_d}^2 = \frac{E[n_0^2(t_0)]}{K_\omega^2} \tag{12.1-15}$$

In summary, we have required three criteria be satisfied for a filter to be optimum. (1) It must give $s_0(t_0) = 0$ when there is no error in our initial estimate of ω_d, that is, when $\omega_{d0} = \omega_d$. (2) It must produce a signal response $s_0(t_0)$ that is proportional to $\omega_{d0} - \omega_d$ for $|\omega_{d0} - \omega_d|$ small. (3) It must provide the smallest variance of Doppler frequency measurements by minimizing $\sigma_{\hat{\omega}_d}^2$ of (12.1-15). We ultimately show that all three of these requirements are satisfied.

Optimum Filter's Accuracy

First, we evaluate (12.1-15) and then minimize $\sigma_{\hat{\omega}_d}^2$. To determine K_ω, we first find $s_0(t)$ from (12.1-9). By substitution of (12.1-3), we obtain

$$\begin{aligned} s_0(t) &= \frac{1}{2}\text{Re}\left\{\int_{-\infty}^{\infty} \psi_r(\xi) z(t - \xi)\, d\xi\right\} \\ &= \frac{\alpha}{2}\text{Re}\left\{\int_{-\infty}^{\infty} \psi(\xi - \tau_R) z(t - \xi) e^{j\omega_d(\xi - \tau_R)}\, d\xi\right\} \end{aligned} \tag{12.1-16}$$

On setting $t = t_0$, the time at which the slope K_ω is defined (the time at which the optimum estimate of Doppler frequency is to occur), we differentiate according to (12.1-13) and get

$$K_\omega = \frac{-\alpha}{2}\text{Re}\left\{\int_{-\infty}^{\infty} j(\xi - \tau_R)\psi(\xi - \tau_R) z(t_0 - \xi) e^{j\omega_{d0}(\xi - \tau_R)}\, d\xi\right\} \tag{12.1-17}$$

To best use this expression, we convert it to an integral in the frequency variable.

First, we replace terms in (12.1-17) by their inverse Fourier transforms, making use of the following transform pairs

$$j(\xi - \tau_R)\psi(\xi - \tau_R) \leftrightarrow -\frac{d\Psi(\omega)}{d\omega} e^{-j\omega\tau_R} \qquad (12.1\text{-}18)$$

$$z(t - \xi) \leftrightarrow Z(\eta) e^{-j\eta\xi} \qquad (12.1\text{-}19)$$

where $\Psi(\omega)$ is the Fourier transform of $\psi(t)$. Thus (12.1-17) becomes

$$
\begin{aligned}
K_\omega &= \frac{\alpha}{2} \mathrm{Re} \left\{ \int_{-\infty}^{\infty} \frac{1}{2\pi} \int_{-\infty}^{\infty} \frac{d\Psi(\omega)}{d\omega} e^{j\omega(\xi - \tau_R)} d\omega \right. \\
&\qquad \cdot \frac{1}{2\pi} \int_{-\infty}^{\infty} Z(\eta) e^{j\eta(t_0 - \xi)} d\eta \, e^{j\omega_{do}(\xi - \tau_R)} d\xi \Bigg\} \\
&= \frac{\alpha}{2} \mathrm{Re} \left\{ \frac{1}{2\pi} \int_{-\infty}^{\infty} \frac{d\Psi(\omega)}{d\omega} \frac{1}{2\pi} \int_{-\infty}^{\infty} Z(\eta) \int_{-\infty}^{\infty} e^{j(\omega - \eta + \omega_{do})\xi} d\xi \right. \\
&\qquad \cdot e^{j\eta t_0} d\eta \, e^{-j\omega\tau_R} d\omega \, e^{-j\omega_{do}\tau_R} \Bigg\}
\end{aligned}
\qquad (12.1\text{-}20)
$$

The third (inner) integral in the last right-side form of (12.1-20) is recognized as $2\pi\delta(\omega - \eta + \omega_{do})$ from (A.3-10). The presence of the impulse then allows the integral over η to be immediately evaluated by the definition of the impulse function using (A.3-1). After a simple variable change K_ω is expressed as

$$K_\omega = \frac{\alpha}{2} \mathrm{Re} \left\{ \frac{1}{2\pi} \int_{-\infty}^{\infty} \frac{d\Psi(\omega - \omega_{do})}{d\omega} Z(\omega) e^{j\omega(t_0 - \tau_R)} d\omega \right\} \qquad (12.1\text{-}21)$$

The next quantity needed in (12.1-15) is $E[n_0^2(t_0)]$, which is the average power in the output noise at time t_0. For the assumed stationary input noise, this output noise power, denoted by N_0, is the same for all time. N_0 is given by the several equivalent forms as follows:

$$
\begin{aligned}
N_0 &= \frac{1}{2\pi} \int_{-\infty}^{\infty} \mathscr{S}_{NN}(\omega) |H(\omega)|^2 \, d\omega = \frac{1}{\pi} \int_{0}^{\infty} \mathscr{S}_{NN}(\omega) |H(\omega)|^2 \, d\omega \\
&= \frac{1}{2} \frac{1}{2\pi} \int_{0}^{\infty} \mathscr{S}_{NN}(\omega) |2H(\omega)|^2 \, d\omega = \frac{1}{2} \frac{1}{2\pi} \int_{-\infty}^{\infty} \mathscr{S}_{NN}(\omega) |2U(\omega)H(\omega)|^2 \, d\omega \\
&= \frac{1}{2} \frac{1}{2\pi} \int_{-\infty}^{\infty} \mathscr{S}_{NN}(\omega) |Z(\omega)|^2 \, d\omega
\end{aligned}
\qquad (12.1\text{-}22)
$$

On substitution of (12.1-22) and (12.1-21) into (12.1-15), we have

$$\sigma_{\hat{\omega}_d}^2 = \frac{\dfrac{1}{2\pi} \displaystyle\int_{-\infty}^{\infty} \mathscr{S}_{NN}(\omega)|Z(\omega)|^2 \, d\omega}{\dfrac{\alpha^2}{2}\left[\mathrm{Re}\left\{ \dfrac{1}{2\pi}\displaystyle\int_{-\infty}^{\infty} \dfrac{d\Psi(\omega - \omega_{d0})}{d\omega} Z(\omega) e^{j\omega(t_0 - \tau_R)} \, d\omega \right\} \right]^2} \qquad (12.1\text{-}23)$$

We will minimize (12.1-23) by use of Schwarz's inequality. Part of the work will set the stage for finding the optimum filter in Section 12.2.

The form of Schwarz's inequality that we need says that (Thomas, 1969, app. E, and Problem 12.1-4)

$$\left\{ \mathrm{Re}\left[\int_a^b A(\omega) B(\omega) \, d\omega \right] \right\}^2 \le \int_a^b |A(\omega)|^2 \, d\omega \int_a^b |B(\omega)|^2 \, d\omega \qquad (12.1\text{-}24)$$

where a and b are real constants ($a = -\infty$ and $b = \infty$ are allowed), and $A(\omega)$ and $B(\omega)$, in general, may be complex functions of the real variable ω. The equality in (12.1-24) is true only if

$$A(\omega) = CB^*(\omega) \qquad (12.1\text{-}25)$$

where C is an arbitrary nonzero real constant.

We apply (12.1-24) to our problem by defining

$$A(\omega) = \frac{1}{\sqrt{2\pi}} \sqrt{\mathscr{S}_{NN}(\omega)} \, Z(\omega) \qquad (12.1\text{-}26)$$

$$B(\omega) = \frac{1}{\sqrt{2\pi}} \frac{d\Psi(\omega - \omega_{d0})}{d\omega} \frac{e^{j\omega(t_0 - \tau_R)}}{\sqrt{\mathscr{S}_{NN}(\omega)}} \qquad (12.1\text{-}27)$$

and using these quantities in (12.1-23). The resulting inequality allows (12.1-23) to be written as

$$\sigma_{\hat{\omega}_d}^2 \ge \frac{1}{\dfrac{\alpha^2}{4\pi}\displaystyle\int_{-\infty}^{\infty} \dfrac{1}{\mathscr{S}_{NN}(\omega)}\left| \dfrac{d\Psi(\omega - \omega_{d0})}{d\omega} \right|^2 d\omega} = \sigma_{\hat{\omega}_d(\min)}^2 \qquad (12.1\text{-}28)$$

where $\sigma_{\hat{\omega}_d(\min)}^2$ denotes the smallest (bounding) value of $\sigma_{\hat{\omega}_d}^2$.

The right side of (12.1-28) defines the smallest variance of error achievable by any linear filter in estimating ω_d; it is a lower bound on error variance. It is to be noted that this bound is not dependent on t_0 or τ_R. In Section 12.2 we use (12.1-25) through (12.1-27) to define the optimum filter that corresponds to the equality in (12.1-28), that is, the filter that produces the lower bound on variance. Prior to that work we demonstrate for white noise that the variance lower bound in (12.1-28) is equal to the Cramér-Rao bound developed in Chapter 10.

Accuracy for White Noise

For white input noise the power spectrum is constant: $\mathscr{S}_{NN}(\omega) = \mathscr{N}_0/2$, $-\infty < \omega < \infty$, where $\mathscr{N}_0$ is a positive constant. The smallest estimation error variance from (12.1-28) becomes

$$\sigma^2_{\hat{\omega}_d(\min)} = \frac{1}{\dfrac{\alpha^2}{\mathscr{N}_0 2\pi} \displaystyle\int_{-\infty}^{\infty} \left| \dfrac{d\Psi(\omega - \omega_{d0})}{d\omega} \right|^2 d\omega} \tag{12.1-29}$$

On using the Fourier transform pair

$$-jt\psi(t)e^{j\omega_{d0}t} \leftrightarrow \frac{d\Psi(\omega - \omega_{d0})}{d\omega} \tag{12.1-30}$$

in Parseval's theorem, we can write (12.1-29) as

$$\sigma^2_{\hat{\omega}_d(\min)} = \frac{1}{\mathscr{R}[\tau^2_{\psi,\mathrm{rms}} + (\bar{t}_\psi)^2]} = \frac{1}{\mathscr{R}[\tau^2_{g,\mathrm{rms}} + (\bar{t}_g)^2]} \tag{12.1-31}$$

(see Problem 12.1-5) where $\bar{t}_\psi$ and $\bar{t}_g$ are mean-times defined by (6.4-4) and Problem 6.4-2, respectively, and

$$\mathscr{R} = \frac{2E_\mathrm{r}}{\mathscr{N}_0} \tag{12.1-32}$$

$$\tau^2_{\psi,\mathrm{rms}} = \frac{\displaystyle\int_{-\infty}^{\infty} (t - \bar{t}_\psi)^2 |\psi(t)|^2 \, dt}{\displaystyle\int_{-\infty}^{\infty} |\psi(t)|^2 \, dt} \tag{12.1-33}$$

$$\tau^2_{g,\mathrm{rms}} = \frac{\displaystyle\int_{-\infty}^{\infty} (t - \bar{t}_g)^2 |g(t)|^2 \, dt}{\displaystyle\int_{-\infty}^{\infty} |g(t)|^2 \, dt} \tag{12.1-34}$$

$$E_\mathrm{r} = \frac{\alpha^2}{2} E_\psi = \frac{\alpha^2}{2} \int_{-\infty}^{\infty} |\psi(t)|^2 \, dt \tag{12.1-35}$$

In most radars the time origin is defined such that the mean-times $\bar{t}_\psi$ and $\bar{t}_g$ are zero.[2]

Our principal result is (12.1-31). The last right-side form derives from $\bar{t}_\psi = \bar{t}_g$ and $\tau_{\psi,\mathrm{rms}} = \tau_{g,\mathrm{rms}}$, since $|\psi(t)| = |g(t)|$. On comparing (12.1-31) with (10.3-15), we find that the bounding variance of measurement error is equal to the Cramér-Rao bound if $\mathscr{R}$ here equals $N\mathscr{R}_\mathrm{p}$ in (10.3-15). If $\psi(t)$ represents a *single* pulse, then $N = 1$ in (10.3-15) and $\mathscr{R} = \mathscr{R}_\mathrm{p}$ in (12.1-31), and the two results are equal. If $\psi(t)$ is a representation of N coherently processed pulses, then E_r in (12.1-32) is N times that of one pulse such that $\mathscr{R} = N\mathscr{R}_\mathrm{p}$ in (12.1-31), and the two results are again equal.

---

[2] The time origin can always be *selected* such that mean-times are zero.

Example 12.1-1   We find $\sigma^2_{\hat{\omega}_d(\min)}$ for a single constant-frequency transmitted pulse with a rectangular envelope of duration T centered on the origin and having a peak positive amplitude A. Thus

$$|\psi(t)| = |g(t)| = A \operatorname{rect}\left(\frac{t}{T}\right) \tag{1}$$

$$\int_{-\infty}^{\infty} |\psi(t)|^2 \, dt = A^2 \int_{-T/2}^{T/2} dt = A^2 T \tag{2}$$

$$\int_{-\infty}^{\infty} t|\psi(t)|^2 \, dt = A^2 \int_{-T/2}^{T/2} t\,dt = 0 \tag{3}$$

$$\int_{-\infty}^{\infty} t^2|\psi(t)|^2 \, dt = A^2 \int_{-T/2}^{T/2} t^2\,dt = \frac{A^2 T^3}{12} \tag{4}$$

From (3) we have $\bar{t}_\psi = 0$. From (12.1-33), $\tau^2_{\psi,\text{rms}} = T^2/12$. From (12.1-31), $\sigma^2_{\hat{\omega}_d(\min)} = 12/\mathscr{R}T^2$.

12.2   OPTIMUM FILTER FOR DOPPLER MEASUREMENTS

The optimum filter is that which measures the target's Doppler angular frequency ω_d with the smallest variance of measurement error. This smallest error was shown to be given by the right side of (12.1-28). The actual measurement variance equals the smallest value only when the equality in (12.1-28) holds, that is, when (12.1-25) is true.

Optimum Filter for Nonwhite Noise

So far we know that the optimum filter must derive from solving (12.1-25) for $Z(\omega) = Z_{\text{opt}}(\omega)$ when (12.1-26) and (12.1-27) are substituted. The result is

$$Z_{\text{opt}}(\omega) = C\frac{d\Psi^*(\omega - \omega_{d0})}{d\omega} \frac{e^{-j\omega(t_0 - \tau_R)}}{\mathscr{S}_{NN}(\omega)} \tag{12.2-1}$$

By referring to (12.1-30) and comparing this result with the matched filter expression of (6.6-11), we can interpret (12.2-1) as a matched filter for nonwhite noise plus a received signal that is $\psi(t)$ delayed to τ_R, Doppler shifted to ω_{d0} (the value to which the receiver is tuned), and multiplied by $-j(t - \tau_R)$. The quantity t_0 in (12.2-1) only determines the time at which the "matched filter" produces its maximum response. In general, t_0 can be any value we wish, but practical considerations limit t_0 to $t_0 \geq \tau_R$ by use of receiver delay.[3] If we do not concern ourselves with the effect of choosing t_0 on the filter's realizability, we may choose $t_0 = \tau_R$. For this case

$$Z_{\text{opt}}(\omega) = C\frac{d\Psi^*(\omega - \omega_{d0})}{d\omega} \frac{1}{\mathscr{S}_{NN}(\omega)}, \qquad t_0 = \tau_R \tag{12.2-2}$$

---

[3] For causality t_0 must be at least larger than τ_R plus the duration that the transmitted waveform is nonzero for $t > 0$.

Optimum Filter for White Noise

For white noise $\mathcal{S}_{NN}(\omega) = \mathcal{N}_0/2$, for $-\infty < \omega < \infty$, and (12.2-1) becomes

$$Z_{\text{opt}}(\omega) = \frac{2C}{\mathcal{N}_0} \frac{d\Psi^*(\omega - \omega_{d0})}{d\omega} e^{-j\omega(t_0 - \tau_R)} \tag{12.2-3}$$

The transfer function $H_{\text{opt}}(\omega)$ of the optimum real filter is found by using (12.2-3) in (12.1-7).

The impulse response of the optimum analytic filter is the inverse Fourier transform of $Z_{\text{opt}}(\omega)$. By using several properties of Fourier transforms (time and frequency shifting, conjugation, and frequency differentiation), from Appendix B we obtain

$$z_{\text{opt}}(t) = \frac{-2C}{\mathcal{N}_0} j(t - t_0 + \tau_R)\psi^*(-t + t_0 - \tau_R) e^{j\omega_{d0}(t - t_0 + \tau_R)} \tag{12.2-4}$$

The corresponding real filter's impulse response $h_{\text{opt}}(t)$ can be found from (12.1-6) using (12.2-4).

Again, if we are not too concerned about filter causality, we can select t_0, the time of the sample to be used in estimating ω_d, equal to τ_R. For this case

$$Z_{\text{opt}}(\omega) = \frac{2C}{\mathcal{N}_0} \frac{d\Psi^*(\omega - \omega_{d0})}{d\omega}, \qquad t_0 = \tau_R \tag{12.2-5}$$

$$z_{\text{opt}}(t) = \frac{2C}{\mathcal{N}_0}(-jt)\psi^*(-t) e^{j\omega_{d0}t}, \qquad t_0 = \tau_R \tag{12.2-6}$$

Recall that the optimum filter was required to satisfy three conditions: (1) It must give $s_0(t_0) = 0$ when $\omega_{d0} = \omega_d$; (2) it must produce a signal response $s_0(t_0)$ that is proportional to $\omega_{d0} - \omega_d$ for $|\omega_{d0} - \omega_d|$ small; (3) it must provide the smallest variance of noise error in its Doppler frequency measurements. Requirement 3 has previously been demonstrated. We next show that requirements 1 and 2 are also satisfied. The signal response for any time and any filter is given by (12.1-16). For the optimum filter we substitute (12.2-4) for $z(t)$,

$$s_0(t) = \frac{\alpha C}{\mathcal{N}_0} \text{Re}\left\{ j \int_{-\infty}^{\infty} \xi\psi^*(\xi)\psi(\xi + t - t_0) e^{-j(\omega_{d0} - \omega_d)\xi} d\xi\, e^{j\omega_d(t - t_0)} \right\} \tag{12.2-7}$$

for optimum measurements this response at $t = t_0$ is needed

$$s_0(t_0) = \frac{\alpha C}{\mathcal{N}_0} \text{Re}\left\{ j \int_{-\infty}^{\infty} \xi|\psi(\xi)|^2 e^{-j(\omega_{d0} - \omega_d)\xi} d\xi \right\}$$

$$= \frac{\alpha C}{\mathcal{N}_0} \int_{-\infty}^{\infty} \xi|\psi(\xi)|^2 \sin[(\omega_{d0} - \omega_d)\xi] d\xi \tag{12.2-8}$$

Since the last form of (12.2-8) is zero when $\omega_{d0} = \omega_d$, requirement 1 is satisfied. Requirement 2 can be seen to be satisfied in two ways. First (12.2-8) is an odd function of $\omega_{d0} - \omega_d$, which means that $s_0(t_0)$ is approximately a linear function of $(\omega_{d0} - \omega_d)$ for $|\omega_{d0} - \omega_d|$ small. Second, if $|\omega_{d0} - \omega_d|$ is small, the sine function can be replaced by its argument giving $s_0(t_0) \approx K_\omega(\omega_{d0} - \omega_d)$, where

$$K_\omega = \frac{\alpha C}{\mathcal{N}_0} \int_{-\infty}^{\infty} \xi^2 |\psi(\xi)|^2 \, d\xi \qquad (12.2\text{-}9)$$

is the slope of $s_0(t_0)$ when $\omega_{d0} = \omega_d$. That (12.2-9) is true can also be shown by formally differentiating (12.2-8) according to (12.1-13). (See Problem 12.2-1):

By using (12.1-32) through (12.1-35) another form for K_ω can be developed as follows (see Problem 12.2-2):

$$K_\omega = \frac{C}{\alpha} \mathcal{R}[\tau_{\psi,\,\mathrm{rms}}^2 + (\bar{t}_\psi)^2] = \frac{C}{\alpha} \mathcal{R}[\tau_{g,\,\mathrm{rms}}^2 + (\bar{t}_g)^2] \qquad (12.2\text{-}10)$$

12.3   SOME PRACTICAL CONSIDERATIONS

For a given transmitted signal for which $\Psi(\omega)$ is known, it may not always be possible for the receiver to implement the optimum filter of (12.2-3) exactly. Practical departures can arise from at least two effects. In one, it may be difficult to implement the exact form of $\Psi(\omega)$ with a realizable filter. In the second, the required derivative with frequency may not be exact. We next consider these two practical effects.

Effect of Filter Mismatch on Accuracy

Let $Z(\omega)$ be any practical filter transfer function as represented in analytic form. $Z(\omega)$ is not necessarily the optimum function of (12.2-3). The variance of estimation error is given, in general, by (12.1-23). For white noise we write the expression as

$$\sigma_{\hat{\omega}_d}^2 = \frac{\displaystyle\int_{-\infty}^{\infty} |Z(\omega)|^2 \, d\omega}{\dfrac{\alpha^2}{2\pi \mathcal{N}_0} \left[\mathrm{Re} \left\{ \displaystyle\int_{-\infty}^{\infty} \frac{d\Psi(\omega - \omega_{d0})}{d\omega} Z(\omega) e^{j\omega(t_0 - \tau_R)} \, d\omega \right\} \right]^2} \qquad (12.3\text{-}1)$$

When $Z(\omega)$ is not the optimum filter, the variance $\sigma_{\hat{\omega}_d}^2$ will be larger than that for the optimum filter, as given by (12.1-29). The two variances may be related as

$$\sigma_{\hat{\omega}_d}^2 = \frac{1}{\eta_\omega^2} \sigma_{\hat{\omega}_d(\min)}^2 \qquad (12.3\text{-}2)$$

where we define a *filter efficiency factor* η_ω for the mismatched filter $Z(\omega)$ by

$$\eta_\omega^2 = \frac{\left[\mathrm{Re} \left\{ \displaystyle\int_{-\infty}^{\infty} \frac{d\Psi(\omega - \omega_{d0})}{d\omega} Z(\omega) e^{j\omega(t_0 - \tau_R)} \, d\omega \right\} \right]^2}{\displaystyle\int_{-\infty}^{\infty} |Z(\omega)|^2 \, d\omega \int_{-\infty}^{\infty} \left| \frac{d\Psi(\omega - \omega_{d0})}{d\omega} \right|^2 \, d\omega} \qquad (12.3\text{-}3)$$

If casuality is not of concern, (12.3-3) can be simplified by setting t_0, the time the system performs the Doppler measurement, equal to τ_R, the target's true delay time.

Our principal result is (12.3-2). It shows how the measurement variance $(\sigma_{\hat{\omega}_d}^2)$ of an arbitrary filter is related to the variance $[\sigma_{\hat{\omega}_d(\min)}^2]$ achieved by the optimum filter. The former is larger than the latter by a degradation factor $1/\eta_\omega^2$ where η_ω is the filter efficiency factor of the arbitrary filter. We take an example to illustrate a mismatched filter.

Example 12.3-1 We consider a filter that has the same "shape" as required in the optimum filter but is mismatched in bandwidth. Let the spectrum of the analytic representation of the transmitted signal have the form[4]

$$\Psi(\omega) = e^{-4\ln(\sqrt{2})(\omega - \omega_0)^2/W_\Psi^2}$$

where ω_0 is the carrier's angular frequency and W_Ψ is the 3.01-dB bandwidth. The spectrum of the actual filter will assume the same form as $\Psi(\omega)$ for the transmitted signal but will have a different 3.01-dB bandwidth of W_z. From (12.2-3) with $t_0 = \tau_R$ the "optimum" receiver filter's transfer function is

$$Z(\omega) = \frac{2C}{\mathcal{N}_0} \left[\frac{-8\ln(\sqrt{2})}{W_z^2} \right] (\omega - \omega_0 - \omega_{d0}) e^{-4\ln(\sqrt{2})(\omega - \omega_0 - \omega_{d0})^2/W_z^2}$$

On solving the integrals in (12.3-3), using the above two expressions and $t_0 = \tau_R$, it is found (Problem 12.3-3) that

$$\eta_\omega^2 = \left[\frac{2(W_\Psi/W_z)}{1 + (W_\Psi/W_z)^2} \right]^3$$

A plot of η_ω versus W_Ψ/W_z is shown in Fig. 12.3-1. It is to be noted that the efficiency factor is 1.0 when $W_\Psi = W_z$, that is, when no mismatch is present. This figure is similar to a plot in Barton (1988, p. 442), which is also based on a gaussian-shaped filter with bandwidth mismatch.

Frequency Discriminator Approximation

An important practical consideration is how to realize the derivative required in the optimum filter's characteristic of (12.2-3). We describe a practical approximation based on the parallel combination of two filters as depicted in Fig. 12.3-2. Here $Z_d(\omega)$ is an arbitrary function for the moment, and ω_q is a positive constant that is subsequently taken to be small relative to the bandwidth of $Z_d(\omega)$. $Z_\Delta(\omega)$ is the transfer function of the parallel combination; it is given by

$$Z_\Delta(\omega) = Z_d(\omega + \omega_q) - Z_d(\omega - \omega_q) \tag{12.3-4}$$

---

[4] Since $\Psi(\omega) \neq 0$ for $\omega < 0$, this spectrum is not strictly one of an analytic signal. However, since $\omega_0 \gg W_\Psi$ in most systems, the contributions for negative frequencies can reasonably be approximated as zero.

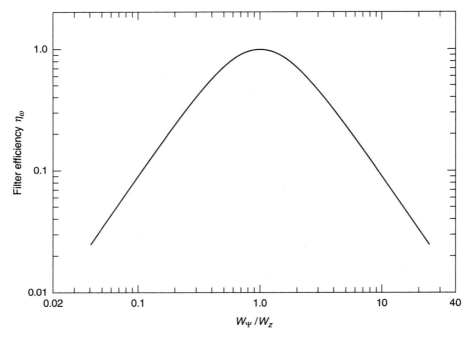

Figure 12.3-1 Filter efficiency for a gaussian filter mismatched in bandwidth for Doppler measurement using a gaussian signal.

We will also be interested later in the sum of the two functions, denoted by $Z_\Sigma(\omega)$,

$$Z_\Sigma(\omega) = Z_d(\omega + \omega_q) + Z_d(\omega - \omega_q) \tag{12.3-5}$$

Next we treat $Z_d(\omega + \omega_q)$ and $Z_d(\omega - \omega_q)$ as functions of the variable ω_q and expand them in Taylor series about the point $\omega_q = 0$:

$$Z_d(\omega + \omega_q) = [Z_d(\omega + \omega_q)|_{\omega_q = 0}] + \left[\frac{dZ_d(\omega + \omega_q)}{d\omega_q}\bigg|_{\omega_q = 0}\right]\frac{\omega_q}{1!}$$

$$+ \left[\frac{d^2 Z_d(\omega + \omega_q)}{d\omega_q^2}\bigg|_{\omega_q = 0}\right]\frac{\omega_q^2}{2!} + \cdots$$

$$= [Z_d(\omega)] + \left[\frac{dZ_d(\omega)}{d\omega}\right]\omega_q + \left[\frac{d^2 Z_d(\omega)}{d\omega^2}\right]\frac{\omega_q^2}{2} + \cdots \tag{12.3-6}$$

$$Z_d(\omega - \omega_q) = [Z_d(\omega - \omega_q)|_{\omega_q = 0}] + \left[\frac{dZ_d(\omega - \omega_q)}{d\omega_q}\bigg|_{\omega_q = 0}\right]\frac{\omega_q}{1!}$$

$$+ \left[\frac{d^2 Z_d(\omega - \omega_q)}{d\omega_q^2}\bigg|_{\omega_q = 0}\right]\frac{\omega_q^2}{2!} + \cdots$$

$$= [Z_d(\omega)] - \left[\frac{dZ_d(\omega)}{d\omega}\right]\omega_q + \left[\frac{d^2 Z_d(\omega)}{d\omega^2}\right]\frac{\omega_q^2}{2} - \cdots \tag{12.3-7}$$

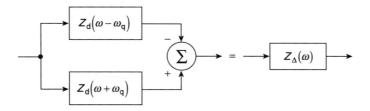

Figure 12.3-2 An approximately optimum filter for Doppler frequency measurements when $Z_d(\omega)$ is chosen according to (12.3-10).

If ω_q is small such that cubic and higher terms can be ignored in the difference of (12.3-4), we have

$$Z_\Delta(\omega) \approx 2\omega_q \frac{dZ_d(\omega)}{d\omega} \tag{12.3-8}$$

Similarly, if quadratic and higher terms are neglected in the sum of (12.3-5), we have

$$Z_\Sigma(\omega) \approx 2Z_d(\omega) \tag{12.3-9}$$

Because $Z_d(\omega)$ is arbitrary, we may ideally choose it to be a matched filter for the transmitted signal shifted to a Doppler frequency ω_{d0}. Thus we choose

$$Z_d(\omega) = \frac{2C}{\mathcal{N}_0} \Psi^*(\omega - \omega_{d0}) \tag{12.3-10}$$

On substituting (12.3-10) into (12.3-8), we have

$$Z_\Delta(\omega) \approx 2\omega_q \frac{2C}{\mathcal{N}_0} \frac{d\Psi^*(\omega - \omega_{d0})}{d\omega} = 2\omega_q Z_{opt}(\omega) e^{j\omega(t_0 - \tau_R)} \tag{12.3-11}$$

This last form derives from (12.2-3). For the sum of (12.3-9) we have

$$Z_\Sigma(\omega) \approx 2 \frac{2C}{\mathcal{N}_0} \Psi^*(\omega - \omega_{d0}) \tag{12.3-12}$$

The preceding developments have shown that two filters, each matched to the transmited signal shifted to a Doppler frequency ω_{d0}, can be "offset-tuned" by amounts $-\omega_q$ for one and $+\omega_q$ for the other and used in the circuit of Fig. 12.3-2 to approximate the optimum Doppler measurement filter. The approximation becomes better as ω_q is made smaller. However, as ω_q changes, the shape of the discriminator's characteristic changes and the slope at crossover is affected. To demonstrate, at least to some extent, these effects, we take two illustrative examples; both assume the same signal spectrum. The first develops the ideal response $Z_{opt}(\omega)$, while the second demonstrates the approximation to the ideal.

Example 12.3-2   Assume that[5]

$$\Psi(\omega) = \begin{cases} \cos\left[\dfrac{\pi}{2W}(\omega - \omega_0)\right], & -W \le (\omega - \omega_0) \le W \\ 0, & \text{elsewhere} \end{cases}$$

where $W > 0$ is a constant (half the spectral extent). After substitution into (12.2-5), we have

$$\frac{\mathcal{N}_0 W}{\pi C} Z_{\text{opt}}(\omega) = \begin{cases} -\sin\left[\dfrac{\pi}{2W}(\omega - \omega_0 - \omega_{d0})\right], & -W \le (\omega - \omega_0 - \omega_{d0}) \le W \\ 0, & \text{elsewhere} \end{cases}$$

This function is plotted in Fig. 12.3-3. The slope of $Z_{\text{opt}}(\omega)$ at the crossover point is $-\pi^2 C/(2\mathcal{N}_0 W^2)$.

Example 12.3-3   Here we use $\Psi(\omega)$ of Example 12.3-2 in (12.3-10) to define $Z_d(\omega)$:

$$Z_d(\omega) = \begin{cases} \dfrac{2C}{\mathcal{N}_0} \cos\left[\dfrac{\pi}{2W}(\omega - \omega_0 - \omega_{d0})\right], & -W \le (\omega - \omega_0 - \omega_{d0}) \le W \\ 0, & \text{elsewhere} \end{cases}$$

By direct substitution into (12.3-4), we obtain

$$\frac{\mathcal{N}_0}{2C} Z_d(\omega)$$

$$= \begin{cases} 0, & (\omega - \omega_0 - \omega_{d0}) \le -\omega_q - W \\ \cos\left[\dfrac{\pi}{2W}(\omega - \omega_0 - \omega_{d0} + \omega_q)\right], & -\omega_q - W < (\omega - \omega_0 - \omega_{d0}) \le \omega_q - W \\ -2\sin\left(\dfrac{\pi\omega_q}{2W}\right)\sin\left[\dfrac{\pi}{2W}(\omega - \omega_0 - \omega_{d0})\right], & \omega_q - W < (\omega - \omega_0 - \omega_{d0}) \le -\omega_q + W \\ -\cos\left[\dfrac{\pi}{2W}(\omega - \omega_0 - \omega_{d0} - \omega_q)\right], & -\omega_q + W < (\omega - \omega_0 - \omega_{d0}) \le \omega_q + W \\ 0 & \omega_q + W < (\omega - \omega_0 - \omega_{d0}) \end{cases}$$

$$(1)$$

---

[5] This spectrum choice does not necessarily correspond to a practical signal $\psi(t)$, but it is simple and convenient in illustrating the points of interest here.

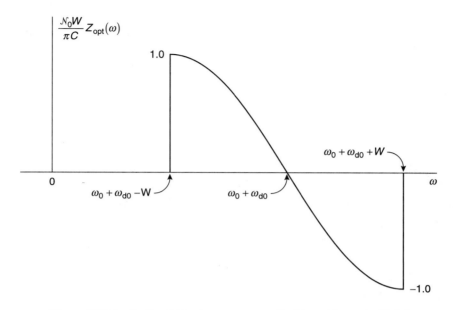

Figure 12.3-3 Optimal filter's response applicable to Example 12.3-2.

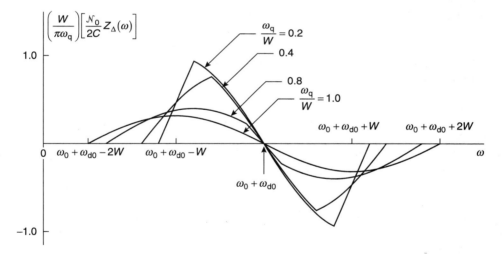

Figure 12.3-4 Approximation to optimum filter's response applicable to Example 12.3-3.

Because $Z_\Delta(\omega)$ and $Z_{opt}(\omega)$ are related through (12.3-11), we may write

$$\left(\frac{W}{\pi\omega_q}\right)\left[\frac{\mathcal{N}_0}{2C}Z_\Delta(\omega)\right] \approx \frac{\mathcal{N}_0 W}{\pi C}Z_{opt}(\omega) \tag{2}$$

The left side of (2) is plotted in Fig. 12.3-4 using (1) for several values of ω_q/W; these plots can be considered the approximations to the right side of (2). The right side is the response of the optimum filter; it is plotted in Fig. 12.3-3. Thus the two figures can be compared directly. Some observations are in order. As ω_q becomes small relative to W, the approximation approaches the ideal, and the two plots have the same slope at crossover as $\omega_q \to 0$. For increasing ω_q there is a decrease in the approximately linear region around crossover, and the slope decreases. The slope decrease is not too excessive, however, until ω_q approaches W. When $\omega_q = W$, the slope of the approximate filter is $1/\pi$ times the slope of the optimum filter.

12.4   PRACTICAL NONCOHERENT IMPLEMENTATIONS FOR DOPPLER MEASUREMENT

Here we consider some practical ways of implementing near-optimum systems for measurement of Doppler frequency ω_d when using one or more target pulses processed noncoherently.

Single-Pulse Measurement—Initial Estimate Available

For measurement of ω_d using a single pulse, the optimum system should theoretically use the optimum filter of (12.2-3) and sample the response at $t = t_0 = \tau_R$. However, recall that this filter was derived assuming that only ω_d was unknown but a reasonably close initial estimate (ω_{d0}) of ω_d was available. From a practical standpoint the *amplitude* of the received target signal is not really known. In fact it can fluctuate from pulse to pulse or it can be steady. This is an extreme range of possible amplitudes that must be anticipated and compensated where needed. A good compensation amounts to converting (normalizing) the optimum filter's response to be independent of target response amplitude.

A second practical problem is that the output of the optimum filter of (12.2-3) is bandpass; that is, it involves a carrier. The carrier needs to be removed in most systems so that measurements can be based on more easily processed video (baseband) waveforms.

A third practical problem arises from the need to build the optimum filter of (12.2-3) at a frequency ω_{d0}. Once the initial estimate of ω_d is corrected by use of a (current) pulse, ω_{d0} may change. For physically designed filters it is not easy to make them variable as ω_{d0} changes. Indeed in some cases the initial estimate ω_{d0} may not be available; this case is discussed in the following subsection.

A method of overcoming all the preceding practical problems is shown in Fig. 12.4-1; it assumes that a close intial estimate ω_{d0} of ω_d is available. First, instead of designing an optimum filter for a center frequency of $\omega_0 + \omega_{d0}$, a mixer is used to reduce the input frequency to an *intermediate frequency* (or IF) ω_{IF} with the nominal Doppler frequency ω_{d0} removed. This operation allows the optimum filter to be centered at a fixed frequency ω_{IF} rather than at $\omega_0 + \omega_{d0}$. The offset matched filters that approximate the derivative in the optimum filter are centered at frequencies $\omega_{IF} + \omega_q$ and $\omega_{IF} - \omega_q$. The signal response s_Δ in Fig. 12.4-1 becomes the approximate equivalent of s_0 for the optimum filter. As ω_{d0} changes, it is only necessary to change the frequency $\omega_0 - \omega_{IF} + \omega_{d0}$ of the "local oscillator," which is generated by a voltage-controlled oscillator (VCO) using a baseband designation for the proper value of ω_{d0}.

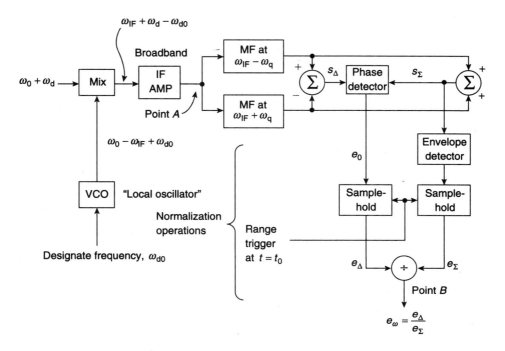

Figure 12.4-1 A practical system for measurement of Doppler angular frequency ω_d on one pulse.

For the moment we ignore noises and discuss only the signals shown in Fig. 12.4-1. To remove the dependence of s_Δ on the target's carrier phase, we may form a second response s_Σ that makes use of (12.3-12). At the time of interest $(t = t_0 = \tau_R)$ for Doppler measurements, s_Σ will reach a peak amplitude and be nearly constant in amplitude for times near τ_R; it will have the same phase as s_Δ. Thus the phase detector can be used to remove the carrier phase of s_Δ and produce a response proportional to the amplitude of s_Δ. If $\psi_\Delta(t)$ and $\psi_\Sigma(t)$ are the analytic signal representations of $s_\Delta(t)$ and $s_\Sigma(t)$, respectively, the phase detector's real output can be shown to be $\frac{1}{2} \text{Re}[\psi_\Delta(t)\psi_\Sigma^*(t)]$ (see Problem 12.4-1). At time $t = \tau_R$ this response is

$$e_\Delta(\tau_R) = \tfrac{1}{2} \text{Re}\left[\psi_\Delta(\tau_R)\psi_\Sigma^*(\tau_R)\right] \tag{12.4-1}$$

Finally, to remove the dependence of $e_\Delta(\tau_R)$ on the target's amplitude (through parameter α), we normalize $e_\Delta(\tau_R)$ by dividing it by the square of the amplitude of $s_\Sigma(t)$ at time $t = \tau_R$.[6] In terms of the analytic signal representations, we desire to create $e_\Sigma = |\psi_\Sigma(\tau_R)|^2$, so

$$e_\omega = \frac{e_\Delta}{e_\Sigma} = \frac{1}{2}\text{Re}\left\{\frac{\psi_\Delta(\tau_R)\psi_\Sigma^*(\tau_R)}{|\psi_\Sigma(\tau_R)|^2}\right\} = \frac{1}{2}\text{Re}\left[\frac{\psi_\Delta(\tau_R)}{\psi_\Sigma(\tau_R)}\right] \tag{12.4-2}$$

---

[6] Since $e_\Delta(\tau_R)$ is proportional to the product of $s_\Delta(\tau_R)$ and $s_\Sigma(\tau_R)$ due to the action of the (typical) phase detector, normalization of the product requires the square of the magnitude of $s_\Sigma(\tau_R)$. The operation is equivalent to normalization of both $s_\Delta(\tau_R)$ and $s_\Sigma(\tau_R)$ prior to the product by the peak amplitude of $s_\Sigma(\tau_R)$.

from (12.4-1). The division is best done by digital means on samples of $e_\Delta(t)$ and $e_\Sigma(t)$ taken at time $t = \tau_R$. An analog method of normalization is also possible (see Problem 12.4-2), but it is difficult for this technique to provide the desired "instantaneous" normalization provided by the digital approach.

In the practical system of Fig. 12.4-1, there are noises added to the various signals shown. An analysis of measurement error based on estimates using the phase detector's output (no normalization) is a bit lengthy, but when (12.1-15) is applied, we find that

$$\sigma_{\hat{\omega}_d}^2 = \frac{2 + \mathscr{R}_p}{\mathscr{R}_p^2 \tau_{\psi,\text{rms}}^2} = \frac{L_x}{\mathscr{R}_p \tau_{\psi,\text{rms}}^2} = \frac{L_x}{\mathscr{R}_p \tau_{g,\text{rms}}^2} \tag{12.4-3}$$

where $\mathscr{R}_p$ is defined by (12.1-32) with E_r applicable to a single pulse and L_x is a phase detector loss defined by

$$L_x = \frac{2 + \mathscr{R}_p}{\mathscr{R}_p} = 1 + \frac{2}{\mathscr{R}_p} \tag{12.4-4}$$

Except for the detector's loss the practical system's performance equals that of the optimum system of (12.1-31) when $\bar{t}_\psi = 0$, the usual case. For high signal-to-noise ratios the effect of the normalization on noise performance is small and (12.4-3) can be assumed to apply.

Single-Pulse Measurement—No Initial Doppler Estimate

The practical system of Fig. 12.4-1 assumes that an initial estimate ω_{d0} of the true Doppler angular frequency is available. The value of ω_{d0} is presumed close enough to ω_d that the difference $\omega_{d0} - \omega_d$ is small enough to fall within the linear response region of the frequency discriminator. For practical discriminators this difference should have a magnitude not exceeding about half the width of the uncertainty function's cut along the Doppler axis. For a single pulse with a rectangular envelope of duration T, this width is $2\pi/T$ rad/s, which equals the frequency resolution constant, Ω_{res}, of the transmitted signal. For pulses with nonconstant, or tapered, envelopes, Ω_{res} may also equal $2\pi/T$, depending on how duration T is defined (see Problems 12.4-3–12.4-7). The most common definitions of duration T tend to lead to a value of Ω_{res} a bit less than $2\pi/T$ (see Problems 12.4-8–12.4-12). With these qualifying observations in mind, we observe that the initial Doppler estimate should be sufficiently accurate that $|\omega_{d0} - \omega_d|$ should not exceed approximately half the Doppler resolution constant, Ω_{res}, of the system [see (8.2-10)]. If the accuracy of ω_{d0} is not good enough, or if no initial estimate is available, another approach is required.

When no initial Doppler estimate is available but the radar has established range measurements so that target delay is known, a bank of single-pulse matched filters operating in parallel can be used to obtain a single-pulse estimate of ω_d. One possible filter bank is illustrated in Fig. 12.4-2a. Here $2M$ filters, separated in angular frequency between pairs by Ω_{res},[7] are centered on the IF frequency. Filter m has center frequency

---

[7] Again we emphasize the separation should equal about the width of the uncertainty function cut along the Doppler axis, to be more precise. For the important case of a constant-amplitude pulse, this width equals Ω_{res}.

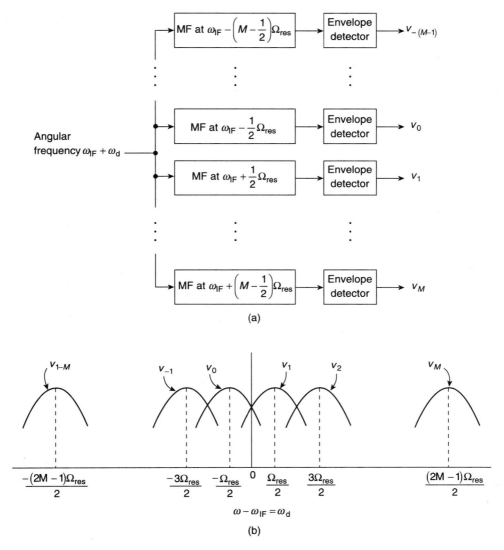

Figure 12.4-2   A Doppler filter bank (*a*) and its responses (*b*) for single-pulse measurement of Doppler frequency.

$\omega_{IF} + (m - \frac{1}{2})\Omega_{res}$ for $(1 - M) \le m \le M$. The responses of the filters are shown in panel *b*. In operation, the target's pulse will excite the largest response in the filter most closely tuned to its true Doppler frequency ω_d when the responses are observed at the target's delay time. For a largest response in filter m, the system forms a single-pulse Doppler estimate $\omega_{do} = (m - \frac{1}{2})\Omega_{res}$. This estimate is only precise to about $\pm \Omega_{res}/2$. However, if the two center filters are used to form the discriminator in Fig. 12.4-1 (this assumes that $\omega_q = \Omega_{res}/2$), the estimate should be adequate to designate the local oscillator such that later target pulses can be measured using the system of Fig. 12.4-1. Clearly the allowable maximum value of ω_d depends on the

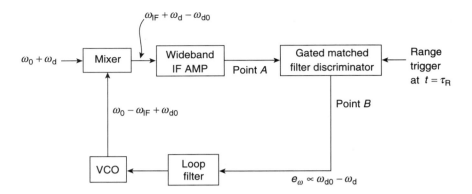

Figure 12.4-3 A loop for continuous measurement (tracking) of Doppler frequency.

number of filters; for $-|\omega_{d(max)}| \leq \omega_d \leq |\omega_{d(max)}|$ we require that

$$|\omega_{d(max)}| \leq M\Omega_{res} \qquad (12.4\text{-}5)$$

If the offset $\omega_q = \Omega_{res}/2$ is deemed too large to give good discriminator performance, the more optimum filters in Fig. 12.4-1 can be used to form the center filter (at center frequency ω_{IF} when the initial Doppler estimate is zero) of a bank of $(2M + 1)$ filters. The side filters are now at center frequencies $\omega_{IF} + m\Omega_{res}$, for $m = \pm 1$, $\pm 2, \ldots, \pm M$.

With a proper scale constant, the envelope detected version of s_Σ in Fig. 12.4-1 is used within the filter bank when determining which filter gives the largest response on the first target pulse. This system is possible because s_Σ is a matched filter response for zero Doppler according to (12.3-12) when ω_{d0} is set to zero. For this arrangement the largest Doppler magnitude must satisfy $|\omega_{d(max)}| \leq (M + \frac{1}{2})\Omega_{res}$.

Doppler Tracking by Continuous Measurements

If the single-pulse Doppler measurements of Fig. 12.4-1 are averaged in a feedback loop closed on the voltage-controlled oscillator (VCO), Doppler frequency can be continuously "tracked." The feedback (servo) loop with a closed-loop low-pass effectively bandwidth W_L (rad/s) will effectively average

$$n_{eff} = \frac{\pi}{W_L T_R} \qquad (12.4\text{-}6)$$

pulses (Barton, 1964, p. 394) in a noncoherent manner. The applicable system is sketched in Fig. 12.4-3 where the discriminator between points A and B is the same as in Fig. 12.4-1 from point A clockwise to point B. The system's action is to adjust the VCO's output frequency until ω_{d0} is nearly equal to ω_d. To see the action, assume that $\omega_d > \omega_{d0}$, so e_ω is a negative voltage. The VCO responds by *raising* its output frequency,[8] so

---

[8] The changes in the VCO's output frequency from $\omega_0 - \omega_{IF}$ must be proportional to the negative of its input control voltage if the discriminator's response is proportional to $\omega_{d0} - \omega_d$.

ω_{d0} approaches ω_d. If the loop's low-pass filter is simple with finite gain at dc, the loop is a simple frequency-locked loop where ω_{d0} can be made to equal ω_d within a small, but finite, error.

In principle, for a nonaccelerating target, the bandwidth W_L of the closed loop of Fig. 12.4-3 can be made very small, so n_{eff} can be large. In practice, there is a limit because of target dynamics (acceleration and higher rates) and other practical factors. For a properly functioning loop the Doppler measurement error's variance is

$$\sigma_{\hat{\omega}_d}^2 = \frac{1}{n_{eff}} \sigma_{\hat{\omega}_d}^2 (1 \text{ pulse}) = \frac{W_L T_R L_x}{\pi \mathscr{R}_p \tau_{\psi,rms}^2} = \frac{W_L T_R (2 + \mathscr{R}_p)}{\pi \mathscr{R}_p^2 \tau_{\psi,rms}^2} \qquad (12.4\text{-}7)$$

Finally we note that correlator versions of the optimum Doppler processor are known as described in Barton and Ward (1969). Also described are other practical details such as processing losses, applications to scanning radars, and other details on mismatch losses.

12.5 OPTIMUM COHERENT DOPPLER MEASUREMENTS

If the phase properties of the system, medium, and target are stable over an N-pulse interval, the N pulses can be coherently processed in the receiver. Coherent processing yields improved Doppler frequency measurement performance compared to non-coherent processing. System stability requires both transmitter and receiver operations be phase-stable.

The optimum-system results derived earlier can be applied to coherent processing if we interpret the sequence of N pulses as a single waveform.

Properties of N Coherent Pulses

We examine the temporal and spectral characteristics of a train of N coherent transmitted pulses. For convenience we center the origin on the pulse group as shown in Fig. 12.5-1. For an odd number N of pulses, one pulse is centered at the origin such that its mean time is zero (see Chapter 6), as shown in panel a, and the separation between adjacent pulse pairs is T_R seconds. For N even, the two central pulses are displaced from the origin so that their mean times are $\pm T_R/2$, as shown in panel b. All pulses are assumed to have identical amplitude and phase modulations $a(t)$ and $\theta(t)$, respectively. The transmitted waveform $s(t)$ can be written as

$$s(t) = \sum_{n=0}^{N-1} a(t + \bar{T} - nT_R) \cos[\omega_0 t + \theta(t + \bar{T} - nT_R) + \phi_0] \qquad (12.5\text{-}1)$$

where

$$\bar{T} = \frac{(N-1)T_R}{2} \qquad (12.5\text{-}2)$$

ω_0 and ϕ_0 are the carrier's angular frequency and phase angle, respectively.

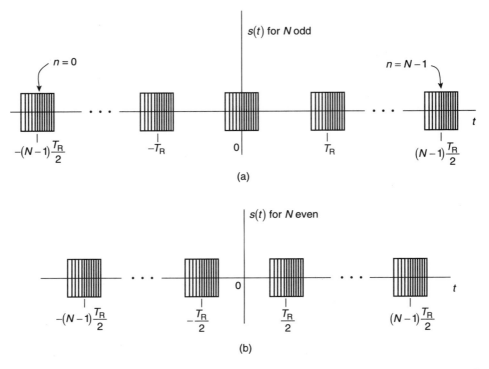

Figure 12.5-1 A transmitted waveform having N pulses: (a) when N is odd, and (b) when N is even. Sketches assume rectangular pulse envelopes only for ease of illustration.

For later purposes the signal $s(t)$ is better represented by its analytic form $\psi(t)$. From Chapter 6 we have

$$s(t) = \text{Re}[\psi(t)] = \text{Re}\left\{ \sum_{n=0}^{N-1} a(t + \bar{T} - nT_R) e^{j\theta(t + \bar{T} - nT_R) + j\omega_0 t + j\phi_0} \right\}$$

$$= \text{Re}\left\{ \left[\sum_{n=0}^{N-1} g(t + \bar{T} - nT_R) \right] e^{j\omega_0 t + j\phi_0} \right\}$$

$$= \text{Re}\left\{ g_N(t) e^{j\omega_0 t + j\phi_0} \right\} \tag{12.5-3}$$

where

$$\psi(t) = g_N(t) e^{j\omega_0 t + j\phi_0} \tag{12.5-4}$$

$$g_N(t) = \sum_{n=0}^{N-1} g(t + \bar{T} - nT_R) \tag{12.5-5}$$

$$g(t) = a(t) e^{j\theta(t)} \tag{12.5-6}$$

The function $g_N(t)$ can be interpreted as the complex envelope of the transmitted signal when N pulses are used but treated as a single modulating waveform.

The Fourier transform of $\psi(t)$, denoted by $\Psi(\omega)$, is the spectrum of the analytic form of the transmitted signal. In Problem 12.5-1 it is found to be

$$\Psi(\omega) = G(\omega - \omega_0)e^{j\phi_0}G_L(\omega) \tag{12.5-7}$$

where we define

$$G_L(\omega) = \frac{\sin[N(\omega - \omega_0)T_R/2]}{\sin[(\omega - \omega_0)T_R/2]} \tag{12.5-8}$$

A sketch of $\Psi(\omega)$ less the exponential factor is shown in Fig. 12.5-2. The modulus is proportional to the factor $G(\omega - \omega_0)$, which is due to the complex envelope of a single typical pulse. If only one pulse were used, this factor times $\exp(j\phi_0)$ would be the full spectrum $\Psi(\omega)$. The presence of the train of pulses causes the spectrum to grow in amplitude by the factor N and break into "lines" at multiples of $\omega_R = 2\pi/T_R$ on both sides of ω_0. The width of each line between points 3.92 dB below the line's maximum amplitude is $2\pi/(NT_R)$ for large N. These lines are mainly controlled by the spectrum's factor $G_L(\omega)$. For N an odd integer, these lines all have a positive amplitude as

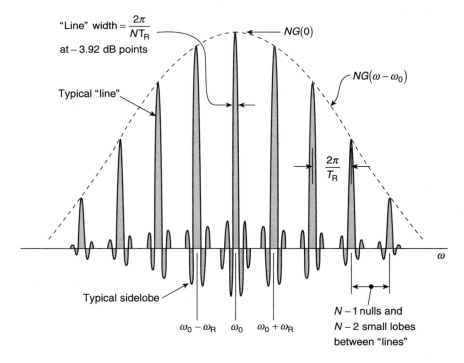

Figure 12.5-2 Example illustration of $\Psi(\omega)$ showing lines at multiples of $2\pi/T_R$ about the carrier frequency ω_0.

sketched in the figure.[9] For N an even integer, lines at $\omega_0 \pm k\omega_R, k = 0, 2, 4, \ldots$, have positive amplitudes, while those for $k = 1, 3, 5, \ldots$, are negative. Sidelobes exist around each line; there are $N - 2$ sidelobes and $N - 1$ nulls between pairs of lines. Sidelobe peak amplitudes decrease for positions away from a line with the smallest ones nearest to midway between two lines. The largest sidelobes are the first on each side of a line, and their peak amplitudes are aproximately 13.26 dB below the line's peak amplitude. There is some tapering effect on lines and sidelobes in $\Psi(\omega)$ due to $G(\omega - \omega_0)$, but this effect is small for N large and lines not near the null regions of $G(\omega - \omega_0)$.

For the waveform of (12.5-4), the received waveform is given by (12.1-3). Its spectrum is

$$\Psi_r(\omega) = \alpha \Psi(\omega - \omega_d) e^{-j\omega\tau_R} \tag{12.5-9}$$

Except for the scale constant α due to the radar equation, and the phase shift due to target delay, the received signal's spectrum is just a frequency-shifted version of the transmitted signal. They both have the same shape.

Optimum Filter for N Pulses

We now define $\psi_1(t)$ and $\Psi_1(\omega)$ as the respective quantities $\psi(t)$ and $\Psi(\omega)$ when $N = 1$. By Fourier transformation of (12.5-4) for $N = 1$, we have

$$\Psi_1(\omega) = G(\omega - \omega_0) e^{j\phi_0} \tag{12.5-10}$$

Equation (12.5-7) becomes

$$\Psi(\omega) = \Psi_1(\omega) G_L(\omega) \tag{12.5-11}$$

In other words, the spectrum of the N-pulse coherent signal is the product of the spectrum of a 1-pulse signal and $G_L(\omega)$, the factor that generates the "line" behavior of $\Psi(\omega)$.

Because we have represented the N-pulse train as a single waveform, the optimum filter defined by (12.2-3) still applies except now with $\Psi(\omega)$ given by (12.5-11). Since $G_L(\omega)$ is real, we have

$$Z_{opt}(\omega) = \frac{2C}{\mathcal{N}_0} \frac{d[\Psi_1^*(\omega - \omega_{d0}) G_L(\omega - \omega_{d0})]}{d\omega} e^{-j\omega(t_0 - \tau_R)} \tag{12.5-12}$$

As before, C is the arbitrary real scale constant of the filter, $\mathcal{N}_0/2$ is the input white noise's power density, and t_0 is the time at which the Doppler frequency measurement is made.

Implementation methods for the filter of (12.5-12) are discussed in Section 12.6.

---

[9] Positive or negative values are quoted for the factor $G_L(\omega)$. The polarity of a line in the whole spectrum $\Psi(\omega)$ is also affected by the polarity of $G(\omega - \omega_0)$ and the exponential phase factor.

Accuracy of Optimum Filter

Measurement accuracy when using the optimum filter of (12.5-12) is obtained from (12.1-31) when $\mathscr{R}$, $\tau_{\psi,\text{rms}}^2$, and $\bar{t}_\psi$ are determined from the signal defined by $\Psi(\omega)$ in (12.5-11). From Problem 12.5-6 we find that

$$\mathscr{R} = N\mathscr{R}_\text{p} = N\frac{2E_\text{r}}{\mathscr{N}_0} \tag{12.5-13}$$

where $\mathscr{R}_\text{p} = 2E_\text{r}/\mathscr{N}_0$ and E_r is the energy in one of the N received pulses. For $\tau_{\psi,\text{rms}}^2$ it is found, assuming $\bar{t}_\psi = 0$ (Problem 12.5-7), that

$$\tau_{\psi,\text{rms}}^2(N\text{ pulses}) = \tau_{g,\text{rms}}^2(1\text{ pulse}) + \frac{(N^2 - 1)T_\text{R}^2}{12} \tag{12.5-14}$$

Thus

$$\sigma_{\hat{\omega}_\text{d}(\text{min})}^2 = \frac{1}{N\mathscr{R}_\text{p}\left\{\tau_{\psi,\text{rms}}^2(1\text{ pulse}) + [(N^2 - 1)T_\text{R}^2/12]\right\}} \tag{12.5-15}$$

In nearly all cases of interest $\tau_{\psi,\text{rms}}$ for one pulse is much smaller than $T_\text{R}/2$ so the second term in the denominator of (12.5-15) dominates, even when N is as small as 2. Thus, to a good accuracy,

$$\sigma_{\hat{\omega}_\text{d}(\text{min})}^2 \approx \frac{12}{N\mathscr{R}_\text{p}(N^2 - 1)T_\text{R}^2}, \qquad N \geq 2 \tag{12.5-16}$$

Another useful form for (12.5-16) is

$$\sigma_{\hat{\omega}_\text{d}(\text{min})}^2 \approx \frac{12}{\mathscr{R}T_N^2}\left(\frac{N-1}{N+1}\right), \qquad N \geq 2 \tag{12.5-17}$$

where $\mathscr{R} = N\mathscr{R}_\text{p}$ and

$$T_N = (N-1)T_\text{R} \tag{12.5-18}$$

is the total span of time over which pulses are transmitted.[10]

Example 12.5-1 We assume rectangular transmitted pulses of duration T and no frequency modulation. We compare $\sigma_{\hat{\omega}_\text{d}(\text{min})}^2$ when N coherent pulses are used to that

---

[10] If individual pulse durations are T, the total span is actually $(N-1)T_\text{R} + T$, but since $T_\text{R} \gg T$, we neglect the small contribution from pulse duration T.

when N pulses are noncoherently processed. Here $\tau^2_{\psi,\mathrm{rms}}$ (1 pulse) $= T^2/12$ from Example 12.1-1. From (12.5-15) and Example 12.1-1 with N pulses integrated,

$$\frac{\sigma^2_{\hat{\omega}_d(\min)}(N \text{ pulses, noncoherent})}{\sigma^2_{\hat{\omega}_d(\min)}(N \text{ pulses, coherent})} = 1 + (N^2 - 1)\left(\frac{T_\mathrm{R}}{T}\right)^2$$

By assuming some numerical values for T_R and T, such that $T_\mathrm{R}/T = 10$, we continue the example. The ratio becomes $1 + 100(N^2 - 1)$ which equals 1, 301, and 801 for $N = 1, 2$, and 3, respectively. Clearly the improvement offered by coherent processing increases rapidly compared to noncoherent processing as either or both N and T_R/T increase.

12.6 PRACTICAL COHERENT IMPLEMENTATIONS FOR DOPPLER MEASUREMENT

The practical implementations described in Section 12.4 for processing noncoherent pulses have their counterparts when using N coherent pulses treated as a single waveform. We subsequently develop these implementations. However, we first need to discuss a special filter called a *transversal filter* because it is needed in the various implementations.

Transversal Filter

A transversal filter consists of a cascade of $N - 1$ delays of T_R seconds each as shown in Fig. 12.6-1. At the output of the cascade and at each input of a delay element (called taps), the signal is passed through a phase shift[11] of β_n, for $n = 0, 1, 2, \ldots, (N - 1)$. These phase shifters are assumed to be broadband, so phases are constants with frequency. For our current purposes we assume that

$$\beta_n = \omega_x(nT_\mathrm{R} - \bar{T}), \qquad n = 0, 1, \ldots, (N - 1) \tag{12.6-1}$$

where ω_x is an arbitrary (selectable) angular frequency constant and $\bar{T}$ is given by (12.5-2). For this choice of phase shifts the transfer function, denoted by $H_\mathrm{L}(\omega; \omega_x)$, of the filter is found to be (Problem 12.6-1)

$$H_\mathrm{L}(\omega; \omega_x) = \frac{\sin\left[N(\omega - \omega_x)T_\mathrm{R}/2\right]}{\sin\left[(\omega - \omega_x)T_\mathrm{R}/2\right]} e^{-j\omega\bar{T}} \tag{12.6-2}$$

The value of the transversal filter resides in its ability to realize the function $G_\mathrm{L}(\omega)$ of (12.5-8) by choice of $\omega_x = \omega_0$ [the factor $\exp(-j\omega\bar{T})$ contributes nothing to the shape of $G_\mathrm{L}(\omega)$; it only adds delay]. More important, as we will see, other choices of

---

[11] A more general transversal filter uses "weighting coefficients" that are complex numbers representing both amplitude and phase variations with n. These are sometimes called *comb filters*, and they can have other names as well. Such filters are useful in *moving target indication* (or MTI) systems.

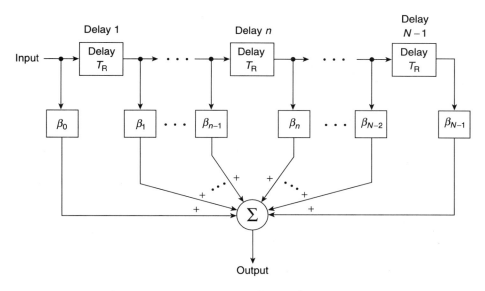

Figure 12.6-1   A transversal filter having one output.

ω_x will allow the function $G_L(\omega)$ to be shifted to other center frequencies as needed by the optimum filter of (12.5-12).

Optimum Filter

To implement the optimum filter, we first define a "receiver" transfer function $Z_{rec}(\omega)$ for a parallel pair of filters, as shown in Fig. 12.3-2. Each path consists of a matched filter for a single pulse in cascade with a transversal filter, both centered at the carrier plus the available Doppler angular frequency estimate ω_{d0}. However, in one path the cascade is shifted in angular frequency by $+\omega_q$, while the other is shifted by $-\omega_q$. Thus

$$Z_{rec}(\omega) = \frac{2C}{\mathcal{N}_0} \Psi_1^*(\omega - \omega_{d0} + \omega_q) H_L(\omega; \omega_x = \omega_0 + \omega_{d0} - \omega_q)$$

$$-\frac{2C}{\mathcal{N}_0} \Psi_1^*(\omega - \omega_{d0} - \omega_q) H_L(\omega; \omega_x = \omega_0 + \omega_{d0} + \omega_q) \quad (12.6\text{-}3)$$

On using (12.6-2) and (12.5-8), we have

$$Z_{rec}(\omega) = \frac{2C}{\mathcal{N}_0} \{\Psi_1^*(\omega - \omega_{d0} + \omega_q) G_L(\omega - \omega_{d0} + \omega_q)$$

$$- \Psi_1^*(\omega - \omega_{d0} - \omega_q) G_L(\omega - \omega_{d0} - \omega_q)\} e^{-j\omega T} \quad (12.6\text{-}4)$$

Next we apply (12.3-8) to approximate the quantity within the braces in (12.6-4) by the derivative

$$Z_{\text{rec}}(\omega) \approx 2\omega_q \frac{2C}{\mathcal{N}_0} \frac{d}{d\omega} \{\Psi_1^*(\omega - \omega_{d0}) G_L(\omega - \omega_{d0})\} e^{-j\omega T} \qquad (12.6\text{-}5)$$

Finally we relate $Z_{\text{rec}}(\omega)$ to the optimum filter through (12.5-12) to obtain

$$Z_{\text{opt}}(\omega) \approx \frac{1}{2\omega_q} Z_{\text{rec}}(\omega) e^{-j\omega(t_0 - \tau_R - \bar{T})} \qquad (12.6\text{-}6)$$

This result shows that the "receiver" function $Z_{\text{rec}}(\omega)$ produces the optimum filter within a scale constant $(1/2\omega_q)$ and phase shift factor $\{\exp[-j\omega(t_0 - \tau_R - \bar{T})]\}$.

Use of the optimum filter requires that the intial Doppler estimate ω_{d0} be accurate enough such that $\omega_{d0} - \omega_d$, the error between ω_{d0} and the true Doppler value ω_d, be within the response region of the filter. Even if this condition is true, ω_d can change with time such that its value drifts outside the linear range of the filter, making a fixed-filter design usually impractical.

Single Waveform Measurements with Initial Estimate

A better approach is to recenter the optimum filter to an intermediate frequency, ω_{IF}, as sketched in Fig. 12.4-1. The filter now uses a fixed center frequency, and the current Doppler estimate (that may vary with time) is removed using the mixer. The requirement that ω_{d0} be close enough to ω_d such that the IF signal is within the linear response region of the filter still remains, however. Generally, this constraint means that $|\omega_{d0} - \omega_d|$ should be not more than about half the width of the uncertainty function of the transmitted waveform as measured along the Doppler axis (the cut for $\tau = 0$).

For the usual case where $s(t)$ consists of N coherent pulses of the same form, it is known that the width of the uncertainty function is $2\pi/NT_R$ rad/s from (7.6-14). Therefore the initial estimate ω_{d0} must satisfy $|\omega_{d0} - \omega_d| < \pi/NT_R$.

Doppler Measurement with Poor or No Initial Estimate

Use of an N-pulse coherent processor to measure a target's Doppler frequency requires that an accurate initial Doppler estimate be available. If no estimate, or an estimate with poor accuracy, is available, then other steps must be taken to obtain an initial estimate of suitable accuracy. For discussion purposes we will consider two regions of estimate accuracy.

In the first region we presume that Doppler frequency is either not known or is known to very poor accuracy. For example $|\omega_d|$ may be unknown but is known to be larger than half the width of the uncertainty function of a single-pulse waveform (which is about half of $2\pi/T$ for a pulse duration T). Here it becomes necessary to use methods such as differentiation of range data; parallel bank of 1-pulse filters, or 1-pulse bank in conjunction with a 1-pulse discriminator to improve 1-pulse accuracy; or use prior knowledge (e.g., knowing the radial speed of a target satellite that rises over the horizon) to develop an initial estimate of Doppler frequency (see also Barton, 1988, p. 445). A combination of these methods may even be necessary. Regardless of the method chosen we will assume that it either gives the full accuracy required to proceed to use the coherent N-pulse processor for final measurements or gives a Doppler estimate with an error $|\omega_{d0} - \omega_d| \leq \pi/T$ such that the estimate serves only as a refinement to allow a second step to generate the final initial Doppler estimate.

The second region of estimation accuracy presumes Doppler frequency is known to an error not exceeding $\pm \pi/T$. In this region a bank of coherent N-pulse matched filters may be used. An examination of the uncertainty function of these filters [see (7.6-14) and Fig. 7.6-4] shows ambiguities in Doppler frequency at multiples of $2\pi/T_R$. This fact means that the filter bank is unambiguous only for $|\omega_{d0} - \omega_d| < \pi/T_R$. The filter bank can give an estimate of Doppler frequency to an error of not more than about $\pm \pi/NT_R$ (the half-width of the uncertainty function) but some form of ambiguity resolution is needed when $|\omega_{d0} - \omega_d| \geq \pi/T_R$. Resolution methods do exist, and one method relies on varying pulse repetition frequency (see Skolnik, 1980; Barton, 1988, p. 445; Edde, 1993, sec. 6–4). When $|\omega_{d0} - \omega_d| < \pi/T_R$ the coherent N-pulse filter bank is unambiguous. We next consider one way of implementing the coherent N-pulse filter bank.

Technically a filter bank of coherent N-pulse matched filters requires each filter to be a cascade of a 1-pulse matched filter and the function $G_L(\omega)$ from (12.5-11); the cascade is shifted to the center frequency of the specific filter in the bank. However, for the Doppler accuracy we have assumed when using this bank, which is $\pm \pi/T$ or better, the frequency shifts in the single-pulse matched filters are relatively small such that a practical realization can usually use only one such filter. Thus the N-pulse bank consists approximately of one 1-pulse matched filter at the center frequency of the bank (which is the IF frequency) in cascade with a bank of filters consisting of frequency-shifted replicas of the function $G_L(\omega)$[12] of (12.5-11). Because $G_L(\omega)$ is periodic with period $2\pi/T_R$, at most N parallel filters are needed to achieve full-spectrum coverage.

By using a matrix of phase shifters, a transversal filter can synthesize the necessary bank of filters for the function $G_L(\omega)$ as shown in Fig. 12.6-2. If the phase shifts are selected as

$$\beta_{nm} = \left[\omega_{IF} + m\frac{2\pi}{NT_R} \right](nT_R - \bar{T}), \qquad n = 0, 1, 2, \ldots, (N-1), m = 0 \pm 1, \ldots, \pm M$$

$$(12.6\text{-}7)$$

where

$$M = \begin{cases} \dfrac{N-1}{2}, & N \text{ odd} \\[2mm] \dfrac{N-2}{2}, & N \text{ even} \end{cases} \qquad (12.6\text{-}8)$$

---

[12] For a bank of filters about the carrier ω_0, the function $G_L(\omega)$ is shifted by multiples of $2\pi/NT_R$; for our case we replace ω_0 by ω_{IF} to obtain the function shifted by multiples of $2\pi/NT_R$ about ω_{IF}.

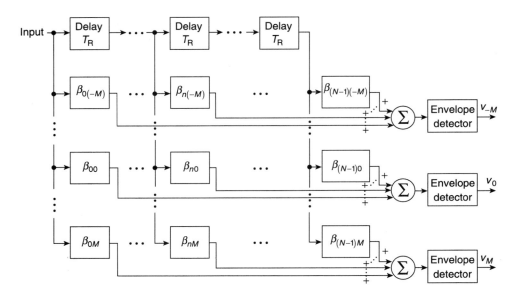

Figure 12.6-2   Transversal filter in a form that implements a bank of coherent N-pulse filters.

then the transfer function of output m (prior to the envelope detector) is (see Problem 12.6-2)

$$H_{\mathrm{L}}\left(\omega;\omega_{\mathrm{IF}}+m\frac{2\pi}{NT_{\mathrm{R}}}\right)=\frac{\sin\left[N(\omega-\omega_{\mathrm{IF}}-m2\pi/NT_{\mathrm{R}})T_{\mathrm{R}}/2\right]}{\sin\left[(\omega-\omega_{\mathrm{IF}}-m2\pi/NT_{\mathrm{R}})T_{\mathrm{R}}/2\right]}e^{-j\omega\bar{T}} \qquad (12.6\text{-}9)$$

where $\bar{T}$ is given by (12.5-2). Output m corresponds to a filter centered at frequency $\omega_{\mathrm{IF}}+(m2\pi/NT_{\mathrm{R}})$; if its response (from envelope detector) is the largest of all the filters at the target's delay, then the refined Doppler estimate is $m2\pi/NT_{\mathrm{R}}$ and its error is at most about $\pm\pi/NT_{\mathrm{R}}$.

We summarize the above discussions. If some form of preliminary Doppler estimate is available (range data differentiation, bank of 1-pulse matched filters, etc.) So that Doppler frequency is known to an error of not more than about $\pm\pi/T$, then a bank of coherent N-pulse matched filters can be used to develop a final estimate of Doppler frequency to an error not more than $\pm\pi/NT_{\mathrm{R}}$, as needed to designate the precision coherent N-pulse measurement system. If the preliminary estimate's error is not more than $\pm\pi/T_{\mathrm{R}}$, the final estimate is unambiguous; if not, ambiguity resolution must be used. The bank of N-pulse matched filters is approximated as one single-pulse matched filter in cascade with the function defined by (12.6-9).

The advantages of the coherent N-pulse matched filter bank are that it requires no range gating and operates at the full bandwidth of the target pulses. It responds equally therefore to all targets that create responses at their respective times of occurrence. Since each filter in the bank is approximately a matched filter, it makes full use of all energy in the lines of the N-pulse spectrum. Disadvantages center mainly around the practical limitations of realizing the delays, and various feedback schemes have been devised to use only a single delay. It is also possible to synthesize the transversal Doppler filter bank by digital methods (see Schleher, 1991, sec. 2.3.3); when N is a power of two, the digital approach is simplified through use of the fast Fourier transform (FFT).

Doppler Tracking

A coherent N-pulse processor can also be used for continuous Doppler frequency measurement, a procedure that we have called *tracking*. The system for tracking with N-pulse waveforms is an extension of tracking with single pulses. The tracking loop is again as given in Fig. 12.4-3 with the circuit between points A and B defined as in Fig. 12.4-1 except for one change. The change is in the two frequency-offset matched filters, which now are N-pulse coherent filters.

Measurement error variance is a direct extension of (12.4-7) with the variance for a 1-pulse processor replaced by that of an N-pulse processor and n_{eff} properly interpreted. If W_L is the closed-loop bandwidth (rad/s), as before, and pulses are continuously transmitted with period T_R, then $n_{eff} = \pi/(W_L N T_R)$. If transmission consists of a group (burst) of N pulses each of duration T every T_R seconds, followed by other groups (bursts) every T_{R0} seconds (where $T_{R0} \geq N T_R$), then $n_{eff} = \pi/(W_L T_{R0})$.

12.7 FILTER MISMATCH AND FINE-LINE MEASUREMENTS

When attempting to make Doppler frequency measurements using coherent N-pulse processing, a reasonable question to ask is: Can the rather complicated transversal (comb) filter needed in the optimum system be simplified or eliminated? The answer is yes, but it requires that we consider a mismatched (nonoptimum) filter. In this section we first summarize the effects of using a mismatched system and then discuss a simplified receiver that uses a filter that responds to only one line in the received signal's spectrum.

Effects of Filter Mismatch

In Section 12.3 the ratio of the variance of Doppler frequency measurement error of the optimum filter to that of an arbitrary filter was defined as the square of a filter efficiency factor η_ω. For white input noise this factor was given by (12.3-3). The expression remains valid for the coherent N-pulse system if $\Psi(\omega)$ is defined by (12.5-7).

Equation (12.3-3) is useful when quantities are specified in the frequency domain and when the integrals can be solved. However, it is sometimes more convenient in integral solutions to have a form defined in the time domain. This form is found in Problem 12.7-1 to be

$$\eta_\omega^2 = \frac{\left[\mathrm{Re}\left\{ -j \int_{-\infty}^{\infty} t\psi(t) z(t_0 - \tau_R - t) e^{j\omega_{d0} t}\, dt \right\} \right]^2}{\int_{-\infty}^{\infty} |z(t)|^2\, dt \int_{-\infty}^{\infty} t^2 |\psi(t)|^2\, dt} \tag{12.7-1}$$

where $z(t)$ is the inverse Fourier transform of $Z(\omega)$. The filter efficiency factor η_ω is developed below to evaluate a simple "fine line" filter to replace the comb (or transversal) filter needed in the optimum system.

Use of an Ungated Fine-Line Filter

We now examine the use of a simple mismatched filter for making N-pulse coherent Doppler measurements. The assumed filter has only the central "tooth" of the comb filter function $G_L(\omega - \omega_{d0})$ of (12.5-12). The shape of the assumed filter is the same as the shape of the central tooth (see Problem 12.7-2). Thus we will assume that the mismatched filter is defined by

$$Z(\omega) = \frac{2C}{\mathcal{N}_0} \Psi_1^*(\omega - \omega_{d0}) \frac{d}{d\omega} \left\{ \frac{\sin[N(\omega - \omega_0 - \omega_{d0})T_R/2]}{N(\omega - \omega_0 - \omega_{d0})T_R/2} \right\} e^{-j\omega(t_0 - \tau_R)} \quad (12.7\text{-}2)$$

This function consists of a single-pulse matched filter at the expected Doppler frequency followed by a "discriminator," or derivative, function based on a fine-line filter with a $\sin(x)/x$ shape. The inverse Fourier transform (impulse response) of the filter of (12.7-2) can be shown to be (see Problem 12.7-3).

$$z(t) = \frac{-2Cj}{\mathcal{N}_0 N T_R} e^{j\omega_{d0}(t - t_0 + \tau_R)} \int_{-\infty}^{\infty} \xi \psi_1^*(\xi - t + t_0 - \tau_R) \operatorname{rect}\left(\frac{\xi}{N T_R}\right) e^{j\omega_0 \xi} d\xi$$

$$(12.7\text{-}3)$$

Although the development is somewhat lengthy, (12.7-1) can be evaluated for $z(t)$ given by (12.7-3), $\psi(t)$ given by (12.5-4), and for $\Psi_1(\omega)$ broadband relative to $2\pi/T_R$. The result is (see Problems 12.7-4–12.7-7)

$$\eta_\omega^2 = \frac{\sigma_{\hat{\omega}_d(\min)}^2}{\sigma_{\hat{\omega}_d}^2} \approx \frac{12\left[\operatorname{Re}\left\{ \int_{-\infty}^{\infty} g^*(t)\,dt \int_{-\infty}^{\infty} \xi^2 g(\xi)\,d\xi + \frac{T_R^2(N^2 - 1)}{12} \left| \int_{-\infty}^{\infty} g(\xi)\,d\xi \right|^2 \right\} \right]^2}{N^2 T_R^3 \left| \int_{-\infty}^{\infty} g(t)\,dt \right|^2 \left[\int_{-\infty}^{\infty} \xi^2 |g(\xi)|^2\,d\xi + \frac{T_R^2(N^2 - 1)}{12} \int_{-\infty}^{\infty} |g(\xi)|^2\,d\xi \right]}$$

$$(12.7\text{-}4)$$

As usual, $g(t)$ is the complex envelope of a single typical pulse in the group of N coherent pulses. This result applies to a white noise channel when the receiver's transfer function is defined by (12.7-2) and any shaped transmitted pulse is allowed.

To see an application of (12.7-4), we take an example of N coherent, constant-frequency, constant-amplitude, pulses, each of duration T.

Example 12.7-1   Assume that the complex envelope of each pulse in the N-pulse burst is

$$g(t) = A \operatorname{rect}\left(\frac{t}{T}\right) = g^*(t)$$

where A and T are positive constants. The integrals required in (12.7-4) evaluate to

$$\int_{-\infty}^{\infty} g(\xi)\, d\xi = AT, \quad \int_{-\infty}^{\infty} \xi^2 |g(\xi)|^2\, d\xi = \frac{A^2 T^3}{12}$$

$$\int_{-\infty}^{\infty} \xi^2 g(\xi)\, d\xi = \frac{AT^3}{12}, \quad \int_{-\infty}^{\infty} |g(\xi)|^2\, d\xi = A^2 T$$

which gives

$$\sigma_{\hat{\omega}_d}^2 \approx \sigma_{\hat{\omega}_d(\min)}^2 \frac{N^2}{(T/T_R)[(T/T_R)^2 + (N^2 - 1)]}, \qquad N \geq 2$$

For the usual case in pulsed radar where $T \ll T_R$, this result reduces to

$$\sigma_{\hat{\omega}_d}^2 \approx \sigma_{\hat{\omega}_d(\min)}^2 \frac{N^2}{(T/T_R)(N^2 - 1)}, \qquad N \geq 2, \quad T \ll T_R$$

The results of Example 12.7-1 show that the ungated fine-line filter for constant-frequency pulses generally gives a measurement error variance significantly larger than that of the optimum filter due to the factor $1/(T/T_R)$. For N larger than about 3, the factor $N^2/(N^2 - 1)$ is nearly unity, and its effect can be ignored. It can be shown (Problem 12.7-10) that the energy in the central line of the spectrum of $\psi(t)$ equals (T/T_R) times the energy in all lines (total energy). Thus the increase in error variance can be associated with the decrease in effective signal energy available to the fine-line filter.

The results produced by (12.7-4) obviously vary with the type of transmitted pulse used. As a second example, assume linear FM pulses for which

$$g(t) = A \operatorname{rect}\left(\frac{t}{T}\right) e^{jDt^2/(2T^2)} \tag{12.7-5}$$

where A and T are positive constants (amplitude and duration, respectively) and D is defined by

$$D = \Delta\omega T = 2\pi\Delta f T \tag{12.7-6}$$

where Δf is the frequency sweep (Hz) over the pulse's duration T. For this case the needed integrals evaluate to (see Problems 12.7-11 and 12.7-12)

$$\int_{-\infty}^{\infty} g(\xi)\, d\xi = 2AT \sqrt{\frac{\pi}{D}} \left[C\left(\sqrt{\frac{D}{4\pi}}\right) + jS\left(\sqrt{\frac{D}{4\pi}}\right) \right] \tag{12.7-7}$$

$$\int_{-\infty}^{\infty} |g(\xi)|^2\, d\xi = A^2 T \tag{12.7-8}$$

$$\int_{-\infty}^{\infty} \xi^2 g(\xi)\, d\xi = \left(\frac{AT^3}{12}\right) {}_1F_1\left(\frac{3}{2}; \frac{5}{2}; j\frac{D}{8}\right)$$ (12.7-9)

$$\int_{-\infty}^{\infty} \xi^2 |g(\xi)|^2\, d\xi = \frac{A^2 T^3}{12}$$ (12.7-10)

Here $C(\cdot)$ and $S(\cdot)$ are the Fresnel integrals of (7.2-16) and (7.2-17). The function ${}_1F_1(a; b; z)$ is known as the *confluent hypergeometric function* defined by[13]

$$ {}_1F_1(a; b; z) = \sum_{n=0}^{\infty} \frac{(a)_n}{(b)_n} \frac{z^n}{n!}$$ (12.7-11)

where a and b are real numbers, z can be complex, and

$$(a)_n = a(a + 1)(a + 2) \ldots (a + n - 1), \qquad (a)_0 = 1$$ (12.7-12)

${}_1F_1(a; b; z)$ is a convergent series for all values of a, b, and z provided that a and b do not equal negative integers. Other properties are summarized by Abramowitz and Stegun (1964, p. 504).

On using the above integrals in (12.7-4), the error variance is found to be (Problem 12.7-13)

$$\sigma_{\hat{\omega}_d}^2 = \sigma_{\hat{\omega}_d(\min)}^2 \frac{N^2 |C(\sqrt{D/4\pi}) + jS(\sqrt{D/4\pi})|^2 [(T/T_R)^2 + (N^2 - 1)]}{(T/T_R)^5 \left\langle \mathrm{Re}\left\{ [C(\sqrt{D/4\pi}) - jS(\sqrt{D/4\pi})] {}_1F_1(3/2; 5/2; jD/8) \right. \right.}$$

$$\left. \left. + 2\sqrt{\frac{\pi}{D}}(N^2 - 1)\left(\frac{T_R}{T}\right)^2 \left| C\left(\sqrt{\frac{D}{4\pi}}\right) + jS\left(\sqrt{\frac{D}{4\pi}}\right) \right|^2 \right\}\right\rangle^2$$ (12.7-13)

for $N \geq 2$. For $D \to 1$ (small D) this result reduces to that of Example 12.7-1 for the N-pulse coherent train of constant-frequency, constant-amplitude pulses, as it should. For $D \to \infty$ (large D) it is found to be

$$\sigma_{\hat{\omega}_d}^2 = \sigma_{\hat{\omega}_d(\min)}^2 \frac{N^2}{\left(\frac{1/\Delta f}{T_R}\right)(N^2 - 1)}, \qquad N \geq 2, \quad D = \Delta\omega T \gg 1$$ (12.7-14)

Since $1/\Delta f$ is the duration of the compressed pulse, the form of (12.7-14) is identical to the final form in Example 12.7-1 for constant-frequency pulses. As before, the factor

---

[13] Other notations for the confluent hypergeometric function are also found. Some are $F(a; b; z)$ (Korn and Korn, 1961, p. 253, where it is called *Kummer's* confluent hypergeometric function), $\Phi(a, b; z)$ (Gradshteyn and Ryzhik, 1965, p. 1058, where it is called the *degenerate hypergeometric function*), and $M(a, b, z)$ (Abramowitz and Stegun, 1964, p. 504, where it is called *Kummer's function*). The symbol ${}_1F_1(a; b; z)$ is Pochhammer's notation for the more general hypergeometric function ${}_mF_n(a_1, a_2, \ldots, a_m; b_1, b_2, \ldots, b_n; z)$ defined in Korn and Korn (1961, p. 253), when $m = 1$ and $n = 1$.

$(1/\Delta f)/T_R$ represents a loss factor in signal energy; it is the ratio of energy in the central spectral line divided by the total energy in all lines.

Gated Fine-Line Filters

The reduction in effective signal energy that occurs when using a fine-line filter can be offset by a reduction in the effective noise level if gating is used. When gating is present, we refer to the system as a gated fine-line-filter.

Although the preceding discussions presumed a filter realized at a center frequency $\omega_0 + \omega_{d0}$ for purposes of simple exposition of principles, a practical design would use a mixer (for removal of the Doppler estimate), a gate (to remove the loss in performance that occurs without gating), and a filter fixed at an IF frequency. Figure 12.7-1 gives the block diagram of one possible system. For simplicity, only one mixer and one IF frequency are shown, although practical systems may use two or more mixers and IF frequencies.[14]

The gate in Fig. 12.7-1 is shown after the single-pulse matched filter. For a constant-frequency constant-amplitude pulse the gate can be either before or after the filter. If before, the gate's width τ_g should be equal or slightly larger than the pulse's duration T. If after, its width should be about $4T/3$ or slightly larger. These choices compensate for the loss in signal energy (due to the fine-line filter) by reducing the effective noise level at the gate's output such that the system gives a measurement error variance of $\sigma^2_{\hat{\omega}_d(min)}$ as given by (12.5-15) or (12.5-16). This action of the gate is discussed in Problems 12.7-15 and 12.7-16.

When pulse compression is used, the proper gate position in Fig. 12.7-1 is, as shown, after the single-pulse matched filter. A gate width of $\tau_g = 1/\Delta f$, or slightly larger, will compensate for the energy loss in the ungated filter. This gate width corresponds to a system measurement error variance as given in (12.5-15) or (12.5-16). Problem 12.7-17 develops the effective improvement in the post–gate noise level.

In the preceding two paragraphs the gate durations were suggested to be a bit larger than the ideal values. The reason is that some allowances must be made in a practical system for target motion during a pulse interval and for jitter in gate timing due to noise and other effects. With the chosen gate durations the system error variance became $\sigma^2_{\hat{\omega}_d(min)}$, which is that of the optimum system. The reader may wonder how this seriously mismatched system can give a performance equal to that of the optimum system. The answer hinges on the fact that the optimum system belongs to a class of *ungated* receivers, so the use of gating was not part of our earlier definition of optimum.

Following the gate in Fig. 12.7-1 are two fine-line filters in parallel. Their difference of transfer functions is approximately the derivative required in (12.7-2) for filter transfer functions

$$Z_d(\omega \pm \omega_q) = \frac{\sin[N(\omega - \omega_{IF} \pm \omega_q)T_R/2]}{N(\omega - \omega_{IF} \pm \omega_q)T_R/2} \qquad (12.7\text{-}15)$$

---

[14] The reasons are mainly practical, such as placing high-gain circuits at lower frequencies where they are more easily realized, using frequencies low enough to realize the narrow-band fine-line filter, and choosing IF frequencies to aid in removing undesired mixer products and spurious signals.

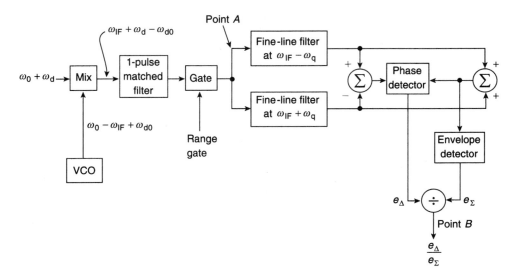

Figure 12.7-1 A practical system for measurement of Doppler angular frequency using a gated fine-line filter.

[see (12.3-8) and Fig. 12.3-2]. The sum is the "reference" output [see (12.3-9)]; when used with the "difference" output in the phase detector, the Doppler "error" voltage is generated. Finally the envelope detector and "error" outputs are sampled at the end of the N-pulse interval at the target's time of occurrence to provide values e_Δ and e_Σ. The divider (best formed digitally) produces the normalized ratio e_Δ/e_Σ. These operations are analogous to those of Fig. 12.4-1 between points A and B.

By connecting point B in Fig. 12.7-1 to the input of the VCO through a suitable low-pass loop filter, the system becomes a Doppler tracker that continuously measures Doppler frequency.

A main advantage of the gated fine-line system is its ability to give the same error variance as the optimum system but without the complicated transversal (comb) filter. Its main disadvantage is that it can function with only one target because of gating. If more than one target is to be processed, parallel gates are needed (one for each target), and all functions after the gate must be duplicated for each target. This disadvantage was not present in the optimum system (ungated) which responded with full bandwidth to all target signals.

PROBLEMS

12.1-1 A monostatic radar at 6.8 GHz must work with fast aircraft targets with speeds up to 750 m/s (about 1678 mi/h). Doppler frequency is measured on each pulse using a frequency discriminator with a slope of $K_\omega = -(1.35/\pi)10^{-5}$ volts per rad/s of Doppler angular frequency. What correction voltage occurs for a maximum-speed target if $\omega_{d0} = 0$?

12.1-2 If the linear range of the frequency discriminator in Problem 12.1-1 occurs for $-2.9(10^5) \le \omega_d \le 2.9(10^5)$ rad/s, what maximum target speed will be allowed while operating in the linear range?

12.1-3 A radar receiver uses a frequency discriminator to make single-pulse Doppler frequency measurements when $\omega_{d0} = 0$. If the discriminator must produce voltages of $\pm V$ when target speeds are $\mp |v|_{\text{max}}$, derive an expression for the discriminator's slope K_ω that is a function of ω_0, V, and $|v|_{\text{max}}$. Evaluate your result for $V = 5$ volts, $|v|_{\text{max}} = 2000$ m/s, and $\omega_0/2\pi = 9.7$ GHz.

★12.1-4 Prove Schwarz's inequality of (12.1-24) and show that the equality occurs only when (12.1-25) is true. [*Hint*: Expand the function $\int_{-\infty}^{\infty} |B(\omega) - \alpha A^*(\omega)|^2 \, d\omega \ge 0$ for α an arbitrary real, but nonzero, variable, and examine when the quadratic function of α must be nonnegative.]

12.1-5 Show that (12.1-31) results from (12.1-29) when (12.1-30) is used.

12.1-6 A radar uses an optimum Doppler frequency estimator so that (12.1-31) applies. If the radar's frequency is 5 GHz and a pulse for which $\tau_{\psi,\text{rms}} = 1.0 \, \mu\text{s}$ is used, what value of $\mathcal{R}$ is required to produce a Doppler estimate $\hat{\omega}_d$ with an rms error of not more than 25% of the maximum Doppler frequency $|\omega_d|_{\text{max}}$? Assume that the maximum target speed is 1200 m/s and $\bar{t}_\psi = 0$.

12.1-7 Work Problem 12.1-6 except assume that the radar frequency is 10 GHz, the maximum target speed is 800 m/s, and the rms error is not more than 10% of $|\omega_d|_{\text{max}}$.

12.1-8 Find $\tau_{g,\text{rms}}$ for the complex envelope of Fig. P11.7-5 and use the result in (12.1-31) to obtain an expression for $\sigma^2_{\hat{\omega}_d(\text{min})}$. Mean time $\bar{t}_g$ is zero for this envelope.

12.1-9 Work Problem 12.1-8 except for a triangular pulse of the form of Fig. 6.9-1a with A^2T replaced here by A.

12.1-10 Work Problem 12.1-8 except assume that $g(t) = A \exp[-4t^2 \ln(\sqrt{2})/T^2]$. Here T is the 3.01-dB pulsewidth.

12.2-1 Formally differentiate (12.2-8) according to (12.1-13), and show that (12.2-9) is true.

12.2-2 Use (12.1-32) through (12.1-35) to show that (12.2-10) is true.

12.2-3 The complex envelope of the transmitted pulse in a monostatic radar with white input noise is

$$g(t) = A \operatorname{rect}\left(\frac{t}{T}\right)$$

where T is the pulse's duration. Find the transfer function of the optimum Doppler measurement filter in (a) analytic, and (b) real form. Assume that $t_0 = \tau_R$.

12.2-4 A radar transmits a pulse for which

$$g(t) = A \operatorname{rect}\left(\frac{t}{2T}\right)\cos\left(\frac{\pi t}{2T}\right)$$

where A and T are positive constants. Find $Z_{opt}(\omega)$ for the optimum Doppler frequency measurement filter for white noise. Sketch $Z_{opt}(\omega)$.

12.3-1 A radar uses a filter mismatched only in bandwidth to measure Doppler frequency. If the filter's bandwidth is 80% larger than it should be when matched to a signal with a gaussian-shaped spectrum, what is the filter's efficiency factor?

12.3-2 Work Problem 13.3-1 except for a bandwidth 60% less than that for an optimum system.

★**12.3-3** Show that the third equation in Example 12.3-1 is true when the first two equations are used in (12.3-3). Assume that $t_0 = \tau_R$.

12.4-1 Assume that the phase detector of Fig. 12.4-1 can be modeled as an analog product device with the product output being passed through a low-pass filter that removes second harmonic products (terms at nominal frequency $2\omega_0$). If inputs are modeled as $s_\Delta(t) = a_\Delta(t)\cos[\omega_0 t + \phi_0 + \theta_\Delta(t)]$ and $s_\Sigma(t) = a_\Sigma(t)\cos[\omega_0 t + \phi_0 + \theta_\Sigma(t)]$, show that the use of analytic representations gives a real output $e_0(t) = \frac{1}{2}\operatorname{Re}[\psi_\Delta(t)\psi_\Sigma^*(t)]$.

12.4-2 Let the sum and difference output signals, s_Σ and s_Δ, respectively, in Fig. 12.4-1 be processed according to Fig. P12.4-2 where all circuits are broadband such that operations occur "instantaneously." The phase detector operates as in Problem 12.4-1 to produce a real response e_ω at the target's delay time τ_R. The gain control loop (envelope detector, difference junction, and amplifier) acts to produce a constant peak signal in the lower channel by adjusting the voltage gains of identical amplifiers to $A_0 = V_0/|\psi_\Sigma(\tau_R)|$. Show that the circuit normalizes the difference signal such that

$$e_\omega(\tau_R) = \frac{V_0^2}{2}\operatorname{Re}\left[\frac{\psi_\Delta(\tau_R)}{\psi_\Sigma(\tau_R)}\right]$$

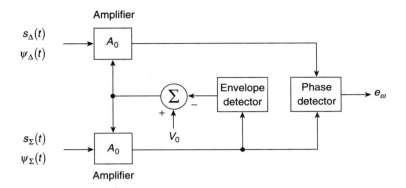

Figure P12.4-2

12.4-3 Assume that a radar transmits a pulse defined by the complex envelope

$$g(t) = Ae^{-(\alpha t/T)^2}$$

where $A > 0$ is a constant and T is defined as the pulsewidth. (a) Find the value of constant α such that $\Omega_{\mathrm{res}} = 2\pi/T$. (b) What width of $g(t)$ does T define (the width between what dB points)?

12.4-4 Work Problem 12.4-3 except for the pulse

$$g(t) = A \, \mathrm{Sa}\left(\frac{\alpha t}{T}\right)$$

12.4-5 Work Problem 12.4-3 except for the pulse

$$g(t) = A \, \mathrm{tri}\left(\frac{\alpha t}{T}\right)$$

12.4-6 Work Problem 12.4-3 except for the pulse

$$g(t) = A \cos\left(\frac{\alpha t}{T}\right) \mathrm{rect}\left(\frac{\alpha t}{\pi T}\right)$$

12.4-7 Work Problem 12.4-3 except for the pulse

$$g(t) = A\left[1 - \left(\frac{\alpha t}{T}\right)^2\right] \mathrm{rect}\left(\frac{\alpha t}{2T}\right)$$

12.4-8 The complex envelope of a radar pulse is often defined by

$$g(t) = Ae^{-1.382(t/T)^2}$$

where T is the duration between points 3 dB down from the peak. Find Ω_{res} for this waveform, and note that it is less than $2\pi/T$, which applies to a rectangular pulse of duration T.

12.4-9 Work Problem 12.4-8 except for the waveform

$$g(t) = A\,\mathrm{Sa}\left(\frac{0.885\pi t}{T}\right)$$

12.4-10 Work Problem 12.4-8 except for the waveform

$$g(t) = A\,\mathrm{tri}\left(\frac{0.5842t}{T}\right)$$

12.4-11 Work Problem 12.4-8 except for the waveform

$$g(t) = A\cos\left(\frac{\pi t}{2T}\right)\mathrm{rect}\left(\frac{t}{2T}\right)$$

12.4-12 Work Problem 12.4-8 except for the waveform

$$g(t) = A\left[1 - \left(\frac{1.081t}{T}\right)^2\right]\mathrm{rect}\left(\frac{1.081t}{2T}\right)$$

12.4-13 A radar at 12 GHz is to measure objects with speeds up to $v_{max} = 700$ m/s when using a single constant-frequency rectangular pulse of duration 250 μs. (a) What maximum Doppler frequency can be expected? (b) At what minimum target range can the radar operate? (c) How many filters must be used in a Doppler filter bank?

12.4-14 Work Problem 12.4-13 except assume that $v_{max} = 350$ m/s, the radar's frequency is 6 GHz, and pulse duration is 500 μs.

12.4-15 A radar uses constant-frequency 5-μs rectangular pulses at a 500 pulse/s rate. Single-pulse energy-to-noise density ratio is $\mathscr{R}_p = 26$. Pulses are integrated noncoherently in a Doppler tracking loop having $W_L/2\pi = 12.5$ Hz. (a) What single-pulse Doppler rms measurement error is possible? (b) How many pulses are effectively integrated in the loop? (c) What rms Doppler tracking error is produced by the loop?

12.5-1 Show that (12.5-7) is the spectrum of the signal defined by (12.5-4).

12.5-2 Assume that a radar transmits a basic pulse envelope $g(t)$ such that its spectrum is approximated over its central region by

$$G(\omega) \approx \cos\left(\frac{\pi\omega}{12\omega_R}\right), \qquad -6\omega_R \le \omega \le 6\omega_R$$

Sketch the shape of $\Psi(\omega)$, assuming a 6-pulse burst of pulses at a rate $\omega_R = 2\pi f_R$, and assume that $\phi_0 = 0$.

12.5-3 Work Problem 12.5-2 except assume 21 pulses and that

$$G(\omega) \approx \cos\left(\frac{\pi\omega}{42\omega_R}\right), \qquad -21\omega_R \leq \omega \leq 21\omega_R$$

12.5-4 A radar uses the highest pulse rate f_R that will allow unambiguous range measurements to 16 km. Its carrier frequency f_0 is chosen to give unambiguous Doppler frequency measurements. Constant-frequency rectangular 2-μs pulses are used, and the targets may have speeds up to 1000 m/s. (a) What is f_R? (b) What is f_0? (c) If pulses can be coherently processed for 5 ms, how many are processed? (d) What is the width of a "line" in the signal being processed for 5 ms?

12.5-5 Generalize Problem 12.5-4 [part (b) only] by finding an expression for radar frequency f_0 that gives unambiguous Doppler and range measurements for any specified unambiguous range R. Assume that pulse duration is T and that v_{max} is the maximum speed of the target.

12.5-6 Show that (12.5-13) is true.

$\star$**12.5-7** Show that (12.5-14) is true.

12.5-8 A radar uses a typical pulse for which $\tau_{\psi,rms} = 1.0$ μs. Pulse transmission rate is 10^4 pulses/s. How much larger is the single-pulse rms Doppler measurement error than that for N pulses if $N = 2, 4$, and 6?

12.5-9 Use (12.5-16) and find an expression for the required minimum value of $\mathscr{R}_p$ to reduce the Doppler error variance to not more than $\omega_R^2/1000$. Repeat for $\omega_R^2/10^4$.

12.5-10 A radar coherently processes N pulses to realize an rms Doppler measurement error of not more than 5% of the pulse rate when $\mathscr{R}_p = 15$. What is the smallest allowable value of N?

12.5-11 Work Problem 12.5-10 except assume an error of not more than 1% of the pulse rate.

12.6-1 If the phase shifts of Fig. 12.6-1 are defined by (12.6-1), show that (12.6-2) is true.

12.6-2 Generalize Problem 12.6-1 by showing that (12.6-9) results when the transversal filter's phases are given by (12.6-7).

12.6-3 A radar has a continuous pulse rate of 1800 pulses per second. An optimum 5-pulse coherent Doppler measurement filter is used, and continuous tracking occurs through a closed-loop tracker for which $W_L/2\pi = 4$ Hz. How many 5-pulse bursts are being averaged (integrated) by the loop?

12.7-1 Show that (12.3-3) can be converted to the form of (12.7-1).

★**12.7-2** By Fourier transformation of (12.5-5) the spectrum of $g_N(t)$, denoted as $G_N(\omega)$, is found to be

$$G_N(\omega) = G(\omega)\frac{\sin(N\omega T_R/2)}{\sin(\omega T_R/2)}$$

Show that $G_N(\omega)$ can also be put in the form

$$G_N(\omega) = N\sum_{k=-\infty}^{\infty}(-1)^{(N-1)k}G\left(k\frac{2\pi}{T_R}\right)\mathrm{Sa}\left[N\left(\omega - k\frac{2\pi}{T_R}\right)\frac{T_R}{2}\right]$$

[*Hint*: Represent $g_N(t)$ by a complex Fourier series with period from $-NT_R/2$ to $+NT_R/2$, find the series' coefficients in terms of values of $G_N(\omega)$ at specific values of ω, and then evaluate $G_N(\omega)$ by Fourier transformation of the series representing $g_N(t)$.]

★**12.7-3** Derive (12.7-3) by inverse Fourier transformation of (12.7-2).

★**12.7-4** Assume that $\Psi_1^*(\omega - \omega_{do})$ in (12.7-2) is wideband relative to the derivative term so that it is approximately constant at $\Psi_1^*(\omega_0)$, and show that

$$z(t) \approx \frac{-2Cj}{\mathcal{N}_0 NT_R}\Psi_1^*(\omega_0)(t - t_0 + \tau_R)\mathrm{rect}\left(\frac{t - t_0 + \tau_R}{NT_R}\right)e^{j(\omega_0 + \omega_{do})(t - t_0 + \tau_R)}$$

so that

$$\int_{-\infty}^{\infty}|z(t)|^2\,dt \approx \frac{C^2 NT_R}{3\mathcal{N}_0^2}|\Psi_1^*(\omega_0)|^2$$

which is needed to put (12.7-1) in the form of (12.7-4).

★**12.7-5** Show that the second denominator factor in (12.7-1) evaluates to

$$\int_{-\infty}^{\infty}t^2|\psi(t)|^2\,dt = N\left[\int_{-\infty}^{\infty}\xi^2|g(\xi)|^2\,d\xi + \frac{(N^2-1)T_R^2}{12}\int_{-\infty}^{\infty}|g(\xi)|^2\,d\xi\right]$$

★**12.7-6** Use the approximation of Problem 12.7-4 for $z(t)$ to show that the integral of the numerator of (12.7-1) evaluates to

$$\int_{-\infty}^{\infty}t\psi(t)z(t_0 - \tau_R - t)e^{j\omega_{do}t}\,dt$$

$$\approx \frac{2Cj}{\mathcal{N}_0 T_R}e^{j\phi_0}\Psi_1^*(\omega_0)\left[\int_{-\infty}^{\infty}\xi^2 g(\xi)\,d\xi + \frac{(N^2-1)T_R^2}{12}\int_{-\infty}^{\infty}g(\xi)\,d\xi\right]$$

12.7-7 Combine the results of Problems 12.7-4 through 12.7-6 to show that (12.7-1) can be approximated by (12.7-4). [*Hint*: Evaluate $\Psi_1^*(\omega_0)$ by the area property of Fourier transforms.]

12.7-8 A radar processes $N = 12$ constant-amplitude, constant-frequency pulses for Doppler frequency measurement. The mismatched fine-line filter defined by (12.7-3) is used. For a 3-μs pulse duration and a 1400-Hz pulse rate, what is the filter's efficiency?

12.7-9 Work Problem 12.7-8 except for $N = 4$ pulses and 50-μs pulses.

12.7-10 Show that the energy in the central line of the spectrum of $\psi(t)$ given by (12.5-4) is T/T_R times the total energy in all lines. Assume constant-frequency pulses of constant amplitude and duration T. [*Hint*: Use the represenation for $G_N(\omega)$ given in Problem 12.7-2 to find the energy of the central line.]

12.7-11 For the linear FM pulse defined by (12.7-5), show that (12.7-7) and (12.7-8) are true.

12.7-12 Work Problem 12.7-11 except show that (12.7-9) and (12.7-10) are true. [*Hint*: Use the known integrals

$$\int_0^1 x^{\mu-1} \sin(ax)\, dx = \frac{-j}{2\mu}[_1F_1(\mu;\mu+1;ja) - {}_1F_1(\mu;\mu+1;-ja)]$$

for $a > 0$ and $\mathrm{Re}(\mu) > -1$ and

$$\int_0^1 x^{\mu-1} \cos(ax)\, dx = \frac{1}{2\mu}[_1F_1(\mu;\mu+1;ja) + {}_1F_1(\mu;\mu+1;-ja)]$$

for $a > 0$ and $\mathrm{Re}(\mu) > 0$ (Gradshteyn and Ryzhik, 1965, pp. 420–421).]

12.7-13 For the linear FM pulse use (12.7-7) through (12.7-10) in (12.7-4) and show that (12.7-13) results.

12.7-14 A radar transmits pulses every 400 μs that have a linear FM sweep of 1.6 MHz. The receiver filter is an ungated fine-line filter, as defined by (12.7-2). Find the range that the filter's efficiency η_ω can have as N varies from 2 to ∞. Note that N is not an effective way of controlling η_ω. How else can one effectively control η_ω?

★12.7-15 Figure P12.7-15 shows a single-pulse matched filter followed by a gate with periodic pulses each of duration τ_g. Show that the power spectrum of the random process $N_0(t)$ is

$$\mathscr{S}_{N_0 N_0}(\omega) = \frac{\mathscr{N}_0}{2}\left(\frac{\tau_g}{T_R}\right)\frac{\tau_g}{2\pi}\int_{-\infty}^{\infty} |\Psi_1(\xi)|^2\, \mathrm{Sa}^2\left[\frac{(\omega-\xi)\tau_g}{2}\right]d\xi$$

[*Hint*: Represent the gates by a complex Fourier series, compute the time average of the autocorrelation function of $N_0(t)$ assuming that the process $N_i(t)$ is stationary, use the known series

$$\sum_{n=1}^{\infty} \frac{\cos(nx)}{n^2} = \frac{\pi^2}{6} - \frac{\pi}{2}x + \frac{1}{4}x^2, \qquad 0 \le x \le 2\pi$$

(Gradshteyn and Ryzhik, 1965, p. 39), and Fourier transform the time-averaged autocorrelation function.]

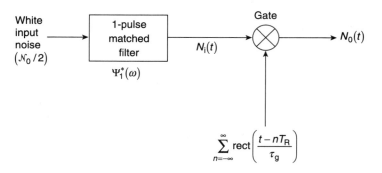

Figure P12.7-15

★**12.7-16** Assume that the power spectrum of Problem 12.7-15 does not change greatly for a small band of frequencies near ω_0. Evaluate $\mathscr{S}_{N_oN_o}(\omega_0)$ for a constant-frequency rectangular pulse of duration T, and show that the value compared to the value with no gating is

$$\frac{\mathscr{S}_{N_oN_o}(\omega_0)(\text{gating})}{\mathscr{S}_{N_oN_o}(\omega_0)(\text{no gating})} = \frac{1}{3}\left(\frac{T}{T_R}\right)\begin{cases} 3\left(\dfrac{\tau_g}{T}\right) - 1, & 1 \le \dfrac{\tau_g}{T} \\[2ex] \left(\dfrac{\tau_g}{T}\right)^2\left[3 - \dfrac{\tau_g}{T}\right], & 0 \le \dfrac{\tau_g}{T} \le 1 \end{cases}$$

[*Hint*: Use the known integral

$$\int_0^\infty \frac{\sin^2(ax)\sin^2(bx)}{x^4}\,dx = \begin{cases} \dfrac{a^2\pi}{6}(3b - a), & 0 \le a \le b \\[2ex] \dfrac{b^2\pi}{6}(3a - b), & 0 \le b \le a \end{cases}$$

(Gradshteyn and Ryzhik, 1965, p. 451).]

★**12.7-17** Work Problem 12.7-16 except assume a linear FM pulse with a large pulse compression ratio such that the approximation of (7.2-23) holds. The final result to be obtained is

$$\frac{\mathscr{S}_{N_oN_o}(\omega_0)(\text{gating})}{\mathscr{S}_{N_oN_o}(\omega_0)(\text{no gating})} \approx \frac{\tau_g}{T_R}$$

13

ANGLE MEASUREMENT AND TRACKING BY CONICAL SCAN

This chapter is the first of two that focus on methods of measuring the spatial angles that define a target's position in the space surrounding a radar. The continuous measurement of these angles is called *tracking*, much the same as tracking was defined for delay and Doppler measurement described earlier.

Although several methods have been devised for angle measurements, we will concentrate on two of the most classic in this and the following chapter. In Chapter 14 the angle measurement method called *monopulse* allows target angles to be determined on each and every pulse, at least in principle. In this chapter the angle measurement method, called *conical scan*, is briefly introduced. Conical scan uses a sequence of pulses over time to develop a measurement of the target's angles. It is not so powerful and modern a method as monopulse. It is rarely used in current systems, but it is important for some of the concepts involved and for its historical importance. Conical scan was developed early in the history of radar and most of the early forms of tracking radar used the method.

Conical scan is described briefly in most radar textbooks. Some detailed analysis has been given by Damonte and Stoddard (1956), and part of our discussion in Section 13.5 is based on their work.

13.1 GEOMETRY AND SYSTEM DEFINITION

A conical scan radar typically uses one antenna pattern (beam) of approximately "pencil-beam" shape. Transmission and reception of pulses are through this single pattern. The antenna pattern's "nose" is the direction in space, called the *beam's direction*, where the maximum gain occurs. Two other spacial directions are of interest. One, called the *reference* (or *boresight*) *direction*, defines the *boresight axis*. In accordance with the notation of Chapter 1, x_R, y_R, and z_R define the rectangular coordinate system having the radar at it's origin. In this system the boresight direction is defined by an azimuth angle A_b and an elevation angle E_b. The beam's direction is offset by a *beam squint angle*, θ_q, from the boresight direction. These relationships are shown in Fig. 13.1-1, where pattern sidelobes are omitted for simplicity.

The second direction of interest is that of the target. It is assumed to be squinted away from the boresight direction by an angle θ_T, as shown in Fig. 13.1-1. For proper system operation $\theta_T < \theta_q$ is necessary, as will become obvious as our discussions develop. Finally, for a given beam offset θ_q and given target offset θ_T, we define θ_s as the angle from the line in the beam's direction to the line in the target's direction.

Conical scan operation requires revolving the beam's direction around the boresight direction with time. The *rotation*, or *scan*, rate is denoted by ω_s. The scan rate is usually slow enough that many target pulses are returned during each scan period. These pulses are modulated in amplitude because the target's angle θ_s varies with time as the beam's direction changes. The modulation contains adequate information to allow measurement of the target in two orthogonal angles relative to the boresight direction when proper signal processing (to be described) is used. Scanning typically involves keeping θ_q constant.

Scanning can be realized in two ways. In one, the antenna pattern is *nutated* around the boresight axis. With nutation the pattern shape is fixed, and polarization is constant with scan time. Only the *direction* of the pattern changes. In the second scan method the pattern is *rotated* in space once each scan period; this form of scanning does not preserve polarization, since the polarization rotates in space along with the pattern. Because most conical scan systems utilize aperture-type antennas, pattern nutation or rotation will mainly be determined by the type of feed used.

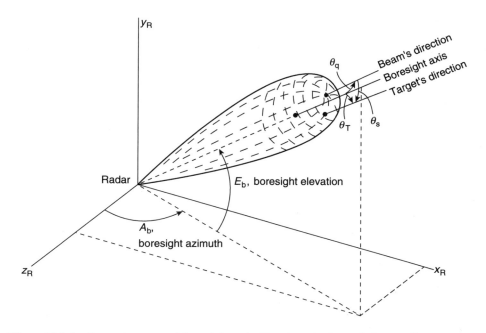

Figure 13.1-1   Beam, target, and boresight axis directions in the radar's coordinate system.

Geometry

To understand what spatial target angles are measured in conical scan, we define an x, y, z coordinate system where z is the boresight direction, with axis x in the x_R, z_R plane of Fig. 13.1-1. The aperture then is in the x, y plane. In the x, y, z system let the target be a point on a sphere of radius r; the target's direction is defined by angles ψ_x, ψ_y, and $\psi_z = \theta_T$ from the x, y, and z axes, respectively. The complements of ψ_x and ψ_y are defined by

$$\theta_x = \frac{\pi}{2} - \psi_x \tag{13.1-1}$$

$$\theta_y = \frac{\pi}{2} - \psi_y \tag{13.1-2}$$

respectively, so

$$\cos(\psi_x) = \sin(\theta_x) \tag{13.1-3}$$

$$\cos(\psi_y) = \sin(\theta_y) \tag{13.1-4}$$

The target is at point T in Fig. 13.1-2a, which shows the applicable geometry. Angles θ_x and θ_y are the angles measured in conical scan; they are the angles of the target's direction from the y, z and x, z planes, respectively.

Figure 13.1-2a also defines a point B on the surface of the sphere of radius r as determined by the beam's direction. The distance D between point B and T is the chord of a great circle segment of length $r\theta_s$. From the geometry we have

$$D^2 = 4r^2 \sin^2\left(\frac{\theta_s}{2}\right) = D_{xy}^2 + r^2[\cos(\theta_q) - \cos(\theta_T)]^2 \tag{13.1-5}$$

For *small* angles θ_q and θ_T, as are typical, we have $D \approx D_{xy}$, and points B and T fall approximately in a plane perpendicular to the boresight axis. As a consequence of the small-angle assumption, the relationships between various angles are found from distances projected into any plane parallel to the x, y plane. Since each distance has a common factor, the "angle" geometry of Fig. 13.1-2b applies. Angles ϕ_s and ϕ_T are defined as those made by the projections of points B and T, respectively, relative to axis x. We call ϕ_s or ϕ_T the *rotation* (or *orientation*) angle of the beam or target. From this geometry a very important relationship for θ_s is found by the law of cosines:

$$\theta_s^2 = \theta_q^2 + \theta_T^2 - 2\theta_q\theta_T \cos(\phi_s - \phi_T) \tag{13.1-6}$$

In the remainder of this chapter we will assume the antenna pattern to be of the pencil-beam type. Its one-way power, or two-way voltage, radiation pattern is assumed to have rotational symmetry so that it is a function of θ_s, only, denoted by $P(\theta_s)$. Although conical scan can operate in a passive, receive-only mode, we discuss only the active system. For other details on passive operation, the reader is referred to Barton and Ward (1969) and Barton (1988).

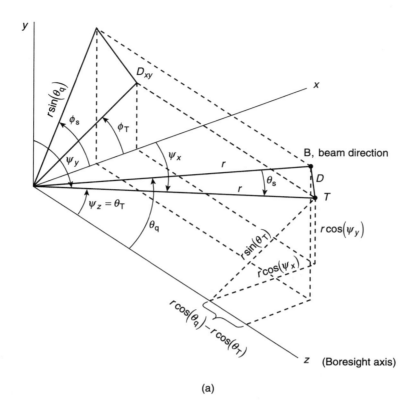

(a)

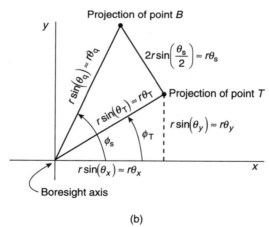

(b)

Figure 13.1-2   Geometries of beam and target directions: (*a*) in *x*, *y*, *z* coordinates, and (*b*) in the *x*, *y* plane.

Conical Scan System

The functions necessary in a conical scan system are defined in Fig. 13.1-3. All functions following the matched filter are for detection purposes. The envelope

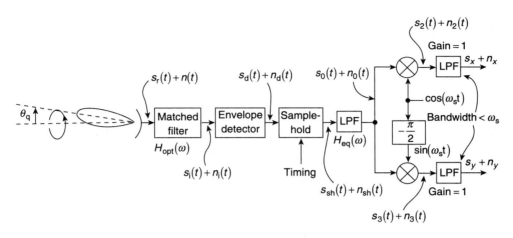

Figure 13.1-3   Block diagram of functions of a conical scan system.

detector removes the carrier's phase. The sample-hold function samples each returned pulse and "holds" the sample value until the time of the next target pulse. This function plus the equalizer (LPF) provide smoothing of the scan modulation of the pulse envelopes. Finally each product and output LPF act as a synchronous detector to determine signal voltages (s_x and s_y) that are proportional to the angles θ_x and θ_y, respectively, which are to be measured. All of these functions are analyzed for both signal and noise responses in the following two sections. Not shown in Fig. 13.1-3 is a duplexer that allows the transmitter to feed the antenna for transmission; other practical devices such as mixers and amplifiers are also omitted for simplicity, since they have little effect on performance.

13.2   SIGNAL ANALYSIS

As usual, we let τ_R and T_R represent target delay and the pulse repetition period, respectively. We model the transmitted and received pulse sequences as (approximately) periodic. Next let $s_{mf}(t)$ represent the response of the matched filter to a target pulse that would be received at delay τ_R *if the antenna pattern's maximum were to point directly at the target.* This condition corresponds to $\theta_s = 0$. For any other value of θ_s, the received pulse is reduced in amplitude by the two-way voltage gain of the pattern. Thus for short-duration pulses

$$s_i(t) = \left\{ \sum_{k=-\infty}^{\infty} s_{mf}(t - kT_R) \right\} P[\theta_s(t)] \tag{13.2-1}$$

where $P(\theta_s)$ is the two-way voltage, or one-way power, radiation pattern[1] of the antenna.

---

[1] Recall that the radiation pattern $P(\theta, \phi)$ of an antenna is the normalized version of the radiation intensity pattern $\mathscr{P}(\theta, \phi)$ obtained by dividing $\mathscr{P}(\theta, \phi)$ by its maximum value; thus $P(0) = 1$ (see Section 3.4).

Since $P(\theta_s)$ is a nonnegative quantity for angles of interest, and pulses are generally short compared to T_R, we have

$$s_d(t) = \left\{ \sum_{k=-\infty}^{\infty} |s_{mf}(t - kT_R)| \right\} P[\theta_s(t)] \tag{13.2-2}$$

where a reasonably large received signal-to-noise ratio and a "linear" envelope detector have been assumed. The signal $s_d(t)$ can be viewed as a "message" function $P[\theta_s(t)]$ being "sampled" by the periodic function within the braces in (13.2-2). Sampling theory (Peebles, 1976, ch. 7, 1987, ch. 2) indicates that an efficient (high-gain) method of recovering the "message" $P(\theta_s)$ is to use a zero-order sample-hold circuit in cascade with an equalizing filter as shown in Fig. 13.1-3. The sample-hold device takes a (nearly instantaneous) sample of the target's pulse in each pulse interval at time $t = \tau_R + kT_R$ and "holds" this value until the next sample time. Thus

$$s_{sh}(t) = |s_{mf}(\tau_R)| \sum_{k=-\infty}^{\infty} P[\theta_s(\tau_R + kT_R)] \, \text{rect}\left[\frac{t - \tau_R - kT_R - (T_R/2)}{T_R} \right] \tag{13.2-3}$$

where it has been assumed that $s_{mf}(\tau_R + kT_R) = s_{mf}(\tau_R)$. This assumption is true if the target's cross section is constant with time. The spectrum of $s_{sh}(t)$, denoted by $S_{sh}(\omega)$, is

$$S_{sh}(\omega) = |s_{mf}(\tau_R)| \frac{1}{T_R} H_{sh}(\omega) \sum_{k=-\infty}^{\infty} F(\omega - k\omega_R)e^{-jk\omega_R \tau_R} \tag{13.2-4}$$

where $\omega_R = 2\pi/T_R$ and

$$
\begin{aligned}
H_{sh}(\omega) &= \mathscr{F}\left\{ \text{rect}\left[\frac{t - (T_R/2)}{T_R} \right] \right\} \\
&= T_R \, \text{Sa}\left(\frac{\omega T_R}{2} \right)e^{-j\omega T_R/2}
\end{aligned}
\tag{13.2-5}
$$

Here $F(\omega)$ denotes the Fourier transform of $P[\theta_s(t)]$, whatever it is. Proofs of (13.2-3) and (13.2-4) are given in Problem 13.2-1.

Next we digress to obtain $|s_{mf}(\tau_R)|$. If we let $s(t)$ denote the transmitted pulse in the central pulse interval (for $k = 0$), and denote its Fourier transform by $S(\omega)$, the corresponding received pulse is $\alpha s(t - \tau_R)$, where α is a constant determined through the radar equation. The received pulse's transform is $\alpha S(\omega) \exp(-j\omega \tau_R)$. The matched filter to this pulse in white noise, $n(t)$, is defined at time $t_0 = \tau_R$ (see Section 6.7) by[2]

$$H_{opt}(\omega) = \frac{2C\alpha}{\mathscr{N}_0} S^*(\omega) \tag{13.2-6}$$

---

[2] The constant C is real and nonzero but is otherwise arbitrary.

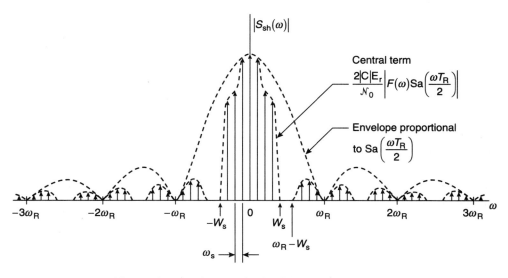

Figure 13.2-1 The magnitude of the spectrum $S_{sh}(\omega)$.

From the inverse Fourier transform of the spectrum of the matched filter's output signal evaluated at the sample time $t = \tau_R$, we have

$$|S_{mf}(\tau_R)| = \left| \frac{2C}{\mathcal{N}_0} \frac{1}{2\pi} \int_{-\infty}^{\infty} |\alpha S(\omega)|^2 \, d\omega \right| = \frac{2|C|}{\mathcal{N}_0} E_r \qquad (13.2\text{-}7)$$

where E_r is the energy in a single received pulse.

We now return to (13.2-4) and use (13.2-7) and (13.2-5) to write

$$S_{sh}(\omega) = \frac{2|C|E_r}{\mathcal{N}_0} \operatorname{Sa}\left(\frac{\omega T_R}{2}\right) e^{-j\omega T_R/2} \sum_{k=-\infty}^{\infty} F(\omega - k\omega_R) e^{-jk\omega_R \tau_R} \qquad (13.2\text{-}8)$$

A rough sketch of $|S_{sh}(\omega)|$ is shown in Fig. 13.2-1;[3] it assumes that ω_R is larger than twice the spectral extent of $P[\theta_s(t)]$.[4] This condition is required by sampling theory if the "message" $P[\theta_s(t)]$ is to be recoverable from its samples with no distortion.

It is clear from Fig. 13.2-1 that the central term would be proportional to $F(\omega)$ if it were not for the distortion due to the factor $\operatorname{Sa}(\omega T_R/2)$. However, the effect is removed

---

[3] Since $P(\theta_s)$ is periodic, its spectrum $F(\omega)$ will contain impulses at multiples of ω_s around $\omega = 0$. Figure 13.2-1 illustrates $F(\omega)$ as three nonzero impulses on each side of $\omega = 0$, so replica spectra at multiples of ω_R have three impulses on each side of $k\omega_R$. These replicas are highly distorted due to the nulls in $\operatorname{Sa}(\omega T_R/2)$ at $\omega = k\omega_R$, $k = \pm 1, \pm 2, \dots$.
[4] Spectral extent is the largest nonnegligible frequency in the spectrum of $P[\theta_s(t)]$; it is W_s in Fig. 13.2.1.

if the equalizing LPF in Fig. 13.1-3 has a transfer function

$$
H_{eq}(\omega) = \begin{cases} \dfrac{T_R}{H_{sh}(\omega)} = \dfrac{e^{j\omega(T_R/2)}}{Sa(\omega T_R/2)}, & |\omega| \leq W_s \\[2mm] \text{arbitrary,} & W_s < |\omega| < \omega_R - W_s \\[2mm] 0, & \omega_R - W_s \leq |\omega| \end{cases} \tag{13.2-9}
$$

Since $W_s \ll \omega_R$ in most practical cases, there is no problem approximating $H_{eq}(\omega)$ in the band $|\omega| \leq W_s$. The region from W_s to $\omega_R - W_s$ allows the filter's transfer function to "roll off" to a negligible level at frequency $\omega_R - W_s$ in order to remove the "sampling replicas" that start to occur at $\omega_R - W_s$. The spectrum of the equalizing filter's response, denoted by $S_0(\omega)$ is

$$
S_0(\omega) = \frac{2|C|E_r}{\mathcal{N}_0} F(\omega) \tag{13.2-10}
$$

from (13.2-8) and (13.2-9). By inverse Fourier transformation of (13.2-10), we have

$$
s_0(t) = \frac{2|C|E_r}{\mathcal{N}_0} P[\theta_s(t)] \tag{13.2-11}
$$

Next we observe that the "message" $P[\theta_s(t)]$ is a periodic function of time because scanning is periodic at the scan rate ω_s. As a consequence $P[\theta_s(t)]$ can be represented by the Fourier series

$$
P[\theta_s(t)] = \frac{a_0}{2} + \sum_{n=1}^{\infty} a_n \cos(n\omega_s t) + \sum_{n=1}^{\infty} b_n \sin(n\omega_s t) \tag{13.2-12}
$$

where a_0, a_n, and b_n are the Fourier series coefficients. On multiplying $s_0(t)$ by $\cos(\omega_s t)$ as implied in Fig. 13.1-3, we have

$$
s_2(t) = \frac{2|C|E_r}{\mathcal{N}_0} \left(\frac{a_0}{2}\cos(\omega_s t) + \sum_{n=1}^{\infty} \frac{a_n}{2}\{\cos[(n+1)\omega_s t] + \cos[(n-1)\omega_s t]\} \right.
$$
$$
\left. + \sum_{n=1}^{\infty} \frac{b_n}{2}\{\sin[(n+1)\omega_s t] + \sin[(n-1)\omega_s t]\} \right) \tag{13.2-13}
$$

The output LPF is designed to pass only the dc term in (13.2-13) while rejecting all harmonics of ω_s. Thus

$$
s_x = \frac{|C|E_r a_1}{\mathcal{N}_0} \tag{13.2-14}
$$

A similar development for the lower output path gives

$$
s_y = \frac{|C|E_r b_1}{\mathcal{N}_0} \tag{13.2-15}
$$

From (B.1-3) and (B.1-4) the two required Fourier series coefficients are

$$a_1 = \frac{2}{T_s} \int_{-T_s/2}^{T_s/2} P[\theta_s(t)] \cos(\omega_s t)\, dt \qquad (13.2\text{-}16)$$

$$b_1 = \frac{2}{T_s} \int_{-T_s/2}^{T_s/2} P[\theta_s(t)] \sin(\omega_s t)\, dt \qquad (13.2\text{-}17)$$

where

$$T_s = \frac{2\pi}{\omega_s} \qquad (13.2\text{-}18)$$

and $\theta_s(t)$ is given by (13.1-6).

We use (13.2-14) and (13.2-15) in Section 13.4 to determine the accuracy of measurements.

13.3 NOISE ANALYSIS

We now analyze the system of Fig. 13.1-3 to determine its noise responses n_x and n_y. From use of (13.2-6) the power density spectrum, denoted by $\mathscr{S}_{N_iN_i}(\omega)$, of the noise $n_i(t)$ is

$$\mathscr{S}_{N_iN_i}(\omega) = \frac{\mathscr{N}_0}{2} |H_{\text{opt}}(\omega)|^2 = \frac{2C^2\alpha^2}{\mathscr{N}_0} |S(\omega)|^2 \qquad (13.3\text{-}1)$$

where the input noise $n(t)$ is assumed to be white with power density $\mathscr{N}_0/2$ for $-\infty < \omega < \infty$.

Now because the noise $n_d(t)$ from the envelope detector is to be sampled, we need be concerned only with the representation of $n_i(t)$ and $n_d(t)$ at times near the sample times. At these times the signal $s_i(t)$ will reach its maximum amplitude (denoted by A_i) and can be represented as a cosine function of time. The noise $n_i(t)$ is bandpass, and we can use its quadrature representation [see (C.8-2)]. Thus, for times near the sample time τ_R,

$$s_i(t) + n_i(t)|_{t \text{ near } \tau_R} \approx A_i \cos(\omega_0 t) + N_{Ii}(t)\cos(\omega_0 t)$$

$$- N_{Qi}(t)\sin(\omega_0 t) \qquad (13.3\text{-}2)$$

$$= R_d(t)\cos(\omega_0 t + \phi_d)$$

where $N_{Ii}(t)$ and $N_{Qi}(t)$ are the baseband quadrature components of noise that represent $n_i(t)$ and

$$R_d(t) = A_i \sqrt{1 + \frac{2N_{Ii}(t)}{A_i} + \frac{N_{Ii}^2(t) + N_{Qi}^2(t)}{A_i^2}} \qquad (13.3\text{-}3)$$

$$\approx A_i + N_{Ii}(t)$$

The angle ϕ_d depends on A_i, $N_{Ii}(t)$, and $N_{Qi}(t)$ but is of no interest because the envelope detector responds only to the envelope function, denoted by $R_d(t)$. In writing the last form of (13.3-3) it is assumed that $A_i \gg |N_{Ii}(t)|$ or $|N_{Qi}(t)|$ *most of the time*. This assumption is true if signal-to-noise ratio at the matched filter's output is large.

The logic leading to (13.3-3) means that the noise $n_d(t)$ is given by

$$n_d(t) \approx N_{Ii}(t) \tag{13.3-4}$$

for times near τ_R. From (C.8-13) the power density spectrum of $N_{Ii}(t)$, denoted by $\mathscr{S}_{N_{Ii}N_{Ii}}(\omega)$ is

$$\mathscr{S}_{N_{Ii}N_{Ii}}(\omega) \approx L_p \left\{ \frac{2C^2\alpha^2}{\mathscr{N}_0} [\,|S(\omega - \omega_0)|^2 + |S(\omega + \omega_0)|^2\,] \right\} \tag{13.3-5}$$

where $L_p\{\cdot\}$ represents taking only the baseband part of the quantity in the braces. The power of this noise is given by the power formula (C.6-5). This power is that present on the sample of $n_d(t)$ at time τ_R; it is also the same for other samples at times $\tau_R + kT_R$ by means of similar developments. On denoting the sample noise power by $\sigma^2_{N_d}$ we have (Problem 13.3-1)

$$\sigma^2_{N_d} = \frac{1}{2\pi} \int_{-\infty}^{\infty} \mathscr{S}_{N_{Ii}N_{Ii}}(\omega)\, d\omega = \frac{2C^2}{\mathscr{N}_0} E_r \tag{13.3-6}$$

when (13.3-5) is used.

Next we examine the sampled and held noise of Fig. 13.1-3; it can be written as

$$n_{sh}(t) = \sum_{k=-\infty}^{\infty} n_d(\tau_R + kT_R)\mathrm{rect}\left[\frac{t - \tau_R - kT_R - (T_R/2)}{T_R} \right] \tag{13.3-7}$$

Although the developments are somewhat detailed, the procedure to reduce (13.3-7) is to (1) form the autocorrelation function of $n_{sh}(t)$ using (13.3-6) to evaluate the expectations needed (Problem 13.3-2) and (2) form the time average of the autocorrelation function, denoted by $R_{N_{sh}N_{sh}}(\tau)$. This time average is found in Problem 13.3-3 to be

$$R_{N_{sh}N_{sh}}(\tau) = \frac{2C^2 E_r}{\mathscr{N}_0} \mathrm{tri}\left(\frac{\tau}{T_R} \right) \tag{13.3-8}$$

which has the Fourier transform [power spectrum $\mathscr{S}_{N_{sh}N_{sh}}(\omega)$]

$$\mathscr{S}_{N_{sh}N_{sh}}(\omega) = \frac{2C^2 E_r}{\mathscr{N}_0} T_R \mathrm{Sa}^2\left(\frac{\omega T_R}{2} \right) \tag{13.3-9}$$

By using (13.2-9) with (13.3-9) we can derive the power spectrum of $n_0(t)$ in Fig. 13.1-3, denoted by $\mathscr{S}_{N_0 N_0}(\omega)$

$$\mathscr{S}_{N_0 N_0}(\omega) = \mathscr{S}_{N_{sh} N_{sh}}(\omega)|H_{eq}(\omega)|^2 \approx \frac{2C^2 E_r T_R}{\mathscr{N}_0} \tag{13.3-10}$$

This result shows that the power density of $n_0(t)$ is approximately constant (for frequencies up to at least W_s).

The response of a product device to noise when one input is a sinusoid is known (see Peebles, 1993, pp. 218–221). The output noise's power density spectrum is the same for both sine and cosine product signals, so

$$\mathscr{S}_{N_2 N_2}(\omega) = \mathscr{S}_{N_3 N_3}(\omega) \approx \tfrac{1}{4}[\mathscr{S}_{N_0 N_0}(\omega - \omega_s) + \mathscr{S}_{N_0 N_0}(\omega + \omega_s)]$$

$$\approx \frac{C^2 E_r T_R}{\mathscr{N}_0} \tag{13.3-11}$$

for $|\omega|$ as large as W_s [or larger if $H_{eq}(\omega)$ approximates $T_R/H_{sh}(\omega)$ for $|\omega| > W_s$ according to (13.2-9)].

Ultimately we desire the powers of the noises n_x and n_y in Fig. 13.1-3. If the output low-pass filters have a transfer function $H_{LPF}(\omega)$, these powers are equal and given by

$$\overline{n_x^2} = \overline{n_y^2} = \frac{1}{2\pi} \int_{-\infty}^{\infty} \frac{C^2 E_r T_R}{\mathscr{N}_0} |H_{LPF}(\omega)|^2 \, d\omega = \frac{C^2 E_r T_R W_{N,LPF}}{\pi \mathscr{N}_0} \tag{13.3-12}$$

where

$$W_{N,LPF} = \frac{\int_0^{\infty} |H_{LPF}(\omega)|^2 \, d\omega}{|H_{LPF}(0)|^2} \tag{13.3-13}$$

is the *noise bandwidth* (rad/s) of the low-pass filter. In writing (13.3-12) the bandwidth of the low-pass filter is typically small relative to the bandwidth of the broader bandwidth noises $n_2(t)$ and $n_3(t)$ so that limits can be approximated as $-\infty$ to $+\infty$. In writing (13.3-12), we have used the fact that $|H_{LPF}(0)| = 1$, by assumption; if this is not a true condition, then the last right-side form of (13.3-12) must have an added factor $|H_{LPF}(0)|^2$.

13.4 ACCURACY

The radar works with signals s_x and s_y in Fig. 13.1-3 to determine angles θ_x and θ_y as defined in Fig. 13.1-2b. For a target exactly on boresight such that $\theta_x = 0$ and $\theta_y = 0$, or equivalently $\theta_T = 0$, (13.1-6) shows that $\theta_s(t) = \theta_q$ (a constant). As a consequence both a_1 and b_1 are zero from (13.2-16) and (13.2-17). Hence, from (13.2-14) and (13.2-15), $s_x = 0$ and $s_y = 0$ for a target on boresight. For small off-boresight target angles θ_T, the "error" signals s_x and s_y will increase with θ_x and θ_y, respectively, as shown in Fig. 13.4-1.

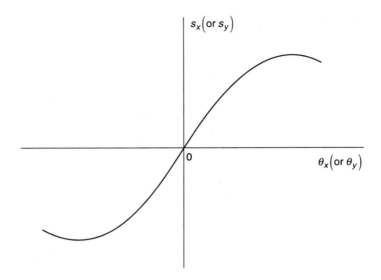

Figure 13.4-1 Radar response voltage versus target angle θ_x (or θ_y) from boresight.

In a manner similar to that already used in our discussions on range and Doppler frequency measurements, the variance of measurement of θ_x, denoted by $\sigma_{\hat{\theta}_x}^2$, is

$$\sigma_{\hat{\theta}_x}^2 = \frac{\overline{n_x^2}}{K_{\theta_x}^2} \tag{13.4-1}$$

where $\overline{n_x^2}$ is the power in the noise n_x and K_{θ_x} is the slope of the voltage response of Fig. 13.4-1 at the origin. Similarly, for measurement of θ_y,

$$\sigma_{\hat{\theta}_y}^2 = \frac{\overline{n_y^2}}{K_{\theta_y}^2} \tag{13.4-2}$$

The slopes are defined as those on boresight (origin) by

$$K_{\theta_x} = \left.\frac{\partial s_x}{\partial \theta_x}\right|_{\theta_T = 0} = \frac{|C|E_r}{\mathcal{N}_0}\left(\left.\frac{\partial a_1}{\partial \theta_x}\right|_{\theta_T = 0}\right) \tag{13.4-3}$$

$$K_{\theta_y} = \left.\frac{\partial s_y}{\partial \theta_y}\right|_{\theta_T = 0} = \frac{|C|E_r}{\mathcal{N}_0}\left(\left.\frac{\partial b_1}{\partial \theta_y}\right|_{\theta_T = 0}\right) \tag{13.4-4}$$

after using (13.2-14) and (13.2-15).

Next, the noise powers are substituted from (13.3-12) into (13.4-1) and (13.4-2) to get

$$\sigma_{\theta_x}^2 = \cfrac{2}{\mathscr{R}_p \cfrac{\pi}{W_{N,LPF} T_R} \left(\cfrac{\partial a_1}{\partial \theta_x}\bigg|_{\theta_T = 0}\right)^2} \qquad (13.4\text{-}5)$$

$$\sigma_{\theta_y}^2 = \cfrac{2}{\mathscr{R}_p \cfrac{\pi}{W_{N,LPF} T_R} \left(\cfrac{\partial b_1}{\partial \theta_y}\bigg|_{\theta_T = 0}\right)^2} \qquad (13.4\text{-}6)$$

where

$$\mathscr{R}_p = \frac{2E_r}{\mathscr{N}_0} \qquad (13.4\text{-}7)$$

is the single-pulse peak signal-to-noise ratio at the matched filter's output.[5]

To further develop the derivatives required in (13.4-5) and (13.4-6), we note that

$$\theta_x = \theta_T \cos(\phi_T) \qquad (13.4\text{-}8)$$

$$\theta_y = \theta_T \sin(\phi_T) \qquad (13.4\text{-}9)$$

from Fig. 13.1-2*b*. Also we expand (13.1-6) to get

$$\begin{aligned}
\theta_s^2 &= \theta_q^2 + \theta_T^2 - 2\theta_q\theta_T[\cos(\phi_T)\cos(\omega_s t) + \sin(\phi_T)\sin(\omega_s t)] \\
&= \theta_q^2 + \theta_x^2 + \theta_y^2 - 2\theta_q[\theta_x \cos(\omega_s t) + \theta_y \sin(\omega_s t)]
\end{aligned} \qquad (13.4\text{-}10)$$

These allow us to write

$$\begin{aligned}
\frac{\partial P(\theta_s)}{\partial \theta_x}\bigg|_{\theta_T = 0} &= \frac{\partial P(\theta_s)}{\partial \theta_s} \frac{\partial \theta_s}{\partial \theta_x}\bigg|_{\theta_T = 0} = \frac{\partial P(\theta_s)}{\partial \theta_s}\bigg|_{\theta_s = \theta_q} \frac{\partial \theta_s}{\partial \theta_x}\bigg|_{\theta_T = 0} \\
&= \frac{\partial P(\theta_s)}{\partial \theta_s}\bigg|_{\theta_s = \theta_q} [-\cos(\omega_s t)]
\end{aligned} \qquad (13.4\text{-}11)$$

$$\frac{\partial P(\theta_s)}{\partial \theta_y}\bigg|_{\theta_T = 0} = \frac{\partial P(\theta_s)}{\partial \theta_s}\bigg|_{\theta_s = \theta_q} [-\sin(\omega_s t)] \qquad (13.4\text{-}12)$$

[5] Recall that E_r corresponds to the beam pointing directly at the target, so $\mathscr{R}_p$ is that which would occur assuming that this condition is true.

These results, when used with (13.2-16) and (13.2-17), give

$$
\begin{aligned}
\left.\frac{\partial a_1}{\partial \theta_x}\right|_{\theta_T = 0} &= \frac{2}{T_s} \int_{-T_s/2}^{T_s/2} \left.\frac{\partial P(\theta_s)}{\partial \theta_x}\right|_{\theta_T = 0} \cos(\omega_s t)\, dt \\
&= \frac{2}{T_s} \int_{-T_s/2}^{T_s/2} \left[\left.\frac{\partial P(\theta_s)}{\partial \theta_x}\right|_{\theta_s = \theta_q}\right] [-\cos^2(\omega_s t)]\, dt \\
&= \left.\frac{\partial P(\theta_s)}{\partial \theta_s}\right|_{\theta_s = \theta_q} \left\{ -\frac{1}{T_s} \int_{-T_s/2}^{T_s/2} [1 + \cos(2\omega_s t)]\, dt \right\} \\
&= -\left.\frac{\partial P(\theta_s)}{\partial \theta_s}\right|_{\theta_s = \theta_q}
\end{aligned}
\tag{13.4-13}
$$

$$
\left.\frac{\partial b_1}{\partial \theta_y}\right|_{\theta_T = 0} = -\left.\frac{\partial P(\theta_s)}{\partial \theta_s}\right|_{\theta_s = \theta_q} = \left.\frac{\partial a_1}{\partial \theta_x}\right|_{\theta_T = 0}
\tag{13.4-14}
$$

We may now put (13.4-5) and (13.4-6) into a final form. First, we recognize the effective number of pulses integrated by the output LPF to be

$$
n_{\mathrm{eff}} = \frac{\pi}{W_{N,\mathrm{LPF}} T_R}
\tag{13.4-15}
$$

[similar to (11.7-10)]. Next, we use (13.4-14) to write

$$
\sigma_{\hat{\theta}_x}^2 = \sigma_{\hat{\theta}_y}^2 = \frac{2}{\mathscr{R}_p n_{\mathrm{eff}} \left(-\left.\dfrac{\partial P(\theta_s)}{\partial \theta_s}\right|_{\theta_s = \theta_q}\right)^2}
\tag{13.4-16}
$$

Finally, we define a normalized two-way slope constant k_{s2} by

$$
\frac{k_{s2}}{\sqrt{L_{k2}}} = \theta_B \left[-\left.\frac{\partial P(\theta_s)}{\partial \theta_s}\right|_{\theta_s = \theta_q} \right]
\tag{13.4-17}
$$

where $\sqrt{L_{k2}}$ is the two-way "loss" in voltage gain (a number greater than one) and θ_B is the 3-db beamwidth of the one-way power pattern $P(\theta_s)$. L_{k2} represents the two-way loss in received signal *power* that occurs on boresight due to the fact that the antenna pattern's boresight gain is less than at the beam's maximum. L_{k2} is called the *crossover loss*. Thus

$$
L_{k2} = \frac{1}{P^2(\theta_q)}
\tag{13.4-18}
$$

$$
k_{s2} = \frac{\theta_B}{P(\theta_q)} \left[-\left.\frac{\partial P(\theta_s)}{\partial \theta_s}\right|_{\theta_s = \theta_q} \right]
\tag{13.4-19}
$$

so

$$
\sigma_{\hat{\theta}_x}^2 = \sigma_{\hat{\theta}_y}^2 = \frac{2L_{k2}\theta_B^2}{\mathscr{R}_p n_{\mathrm{eff}} k_{s2}^2}
\tag{13.4-20}
$$

Our principal result is (13.4-20), which gives the variance of measurement errors. The expression is valid for any shape of pencil-beam antenna pattern when the radar operates monostatically (active or two-way case), and it is in agreement with Barton (1988, p. 387) and Barton and Ward (1969, p. 34). To develop (13.4-20) further requires specification of a pattern. In the following section we take an example of a gaussian pattern.

The case of continuous measurement, or tracking, of an angle θ_x or θ_y is easily handled. If $W_{N,\mathrm{LPF}}$ is interpreted as including an angle tracking loop, then $W_{N,\mathrm{LPF}}$ is the overall angle tracking loop's closed-loop noise bandwidth, including any low-pass filter at the output of Fig. 13.1-3. The effective number of pulses n_{eff} becomes the number of pulses effectively integrated by the tracking system.

13.5 EXAMPLE OF A GAUSSIAN PATTERN

The gaussian function is one that can be solved, and it is a reasonable approximation to other patterns having other shapes, at least over their 3-dB beamwidths. We consider this pattern in this section as an example of the previous developments. We give first a summary of the most important results using the general developments of Section 13.4 and then give as alternative development for the specific pattern.[6] In each case we assume the two-way voltage (one-way power) pattern to be

$$P(\theta_s) = e^{-a^2\theta_s^2} \tag{13.5-1}$$

where

$$a^2 = \frac{4\ln(2)}{\theta_B^2} \approx \frac{2.773}{\theta_B^2} \tag{13.5-2}$$

where θ_B is the 3-dB beamwidth of $P(\theta_s)$ treated as a one-way power pattern [i.e., where $P(\theta_B/2) = 0.5$].

Most Important Results

The most important quantities of interest are the pattern derivative of (13.4-16), the crossover loss of (13.4-18), and the normalized slope constant of (13.4-19). On using (13.5-1), these quantities become

$$\left[-\frac{\partial P(\theta_s)}{\partial \theta_s}\bigg|_{\theta_s=\theta_q}\right] = \frac{k_{s2}}{\theta_B\sqrt{L_{k2}}} = 8\ln(2)\frac{\theta_q}{\theta_B^2}e^{-4\ln(2)(\theta_q/\theta_B)^2} \tag{13.5-3}$$

$$L_{k2} = e^{8\ln(2)(\theta_q/\theta_B)^2} \tag{13.5-4}$$

$$k_{s2} = 8\ln(2)\left(\frac{\theta_q}{\theta_B}\right) \tag{13.5-5}$$

---

[6] Those interested in the gaussian pattern only as an example may ignore the alternative development, which serves only as a verification of the more general developments.

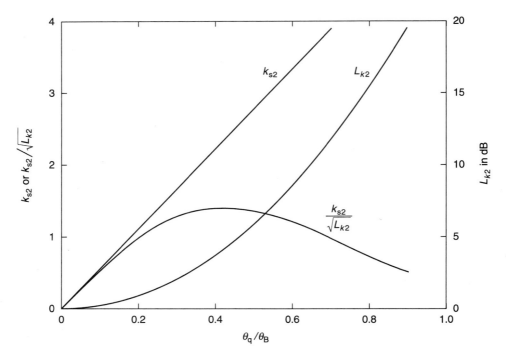

Figure 13.5-1   Quantities relating to using a gaussian pattern in conical scan.

These quantities are plotted in Fig. 13.5-1. From (13.4-16) and (13.4-17) it is clear that the best performance occurs when the quantity $k_{s2}/\sqrt{L_{k2}}$ is largest. The maximum is 1.428 when $(\theta_q/\theta_B) = [8\ln(2)]^{-1/2}$. We refer to this case as one of maximum slope.

With maximum slope the crossover loss is 4.34 dB. This is a reasonably large loss that reduces the system's ability to perform range measurements. Sometimes a compromise is chosen (Skolnik, 1980, p. 159) by selecting θ_q/θ_B near 0.3 where crossover loss is only about 2.17 dB. This "system" optimum increases the angle error variance by about 21.4% compared to the maximum slope case. It can be shown that these variances are (Problems 13.5-8 and 13.5-9)

$$\sigma_{\hat{\theta}_x}^2 = \sigma_{\hat{\theta}_y}^2 \approx \begin{cases} \dfrac{0.980\theta_B^2}{\mathcal{R}_p n_{\text{eff}}}, & \text{for maximum slope} \\[3mm] \dfrac{1.190\theta_B^2}{\mathcal{R}_p n_{\text{eff}}}, & \text{for "system" optimum} \end{cases} \tag{13.5-6}$$

Alternative Development

The preceding results for the gaussian pattern of (13.5-1) can also be shown using an alternative procedure. On substituting (13.1-6) into (13.5-1), we have

$$P(\theta_s) = e^{-a^2[\theta_q^2 + \theta_T^2 - 2\theta_q\theta_T\cos(\omega_s t - \phi_T)]} \tag{13.5-7}$$

But since (Abramowitz and Stegun, 1964, p. 376)

$$e^{x\cos(z)} = I_0(x) + 2 \sum_{n=1}^{\infty} I_n(x)\cos(nz) \tag{13.5-8}$$

(13.5-7) can be expanded as

$$P(\theta_s) = e^{-a^2(\theta_q^2 + \theta_T^2)} \bigg\{ I_0(2a^2\theta_q\theta_T)$$

$$+ 2\sum_{n=1}^{\infty} [I_n(2a^2\theta_q\theta_T)\cos(n\phi_T)]\cos(n\omega_s t) \tag{13.5-9}$$

$$+ 2\sum_{n=1}^{\infty} [I_n(2a^2\theta_q\theta_T)\sin(n\phi_T)]\sin(n\omega_s t) \bigg\}$$

If we recognize that $P(\theta_s)$ is a periodic function with fundamental angular frequency ω_s (because scanning is periodic making θ_s periodic), then (13.5-9) is recognized as the Fourier series of $P(\theta_s)$. In particular, we are mainly interested in coefficients a_1 and b_1 [see (13.2-12)] which are

$$a_1 = K_q\cos(\phi_T) \tag{13.5-10}$$

$$b_1 = K_q\sin(\phi_T) \tag{13.5-11}$$

where we define K_q by

$$K_q = 2e^{-a^2(\theta_q^2 + \theta_T^2)} I_1(2a^2\theta_q\theta_T) \tag{13.5-12}$$

Figure 13.5-2 shows the behavior of K_q with θ_T/θ_B for various beam squint angles θ_q/θ_B. The plots agree with those of Skolnik (1980, p. 159), who plots $K_q/2$. The maximum slope of K_q at the origin occurs when $(\theta_q/\theta_B) = [8\ln(2)]^{-1/2}$.

For θ_T small we use the approximation $I_1(x) \approx x/2$, $|x|$, small, to write K_q as

$$K_q \approx 2a^2\theta_q\theta_T e^{-a^2\theta_q^2}, \qquad \theta_T \text{ small} \tag{13.5-13}$$

With this approximation we write (13.5-10) and (13.5-11) as

$$\left.\begin{matrix} a_1 \\ b_1 \end{matrix}\right\} \approx 2a^2\theta_q e^{-a^2\theta_q^2} \begin{cases} \theta_T\cos(\phi_T) \\ \theta_T\sin(\phi_T) \end{cases} = 2a^2\theta_q e^{-a^2\theta_q^2} \begin{cases} \theta_x \\ \theta_y \end{cases} \tag{13.5-14}$$

From (13.2-14) and (13.2-15) we obtain

$$\left.\begin{matrix} s_x \\ s_y \end{matrix}\right\} = \frac{|C|E_r}{\mathcal{N}_0} \begin{Bmatrix} a_1 \\ b_1 \end{Bmatrix} \approx \frac{|C|E_r}{\mathcal{N}_0} 2a^2\theta_q e^{-a^2\theta_q^2} \begin{cases} \theta_x \\ \theta_y \end{cases} \tag{13.5-15}$$

which shows that s_x and s_y are proprtional to θ_x and θ_y, respectively, as they should for small θ_T.

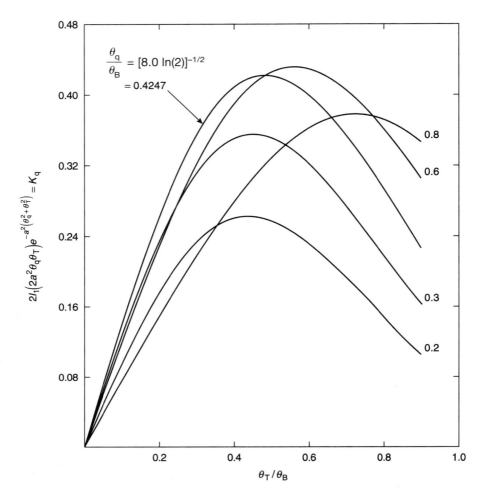

Figure 13.5-2 K_q versus θ_T/θ_B for various θ_q/θ_B and a gaussian pattern.

The conclusion of the alternative development follows the use of (13.5-15) in calculating K_{θ_x} and K_{θ_y} from (13.4-3) and (13.4-4), respectively. We have

$$K_{\theta_x} = K_{\theta_y} = \frac{|C|E_r}{\mathcal{N}_0} 2a^2 \theta_q e^{-a^2\theta_q^2} \tag{13.5-16}$$

When this result is used in (13.4-1) and (13.4-2) with (13.3-12), we have

$$\sigma_{\hat{\theta}_x}^2 = \sigma_{\hat{\theta}_y}^2 = \frac{2}{\dfrac{2E_r}{\mathcal{N}_0} \dfrac{\pi}{W_{N,\mathrm{LPF}} T_R} \left[8\ln(2) \dfrac{\theta_q}{\theta_B^2} e^{-4\ln(2)(\theta_q/\theta_B)^2} \right]^2} \tag{13.5-17}$$

Finally we use (13.4-7), (13.4-15), and (13.5-3), which is true for the gaussian pattern, to write (13.5-17) exactly as (13.4-20). Thus the alternative development gives the same error variance results as with our general procedure.

PROBLEMS

13.1-1 Assume that the two-way voltage pattern in a conical scan radar is

$$P(\theta_s) = e^{-4\ln(2)(\theta_s/\theta_B)^2}$$

where θ_B is the 3-dB beamwidth. (a) Assume that $\theta_T = 0.3\theta_B$, $\theta_q = 0.4\theta_B$, and $\phi_T = 0$, and plot $P(\theta_s)$ versus $\omega_s t$ for one period of $\omega_s t$. (b) Repeat part (a) for $\theta_T = 0.15\theta_B$. The plotted functions represent the envelope of pulses received by the system.

13.2-1 Define

$$h_{sh}(t) = \text{rect}\left[\frac{t - (T_R/2)}{T_R}\right]$$

and show that Fig. P13.2-1 is equivalent to the sample-hold operation in Fig. 13.1-3 so that its output is given by (13.2-3). Also use Fig. P13.2-1 to prove that (13.2-4) is true.

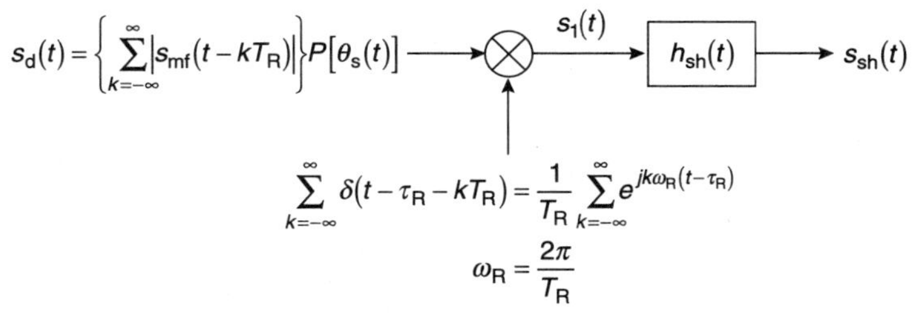

Figure P13.2-1

⋆13.2-2 Assume that the antenna pattern of a conical scan system is adequately approximated over its main beam by

$$P(\theta_s) = \cos\left[\frac{4\pi}{3}\left(\frac{\theta_s}{\theta_B}\right)^2\right]$$

where θ_B is the 3-dB beamwidth. (a) Determine a_1 and b_1 for this pattern. (b) Evaluate a_1 and b_1 for θ_T small and write the results in terms of θ_x and θ_y.

⋆13.2-3 Work Problem 13.2-2 except assume the pattern

$$P(\theta_s) = \cos^2\left[\pi\left(\frac{\theta_s}{\theta_B}\right)^2\right]$$

★13.2-4 Define K_q as $a_1/\cos(\phi_T)$. For a particular antenna pattern it is known that

$$K_q = 2J_1(2a^2\theta_q\theta_T)\sin[a^2(\theta_q^2 + \theta_T^2)]$$

where $a^2 = 4\pi/(3\theta_B^2)$ and $J_1(x)$ is the Bessel function of the first kind of order one. Plot K_q versus θ_T/θ_B for $\theta_q/\theta_B = 0.1$ to 0.7 in steps of 0.1. Compare the results to those of Fig. 13.5-2 for a gaussian pattern. (*Hint*: Use the approximation $J_1(x) \approx x[0.5 - 0.56249985(x/3)^2 + 0.21093573(x/3)^4 - 0.03954289(x/3)^6 + 0.00443319(x/3)^8 - 0.00031761(x/3)^{10} + 0.00001109(x/3)^{12} + \varepsilon]$ where $|\varepsilon| < 1.3(10^{-8})$ when $-3 \le x \le 3$. See Abramowitz and Stegun (eds., 1964, p. 370).)

13.2-5 For K_q as given in Problem 13.2-4 find the value of θ_q/θ_B that maximizes $[\partial K_q/\partial\theta_T]|_{\theta_T = 0}$.

★13.2-6 Work Problem 13.2-4 except assume that

$$K_q = J_1\left(\frac{4\pi\theta_q\theta_T}{\theta_B^2}\right)\sin\left[2\pi\frac{(\theta_q^2 + \theta_T^2)}{\theta_B^2}\right]$$

13.2-7 For K_q as given in Problem 13.2-6 find the value of θ_q/θ_B that maximizes $[\partial K_q/\partial\theta_T]|_{\theta_T = 0}$.

13.3-1 Use (13.3-5) and show that (13.3-6) is true.

★13.3-2 Use (13.3-7) and show that the autocorrelation function of $n_{sh}(t)$ depends on time and is given by

$$R_{N_{sh}N_{sh}}(t, t + \tau) = \frac{2C^2 E_r}{\mathcal{N}_0} \sum_{k=-\infty}^{\infty} \text{rect}\left[\frac{t - \tau_R - kT_R - (T_R/2)}{T_R}\right]$$

$$\cdot \text{rect}\left[\frac{t + \tau - \tau_R - kT_R - (T_R/2)}{T_R}\right]$$

★13.3-3 Show that the time average of the autocorrelation function of Problem 13.3-2, as defined by (C.5-19), is given by (13.3-8).

13.3-4 Discuss how terms in (13.3-3) can be *roughly* replaced by the peak signal power-to-average noise power ratio, denoted by (S_i/N_i), at the matched filter's output. Using your rough replacement, evaluate the relative magnitudes of the three terms for $(S_i/N_i) = 8$ (9 dB), 16, 32, 64, and 128 (21 dB).

13.3-5 Suppose the low-pass filter at the outputs in Fig. 13.1-3 is defined by

$$|H_{\text{LPF}}(\omega)|^2 = \frac{1}{1 + (\omega/W_c)^{2n}}, \quad n = 1, 2, \ldots$$

where W_c is a positive constant. Find the noise bandwidth $W_{N,\text{LPF}}$ for any n.

13.3-6 Work Problem 13.3-5 except assume that

$$|H_{LPF}(\omega)|^2 = \frac{1}{[1 + (\omega/W_c)^2]^n}, \qquad n = 2, 3, 4 \ldots$$

13.4-1 Assume that $P(\theta_s)$ for an antenna is defined by

$$P(\theta_s) = \left[\frac{2J_1(3.227\theta_s/\theta_B)}{(3.227\theta_s/\theta_B)} \right]^2$$

where θ_B (rad) is the -3-dB beamwidth of $P(\theta_s)$ treated as a one-way power pattern. (a) Find $k_{s2}/\sqrt{L_{k2}}$ as defined by (13.4-17). (b) Find $\sqrt{L_{k2}}$. (c) Use your results to find k_{s2}. Plot the three results versus θ_q/θ_B, and compare with Fig. 13.5-1.

13.4-2 Work Problem 13.4-1 except assume that

$$P(\theta_s) = \cos^2\left[\frac{\pi}{2}\left(\frac{\theta_s}{\theta_B} \right) \right]$$

13.4-3 Work Problem 13.4-1 except assume that

$$P(\theta_s) = \cos^4\left[0.7281 \frac{\pi}{2}\left(\frac{\theta_s}{\theta_B} \right) \right]$$

13.4-4 A generalization of the antenna patterns of Problems 13.4-2 and 13.4-3 is

$$P(\theta_s) = \cos^{2N}\left(\frac{a\theta_s}{\theta_B} \right), \qquad N = 1, 2, \ldots$$

Find (a) constant a so that $P(\theta_B/2) = 0.5$, (b) $\sqrt{L_{k2}}$, (c) $k_{s2}/\sqrt{L_{k2}}$, and (d) the value of θ_q/θ_B that maximizes $k_{s2}/\sqrt{L_{k2}}$. (e) Plot the maximum value found in part (d) versus N for N up to 10.

13.4-5 A conical scan radar integrates 26 pulses effectively where $\mathscr{R}_p = 9$ (or 9.54 dB). If the system is able to measure each spatial angle to an rms error of 0.05 times the 3-dB beamwidth of $P(\theta_s)$, what is $\partial P(\theta_s)/\partial \theta_s|_{\theta_s = \theta_q}$?

13.5-1 A conical scan radar uses a pattern that is approximately gaussian. If θ_q/θ_B is 0.35, how much lower is the slope constant $k_{s2}/\sqrt{L_{k2}}$ as compared to θ_q/θ_B that gives the maximum slope?

13.5-2 What is the crossover loss L_{k2} of a conical scan radar having a gaussian beam when $\theta_q = 0.35\theta_B$?

13.5-3 Work Problem 13.5-2 except assume that $\theta_q = 0.60\theta_B$.

13.5-4 A conical scan radar has a gaussian pattern with a squint angle θ_q that is the arithmetic mean of the squint angles for maximum error slope and the "system optimum." (a) What is the value of θ_q? (b) What is the resulting error slope $k_{s2}/\sqrt{L_{k2}}$? (c) What is the crossover loss L_{k2}?

13.5-5 Work Problem 13.5-4 except for a beam squint angle that is the geometric mean of the squint angles for maximum slope and "system optimum."

13.5-6 The gaussian pattern of a conical scan system has a one-way 3-dB beamwidth of 6 degrees. The beam squint angle is set for a "system optimum." (a) If $\mathcal{R}_p = 8$ is available, what effective number of pulses must be integrated to achieve an rms angle error of 0.2 degrees? (b) If the angle tracking loops are to have noise bandwidths of 10 Hz, what pulse rate is required?

13.5-7 Work Problem 13.5-6 except assume a 4-degree beamwidth, $\mathcal{R}_p = 23$, and an rms error of 0.1 degree.

13.5-8 Show that (13.5-6), for the case of maximum slope, is true.

13.5-9 Show that (13.5-6), for the case of "system optimum," is true.

13.5-10 A conical scan system is to produce an rms tracking error of $0.1\theta_B$. If the antenna has approximately a gaussian pattern and θ_q is set for a "system optimum," plot curves of the required value of $\mathcal{R}_p$ versus $100 \le f_R$ for values of $W_{N,LPF}/2\pi$ of 5, 10, 20, and 40 Hz.

14

ANGLE MEASUREMENT AND TRACKING BY MONOPULSE

By current technology the most-used, effective, and attractive method for deriving information about the spatial angles of a target is called *monopulse*. The name derives from the fact that the method is theoretically capable of obtaining estimates of a target's angles by using only one (*mono*) pulse. Monopulse is richly discussed in the literature.[1] In this chapter we discuss some of the fundamental theoretical concepts on which monopulse is based. We begin by developing the optimum monopulse system and its performance in angle measurement. Then we discuss various practical aspects of monopulse, give some realizations, and some extensions.

All monopulse systems break down into two basic parts. One is the antenna and its associated microwave elements (that form what we will call the *comparator*), and the second is a signal processor appropriate for the type of antenna patterns used. There are three types of monopulse of interest. Two are important enough to have names. One, called *amplitude-sensing monopulse* (or just *amplitude monopulse*) uses only the amplitude characteristic (magnitude) of its antenna patterns (more than one pattern is required in the system). The phase characteristics of the patterns are as nearly equal and independent of target angles as design allows.

The second type of monopulse is called *phase sensing monopulse* (or just *phase monopulse*). It uses only the (different) phase characteristics of the patterns that have as nearly identical amplitude characteristics as possible. This type is less popular than amplitude monopulse but is sometimes used.

The third type of monopulse does not carry any particular name. It could be called *general monopulse* or *amplitude and phase monopulse*, or *hybrid monopulse*, because the system uses both the amplitude and phase characteristics of the system's patterns. This

---

[1] A few references that can provide more details and other references are Edde (1993), Skolnik (ed., 1990), Barton (1988), Levanon (1988), Eaves and Reedy (eds., 1987), Leonov and Fomichev (1986), Sherman (1984), Skolnik (1980), Barton (ed., 1977), this book is a collection of important papers on monopulse up to the time, especially the paper by G. M. Kirkpatrick which is a reprint of a General Electric research report, "Final Engineering Report on Angular Accuracy Improvement," dated 1 August 1952), and Rhodes (1959).

system has been studied in the literature and is discussed in this chapter. However, it has received little practical attention.

Finally we briefly discuss conopulse radar. This is also a hybrid system; it is part *con*ical scan and part mon*opulse*.

14.1   SOME PRELIMINARY DEFINITIONS

In this section we give some preliminary definitions that link the antenna details of Chapter 3 to the problem of angle measurement. These details relate to the vector representation of patterns and a definition of a one-way vector voltage pattern. Although the material should prove enlightening to the more academic reader, it can be bypassed by the more practical reader. In the latter case, if geometry (Fig. 14.1-1), the system's structure (Fig. 14.1-2), and the definitions of the voltage pattern, output voltage, and output noise power are understood [through (14.1-12), (14.1-20), and (14.1-21)], then the reader can go directly to Section 14.2 without too great a loss.

We consider a monostatic radar that transmits its pulses through a transmitting antenna pattern having its maximum gain in the direction of the reference axis (the z axis, or boresight axis) as shown in Fig. 14.1-1. The radar is at the origin of the x, y, z coordinates. The transmitting antenna's aperture is in the x, y plane. A target of interest is located at point T, which is near the reference axis. In radar-centered spherical coordinates, the target is located at a point (R, θ, ϕ). Since point T is near the reference axis, θ can be taken as a small angle. This fact allows angles θ_x and θ_y, which

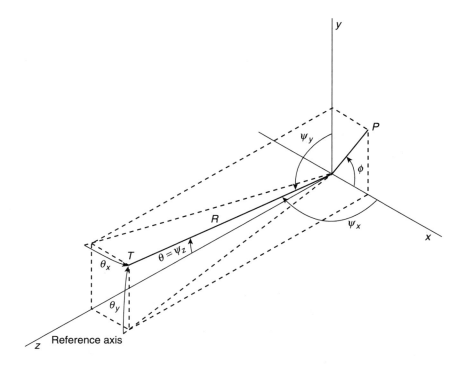

Figure 14.1-1   Geometry of important angles in monopulse radar.

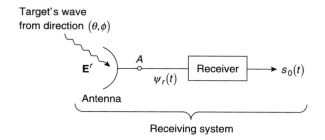

Figure 14.1-2 General form of a monopulse receiving system.

are the angles of the target's direction from the y, z and x, z planes, respectively, to be considered small. Thus from the figure, (3.11-4), and (3.11-5),

$$\theta_x \approx \sin(\theta_x) = \cos(\psi_x) = \sin(\theta)\cos(\phi) \qquad (14.1\text{-}1)$$

$$\theta_y \approx \sin(\theta_y) = \cos(\psi_y) = \sin(\theta)\sin(\phi) \qquad (14.1\text{-}2)$$

It is the angles θ_x and θ_y that the radar typically measures. If it is in a tracking mode, the system uses these measurements to move the reference (z) axis to follow the target's position such as to drive the measured values of θ_x and θ_y to zero. Angles θ_x and θ_y are sometimes called *error angles*.

The transmitting antenna's beamwidth is assumed large enough relative to the largest values of $|\theta_x|$ and $|\theta_y|$ so that the field intensity of the wave at the target may be approximated as constant.

The transmitted wave strikes the target which scatters energy towards the receiving antenna in accordance with a radar cross section σ. The scattered wave arrives at the receiving antenna. In the radar's coordinates, the scattered wave is the sum of two component waves (see Section 5.1). One has a polarization matched to the receiving antenna, and this component is that to which the receiving antenna responds by generating an available power determined by cross section σ and the radar equation. We denote this wave component by $\mathbf{E}^r$, where

$$\mathbf{E}^r = E_\theta \hat{a}_\theta + E_\phi \hat{a}_\phi \qquad (14.1\text{-}3)$$

Here the wave's complex electric field components in directions θ and ϕ are E_θ and E_ϕ, respectively, while $\hat{a}_\theta$ and $\hat{a}_\phi$ and are unit vectors in these respective directions. The radar does not respond to the orthogonal (second) component of the scattered wave.[2]

The received electric field $\mathbf{E}^r$ from the target at direction (θ, ϕ) induces a voltage into the receive antenna. This induced voltage (described below) is affected by the receiving antenna's characteristics and emerges at the antenna's output terminals (point A in

---

[2] From another viewpoint, given a receiving antenna with a prescribed polarization, it will only respond to the component of the scattered wave that is matched to the antenna's polarization. The available power in the response component is that for which σ is defined.

the system defined in Fig. 14.1-2). For a complex transmitted waveform $\psi(t)$, this complex voltage is denoted as $\psi_r(t)$, and its definition assumes that the receiver is impedance-matched to the antenna at its output terminals.

The receiving system of Fig. 14.1-2 consists of the antenna, with an unspecified pattern at this point, and a receiver with an unspecified transfer function. The optimization problem will lead to a solution for both of these components, that is, an optimum antenna pattern and an optimum transfer function for the receiver.

Received Signal

Because the signal received from the target is directly related to the antenna's characteristics, we will need to borrow some material from Chapter 3. Several steps are needed to show how the antenna develops the received signal.

From Section 3.4 and (3.4-24), in particular, we know that the open-circuit complex voltage, denoted by $\psi_{\mathrm{ant}}(t)$, induced into the receiving antenna by the received field $\mathbf{E}^r$ (defined in the radar's spherical coordinates) is

$$\psi_{\mathrm{ant}}(t) = \mathbf{E}^r \cdot \mathbf{h} = E_\theta h_\theta + E_\phi h_\phi \tag{14.1-4}$$

where $\mathbf{h}$ is the (vector) effective length of the receiving antenna given by

$$\mathbf{h} = h_\theta \hat{a}_\theta + h_\phi \hat{a}_\phi \tag{14.1-5}$$

The components h_θ and h_ϕ are defined by (3.4-22) and (3.4-23). However, before substituting for h_θ and h_ϕ to reduce (14.1-4), we observe that the definitions for h_θ and h_ϕ in Chapter 3 apply to any aperture antenna. For our purposes it is useful to adopt specific notation for the monopulse antenna. In particular, $F_x(\theta, \phi)$, $F_y(\theta, \phi)$, $\mathbf{F}_a(\theta, \phi)$, $E_{ax}(\xi, \zeta)$, $E_{ay}(\xi, \zeta)$, and $E_a(\xi, \zeta)$ of Chapter 3 will now be replaced, respectively, by $F_{\Delta, x}(\theta, \phi)$, $F_{\Delta, y}(\theta, \phi)$, $\mathbf{F}_{\Delta, a}(\theta, \phi)$, $E_{\Delta, x}(\xi, \zeta)$, $E_{\Delta, y}(\xi, \zeta)$, and $\mathbf{E}_\Delta(\xi, \zeta)$ all with exactly the same meanings. Use of the subscript Δ is to facilitate remembering that we are dealing with the monopulse (error) pattern. On substituting h_θ and h_ϕ, (14.1-4) becomes

$$\psi_{\mathrm{ant}}(t) = \frac{(2\pi)^2 [1 + \cos(\theta)]}{\eta I} \{ [E_\theta \cos(\phi) - E_\phi \sin(\phi)] F_{\Delta, x}(\theta, \phi)$$

$$+ [E_\theta \sin(\phi) + E_\phi \cos(\phi)] F_{\Delta, y}(\theta, \phi) \} \tag{14.1-6}$$

$$\approx \frac{2(2\pi)^2}{\eta I} [\mathbf{E}^r \cdot \mathbf{F}_{\Delta, a}(\theta, \phi)]$$

In (14.1-6), $\mathbf{F}_{\Delta, a}(\theta, \phi)$ is the vector angular spectrum of (3.2-9); its components are the angular spectra $F_{\Delta, x}(\theta, \phi)$ and $F_{\Delta, y}(\theta, \phi)$ defined by (3.1-15) and (3.1-16). Parameter η is the intrinsic impedance of the medium defined in (3.1-5), and I is, in general, a complex current that would be necessary at the antenna's terminals (point A in Fig. 14.1-2) if it were used as a radiating antenna that produces a radiated power P_{rad}.[3] For our purposes I can be taken as real because only one terminal pair is involved, and the location of the terminals can always be chosen to make I real.

A further short discussion of (14.1-6) is needed. The first right-side form is just the direct substitution of expressions for h_θ and h_ϕ into the second right-side form of (14.1-4). To understand the second right-side form of (14.1-6), we first represent $\mathbf{E}^r$ in the radar's *rectangular coordinates*. For the small-θ assumption we have

$$\begin{aligned}
\mathbf{E}^r &= E_x \hat{a}_x + E_y \hat{a}_y \\
&\approx [E_\theta \cos(\phi) - E_\phi \sin(\phi)]\hat{a}_x + [E_\theta \sin(\phi) + E_\phi \cos(\phi)]\hat{a}_y
\end{aligned} \tag{14.1-7}$$

where

$$E_x \approx E_\theta \cos(\phi) - E_\phi \sin(\phi) \tag{14.1-8}$$

$$E_y \approx E_\theta \sin(\phi) + E_\phi \cos(\phi) \tag{14.1-9}$$

are the components of $\mathbf{E}^r$ in rectangular coordinates. Since $\mathbf{F}_{\Delta,a}(\theta, \phi)$ is a vector quantity in rectangular coordinates, the last form of (14.1-6) results.

To put (14.1-6) in a more useful form, we next *define* what we will call the one-way vector voltage pattern[4] of the receiving antenna. The pattern is denoted by $\mathbf{F}_\Delta(\theta, \phi)$ and defined such that its squared magnitude equals the directive gain of the antenna, as defined by (3.4-5). It can be shown that (see Problem 14.1-4)

$$\mathbf{F}_\Delta(\theta, \phi) = \sqrt{\frac{2(2\pi)^5}{\eta\lambda^2|I|^2 R_{rr}}} \left[\frac{1 + \cos(\theta)}{2}\right] \mathbf{F}_{\Delta,a}(\theta, \phi) \tag{14.1-10}$$

where R_{rr} is the radiation resistance of the receiving antenna. On substitution of (14.1-10) into (14.1-6), we have

$$\psi_{ant}(t) = \sqrt{\frac{\lambda^2 R_{rr}}{\pi\eta}} [\mathbf{E}^r \cdot \mathbf{F}_\Delta(\theta, \phi)] \tag{14.1-11}$$

where $\mathbf{E}^r$ is given by (14.1-7), and (for θ small) it can be shown that (see Problem 14.1-5)

$$\mathbf{F}_\Delta(\theta, \phi) = \sqrt{\frac{4\pi}{\lambda^2}} \frac{\int_A \int \mathbf{E}_\Delta(\xi, \zeta) e^{jk\xi\theta_x + jk\zeta\theta_y} d\xi\, d\zeta}{\sqrt{\int_A \int |\mathbf{E}_\Delta(\xi, \zeta)|^2 d\xi\, d\zeta}}, \quad \theta \text{ small} \tag{14.1-12}$$

Here $\mathbf{E}_\Delta(\xi, \zeta)$ is the current notation for the vector aperture distribution function $\mathbf{E}_a(\xi, \zeta)$ of (3.1-2). The problem of choosing an optimum receiving pattern for

---

[3] P_{rad} is power defined by averaging over one cycle of the carrier's frequency.
[4] This pattern is necessarily a complex vector as long as we allow an aperture with arbitrary polarization, as we have consistently done. For a constant polarization over the aperture, the pattern can be defined by a suitable complex scalar function.

monopulse reduces to choosing an optimum aperture distribution for the receiving antenna. We observe that (14.1-12), which is a function of θ and ϕ in general, reduces to a function of θ_x and θ_y when θ is a small angle.

Next we use the equivalent circuit of the receiving antenna, given in Fig. 3.4-3, to develop the voltage $\psi_r(t)$ that occurs at the antenna's output.[5] For a matched-impedance (receiver) load on the antenna, $\psi_r(t)$ becomes

$$\psi_r(t) = \frac{Z_{\Delta r}^*}{2L_{rr}R_{rr}}\psi_{ant}(t) = \frac{Z_{\Delta r}^*}{L_{rr}}\sqrt{\frac{\lambda^2}{4\pi\eta R_{rr}}}[\mathbf{E}^r \cdot \mathbf{F}_\Delta(\theta, \phi)] \qquad (14.1\text{-}13)$$

Here the ohmic loss on reception is defined by

$$L_{rr} = \frac{R_{rr} + R_{Lr}}{R_{rr}} \qquad (14.1\text{-}14)$$

R_{Lr} is the (ohmic) loss resistance, and $Z_{\Delta r} = R_{rr} + R_{Lr} + jX_{Ar}$ is the antenna's impedance.

The final step in developing a useful form for $\psi_r(t)$ is to recognize that all terms in (14.1-13) are constants with respect to time except $\mathbf{E}^r$. $\mathbf{E}^r$ is proportional to the transmitted waveform delayed and Doppler-shifted. We may then define a new vector of proportionality, $\boldsymbol{\alpha}$, by

$$\boldsymbol{\alpha} = \alpha_x\hat{a}_x + \alpha_y\hat{a}_y \qquad (14.1\text{-}15)$$

such that

$$\frac{Z_{\Delta r}^*}{L_{rr}}\sqrt{\frac{\lambda^2}{4\pi\eta R_{rr}}}\mathbf{E}^r = \boldsymbol{\alpha}\psi(t - \tau_R)e^{j\omega_d(t - \tau_R)} \qquad (14.1\text{-}16)$$

Equation (14.1-13) can now be written in the very useful form

$$\psi_r(t) = [\boldsymbol{\alpha} \cdot \mathbf{F}_\Delta(\theta, \phi)]\psi(t - \tau_R)e^{j\omega_d(t - \tau_R)} \qquad (14.1\text{-}17)$$

As usual τ_R and ω_d are the target's delay and angular Doppler frequency, respectively.

For later purposes it is convenient to find $|\psi_r(t)|^2$. Because the received wave $\mathbf{E}^r$ is polarization-matched to the receiving antenna, the magnitude-squared of the dot product in (14.1-4) is given by $|\mathbf{E}^r|^2 |\mathbf{h}|^2$. By retracing the above developments leading to (14.1-17), it is found that

$$|\psi_r(t)|^2 = |\boldsymbol{\alpha}|^2 |\psi(t - \tau_R)|^2 |\mathbf{F}_\Delta(\theta, \phi)|^2 \qquad (14.1\text{-}18)$$

---

[5] Quantities v_r, R_r, R_L, and X_A in Fig. 3.4-3, which apply to any antenna, are replaced for the receiving antenna by $\psi_{ant}(t)$, R_{rr}, R_{Lr}, and X_{Ar}, respectively, for present purposes. For a matched receiver $Z_L = R_{rr} + R_{Lr} - jX_{Ar}$ is assumed, which leads to (14.1-13).

Output Signal and Noise

From Section 6.5 we know that the real response, denoted by $s_0(t)$, of an analytic network, with transfer function $Z(\omega)$ and impulse response $z(t)$, when excited at its input by an analytic signal $\psi_r(t)$, is half the real part of the convolution of $\psi_r(t)$ and $z(t)$. For $\psi_r(t)$ given by (14.1-17), we have

$$s_0(t) = \tfrac{1}{2}\,\mathrm{Re}\left\{[\boldsymbol{\alpha}\cdot\mathbf{F}_\Delta(\theta,\phi)]\int_{-\infty}^{\infty}\psi(\xi)z(t-\tau_R-\xi)e^{j\omega_d\xi}\,d\xi\right\} \qquad (14.1\text{-}19)$$

Alternatively, in terms of frequency-domain functions

$$s_0(t) = \tfrac{1}{2}\,\mathrm{Re}\left\{[\boldsymbol{\alpha}\cdot\mathbf{F}_\Delta(\theta,\phi)]\frac{1}{2\pi}\int_{-\infty}^{\infty}\Psi(\omega-\omega_d)Z(\omega)e^{j\omega(t-\tau_R)}\,d\omega\right\} \qquad (14.1\text{-}20)$$

We use these expressions in the next section to find the optimum monopulse system.

For noise, let $N(t)$ be a zero-mean wide-sense stationary random process that represents the real noise at the input (point A) in Fig. 14.1-2. If the analytic representation of the receiver's transfer function is $Z(\omega)$, the output average noise power, denoted N_0, is readily found to be [see (12.1-22)]

$$N_0 = \frac{1}{2}\frac{1}{2\pi}\int_{-\infty}^{\infty}\mathscr{S}_{NN}(\omega)|Z(\omega)|^2\,d\omega \qquad (14.1\text{-}21)$$

Here $\mathscr{S}_{NN}(\omega)$ is the power density spectrum of $N(t)$. We use this noise power in the section to follow.

14.2   OPTIMUM MONOPULSE SYSTEM

We next apply the preceding preliminary definitions to the problem of finding the optimum monopulse signal processor and its performance.

Definition of Optimality

Figure 14.1-2 defines the system. We presume that the radar has already detected the presence of a target of interest and range, and Doppler frequency measurements have been established such that τ_R and ω_d may be assumed known. Similarly the radar is assumed to have made an estimate (measurement) of the target pulse's amplitude so that $\boldsymbol{\alpha}$ is known. All these measurements may be, and typically are, made in another receiver path in parallel with that in Fig. 14.1-2.[6] The principal purpose of the monopulse processor is to observe the output $s_0(t)$ at the time of the target's pulse, denoted by t_0, and use the observation to form an estimate (measurement) of one of

---

[6] As we proceed, it will be seen that the parallel signal processor may even use a different antenna pattern, although all patterns may be generated simultaneously by the same aperture.

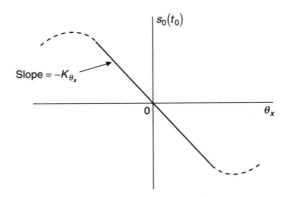

Figure 14.2-1   Angle discriminator characteristic of a monopulse system.

the target's angles, θ_x or θ_y. A second monopulse processor in parallel, that uses the same concepts, would then measure the other target angle, θ_y or θ_x. Since the angle measurement is made using only one pulse, the system is monopulse.

As with previous range and Doppler frequency measurement systems, the optimum system will be that which satisfies three requirements. First, for a target on the reference axis where θ, θ_x, and θ_y are all zero, we ask that $s_0(t_0)$ be zero. A zero response at the desired measurement time, t_0, simply indicates that the reference axis points toward the target. Second, the response $s_0(t_0)$ must be proportional to target angle θ_x,[7] as depicted in Fig. 14.2-1. In the linear region around the origin when there is no output noise, $s_0(t_0) = - K_{\theta_x}\theta_x$, so the measurement of θ_x is the negative ratio of the response $s_0(t_0)$ to the slope constant K_{θ_x} (which is known). When output noise $n_0(t_0)$ occurs, the response becomes $s_0(t_0) + n_0(t_0) = - K_{\theta_x}\hat{\theta}_x$, where $\hat{\theta}_x$ is now a measurement with noise. Clearly this expression gives $\hat{\theta}_x = - [s_0(t_0) + n_0(t_0)]/K_{\theta_x} = \theta_x - [n_0(t_0)/K_{\theta_x}]$, so noise causes an error in measurement of

$$\Delta\theta_x = \hat{\theta}_x - \theta_x = - \frac{n_0(t_0)}{K_{\theta_x}} \tag{14.2-1}$$

The mean-squared error $E[(\Delta\theta_x)^2]$, which is the variance in the measurement of θ_x, denoted by $\sigma_{\hat{\theta}_x}^2$, is

$$\sigma_{\hat{\theta}_x}^2 = \frac{E[N_0^2(t_0)]}{K_{\theta_x}^2} = \frac{N_0}{K_{\theta_x}^2} \tag{14.2-2}$$

Here $N_0(t)$ is a random process representing $n_0(t_0)$. The power in $N_0(t)$ at any time is N_0 as given by (14.1-21).

Our third requirement of the optimum system is that it must produce the smallest value of $\sigma_{\hat{\theta}}^2$ of any possible linear system.

---

[7] We will examine the monopulse processor (channel) that estimates θ_x. The channel for estimation of θ_y uses identical procedures and gives similar results, although the notation is different.

From Fig. 14.2-1 we have

$$K_{\theta_x} = -\left.\frac{\partial s_0(t_0)}{\partial \theta_x}\right|_{\theta_x = 0, \theta_y = 0} = -\left.\frac{\partial s_0(t_0)}{\partial \theta_x}\right|_{\theta = 0} \tag{14.2-3}$$

The required derivative is found from (14.1-20) by noting that only $\mathbf{F}_\Delta(\theta, \phi)$ is a function of θ_x through (14.1-12). Thus (14.2-2) becomes

$$\sigma_{\theta_x}^2 = \frac{\dfrac{1}{2\pi} \displaystyle\int_{-\infty}^{\infty} \mathscr{S}_{NN}(\omega)|Z(\omega)|^2 d\omega}{\dfrac{1}{2}\left\langle \mathrm{Re}\left\{\left[\boldsymbol{\alpha} \cdot \left.\dfrac{\partial \mathbf{F}_\Delta(\theta, \phi)}{\partial \theta_x}\right|_{\theta_x = 0, \theta_y = 0}\right]\dfrac{1}{2\pi}\displaystyle\int_{-\infty}^{\infty} \Psi(\omega - \omega_d)Z(\omega)e^{j\omega(t_0 - \tau_R)}d\omega\right\}\right\rangle^2} \tag{14.2-4}$$

The optimum monopulse signal processor will consist of the pattern $\mathbf{F}_\Delta(\theta, \phi)$ and receiver (filter) function $Z(\omega)$ that minimize $\sigma_{\theta_x}^2$ of (14.2-4).

Optimum System for Nonwhite Noise

To minimize (14.2-4) we use Schwarz's inequality of (12.1-24) in the denominator. Define a complex constant with respect to ω by D, where

$$D = \boldsymbol{\alpha} \cdot \left.\frac{\partial \mathbf{F}_\Delta(\theta, \phi)}{\partial \theta_x}\right|_{\theta_x = 0, \theta_y = 0} = |D|e^{j\phi_D} \tag{14.2-5}$$

and choose[8]

$$A(\omega) = \frac{\sqrt{D}}{\sqrt{2\pi}}\sqrt{\mathscr{S}_{NN}(\omega)}Z(\omega) \tag{14.2-6}$$

$$B(\omega) = \frac{\sqrt{D}}{\sqrt{2\pi}}\frac{\Psi(\omega - \omega_d)}{\sqrt{\mathscr{S}_{NN}(\omega)}}e^{j\omega(t_0 - \tau_R)} \tag{14.2-7}$$

Equation (14.2-4) reduces to

$$\sigma_{\theta_x}^2 \geq \frac{1}{\dfrac{1}{2}\left|\boldsymbol{\alpha} \cdot \left.\dfrac{\partial \mathbf{F}_\Delta(\theta, \phi)}{\partial \theta_x}\right|_{\theta_x = 0, \theta_y = 0}\right|^2 \dfrac{1}{2\pi}\displaystyle\int_{-\infty}^{\infty}\dfrac{|\Psi(\omega - \omega_d)|^2}{\mathscr{S}_{NN}(\omega)}d\omega} \tag{14.2-8}$$

---

[8] D may be associated with $A(\omega)$ or $B(\omega)$ or partially with each, as done here. Because of the arbitrary constant resulting from Schwarz's inequality, the magnitude of D is unimportant. The effect of the phase of D is the same for all three associations.

The equality in (14.2-8) holds only if $A(\omega) = CB^*(\omega)$, where C is an arbitrary real nonzero constant. From (14.2-6) and (14.2-7) we solve this condition to get

$$Z(\omega) = Z_{\text{opt}}(\omega) = C\frac{\Psi^*(\omega - \omega_d)}{\mathscr{S}_{NN}(\omega)}e^{-j\omega(t_0 - \tau_R) - j\phi_D} \tag{14.2-9}$$

This expression defines the optimum monopulse receiver. Except for the phase factor $\exp(-j\phi_D)$ it is just the transfer function of an analytic matched filter for the delayed and Doppler-shifted target signal. The phase factor only implies that the optimum receiver removes any phase present on the output waveform from the antenna.

Our new knowledge of the form of the optimum receiver has no bearing on the optimum antenna, which is found by further minimization of (14.2-8) by maximization of the denominator's dot product. The dot product of two complex vectors will have a maximum magnitude equal to the product of the two magnitudes when the two vectors spatially align and the phase of one is the conjugate of the other. This condition was previously used for $\boldsymbol{\alpha}$ and $\mathbf{F}_\Delta(\theta, \phi)$ which lead to (14.1-18). The requirement for $|\boldsymbol{\alpha} \cdot \mathbf{F}_\Delta(\theta, \phi)|^2 = |\boldsymbol{\alpha}|^2|\mathbf{F}_\Delta(\theta, \phi)|^2$ to be true becomes (see Problem 14.2-6)

$$\frac{\alpha_y}{\alpha_x} = \frac{F^*_{\Delta,y}(\theta, \phi)}{F^*_{\Delta,x}(\theta, \phi)} = \frac{\int_A\int E^*_{\Delta,y}(\xi, \zeta)e^{-jk\xi\theta_x - jk\zeta\theta_y}d\xi d\zeta}{\int_A\int E^*_{\Delta,x}(\xi, \zeta)e^{-jk\xi\theta_x - jk\zeta\theta_y}d\xi d\zeta} \tag{14.2-10}$$

from (14.1-12).

Similarly, by differentiation of (14.1-12),

$$\frac{\partial \mathbf{F}_\Delta(\theta, \phi)}{\partial \theta_x} = \sqrt{\frac{4\pi}{\lambda^2}}\frac{\int_A\int jk\xi\, \mathbf{E}_\Delta(\xi, \zeta)e^{jk\xi\theta_x + jk\zeta\theta_y}d\xi d\zeta}{\sqrt{\int_A\int |\mathbf{E}_\Delta(\xi, \zeta)|^2 d\xi d\zeta}} \tag{14.2-11}$$

We apply this result in asking that

$$\left|\boldsymbol{\alpha} \cdot \frac{\partial \mathbf{F}_\Delta(\theta, \phi)}{\partial \theta_x}\right|^2_{\theta_x = 0, \theta_y = 0} = |\boldsymbol{\alpha}|^2\left|\frac{\partial \mathbf{F}_\Delta(\theta, \phi)}{\partial \theta_x}\right|^2_{\theta_x = 0, \theta_y = 0} \tag{14.2-12}$$

be true, which requires that (see Problem 14.2-7)

$$\frac{\alpha_y}{\alpha_x} = \frac{-\int_A\int jk\xi E^*_{\Delta,y}(\xi, \zeta)d\xi d\zeta}{-\int_A\int jk\xi E^*_{\Delta,x}(\xi, \zeta)d\xi d\zeta} \tag{14.2-13}$$

By equating (14.2-10) for $\theta \to 0$ with (4.2-13), we find that

$$\frac{\int_A\int \xi E_{\Delta,y}(\xi, \zeta)d\xi d\zeta}{\int_A\int E_{\Delta,y}(\xi, \zeta)d\xi d\zeta} = \frac{\int_A\int \xi E_{\Delta,x}(\xi, \zeta)d\xi d\zeta}{\int_A\int E_{\Delta,x}(\xi, \zeta)d\xi d\zeta} \tag{14.2-14}$$

must be satisfied for (14.2-12) to be true. Note that constant-polarization apertures, where Q_Δ is a constant and $E_{\Delta,y}(\xi, \zeta) = Q_\Delta E_{\Delta,x}(\xi, \zeta)$ for all (ξ, ζ), satisfy (14.2-14) for all polarizations.

In the following work we will assume that the receiving pattern originates from a constant-polarization aperture or its aperture distribution satisfies (14.2-14) such that the minimum value of (14.2-8) becomes

$$\sigma_{\theta_x}^2 \geq \frac{1}{\frac{1}{2}|\alpha|^2 \left|\frac{\partial \mathbf{F}_\Delta(\theta, \phi)}{\partial \theta_x}\right|_{\theta_x = 0, \theta_y = 0}^2 \frac{1}{2\pi}\int_{-\infty}^{\infty} \frac{|\Psi(\omega - \omega_d)|^2}{\mathscr{S}_{NN}(\omega)} d\omega} \tag{14.2-15}$$

We use (14.1-12) and apply Schwarz's inequality to (14.2-15) to further minimize the error variance:

$$\left|\frac{\partial \mathbf{F}_\Delta(\theta, \phi)}{\partial \theta_x}\right|_{\theta_x = 0, \theta_y = 0}^2 = \frac{4\pi}{\lambda^2} \frac{|\int_A \int jk\xi \mathbf{E}_\Delta(\xi, \zeta) d\xi d\zeta|^2}{\int_A \int |\mathbf{E}_\Delta(\xi, \zeta)|^2 d\xi d\zeta}$$

$$\leq \frac{4\pi}{\lambda^2} \frac{\int_A \int |jk\xi|^2 d\xi d\zeta \int_A \int |\mathbf{E}_\Delta(\xi, \zeta)|^2 d\xi d\zeta}{\int_A \int |\mathbf{E}_\Delta(\xi, \zeta)|^2 d\xi d\zeta}$$

$$= \frac{4\pi}{\lambda^2} \frac{1}{\lambda^2} \int_A \int (2\pi\xi)^2 d\xi d\zeta = \frac{4\pi A}{\lambda^4} L_{0,x}^2 \tag{14.2-16}$$

Here A is the aperture's physical area, and

$$L_{0,x}^2 = \frac{1}{A} \int_A \int (2\pi\xi)^2 d\xi d\zeta \tag{14.2-17}$$

is called the *mean-squared length* of the optimum aperture for the x direction. Equation (14.2-15) reduces to

$$\sigma_{\theta_x}^2 \geq \frac{\lambda^2}{|\alpha|^2 \frac{A}{\lambda^2} \int_{-\infty}^{\infty} \frac{|\Psi(\omega - \omega_d)|^2}{\mathscr{S}_{NN}(\omega)} d\omega L_{0,x}^2} \tag{14.2-18}$$

The equality, or minimum value, of (14.2-18) holds only when

$$\mathbf{E}_\Delta(\xi, \zeta) = \frac{-j2\pi\xi}{\lambda} \mathbf{C} \tag{14.2-19}$$

where

$$\mathbf{C} = C_x \hat{a}_x + C_y \hat{a}_y \tag{14.2-20}$$

is a vector with real nonzero arbitrary constant components C_x and C_y (see Problem 14.2-14). We note that (14.2-19) defines the optimum receiving antenna through (14.1-12); it also satisfies the condition (14.2-14), since it must for the equality in (14.2-15) to hold.

Our optimization is essentially complete. Because the various steps were lengthy, a summary is in order. The optimum monopulse antenna pattern is given by (14.1-12) when using the vector aperture distribution of (14.2-19), which is seen to be a linear odd function of ξ. The optimum receiver's transfer function is that of a matched filter as defined by (14.2-9). When these two optimum functions are used, the variance of error in measuring θ_x is given by the equality in (14.2-18).

Some refinements of our results are possible, as developed in the following subsection for white noise. It should be noted that we have examined only the optimum system for measuring θ_x. An identical procedure applies to the measurement of θ_y. Practical systems usually operate the two receiving systems independently and in parallel, one to measure θ_x and one for θ_y. When the "channel" used for range, Doppler, and amplitude measurements is added, the usual practical monopulse system has at least three channels (Peebles, 1969) operating (more-or-less) independently. However, the antenna patterns associated with the three channels can be, and often are, generated by the same aperture.

Optimum System for White Noise

With white noise where $\mathscr{S}_{NN}(\omega) = \mathscr{N}_0/2$, $-\infty < \omega < \infty$, with $\mathscr{N}_0$ a positive constant, the antenna is still defined by the distribution of (14.2-19). The filter of (14.2-9) becomes a white noise matched filter

$$Z_{\text{opt}}(\omega) = \frac{2C}{\mathscr{N}_0} \Psi^*(\omega - \omega_d)^{-j\omega(t_0 - \tau_R) - j\phi_D} \tag{14.2-21}$$

and measurement minimum variance, denoted by $\sigma^2_{\theta_x(\text{min})}$, reduces to

$$\sigma^2_{\theta_x(\text{min})} = \frac{\lambda^2}{|\alpha|^2 \dfrac{4\pi A}{\lambda^2} \dfrac{E_\psi}{\mathscr{N}_0} L^2_{0,x}} = \frac{\lambda^2}{|\alpha|^2 \dfrac{4\pi A}{\lambda^2} \dfrac{2E_s}{\mathscr{N}_0} L^2_{0,x}} \tag{14.2-22}$$

[from (14.2-18)]. Here E_s and E_ψ are the energies in the real transmitted signal and its analytic representation, respectively.

It can be shown that $4\pi A/\lambda^2 = G_{D0}$ is the directivity of a uniformly illuminated aperture with area A. On noting that (14.1-18) applies to any aperture, even the uniform one, it is clear that the energy received by such an aperture, denoted by E_{ψ_r}, would be

$$E_{\psi_r} = |\alpha|^2 E_\psi G_{D0} = |\alpha|^2 2E_s G_{D0} = 2E_r \tag{14.2-23}$$

Here E_r is the real energy received by a uniformly illuminated aperture. This relationship allows (14.2-22) to be expressed as

$$\sigma^2_{\theta_x(\text{min})} = \frac{\lambda^2}{\mathscr{R}_{p0} L^2_{0,x}} \tag{14.2-24}$$

where

$$\mathscr{R}_{p0} = \frac{E_{\psi_r}}{\mathscr{N}_0} = \frac{2E_r}{\mathscr{N}_0} \qquad (14.2\text{-}25)$$

is the energy to noise density ratio that would be produced by a uniformly illuminated receiving aperture.

The result (14.2-24) agrees with that of Barton and Ward (1969, p. 22). Another form for (14.2-24) uses the maximum slope associated with the aperture, denoted by $K_{0,x}$ and given by (see Problem 14.2-16)

$$K_{0,x}^2 = \frac{L_{0,x}^2}{\lambda^2} = \frac{1}{G_{D0}} \left| \frac{\partial \mathbf{F}_\Delta(\theta, \phi)}{\partial \theta_x} \right|_{\theta=0}^2 \bigg|_{\max} \qquad (14.2\text{-}26)$$

It is

$$\sigma_{\hat{\theta}_x(\min)}^2 = \frac{1}{\mathscr{R}_{p0} K_{0,x}^2} \qquad (14.2\text{-}27)$$

Example 14.2-1 We find $L_{0,x}$ for a circular aperture of diameter $D = 0.86$ m. From (14.2-17) with $\xi^2 + \zeta^2 = (D/2)^2$ on the aperture's edge,

$$L_{0,x}^2 = \frac{1}{\pi(D/2)^2} \int_{\zeta=-D/2}^{D/2} \int_{\xi=-\sqrt{(D/2)^2-\zeta^2}}^{\sqrt{(D/2)^2-\zeta^2}} (2\pi\xi)^2 \, d\xi d\zeta$$

$$= \frac{64\pi}{3D^2} \int_0^{D/2} \left[\left(\frac{D}{2}\right)^2 - \zeta^2 \right]^{3/2} d\zeta = \frac{64\pi}{3D^2} \frac{3\pi D^4}{256} = \left(\frac{\pi D}{2}\right)^2$$

Thus $L_{0,x} = \pi D/2 = \pi(0.86/2) = 1.3509$ m.

Example 14.2-2 Continue Example 14.2-1 for a radar at 7 GHz and $\mathscr{R}_{p0} = 14.3$ to find $\sigma_{\hat{\theta}_x(\min)}$. Since $\lambda = c/f = 3(10^8)/7(10^9)$ m, then

$$\sigma_{\hat{\theta}_x(\min)}^2 = \frac{\lambda^2}{\mathscr{R}_{p0} L_{0,x}^2} = \frac{(3/70)^2}{14.3(1.3509)^2} = 0.7038(10^{-4}) \text{ rad}^2$$

and $\sigma_{\hat{\theta}_x(\min)} = 0.008389$ rad $= 0.4807$ degrees. From (3.3-33) this aperture can have a minimum beamwidth of $\theta_B^\circ = 58.86\lambda/D = 58.86(3/70)/0.86 = 2.9332$ degrees so the accuracy is about $(0.4807/2.9332)100 = 16.39\%$ of the beamwidth.

14.3 ANTENNA MISMATCH IN MONOPULSE

It is of interest to examine the performance of a "monopulse" system that does not use the optimum antenna defined by (14.2-26) and (14.2-17). We refer to this case as *antenna mismatch*. However, we continue to assume the use of the optimum (matched filter) receiver.

Slope Parameters

Let $\mathbf{F}_\Delta(\theta, \phi)$ be the pattern of an arbitrary receiving antenna and define an "*error pattern slope parameter*," $K_x{}^9$ for the x direction by

$$K_x^2 = \frac{1}{G_{D0}} \left| \frac{\partial \mathbf{F}_\Delta(\theta, \phi)}{\partial \theta_x} \right|^2_{\theta=0} = \left(\frac{2\pi}{\lambda} \right)^2 \frac{|\int_A \int \xi \mathbf{E}_\Delta(\xi, \zeta) d\xi d\zeta|^2}{A \int_A \int |\mathbf{E}_\Delta(\xi, \zeta)|^2 d\xi d\zeta} \tag{14.3-1}$$

On retracing the steps following (14.2-8), except for white noise, we have

$$\sigma_{\theta_x}^2 = \frac{1}{\mathscr{R}_{p0} K_x^2} \tag{14.3-2}$$

This expression is the same as (14.2-27) except that K_x has a smaller value than $K_{0,x}$ because the antenna is not optimum.

To readily see that $K_x \leq K_{0,x}$, we can apply Schwarz's inequality to the numerator of (14.3-1):

$$K_x^2 \leq \left(\frac{2\pi}{\lambda} \right)^2 \frac{\int_A \int \xi^2 d\xi d\zeta \int_A \int |\mathbf{E}_\Delta(\xi, \zeta)|^2 d\xi d\zeta}{A \int_A \int |\mathbf{E}_\Delta(\xi, \zeta)|^2 d\xi d\zeta} = \frac{1}{\lambda^2} L_{0,x}^2 = K_{0,x}^2 \tag{14.3-3}$$

In terms of K_x and $K_{0,x}$, (14.3-2) becomes

$$\sigma_{\theta_x}^2 = \frac{1}{\mathscr{R}_{p0} K_x^2} \geq \frac{1}{\mathscr{R}_{p0} K_{0,x}^2} = \sigma_{\theta_x(\min)}^2 \tag{14.3-4}$$

Other Slope Constants

From (14.3-4) we have

$$\frac{\sigma_{\theta_x(\min)}^2}{\sigma_{\theta_x}^2} = \frac{K_x^2}{K_{0,x}^2} = \eta_{\theta_x}^2 \tag{14.3-5}$$

where

$$\eta_{\theta_x} = \frac{K_x}{K_{0,x}} \tag{14.3-6}$$

is here called the *angle efficiency factor* by analogy with the filter efficiency factor of (12.3-2) for Doppler measurement. The ratio $K_x/K_{0,x}$ has been called the *difference slope ratio* by Hannan (1961).

Barton (1964, 1988) has defined a useful slope parameter. It is denoted here by $k_{m,x}$ as it relates to the measurement of θ_x (Barton uses the notation k_m; however, the

---

[9] K_x is called by other names and given other symbols in some literature. For example, it is denoted by K and called the *relative difference slope* by Barton and Ward (1969, p. 18) and by Hannan (1961, p. 453). We prefer not to use the "difference" notation, since some monopulse systems do not have "difference" patterns but all have an "error" pattern for each coordinate θ_x and θ_y.

parameter can differ for the two channels measuring θ_x and θ_y. We attempt to be more specific by choice of subscripts.). Barton's parameter is defined in such a way as to form a link between the error channel (receiving system for measuring θ_x) and the reference channel (receiving system for measuring range, Doppler, and signal amplitude).

Let the reference channel's pattern have a directivity denoted by $G_{D\Sigma}$ and a 3-dB one-way beamwidth θ_B(rad) in the direction of θ_x.[10] From (14.3-1) we develop

$$K_x^2 = \frac{G_{D\Sigma}\theta_B^2}{G_{D\Sigma}\theta_B^2} \frac{1}{G_{D0}} \left|\frac{\partial \mathbf{F}_\Delta(\theta, \phi)}{\partial \theta_x}\right|^2_{\theta=0} = \frac{G_{D\Sigma}}{\theta_B^2 G_{D0}} \frac{\theta_B^2}{G_{D\Sigma}} \left|\frac{\partial \mathbf{F}_\Delta(\theta, \phi)}{\partial \theta_x}\right|^2_{\theta=0}$$

$$= \frac{G_{D\Sigma}}{\theta_B^2 G_{D0}} k_{m,x}^2 \qquad (14.3\text{-}7)$$

where

$$k_{m,x}^2 = \frac{\theta_B^2}{G_{D\Sigma}} \left|\frac{\partial \mathbf{F}_\Delta(\theta, \phi)}{\partial \theta_x}\right|^2_{\theta=0} = \frac{\theta_B^2 G_{D0}}{G_{D\Sigma}} K_x^2 \qquad (14.3\text{-}8)$$

which is valid for all reference channel antenna patterns.

On using (14.3-7) in (14.3-2), we may rewrite the measurement error variance as

$$\sigma_{\hat{\theta}_x}^2 = \frac{\theta_B^2}{\mathscr{R}_{p0}(G_{D\Sigma}/G_{D0})k_{m,x}^2} = \frac{\theta_B^2}{\mathscr{R}_\Sigma k_{m,x}^2} \qquad (14.3\text{-}9)$$

where

$$\mathscr{R}_\Sigma = \frac{G_{D\Sigma}}{G_{D0}} \mathscr{R}_{p0} \qquad (14.3\text{-}10)$$

To interpret the meaning of $\mathscr{R}_\Sigma$, recall that $\mathscr{R}_{p0}$ was the energy to noise density ratio that would be produced by the receiving pattern of a uniformly illuminated aperture with directivity G_{D0}. The factor $G_{D\Sigma}/G_{D0}$ of (14.3-10) means that $\mathscr{R}_\Sigma$ is the energy to noise density ratio of the reference channel which has directivity $G_{D\Sigma}$. The ratio $G_{D\Sigma}/G_{D0}$ is also recognized as the aperture efficiency of the reference channel's antenna pattern.

Error Pattern as Reference Pattern Derivative

In real monopulse systems the pattern of the error channel measuring θ_x resembles the derivative of the reference channel's pattern. It is of some academic interest to determine conditions under which the relationship can be true. Toward this purpose recognize that (14.1-12), which defines the pattern of the error channel, is actually

---

[10] To be consistent with Chapter 3, the 3-dB one-way beamwidth in the direction of θ_y is denoted by ϕ_B (rad).

general in that it can be applied to any pattern with suitable notation changes. For present purposes, let $\mathbf{F}_\Sigma(\theta, \phi)$ denote the one-way vector voltage pattern of the reference channel, defined for a vector aperture field distribution denoted by $\mathbf{E}_\Sigma(\xi, \zeta)$.

If we ask the error pattern to be proportional to the derivative of the reference pattern according to

$$\mathbf{F}_\Delta(\theta, \phi) = Q \frac{\partial \mathbf{F}_\Sigma(\theta, \phi)}{\partial \theta_x} \tag{14.3-11}$$

where Q is the proportionality constant (assumed real but not yet known), then (14.1-12) gives

$$\begin{aligned} \mathbf{F}_\Delta(\theta, \phi) &= \sqrt{\frac{4\pi}{\lambda^2}} \frac{\int_A \int \mathbf{E}_\Delta(\xi, \zeta) e^{jk\xi\theta_x + jk\zeta\theta_y} d\xi d\zeta}{\sqrt{\int_A \int |\mathbf{E}_\Delta(\xi, \zeta)|^2 d\xi d\zeta}} \\ &= Q \sqrt{\frac{4\pi}{\lambda^2}} \frac{\int_A \int (jk\xi) \mathbf{E}_\Sigma(\xi, \zeta) e^{jk\xi\theta_x + jk\zeta\theta_y} d\xi d\zeta}{\sqrt{\int_A \int |\mathbf{E}_\Sigma(\xi, \zeta)|^2 d\xi d\zeta}} \end{aligned} \tag{14.3-12}$$

Now suppose that the same aperture area is used to generate the two patterns.[11] Then, if, as a radiator, one pattern radiates an amount of power, it is reasonable to expect the other to radiate the same power when both are to be excited by the same incident power. For equal radiated powers the denominator integrals in (14.3-12) must be equal, so the expression is true if

$$\mathbf{E}_\Delta(\xi, \zeta) = Q jk\xi \mathbf{E}_\Sigma(\xi, \zeta) \tag{14.3-13}$$

Finally, on integrating the magnitude squared of both sides of (14.3-13), and again invoking the equal-power constraint, we find that (14.3-11) is true if

$$Q^2 = \frac{\lambda^2}{L_{\Sigma, x}^2} \tag{14.3-14}$$

where

$$L_{\Sigma, x}^2 = \frac{\int_A \int (2\pi\xi)^2 |\mathbf{E}_\Sigma(\xi, \zeta)|^2 d\xi d\zeta}{\int_A \int |\mathbf{E}_\Sigma(\xi, \zeta)|^2 d\xi d\zeta} \tag{14.3-15}$$

We will call $L_{\Sigma, x}$ the *rms length* of the reference pattern's aperture for the x direction. $L_{\Sigma, x}$ has also been called the aperture's *effective width* (Skolnik, 1962). On substitution of Q into (14.3-11) and (14.3-13), we see that the error pattern is the derivative of the reference pattern according to

$$\mathbf{F}_\Delta(\theta, \phi) = \frac{\lambda}{L_{\Sigma, x}} \frac{\partial \mathbf{F}_\Sigma(\theta, \phi)}{\partial \theta_x} \tag{14.3-16}$$

---

[11] We will see later that this condition is possible.

provided that the error pattern's aperture distribution is related to that of the reference pattern according to

$$\mathbf{E}_\Delta(\xi, \zeta) = \frac{j2\pi}{L_{\Sigma,x}} \xi \mathbf{E}_\Sigma(\xi, \zeta) \tag{14.3-17}$$

The error pattern's distribution is therefore a linear odd function times the reference pattern's distribution.

Finally we obtain K_x from (14.3-1) when using (14.3-17):

$$K_x^2 = \left(\frac{2\pi}{\lambda}\right)^2 \frac{|\iint_A \xi^2 \mathbf{E}_\Sigma(\xi, \zeta) d\xi d\zeta|^2}{A \iint_A \xi^2 |\mathbf{E}_\Sigma(\xi, \zeta)|^2 d\xi d\zeta} \tag{14.3-18}$$

Example 14.3-1 As an example of patterns that have a derivative relationship, consider a rectangular aperture of lengths D_x and D_y in directions x and y, respectively. The aperture is centered on the origin, and its aperture distribution is presumed factorable with electric field only in the x direction. We define an example distribution, find the patterns $\mathbf{F}_\Sigma(\theta, \phi)$ and $\mathbf{F}_\Delta(\theta, \phi)$, and then find $L_{\Sigma,x}^2$. Begin by assuming the distribution

$$\mathbf{E}_\Sigma(\xi, \zeta) = \hat{a}_x E_0 \left[1 - \left(\frac{2\xi}{D_x}\right)^2\right] \text{rect}\left(\frac{\xi}{D_x}\right) E(\zeta) \text{rect}\left(\frac{\zeta}{D_y}\right) \tag{1}$$

where E_0 is a real positive constant and $E(\zeta)$ is the distribution's factor for the y direction.

Pattern $\mathbf{F}_\Sigma(\theta, \phi)$ derives from (14.1-12) using (1):

$$\mathbf{F}_\Sigma(\theta, \phi) = \sqrt{\frac{4\pi}{\lambda^2}} \frac{\hat{a}_x E_0 \int_{-D_x/2}^{D_x/2} [1 - (2\xi/D_x)^2] e^{jk\theta_x\xi} d\xi \int_{-D_y/2}^{D_y/2} E(\zeta) e^{jk\theta_y\zeta} d\zeta}{\sqrt{E_0^2 \int_{-D_x/2}^{D_x/2} [1 - (2\xi/D_x)^2]^2 d\xi \int_{-D_y/2}^{D_y/2} |E(\zeta)|^2 d\zeta}} \tag{2}$$

The integrals involving ξ are solved from known integrals in Appendix D. By defining

$$F_g(\theta_y) = \frac{\int_{-D_y/2}^{D_y/2} E(\zeta) e^{jk\theta_y\zeta} d\zeta}{\sqrt{D_y \int_{-D_y/2}^{D_y/2} |E(\zeta)|^2 d\zeta}}$$

and

$$w = \frac{k\theta_x D_x}{2} = \frac{\pi D_x \theta_x}{\lambda}$$

we obtain (see Problem 14.3-15)

$$\mathbf{F}_\Sigma(\theta, \phi) = \hat{a}_x \sqrt{\frac{10\pi D_x D_y}{3\lambda^2}} \frac{3}{w^2} \left[\frac{\sin(w)}{w} - \cos(w)\right] F_g(\theta_y) \tag{3}$$

To find $\mathbf{F}_\Delta(\theta, \phi)$, we again use (14.1-12) except with $\mathbf{E}_\Delta(\xi, \zeta)$ defined by (14.3-13):

$$\mathbf{F}_\Delta(\theta, \phi) = \sqrt{\frac{4\pi}{\lambda^2} \frac{j\hat{a}_x E_0 \int_{-D_x/2}^{D_x/2} \xi[1 - (2\xi/D_x)^2] e^{jk\theta_x\xi} d\xi \int_{-D_y/2}^{D_y/2} E(\zeta) e^{jk\theta_y\zeta} d\zeta}{\sqrt{E_0^2 \int_{-D_x/2}^{D_x/2} \xi^2 [1 - (2\xi/D_x)^2]^2 d\xi \int_{-D_y/2}^{D_y/2} |E(\zeta)|^2 d\zeta}}} \tag{4}$$

The necessary integrals exist in Appendix D. The evaluation gives (see Problem 14.3-16)

$$\mathbf{F}_\Delta(\theta, \phi) = -\hat{a}_x \sqrt{\frac{10\pi D_x D_y}{3\lambda^2}} \frac{3\sqrt{63}}{w^3} \left[\frac{\sin(w)}{w} - \cos(w) - \frac{w}{3}\sin(w)\right] F_g(\theta_y) \tag{5}$$

Plots of $\mathbf{F}_\Sigma(\theta, \phi)$ and $\mathbf{F}_\Delta(\theta, \phi)$ normalized (divided) by $\hat{a}_x \sqrt{10\pi D_x D_y/(3\lambda^2)}$ are shown in Fig. 14.3-1 along with some important parameters.

We can find $L_{\Sigma,x}$ from (14.3-15) or (14.3-16). We demonstrate the latter. The former can be calculated as a check if the reader is so inclined. We start by finding the derivative

$$\frac{\partial \mathbf{F}_\Sigma(\theta, \phi)}{\partial \theta_x} = \hat{a}_x \sqrt{\frac{10\pi D_x D_y}{3\lambda^2}} 3F_g(\theta_y) \left[\frac{w^3\cos(w) - 3w^2\sin(w)}{w^6}\right.$$
$$\left. + \frac{w^2\sin(w) + 2w\cos(w)}{w^4}\right] \frac{dw}{d\theta_x} \tag{6}$$

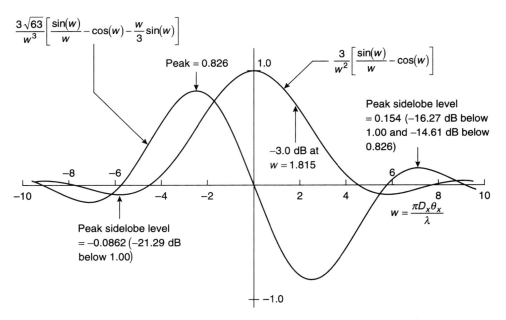

Figure 14.3-1   Normalized plots of $\mathbf{F}_\Sigma(\theta, \phi)$ and $\mathbf{F}_\Delta(\theta, \phi)$ that are applicable to the rectangular aperture of Example 14.3-1.

from (3). After grouping terms in (6) and forming the ratio required by (14.3-16), we get

$$L_{\Sigma, x} = \frac{3\lambda}{\sqrt{63}} \frac{dw}{d\theta_x} = \frac{\pi D_x}{\sqrt{7}} \approx 1.187 D_x$$

Other parameters, such as K_x and η_{θ_x}, can also be found for this example aperture (see Problems 14.3-17 and 14.3-18).

Another example of patterns with a derivative relationship is given in the problems (see Problems 14.3-19 and 14.3-20).

Error Distribution as Reference Distribution Derivative

In the preceding discussion we found that an error pattern that was proportional to the derivative of the reference pattern occurred when the error pattern's aperture distribution was the product of the reference pattern's distribution and a linear odd function of ξ (when measuring θ_x). A sort of reverse situation occurs when we require the error pattern to equal the product of the reference pattern times a linear odd function of θ_x. We will examine this problem and show that its solution requires the error pattern's distribution to be proportional to the derivative of the reference pattern's distribution. We will assume that both the reference and error patterns radiate the same amounts of power when excited by the same amounts of incident power.

A vector voltage pattern and the vector aperture distribution that produces the pattern are related through (14.1-12). For a reference pattern $\mathbf{F}_\Sigma(\theta, \phi)$ and error pattern $\mathbf{F}_\Delta(\theta, \phi)$ generated, respectively, by distributions $\mathbf{E}_\Sigma(\xi, \zeta)$ and $\mathbf{E}_\Delta(\xi, \zeta)$, we will require that

$$\frac{\mathbf{F}_\Delta(\theta, \phi)}{\mathbf{F}_\Sigma(\theta, \phi)} = \frac{\int_A \int \mathbf{E}_\Delta(\xi, \zeta) e^{jk\theta_x\xi + jk\theta_y\zeta} d\xi d\zeta}{\int_A \int \mathbf{E}_\Sigma(\xi, \zeta) e^{jk\theta_x\xi + jk\theta_y\zeta} d\xi d\zeta} = P_x \theta_x \tag{14.3-19}$$

Here P_x is the (real) constant or proportionality for the linear odd function of θ_x (which is $P_x\theta_x$). The ratio of vectors is to be interpreted as one being proportional to the other. In writing (14.3-19) the denominators of (14.1-12) that define $\mathbf{F}_\Sigma(\theta, \phi)$ and $\mathbf{F}_\Delta(\theta, \phi)$ are equal due to our equal-power constraint.

Next define u_1, u_2, and $\mathbf{q}_\Sigma(u_1, u_2)$ by

$$u_1 = k\theta_x \tag{14.3-20}$$

$$u_2 = k\theta_y \tag{14.3-21}$$

$$\mathbf{q}_\Sigma(u_1, u_2) = \frac{1}{(2\pi)^2} \int_A \int \mathbf{E}_\Sigma(\xi, \zeta) e^{ju_1\xi + ju_2\zeta} d\xi d\zeta \tag{14.3-22}$$

Since (14.3-22) is a two-dimensional inverse Fourier transform,

$$\mathbf{q}_\Sigma(u_1, u_2) \leftrightarrow \mathbf{E}_\Sigma(\xi, \zeta) \tag{14.3-23}$$

From the "frequency" differentiation property of Fourier transforms,

$$-ju_1 \mathbf{q}_\Sigma(u_1, u_2) = -jk\theta_x \mathbf{q}_\Sigma(u_1, u_2) \leftrightarrow \frac{\partial \mathbf{E}_\Sigma(\xi, \zeta)}{\partial \xi} \tag{14.3-24}$$

By using (14.3-24), we can write (14.3-19) as

$$\int_A \int \mathbf{E}_\Delta(\xi, \zeta) e^{jk\theta_x\xi + jk\theta_y\zeta} d\xi d\zeta = j2\pi\lambda P_x(-ju_1)\mathbf{q}_\Sigma(u_1, u_2)$$

$$= j\frac{\lambda P_x}{2\pi} \int_A \int \frac{\partial \mathbf{E}_\Sigma(\xi, \zeta)}{\partial \xi} e^{jk\theta_x\xi + jk\theta_y\zeta} d\xi d\zeta \tag{14.3-25}$$

This condition is guaranteed for any θ_x and θ_y and either $\mathbf{E}_\Delta(\xi, \zeta)$ or $\mathbf{E}_\Sigma(\xi, \zeta)$ treated as arbitrary if the two functions are related by

$$\mathbf{E}_\Delta(\xi, \zeta) = \frac{jP_x\lambda}{2\pi} \frac{\partial \mathbf{E}_\Sigma(\xi, \zeta)}{\partial \xi} \tag{14.3-26}$$

To evaluate P_x, we form the magnitude squared of both sides of (14.3-26) and integrate over the aperture's area. On invoking the equal-power constraint again, we have

$$P_x^2 = \left(\frac{2\pi}{\lambda}\right)^2 \frac{\int_A \int |\mathbf{E}_\Sigma(\xi, \zeta)|^2 d\xi d\zeta}{\int_A \int |\partial \mathbf{E}_\Sigma(\xi, \zeta)/\partial \xi|^2 d\xi d\zeta} \tag{14.3-27}$$

On recognizing that $\mathbf{F}_\Delta(\theta, \phi)$ can be taken as equivalent to $\mathbf{F}_\Delta(\theta, \phi)$ in (14.3-1), we use (14.3-26) to find the error pattern's slope parameter

$$K_x^2 = \left(\frac{2\pi}{\lambda}\right)^2 \frac{|\int_A \int \xi[\partial \mathbf{E}_\Sigma(\xi, \zeta)/\partial \xi] d\xi d\zeta|^2}{A\int_A \int |\partial \mathbf{E}_\Sigma(\xi, \zeta)/\partial \xi|^2 d\xi d\zeta} \tag{14.3-28}$$

In summary, we have shown that the ratio of an odd pattern to an even pattern will be proportional to θ_x according to (14.3-19), provided that the patterns have aperture distributions that satisfy (14.3-26), with P_x given by (14.3-27). In essence, one pattern, either $\mathbf{F}_\Sigma(\theta, \phi)$ or $\mathbf{F}_\Delta(\theta, \phi)$, can be arbitrary but the other is not. Our analysis is a generalization of the same problem treated by Kirkpatrick (1952) who limited his work to rectangular apertures with factorable, scalar, patterns. Kirkpatrick's procedures produced *two* conditions for (14.3-19) to be true. One was essentially the same as (14.3-26). The second was that $\mathbf{E}_\Sigma(\xi, \zeta)$ should be zero on the aperture's edges, a condition not found to be theoretically required here. As a practical matter, the condition does prevent unrealizable impulses from occurring in $\mathbf{E}_\Delta(\xi, \zeta)$ on the aperture's edges.

Example 14.3-2   As an example of patterns that form a linear pattern ratio, we again consider the origin-centered rectangular aperture of Example 14.3-1 with side lengths D_x and D_y in x and y directions, respectively. We again assume a factorable aperture distribution defined by

$$\mathbf{E}_\Sigma(\xi, \zeta) = \hat{a}_x E_0 \left[1 - \left(\frac{2\xi}{D_x} \right)^2 \right] \text{rect}\left(\frac{\xi}{D_x} \right) E(\zeta) \, \text{rect}\left(\frac{\zeta}{D_y} \right)$$

where E_0 and $E(\zeta)$ were previously defined. The corresponding pattern was also determined previously:

$$\mathbf{F}_\Sigma(\theta, \phi) = \hat{a}_x \sqrt{\frac{10\pi D_x D_y}{3\lambda^2}} \frac{3}{w^2} \left[\frac{\sin(w)}{w} - \cos(w) \right] F_g(\theta_y) \tag{1}$$

where

$$w = \frac{\pi D_x \theta_x}{\lambda}$$

$$F_g(\theta_y) = \frac{\int_{-D_y/2}^{D_y/2} E(\zeta) e^{jk\theta_y \zeta} d\zeta}{\sqrt{D_y \int_{-D_y/2}^{D_y/2} |E(\zeta)|^2 \, d\zeta}}$$

Our example will find $\mathbf{F}_\Delta(\theta, \phi)$ and form the ratio of patterns to demonstrate that (14.3-19) is true. Finally P_x will be found.

From (14.3-26) the odd pattern's distribution must be

$$\mathbf{E}_\Delta(\xi, \zeta) = \left(\frac{jP_x\lambda}{2\pi} \right) \frac{\partial \mathbf{E}_\Sigma(\xi, \zeta)}{\partial \xi}$$

$$= \left(\frac{jP_x\lambda}{2\pi} \right) \hat{a}_x \left(\frac{-8E_0}{D_x^2} \right) \xi \, \text{rect}\left(\frac{\xi}{D_x} \right) E(\zeta) \, \text{rect}\left(\frac{\zeta}{D_y} \right)$$

The odd pattern derives from (14.1-12)

$$\mathbf{F}_\Delta(\theta, \phi) = \sqrt{\frac{4\pi}{\lambda^2}} \frac{\hat{a}_x(-8jP_x\lambda E_0/2\pi D_x^2) \int_{-D_x/2}^{D_x/2} \xi e^{jk\theta_x \xi} d\xi \int_{-D_y/2}^{D_y/2} E(\zeta) e^{jk\theta_y \zeta} d\zeta}{\sqrt{|-8jP_x\lambda E_0/2\pi D_x^2|^2 \int_{-D_x/2}^{D_x/2} \xi^2 \, d\xi \int_{-D_y/2}^{D_y/2} |E(\zeta)|^2 \, d\zeta}}$$

On using integrals from Appendix D, this expression becomes

$$\mathbf{F}_\Delta(\theta, \phi) = \hat{a}_x \sqrt{\frac{10\pi D_x D_y}{3\lambda^2}} \sqrt{\frac{2}{5}} \frac{3}{w} \left[\frac{\sin(w)}{w} - \cos(w) \right] F_g(\theta_y) \tag{2}$$

When the ratio of $\mathbf{F}_\Delta(\theta, \phi)$ to $\mathbf{F}_\Sigma(\theta, \phi)$ is formed, we get

$$\frac{\mathbf{F}_\Delta(\theta, \phi)}{\mathbf{F}_\Sigma(\theta, \phi)} = \sqrt{\frac{2}{5}}\, w = \frac{2\pi D_x}{\sqrt{10\lambda}}\, \theta_x$$

so

$$P_x = \frac{2\pi D_x}{\sqrt{10\lambda}} \tag{3}$$

Plots of $\mathbf{F}_\Sigma(\theta, \phi)$ and $\mathbf{F}_\Delta(\theta, \phi)$ normalized (divided) by $\hat{a}_x\sqrt{10\pi D_x D_y/(3\lambda^2)}$ are shown in Fig. 14.3-2 along with some useful parameters.

Other examples of patterns with linear pattern ratios are given in the problems (see Problems 14.3-19 and 14.3-22 for one example; another is given in Problems 14.3-23 and 14.3-24). On comparing data of these two examples and that of Example 14.3-2, it is found that as the distribution taper of the reference pattern increases, the error slope decreases while the error pattern sidelobe level improves (sidelobes decrease). This behavior was similar to that observed in Chapter 3 where directivity/sidelobe behavior was examined for even patterns. For even (reference) patterns we found that the Taylor distributions were a practical compromise that achieved highest directivity for a specified sidelobe level. For error (odd) patterns there is a similar Taylor-like solution that gives largest practical error slope for a specified error pattern sidelobe level. It was developed by Bayliss (1968) for circular apertures and line sources. Elliott (1976) generalized the work for lines sources with specified variations in sidelobes (see also Elliott, 1981).

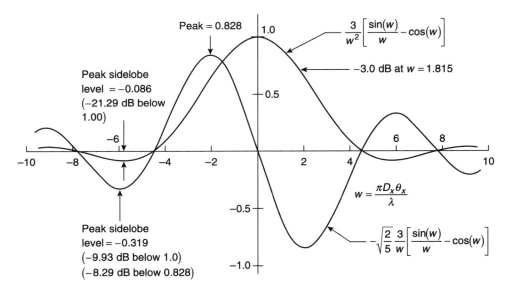

Figure 14.3-2   Normalized plots of $\mathbf{F}_\Sigma(\theta, \phi)$ and $\mathbf{F}_\Delta(\theta, \phi)$ that are applicale to the rectangular aperture of Example 14.3-2.

14.4 PRACTICAL MONOPULSE CONCEPTS

In Section 14.2 we *defined* an optimum monopulse system as one that satisfies three conditions:

1. The error channel's real output voltage due to the target, at the time of the target, must be zero if the target's direction is that of the reference axis.
2. The error channel's real output voltage due to the target, at the time of the target, must be proportional to the target's error angle θ_x from the reference axis, with the proportionality constant being known.
3. For additive input noise the system must provide the smallest possible variance of error in the measurement of θ_x of all linear systems.

Implicit in these conditions (especially condition 2) is that the system must allow an independent measurement of θ_x on each pulse (monopulse).

Conditions 1 and 2 relate to the noise-free response of the system. Condition 3 admits the effects of noise. Although our definition is helpful in obtaining an optimum system, there seems to be no widely accepted definition of monopulse, especially when practical systems are considered.

Kirkpatrick's Monopulse

For example, in an important early work Kirkpatrick (1952) analyzed monopulse systems. Although he did not formally define monopulse, he observed that if the error channel's response signal (no noise case) had certain characteristics it could be used to measure the target's angle (θ_x or θ_y). He called the response signal an *electrical correction signal* (or ECS) and noted three characteristics:

1. ECS is not a function of target range.
2. ECS is not a function of target cross section.
3. ECS is proportional to the target's angle (θ_x or θ_y).

The first two conditions can be interpreted more generally by requiring that the ECS not be a function of the received target signal's amplitude (i.e., not dependent on parameters in the radar equation); they are equivalent to the optimum system's requirement 2 that the proportionality constant be known. In essence, Kirkpatrick's three conditions are equivalent to conditions 1 and 2 for the optimum system.

Although Kirkpatrick analyzed monopulse systems with noise, these analyses were not part of a larger effort to define monopulse through the behavior of the ECS with noise.

Rhodes's Monopulse

In another important early work, Rhodes (1959) developed a unified theory in which he defined all known forms of monopulse system through three postulates (Rhodes, 1959, pp. 21–22):

1. Monopulse angle information appears in the form of a (sensing) ratio.

2. The sensing ratio for a positive angle of arrival is the inverse of the ratio for an equal negative angle.[12]

3. The angle-output function is an odd, real function of the angle of arrival.

As in Kirkpatrick's work, Rhodes's definition does not incorporate noise into the definition of monopulse. This fact does not invalidate any of the approaches; it only emphasizes that monopulse can be discussed through its antenna–signal processor operations alone, or noise can be included. Indeed, there are many variations in antenna realizations and signal processing that may be called monopulse. Rhodes's work has proved useful in defining types and classes of some of these variations. Discussion of the classes will be given later in this chapter (see Section 14.7) after some forms of signal processors have been introduced.

Types of Monopulse

For most practical systems it is adequate to define two types of monopulse: *amplitude sensing* (or just *amplitude monopulse*) and *phase sensing* (or just *phase monopulse*). In amplitude sensing, the angle information is derived from two antenna patterns that have different amplitude characteristics but identical (or nearly identical) phase characteristics. In phase sensing, the pattern's amplitudes are the same, but different phase behavior's produce angle information. A third type, which might be called either *general monopulse* or *amplitude and phase sensing*, can also be defined. It is probably the least used but this type has been considered (Peebles and Berkowitz, 1968). General monopulse is briefly described in Section 14.8.

Practical Pattern Representation

For the remainder of this chapter it will not be necessary to consider the detailed relationship between vector pattern representations and their vector aperture distributions. It will be adequate to discuss patterns using a complex scalar representation instead of the complex vector representation. To see that this fact is true, imagine that a target presents a constant cross section over all angles θ_x and θ_y of interest in the receiving patterns. If all other parameters in the radar equation are also constant, the complex voltage produced at the output of a pattern is constant except for the effect of the pattern itself as θ_x and θ_y vary. This behavior can be described by using a complex scalar one-way voltage pattern, and the polarization (vector) behavior of the pattern can be ignored. From another viewpoint, the pattern responds only to the wave's polarization component that is a match to the antenna's polarization, so there is some complex scalar "pattern" that describes this response.

Thus we represent an arbitrary one-way complex receiving voltage pattern by the notation $p(\theta_x, \theta_y)$ and define it by

$$p(\theta_x, \theta_y) = f(\theta_x, \theta_y) e^{j\beta(\theta_x, \theta_y)} \tag{14.4-1}$$

---

[12] Inverse is defined as in the theory of groups; there can be both additive and multiplicative ratios. In the multiplicative case the inverse of a group element is its reciprocal, the identity element being unity. In the additive case the inverse of a group element is its negative, the identity element being zero. The identity element physically represents the value of the sensing ratio on the reference axis (Rhodes, 1959, p. 22).

Here $f(\theta_x, \theta_y)$ is the amplitude of the pattern and $\beta(\theta_x, \theta_y)$ is its phase.[13] In some discussions involving measurements in one coordinate, the explicit dependence on the other coordinate's angle is omitted for convenience, and we use the notation

$$p(\theta_x) = f(\theta_x)e^{j\beta(\theta_x)} \tag{14.4-2}$$

for measurement of θ_x.

Angle-Sensing Ratios

In describing practical monopulse systems, we introduce the angle-sensing ratios of Rhodes (1959). As previously noted, angle information resides in the ratio of a pair of patterns in order for the angle measurement to be independent of target amplitude. For measurement of θ_x (a similar development applies to the measurement of θ_y) let $p(\theta_x)$ represent some arbitrary pattern. We define the *multiplicative angle-sensing ratio*, denoted by $r_m(\theta_x)$, as

$$r_m(\theta_x) = \frac{p(\theta_x)}{p(-\theta_x)} \tag{14.4-3}$$

(Rhodes, 1959, p. 24).

A second ratio, called the *additive angle-sensing ratio*, denoted by $r_a(\theta_x)$, is defined by

$$r_a(\theta_x) = \frac{p_o(\theta_x)}{p_e(\theta_x)} \tag{14.4-4}$$

where $p_o(\theta_x)$ and $p_e(\theta_x)$ are arbitrary odd and even functions of θ_x, respectively (Rhodes, 1959, p. 24). In a practical application $p_o(\theta_x)$ and $p_e(\theta_x)$ can be taken as the odd and even parts of $p(\theta_x)$ according to

$$p_o(\theta_x) = \tfrac{1}{2}[p(\theta_x) - p(-\theta_x)] \tag{14.4-5}$$

$$p_e(\theta_x) = \tfrac{1}{2}[p(\theta_x) + p(-\theta_x)] \tag{14.4-6}$$

With this interpretation (14.4-4) becomes

$$r_a(\theta_x) = \frac{p(\theta_x) - p(-\theta_x)}{p(\theta_x) + p(-\theta_x)} = \frac{r_m(\theta_x) - 1}{r_m(\theta_x) + 1} \tag{14.4-7}$$

from use of (14.4-3). Thus $r_a(\theta_x)$ and $r_m(\theta_x)$ are not independent, but they are related through the bilinear transformation, which is the last form in (14.4-7). We later show that the outputs of some monopulse systems are proportional to a suitable *function* of either $r_a(\theta_x)$ or $r_m(\theta_x)$.

---

[13] Phase angles $+\pi$ or $-\pi$ that represent negative amplitudes are presumed part of the definition of $f(\theta_x, \theta_y)$ when it is desirable to do so.

14.5 AMPLITUDE-SENSING MONOPULSE

Consider the problem of measuring a target's error angle θ_x. We need not consider measurement of θ_y because the applicable procedures will be the same as for θ_x. Suppose that by some means yet to be introduced, an antenna produces a one-way voltage pattern squinted slightly away from the reference axis. This pattern is shown in Fig. 14.5-1a as $p(\theta_x)$. A second pattern squinted in the opposite direction by an equal, but negative, amount is shown as $p(-\theta_x)$. Both patterns are defined through (14.4-2), but in amplitude-sensing monopulse their amplitudes are different while their phases are the same (often can be taken as zero), as indicated in the figure.

Multiplicative Sensing Ratio

If identical linear signal processors are attached to the pattern outputs of Fig. 14.5-1a, the ratio of their output complex voltages is just the ratio of the patterns. The ratio is the multiplicative sensing ratio of (14.4-3) except that the phase factors cancel

$$r_m(\theta_x) = \frac{f(\theta_x)}{f(-\theta_x)} \tag{14.5-1}$$

This function is sketched in panel b. From the plot of $r_m(\theta_x)$, it may appear that the multiplicative sensing ratio does not behave properly for a measurement of θ_x, since this function is *not* proportional to θ_x. Recall, however, that our conditions for monopulse do not say that the sensing ratio itself must be proportional to θ_x. It is the receiving system's final output voltage that must be proportional to θ_x. This final output can be the result of some *function of the sensing ratio*. Suppose that here, for example, we form the logarithm (to any base) of $r_m(\theta_x)$, as shown in panel c. For small θ_x it is clear that $\log[r_m(\theta_x)]$ is proportional to θ_x for $|\theta_x|$ small.

Realization of the logarithmic function is equivalent to choosing a suitable angle detector (signal processor) to work with the two signals produced by the patterns of Fig. 14.5-1a. If we adopt the notation of Fig. 14.5-2a to represent an antenna that generates the two amplitude patterns, a practical system is shown in panel b. The log-IF amplifiers are assumed identical and produce IF output voltages that have amplitudes that are the logarithms of the amplitudes of their input IF voltages. If A, B, P, Q, and R in panel b represent complex voltages, in general, we have

$$R = \text{complex response} = P - Q = \log(|A|) - \log(|B|)$$
$$= \log\left[\left|\frac{A}{B}\right|\right] = \log\left[\frac{f(\theta_x)}{f(-\theta_x)}\right] = \log[r_m(\theta_x)] \tag{14.5-2}$$

$$s_0 = \text{Re}[R] = R = \log[r_m(\theta_x)] \tag{14.5-3}$$

Equation (14.5-3) and the last two forms of (14.5-2) are restricted to values of θ_x near the origin, since the expressions assume that $r_m(\theta_x)$ is positive.

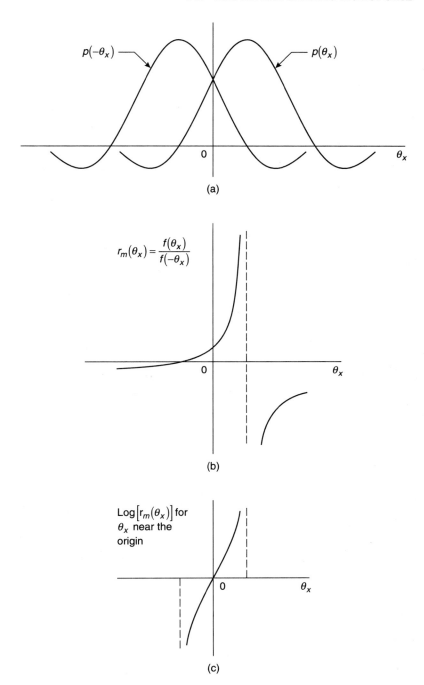

Figure 14.5-1 Patterns (*a*), their ratio (*b*), and the logarithm of their ratio (*c*), applicable to amplitude-sensing monopulse.

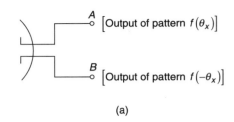

(a)

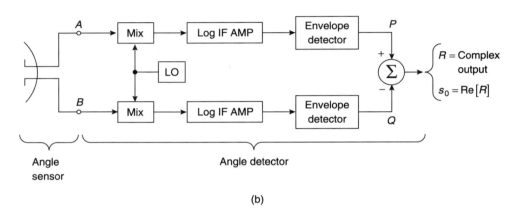

(b)

Figure 14.5-2   (*a*) Notation for an antenna that produces two amplitude patterns, and (*b*) the angle sensor and its angle detector to produce an amplitude monopulse system using multiplicative sensing.

From (14.5-3) with small values of $|\theta_x|$ and the natural logarithm assumed, we have $s_0 = -K_x\theta_x$, where

$$K_x = -\frac{\partial s_0}{\partial \theta_x}\bigg|_{\theta_x=0} = \frac{-2}{f(0)}\frac{\partial f(\theta_x)}{\partial \theta_x}\bigg|_{\theta_x=0} \qquad (14.5\text{-}4)$$

Thus a measurement of θ_x can be taken as the ratio of the value of s_0 measured at the time of the target to the known constant $-K_x$. The measurements may be taken for every pulse so the system is monopulse.

Additive Sensing Ratio

A microwave element called a *hybrid T* (also *hybrid junction, magic T*, or just *hybrid*) is a 4-port device shown symbolically in Fig. 14.5-3*a*. It is able to provide complex voltages (*C* and *D*) that are proportional to the sum and difference of the complex input voltages *A* and *B*, as shown. A *generalized hybrid* is shown in panel *b*; it depends on a selectable constant ϕ_0. When $\phi_0 = -\pi/4$ rad, the generalized hybrid becomes a hybrid *T*. Also shown in panel *c* is a device called a *short-slot coupler*. It is useful in some implementations and it can be modified slightly to become a hybrid *T* (see Problem 14.5-3).

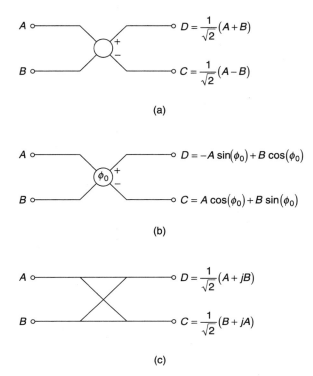

$$D = \frac{1}{\sqrt{2}}(A+B)$$

$$C = \frac{1}{\sqrt{2}}(A-B)$$

(a)

$$D = -A\sin(\phi_0) + B\cos(\phi_0)$$

$$C = A\cos(\phi_0) + B\sin(\phi_0)$$

(b)

$$D = \frac{1}{\sqrt{2}}(A+jB)$$

$$C = \frac{1}{\sqrt{2}}(B+jA)$$

(c)

Figure 14.5-3 Symbols representing microwave components: (*a*) Hybrid *T*, (*b*) generalized hybrid, and (*c*) short-slot coupler.

If amplitude monopulse patterns $f(\theta_x)$ and $f(-\theta_x)$ provide the inputs A and B to a hybrid T, as shown in Fig. 14.5-4*a*, then we have

$$C = \frac{1}{\sqrt{2}}[f(\theta_x) - f(-\theta_x)] \qquad (14.5\text{-}5)$$

$$D = \frac{1}{\sqrt{2}}[f(\theta_x) + f(-\theta_x)] \qquad (14.5\text{-}6)$$

Functions C and D are sketched in panel *b*. Their ratio C/D, as shown in panel *c*, is the additive angle sensing ratio of (14.4-4) because C and D are clearly proportional to the odd and even parts of $f(\theta_x)$ from (14.4-5) and (14.4-6). Thus

$$r_a(\theta_x) = \frac{C}{D} = \frac{f(\theta_x) - f(-\theta_x)}{f(\theta_x) + f(-\theta_x)} = \frac{f_o(\theta_x)}{f_e(\theta_x)} \qquad (14.5\text{-}7)$$

Because $r_a(\theta_x)$ is a real odd function of θ_x it can be used directly to measure θ_x. It is only necessary to add a suitable angle detector to process the signals at the outputs of the hybrid T and form the sensing ratio. One such detector is added as illustrated in Fig. 14.5-5. Some discussion is needed.

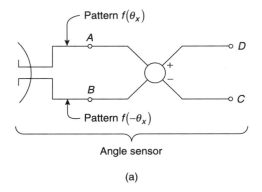

(a)

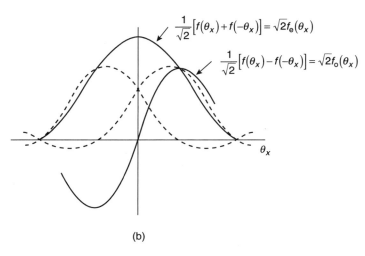

(b)

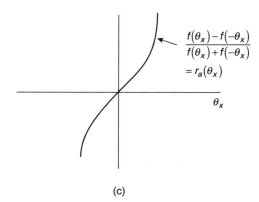

(c)

Figure 14.5-4   Angle sensor (*a*), patterns (*b*), and the additive sensing ratio (*c*) used in amplitude-sensing monopulse.

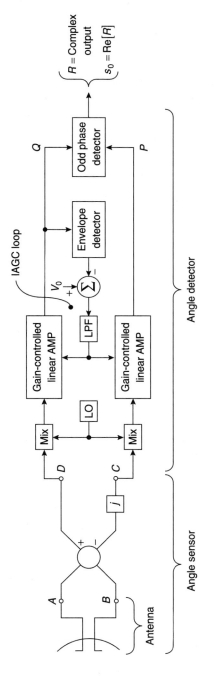

Figure 14.5-5   Amplitude-sensing monopulse receiving system to measure angle θ_x.

587

The j in the hybrid T represents a phase shift of $\pi/2$; it has been added to be consistent with the work of Rhodes (1959). The phase shift could be omitted but a different phase detector would be needed (these are called phase comparators by Rhodes). Complex-signal representations for two principal forms of phase detector are illustrated in Fig. 14.5-6. They represent real detectors that can be modeled as a real 4-quadrant analog product device followed by a low-pass filter to remove bandpass terms from the response.[14] The only difference between the even and odd detectors is the phase shift of $-\pi/2$ represented by $-j$ in the odd device. For equal-phase complex inputs P and Q the phase shift causes the real output s_0 to be zero in the odd detector and maximum in the even device. Phase detectors are also variously known as *product detectors*, *synchronous detectors*, *phase-sensitive detectors*, and *dot-product detectors* (see Sherman, 1984, p. 165).

The purpose of the IAGC loop in Fig. 14.5-5 is normalization, where the I stands for "instantaneous." If C, D, P, and Q represent complex voltages, the action of the loop is to force the amplitude of Q to equal some constant V_0. Because the two IF amplifiers are designed to be identical, their voltage gains are $Q/D = V_0/|D|$, so

$$Q = V_0 \frac{D}{|D|} \tag{14.5-8}$$

$$P = V_0 \frac{C}{|D|} \tag{14.5-9}$$

On using these expressions with the phase detector (Fig. 14.5-6b), the complex response voltage R is

$$R = \frac{-j}{2}PQ^* = \frac{-jV_0^2}{2}\frac{CD^*}{|D|^2} = \frac{-jV_0^2}{2}\frac{C}{D}$$
$$= \frac{-jV_0^2}{2}\frac{j[f(\theta_x) - f(-\theta_x)]}{[f(\theta_x) + f(-\theta_x)]} = \frac{V_0^2}{2}r_a(\theta_x) = s_0 \tag{14.5-10}$$

The last form of (14.5-10) results because R is real.

Ideally the IAGC loop would sample the target pulse and instantaneously adjust the gain so that voltage Q has amplitude V_0. This adjustment is made for each and every pulse. In practical systems, however, the loop may average over several pulses. If the target returns are constant, or fluctuate slowly compared to the loop's averaging time, there is little degradation of performance, and pulse-to-pulse measurements are nearly independent of the target's amplitude. For fast target fluctuations there may be some loss in performance unless some form of "fast" normalization is added.[15]

The amplitude monopulse system of Fig. 14.5-5 is a very common form sometimes called *sum-and-difference monopulse* because of the form of $r_a(\theta_x)$.

---

[14] Some real phase detectors do not respond to the amplitudes of one (or both) of their inputs, as does the product device assumed here. These forms can be modeled by adding a hard (symmetric) limiter to the forms of Fig. 14.5-6 at each affected input.

[15] For example, P and Q may be sampled at the target's time, digitally normalized, and the action of the phase detector performed digitally.

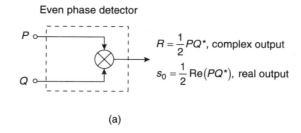

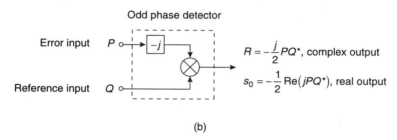

Figure 14.5-6 Symbols and responses representing (a) even and (b) odd phase detectors.

Pattern Realizations

We will next discuss some basic methods for achieving the patterns required in amplitude sensing monopulse. We mainly describe the generation of patterns using a single aperture of the reflector (or lens) type. All the concepts have a counterpart with the phased array.

When a reflector is illuminated by a waveguide horn or other radiator at the reflector's focus,[16] a pattern is generated. If the horn is offset from the focus in the x direction by a small amount, the pattern will shift (squint) away from the reference axis. For two similar horns opposingly displaced from the focus, two patterns are generated as illustrated in Fig. 14.5-7. Beams A and B (no sidelobes are shown for simplicity) correspond to patterns $f(\theta_x)$ and $f(-\theta_x)$, respectively, in preceding work.

For measurement in one coordinate (θ_x) only, the two patterns of Fig. 14.5-7 would be adequate. However, most practical systems measure both target angles θ_x and θ_y. Three patterns are required, one reference and two error. The 2-horn feed can be extended to a 4-horn feed as shown in Fig. 14.5-8a to create the three patterns. The main beams produced by the four horns are shown in panel b.

The hybrid network that produces the three desired patterns using the outputs A, B, C, and D from the four beams, is shown in Fig. 14.5-9a; the network is usually called a *comparator*. The reference pattern derives by summing all four beams, an action that often results in the name *sum pattern*. The output of the reference (or sum) pattern is denoted by Σ in the figure. Error pattern outputs, denoted by outputs Δ_x and Δ_y in the figure, are differences between sums of pairs of beams, which usually gives rise to the

---

[16] It may be helpful to envision the reflector as a paraboloid, although other shapes, such as spherical, are also possible.

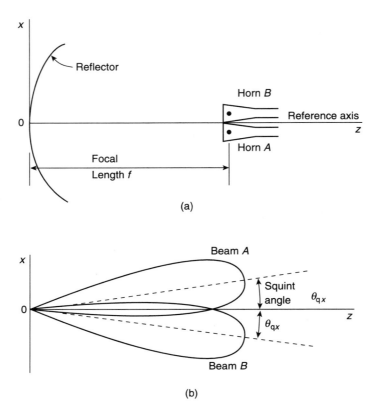

Figure 14.5-7 Two-horn method of generating two simultaneous patterns for amplitude monopulse. (*a*) Feed arrangement, and (*b*) the generated patterns' main lobes (sidelobes omitted for clarity).

name *difference patterns*. Appropriate addition or subtraction of beams is defined by the signs shown in panel *b*. The output $\Delta_{x,y}$ is called the *quadrupolar signal*; it is the difference of the two sums of diagonally positioned beams and is not used. Although it has been known for a long time that practical systems did not need to use the quadrupolar signal, Peebles and Berkowitz (1968) showed, for factorable patterns, that it is theoretically unnecessary to monopulse. The system using 4-horn illumination for pattern generation is sometimes called *sum-and-difference monopulse*.

The 4-horn feed is simple and easy to understand. It is a good learning tool for first exposure by readers. However, its performance is poor relative to other feed structures that have been devised, and it is seldom used in current-day systems. In the 4-horn feed each horn is typically excited in only one waveguide mode. If horn size is chosen to optimize the reference pattern (large gain, low sidelobes), the error patterns are poor (low error slope and high sidelobes). If the error patterns are "optimized" the reference pattern is poor. In other words, it is not possible to optimize *both* error and reference patterns with the simple 4-horn feed.

It has been found that either by allowing the incoming wave to excite higher modes in the horns or by using more horns, or both, many design options become available.

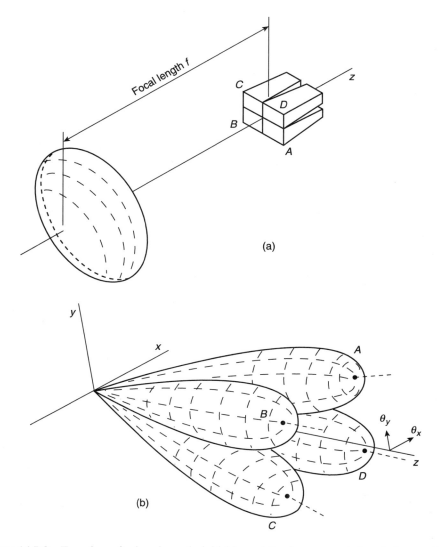

Figure 14.5-8 Four-horn feed and paraboloid (*a*), and the four generated main beams (*b*) in amplitude-sensing monopulse.

Leonov and Fomichev (1986, pp. 33–34) describe a feed with one horn that uses six modes. At the other extreme, Hannan (I and II, 1961) has described a feed with 12 single-mode horns that gives good control over all three monopulse patterns. The horn arrangement, comparator network, and sums used to generate Σ, Δ_x, and Δ_y are shown in Fig. 14.5-10*a*, *b*, and *c*, respectively.

   Hannan (I and II, 1961) and Hannan and Loth (1961) have described a relatively simple near-optimum feed that is between the complexity extremes of the 1-horn-six-mode feed and the 12-horn–one-mode feed. It uses four stacked horns in one plane and multi-modes in the other plane (three modes per horn) to achieve independent control of the reference and two error patterns. The necessary comparator requires

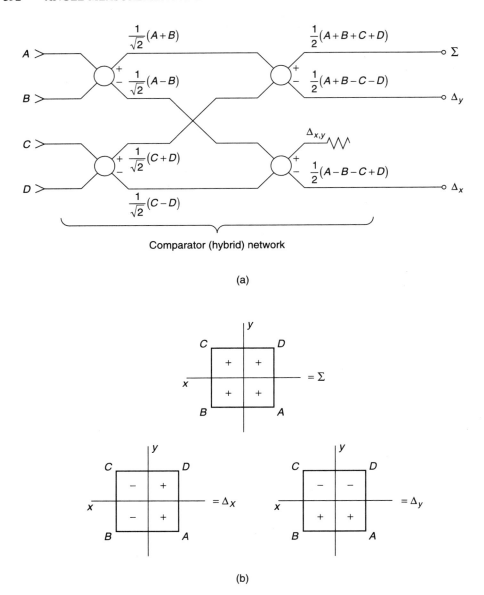

Figure 14.5-9 (*a*) Comparator and four horns of a 4-horn feed in amplitude-sensing monopulse. (*b*) The signs of addition of horn outputs by the comparator.

only eight hybrid T elements, which is a significant reduction compared with the 12-horn feed.

Four-Horn Difference Patterns as Derivative Pattern Approximations

It is of academic interest to show that the difference (error) patterns generated by the 4-horn feed method of Figs. 14.5-8 and 14.5-9 are approximately proportional to the

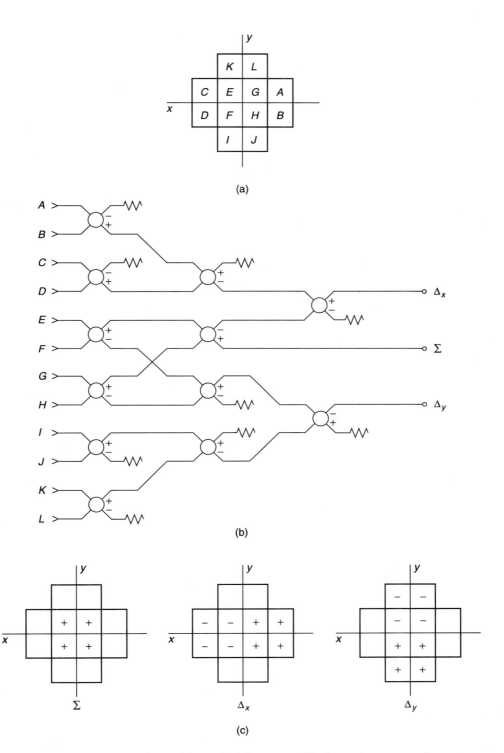

Figure 14.5-10 (*a*) Horns in a 12-horn feed for an amplitude-sensing monopulse system; view is of the horn mouths from the reflector. (*b*) Horns and comparator for the 12-horn feed. (*c*) The signs of addition of horn outputs by the comparator; view is as in (*a*).

derivatives of the sum pattern. For a basic aperture pattern $f(\theta_x, \theta_y)$ corresponding to a horn on axis (at the focus), the squinted patterns become

$$A = f(\theta_x - \theta_{qx}, \theta_y - \theta_{qy}) \tag{14.5-11}$$

$$B = f(\theta_x + \theta_{qx}, \theta_y - \theta_{qy}) \tag{14.5-12}$$

$$C = f(\theta_x + \theta_{qx}, \theta_y + \theta_{qy}) \tag{14.5-13}$$

$$D = f(\theta_x - \theta_{qx}, \theta_y + \theta_{qy}) \tag{14.5-14}$$

Here θ_{qx} and θ_{qy} are the squint angles in respective directions x and y. The patterns represented by the outputs of the hybrid networks may be defined as

$$f_\Sigma(\theta_x, \theta_y) = \tfrac{1}{2}(A + B + C + D) \tag{14.5-15}$$

$$f_{\Delta_x}(\theta_x, \theta_y) = \tfrac{1}{2}(A + D - B - C) \tag{14.5-16}$$

$$f_{\Delta_y}(\theta_x, \theta_y) = \tfrac{1}{2}(A + B - C - D) \tag{14.5-17}$$

The two-dimensional Taylor's series of a function $h(u, v)$ about the origin is (Wylie, 1951, p. 604)

$$h(u, v) = \sum_{m=0}^{\infty} \sum_{n=0}^{m} \alpha_{mn} u^{m-n} v^n \tag{14.5-18}$$

where

$$\alpha_{mn} = \binom{m}{n} \frac{1}{m!} \frac{\partial^m h(u, v)}{\partial u^{m-n} \partial v^n} \bigg|_{u=0, v=0} \tag{14.5-19}$$

$$\binom{m}{n} = \frac{m!}{n!(m-n)!} \tag{14.5-20}$$

If we identify the function h with f_Σ and let u and v represent θ_{qx} and θ_{qy}, respectively, the series expansion of $f_\Sigma(\theta_x, \theta_y)$ becomes

$$f_\Sigma(\theta_x, \theta_y) \approx 2f(\theta_x, \theta_y) \tag{14.5-21}$$

for small values of θ_{qx} and θ_{qy}. Similarly, when h is identified with f_{Δ_x} and f_{Δ_y}, we obtain

$$f_{\Delta_x}(\theta_x, \theta_y) \approx -2\theta_{qx} \frac{\partial f(\theta_x, \theta_y)}{\partial \theta_x} \approx -\theta_{qx} \frac{\partial f_\Sigma(\theta_x, \theta_y)}{\partial \theta_x} \tag{14.5-22}$$

$$f_{\Delta_y}(\theta_x, \theta_y) \approx -2\theta_{qy} \frac{\partial f(\theta_x, \theta_y)}{\partial \theta_y} \approx -\theta_{qy} \frac{\partial f_\Sigma(\theta_x, \theta_y)}{\partial \theta_y} \tag{14.5-23}$$

Equations (14.5-21) through (14.5-23) show that, when the squint angles are small, the error patterns are approximately proportional to the derivative of the sum pattern. Therefore 4-horn sum-and-difference amplitude-sensing monopulse is approximately equivalent to the derivative relationship of (14.3-11) with Q equal to $-\theta_{qx}$ or $-\theta_{qy}$ for x and y directions, respectively.

Overall Amplitude-Sensing Monopulse System

Figure 14.5-5 was given as a typical realization of the reference and one error channel in an amplitude monopulse system based on the additive sensing ratio. We add the second error channel and a few other practical functions needed to realize an overall system, as given in Fig. 14.5-11. This system is probably the most common one for monopulse realizations. Here a single aperture (antenna) with an appropriate feed, and its necessary hybrids generates the reference and two error patterns. The transmitter transmits through the reference pattern by sharing the Σ port with the receiver. During the transmitted pulse the *duplexer* (DUP) connects the antenna to the transmitter. At all other times the duplexer connects the antenna to the receiver. In essence, the duplexer acts as a rapidly controlled switch. Ideally the duplexer should have no loss, but in reality it has losses related to the devices needed for its realization. These devices are typically gas discharge tubes, sections of waveguide, circulators, limiters, and some solid-state devices (see Skolnik, ed. 1990, p. 4.4), although not all these devices are necessarily present in any one duplexer design.

For reception the signals $j\Delta_x$, $j\Delta_y$, and Σ are connected to three identical channels that are gain and phase matched up to the range gate outputs. These channels are shown each with one down-converting mixer stage to produce a single IF frequency, but some systems use two or more IF frequencies. Most of the receiving system's gain is in the IF circuits. Much of this gain is gain controlled via the AGC loop in the reference channel. As noted previously, the AGC action establishes normalization in the three channels. Since AGC can cause many tens of dB in gain change, it places tight requirements on the amplifiers to maintain gain and phase match over the large dynamic range.

The system of Fig. 14.5-11 is shown with single-pulse matched filters (MF) and range gates in the IF circuits. This choice is a good one if only one target is being processed. However, if several targets are to be simultaneously processed (but having close enough Doppler frequencies that the MF is usable for all), it may be preferable to have the range gating follow the detectors (gate at video). Here a parallel bank of gates, one for each target, would be needed; it would also be necessary to modify the normalization (AGC) operation. For example, the AGC common to all three channels could be derived from the largest target, an average of two or more of the targets, or some other method. On top of this normalization, individual normalization would also be needed, perhaps by digital methods using samples from the envelope detector's output.

Clearly other architectures are possible. Peebles and Wang (1982) have studied four of the most important forms. These forms were called systems 1, 2, 3, and 4 as defined in Table 14.5-1. Analysis has found that the accuracy expression for measuring θ_x (or θ_y) has the same form for all four systems:

$$\sigma_{\theta_x}^2 = \sigma_{\theta_x(\min)}^2 R_A \left[1 + \frac{R_B}{(S_1/N)} \right] \tag{14.5-24}$$

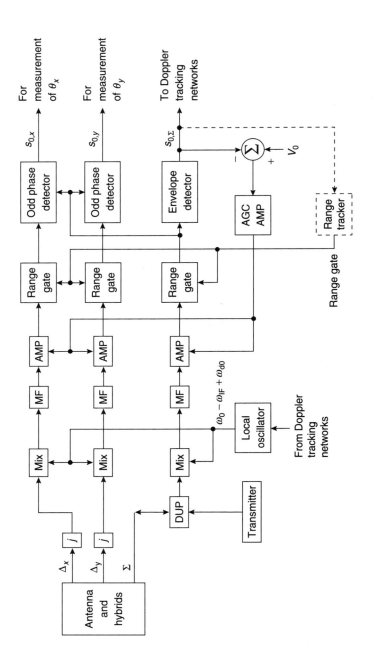

Figure 14.5-11 A practical amplitude-sensing monopulse receiving system.

TABLE 14.5-1 Matched filter and range gate locations for four system definitions.

System	Matched Filter at		Range Gate at	
	IF	Video	IF	Video
1	X			X
2	X		X	
3		X		X
4		X	X	

Here R_A and R_B are constants for a specific set of system parameters (pulse width, range gate width, etc.). The quantity (S_1/N) is the single-pulse signal-to-noise ratio given by

$$\left(\frac{S_1}{N}\right) = \frac{1}{2}\mathscr{R}_{p0} \tag{14.5-25}$$

where $\mathscr{R}_{p0}$ is defined by (14.2-25), and $\sigma^2_{\theta_x(min)}$ is given by (14.2-24) or 14.2-27).

The relative performance of the four systems is determined by the relative behavior of R_A and R_B. Peebles and Wang (1982) have given general formulas for R_A and R_B for all four systems for any shaped transmitted pulse and arbitrary range gate duration τ_g. Although the formulas may be solved by computer, they are too complicated to reproduce here. Instead, we try to compare the four systems in a general way for rectangular constant-frequency transmitted pulses of duration τ_p. We take two important cases.

First, assume that (S_1/N) is large enough so that the term of (14.5-24) involving R_B can be neglected. When $\tau_g/\tau_p \geq 1$, system, 4 gives the best performance,[17] but the other three systems are also close (within 1 dB). For $\tau_g/\tau_p < 1$, performance of systems 1, 2, and 3 are about the same (within about 0.5 dB), while that of system 4 is poor. In a realistic system $\tau_g/\tau_p \geq 1$ would usually be selected for various practical reasons.

Second, if (S_1/N) is small enough so that the term involving R_B in (14.5-24) is not negligible, the curves given by Peebles and Wang (1982) would be used for a careful comparison. However, a broad-based comparison of R_B, assuming that values of R_A are roughly equal, shows that systems 1 and 2 are equivalent with accuracy better than system 3, which is better than system 4 for $\tau_g/\tau_p > 1.3$. For $1 < \tau_g/\tau_p \leq 1.3$, systems 3 and 4 have nearly the same accuracy and the accuracies of systems 1 and 2 are the same, with the latter pair having the better accuracy. For $\tau_g/\tau_p < 1$, the comparison is very difficult and usually unnecessary for this choice of gate width is rarely used. The accuracies of systems 1 and 2 are equal, and in most cases this accuracy is better than system 3, which, in turn, is better than system 4. Plots of R_A and R_B for systems 1 and 2 are shown in Fig. 14.5-12 where $R_A = 99/80$ for $\tau_g/\tau_p \geq 2$.

Another practical monopulse architecture uses a limiter in the reference channel placed at the output of the IF amplifiers. Peebles (1985) analyzed system 1 (above)

---

[17] R_A and R_B for systems 3 and 4 depend on the bandwidth of the IF amplifier in Fig. 14.5-11 when the matched filter is at video (baseband). Our conclusions assume this bandwidth is large relative to the signal's bandwidth.

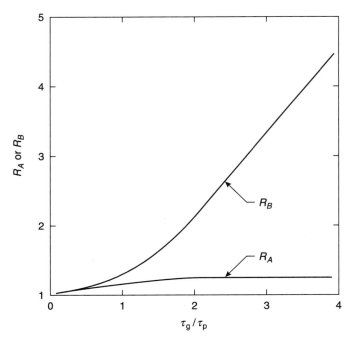

Figure 14.5-12 R_A and R_B for either system 1 or system 2. Plots assume a constant-frequency rectangular pulse of duration τ_p and a rectangular gate of duration τ_g. (From Peebles and Wang, © 1982 IEEE, with permission.)

with a limiter added. Accuracy was again found to be given by (14.5-24) and general formulas were given for R_A and R_B. Unfortunately, these formulas are difficult to evaluate, especially that for R_B.[18] For rectangular, constant-frequency pulses of duration τ_p, R_A has been evaluated by various means to give the curves of Fig. 14.5-13. The limiting curve for $(S_1/N) \to \infty$ is given by

$$
R_A|_{(S_1/N) \to \infty} = \begin{cases} 5/3, & 2\tau_p < \tau_g \\[2mm] \dfrac{16[3(\tau_g/\tau_p) - 1]}{3(\tau_g/\tau_p)^2[4 - (\tau_g/\tau_p)]^2}, & \tau_p < \tau_g \leq 2\tau_p \\[2mm] \dfrac{16[3 - (\tau_g/\tau_p)]}{3[4 - (\tau_g/\tau_p)]^2}, & 0 < \tau_g \leq \tau_p \end{cases} \qquad (14.5\text{-}26)
$$

It is interesting to note that system 1 gives a measurement variance 99/80 times as large as the theoretical minimum (or 0.93 dB larger) when phase detectors are not saturated (i.e., they are equivalent to product devices followed by low-pass filters).

---

[18] Even computer evaluation of R_B was not attempted due to the considerable computation time and cost involved.

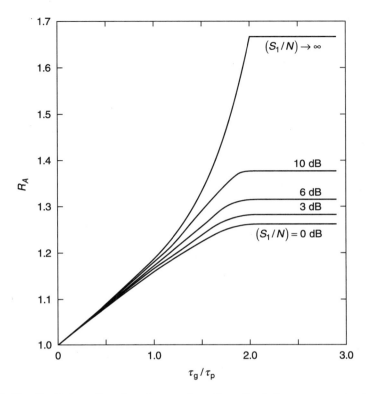

Figure 14.5-13 Coefficient R_A plotted as a function of relative range gate duration for constant-frequency rectangular transmitted pulses. (From Peebles, © 1985 IEEE, with permission.)

When a limiter is added, or used to represent a phased detector saturated at its reference input, the variance is 5/3 times the minimum (or 2.22 dB larger). We conclude that operation with limiting, or reference-side saturated phase detectors, degrades the system by $2.22 - 0.93 = 1.29$ dB.

14.6 PHASE-SENSING MONOPULSE

Phase-sensing monopulse (or phase monopulse) corresponds to using patterns having the same amplitudes but different phases. This form of monopulse is not as widely used as amplitude monopulse, but practical systems have been built. The phases are typically developed from offsetting separate apertures in space. We will describe the basic ideas assuming measurements in only one angle (θ_x). At the end of the section we describe a full practical system for two-coordinate measurements.

We begin by assuming two identical apertures are displaced from each other as shown in Fig. 14.6-1. The origin, referred to as the *phase center* of the *system*, is half way between the two antennas. The phase centers of the antennas are assumed located in their geometric center.[19] These definitions allow the aperture phase centers to be displaced by distances $d_x/2$ (aperture A) and $-d_x/2$ (aperture B) for a total separation d_x.

In Fig. 14.6-1 a wave from a target is shown arriving from an arbitrary direction (θ, ϕ) in spherical coordinates. The target's direction (origin-target line) makes an angle ψ_x from the x axis. Since the target's wave arrives first at aperture A (for $0 < \theta \leq \pi/2$ and $-\pi/2 < \phi < \pi/2$), its phase is advanced relative to when the wave arrives at the system's phase center. This phase advance equals $2\pi/\lambda$ times the projected length of the phase center of aperture A onto the origin-target line. This length is $(d_x/2)\cos(\psi_x) = (d_x/2)\cos[(\pi/2) - \theta_x] = (d_x/2)\sin(\theta_x)$, so the phase advance is $(\pi d_x/\lambda)\sin(\theta_x)$. The wave arrives at the phase center of aperture B after it passes the system's phase center. The phase associated with aperture B is therefore retarded by the same amount that aperture A was advanced. The two patterns, denoted by $p_A(\theta_x, \theta_y)$ and $p_B(\theta_x, \theta_y)$, may be written in the forms

$$p_A(\theta_x, \theta_y) = f(\theta_x, \theta_y)e^{j\beta(\theta_x, \theta_y)} \tag{14.6-1}$$

$$p_B(\theta_x, \theta_y) = f(\theta_x, \theta_y)e^{-j\beta(\theta_x, \theta_y)} \tag{14.6-2}$$

where

$$\beta(\theta_x, \theta_y) = \frac{\pi d_x}{\lambda}\sin(\theta_x) \tag{14.6-3}$$

As with amplitude monopulse, we consider both multiplicative and additive angle sensing ratios for phase monopulse.

Multiplicative Sensing Ratio

We associate the general pattern of (14.4-1) with $p_A(\theta_x, \theta_y)$. Then $p(-\theta_x) = p_A(-\theta_x, \theta_y) = f(-\theta_x, \theta_y)\exp[j\beta(-\theta_x, \theta_y)] = f(\theta_x, \theta_y)\exp[-j\beta(\theta_x, \theta_y)] = p_B(\theta_x, \theta_y)$. The last form derives from (14.6-3) and (14.6-2) and the fact that the amplitude patterns are required to have even symmetry in θ_x and θ_y. The multiplicative angle-sensing ratio can now be written as

$$r_m(\theta_x) = \frac{p(\theta_x)}{p(-\theta_x)} = \frac{p_A(\theta_x, \theta_y)}{p_B(\theta_x, \theta_y)} = \frac{e^{j\beta(\theta_x, \theta_y)}}{e^{-j\beta(\theta_x, \theta_y)}} = e^{j2\beta(\theta_x, \theta_y)} \tag{14.6-4}$$

By use of a suitable angle detector, the outputs of the patterns generated by the apertures of Fig. 14.6-1 can fully utilize the angle information contained in $r_m(\theta_x)$. Figure 14.6-2b illustrates one suitable detector, based on using the symbolic notation of panel a to represent the two antennas, and the odd phase detector defined in Fig. 14.5-6b. If A, B, P, Q, and R are interpreted as complex voltages, and V_0 is a

---

[19] This condition is true for nearly all aperture distributions that define the pattern.

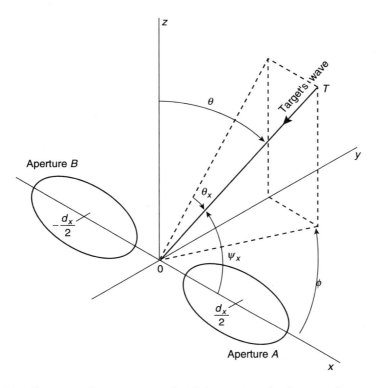

Figure 14.6-1 Geometry of two apertures that define patterns in phase-sensing monopulse for measurements of θ_x.

constant representing the amplitudes of both P and Q, then

$$R = \frac{-j}{2} PQ^* = \frac{-j}{2} V_0 \frac{A}{|A|} V_0 \frac{B^*}{|B|} = \frac{-jV_0^2}{2} \frac{p_A(\theta_x, \theta_y)}{|p_A(\theta_x, \theta_y)|} \frac{p_B^*(\theta_x, \theta_y)}{|p_B(\theta_x, \theta_y)|}$$

$$= \frac{-jV_0^2}{2} e^{2j\beta(\theta_x, \theta_y)} = \frac{-jV_0^2}{2} r_m(\theta_x) \qquad (14.6\text{-}5)$$

from (14.6-4). It remains to show that the real output is proportional to θ_x.
The real response is

$$s_0 = \text{Re}(R) = \text{Re}\left[\frac{-jV_0^2}{2} e^{j2\beta(\theta_x, \theta_y)}\right] = \frac{V_0^2}{2} \sin\left[\frac{2\pi d_x}{\lambda} \sin(\theta_x)\right]$$

$$\approx \frac{V_0^2}{2} \sin\left[\frac{2\pi d_x}{\lambda} \theta_x\right], \qquad |\theta_x| \text{ small} \qquad (14.6\text{-}6)$$

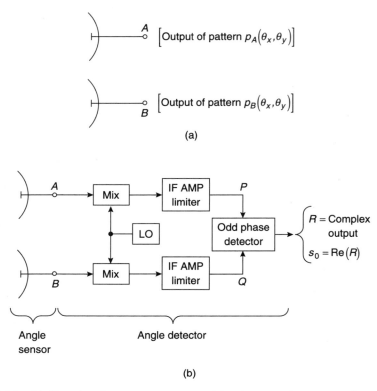

Figure 14.6-2 (a) Notation for two antennas used in phase-sensing monopulse, and (b) the angle sensor and its angle detector to produce a phase monopulse system using multiplicative sensing.

For θ_x small enough for $2\pi d_x \theta_x / \lambda$ to be small, s_0 is seen to be proportional to θ_x according to $s_0 \approx -K_x \theta_x$, where

$$K_x = -\frac{\partial s_0}{\partial \theta_x}\bigg|_{\theta_x = 0, \, \theta_y = 0} = -\frac{V_0^2 \pi d_x}{\lambda} \tag{14.6-7}$$

is a known constant. As usual, θ_x is measured as the ratio $-s_0/K_x$, taken on each pulse to achieve monopulse operation.

Additive Sensing Ratio

The additive sensing ratio for measurement of angle θ_x is defined, in general, by

$$r_a(\theta_x) = \frac{p_o(\theta_x)}{p_e(\theta_x)} = \frac{p(\theta_x) - p(-\theta_x)}{p(\theta_x) + p(-\theta_x)} \tag{14.6-8}$$

Where $p_o(\theta_x)$ and $p_e(\theta_x)$ are arbitrary odd and even patterns, respectively, in the variable θ_x. The last form in (14.6-8) derives from using the odd and even parts of a single arbitrary pattern $p(\theta_x)$ to satisfy the middle form. As above for multiplicative sensing, we associate $p(\theta_x)$ and $p(-\theta_x)$ with $p_A(\theta_x, \theta_y)$ and $p_B(\theta_x, \theta_y)$, respectively, and obtain

$$r_a(\theta_x) = \frac{p_A(\theta_x, \theta_y) - p_B(\theta_x, \theta_y)}{p_A(\theta_x, \theta_y) + p_B(\theta_x, \theta_y)} = \frac{e^{j2\beta(\theta_x, \theta_y)} - 1}{e^{j2\beta(\theta_x, \theta_y)} + 1} \tag{14.6-9}$$

We have used (14.6-1) and (14.6-2) to write the last form in (14.6-9).

Two observations will complete our work. First, a hybrid T can be used to generate the sum and difference of patterns as required in (14.6-9). Second, we already have seen that the angle detector of Fig. 14.5-5 produces the proper error channel response for a sum and difference of patterns derived from the hybrid T. Thus the complete receiving system's block diagram appears in Fig. 14.6-3. It can be shown (see Problem 14.6-3) that the complex response is

$$R = \frac{-jV_0^2}{2} r_a(\theta_x) = \frac{V_0^2}{2} \tan\left[\frac{\pi d_x}{\lambda} \sin(\theta_x)\right]$$

$$\approx \frac{V_0^2}{2} \tan\left[\frac{\pi d_x}{\lambda} \theta_x\right], \qquad |\theta_x| \text{ small} \tag{14.6-10}$$

The real response is the same since R is real.

For $\pi d_x \theta_x / \lambda$, small, the response is $s_0 \approx -K_x \theta_x$ where the known constant of proportionality is

$$K_x = -\left.\frac{\partial s_0}{\partial \theta_x}\right|_{\theta_x = 0, \theta_y = 0} = -\frac{V_0^2 \pi d_x}{2\lambda} \tag{14.6-11}$$

Overall Phase-Sensing Monopulse System

A typical phase-sensing monopulse system for making measurements of both θ_x and θ_y uses four apertures displaced from the system's phase center (origin), as shown in Fig. 14.6-4a. A target's wave is shown arriving from direction (θ, ϕ). The phase shift in a given aperture, due to target direction, is $2\pi/\lambda$ times the projection (dot product) of the vector position of the aperture's phase center onto a unit vector in the target's direction. It can be shown that these phase shifts are (Problem 14.6-4)

$$\beta_A(\theta_x, \theta_y) = \beta_x(\theta_x) + \beta_y(\theta_y) \tag{14.6-12a}$$

$$\beta_B(\theta_x, \theta_y) = -\beta_x(\theta_x) + \beta_y(\theta_y) \tag{14.6-12b}$$

$$\beta_C(\theta_x, \theta_y) = -\beta_x(\theta_x) - \beta_y(\theta_y) \tag{14.6-12c}$$

$$\beta_D(\theta_x, \theta_y) = \beta_x(\theta_x) - \beta_y(\theta_y) \tag{14.6-12d}$$

where

$$\beta_x(\theta_x) = \frac{\pi d_x}{\lambda} \sin(\theta_x) \tag{14.6-13}$$

$$\beta_y(\theta_y) = \frac{\pi d_y}{\lambda} \sin(\theta_y) \tag{14.6-14}$$

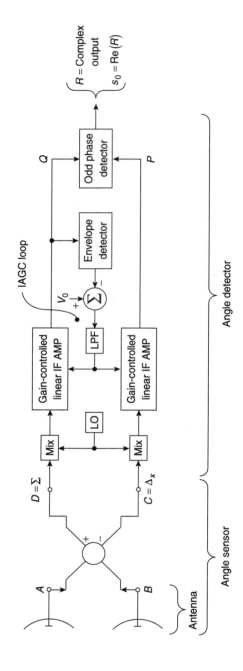

Figure 14.6-3 Phase-sensing monopulse receiving system to measure angle θ_x.

604

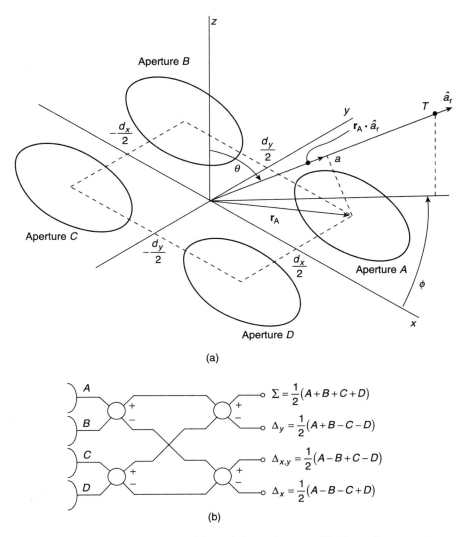

$$\Sigma = \frac{1}{2}(A+B+C+D)$$

$$\Delta_y = \frac{1}{2}(A+B-C-D)$$

$$\Delta_{x,y} = \frac{1}{2}(A-B+C-D)$$

$$\Delta_x = \frac{1}{2}(A-B-C+D)$$

(b)

Figure 14.6-4 Four-aperture antenna (a), and the angle sensor (b), for a phase-sensing mono-pulse system to measure θ_x and θ_y.

For identical pattern amplitudes, denoted by $f(\theta_x, \theta_y)$, the four patterns become

$$p_A(\theta_x, \theta_y) = f(\theta_x, \theta_y)\, e^{j\beta_x(\theta_x) + j\beta_y(\theta_y)} \tag{14.6-15a}$$

$$p_B(\theta_x, \theta_y) = f(\theta_x, \theta_y)\, e^{-j\beta_x(\theta_x) + j\beta_y(\theta_y)} \tag{14.6-15b}$$

$$p_C(\theta_x, \theta_y) = f(\theta_x, \theta_y)\, e^{-j\beta_x(\theta_x) - j\beta_y(\theta_y)} \tag{14.6-15c}$$

$$p_D(\theta_x, \theta_y) = f(\theta_x, \theta_y)\, e^{j\beta_x(\theta_x) - j\beta_y(\theta_y)} \tag{14.6-15d}$$

An overall phase-sensing monopulse system can be implemented several ways. We choose to illustrate the merging of two sets of single-coordinate systems, as in Fig. 14.6-3. The key is to use the comparator shown in Fig. 14.6-4b to generate the patterns represented by Σ, Δ_x, Δ_y, and $\Delta_{x,y}$. These are found to be (Problem 14.6-5)

$$\Sigma = 2f(\theta_x, \theta_y)\cos[\beta_x(\theta_x)]\cos[\beta_y(\theta_y)] \tag{14.6-16a}$$

$$\Delta_x = j2f(\theta_x, \theta_y)\sin[\beta_x(\theta_x)]\cos[\beta_y(\theta_y)] \tag{14.6-16b}$$

$$\Delta_y = j2f(\theta_x, \theta_y)\cos[\beta_x(\theta_x)]\sin[\beta_y(\theta_y)] \tag{14.6-16c}$$

$$\Delta_{x,y} = -2f(\theta_x, \theta_y)\sin[\beta_x(\theta_x)]\sin[\beta_y(\theta_y)] \tag{14.6-16d}$$

The overall two-coordinate phase-sensing monopulse system is exactly the same as in Fig. 14.5-11 except the patterns denoted by $\Sigma, j\Delta_x$, and $j\Delta_y$ are replaced by Σ, Δ_x, and Δ_y, respectively, as defined by (14.6-16). In other words, the outputs of the comparator of Fig. 14.6-4b are fed directly to the mixers in Fig. 14.5-11 (no factor j). It can be shown that the real error output signals, denoted by $s_{0,x}$ and $s_{0,y}$ for x- and y-coordinate error channels, are

$$s_{0,x} = \frac{V_0^2}{2}\tan[\beta_x(\theta_x)] = \frac{V_0^2}{2}\tan\left[\frac{\pi d_x}{\lambda}\sin(\theta_x)\right]$$

$$\approx \frac{V_0^2}{2}\tan\left[\frac{\pi d_x}{\lambda}\theta_x\right], \qquad |\theta_x| \text{ small} \tag{14.6-17}$$

$$s_{0,y} = \frac{V_0^2}{2}\tan[\beta_y(\theta_y)] = \frac{V_0^2}{2}\tan\left[\frac{\pi d_y}{\lambda}\sin(\theta_y)\right]$$

$$\approx \frac{V_0^2}{2}\tan\left[\frac{\pi d_y}{\lambda}\theta_y\right], \qquad |\theta_y| \text{ small} \tag{14.6-18}$$

The slope constants K_{θ_x} and K_{θ_y} for the x and y error channels, respectively, are found to be

$$K_{\theta_x} = -\frac{V_0^2 \pi d_x}{2\lambda} \tag{14.6-19}$$

$$K_{\theta_y} = -\frac{V_0^2 \pi d_y}{2\lambda} \tag{14.6-20}$$

when $\pi d_x \theta_x/\lambda$ and $\pi d_y \theta_y/\lambda$ are small in (14.6-17) and (14.6-18), respectively.

14.7 RHODES'S OTHER FORMS OF MONOPULSE

Rhodes (1959) categorized a number of monopulse systems according to classes and types. His *types* were defined in Section 14.4 as either amplitude (A) or phase (P).

Monopulse Classifications

Monopulse systems were also categorized into three *classifications*. These were called I, II, and III, according to the form of angle detector used. Thus Rhodes's monopulse types were based on the angle sensor used, while classification is defined by the angle detector.

Rhodes's Eight Types and Classes of Monopulse

Figure 14.7-1 shows eight monopulse configurations according to type and class. The simplified diagrams are derived from Rhodes (1959, fig. 3.12, p. 57), but the notation has been changed to that used here. All adjacent pairs of forms within a given classification (including diagonal pairs in class I) are equivalent (Rhodes, 1959, p. 57).

From Fig. 14.7-1 the reader may recognize the following correspondences between class-type with previously described one-coordinate systems:

$I_P - P$: Fig. 14.6-3

$III - P$: Fig. 14.6-2*b*

$I_A - A$: Fig. 14.5-5

$II - A$: Fig. 14.5-2*b*.

The behaviors of some of the other configurations are left for reader exercises (see Problems 14.7-1–14.7-4).

14.8 GENERAL AMPLITUDE AND PHASE MONOPULSE

Up to this point all of our practical discussions have centered on either amplitude- or phase-sensing systems. In this section we take a brief look at the more general problem of amplitude *and* phase (general) monopulse where the patterns may be arbitrary complex functions of target angles θ_x and θ_y.

We approach the general problem by using the method of Peebles and Berkowitz (1968), who examined the problem of general monopulse when multiple targets were present that were unresolved in all dimensions of range, Doppler frequency, and two spatial angles. However, our interest is mainly on the one-target problem, and the multiple-target discussion is limited to only that which will give a flavor for the more general problem. The two-target case has been applied to multipath reduction (Peebles, 1971).

Pattern Representation by Taylor's Expansions

Our approach centers on representing an arbitrary number of patterns by a two-dimensional Taylor's series expansion for each pattern. The polynomial representations are truncated to preserve terms of degree K or less. The series-represented patterns lead to voltage representations using each pattern. The procedures produce a particular number of equations of degree K and a particular number of target unknown variables. Their solution gives the required number of patterns and value of K necessary for a solution of the target unknown variables, all for a given number (N) of targets.

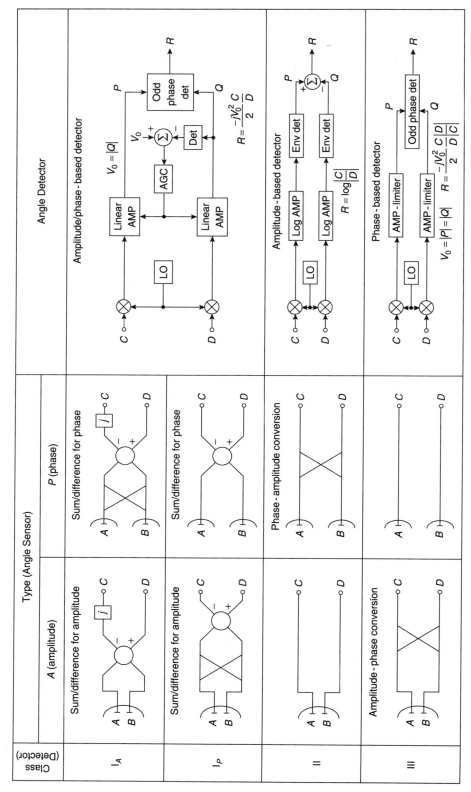

Figure 14.7-1 Rhodes's eight monopulse configurations according to type (the angle sensor used) and class (the angle detector used). (Adapted from **Rhodes**, 1959, with permission).

Let M arbitrary patterns $p_m(\theta_x, \theta_y)$, $m = 1, 2, \ldots, M$, be represented by a Taylor's series expansion with the series truncated to delete terms of degree larger than K. Then (Wylie, 1951)

$$p_m(\theta_x, \theta_y) = \sum_{k=0}^{K} \sum_{l=0}^{k} \alpha_{mkl} \theta_x^{k-l} \theta_y^l, \qquad m = 1, 2, \ldots, M \qquad (14.8\text{-}1)$$

where α_{mkl} is coefficient kl for pattern m given by

$$\alpha_{mkl} = \binom{k}{l} \frac{1}{k!} \frac{\partial^k p_m(\theta_x, \theta_y)}{\partial \theta_x^{k-l} \partial \theta_y^l} \Bigg|_{\theta_x = 0, \theta_y = 0} \qquad (14.8\text{-}2)$$

$$\binom{k}{l} = \frac{k!}{(k-l)!\, l!} \qquad (14.8\text{-}3)$$

If N targets are unresolved, each pattern's voltage response, denoted by e_m, is the sum of N responses, one for each target. These responses all occur at the same delay due to the assumption that the targets are unresolved in range. In general, the receiving system can observe (measure) the voltages e_m at the delay time, but has no other information available except knowledge of its own patterns through knowledge of the coeffieients α_{mkl}. Since there are four unknowns for each target (signal amplitude and phase, and two angles), the system has M complex equations representing the voltages e_m, and $4N$ real unknown parameters to determine.

Denote the unknown angles by θ_{xn} and θ_{yn}, and let A_n be a complex voltage representing the unknown amplitude ($|A_n|$) and phase (angle of A_n), all for target n.

The pattern responses have the form

$$e_m = \sum_{n=1}^{N} A_n p_m(\theta_{xn}, \theta_{yn}) = \sum_{n=1}^{N} A_n \sum_{k=0}^{K} \sum_{l=0}^{k} \alpha_{mkl} \theta_{xn}^{k-l} \theta_{yn}^l,$$

$$= \sum_{k=0}^{K} \sum_{l=0}^{k} \alpha_{mkl} Z_{kl}, \qquad m = 1, 2, \ldots, M \qquad (14.8\text{-}4)$$

where we define

$$Z_{kl} = \sum_{n=1}^{N} A_n \theta_{xn}^{k-l} \theta_{yn}^l \qquad (14.8\text{-}5)$$

The general solution for unknowns first recognizes that (14.8-4) is a set of M complex equations in $(K + 1)(K + 2)/2$ complex unknowns Z_{kl}. Since the coefficients α_{mkl} are known from the known patterns, the Z_{kl} can be assumed known from existing methods for solving M equations linear in $(K + 1)(K + 2)/2$ unknowns. The solutions exist only if

$$M \geq \frac{(K + 1)(K + 2)}{2} \qquad (14.8\text{-}6)$$

For the Z_{kl} known, (14.8-5) represents $(K + 1)(K + 2)$ real equations in $4N$ real unknowns so that a solution for all target parameters requires $(K + 1)(K + 2) \geq 4N$, or

$$M \geq \frac{(K + 1)(K + 2)}{2} \geq 2N \tag{14.8-7}$$

Equation (14.8-7) is an important result. It shows that for general monopulse and $N = 1$ target, a pattern representation of degree $K = 1$ for at least $M = 3$ patterns is required. For $N = 2$ targets we require that $K = 2$ and $M = 6$.

By repeating similar procedures, except for factorable patterns, we obtain (Peebles and Berkowitz, 1968, p. 848)

$$M \geq (K + 1)^2 \geq 2N \tag{14.8-8}$$

so $K = 1$ and $M = 4$ for both one and two targets. For the special case of one target using amplitude patterns, the current results lead exactly to the 4-horn monopulse system already described, as noted below.

General Monopulse for One Target

We outline the complete monopulse solution for $N = 1$, $K = 1$, and $M = 3$. From (14.8-4) in matrix notation

$$[e] = [\alpha][Z] \tag{14.8-9}$$

where

$$[e] = \begin{bmatrix} e_1 \\ e_2 \\ e_3 \end{bmatrix}, \quad [Z] = \begin{bmatrix} Z_{00} \\ Z_{10} \\ Z_{11} \end{bmatrix} \tag{14.8-10}$$

$$[\alpha] = \begin{bmatrix} \alpha_{100} & \alpha_{110} & \alpha_{111} \\ \alpha_{200} & \alpha_{210} & \alpha_{211} \\ \alpha_{300} & \alpha_{310} & \alpha_{311} \end{bmatrix} \tag{14.8-11}$$

On inverting (14.8-9) and using (14.8-5), we have the solutions

$$A = Z_{00} = \alpha^{11}e_1 + \alpha^{12}e_2 + \alpha^{13}e_3 \tag{14.8-12}$$

$$\theta_x = \frac{Z_{10}}{Z_{00}} = \frac{\alpha^{21}e_1 + \alpha^{22}e_2 + \alpha^{23}e_3}{\alpha^{11}e_1 + \alpha^{12}e_2 + \alpha^{13}e_3} \tag{14.8-13}$$

$$\theta_y = \frac{Z_{11}}{Z_{00}} = \frac{\alpha^{31}e_1 + \alpha^{32}e_2 + \alpha^{33}e_3}{\alpha^{11}e_1 + \alpha^{12}e_2 + \alpha^{13}e_3} \tag{14.8-14}$$

where α^{kl} represents element kl in the inverse matrix of $[\alpha]$.

Figure 14.8-1 shows the functional operations that make up the receiving system for general one-target monopulse. The matrix operation with summing junctions is equivalent to a comparator. The dividers (shown with n and d to define numerator and denominator, respectively) represent the angle detectors. Clearly the angle information is in the form of a ratio that also provides normalization.

The processor of Fig. 14.8-1 cannot be reduced much farther without specifying patterns. It is of interest to examine the case of amplitude patterns, which has been called *three-beam monopulse* (Peebles, 1969).

Three-Beam Amplitude Monopulse

For amplitude monopulse the three patterns can have any spatial directions, in theory. As a practical arrangement, assume that they are directed to corners of an equilateral triangle as shown in Fig. 14.8-2. The squint angle from the origin is θ_s for each pattern's maximum. Offsets in the directions of θ_x and θ_y are γ_m and β_m, respectively. The cluster of patterns has a "rotation" or "orientation" angle ϕ, as shown. All patterns are assumed symmetrical about their maxima and are identical except for direction. Thus

$$p_m(\theta_x, \theta_y) = f(r_m) \qquad (14.8\text{-}15)$$

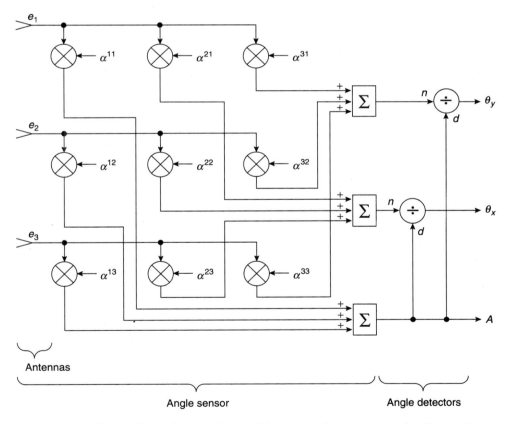

Figure 14.8-1 General form of monopulse receiving system for one target using three patterns.

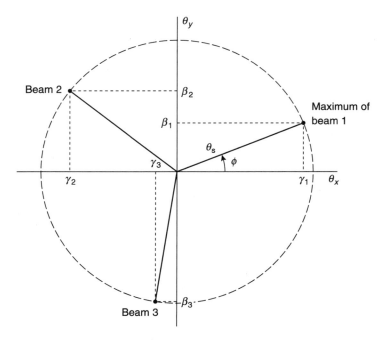

Figure 14.8-2   Directions of the three patterns used in three-beam amplitude monopulse.

where

$$r_m = [(\theta_x - \gamma_m)^2 + (\theta_y - \beta_m)^2]^{1/2}, \qquad m = 1, 2, 3 \qquad (14.8\text{-}16)$$

and $f(r_m)$ defines the basic pattern function.

To show the final form of the three-beam system, define parameters K and G by

$$G = \alpha_{m00} = f(\theta_s) \qquad (14.8\text{-}17)$$

$$K = \frac{\partial f(r_m)}{\partial r_m}\bigg|_{\theta_x = 0, \theta_y = 0} = \frac{\partial f(r_m)}{\partial r_m}\bigg|_{r_m = \theta_s} \qquad (14.8\text{-}18)$$

The matrix $[\alpha]$ can readily be evaluated (Problem 14.8-7), and its inverse found (Problem 14.8-8). From the inverse of (14.8-9) the elements of $[Z]$ are found, and when used in (14.8-5), we obtain

$$Z_{00} = A = \frac{1}{3G}(e_1 + e_2 + e_3) \qquad (14.8\text{-}19)$$

$$Z_{10} = A\theta_x = \frac{2}{3K}\left\{\left[\frac{1}{2}(e_2 + e_3) - e_1\right]\cos(\phi) + \frac{\sqrt{3}}{2}(e_2 - e_3)\sin(\phi)\right\} \qquad (14.8\text{-}20)$$

$$Z_{11} = A\theta_y = \frac{2}{3K}\left\{\left[\frac{1}{2}(e_2 + e_3) - e_1\right]\sin(\phi) - \frac{\sqrt{3}}{2}(e_2 - e_3)\cos(\phi)\right\} \qquad (14.8\text{-}21)$$

Next define Σ, Δ_x, and Δ_y by

$$\Sigma = \frac{1}{\sqrt{3}}(e_1 + e_2 + e_3) \tag{14.8-22}$$

$$\Delta_x = \frac{\sqrt{2}}{\sqrt{3}}\left\{\left[\frac{1}{2}(e_2 + e_3) - e_1\right]\cos(\phi) + \frac{\sqrt{3}}{2}(e_2 - e_3)\sin(\phi)\right\} \tag{14.8-23}$$

$$\Delta_y = \frac{\sqrt{2}}{\sqrt{3}}\left\{\left[\frac{1}{2}(e_2 + e_3) - e_1\right]\sin(\phi) - \frac{\sqrt{3}}{2}(e_2 - e_3)\cos(\phi)\right\} \tag{14.8-24}$$

The parameter solutions become

$$A = \frac{1}{\sqrt{3G}}\Sigma \tag{14.8-25}$$

$$\theta_x = \frac{\sqrt{2G}}{K}\frac{\Delta_x}{\Sigma} \tag{14.8-26}$$

$$\theta_y = \frac{\sqrt{2G}}{K}\frac{\Delta_y}{\Sigma} \tag{14.8-27}$$

Figure 14.8-3 illustrates one method of forming signals Σ, Δ_x, and Δ_y by using the generalized hybrid of Fig. 14.5-3b. The phase shifts $\pm \pi/2$ (the $\pm j$) are present so that the angle detectors can be exactly the same as in Fig. 14.5-11.[20] The final overall system for any orientation angle ϕ of the patterns consists of the comparator of Fig. 14.8-3 and the receiver (angle detectors) of Fig. 14.5-11.

The Cramér-Rao inequality has been used to find the minimum variances of measuring spatial angles, signal amplitude, and phase for the three-beam amplitude monopulse system (Peebles, 1969). We give only the results for angle measurements. Others agree with the bounds given in Chapter 10. It is found that (in current notation)

$$\sigma^2_{\theta_x(\min)} = \frac{\theta_B^2}{K_{m,x}^2 \mathscr{R}_\Sigma} \tag{14.8-28}$$

This result appears similar to (14.3-9). However, there are differences that must be recognized. Here $\mathscr{R}_\Sigma$ is the value of $\mathscr{R}$ produced in the Σ-channel in Fig. 14.8-3, so $\mathscr{R}_\Sigma$ here is directly analogous to $\mathscr{R}_\Sigma$ in (14.3-9). The differences are in θ_B and $K_{m,x}$. Here θ_B is the 3-dB beamwidth of the single pattern $f(r)$, whereas θ_B in (14.3-9) is the 3-dB beamwidth of the full reference channel's pattern, which here would be the sum of the

---

[20] The form of the comparator given in Peebles (1969) has a phase shift of π in the line providing Δ_y. When phase shifts of $\pi/2$ are added to outputs Δ_x and Δ_y there, so that odd phase detectors are used, the comparator of Fig. 14.8-3 results.

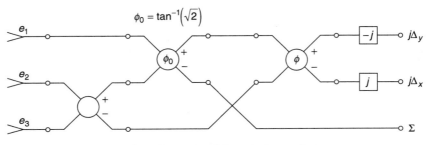

ϕ = orientation angle of triangular beam cluster

Figure 14.8-3 Comparator for use with three-beam amplitude-sensing monopulse.

three patterns. Finally

$$K_{m,x} = \frac{\theta_B K}{\sqrt{2}G} = \frac{\theta_B}{\sqrt{2}f(\theta_s)} \frac{\partial f(r)}{\partial r}\bigg|_{r=\theta_s} = \frac{\theta_B}{\sqrt{2}f(\theta_s)} \frac{\partial f(\theta_s)}{\partial \theta_s} \tag{14.8-29}$$

Which depends on the single basic pattern and not on the full reference and error patterns. For a given pattern $f(r)$, the smallest value of the minimum variance of estimate occurs for the largest value of K and not the largest value of $K_{m,x}$.

Three-Beam Phase Monopulse

In three-beam phase monopulse three patterns are generated by three apertures displaced from the system's phase center (origin) and laying in the x, y plane. Let their phase centers be displaced distances γ_m and β_m (same notation as in the amplitude case but parameters are distances here rather than angles before) along x and y directions, respectively, and each has a distance d_0 from the origin. We again assume that the three aperture phase centers are at corners of an equilateral triangle with cluster orientation ϕ (as in Fig. 14.8-2). Pattern amplitudes are identical and are assumed to be symmetrical as denoted by $f_m(r_m) = f(r_m)$, $m = 1, 2, 3$, and $r_m^2 = \theta_x^2 + \theta_y^2$. Finally we define G and K by

$$G = f(0) \tag{14.8-30}$$

$$K = -j\frac{2\pi d_0 f(0)}{\lambda} \tag{14.8-31}$$

With the preceding definitions it results that the derivations used for three-beam amplitude monopulse again are valid. It is only necessary to use the new values of K and G in previous equations. As a consequence the entire recieving system is the same as for the amplitude system (see Peebles, 1969, p. 55).

General Monopulse for One Target and Factorable Patterns

With one target we require four patterns when they are factorable. Solutions for the target's parameters A, θ_x, and θ_y are known for arbitrary (but factorable) patterns

(Peebles, 1968). For details the reader is referred to the reference. However, it is of interest to note that for amplitude patterns the system is exactly the same as in the 4-horn amplitude-sensing monopulse system of Section 14.5. Perhaps more important is that the analysis showed that the quadrupolar signal, even though it was produced as part of the theory, was not necessary to, or used in, the solutions for the unknown target parameters.

14.9 CONOPULSE—A HYBRID SYSTEM

It is possible to form an angle measurement and tracking system that is a sort of hybrid of the conical scan and monopulse systems. In principle, it has many of the performance properties of monopulse with less complexity, while it has some of the simplicity of conical scan but with less susceptibility to target fluctuations. The system is called *conopulse* which is a contraction for a hybrid *con*ical scan-mon*opulse* system.

Conopulse is similar to conical scan in that it uses beam scanning (rotation) in space. However, *two* patterns are scanned rather than one, and two channels of signal processing are used. Conopulse is also similar to monopulse in that parts of its signal-processing operations produce angle measurements that are independent of target fluctuations. Since angle information depends on both the patterns and their scanning with time, part of the signal processing is similar to conical scan. As a consequence there is a limit on the rate of independent angle measurements in conopulse, as in conical scan.

The fundamentals of conopulse systems are contained in the work of Sakamoto (1975), and Sakamoto and Peebles (1976, pp. 11C-1–11C-3). Details and some history are given in Sakamoto and Peebles (1978). The measurement accuracy of conopulse has been studied by Sakamoto and Peebles (1976, pp. 11D-1–11D-2) and by Peebles and Sakamoto (May 1980, November 1980).

Conopulse Concept

Figure 14.9-1 illustrates the functions needed in an overall conopulse system. The key idea is to scan two independent antenna patterns with time. The scanning action rotates the two patterns together about the reference axis. The pattern maxima are offset from the reference axis by a squint angle θ_q (assumed constant with rotation) and one pattern lags the other by one-half the scan period. Each of the two patterns can be thought of as a conical scan operation. The pattern output voltages both contain sufficient information to measure target angles. However, by proper signal processing the *two* outputs can be used to make angle measurements independent of target fluctuations.[21] The angles being measured are denoted by θ_x and θ_y; these are defined in Fig. 14.1-1.

Analogous to monopulse, the pattern outputs in Fig. 14.9-1 are summed and differenced using a hybrid T to generate reference and error signals, denoted respectively by Σ and Δ. This operation allows the transmitter to utilize the "reference" pattern (proportional to the sum of the two patterns which is maximum on the

---

[21] There are some limitations involved, and they are subsequently defined.

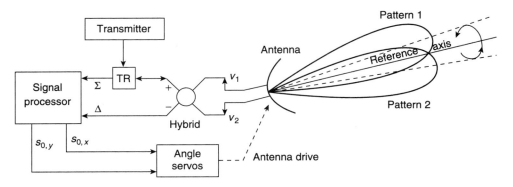

Figure 14.9-1   Basic functions required in a conopulse angle-tracking system.

reference axis) for transmission. Signals Σ and Δ are used in a signal processor (to be defined) to generate error signals, denoted by $s_{0,x}$ and $s_{0,y}$, that are proportional to θ_x and θ_y, respectively. The error signals are used by the angle track circuits to force the reference axis to follow the target in space. The action is to move the reference axis such that $s_{0,x}$ and $s_{0,y}$ are maintained as near zero as noise and other errors allow.

For a mechanically steered antenna, such as a paraboloid or lens, the angle track circuits typically contain motors and gearing to physically move the antenna, thereby moving the reference axis (the same as the boresight axis in this case). For an electronically steered antenna (phased array), the angle track circuits are all electronic. They would then generate the necessary beam-steering drive commands to steer the reference axis.

Realization of Two-Pattern Scanning

One disadvantage of conopulse is the difficulty of realizing the scanning of two patterns simultaneously in space with time. One possible method is illustrated in Fig. 14.9-2. Here two feed radiators (horns or their equivalents) are offset from the focus of a paraboloid to produce the two required squinted patterns. The radiators are physically attached at point C, so both rotate together about the antenna's axis of symmetry. Rotation is made possible by two rotary joints having the same axis of rotation. Rotary joint B is fed by microwave line B along the line of symmetry. Rotary joint A is fed from behind the feed structure by microwave line A which may also serve as a fixed support for the feed. Another support strut is shown dashed; it is physically connected to line A behind the feed. A motor behind rotary joint A drives point C through gears to cause the two radiators to rotate.

In another, perhaps more important, version of Fig. 14.9-2, rotary joint B is moved behind the reflector. In this version line B in front of the reflector rotates with the radiators. The advantage is that the drive motor and gears can be behind the reflector which reduces the focus-located mass and produces a better mechanical design.

At this writing, the structure of Fig. 14.9-2 has not been built or tested. Some studies by O'Brien (1995) have been done on the use of three types of conical horn and a circular dielectric rod as radiators. Although the work is not conclusive, it appears that a circular dielectric rod (James, 1972) can give antenna patterns with good

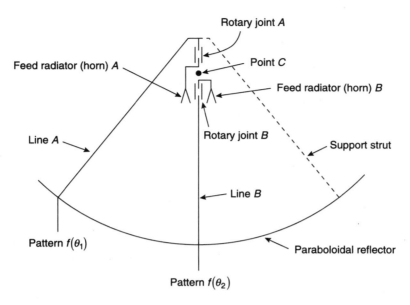

Figure 14.9-2 Feed structure for simultaneous scanning of two offset patterns needed in a conopulse system.

symmetry, narrow beamwidth for a given aperture size, and other reasonably good antenna properties. However, the radiator may have limitations at higher system power levels.

Another method (see Barton, 1988, p. 419) uses an amplitude comparison monopulse antenna with a resolver (realized by the generalized hybrid of Fig. 14.5-3*b* with ϕ_0.varying with time at the scan rate) to generate the equivalent of two scanning patterns.

The conopulse system of Fig. 14.9-1 can be implemented in two basic forms. Each form uses a combination of monopulse and conical scan signal-processing techniques. We name them MOCO, when monopulse methods are first performed on Σ and Δ followed by conical scan techniques, and COMO, when the reverse order of processing Σ and Δ occurs. Due to its better performance the MOCO system would normally be preferred over the COMO variety. Because of this fact we describe only the MOCO system and refer the reader to the literature for a discussion of the COMO system.

MOCO System

The basic elements of a MOCO system are given in Fig. 14.9-3. Subtleties of a realistic system, such as mixing, gating, and filtering, are omitted for clarity of discussion, since they do not alter the basic concepts to be presented. The ratio that generates $R = \Delta/\Sigma$ is performed first. The ratio is equivalent to normalization in a monopulse system (e.g., see Section 14.5). It is in this stage that target fluctuations are removed.

The signal R is a train of video pulses occurring at the pulse rate. The train is periodically amplitude modulated due to scanning of the antenna patterns. In

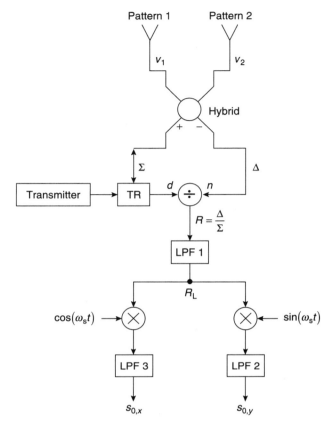

Figure 14.9-3   Functional block diagram of a MOCO form of conopulse system. (Adapted from Sakamoto and Peebles, © 1978 IEEE, with permission.)

practice, the pulse rate is much faster than the rate of amplitude modulation, such that a low-pass filter can be used to smooth out the pulsed waveform. The output, denoted by R_L, is a baseband video waveform proportional to the modulation due to pattern scanning. R_L contains all the information needed to measure angles θ_x and θ_y.

Since scanning is periodic, R_L is closely modeled as periodic. R_L is analogous to the scan-modulated video in a conical scan receiver, and the usual in-phase and quadrature-phase conical scan angle detectors can be used to generate error signals $s_{0,x}$ and $s_{0,y}$. These detectors can be viewed as devices that produce voltages proportional to the cosine and sine coefficients of the fundamental scan frequency term in the Fourier series representation of R_L.

Measurement Error Variance

As in monopulse, the variance of angle error measurements can be found from (14.2-2) when applied to the MOCO system. However, in conopulse the slope constant is the same for measurements of both θ_x and θ_y. This fact means the measurement variances for θ_x and θ_y are equal. The calculation of error variance is involved and will not be developed. The reader is referred to Peebles and Sakamoto (November 1980) for details. Assume that the system's input noise is white, that the target has constant cross section, let the output lowpass filters 2 and 3 in Fig. 14.9-3 have bandwidth W_L (rad/s) which represents the closed-loop bandwidth of any angle-tracking loops, and define a slope constant k_c by

$$k_c = \frac{\theta_B}{f(\theta_q)} \frac{\partial f(\theta_q)}{\partial \theta_q} \tag{14.9-1}$$

Here θ_B is the 3-dB beamwidth of the one-way voltage sum pattern taken along a line of angles between the peaks of the two scanning patterns with the line passing through the origin. The basic form of the scanning patterns is denoted by $f(\theta)$, where θ is the angle from the direction of any one pattern's maximum to the direction of the target. Then it can be shown that

$$\sigma_{\theta_x}^2 = \sigma_{\theta_y}^2 = \sigma_{\theta(\min)}^2 R_A \left[1 + \frac{R_B}{(S_1/N_i)} \right]$$

$$= \sigma_{\theta(\min)}^2 R_A \left[1 + \frac{2R_B}{\mathscr{R}_p} \right] \tag{14.9-2}$$

where

$$\sigma_{\theta(\min)}^2 = \frac{\theta_B^2}{k_c^2 (S_i/N_i)} = \frac{2\theta_B^2}{k_c^2 \mathscr{R}} \tag{14.9-3}$$

$$\mathscr{R}_p = 2 \left(\frac{S_1}{N_i} \right) \tag{14.9-4}$$

$$\mathscr{R} = 2 \left(\frac{S_i}{N_i} \right) \tag{14.9-5}$$

$$\left(\frac{S_i}{N_i} \right) = \frac{\pi}{W_L T_R} \left(\frac{S_1}{N_i} \right) \tag{14.9-6}$$

R_A and R_B are constants that depend on the range gate's width, τ_g, and the form of the received and matched filtered signal. The general equations for R_A and R_B are somewhat complicated but are given by Peebles and Sakamoto (November 1980). For constant-frequency rectangular transmitted pulses of duration τ_p, the parameters R_A and R_B are found to be

$$R_A = \begin{cases} \dfrac{6(120 - 100x + 30x^2 - 3x^3)}{5(144 - 144x + 60x^2 - 12x^3 + x^4)}, & 0 \le x \le 1 \\[2mm] \dfrac{6(-22 + 80x + 10x^2 - 30x^3 + 10x^4 - x^5)}{5x^2(144 - 144x + 60x^2 - 12x^3 + x^4)}, & 1 < x \le 2 \\[2mm] \dfrac{99}{80}, & 2 < x \end{cases} \tag{14.9-7}$$

$$R_B = \begin{cases} \dfrac{20(6 - 4x + x^2)}{120 - 100x + 30x^2 - 3x^3}, & 0 \le x \le 1 \\[3mm] \dfrac{20(-1 + 4x)}{-22 + 80x + 10x^2 - 30x^3 + 10x^4 - x^5}, & 1 < x \le 2 \\[3mm] \dfrac{10}{33}(-1 + 4x), & 2 < x \end{cases} \qquad (14.9\text{-}8)$$

where we define

$$x = \frac{\tau_g}{\tau_p} \qquad (14.9\text{-}9)$$

In the above (S_1/N_i) is the single-pulse signal-to-noise power ratio at the output of the sum channel's matched filter, and (S_i/N_i) is the total signal-to-noise power ratio of (14.9-6).

The variance $\sigma^2_{\theta(\min)}$ of (14.9-3) is the smallest possible measurement variance, as found from the Cramér-Rao bound (Peebles and Sakamoto, May 1980). If compared to monopulse under reasonable conditions (same values of k_c, $\mathscr{R}$, and θ_B), the measurement variance of conopulse is twice that of monopulse, so its performance is not as good as monopulse. It can be shown that $R_A \ge 1$, so the presence of the factor R_A in (14.9-2) indicates that a practical MOCO system has larger variance than the theoretical minimum, even if (S_1/N_i) is large enough for the term involving R_B to be negligible. Presence of the term in R_B only serves to increase the variance.

Example 14.9-1   To attach some numerical significance to (14.9-2), we find the performance degradation relative to ideal when a constant-frequency rectangular pulse is used in a system having $\tau_g = 2.2\tau_p$ and $(S_1/N_i) = 8.6$. From (14.9-7) and (14.9-8), $R_A = 99/80 = 1.2375$, $R_B = (10/33)[4(2.2) - 1] = 78/33 = 2.3636$. Thus

$$\frac{\sigma^2_{\theta_x}}{\sigma^2_{\theta(\min)}} = R_A\left[1 + \frac{R_B}{(S_1/N_i)}\right] = 1.2375\left[1 + \left(\frac{2.3636}{8.6}\right)\right] = 1.5776$$

(or 1.98 dB) is the degradation from ideal. Note that as (S_1/N_i) becomes very large, the degradation equals $R_A = 1.2375$, or 0.93 dB.

PROBLEMS

14.1-1 The reference axis in a radar has azimuth angle $A_b = \pi/8$ rad and elevation angle $E_b = \pi/6$ rad. For a target located at azimuth and elevation angles $A_T = 31\pi/180$ rad and $E_T = 6\pi/45$ rad, respectively, determine the exact angles θ_x and θ_y. You may wish to use some expressions from Chapter 1.

14.1-2 Work Problem 14.1-1 except for $E_b = 0$ and $E_T = -\pi/30$ rad.

14.1-3 (a) Define $|\theta_x|$ as no longer small when $\sin(\theta_x)$ in (14.1-1) departs from its approximation $\sin(\theta_x) \approx \theta_x$ by 10%. If $\phi = 0$, what range of values of θ

correspond to small $|\theta_x|$? (b) What is the effect on your answer if ϕ is not zero?

14.1-4 Use (14.1-10) and show that $|\mathbf{F}_\Delta(\theta, \phi)|^2 = G_D(\theta, \phi)$ where $G_D(\theta, \phi)$ is the directive gain of the pattern.

14.1-5 Show that (14.1-12) is true.

14.2-1 A target is offset from the reference axis by $\theta_x = \pi/150$ rad. It produces and output signal voltage $s_0(t_0) = 1.55$ V. If the noise power at the output is $N_0 = E[N_0^2(t_0)] = 0.15$, what standard deviation of error exists in the measurement of θ_x?

14.2-2 Work Problem 14.2-1 except assume that $\theta_x = -2\pi/325$ rad, $s_0(t_0) = 1.4$ V, and $N_0 = 0.08$.

14.2-3 (a) Replace $\sqrt{D}$ by unity in (14.2-6) and by D in (14.2-7), and show that (14.2-9) is unchanged. (b) Repeat part (a) except use D in (14.2-6) and unity in (14.2-7).

14.2-4 Show that, except for the factor $\exp(-j\phi_D)$, (14.2-9) is a matched filter for a transmitted signal $\psi(t)$ that is delayed by τ_R and Doppler-shifted by ω_d.

14.2-5 For two vectors $\boldsymbol{\alpha} = \alpha_1\hat{a}_1 + \alpha_2\hat{a}_2$ and $\mathbf{F} = F_1\hat{a}_1 + F_2\hat{a}_2$, where α_1, α_2, F_1, and F_2 are complex vector components and $\hat{a}_1$ and $\hat{a}_2$ are orthogonal unit vectors, show that $|\boldsymbol{\alpha} \cdot \mathbf{F}|^2 = |\boldsymbol{\alpha}|^2 |\mathbf{F}|^2$ if the two vectors $\boldsymbol{\alpha}$ and $\mathbf{F}$ spatially align and one is the conjugate of the other, that is, if $(\alpha_2/\alpha_1) = (F_2/F_1)^*$.

14.2-6 Use the result of Problem 14.2-5 to show that (14.2-10) is true.

14.2-7 Use the result of Problem 14.2-5 to show that (14.2-13) is true.

14.2-8 In terms of moments of a function, what interpretation can be given to (14.2-14)?

14.2-9 Find $L_{0,x}$ for a circular aperture of radius R centered on the origin of the x, y plane.

14.2-10 Find $L_{0,x}$ for a rectangular aperture of lengths W (in x direction) and H (in y direction) when centered on the origin in the x, y plane.

14.2-11 For the rectangular aperture of Problem 14.2-10, find the optimum receiving antenna voltage pattern for measuring θ_x. Assume that $|\theta_x|$ is small.

★14.2-12 Use (14.2-19) with (14.1-12) and show that the pattern $\mathbf{F}_\Delta(\theta, \phi)$ of the optimum antenna is given by

$$\mathbf{F}_\Delta(\theta, \phi) = \frac{-\lambda}{L_{0,x}} \frac{\partial \mathbf{F}_\Sigma(\theta, \phi)}{\partial \theta_x}$$

where we define

$$\mathbf{F}_\Sigma(\theta, \phi) = \left[\frac{4\pi}{\lambda^2}\right]^{1/2} \frac{\int_A \int \mathbf{C} e^{jk\theta_x\xi + jk\theta_y\zeta}\, d\xi\, d\zeta}{[\int_A \int |\mathbf{C}|^2\, d\xi\, d\zeta]^{1/2}}$$

Note that the optimum pattern is proportional to the derivative of the pattern $\mathbf{F}_\Sigma(\theta, \phi)$. Discuss the significance of $\mathbf{F}_\Sigma(\theta, \phi)$, including its illumination function and polarization.

14.2-13 Use the results of Problem 14.2-12 to derive the optimum error pattern $\mathbf{F}_\Delta(\theta, \phi)$ for a circular aperture of radius R_0 centered on the origin. [*Hint:* Solve for $\mathbf{F}_\Sigma(\theta, \phi)$ and differentiate; then solve for $L_{0,x}$.]

★14.2-14 Schwarz's inequality can be applied to multiple as well as single integrals (see Korn and Korn, 1961, p. 424). Thus for double integrals

$$\left| \int_A \int g(\xi, \zeta) E(\xi, \zeta) d\xi d\zeta \right|^2 \leq \int_A \int |g(\xi, \zeta)|^2 \, d\xi d\zeta \int_A \int |E(\xi, \zeta)|^2 \, d\xi d\zeta \quad (1)$$

where $g(\xi, \zeta)$ and $E(\xi, \zeta)$ are complex scalar functions of ξ and ζ, and A is the area of integration in the ξ, ζ plane. The equality in (1) will be true if $E(\xi, \zeta) = Cg^*(\xi, \zeta)$, where C is a nonzero arbitrary real constant. Show that if $E(\xi, \zeta)$ is replaced by a complex *vector* function $\mathbf{E}(\xi, \zeta)$, that the requirement for equality in (1) is that $\mathbf{E}(\xi, \zeta) = C g^* (\xi, \zeta)$, where $\mathbf{C}$ is a real vector with constant components.

14.2-15 Show that for use of the optimum receiving antenna pattern and optimum filter, the variance of error in the measurement of θ_x is given by (14.2-22) for white noise, as derived from (14.2-18).

14.2-16 Show that (14.2-26) is true.

14.2-17 A radar uses a circular antenna of radius 15λ that is optimum for measurement of θ_x. What value of $\mathscr{R}_{po}$ is needed if the rms error in measurement is to be 0.2 degree?

14.2-18 Work Problem 14.2-17 except assume a square aperture with a length of 30λ on a side. The aperture is centered on the origin.

14.3-1 For a rectangular aperture of width W (in x direction) and height H (in y direction) centered on the origin, assume that illumination is factorable and only in the x direction according to

$$\mathbf{E}_\Delta(\xi, \zeta) = \hat{a}_x E_{\Delta, x}(\xi) E_{\Delta, y}(\zeta) \operatorname{rect}\left(\frac{\xi}{W}\right) \operatorname{rect}\left(\frac{\zeta}{H}\right) \quad (1)$$

If $E_{\Delta, y}(\zeta) = E_{y0}$, a positive constant, show that

$$K_x = \frac{2\pi}{\lambda \sqrt{W}} \frac{|\int_{-W/2}^{W/2} \xi E_{\Delta, x}(\xi) d\xi|}{[\int_{-W/2}^{W/2} |E_{\Delta, x}(\xi)|^2 d\xi]^{1/2}}$$

14.3-2 Generalize Problem 14.3-1 by showing that

$$K_x = \frac{2\pi}{\lambda \sqrt{W}} \frac{|\int_{-W/2}^{W/2} \xi E_{\Delta, x}(\xi) d\xi|}{[\int_{-W/2}^{W/2} |E_{\Delta, x}(\xi)|^2 d\xi]^{1/2}} \sqrt{\rho_y}$$

where

$$\rho_y = \frac{|\int_{-H/2}^{H/2} E_{\Delta,y}(\zeta)d\zeta|^2}{H \int_{-H/2}^{H/2} |E_{\Delta,y}(\zeta)|^2 d\zeta}$$

is aperture efficiency for the y coordinate and $E_{\Delta,y}(\zeta)$ is arbitrary.

14.3-3 A rectangular aperture has an arbitrary factorable illumination as defined in Problem 14.3-1. Define aperture efficiency ρ_y for the y coordinate as in Problem 14.3-2 and use a similar definition for the x coordinate. Then show that

$$K_x = \frac{2\pi}{\lambda} \sqrt{\rho_x \rho_y} \frac{|\int_{-W/2}^{W/2} \xi E_{\Delta,x}(\xi)d\xi|}{|\int_{-W/2}^{W/2} E_{\Delta,x}(\xi)d\xi|}$$

14.3-4 Show that η_{θ_x} of (14.3-6) is given by

$$\eta_{\theta_x}^2 = \frac{|\int_A \int \xi \mathbf{E}_\Delta(\xi, \zeta)d\xi d\zeta|^2}{\int_A \int |\mathbf{E}_\Delta(\xi, \zeta)|^2 d\xi d\zeta \int_A \int \xi^2 d\xi d\zeta}$$

14.3-5 An optimum rectangular aperture with factorable illumination has constant illumination in the y direction and is a linear function $E_{\Delta,x}(\xi) = \xi E_{x0}$ in the x direction. Use results from Problem 14.3-1 and show that $K_x = K_{0,x} = \pi W/[\lambda(3)^{1/2}]$.

14.3-6 A rectangular aperture as defined in Problem 14.3-1 has the illumination function

$$\mathbf{E}_\Delta(\xi, \zeta) = \hat{a}_x E_{x0} E_{y0} \sin\left(\frac{\pi\xi}{W}\right) \mathrm{rect}\left(\frac{\xi}{W}\right) \mathrm{rect}\left(\frac{\zeta}{H}\right)$$

Where E_{x0} and E_{y0} are real constants. (a) Find K_x. (b) Find η_{θ_x} and compute $20\log(\eta_{\theta_x})$, which represents the performance loss compared to ideal.

14.3-7 Work Problem 14.3-6 except assume that

$$\mathbf{E}_\Delta(\xi, \zeta) = -\hat{a}_x E_{x0} E_{y0} \xi^3 \mathrm{rect}\left(\frac{\xi}{W}\right) \mathrm{rect}\left(\frac{\zeta}{H}\right)$$

14.3-8 Work Problem 14.3-6 except assume that

$$\mathbf{E}_\Delta(\xi, \zeta) = \hat{a}_x E_{x0} E_{y0} \left\{ \mathrm{rect}\left[\frac{\xi - (W/4)}{(W/2)}\right] - \mathrm{rect}\left[\frac{\xi + (W/4)}{(W/2)}\right] \right\}$$

14.3-9 Work Problem 14.3-6 except assume that

$$\mathbf{E}_\Delta(\xi, \zeta) = \hat{a}_x E_{x0} E_{y0} \sin\left(\frac{2\pi\xi}{W}\right) \mathrm{rect}\left(\frac{\xi}{W}\right) \mathrm{rect}\left(\frac{\zeta}{H}\right)$$

14.3-10 Work Problem 14.3-6 except assume that

$$
\mathbf{E_\Delta}(\xi, \zeta) = \begin{cases}
-\hat{a}_x E_{x0} E_{y0} \left(2 + \dfrac{4\xi}{W}\right) \mathrm{rect}\left(\dfrac{\zeta}{H}\right), & \dfrac{-W}{2} \le \xi < \dfrac{-W}{4} \\[2ex]
\hat{a}_x E_{x0} E_{y0} \dfrac{4\xi}{W} \mathrm{rect}\left(\dfrac{\zeta}{H}\right), & \dfrac{-W}{4} \le \xi < \dfrac{W}{4} \\[2ex]
\hat{a}_x E_{x0} E_{y0} \left(2 - \dfrac{4\xi}{W}\right) \mathrm{rect}\left(\dfrac{\zeta}{H}\right), & \dfrac{W}{4} \le \xi < \dfrac{W}{2} \\[2ex]
0, & \dfrac{W}{2} \le \xi < \dfrac{-W}{2}
\end{cases}
$$

14.3-11 Work Problem 14.3-6 except assume that

$$
\mathbf{E_\Delta}(\xi, \zeta) = \hat{a}_x E_{x0} E_{y0} \, \xi \cos\left(\frac{\pi\xi}{W}\right) \mathrm{rect}\left(\frac{\xi}{W}\right) \mathrm{rect}\left(\frac{\zeta}{H}\right)
$$

★14.3-12 Use the second right-side form of (14.3-8) to show that an alternative form for $k_{m,x}$ is

$$
k_{m,x} = \theta_\mathrm{B} \frac{2\pi}{\lambda} \frac{|\int_A \int \xi \mathbf{E_\Delta}(\xi, \zeta)\, d\xi d\zeta|}{|\int_A \int \mathbf{E_\Sigma}(\xi, \zeta)\, d\xi d\zeta|}
$$

Where $\mathbf{E_\Sigma}(\xi, \zeta)$ is the vector aperture illumination function associated with the pattern having directivity $G_{D\Sigma}$. [*Hint*: Assume that both distributions $\mathbf{E_\Delta}(\xi, \zeta)$ and $\mathbf{E_\Sigma}(\xi, \zeta)$ correspond to the same radiated power, if used for radiation, when each is excited by the same incident power.]

14.3-13 A radar uses a mismatched antenna for which $K_x = 37.0$; its angle efficiency factor is 0.816. If a target were to produce $\mathscr{R}_{p0} = 14$ on reception by a uniformly illuminated aperture, what value of standard deviation of error in estimation of θ_x is available in the system? If the antenna were ideal (no mismatch), what would the standard deviation be?

14.3-14 A radar measures θ_x using an error and a reference channel. The 3-dB beamwidth of the reference pattern is 3.4 degrees. What is the value of $k_{m,x}$ if the standard deviation of measurement error is 0.4 degree when $\mathscr{R}_\Sigma = 18$ in the reference channel?

14.3-15 In Example 14.3-1 show that (2) reduces to (3).

14.3-16 In Example 14.3-1 show that (4) reduces to (5).

14.3-17 Find K_x for the antenna of Example 14.3-1.

14.3-18 Find η_{θ_x} for the antenna of Example 14.3-1.

★14.3-19 A radar uses an origin-centered rectangular aperture of width D_x (in x direction) and height D_y (in y direction). The reference pattern's vector illumination function is

$$\mathbf{E}_\Sigma(\xi, \zeta) = \hat{a}_x E_0 \cos\left(\frac{\pi\xi}{D_x}\right) E(\zeta) \operatorname{rect}\left(\frac{\xi}{D_x}\right) \operatorname{rect}\left(\frac{\zeta}{D_y}\right)$$

where E_0 is a real constant and $E(\zeta)$ is an arbitrary function of ζ. Use (14.1-12) to show that the pattern is

$$\mathbf{F}_\Sigma(\theta, \phi) = \hat{a}_x \left[\frac{32 D_x D_y}{\lambda^2 \pi}\right]^{1/2} \left[\frac{(\pi/2)^2 \cos(w)}{(\pi/2)^2 - w^2}\right] F_g(\theta_y)$$

where $w = k\theta_x D_x/2 = \pi\theta_x D_x/\lambda$ and $F_g(\theta_y)$ is defined in Example 14.3-1.

★**14.3-20** For the antenna defined in Problem 14.3-19, the error pattern is to be given by (14.3-16). Find $L_{\Sigma, x}$ and $\mathbf{F}_\Delta(\theta, \phi)$. Plot $\mathbf{F}_\Delta(\theta, \phi)$ and $\mathbf{F}_\Sigma(\theta, \phi)$ normalized (divided) by $\hat{a}_x F_g(\theta_y)[32 D_x D_y/(\lambda^2 \pi)]^{1/2}$ versus $-9.5 \leq w \leq 9.5$.

14.3-21 Use (1) and (2) of Example 14.3-2 with (14.3-19) to show that P_x is given by (3).

14.3-22 A radar uses the antenna defined in Problem 14.3-19 for the reference pattern. The error pattern's distribution is chosen to give a linear pattern ratio according to (14.3-19). (a) Find P_x from (14.3-27). (b) Use P_x and (14.3-19) to find $\mathbf{F}_\Delta(\theta, \phi)$. (c) Plot $\mathbf{F}_\Sigma(\theta, \phi)$ and $\mathbf{F}_\Delta(\theta, \phi)$ normalized (divided) by $\hat{a}_x F_g(\theta_y)[32 D_x D_y/(\pi\lambda^2)]^{1/2}$ for $-9.5 \leq w \leq 9.5$, where $w = k\theta_x D_x/2$.

★**14.3-23** A rectangular aperture of width D_x (in the x direction) and height D_y (in the y direction) is origin centered. A reference pattern's illumination function is

$$\mathbf{E}_\Sigma(\xi, \zeta) = \hat{a}_x E_0 \cos^2\left(\frac{\pi\xi}{D_x}\right) \operatorname{rect}\left(\frac{\xi}{D_x}\right) E(\zeta) \operatorname{rect}\left(\frac{\zeta}{D_y}\right)$$

where E_0 is a real constant and $E(\zeta)$ is an arbitrary function. Use (14.1-12) to show that the reference pattern is

$$\mathbf{F}_\Sigma(\theta, \phi) = \hat{a}_x \left[\frac{8\pi D_x D_y}{3\lambda^2}\right]^{1/2} \left[\frac{\pi^2 \sin(w)}{w(\pi^2 - w^2)}\right] F_g(\theta_y)$$

where $w = \pi\theta_x D_x/\lambda$ and $F_g(\theta_y)$ is defined in Example 14.3-1.

14.3-24 Work Problem 14.3-22 except assume the antenna of Problem 14.3-23 and use the normalization factor a $\hat{a}_x[8\pi D_x D_y/(3\lambda^2)]^{1/2} F_g(\theta_y)$.

14.3-25 Use (14.3-18) to determine K_x for the reference pattern of Problem 14.3-19. This is an example of an error pattern that is proportional to the derivative of the reference pattern.

14.3-26 Use (14.3-28) to determine K_x for the reference pattern of Problem 14.3-19. This is an example of an error pattern distribution that is proportional to the derivative of the reference pattern's distribution.

14.3-27 Work Problem 14.3-26 except use the pattern of Problem 14.3-23.

14.4-1 If an arbitrary pattern $p(\theta_x)$ is to equal the sum of an odd part and an even part, show that these parts are given by (14.4-5) and (14.4-6), respectively.

14.5-1 Discuss how the signals of (14.5-2) and (14.5-3) would change if the "gains" through the mixer-logarithmic amplifiers were K_1 from signal A to P and K_2 from B to Q.

14.5-2 The function K_x in (14.5-4) assumed that the logarithm in (14.5-3) was natural (to base e). For a logarithm to any base b, determine K_x and observe the effect.

14.5-3 Add a phase shift $-j$ after B and prior to the coupler in Fig. 14.5-3c and also after the coupler but before C. Show that the modified short slot coupler is equivalent to a hybrid T by finding outputs C and D.

14.5-4 A practical "even" real phase detector can be modeled as a real analog product device followed by a low-pass filter to remove bandpass components from the output. (a) Define inputs $s_1(t) = a_1 a(t) \cos[\omega_0 t + \theta(t) + \phi_1]$ and $s_2(t) = a_2 a(t) \cos[\omega_0 t + \theta(t) + \phi_2]$, where $a_1, a_2, \omega_0, \phi_1,$ and ϕ_2 are real constants and $a(t)$ and $\theta(t)$ are real functions of t. Determine the real response $s_0(t)$. (b) If $s_1(t)$ and $s_2(t)$ are narrowband and are replaced by their analytic signals, denoted by P and Q, respectively, and the model of Fig. 14.5-6a is used, determine how R and $\text{Re}(R)$ are related to $s_0(t)$ of part (a).

14.5-5 A practical "odd" phase detector can be created from the even detector of Problem 14.5-4 by adding a phase shift of $-\pi/2$ between input $s_1(t)$ and the analog product. (a) For $s_1(t)$ and $s_2(t)$ as given in Problem 14.5-4, find $s_0(t)$. (b) Replace the narrowband signals with their analytic representations, and find R and $\text{Re}(R)$ for the model of Fig. 14.5-6b. How are these related to $s_0(t)$ in part (a)?

14.5-6 Treat A through L in Fig. 14.5-10 as voltages, and find expressions for voltages Σ, Δ_x, and Δ_y in terms of A through L.

14.5-7 A monopulse system is implemented as in Fig. 14.5-11. The antenna is optimum for measurement of θ_x and has an rms length $L_{0,x} = 41\lambda$. On reception $\mathcal{R}_{p0} = 60$ when a constant-frequency rectangular pulse is used. The receiver's range gate is twice the transmitted pulse length. (a) What are R_A and R_B? (b) What is the standard deviation of measurement error?

14.5-8 Work Problem 14.5-7 except assume that $\mathcal{R}_{p0} = 20$ and the phase detectors saturate (limit) on their reference channel inputs. Assume that R_B is the same as in Problem 14.5-7.

14.6-1 A phase-sensing monopulse radar uses two identical rectangular apertures on reception. Their widths in x and y directions are D_x and D_y, respectively. Their basic pattern (if origin centered) is known to be given by

$$f(\theta_x, \theta_y) = \left[\frac{4\pi D_x D_y}{\lambda^2}\right]^{1/2} \text{Sa}\left(\frac{\pi D_x \theta_x}{\lambda}\right) \text{Sa}\left(\frac{\pi D_y \theta_y}{\lambda}\right)$$

The receiving system uses two such antennas displaced as in Fig. 14.6-1 with $d_x = D_x$, and a target on the reference axis produces $\mathcal{R}_{p0} = 18$ in each of the signal channels of Fig. 14.6-2b. What will be the value of $\mathcal{R}_{p0}$ for a target with $\theta_y = 0$ and θ_x equal to the value that produces the first maximum value of s_0?

14.6-2 The system of Problem 14.6-1 must be able to measure θ_x even though θ_y can vary. Assume that the largest that θ_y can be is three times the largest that θ_x can be, with the latter being the value that gives the first maximum of s_0 (due to θ_x variations alone). How should D_y be related to D_x if the reduction in $\mathcal{R}_{p0}$ *due to variations in* θ_y should not exceed 1.0 dB when θ_x is at its largest value?

14.6-3 Show that (14.6-10) defines the complex response R of Fig. 14.6-3.

14.6-4 For the system of Fig. 14.6-4 show that (14.6-12) defines the phase shifts associated with the four apertures due to a wave arriving from direction (θ, ϕ).

14.6-5 Let Σ, Δ_x, Δ_y, and $\Delta_{x,y}$ in Fig. 14.6-4b represent patterns. Similarly let A, B, C, and D represent the respective patterns of (14.6-15). Show that the former patterns are defined by (14.6-16).

14.7-1 (a) For the class I_A, type P, monopulse system in Fig. 14.7-1, find signals C, D, and R in terms of signals A and B. (b) For P-type monopulse signals A and B are proportional to the patterns according to

$$A = Kf(\theta_x)e^{j\beta(\theta_x)}$$
$$B = Kf(\theta_x)e^{-j\beta(\theta_x)}$$

where K is a constant, $f(\theta_x)$ is the amplitude, and $\beta(\theta_x)$ is the phase of the patterns. Use these results to put R in terms of $f(\theta_x)$ and $\beta(\theta_x)$.

14.7-2 (a) For the class I_P, type A, monopulse system in Fig. 14.7-1, find signals C, D, and R in terms of signals A and B. (b) For A-type monopulse signals A and B are proportional to the patterns according to

$$A = Kf(\theta_x)e^{j\beta(\theta_x)}$$
$$B = Kf(-\theta_x)e^{j\beta(\theta_x)}$$

where K is a constant, $f(\theta_x)$ and $f(-\theta_x)$ are amplitudes, and $\beta(\theta_x)$ is the phase of the patterns. Use these results to put R in terms of $f(\theta_x)$, $f(-\theta_x)$, and $\beta(\theta_x)$.

14.7-3 Work Problem 14.7-1 except for the class II, type P system.

14.7-4 Work Problem 14.7-2 except for the class III, type A system.

14.8-1 Explain why (14.8-4) represents $(K + 1)(K + 2)/2$ complex unknowns Z_{kl}.

14.8-2 By using the work in Section 14.8, determine how many beams are needed to synthesize a general monopulse system to measure amplitude, phase, and two spatial angles of three unresolved target signals. What degree of polynomial representation of beams is necessary?

14.8-3 Work Problem 14.8-2 except for four targets.

14.8-4 Work Problem 14.8-2 except assume that the beams are factorable.

14.8-5 Work Problem 14.8-2 except for four targets and factorable beams.

14.8-6 Three one-way voltage patterns have the general representation of (14.4-1):

$$p_m(\theta_x, \theta_y) = f_m(\theta_x, \theta_y) \, e^{j\beta_m(\theta_x, \theta_y)}, \qquad m = 1, 2, 3$$

where $f_m(\theta_x, \theta_y)$ is the amplitude and $\beta_m(\theta_x, \theta_y)$ is the phase of the pattern. Find the most general representation of the elements of the matrix $[\alpha]$ of (14.8-11).

14.8-7 For three-beam amplitude monopulse, where (14.8-15) through (14.8-18) apply, show that the matrix $[\alpha]$ defined in (14.8-11) is

$$[\alpha] = \begin{bmatrix} G & -K\cos(\phi) & -K\sin(\phi) \\ G & K\cos[\phi - (\pi/3)] & K\sin[\phi - (\pi/3)] \\ G & K\cos[\phi + (\pi/3)] & K\sin[\phi + (\pi/3)] \end{bmatrix}$$

14.8-8 Find the inverse of the matrix $[\alpha]$ of Problem 14.8-7. Use the inverse to show that (14.8-19) through (14.8-21) are true.

14.8-9 For the comparator of Fig. 14.8-3, show that the outputs shown are true for Σ, Δ_x, and Δ_y given by (14.8-22) through (14.8-24).

14.8-10 Assume that the pattern used in a three-beam amplitude monopulse radar has the form $f(r) = 2J_1(r)/r$, where $J_1(r)$ is the Bessel function of the first kind of order one. (a) Find the squint angle $r = \theta_s$ that makes K maximum. (b) For the optimum squint angle find $f(\theta_s)$ and θ_B. (c) Find $K_{m,x}$. (d) Find the value of $\mathcal{R}_\Sigma$ needed to make the minimum standard deviation of measurement error not larger than $0.15\theta_B$.

14.8-11 Work Problem 14.8-10 except assume the pattern $f(r) = \exp(-r^2)$.

14.9-1 A MOCO form of conopulse radar uses constant-frequency rectangular transmitted pulses of duration 2 μs. In the receiver the range gate's duration is 3 μs. Find R_A and R_B.

14.9-2 Work Problem 14.9-1 except change τ_g to 3.5 μs.

15

DIGITAL SIGNAL PROCESSING IN RADAR

Profound changes in radar systems have taken place in recent years due to the use of the digital computer. Not too many years ago the computer was used mainly as a low-frequency data processor. In recent times advances in computer speed and memory, as well as the development of efficient *digital signal-processing* (DSP) techniques, have allowed the computer to perform real-time processing. The trend toward "all digital" has resulted in an ever-increasing number of analog functions being replaced by digital ones.

Because the digital computer is now such an important part of radar, DSP has become a basic principle on which radar is dependent and it must be included in the scope of any book covering these principles. In this chapter we present the most important aspects of DSP.[1] It is important to observe that the use of digital processing is not *necessary* to a radar that, in its origins, was analog. DSP is used because of its advantages, which include reductions in system cost, size, and weight and increases in reliability, flexibility, and performance. Its discussion has been placed in this last chapter because the writer believes that the earlier radar topics are most easily learned when developed in "analog" terms.

In principle, nearly every analog radar signal-processing function can be replaced by some equivalent digital operation. Practical limitations prevent complete replacement at this time, but there is always the promise of an "all digital" radar for the future.

---

[1] The topic of DSP is large and well developed. For additional detail beyond the scope of this chapter, and for other references, the reader is referred to some of the recent books as a start: Kamen and Heck (1997), Oppenheim and Willsky (1997), Shenoi (1995), Taylor (1994), Ziemer et al. (1993), Kamen (1990), Soliman and Srinath (1990), Oppenheim and Schafer (1989), Gabel and Roberts (1987), Cadzow (1987), and Ludeman (1986).

15.1 FUNDAMENTAL CONCEPTS

Two things come into play when using digital methods in radar, the digital processor, which we will just call the computer, and signal-processing techniques. The former (computer plus some peripheral equipment) actually does the operations specified by the latter. We will not be too interested in *how* the computer does its operations. Rather, we concentrate on signal-processing techniques, and comment on the computer's functions only when it is helpful to do so. We are able to separate the two things because the principal computer operations (addition, subtraction, multiplication, division, and delay) show up only as mathematical operations in the signal-processing techniques.

Digital signal processing is to be broadly defined here as including all functions needed to take an analog signal, convert it to a suitable digital form, operate on it digitally to create a desired response, and then (if required) convert the response to a representative analog signal. For example, we might convert a received pulse from a target to digital form, implement a digital version of a single-pulse matched filter, and then convert back to the analog form of the matched filter's response. In another example, the N-pulse matched filter needed in the Doppler measurement systems of Chapter 12 could be implemented. Three operations are most fundamental to our problem: (1) analog-to-digital (A/D) conversion of signals, (2) digital-to-digital (D/D) signal processing, and (3) digital-to-analog (D/A) conversion of signals when desired.

A/D Conversion

Let $x(t)$ represent some arbitrary (real or complex) waveform that has continuous values over continuous time. It is normally called a *continuous-time signal*. Because digital computers can only work with discrete data, it is necessary to sample $x(t)$ at defined times to obtain the computer's data. Although it is possible to use nonperiodic sampling, periodic sampling is the most practical and is the most common form.

We discuss only periodic sampling where samples occur at times $t_n = nT_s$, $n = 0, \pm 1, \pm 2, \ldots$. Here T_s is called the *sampling period* and $1/T_s$ is known as the *sampling frequency* or *sampling rate*. The sampled version of $x(t)$ consists of a sequence of sample values $x(nT_s)$ that occur at discrete values of time. The sequence is called a *discrete-time signal*. Mathematically a sequence is defined by the contents of braces, so the discrete-time signal would be formally represented as $\{x(nT_s)\}$. However, such formalism is not needed in our work, so we refer to the sampled signal as just $x(nT_s)$ and think of it as being dependent on the integer index (variable) n.

After sampling, an A/D converter normally does two other functions: quantization and coding. These functions are necessary to convert the signal samples to a form that is useful for the computer's operation. They are *not* needed for the understanding of DSP, since they relate only to the *realization* of the DSP operations. Although we briefly mention quantization and coding, the reader should keep in mind the separation between computer functions and signal-processing techniques.

The digital computer must represent any number by a group of binary digits. If N_b is the number of binary digits, and each digit can take on only two values (binary numbers 0 or 1), then by various rearrangement of the N_b values there can be only 2^{N_b} numbers represented by the N_b digits. The purpose of the quantizer is to take each sample $x(nT_s)$ and "round off" the exact value to the nearest of 2^{N_b} values for computer use. This operation constitutes quantization. Coding consists of the assignment of the binary digit sequences to properly represent the quantizer's discrete values.

Clearly, there are practical problems that can arise. It is necessary that the discrete values span the total amplitude range of $x(t)$. If samples exceed the quantizer's range the quantizer is said to *amplitude overload*. We shall not need additional practical detail on quantizers or coders; some is given in Peebles (1987) for those interested in more detail and other references.

D/D Processing

The digital-to digital operation represents the computer processing of the digitally represented sequence of signal samples. These operations are physically done within the computer according to programmed DSP instructions designed to produce a desired response. In some cases the response might represent the receiver's output signal from the matched filter, or it might represent one of several signals from a filter bank used in Doppler processing.

Regardless of the form of the programmed DSP instructions, they constitute digital signal-processing operations. We are able to describe these operations independently of the computer hardware, and they may be thought of as a *discrete-time processor* or *discrete-time system*.

The boundary lines between the A/D converter and the D/D processor are not well defined. Sometimes the coding operations described above as part of A/D conversion might actually be done by the computer (in the D/D processor). The distinction is in the *definition* of the A/D converter. Our definition chooses to place the encoder in the A/D converter. Similarly the decoding operations of the D/A converter described below might actually reside in the D/D processor.

D/A Conversion

In many systems the outputs of the D/D processor are used directly. For example if range, Doppler, and angle measurement systems are highly computerized, the output data of target range, Doppler, spatial angles, and other data might be used digitally at other places in the system or generated directly in digital form (e.g., base 10 digital registers). In such cases there is no need for the D/A converter.

D/A conversion becomes necessary only when an analog response is needed (e.g., for some meters and displays). The functions performed by a D/A converter are the inverse functions of those done in A/D conversion. First, the computer's codewords are decoded; that is, they are converted to the appropriate level of the 2^{N_b} possible levels. Next, comes inverse quantization where the levels are converted to equivalent signal amplitudes (discrete values).[2] Finally, the discrete signal amplitudes (samples) are properly filtered to obtain an analog waveform.

Practical Considerations

There are two very fundamental practical problems that need to be resolved. First, we might ask, how can samples of a signal represent a continuous signal when the

---

[2] In practice, inverse coding and inverse quantization are often done in one operation.

continuous signal can have *any* values between samples? The answer is, it cannot in general. How, then, can we use sampled signals at all? Fortunately there is a very basic theorem, called the *sampling theorem*, that guarantees that a signal $x(t)$ can be represented by its periodic samples $x(nT_s)$, and recovered from its samples *without error* over *all time*, if certain conditions are satisfied. These conditions are almost always satisfied, at least to an acceptable practical approximation, by all real-world waveforms. The sampling theorem is so fundamental that we discuss its details in the next section.

A second practical problem arises from quantization. The rounding of exact samples to the nearest of a set of discrete amplitudes leads to an error in signal recovery that is called *quantization error* (see Peebles, 1987, ch. 3). The power that exists in the quantization error, denoted by N_q, can be considered a "noise" that interferes with the true recovery of the signal being sampled.

A loss in performance due to quantization can be reflected in the ratio of the recovered signal power (S_0) to quantization noise power. For a uniform quantizer with no amplitude overload, that operates with maximum input signal amplitude, the largest value of (S_0/N_q) is

$$\left(\frac{S_0}{N_q}\right)_{max} = \frac{3(2^{2N_b})}{K_{cr}^2} \qquad (15.1\text{-}1)$$

where

$$K_{cr}^2 = \frac{|x(t)|_{max}^2}{\overline{x^2(t)}} \qquad (15.1\text{-}2)$$

K_{cr} is called the *crest factor* of $x(t)$ which is presumed to have a maximum magnitude $|x(t)|_{max}$. The power in $x(t)$ is $\overline{x^2(t)}$. The crest factor is purely a shape characteristic of the signal $x(t)$. From (15.1-1) it is clear that performance increases as N_b increases. Practical systems choose N_b large enough that $(S_0/N_q)_{max}$ is acceptable. Overload and other factors decrease (S_0/N_q) below the value of (15.1-1).

Example 15.1-1   We find the performance increase in a digital system that occurs when the number of binary digits (N_b) is increased by one. From (15.1-1) expressed in decibels, $[(S_0/N_q)_{max}]_{dB} = (3)_{dB} + 2N_b(2)_{dB} - (K_{cr}^2)_{dB} = 4.7712 + 6.0206N_b - (K_{cr}^2)_{dB}$. Thus, as N_b increases by one, performance increases by 6.02 dB.

15.2   SAMPLING THEOREMS

In this section we discuss various sampling theorems. However, we develop only the sampling theorem for baseband (also called low-pass) signals in some detail, since it is the most important one of many sampling theorems. Because the theorem applies to analog continuous-time signals, we present this form and later show its relationship to the discrete-time signal.

Baseband Sampling Theorem

Theorem Let $x(t)$ be a real or complex baseband[3] waveform that is band-limited such that its Fourier transform (spectrum) is nonzero only over a frequency band $-W_x \leq \omega \leq W_x$, where $W_x > 0$ is a constant; W_x is the highest nonzero frequency present in $x(t)$. W_x is called the *spectral extent* of $x(t)$. The signal $x(t)$ can be completely specified (recovered) without error from its periodic samples taken at times nT_s, $n = 0, \pm 1, \pm 2, \ldots$, provided that the sample rate satisfies

$$\omega_s = \frac{2\pi}{T_s} > 2W_x \quad \text{(rad/s)} \tag{15.2-1}$$

The quantity $2W_x$ is the lowest bound on the sampling rate, and it is called the *Nyquist rate*.

We prove the sampling theorem with the help of Fig. 15.2-1. The signal $x_s(t)$ is called the ideally sampled version of $x(t)$ when it equals $x(t)$ multiplied by the periodic impulse train as shown. It is given by

$$x_s(t) = x(t) \sum_{n=-\infty}^{\infty} \delta(t - nT_s) \tag{15.2-2a}$$

$$= \sum_{n=-\infty}^{\infty} x(nT_s)\delta(t - nT_s) \tag{15.2-2b}$$

$$= \frac{1}{T_s} \sum_{n=-\infty}^{\infty} x(t)e^{jn\omega_s t} \tag{15.2-2c}$$

Equation (15.2-2b) derives from use of property (A.3-24) of impulses with (15.2-2a). Similarly (15.2-2c) derives from (15.2-2a) by using a Fourier series representation of the impulse train (Problem 15.2-3). Fourier transformation of (15.2-2c) gives the spectrum, denoted by $X_s(\omega)$, of $x_s(t)$:

$$X_s(\omega) = \frac{1}{T_s} \sum_{n=-\infty}^{\infty} X(\omega - n\omega_s) \tag{15.2-3}$$

Here $X(\omega)$ is the Fourier transform (spectrum) of $x(t)$. Thus the sampled signal's spectrum is a sum of scaled (by $1/T_s$) replicas of the spectrum of the unsampled signal displaced to all frequencies $n\omega_s$, as sketched in Fig. 15.2-2.

Now, if the ideal filter of Fig. 15.2-1 has the transfer function $H(\omega)$ and impulse response $h(t)$ defined by

$$H(\omega) = T_s \text{rect}\left(\frac{\omega}{\omega_s}\right) \tag{15.2-4}$$

$$h(t) = \text{Sa}\left(\frac{\omega_s t}{2}\right) \tag{15.2-5}$$

---

[3] This theorem can also apply to the direct sampling of bandpass signals, but its use indicates the need for high sampling rates which are actually not required, as other sampling theorems prove.

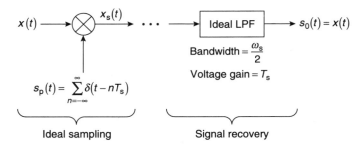

Figure 15.2-1   Circuit models for ideal sampling and signal recovery used in proving the sampling theorem.

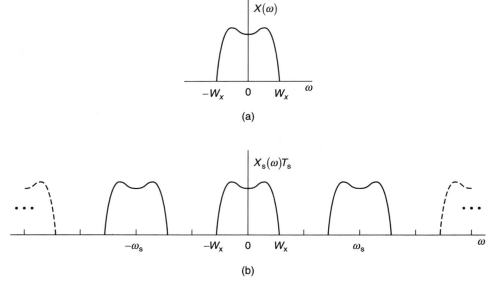

Figure 15.2-2   Sketches of (a) the spectrum $X(\omega)$ of $x(t)$, and (b) the spectrum $X_s(\omega)$ of the sampled signal $x_s(t)$.

it is clear that the filter will pass only the central term in (15.2-3) for $n = 0$ provided that $\omega_s > 2W_x$ is true so that the replica spectra do not overlap. This is the requirement in our theorem for recovery of $x(t)$ from its samples. If ω_s is too small, spectral replicas will overlap, a condition called *aliasing*. Thus for $\omega_s > 2W_x$ the spectrum of the response $s_0(t)$, denoted by $S_0(\omega)$, becomes

$$S_0(\omega) = X_s(\omega)H(\omega) = X(\omega) \qquad (15.2\text{-}6)$$

and so

$$s_0(t) = x(t) \qquad (15.2\text{-}7)$$

In other words, the network of Fig. 15.2-1 creates a sampled version of $x(t)$ and then recovers $x(t)$ *exactly* at its output for all time t.

To see how the ideal filter of Fig. 15.2-1 can reproduce $x(t)$ at times other than sample times, we apply $x_s(t)$, as defined in (15.2-2b), to the filter. Since each impulse excites an impulse response, we get

$$s_0(t) = \sum_{n=-\infty}^{\infty} x(nT_s)h(t - nT_s)$$

$$= \sum_{n=-\infty}^{\infty} x(nT_s)\text{Sa}\left[\frac{\omega_s(t - nT_s)}{2}\right] = x(t) \qquad (15.2\text{-}8)$$

The last equality is true since $s_0(t) = x(t)$ has already been shown. The last form of (15.2-8) shows that $s_0(t)$ consists of sampling functions (this is the reason for the name) added together to produce $x(t)$; each function is associated with one sample, and its peak amplitude equals the sample value. In essence, the sum of the sampling functions provides the exact interpolation (between samples) needed to give $x(t)$. For this reason the sampling function is also known as the *interpolating function*.

Other Sampling Theorems

Theorems exist for sampling of bandpass signals and both baseband and bandpass noise. There are "duality" theorems for sampling the *spectra* of baseband and bandpass deterministic signals and of random signals. There are theorems to handle practical methods of sampling (not the ideal case taken here), higher-order sampling, multidimensional sampling, and a method called quadrature sampling. Most of these theorems are reviewed in Peebles (1987, 1976), and historical references are given.

I and Q Sampling

In radar it is often desirable to replace analog circuits that are high frequency and bandpass by digital processing. In a relatively representative example, it might be desired to replace all components in receiving channels after some point in the IF circuits; perhaps replace the matched filter and all that follows it. If the IF signal is band-limited to a band W_x centered on the IF frequency ω_{IF} and if the signal is considered "low-pass" with highest frequency $\omega_{\text{IF}} + (W_x/2)$, then a sampling frequency of not less than $2\omega_{\text{IF}} + W_x$ seems to be indicated by the baseband sampling theorem. Such a rate for large ω_{IF} can be excessive and is not necessary.

By use of a special interpolating function (see Peebles, 1987, p. 34) developed by Kohlenberg (1953), direct sampling can be used. More often, however, a method called *I and Q sampling*, also called *indirect* or *quadrature sampling*, is used. The two methods are equivalent, but the I and Q method allows the sampling to be less critical (samples are of much lower frequency functions).

I and Q sampling is illustrated in Fig. 15.2-3a. The general form of a received waveform derives from (1.6-21); it is written as

$$x(t) = \alpha a(t - \tau_{\text{R}})\cos\left[(\omega_{\text{IF}} + \omega_{\text{d}})(t - \tau_{\text{R}}) + \theta(t - \tau_{\text{R}}) + \phi_0\right] \qquad (15.2\text{-}9)$$

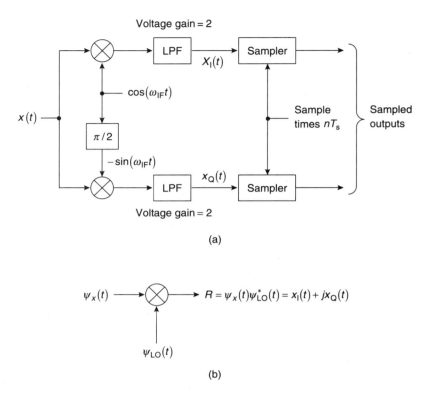

Figure 15.2-3 (a) Functions required in I and Q sampling of a bandpass signal. (b) An equivalent representation of (a) where $x(t)$ and $\cos(\omega_{IF}t)$ are represented by their complex (analytic) forms.

where $a(t)$, ω_d, τ_R, $\theta(t)$, and ϕ_0 were previously defined in Chapter 1, and α is a constant that depends on the radar equation. Analysis of Fig. 15.2-3 shows that

$$x_I(t) = \alpha a(t - \tau_R)\cos\left[\omega_d(t - \tau_R) + \theta(t - \tau_R) + \phi_0 - \omega_{IF}\tau_R\right] \qquad (15.2\text{-}10)$$

$$x_Q(t) = \alpha a(t - \tau_R)\sin\left[\omega_d(t - \tau_R) + \theta(t - \tau_R) + \phi_0 - \omega_{IF}\tau_R\right] \qquad (15.2\text{-}11)$$

Since α, ω_d, ϕ_0, ω_{IF}, and τ_R are constants (over the radar's processing time), and $a(t)$ and $\theta(t)$ are baseband functions, $x_I(t)$ and $x_Q(t)$ are both baseband functions with spectral extent $W_x/2$. They can each be sampled at a rate not less than W_x. The total sample rate is not less than $2W_x$ because two signals must be sampled.[4]

An alternative way of viewing I and Q sampling is to consider complex (analytic) signal representations. If $\psi_x(t)$ is the analytic representation of $x(t)$ in Fig. 15.2-3a and

---

[4]Samples of $x_I(t)$ and $x_Q(t)$ can occur at the same times (representing *two* samples at a rate W_x) or at different times for an average $2W_x$. The quoted sampling rates for $x_I(t)$ and $x_Q(t)$ assume $\omega_d \ll W_x/2$. If not, then sampling must assume spectral extents are $\omega_d + (W_x/2)$.

the analytic representation of $\cos(\omega_{IF} t)$ is denoted by $\psi_{LO}(t)$, then

$$\psi_x(t) = [x_I(t) + jx_Q(t)] e^{j\omega_{IF} t} \qquad (15.2\text{-}12)$$

$$\psi_{LO}(t) = e^{j\omega_{IF} t} \qquad (15.2\text{-}13)$$

The response of the product in panel b is

$$R = \psi_x(t)\psi_{LO}^*(t) = x_I(t) + jx_Q(t) \qquad (15.2\text{-}14)$$

This result means that if samples of $x_I(t)$ and $x_Q(t)$ of Fig. 15.2-3a are treated as a complex number, the result is the same as a complex sample of R of panel b.

The principal conclusion of our developments is that if samples of $x_I(t)$ and $x_Q(t)$ of Fig. 15.2-3a are treated as in-phase (I) and quadrature-phase (Q) components of samples of the response of the processor of Fig. 15.2-3b to the equivalent complex input $\psi_x(t)$, then these complex samples can be processed to form any desirable responses (e.g., matched filter). These facts mean that samples of the complex signal R can be further processed to obtain a complex response having a real part that is the proper response to the processing of the real input $x(t)$.

15.3 DISCRETE-TIME SEQUENCES

In this section we continue our earlier discussions on discrete-time (DT) signals and give some examples. The description of sampled waveforms as DT signals is the first of two main concepts needed to perform digital signal processing (DSP) in radar. The second concept relates to *discrete-time systems* that are discussed in Sections 15.7–15.9.

Basic Definitions

Earlier a discrete-time signal was defined as a *sequence* of sample values of some waveform. For a waveform $x(t)$, sampled at times nT_s with T_s the sampling period and $n = 0, \pm 1, \pm 2, \ldots$, the sample values are $x(nT_s)$. Strictly speaking, the sequence is written as $\{x(nT_s)\}$, but we have noted that short-form notation $x(nT_s)$ is unambiguous and acceptable for all our work. However, we require one additional notation change before proceeding.

In DSP it is typically unnecessary to carry along the constant T_s in our work. A digital computer, which must form the actual processor's implementation, only views the data stream as a function of index n. T_s is unimportant (until one wishes to do D/A conversion, which is *after* the DSP operations). Thus sample values are taken as functions of index n rather than time nT_s. To distinguish this fact, we adopt brackets to enclose the independent variable n for our remaining work. Thus $x[n]$ will denote the *sequence* of sample values that defines a DT signal of $x(nT_s)$.

Technically a *digital signal* (or digital sequence) is one in which both time *and* amplitude are discrete. Quantization of samples of a DT signal produces a digital signal. Similarly a signal processor for which all values are quantized (this includes *all* computers) is a *digital* signal processor (e.g., as opposed to using a name such as DT signal processor). In the real world all sequences and processors (computers) are digital. However, if the quantization noise power is small, as it usually is, we need not distinguish between digital versus DT sequences and systems. Thus in all following work the terms discrete-time and digital are synonymous, and

$$x[n] = x(nT_s), \qquad -\infty < n < \infty \tag{15.3-1}$$

represents either a DT sequence or a digital sequence.

It is typical to represent $x[n]$ graphically as shown in Fig. 15.3-1. The notation seeks to convey that $x[n]$ is not defined for noninteger values of n.

Example 15.3-1   As one example of a discrete-time sequence, we define an *exponential sequence* by

$$x[n] = \begin{cases} Ae^{-n/3}, & n = 0, 1, 2, \ldots \\ 0, & n < 0 \end{cases}$$

where A is a real constant. This sequence is plotted in Fig. 15.3-1 for positive A.

Examples of DT Sequences

A number of DT sequences deserve special definition and names. In most cases these sequences have analogous continuous-time signals. For example, the *unit-impulse sequence* (also called a *unit-pulse sequence, unit sample,* or *Kronecker delta function*), denoted by $\delta[n]$, is defined by

$$\delta[n] = \begin{cases} 1, & n = 0 \\ 0, & n \neq 0 \end{cases} \tag{15.3-2}$$

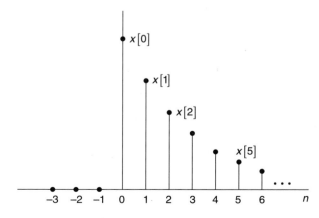

Figure 15.3-1   Graphical notation for a discrete-time sequence.

The unit-impulse sequence has even symmetry, that is, $\delta[-n] = \delta[n]$, and it satisfies the property

$$\sum_{n=-\infty}^{\infty} \delta[n] = 1 \tag{15.3-3}$$

The unit-impulse sequence is analogous to the delta or impulse function $\delta(t)$ for continuous time.

The *unit-step* and *ramp sequences* are defined, respectively, as

$$u[n] = \begin{cases} 1, & n \geq 0 \\ 0, & n < 0 \end{cases} \tag{15.3-4}$$

$$r[n] = \begin{cases} n, & n \geq 0 \\ 0, & n < 0 \end{cases} \tag{15.3-5}$$

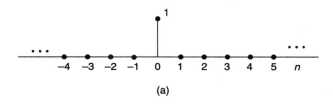

(a)

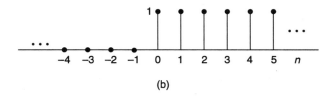

(b)

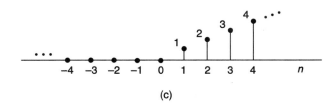

(c)

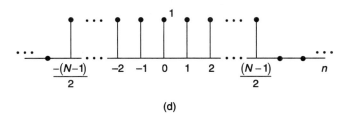

(d)

Figure 15.3-2 Examples of discrete-time sequences. (*a*) The unit-impulse, (*b*) unit-step, (*c*) ramp, and (*d*) retangular sequences.

Let N be an odd positive integer. The *rectangular function* (also called the *rectangular-pulse sequence*), denoted by rect$[k/N]$, is defined by (Kamen and Heck, 1997, p. 23)

$$\text{rect}\left[\frac{k}{N}\right] = \begin{cases} 1, & -\dfrac{N-1}{2} \leq n \leq \dfrac{N-1}{2} \\ 0, & \text{all other } n \end{cases} \tag{15.3-6}$$

The above four sequences are sketched in Fig. 15.3-2.

15.4   PROPERTIES OF DISCRETE-TIME SEQUENCES

As with continuous-time signals, discrete-time sequences have important properties that ease our ability to work with them.

Energy and Average Power

The normalized energy in a DT sequence $x[n]$ over a finite interval $N_1 \leq n \leq N_2$, with $N_1 < N_2$, is defined by[5]

$$\text{Energy in interval } (N_1 \leq n \leq N_2) = \sum_{n=N_1}^{N_2} |x[n]|^2 \tag{15.4-1}$$

The normalized average power is this energy divided by the number of points in the interval

$$\begin{matrix} \text{Average power} \\ \text{in interval} \\ (N_1 \leq n \leq N_2) \end{matrix} = \frac{1}{(N_2 - N_1 + 1)} \sum_{n=N_1}^{N_2} |x[n]|^2 \tag{15.4-2}$$

For the infinite interval let $N_2 = N$ and $N_1 = -N$. We define energy and power, denoted by E_∞ and P_∞ respectively, by

$$E_\infty = \lim_{N \to \infty} \sum_{n=-N}^{N} |x[n]|^2 = \sum_{n=-\infty}^{\infty} |x[n]|^2 \tag{15.4-3}$$

$$P_\infty = \lim_{N \to \infty} \frac{1}{(2N+1)} \sum_{n=-N}^{N} |x[n]|^2 \tag{15.4-4}$$

As with continuous signals there are three categories of sequences defined through energy and power. An *energy sequence* is one with E_∞ finite, so P_∞ is zero. A *power*

---

[5] We use terms energy and power in a customary manner even though energy and power must include the effect of some impedance. See Section 1.5 for further clarification of these normalized quantities.

sequence is one in which P_∞ is finite, so E_∞ is infinite. There are also some sequences in which neither E_∞ nor P_∞ is finite.

Symmetry

A sequence is said to have *conjugate-symmetry* if

$$x[n] = x^*[-n] \tag{15.4-5}$$

It has *conjugate-antisymmetry* if

$$x[n] = -x^*[-n] \tag{15.4-6}$$

If $x[n]$ is a real sequence that satisfies (15.4-5), it is called *even*, while satisfaction of (15.4-6) defines an *odd* sequence. An odd sequence is necessarily zero at $n = 0$.

Any (real or complex) sequence can be decomposed into the sum of a conjugate-symmetric sequence $x_e[n]$ and a conjugate-antisymmetric sequence $x_o[n]$ as follows

$$x[n] = x_e[n] + x_o[n] \tag{15.4-7}$$

The component sequences are generated according to

$$x_e[n] = \tfrac{1}{2}\{x[n] + x^*[-n]\} \tag{15.4-8}$$

$$x_o[n] = \tfrac{1}{2}\{x[n] - x^*[-n]\} \tag{15.4-9}$$

Shifting-Delay

Let $x[n]$ be a DT sequence and q be a positive integer. The DT sequence $x[n-q]$ is the q-step right shift of $x[n]$; this case corresponds to a true delay of $x[n]$ to a later "time" by q units. Similarly $x[n+q]$ represents $x[n]$ shifted left to an earlier "time" by q units; this case corresponds to a negative delay, which is unrealizable.

The representation of a unit impulse shifted by k units is $\delta[n-k]$; it is useful in expressing any sequence $x[n]$ in terms of impulses:

$$x[n] = \sum_{k=-\infty}^{\infty} x[k]\delta[n-k] \tag{15.4-10}$$

The logic leading to (15.4-10) is left to the reader as an exercise.

Example 15.4-1   If we let $x[n] = u[n]$, the unit-step sequence, we can relate $u[n]$ to the impulse by

$$u[n] = \sum_{k=0}^{\infty} \delta[n-k] \tag{1}$$

The reader should sketch (1) and compare to Fig. 15.3-2*b* to confirm its validity.

Analogous to Example 15.4-1 we can also relate the unit-impulse to the unit-step sequence by

$$\delta[n] = u[n] - u[n - 1] \tag{15.4-11}$$

This expression is called the *first backward difference* of the unit-step sequence (see also Problem 15.4-7).

Products and Sums of Sequences

Multiplication of a sequence $x[n]$ by a constant β is defined as the product of each sample value by β; it is the new sequence $\beta x[n]$. The product or sum of two sequences $x[n]$ and $y[n]$ is defined as the sequence of sample-by-sample products or sums, respectively. For example, if $x[n] = 0.1\delta[n - 2] + 0.7\delta[n - 3]$ and $y[n] = 0.3\delta[n] + 0.5\delta[n - 2]$, then the new sequence $w[n] = x[n]y[n] = 0.1(0.5)\delta[n - 2]$.

In another example let $x[n]$ be an arbitrary sequence that we multiply by a single unit-impulse $\delta[n - k]$ to get $x[n]\delta[n - k]$. Now, since $\delta[n - k]$ is defined *only* at $n = k$, the product is a one-term sequence $x[k]\delta[n - k]$. Thus we have

$$x[n]\delta[n - k] = x[k]\delta[n - k] \tag{15.4-12}$$

Sampling

We make use of (15.4-12) to show the sampling property of impulses. Suppose that we form the product of a sequence $x[n]$ as given by (15.4-10) with an impulse $\delta[n - q]$ at $n = q$. We have

$$x[n]\delta[n - q] = \left\{ \sum_{k=-\infty}^{\infty} x[k]\delta[n - k] \right\} \delta[n - q]$$

$$= \sum_{k=-\infty}^{\infty} x[k]\delta[k - q] = x[q] \tag{15.4-13}$$

It is left as a reader's exercise to argue why the two middle forms of (15.4-13) are equal (see Problem 15.4-9). This result is interpreted as the impulse at $n = q$ forming a sample of $x[n]$ at $n = q$.

Periodic Sequences

A sequence $x[n]$ is *periodic* if there exists a positive integer N such that

$$x[n + N] = x[n], \qquad \text{all integers } n \tag{15.4-14}$$

The smallest value of N that satisfies (15.4-14) is called the *fundamental period* of $x[n]$.

For example, consider sampling of the periodic sinusoid $\cos(2\pi t/T_0)$, where T_0 is its period. With samples taken at times nT_s, the sample sequence is $x[n] = \cos(\Omega_0 n)$, where $\Omega_0 = 2\pi T_s/T_0$, and $x[n + N] = \cos(\Omega_0 n + \Omega_0 N)$. For (15.4-14) to be satisfied, we require $\Omega_0 N$ to equal some integer multiple (e.g., q) of 2π. Then $\Omega_0 N = q2\pi$ or

$\Omega_0 = 2\pi q/N$ must be true for some positive integers q and N. A consequence of this condition is that sampling of a periodic continuous-time waveform does not necessarily produce a periodic discrete-time sequence.

Scaling

In continuous-time signals replacing time t by αt with α a real constant scales time and corresponds to a stretching $(\alpha < 1)$ or compression $(\alpha > 1)$ of time. With DT sequences this effect does not necessarily happen, and a totally different sequence can occur. An example helps emphasize this point.

Example 15.4-2   We scale index n in the sequence $x[n] = \cos[\pi n/4]$ by replacing n by $4n$ to get a new sequence $y[n] = x[4n] = \cos[\pi n]$. Since sequences are defined only for integer values of their index, a sketch of these two functions confirms that they are totally different.

15.5   FOURIER TRANSFORMS OF DT SEQUENCES

Our interest now turns to descriptions of discrete-time sequences through their spectral characteristics. For this purpose we develop the *discrete-time Fourier transform* (DTFT) and show its relationship to the spectrum of the continuous-time signal from which the DT sequence was derived. In addition we define the *discrete Fourier Transform* (DFT) and show its relationship to the DTFT.

Discrete-Time Fourier Transform

In discussing the DTFT, we start with the Fourier transform, denoted by $X_c(\omega)$, of the continuous-time signal $x(t)$ from which the DT sequence is defined by sampling. From the sampling theorem we know that (15.2-8) is a valid representation of $x(t)$. Its Fourier transform is

$$X_c(\omega) = \int_{-\infty}^{\infty} x(t)e^{-j\omega t}\,dt$$

$$= \int_{-\infty}^{\infty} \sum_{n=-\infty}^{\infty} x(nT_s)\,\mathrm{Sa}\left[\frac{\omega_s(t-nT_s)}{2}\right]e^{-j\omega t}\,dt \qquad (15.5\text{-}1)$$

By assuming that the integral converges uniformly, the order of integration and summation can be reversed. After a change of variables, we have

$$X_c(\omega) = \frac{2}{\omega_s}\sum_{n=-\infty}^{\infty} x(nT_s)e^{-j\omega T_s n}\int_{-\infty}^{\infty} \mathrm{Sa}(\xi)e^{-j(2\omega/\omega_s)\xi}\,d\xi \qquad (15.5\text{-}2)$$

The integral is recognized as $\pi\,\mathrm{rect}(\omega/\omega_s)$, which gives

$$X_c(\omega) = \mathrm{rect}\left(\frac{\omega T_s}{2\pi}\right)T_s\sum_{n=-\infty}^{\infty} x(nT_s)e^{-j\omega T_s n} \qquad (15.5\text{-}3)$$

We give an important interpretation of (15.5-3) in the next paragraph. First, however, we note that $x(nT_s)$ are the sample values that form the sample sequence $x[n]$. Next, we define a new variable Ω, called the *discrete frequency* (even though it is really an angle), by

$$\Omega = \omega T_s \tag{15.5-4}$$

The sum in (15.5-3), denoted by $X(e^{j\Omega})$,[6] can be written as

$$X(e^{j\Omega}) = \sum_{n=-\infty}^{\infty} x[n] e^{-j\Omega n} \tag{15.5-5}$$

Equation (15.5-5) is the *discrete-time Fourier transform* (DTFT) of the sequence $x[n]$.

A useful interpretation of the meaning of the DTFT is developed by Fourier transformation of the ideally sampled form of the continuous-time signal $x(t)$, as given by (15.2-2b). The transformed left side is just $X_s(\omega)$, the ideally sampled signal's spectrum. The transformed right side is the sum in (15.5-3). After using (15.5-4), it is clear that

$$X_s(\omega) = X(e^{j\Omega})|_{\Omega = \omega T_s} = \frac{1}{T_s} \sum_{n=-\infty}^{\infty} X(\omega - n\omega_s) \tag{15.5-6}$$

where the last right-side term results from substitution of (15.2-3). In the variable ω the DTFT equals $X_s(\omega)$ as sketched in Fig. 15.2-2b. The action of the rectangular function in (15.5-3) is to select the principal term (for $n = 0$) from $X_s(\omega)$ to guarantee the spectrum is that of $x(t)$. An important observation is that $X_s(\omega)$ and the DTFT are periodic with period ω_s in the variable ω. In the variable Ω the DTFT is periodic with period 2π; its behavior is totally specified by its behavior over the central period $-\pi < \Omega < \pi$. We take an example.

Example 15.5-1   We find the DTFT of the rectangular pulse sequence of (15.3-6). From (15.5-5),

$$X(e^{j\Omega}) = \sum_{n=-\infty}^{\infty} \text{rect}\left[\frac{k}{N}\right] e^{-j\Omega n} = \sum_{n=-(N-1)/2}^{(N-1)/2} e^{-j\Omega n}$$

The sum is a well-known series (see Kamen and Heck, 1997, p. 268) that gives

$$X(e^{j\Omega}) = \frac{\sin(N\Omega/2)}{\sin(\Omega/2)}$$

A plot of this expression is shown in Fig. 15.5-1 for $N = 17$. We note that the DTFT does not become zero for some $|\Omega| < \pi$. This fact is a consequence of the

---

[6] The sum is a function of Ω and could be written $X(\Omega)$. However, the variable $e^{j\Omega}$ is commonly used in DSP literature because (15.5-5) becomes the two-sided z-transform of $x[n]$ for the variable $z = e^{j\Omega}$, and much of the spectral analysis of discrete systems is based on z-transforms.

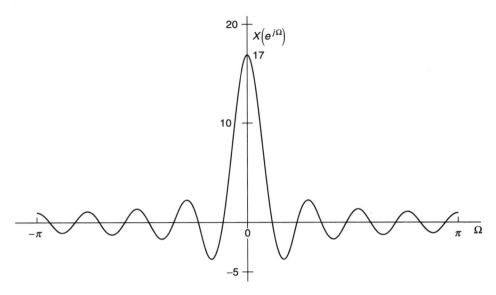

Figure 15.5-1 Plot of DTFT of the rectangular pulse sequence rect $[k/17]$.

continuous-time pulse not being band-limited, which leads to aliasing. The aliasing is most severe for values of Ω near π and $-\pi$. It can be reduced by sampling at a higher rate (increasing N). As N increases, the DTFT approaches a sampling function, more lobes appear, and their amplitudes become smaller for Ω near π and $-\pi$ (and are zero as $N \to \infty$).

Properties of the DTFT

For nearly every property of a continuous-time Fourier transform (Appendix B) there is an analogous property of a DTFT. Tables and details are available in many places (Kamen and Heck, 1997, p. 273; Taylor, 1994, p. 287; Oppenheim et al., 1997, p. 372). We list only several of the more important properties.

In the following a_1 and a_2 are arbitrary constants, N is an integer; and Ω_0 is a real constant; $x[n]$, $x_1[n]$, and $x_2[n]$ are DT sequences having respective DTFTs $X(e^{j\Omega})$, $X_1(e^{j\Omega})$, and $X_2(e^{j\Omega})$. A double-ended arrow implies that the sequence at the left arrowhead has the DTFT listed at the right arrowhead.

Linearity

$$x[n] = a_1 x_1[n] + a_2 x_2[n] \leftrightarrow a_1 X_1(e^{j\Omega}) + a_2 X_2(e^{j\Omega}) = X(e^{j\Omega}) \quad (15.5\text{-}7)$$

Time and frequency shifting

$$x[n - N] \leftrightarrow X(e^{j\Omega})e^{-j\Omega N} \qquad (15.5\text{-}8)$$

$$x[n]e^{j\Omega_0 n} \leftrightarrow X(e^{j\Omega - j\Omega_0}) \qquad (15.5\text{-}9)$$

Conjugation

$$x^*[n] \leftrightarrow X^*(e^{-j\Omega}) \tag{15.5-10}$$

Time reversal

$$x[-n] \leftrightarrow X(e^{-j\Omega}) \tag{15.5-11}$$

Convolution

$$x[n] = \sum_{k=-\infty}^{\infty} x_1[k] x_2[n-k] \leftrightarrow X_1(e^{j\Omega}) X_2(e^{j\Omega}) = X(e^{j\Omega}) \tag{15.5-12}$$

$$x[n] = x_1[n] x_2[n] \leftrightarrow \frac{1}{2\pi} \int_{-\pi}^{\pi} X_1(e^{j\Omega - j\xi}) X_2(e^{j\xi}) \, d\xi = X(e^{j\Omega}) \tag{15.5-13}$$

Inverse Discrete-Time Fourier Transform

If we are given the DTFT $X(e^{j\Omega})$ of some sequence $x[n]$, the sequence can be recovered by the *inverse discrete-time Fourier transform* (IDTFT) defined by

$$x[n] = \frac{1}{2\pi} \int_{-\pi}^{\pi} X(e^{j\Omega}) e^{j\Omega n} \, d\Omega \tag{15.5-14}$$

The IDTFT and the DTFT of (15.5-5) together are called a DTFT pair. The DTFT is often called the *analysis equation*, while the IDTFT is called the *synthesis equation*. Tables of DTFT pairs exist in various books and will be omitted here (see Kamen, 1990, p. 501; Soliman and Srinath, 1990, p. 354).

We outline the procedure for proving (15.5-14) but leave the details as an exercise. Since any DTFT $X(e^{j\Omega})$ is periodic with period 2π, it will have a complex Fourier series. If the series representation is substituted for $X(e^{j\Omega})$ when (15.5-3) is written in the form

$$X_c(\omega) = \text{rect}\left(\frac{\omega T_s}{2\pi}\right) T_s X\left(e^{j\omega T_s}\right) \tag{15.5-15}$$

and the inverse Fourier transform is computed, the resulting function must equal the continuous-time signal $x(t)$. The specific form will be that of (15.2-8), where the Fourier series coefficients must equal $x(nT_s) = x[n]$ as defined by (15.5-14).

Discrete Fourier Transform

For the DTFT to exactly represent the spectrum of a sequence $x[n]$ the sequence must originate from a band-limited signal. Such signals are not typically time limited—that is, values of $x[n]$ are nonzero for $-\infty < n < \infty$—so theoretically an infinite number of samples must be processed by a computer to compute the DTFT.

Obviously no computer can process data over an unlimited time. All practical signals must be limited in time. There are two possibilities.

The first practical signal is one that is already time limited (e.g., a rectangular pulse sequence). Although such signals are not strictly band-limited, if the number of samples (sampling rate) over the finite time is large enough, the aliasing problem is minimized.

The second practical signal derives from truncation of the sample sequence to a finite length, say N. Again, however, if the sampling rate is high enough in relation to the signal's bandwidth, the effects of truncation and aliasing can be reduced.

In both practical cases we must deal with a finite-length sequence. Without loss in generality we may assume $x[n]$ to be nonzero for $0 \leq n \leq N - 1$. Another practical problem arises with the DTFT. It is a continuous function of Ω. Unless this function has an analytic form, it is difficult for a computer to work with it. It is more convenient to discretize the variable Ω. Thus, for various practical reasons, DSP operations are based on finite-length (N) sequences and specification of Ω at a finite number of values, usually the same as N for convenience. The *discrete Fourier transform* (DFT) and *inverse discrete Fourier transform* (IDFT), defined respectively by

$$X[k] = \sum_{n=0}^{N-1} x[n] e^{-j2\pi nk/N}, \qquad k = 0, 1, 2, \ldots, (N-1) \qquad (15.5\text{-}16)$$

$$x[n] = \frac{1}{N} \sum_{k=0}^{N-1} X[k] e^{j2\pi nk/N}, \qquad n = 0, 1, 2, \ldots, (N-1) \qquad (15.5\text{-}17)$$

satisfy all these practical considerations. Here $X[k]$ is a sequence defined by

$$X[k] = X(e^{j\Omega})|_{\Omega = \Omega_k} \qquad (15.5\text{-}18)$$

$$\Omega_k = \omega T_s|_{\omega = 2\pi k/(NT_s)} = 2\pi k/N \qquad (15.5\text{-}19)$$

NT_s is the time interval of $x(t)$ over which the N samples are taken.

Analogous to properties of Fourier transforms of continuous-time signals and DTFTs, there exist properties of DFTs. These are tabulated in many places (Kamen and Heck, 1997, p. 296; Taylor, 1994, p. 293; Oppenheim and Schafer, 1989, p. 547) and will not be given here.

Since the DFT and IDFT have exactly the same form ($x[n]$ in the DFT is equivalent to $X[k]/N$ in the IDFT), similar software can be used to calculate the two functions. Ordinarily either the DFT or IDFT requires N^2 complex multiplications by a computer. However, by use of a special algorithm called the *fast Fourier transform* (FFT), introduced by Cooley and Tukey (1965), the number of multiplications can be reduced to the order of $[N \log_2(N)]/2$, which can be a great savings. For example, with $N = 1024$ we have $N^2 = 1,048,576$ and $[N \log_2(N)]/2 = 5120$ or 99.51% fewer calculations. Software packages such as MATLAB have algorithms to compute a FFT as well as an inverse fast Fourier transform (IFFT).

Relationship of DFT to DTFT

It is of academic interest to show how the DFT is related to the DTFT. For the finite-length discrete-time sequence $x[n]$, where $x[n]$ is nonzero only for

$0 \le n \le (N - 1)$ with N a positive integer, the DTFT from (15.5-5) is

$$X(e^{j\Omega}) = \sum_{n=0}^{N-1} x[n] e^{-j\Omega n} \tag{15.5-20}$$

On comparison with the DFT of (15.5-16), we have

$$X[k] = X(e^{j\Omega})|_{\Omega = 2\pi k/N} = X(e^{j2\pi k/N}) \tag{15.5-21}$$

This expression indicates that the DFT's values $X[k]$ are equal to the DTFT evaluated at the discrete frequency values $\Omega = 2\pi k/N$.

Proof of IDFT

The DFT has been justified by showing its relationship to the DTFT. We now begin with the DFT of (15.5-16) and show that the IDFT is true. On replacing n by m in (15.5-16) and multiplying both sides by $\exp(j2\pi nk/N)$, we have

$$X[k] e^{j2\pi nk/N} = \sum_{m=0}^{N-1} x[m] e^{j2\pi k(n-m)/N} \tag{15.5-22}$$

The sum over k from 0 to $N - 1$ on both sides gives

$$\sum_{k=0}^{N-1} X[k] e^{j2\pi nk/N} = \sum_{m=0}^{N-1} x[m] \sum_{k=0}^{N-1} e^{j2\pi k(n-m)/N} = Nx[n] \tag{15.5-23}$$

The last form of (15.5-23) verifies (15.5-17) and results because the sum over k equals zero for all values of m except for $m = n$ when it equals N (see Problem 15.5-16).

Periodic Sequences and DFTs

The IDFT of (15.5-17) was limited to $n = 0, 1, \ldots, (N - 1)$. If this restriction is removed, then

$$x[n] = \frac{1}{N} \sum_{k=0}^{N-1} X[k] e^{jk(2\pi/N)n} \tag{15.5-24}$$

By replacing n by $n + N$, we find that (15.5-24) gives $x[n] = x[n + N]$, so sequence $x[n]$ is periodic with period N. The right side of (15.5-24) can be interpreted as a finite-term complex Fourier series of $x[n]$ in the variable n with period N. As such, the amplitudes of the series terms at frequencies $k(2\pi/N)$ are recognized as $X[k]/N$. From the DFT of (15.5-16) these amplitudes are $1/N$ times the values of the DFT at the various values of k. Since the DFT is based on a finite sequence for $n = 0, 1, \ldots, (N - 1)$, we arrive at an important conclusion. If a periodic sequence of period N is truncated to its central period for $n = 0, 1, \ldots, (N - 1)$, the truncated sequence's DFT gives values (versus k) that are N times the amplitudes of the frequencies that comprise the complex Fourier series of the periodic sequence.

Zero Padding

In choosing the number (N) of samples of a signal to use in the DFT, it may result that the DFT is not a good approximation to the DTFT. Increasing N for a fixed interval T of samples amounts to sampling faster, which pushes spectral replicas apart and reduces aliasing. If errors are also due to truncation effects, it may be necessary to increase T as well as N. Even after acceptable values of N and T are determined, there are times when a sequence must be modified by *zero padding* (also known as *zero augmentation*). We demonstrate by means of an example.

Example 15.5-2 Consider the rectangular pulse sequence $x[n] = 1$, $n = 0, 1, \ldots,$ $(N_1 - 1)$. A total of N_1 samples spans the pulse's duration. The DTFT of $x[n]$ is found to be

$$X(e^{j\Omega}) = \frac{\sin(N_1\Omega/2)}{\sin(\Omega/2)} e^{-j(N_1 - 1)\Omega/2} \tag{1}$$

As N_1 is increased, $x[n]$ has more samples (is sampled faster), and the DTFT asymptotically approaches the shape of the true spectrum of the rectangular pulse. If the phase factor $\exp[-j(N_1 - 1)\Omega/2]$ is ignored, Fig. 15.5-1 plots the DTFT for $N_1 = 17$.

With $N = N_1$ in (15.5-16) the DFT of $x[n]$ is

$$X[k] = \frac{\sin(\pi k)}{\sin(\pi k/N_1)} e^{-j(N_1 - 1)\pi k/N_1}, \qquad k = 0, 1, \ldots, (N_1 - 1) \tag{2}$$

where (15.5-8) has also been used. From L'Hospital's rule $X[0] = N_1$, but $X[k] = 0$ for all $k = 1, 2, \ldots, (N_1 - 1)$. Thus we find that the DFT is a very poor representation for the spectrum of the sequence because the sample frequencies $\Omega_k = 2\pi k/N_1$ happen to fall at spectral nulls for all but $k = 0$.

To use zero padding, we keep the same sample rate but extend the number of samples beyond the pulse's nonzero interval by assigning zero amplitudes to the extra samples. To form a numerical example, we define $x[n] = 1$, $n = 0, 1, \ldots, N_1 - 1$, and $x[n] = 0$, $n = N_1, N_1 + 1, \ldots, (4N_1 - 1)$. Here $3N_1$ extra sample points (all zero amplitude) are added. The DTFT is still given by (1), as before. The DFT becomes (15.5-16) with $N = 4N_1$, or

$$X[k] = \sum_{n=0}^{4N_1 - 1} x[n] e^{-j2\pi nk/4N_1} = \sum_{n=0}^{N_1 - 1} e^{-j\pi kn/2N_1}$$

$$= \frac{\sin(\pi k/4)}{\sin(\pi k/4N_1)} e^{-j(N_1 - 1)\pi k/4N_1} \tag{3}$$

for $k = 0, 1, \ldots, (4N_1 - 1)$. Equation (3) again gives nulls in the DFT for $k = 4, 8, \ldots, 4(N_1 - 1)$ but gives nonzero values for other k. This means there are now four times as many frequencies being calculated, leading to three nonzero values over each lobe in Fig. 15.5-1 when $N_1 = 17$. The use of zero padding has increased the number of calculated frequencies, which gives a better interpretation of the signal's spectrum shape.

In general, zero padding increases the number of computed frequencies in the DFT, which allows better definition of the sequence's DTFT (see Problem 15.5-17). There are also system applications of zero padding that we discuss in a later section.

15.6 PROPERTIES OF THE DFT

Some special considerations must be taken when dealing with the DFT due to its restriction to finite-length sequences. These considerations mostly result from time shifts. In the following we summarize some of the most important properties of DFTs where a_1 and a_2 are arbitrary constants; N is a positive integer; q is an integer, $x[n]$, $x_1[n]$, and $x_2[n]$ are all sequences of length N with respective DFTs $X[k]$, $X_1[k]$, and $X_2[k]$, all of length N. If some sequence $x_1[n]$ or $x_2[n]$ is of length less than N, it is to be zero-padded to obtain length N. The double-ended arrow implies the quantity on the right head is the DFT of the sequence on the left head. With these definitions the properties are listed without proofs.

Linearity

The DFT of a linear sum of sequences is the sum of the DFTs according to

$$x[n] = a_1 x_1[n] + a_2 x_2[n] \leftrightarrow a_1 X_1[k] + a_2 X_2[k] = X[k] \qquad (15.6\text{-}1)$$

If $x_1[n]$ is of length N_1 and $x_2[n]$ has length N_2, we set N equal to the larger of N_1 or N_2 and zero-pad the other sequence to give length N. The padding is necessary to establish all sequences of the same length.

Time and Frequency Shifting

If time or frequency shifts are attempted with the finite N-point sequences that define the DFT and IDFT, parts of the sequence can move outside the finite region of interest and become undefined. A special form of shift called *circular translation* (or *circular shift*) is necessary. To accomplish circular translation, we repeat the given N-point sequence $x[n]$ periodically with period N to form a periodic sequence. The periodic sequence is next shifted by the desired amount (e.g., q units). The sequence to be used in the DFT (or IDFT) is the portion of the shifted sequence for $n = 0, 1, \ldots, (N-1)$. It can be shown that the shifted sequence is defined by $x[((n-q))_N]$, where

$$((n-q))_N = (n-q) \quad \text{modulo } N \qquad (15.6\text{-}2)$$

For right shifts where q is a positive integer, some values of $n - q$ can be negative; (15.6-2) implies the addition of the smallest multiple of N so that $(n - q + \text{multiple of } N)$ is a number from 0 to $N - 1$. For left shifts where q is a negative integer, we subtract the smallest multiple of N so that the result is a number from 0 to $N - 1$.

Example 15.6-1 Choose $N = 4$ and shift a sequence $x[n]$ right by $q = 2$ units. For $n = 0, 1, \ldots, (N - 1) = 3$, we have

$$((0 - 2))_4 = ((-2))_4 = (-2 + 4) = 2$$
$$((1 - 2))_4 = ((-1))_4 = (-1 + 4) = 3$$
$$((2 - 2))_4 = ((0))_4 = 0$$
$$((3 - 2))_4 = ((1))_4 = 1$$

The sequence values to be used with a DFT for $n = 0, 1, 2, 3$, respectively, are $x[2]$, $x[3]$, $x[0]$, and $x[1]$.

With the above comments in mind, time- and frequency-shifting properties are

$$x[((n - q))_N] \leftrightarrow X[k] e^{-j2\pi kq/N} \tag{15.6-3}$$

$$x[n] e^{j2\pi nq/N} \leftrightarrow X[((k - q))_N] \tag{15.6-4}$$

Other properties are listed without comment.

Conjugation

$$x^*[n] \leftrightarrow X^*[((-k))_N] \tag{15.6-5}$$

$$x^*[((-n))_N] \leftrightarrow X^*[k] \tag{15.6-6}$$

Time Reversal

$$x[((-n))_N] \leftrightarrow X[((-k))_N] \tag{15.6-7}$$

Convolution[7]

$$x[n] = \sum_{m=0}^{N-1} x_1[m] x_2[((n - m))_N] \leftrightarrow X_1[k] X_2[k] = X[k] \tag{15.6-8}$$

$$x[n] = x_1[n] x_2[n] \leftrightarrow \frac{1}{N} \sum_{m=0}^{N-1} X_1[m] X_2[((k - m))_N] = X[k] \tag{15.6-9}$$

15.7 DISCRETE-TIME SYSTEMS

The general form of a DSP operation is to accept some input sequence $x[n]$ and create some desired output sequence $y[n]$ as sketched generally in Fig. 15.7-1. A DSP *system* is defined as the devices that perform the specified operation.

---

[7] These are known as *circular convolutions*.

Figure 15.7-1   A DSP system in general form.

General Comments

A typical system may be the combination of various smaller systems (which we might call system components or elements) through interconnections. Figure 15.7-2 illustrates several connections. In panel *a* two systems are in *cascade* (also called *series*) to form the overall system. In panel *b* two systems are in *parallel* where the overall output is the sum of the two outputs; one of the outputs could also have been added negatively to provide a difference. In panel *a* or *b* there is no path from any output back to any input of the components, which corresponds to a *feed-forward* system. In panel *c* we have a system with various paths from some output back to some input; such a system is said to have *feedback*. Clearly there is an infinite variety of system interconnections to form an overall system.

A discrete time system is said to be *finite dimensional* if for some positive integer N and nonnegative integer M, $y[n]$ is some function only of the N most recent past output values, n, and the current and M most recent past values of the input (Kamen and Heck, 1997). Any other function is infinite dimensional. The function forms an *input-output difference equation*. The system is called *linear* if and only if the input-output difference equation is linear, that is, if it can be written in the form

$$y[n] = - \sum_{i=1}^{N} a_i(n) y[n-i] + \sum_{i=0}^{M} b_i(n) x[n-i] \tag{15.7-1}$$

where $a_i(n)$ and $b_i(n)$ are coefficients that may, in general, depend on n.

In most radar applications of DSP the coefficients in (15.7-1) are constants with respect to n. If (and only if) this is true, such that $a_i(n) = a_i$ and $b_i(n) = b_i$ for all i and n, the system is defined as linear and *time invariant* (abbreviated LTI).[8] LTI systems are sometimes called linear *shift invariant* (LSI). Equation (15.7-1) for the finite-dimensional LTI systems becomes

$$y[n] = - \sum_{i=1}^{N} a_i y[n-i] + \sum_{i=0}^{M} b_i x[n-i] \tag{15.7-2}$$

As a final note we observe that the output of a system defined by (15.7-2) does not depend on any future value of the input. This fact prevents the occurrence of a response prior to the application of a nonzero input, a condition that defines the system as *causal*, provided that the system is at rest (zero initial conditions) prior to occurrence of any input (see Problem 15.7-1).

---

[8] Constant coefficients in the defining difference equation of a DT system are analogous to the constant coefficients of the defining differential equation of a CT system.

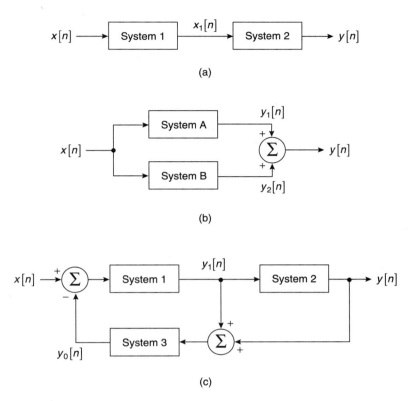

Figure 15.7-2 System interconnection examples. (*a*) Series or cascade, (*b*) parallel, and (*c*) combination system with feedback.

System Descriptions

The various methods of describing (and analyzing) systems broadly fall into two categories. One involves the *sequence* (time) *domain*; the other uses the *transform domain*. In all following work of this chapter, we consider only finite-dimensional LTI DT systems that are initially at rest. As such they must satisfy the difference equation of (15.7-2).

There are three principal techniques used in sequence domain analysis. One involves mathematical solution of a specified difference equation when an analytical solution for $y[n]$ is needed. The theory required parallels that for differential equation solutions in CT systems (recall the *homogeneous* and *particular* solutions). We do not develop these details. For those interested, the book by Hildebrand (1968) forms a good start. On the other hand, if the difference equation is specified, and we are only interested in calculating the response by computer, it is only necessary for the machine to implement the difference equation. Figure 15.7-3 illustrates the operations to be performed, where D represents one unit of delay. The form of panel *a* is called the *direct form I*, since it is a direct realization of (15.7-2).

The *canonic form* (or also called *direct form II*) of Fig. 15.7-3*b* is equivalent to that in panel *a* (Oppenheim and Schafer, 1989, p. 296). We may treat the form of panel *a* as

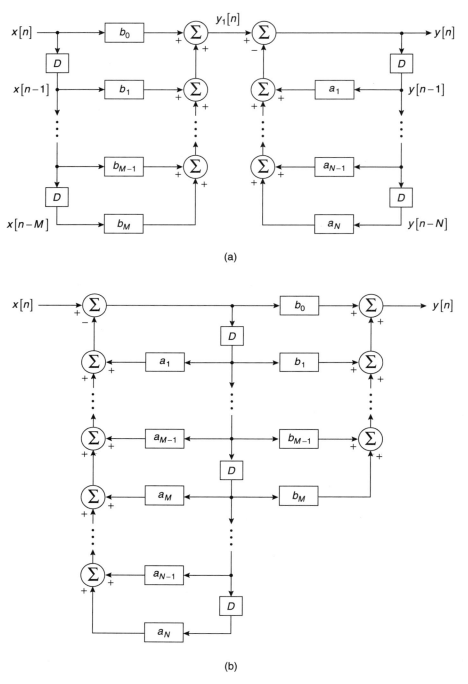

(a)

(b)

Figure 15.7-3 Block diagrams representing an input-output difference equation. (a) Direct form I, and (b) canonic form; the two forms are equivalent.

a cascade of two systems, one from $x[n]$ to $y_1[n]$ and the other from $y_1[n]$ to $y[n]$. For LTI systems the order of cascade is unimportant. On reversing the order, it can be seen that the two delay cascades are identical and can be replaced by one cascade. The result is the canonic form, which is the system that uses the smallest number of delay elements for given values of N and M.

In practice, it is not unusual for the coefficients of (15.7-2) to be determined from a prescribed spectral description of a transfer function derived from DFT or z-transform functions. For example, if the system is to be a digital filter with a specified characteristic, this characteristic can be used to determine the coefficients. Here we have a combination of sequence and transform domain considerations.

A second sequence domain analysis method is based on use of an impulse response to characterize a prescribed system. In this case $y[n]$ is the convolution of $x[n]$, and the impulse response in a manner analogous to CT systems.

A third sequence domain method is to use state variable theory. This theory is beyond the scope of this book and is not discussed further.

In using transform domain analysis the procedure is again analogous to that of a CT system. The DTFT of the response $y[n]$, denoted by $Y(e^{j\Omega})$, is equal to the product of the DTFTs of the input $x[n]$, denoted by $X(e^{j\Omega})$, and the system's impulse response. This latter DTFT gives the *system function* (also called the *transfer function*), usually denoted by $H(e^{j\Omega})$. If the output sequence $y[n]$ is required, the IDTFT of the product $X(e^{j\Omega})H(e^{j\Omega}) = Y(e^{j\Omega})$ is calculated.

15.8   SYSTEM DESCRIPTIONS IN THE SEQUENCE DOMAIN

In this section we briefly summarize the characteristics of LTI systems through sequence domain descriptions.

Impulse Response

If an impulse $\delta[n]$ is applied to a LTI system at rest, the response it evokes is called the *impulse response*, denoted by $h[n]$. The response completely characterizes the system. Because of time invariance, a shifted impulse $\delta[n-m]$ evokes a shifted response $h[n-m]$. Any sequence $x[n]$ can be expressed as

$$x[n] = \sum_{m=-\infty}^{\infty} x[m]\delta[n-m] \tag{15.8-1}$$

This general sequence evokes a general response

$$y[n] = \sum_{m=-\infty}^{\infty} x[m]h[n-m] = \sum_{k=-\infty}^{\infty} x[n-k]h[k] \tag{15.8-2}$$

The last form of (15.8-2) derives from a simple variable change.

Equation (15.8-2) is called a *convolution sum* (or just *convolution*). A star is often used to imply the convolution; the notation is

$$y[n] = \sum_{m=-\infty}^{\infty} x[m]h[n-m] = x[n] \star h[n] \tag{15.8-3}$$

Cascaded and Parallel Systems

Two systems in cascade according to Fig. 15.7-2a with impulse responses $h_1[n]$ and $h_2[n]$ are equivalent to a single system with impulse response $h[n]$ given by

$$h[n] = h_1[n] \star h_2[n] = h_2[n] \star h_1[n] \tag{15.8-4}$$

The last form of (15.8-4) is given only to emphasize that the order of elements in a cascade of LTI elements is arbitrary.

Parallel systems A and B, as defined by Fig. 15.7-2b, are equivalent to a single system with impulse response $h[n]$ given by

$$h[n] = h_A[n] + h_B[n] \tag{15.8-5}$$

where $h_A[n]$ and $h_B[n]$ are the respective impulse responses of systems A and B.

Causality and Linearity

LTI systems at rest are inherently causal and linear, as found from our discussions of (15.7-2). Linearity requires that superposition apply, so if $y_1[n]$ and $y_2[n]$ are respective responses of a system to individual inputs $x_1[n]$ and $x_2[n]$, the response to the linear combination of inputs is

$$y[n] = a_1 y_1[n] + a_2 y_2[n] \tag{15.8-6}$$

for any inputs and any constants a_1 and a_2.

Stability

In a general sense, a system is said to be stable if the application of an input does not cause a divergent response. More specifically, a system is *bounded-input–bounded-output* (BIBO) stable if the response is bounded for every possible bounded input. A system is BIBO stable if and only if

$$\sum_{n=-\infty}^{\infty} |h[n]| < \infty \tag{15.8-7}$$

Invertibility

A system is called invertible if distinct inputs lead to distinct outputs. Such a system has an *inverse system* such that, when placed in cascade with the original system, the cascade's response is the same as the input to the cascade.

Example 15.8-1 An *accumulator* or *summer* is a system such that (see Problem 15.8-1)

$$y[n] = \sum_{k=-\infty}^{n} x[k]$$

The accumulator's inverse system is called a *first backward difference* system (see Problem 15.8-2)

$$w[n] = y[n] - y[n-1]$$

When cascaded the two give $w[n] = x[n]$.

Finite Impulse Response Systems

Digital systems fall into one of two categories. A *finite impulse response system* (FIR) is one with all its elements in a purely feed-forward structure. The system response is obtained from (15.7-2) with $a_i = 0$, $i = 1, 2, \ldots, N$.

$$y[n] = \sum_{i=0}^{M} b_i x[n-i] \tag{15.8-8}$$

If the input is an impulse $x[n] = \delta[n]$, the impulse response is found

$$h[n] = \sum_{i=0}^{M} b_i \delta[n-i] = \begin{cases} b_n, & 0 \le n \le M \\ 0, & \text{all other } n \end{cases} \tag{15.8-9}$$

FIR systems are stable and have a simple structure. However, the systems must usually be of high complexity (large M) to satisfy even elementary modern-day requirements (Taylor, 1994, p. 424).

A FIR system can be implemented with a single shift register with M stages as shown in the left part of Fig. 15.7-3a to generate $y_1[n]$.

Infinite Impulse Response Systems

A system with both feed-forward and feedback connections is called an *infinite impulse response system* (IIR). IIR systems typically require a lower-order design than FIR systems in satisfying modern requirements, but they can have a variety of design problems, including instability. The structure of an IIR system follows the form of Fig. 15.7-3b, where $N \ge M$ is required for a causal system (see Taylor, 1994, p. 442).

Systems for which (15.7-2) applies with $N \ge 1$ are called *recursive systems* because output values can be recursively computed from input and past output values. Recursive systems are usually IIR systems. An example of the complete solution of a simple recursive system is given in Problem 15.7-3.

15.9 SYSTEM DESCRIPTIONS IN THE TRANSFORM DOMAIN

In this section we review the description of digital systems by using the transform domain.

Transforms

The transform-domain description of a DT signal is through the DTFT. For a sequence $x[n]$ its DTFT is

$$X(e^{j\Omega}) = \sum_{n=-\infty}^{\infty} x[n]e^{-j\Omega n} \tag{15.9-1}$$

where Ω is the "frequency-domain" real variable called the discrete frequency.

A more general transform, called the *bilateral* (or *two-sided*) z-transform of the sequence $x[n]$ is defined by

$$X(z) = \sum_{n=-\infty}^{\infty} x[n]z^{-n} \tag{15.9-2}$$

where z is a "frequency-domain" complex variable. In polar form z can be written as

$$z = |z|e^{j\Omega} \tag{15.9-3}$$

where the magnitude and phase of z are $|z|$ and Ω, respectively. By comparing (15.9-1) and (15.9-2), it is clear that the DTFT is the z-transform evaluated on the unit circle $|z| = 1$. That is,

$$X(e^{j\Omega}) = X(z)|_{z=\exp(j\Omega)} \tag{15.9-4}$$

We next discuss how these transforms enter into the descriptions of systems.

System Transfer Function

Crucial to the analysis and synthesis of systems using the transform domain is the concept of a system transfer function. We already know that a LTI system at rest is described by its impulse response, and the system's response to a sequence $x[n]$ is the convolution sum of (15.8-3). On forming the DTFT of (15.8-3), and using the convolution property of (15.5-12), we have

$$Y(e^{j\Omega}) = X(e^{j\Omega})H(e^{j\Omega}) \tag{15.9-5}$$

where $x[n]$, $y[n]$, and $h[n]$ have respective DTFTs $X(e^{j\Omega})$, $Y(e^{j\Omega})$, and $H(e^{j\Omega})$. This latter function

$$H(e^{j\Omega}) = \sum_{n=-\infty}^{\infty} h[n]e^{-j\Omega n} \tag{15.9-6}$$

is called the system *transfer function* (or *frequency response function*).

The transform $H(e^{j\Omega})$ can be likened to the CT transfer function $H(\omega)$. By choice of $H(e^{j\Omega})$ various (digital) filters may be designed. Equation (15.9-5) is an analysis tool that allows us to first find the transform of the filtered sequence. Then, by calculating the IDTFT, we can finally develop the discrete response sequence $y[n]$. When these operations are combined with appropriate A/D and D/A operations, the overall procedures constitute a rather general DSP system.

By transforming (15.7-2) using the time-shifting property of DTFTs, as given by (15.5-8), we obtain

$$H(e^{j\Omega}) = -H(e^{j\Omega}) \sum_{i=1}^{N} a_i e^{-j\Omega i} + \sum_{i=0}^{M} b_i e^{-j\Omega i} \qquad (15.9\text{-}7)$$

or

$$H(e^{j\Omega}) = \frac{\displaystyle\sum_{i=0}^{M} b_i e^{-j\Omega i}}{1 + \displaystyle\sum_{i=1}^{N} a_i e^{-j\Omega i}} \qquad (15.9\text{-}8)$$

Since the coefficients in (15.7-2) and (15.9-8) are the same, the specification of $H(e^{j\Omega})$ in the transform domain determines the difference equation in the sequence domain (a synthesis case). Conversely, specification of the difference equation determines $H(e^{j\Omega})$ (an analysis case).

In terms of the z-transform, (15.9-8) becomes

$$H(z) = \frac{b_0 + b_1 z^{-1} + \cdots + b_M z^{-M}}{1 + a_1 z^{-1} + \cdots + a_N z^{-N}} = b_0 \frac{\displaystyle\prod_{k=1}^{M} (1 - c_k z^{-1})}{\displaystyle\prod_{k=1}^{N} (1 - d_k z^{-1})} \qquad (15.9\text{-}9)$$

The last (factored) form of (15.9-9) recognizes that the numerator and denominator of the middle form are polynomials of degree M and N, respectively, so they have roots at $z = c_k$ and $z = d_k$, respectively. The numerator gives zeros of $H(z)$ at $z = c_k$ and a pole at $z = 0$. The denominator gives poles at $z = d_k$ and a zero at $z = 0$.

Although we will not be concerned with more detail, the design of various digital filters revolves around the behavior of (15.9-8) or (15.9-9). Broadly, the placement of poles and zeros determines a desired transfer function. Having the (filter's) transfer function, we find the impulse response from an IDTFT. The computer then typically processes the data $x[n]$ in the sequence domain for many DSP systems. Some details of digital filters are available in Kamen and Heck (1997), Taylor (1994), Soliman and Srinath (1990), Kamen (1990), Oppenheim and Schafer (1989), and Cadzow (1987). These sources also contain other references.

Use of the FFT in Processing

The FFT (or IFFT) is a very fast algorithm for computing the DFT (or IDFT), and the DFT (or IDFT) is equal or approximately equal to the DTFT (or IDTFT), depending on the signal form. Because of their speed the FFT and IFFT are finding ready applications in radar, so it is important to indicate how these should be used.

We discuss the important case where $h[n]$ has finite length P and the input $x[n]$ has a length very long relative to P (approximated here as infinite). The goal is to compute the system's response $y[n]$ when the FFT version of the DFT is to be used. The direct procedure is to find $y[n]$ as the convolution of $h[n]$ and the full sequence $x[n]$; alternatively, the IFFT of the product of the FFTs of $x[n]$ and $h[n]$ can be used. In both cases a long delay processing time is encountered, and zero padding is needed in the latter.

In another procedure (among others) the sequence $x[n]$ can be subdivided into contiguous subintervals (blocks) each of length $L \geq P$ such that

$$x[n] = \sum_{m=0}^{\infty} x_m[n - mL] \tag{15.9-10}$$

where

$$x_m[n] = \begin{cases} x[n + mL], & 0 \leq n \leq L - 1 \\ 0, & \text{all other } n \end{cases} \tag{15.9-11}$$

is the sequence in subinterval m. From linear convolution it follows that

$$y[n] = x[n] \star h[n] = \sum_{m=0}^{\infty} y_m[n - mL] \tag{15.9-12}$$

where

$$y_m[n] = x_m[n] \star h[n] \tag{15.9-13}$$

Each subinterval's convolution will be of length $L + P - 1$. The responses for adjacent subintervals overlap, but their sum is the correct response. This procedure involves calculating the response as a sum of subinterval convolutions, and it requires no DFT operations. It has the advantage of giving data more rapidly than for full-data convolution. However, the subinterval convolutions can be slower to perform than when the FFT version of the DFT is used.

To use the FFT method, we again compute $y[n]$ for each subinterval and add for the total response. However, now we apply the FFT to each subinterval. First, $x_m[n]$ and $h[n]$ are zero-padded to each have length[9] $N = L + P - 1$. We then calculate $y_m[n]$ as the N-point IFFT of the product of the N-point FFTs of $x_m[n]$ and $h[n]$. This procedure is called *block convolution* (more specifically, it is called the *overlap-add method* of block convolution). For most lengths L and P the FFT procedure is faster than the use of the DFT or direct convolution.

---

[9] In many cases the FFT is implemented with N a power of 2. In this case zeros are added until $N = 2^Q$, where Q is the smallest positive integer such that $L + P - 1 \leq N$.

15.10   SUMMARY AND COMMENTS

This chapter has summarized the most important fundamentals of DSP without the need to involve detailed computer functions. It was initially shown how most practical waveforms can be converted to discrete (digital) form for further processing. The conversion involves principles of sampling and quantization (Sections 15.1 and 15.2). Some details on the description of these discrete sequences were developed using sequence-domain methods (Sections 15.3 and 15.4) and transform- (frequency-) domain methods, including the DTFT, IDTFT, DFT, IDFT, FFT, and IFFT (Sections 15.5 and 15.6).

Having established descriptions of digital sequences and their transforms, the chapter then defined digital systems for the most common case of linear time-invariant systems with zero initial conditions. General sequence- and transform-domain representations were given. In the former domain convolution defines the effect of the system on an input digital signal to create a desired digital response signal. In the latter domain performance is established mainly by the choice of the system's transfer function (that of some desired digital filter in many cases). Sections 15.7 through 15.9 gave these discussions.

We close this chapter by noting that in some cases we may wish to convert a radar's digital response back to an analog waveform. Because these last points do not really involve basic DSP concepts, they were not developed. However, the conversion is done via a digital-to-analog (D/A) converter as briefly touched on in Section 15.1.

This chapter is by no means an exhaustive summary of DSP. For more details the reader should consult the several references given herein.

PROBLEMS

15.1-1 A computer is used in a radar to do DSP operations. It uses $N_b = 8$ digits to express the samples of the envelope of a radar pulse that can have peak voltages from 0 to 3 V. What are the quantized amplitudes with which the computer works if the lowest level is 0 V and the largest is 3 V?

15.1-2 Work Problem 15.1-1 except assume that $N_b = 16$.

15.1-3 Assume that the crest factor is 2.6 for the radar pulse of Problem 15.1-1. What is the largest possible signal power to quantization noise power ratio for the DSP processor?

15.1-4 Work Problem 15.1-3 except assume that $N_b = 16$ digits.

15.2-1 A radar pulse's envelope has a duration of 2 μs, and it can be approximated as having a spectral extent of 4 MHz. What is the minimum rate to be used to sample the envelope for DSP purposes, and how many samples are taken across the pulse's duration?

15.2-2 The processor of Problem 15.2-1 is found to have too much aliasing. Rework the problem except assume that the spectral extent is 8 MHz.

15.2-3 Show that (15.2-2c) derives from (15.2-2a) by use of a Fourier series representation of the impulse train in Fig. 15.2-1.

15.2-4 Show that (15.2-3) derives from Fourier transformation of (15.2-2c).

15.2-5 A periodic baseband signal has a finite number of frequencies in its Fourier series. Discuss how this band-limited signal can be reproduced from a *finite* number of samples taken over one period.

15.2-6 In I and Q sampling the quadrature signals have spectral extent $W_x/2$ if the spectral extent of the bandpass waveform is W_x. Discuss why the factor of $\frac{1}{2}$ occurs.

15.3-1 The first four samples of a signal $x(t)$ taken every second apart are $x(0) = 1.21$, $x(1) = 1.45$, $x(2) = 1.32$, and $x(3) = 0.91$ V. A quantizer is based on $N_b = 4$ digits, $|x(t)|_{max} = 1.5$ V, and $-1.5 \le x(t) \le 1.5$ V. The extreme quantizer's discrete levels are placed at ± 1.5 V. What are the values of the four quantized samples?

15.3-2 A cosine function with angular frequency ω_0 is sampled every T_s seconds to give samples $x(nT_s) = A\cos(\omega_0 nT_s)$, where $A > 0$ is a constant. Sketch the sample sequence $x[n]$ for n spanning one period of $x(t)$ when (a) $\omega_s = 8\omega_0$ and (b) $\omega_s = 7\omega_0$. Observe the effect of varying ω_s.

15.4-1 Find E_∞ and P_∞ for the unit impulse sequence. Is $\delta[n]$ an energy or power sequence?

15.4-2 Work Problem 15.4-1 except for the unit step sequence.

15.4-3 Work Problem 15.4-1 except for the sequence $x[n] = (\frac{1}{2})^n u[n]$.

15.4-4 Work Problem 15.4-1 except for the sequence $x[n] = \exp(jn\pi/8)$.

15.4-5 A sequence is $x[n] = 2n$ for $0 \le n \le 4$ and zero for $4 < n < 0$. Sketch (a) $x[n]$, (b) $x[-n]$, (c) $x_e[n]$, and (d) $x_o[n]$.

15.4-6 Use (1) of Example 15.4-1 to show that

$$u[n] = \sum_{k=-\infty}^{n} \delta[k]$$

15.4-7 Show that $\delta[n+1] = u[n+1] - u[n]$ where the right side is called the *first forward difference* of $u[n]$.

15.4-8 Two sequences are $x[n] = \delta[n] + 2\delta[n-1] + 4\delta[n-3]$ and $y[n] = \delta[n+1] - 3\delta[n-1] + 2\delta[n-3]$. Find sequences (a) $x[n]y[n]$ and (b) $x[n] + y[n]$.

15.4-9 Discuss why the two middle forms in (15.4-13) are equal.

15.5-1 Begin with (15.5-1) and prove that (15.5-3) is true.

15.5-2 Find the DTFT of the sequence $x[n] = \alpha^n u[n]$, where α is a nonzero real constant with $|\alpha| < 1$. Sketch the magnitude of the DTFT. [*Hint*: Use the known sum

$$\sum_{n=N_1}^{N_2} z^n = \frac{z^{N_1} - z^{N_2+1}}{1-z}$$

where N_1 and N_2 are integers with $N_2 > N_1$ and z can be real or complex (Kamen and Heck, 1997, p. 264).]

15.5-3 Reconsider the sequence of Problem 15.5-2 except assume a finite sequence with nonzero values only for $0 \le n \le N$. Find an expression for the DTFT.

15.5-4 Find the DTFT of the exponential sequence

$$x[n] = \begin{cases} e^{-\alpha n}, & 0 \le n \\ 0, & n < 0 \end{cases}$$

where $\alpha > 0$ is a real constant. (*Hint*: Make use of the series given in Problem 15.5-2.)

15.5-5 Find the DTFT of the sequence $x[n] = \alpha^n \cos(2n) u[n]$, where α is a real constant with $|\alpha| < 1$.

15.5-6 Find the DTFT of the sequence $x[n] = \alpha^{|n|}$, where α is a real constant with $|\alpha| < 1$. Determine the maximum and minimum values of the DTFT's magnitude, and find the values of Ω at which they occur.

15.5-7 Find the DTFT of the sequence

$$x[n] = \begin{cases} 1, & n = 0 \\ \frac{2}{3}, & n = \pm 1 \\ \frac{1}{3}, & n = \pm 2 \end{cases}$$

and put it in terms of trigonometric functions. Sketch the DTFT.

★15.5-8 Let the rectangular pulse sequence of Example 15.5-1 extend to $\pm \infty$ by letting $N \to \infty$. Show that the DTFT can be written as

$$X(e^{j\Omega}) = 2\pi \sum_{m=-\infty}^{\infty} \delta(\Omega - m2\pi)$$

where $\delta(\cdot)$ is the ordinary delta function. (*Hint*: As $N \to \infty$, note that the DTFT of the example is first approximated as a sampling function for N large, then becomes an impulse from use of (A.2-6); finally use property (A.3-18) with the fact that the DTFT is periodic.)

15.5-9 Find the DTFT of two sequences: (a) $x[-1] = \frac{1}{2}$, $x[0] = 0$, and $x[1] = \frac{1}{2}$, and (b) $y[-1] = \frac{1}{2}$, $y[0] = 0$, and $y[1] = -\frac{1}{2}$. (c) Use your results to determine the sequence having the DTFT $\exp(j\Omega)$.

15.5-10 Compute the IDTFT for the function $X(e^{j\Omega})$ of Problem 15.5-8, and show that the corresponding sequence is $x[n] = 1$, all n.

15.5-11 Show that (15.5-8) and (15.5-9) are true.

15.5-12 Prove the conjugation property of DTFTs given by (15.5-10).

15.5-13 Prove (15.5-11).

15.5-14 Show that

$$nx[n] \leftrightarrow j\frac{dX(e^{j\Omega})}{d\Omega}$$

is a property of the DTFT (called the *time multiplication property*).

15.5-15 If $X(e^{j\Omega})$ is the DTFT of a sequence $x[n]$, show that *Parseval's theorem*

$$\sum_{n=-\infty}^{\infty} |x[n]|^2 = \frac{1}{2\pi} \int_{-\pi}^{\pi} |X(e^{j\Omega})|^2 \, d\Omega$$

is true. This theorem relates energy to the sequence and transform domains.

⋆15.5-16 Start with (15.5-23) and justify that (15.5-17) is true.

⋆15.5-17 For the sequence

$$x[n] = \begin{cases} e^{-n/2}, & n = 0, 1, \ldots, (N_1 - 1) \\ 0, & \text{all other } n \end{cases}$$

where $N_1 > 0$ is an integer, find (a) the DTFT, (b) the N_1-point DFT, and (c) the N-point DFT that uses zero padding with $N = 4N_1$. Plot the magnitudes of the transforms for $N_1 = 4$, and observe the effects of zero padding.

15.5-18 Generalize the sequence of Problem 15.5-17 by letting

$$x[n] = \begin{cases} e^{-\alpha n}, & n = 0, 1, \ldots, (N_1 - 1) \\ 0, & \text{all other } n \end{cases}$$

where $\alpha > 0$ is a real constant and N_1 may be infinite. (a) For $N_1 = \infty$ find the DTFT. (b) Truncate the infinite sequence to a finite length N_1. Find the DFT of the truncated sequence. (c) Plot the magnitudes of the two transforms for $N_1 = 4$, and 16, all for $\alpha = \frac{1}{2}$, and observe the effect of truncation.

15.6-1 (a) Find the DFT of the sequence $x[0] = 1$, $x[1] = 2$, $x[2] = -1$, and $x[3] = 1$. (b) Calculate the DFT of the sequence shifted later in time by one unit according to the right side of (15.6-3). (c) Use circular translation of the sequence of (a) and calculate the DFT. Does this DFT equal that of part (b) as (15.6-3) promises?

15.7-1 A linear time-invariant system satisfies (15.7-2). Discuss what must be known to solve (15.7-2) if the coefficients are all known but the system is not at rest.

15.7-2 If the response $y[n]$ of a LTI system at rest is equal to the convolution of its impulse response $h[n]$ and the input sequence $x[n]$, show that the DTFT of $y[n]$, denoted by $Y(e^{j\Omega})$, equals $X(e^{j\Omega})H(e^{j\Omega})$, where $X(e^{j\Omega})$ and $H(e^{j\Omega})$ are the DTFTs of $x[n]$ and $h[n]$, respectively.

15.7-3 A LTI system is at rest and is defined by $N = 1$ and $M = 0$ in (15.7-2). Use recursion to find an expression for $y[n]$ for any $n \geq 0$ that is in terms of coefficients a_i and b_i, and the input sequence $x[n]$.

15.8-1 Show that the impulse response of the accumulator of Example 15.8-1 is $h[n] = u[n]$.

15.8-2 Find the impulse response of the first backward difference system defined in Example 15.8-1.

15.8-3 Demonstrate that the accumulator of Example 15.8-1 is unstable.

15.8-4 Use (15.8-9) for the FIR system to find an expression for its DTFT (transfer function) in terms of its coefficients b_i. Note that the structure of a FIR system (from Fig. 15.7-3a with $N = 0$) is the same as a transversal filter of Chapter 12.

APPENDIX A

REVIEW OF THE IMPULSE FUNCTION

The impulse function, also widely known as the delta function, is indispensible in analysis and synthesis of modern radar systems. In this appendix we review the most important definitions and properties of the impulse function and its derivatives as well as the unit-step function (the integral of the impulse function).

A.1 INTRODUCTION

It is common to denote a *unit-impulse function* by the symbol $\delta(t)$, referring to an impulse that "occurs" at $t = 0$. Often the unit-impulse is defined as a "function" that has zero duration, area one, and an infinite "amplitude" at its time of occurrence. Indeed, to graphically represent an impulse function, we typically use an arrow of amplitude one pointing to infinity to imply area one and infinite "amplitude," as shown in Fig. A.1-1a. When the impulse has an area different from one, and occurs at a time other than the origin, we assign a "scale," or proportionality constant, to define its area, and shift it in time, as shown in panel b for an area of 1.5 and an occurrence time of $t = 3$.

Background

In reality $\delta(t)$ is not a function in the ordinary sense. It is best described by a relatively new theory in mathematics called the *theory of distributions*, or sometimes the theory of generalized functions. As pointed out by Lützen (1982, p. 159), the first people to use distributions in mathematics were J. B. J. Fourier in 1822, G. Kirchoff in 1882, and O. Heaviside in 1898, with the latter two actually using the "delta" function. However it was P. A. M. Dirac (1930) who gave the delta function its name, notation and widespread popularity. As time progressed, efforts were made in mathematical areas that eventually helped to lead to adequate and rigorous understanding of impulses and related topics. Notably in 1932 S. Bochner possibly only implicitly used

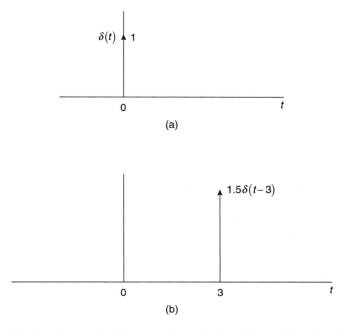

Figure A.1-1 (*a*) A unit-impulse $\delta(t)$, and (*b*) a scaled and time-shifted impulse $1.5\delta(t-3)$ occurring at $t = 3$ seconds.

distributions and S. L. Sobolev in 1935 and 1936 rigorously defined distributions as functionals, while J. G. Mikusiński in 1948 and others defined distributions through sequences.[1]

It remained for Laurent Schwartz (1966), mainly in two classical publications in 1950 and 1951, to rigorously invent the theory of distributions and to show the far-reaching applications of the theory. However, the general theory is difficult for most engineers and applied mathematicians, and efforts followed to reduce the theory to simpler terms. Such simplifications were given by G. Temple (1953, 1955) through the introduction of generalized functions. Further simplification of Temple's work followed in the excellent book by Lighthill (1959). Much of the work discussed below follows Lighthill.

Other early sources containing discussions of distributions or generalized functions are Halperin (1952), Friedman (1956), and Papoulis (1962) who gives a very readable appendix summarizing some of the properties of impulse functions. Finally we note that a number of more recent sources are available that discuss the theory of distributions or generalized functions. The reader is referred to Friedlander (1982), Kanwal (1983), and Al-Gwaiz (1992), to name just three.

Generalized Functions

The fundamental idea behind describing impulse (delta) functions, unit-step functions, their derivatives, as well as other "functions," is to generalize the

---

[1] See Lützen 1982, pp. 159 and 167, where many other early works are cited.

concept of a function to include these "functions." They are then called *generalized functions.*

There are several ways of defining generalized functions. Here we will review and summarize the definition based on use of sequences. We rely heavily on the work of Lighthill (1959) who extended the work of Temple (1953, 1955), who in turn extended the work of Mikusiński. We proceed by listing certain fundamental definitions in the next few paragraphs, and these are then used in Section A.2 to define the impulse function, its derivatives, and the unit-step function. In addition we discuss asymmetrical impulses and unit-steps, and multidimensional impulses. The definitions we require follow (see Lighthill, 1959, pp. 15–19).

1. *Good function.* A good function is one that is everywhere differentiable any number of times and such that it and all its derivatives decrease at most as $|t|^{-N}$ as $|t| \to \infty$ for all N. An example of a good function is $\exp(-t^2)$.

2. *Fairly good function.* A fairly good function is one that is everywhere differentiable any number of times and such that it and all its derivatives decrease at most as $|t|^N$ as $|t| \to \infty$ for some N. Any polynomial is a fairly good function.

3. *Regular sequence.* A sequence $f_m(t)$ of good functions is called a regular sequence if, for any good function $\phi(t)$ whatsoever, the limit

$$\lim_{m \to \infty} \int_{-\infty}^{\infty} f_m(t)\phi(t)dt \qquad (A.1\text{-}1)$$

exists. An example of a regular sequence is $f_m(t) = \exp(-t^2/m^2)$.

4. *Equivalent sequences.* Two regular sequences are called equivalent sequences if, for any good function $\phi(t)$ whatever, the limit in (A.1-1) is the same for each sequence. The two sequences $\exp(-t^2/m^2)$ and $\exp(-t^4/m^4)$ are equivalent.

5. *Generalized function.* A generalized function $f(t)$ is defined as a regular sequence $f_m(t)$ of good functions.

6. *Equality of generalized functions.* Two generalized functions are called equal if the corresponding regular sequences are equivalent.

7. *Integral of generalized function.* The integral of the product $f(t)\phi(t)$, where $f(t)$ is a generalized function and $\phi(t)$ is a good function, is defined as

$$\int_{-\infty}^{\infty} f(t)\phi(t)dt = \lim_{m \to \infty} \int_{-\infty}^{\infty} f_m(t)\phi(t)dt \qquad (A.1\text{-}2)$$

since the limit is the same for all equivalent sequences $f_m(t)$.

8. *Derivatives of generalized functions.* If a generalized function $f(t)$ is defined by a sequence $f_m(t)$, the derivative $df(t)/dt$ is defined by the sequence $df_m(t)/dt$ and

$$\int_{-\infty}^{\infty} \frac{df(t)}{dt}\phi(t)dt = \lim_{m \to \infty} \int_{-\infty}^{\infty} \frac{df_m(t)}{dt}\phi(t)dt \qquad (A.1\text{-}3)$$

for any good function $\phi(t)$.

9. *Sum of generalized functions.* The sum $f(t) + h(t)$ of two generalized functions $f(t)$ and $h(t)$ is defined by the sum $f_m(t) + h_m(t)$ of their defining sequences $f_m(t)$ and $h_m(t)$.

10. *Product with a function.* The product $g(t)f(t)$, where $g(t)$ is a fairly good function and $f(t)$ is a generalized function, is defined by the sequence $g(t)f_m(t)$, where $f_m(t)$ is the defining sequence for $f(t)$.

11. *Linear functional argument.* If $f(at + b)$ is a generalized function of the linear function $at + b$, then $f(at + b)$ is defined by the sequence $f_m(at + b)$.

12. *Fourier transforms of generalized functions.* The Fourier transform $F(\omega)$ of a generalized function $f(t)$ is defined by the sequence $F_m(\omega)$, where $F_m(\omega)$ is the Fourier transform of $f_m(t)$:

$$F(\omega) = \int_{-\infty}^{\infty} f(t)e^{-j\omega t}\,dt = \lim_{m \to \infty} F_m(\omega)$$

$$= \lim_{m \to \infty} \int_{-\infty}^{\infty} f_m(t)e^{-j\omega t}\,dt \qquad \text{(A.1-4)}$$

Similarly the inverse Fourier transform $f(t)$ of a generalized function $F(\omega)$ is defined by the sequence $f_m(t)$ where $f_m(t)$ is the inverse Fourier transform of $F_m(\omega)$, the sequence defining $F(\omega)$:

$$f(t) = \frac{1}{2\pi} \int_{-\infty}^{\infty} F(\omega)e^{j\omega t}\,d\omega = \lim_{m \to \infty} f_m(t)$$

$$= \lim_{m \to \infty} \frac{1}{2\pi} \int_{-\infty}^{\infty} F_m(\omega)e^{j\omega t}\,d\omega \qquad \text{(A.1-5)}$$

The above list of important definitions is not complete, and others exist (Lighthill, 1959). However, our list is sufficient to provide plausible justification for most of the properties to be given without formal proofs in Section A.3.

A.2 BASIC DEFINITIONS

We next apply the above fundamental definitions to the definitions of specific generalized functions, namely, $\delta(t)$, $d^n\delta(t)/dt^n$, and the unit-step function $u(t)$. We also define asymmetrical impulse and step functions as well as multidimensional impulses.

Impulse as a Generalized Function

Consider the sequence

$$f_m(t) = \frac{m}{\sqrt{\pi}} e^{-m^2 t^2} \tag{A.2-1}$$

which has an area of one for all m (Problem A.2-2). Figure A.2-1 sketches $f_m(t)$ which has principal amplitudes mainly in a small region of t about $t = 0$ when m is large. For large enough m, $\phi(t)$ is nearly constant at $\phi(0)$ over the small region of t. Since the area of $f_m(t)$ is one, we have

$$\int_{-\infty}^{\infty} f_m(t)\phi(t)\, dt \approx \phi(0) \int_{-\infty}^{\infty} f_m(t)\, dt = \phi(0), \qquad m \text{ large} \tag{A.2-2}$$

Next we define the generalized (impulse) function $\delta(t)$ by the sequence (A.2-1):

$$\delta(t) = \lim_{m\to\infty} f_m(t) = \lim_{m\to\infty} \frac{m}{\sqrt{\pi}} e^{-m^2 t^2} \tag{A.2-3}$$

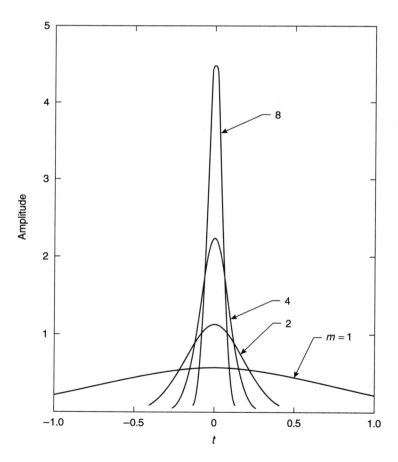

Figure A.2-1   A sequence defining the impulse function.

Thus, from (A.1-2) using (A.2-2),

$$\int_{-\infty}^{\infty} \delta(t)\phi(t)\,dt = \phi(t)|_{t=0} = \phi(0) \tag{A.2-4}$$

This result defines the impulse function using the sequence (A.2-1). From definition 4 we see that $\delta(t)$ can also be defined by any other equivalent sequence. Some such sequences are

$$\delta(t) = \lim_{m \to \infty} \frac{m}{\pi(1 + m^2 t^2)} \tag{A.2-5}$$

$$\delta(t) = \lim_{m \to \infty} \frac{m}{\pi} \frac{\sin(mt)}{(mt)} \tag{A.2-6}$$

$$\delta(t) = \lim_{m \to \infty} \frac{m}{\pi} \left[\frac{\sin(mt)}{mt}\right]^2 \tag{A.2-7}$$

Often a broader class of function $\phi(t)$ is used to define $\delta(t)$ (Papoulis, 1962, p. 271; Friedman, 1956, p. 136; Korn and Korn, 1961, p. 742). In this broader class $\phi(t)$ is an arbitrary function except continuous at the time of the impulse [continuous higher derivatives of $\phi(t)$ are not presumed]. For the broader class, other sequences may be used to define $\delta(t)$. Some such sequences are

$$\delta(t) = \lim_{m \to \infty} m \operatorname{rect}(mt) \tag{A.2-8}$$

$$\delta(t) = \lim_{m \to \infty} m \operatorname{tri}(mt) \tag{A.2-9}$$

$$\delta(t) = \lim_{m \to \infty} \frac{m}{2} e^{-m|t|} \tag{A.2-10}$$

Impulse Derivatives as Generalized Functions

From definition 8 any of the sequences (A.2-1) and (A.2-5) through (A.2-7) can be differentiated to obtain sequences adequate to define $d\delta(t)/dt$. To show the procedures more carefully, consider the integral as expanded by parts

$$\int_{-\infty}^{\infty} \frac{df_m(t)}{dt}\phi(t)\,dt = f_m(t)\phi(t)\Big|_{-\infty}^{\infty} - \int_{-\infty}^{\infty} f_m(t)\frac{d\phi(t)}{dt}\,dt$$

$$= -\int_{-\infty}^{\infty} f_m(t)\frac{d\phi(t)}{dt}\,dt \tag{A.2-11}$$

which follows because $\phi(\pm\infty) = 0$. When (A.2-11) is used in the right side of (A.1-3), we have

$$\int_{-\infty}^{\infty} \frac{df(t)}{dt}\phi(t)\,dt = -\lim_{m \to \infty} \int_{-\infty}^{\infty} f_m(t)\frac{d\phi(t)}{dt}\,dt$$

$$= -\int_{-\infty}^{\infty} f(t)\frac{d\phi(t)}{dt}\,dt \tag{A.2-12}$$

for any generalized function $f(t)$. In particular, if $f(t) = \delta(t)$, (A.2-12) and (A.2-4) give

$$\int_{-\infty}^{\infty} \frac{d\delta(t)}{dt} \phi(t)\, dt = -\int_{-\infty}^{\infty} \delta(t) \frac{d\phi(t)}{dt}\, dt = -\frac{d\phi(t)}{dt}\bigg|_{t=0} \qquad (A.2\text{-}13)$$

for the definition of $d\delta(t)/dt$.

By performing the above procedure n times, we have

$$\int_{-\infty}^{\infty} \frac{d^n f(t)}{dt^n} \phi(t)\, dt = (-1)^n \int_{-\infty}^{\infty} f(t) \frac{d^n \phi(t)}{dt^n}\, dt \qquad (A.2\text{-}14)$$

for any generalized function $f(t)$. In particular, if $f(t) = \delta(t)$, we have

$$\int_{-\infty}^{\infty} \frac{d^n \delta(t)}{dt^n} \phi(t)\, dt = (-1)^n \int_{-\infty}^{\infty} \delta(t) \frac{d^n \phi(t)}{dt^n}\, dt = (-1)^n \frac{d^n \phi(t)}{dt^n}\bigg|_{t=0} \qquad (A.2\text{-}15)$$

for the definition of $d^n \delta(t)/dt^n$.

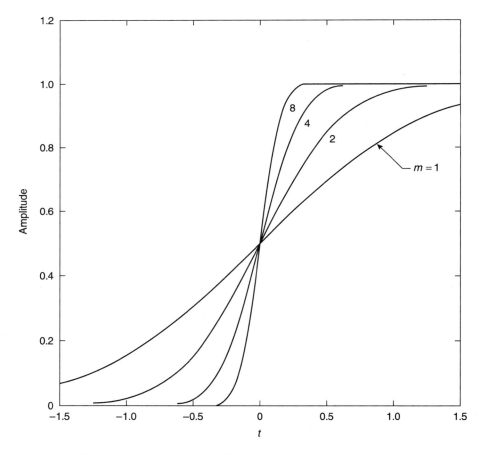

Figure A.2-2   A sequence defining the unit-step function $u(t)$.

Unit-Step as a Generalized Function

Consider the sequence

$$f_m(t) = \tfrac{1}{2} + \tfrac{1}{2}\operatorname{erf}(mt) \qquad (\text{A.2-16})$$

where $\operatorname{erf}(\cdot)$ is the well-known *error function* defined by

$$\operatorname{erf}(x) = \frac{2}{\sqrt{\pi}} \int_0^x e^{-\xi^2}\,d\xi, \qquad x \geq 0 \qquad (\text{A.2-17})$$

$$\operatorname{erf}(-x) = -\operatorname{erf}(x), \qquad x \geq 0 \qquad (\text{A.2-18})$$

This sequence is sketched in Fig. A.2-2. Clearly, as $m \to \infty$, the sequence approaches the unit-step function given by

$$u(t) = \begin{cases} 1, & t > 0 \\ \tfrac{1}{2}, & t = 0 \\ 0, & t < 0 \end{cases} \qquad (\text{A.2-19})$$

Thus we use the sequence to define $u(t)$:

$$u(t) = \lim_{m \to \infty} \left[\tfrac{1}{2} + \tfrac{1}{2}\operatorname{erf}(mt) \right] \qquad (\text{A.2-20})$$

However, the derivative of the right side of (A.2-20) is the sequence on the right side of (A.2-3) which defines $\delta(t)$. Therefore the derivative of the generalized function $u(t)$ is defined by the same sequence as defines $\delta(t)$, so we have symbolically

$$\frac{du(t)}{dt} = \delta(t) \qquad (\text{A.2-21})$$

Asymmetrical Impulse and Unit-Step Functions

Several forms of impulse functions exist. All the preceding work related to $\delta(t)$ called the *symmetrical unit-impulse function*. The corresponding function $u(t)$ of (A.2-21) and (A.2-19) is called a *symmetrical unit-step function*.

Another unit-step function that is important in writing mathematical expressions for probability distribution functions of discrete random variables (see Appendix C) is the *asymmetrical unit-step function* defined by

$$u_-(t) = \begin{cases} 1, & t \geq 0 \\ 0, & t < 0 \end{cases} \qquad (\text{A.2-22})$$

On comparing (A.2-22) with (A.2-19), we see that $u_-(t)$ and $u(t)$ differ only in assignment of the "value" at $t = 0$. If this "value" is assigned to the region $t \leq 0$, a second asymmetrical unit-step function results:

$$u_+(t) = \begin{cases} 1, & t < 0 \\ 0, & t \leq 0 \end{cases} \tag{A.2-23}$$

Although precise definitions exist for asymmetrical unit-impulse functions (see Korn and Korn, 1961, p. 745), we will define them here only through their relationships to the asymmetrical unit-step functions:

$$\delta_-(t) = \frac{du_-(t)}{dt} \tag{A.2-24}$$

$$\delta_+(t) = \frac{du_+(t)}{dt} \tag{A.2-25}$$

The impulse $\delta_-(t)$ is the proper one to be used in defining the probability density function of a discrete random variable (see Appendix C). However, since only the relationship (A.2-24) is needed in the probability work, the "negative" subscripts are usually deleted. The deletion may make it appear that the impulses and steps are symmetrical when they are not, but no damage is incurred because of the limited use.

Multidimensional Impulses

Impulses can be defined for n-dimensional space in various coordinate systems (Korn and Korn, 1961, p. 745; Friedman, 1956, p. 255). For rectangular cartesian coordinates the unit-impulse function $\delta(x, y, z)$ is defined by

$$\int_{-\infty}^{\infty} \int_{-\infty}^{\infty} \int_{-\infty}^{\infty} \delta(x, y, z) \phi(x, y, z)\, dx dy dz = \phi(0, 0, 0) \tag{A.2-26}$$

This expression is equivalent to

$$\int_{-\infty}^{\infty} \int_{-\infty}^{\infty} \int_{-\infty}^{\infty} \delta(x)\, \delta(y)\, (\delta z)\, \phi(x, y, z) dx dy\, dz = \phi(0, 0, 0) \tag{A.2-27}$$

so that the three-dimensional unit impulse is equivalent to the product of three one-dimensional impulses. Symbolically we have

$$\delta(x, y, z) = \delta(x)\, \delta(y)\, \delta(z) \tag{A.2-28}$$

These results are extendable to higher dimensions.

A.3   PROPERTIES OF IMPULSE FUNCTIONS

In this section we list a number of properties of the symmetrical impulse function and its derivatives. No detailed proofs are given. Although some proofs are available through problems at the end of the appendix, most are simply omitted. For the proofs the reader is referred to the literature. However, where it is especially helpful in understanding a particular property, a minimum of development is included. For completeness we begin by restating definitions of $\delta(t)$ and $d^n\delta(t)/dt^n$. In the various properties to follow a, b, t_0, t_1, t_2, ..., t_M (for $M = 1, 2, ...$), and α_1, α_2, ..., α_N (for $N = 1, 2, ...$) are all real constants. Also $g(t)$ is a real function of t, $\phi(t)$ and $\phi_1(t)$, $\phi_2(t)$, ..., $\phi_N(t)$ (for $N = 1, 2, ...$,) are good functions.

Definition of Impulse

$$\int_{-\infty}^{\infty} \delta(t)\phi(t)dt = \phi(0) \tag{A.3-1}$$

Definition of Derivatives of Impulse

For the first derivatives $d\delta(t)/dt$, sometimes called a *unit-doublet*,

$$\int_{-\infty}^{\infty} \frac{d\delta(t)}{dt} \phi(t)\, dt = -\left. \frac{d\phi(t)}{dt}\right|_{t=0} \tag{A.3-2}$$

For derivative n

$$\int_{-\infty}^{\infty} \frac{d^n\delta(t)}{dt^n} \phi(t)\, dt = (-1)^n \left. \frac{d^n\phi(t)}{dt^n}\right|_{t=0} \tag{A.3-3}$$

Linearity

If

$$\phi(t) = \sum_{n=1}^{N} \alpha_n \phi_n(t) \tag{A.3-4}$$

then

$$\int_{-\infty}^{\infty} \delta(t)\phi(t)dt = \sum_{n=1}^{N} \alpha_n \int_{-\infty}^{\infty} \delta(t)\phi_n(t)dt = \sum_{n=1}^{N} \alpha_n \phi_n(0) \tag{A.3-5}$$

Fourier Transforms of Impulses

Let $(\cdot) \leftrightarrow (\cdot)$ denote a Fourier transform pair. The transform of $\delta(t)$ is

$$\int_{-\infty}^{\infty} \delta(t)e^{-j\omega t}dt = e^{-j\omega t}|_{t=0} = 1 \tag{A.3-6}$$

where we identify $\phi(t) = \exp(-j\omega t)$, so

$$\delta(t) \leftrightarrow 1 \tag{A.3-7}$$

Similarly the inverse transform of $\delta(\omega)$ is

$$\frac{1}{2\pi} \int_{-\infty}^{\infty} \delta(\omega) e^{j\omega t} d\omega = \frac{1}{2\pi} e^{j\omega t}\big|_{\omega=0} = \frac{1}{2\pi} \tag{A.3-8}$$

so

$$\frac{1}{2\pi} \leftrightarrow \delta(\omega) \tag{A.3-9}$$

From (A.3-7) and (A.3-9) we have

$$\delta(t) = \frac{1}{2\pi} \int_{-\infty}^{\infty} e^{j\omega t} d\omega \tag{A.3-10}$$

$$\delta(\omega) = \frac{1}{2\pi} \int_{-\infty}^{\infty} e^{-j\omega t} dt \tag{A.3-11}$$

which are useful representations of $\delta(t)$ and $\delta(\omega)$.

Fourier Transforms of Derivatives of Impulses

From use of (A.3-3) with $n = 0, 1, 2, \ldots$,

$$\int_{-\infty}^{\infty} \frac{d^n \delta(t)}{dt^n} e^{-j\omega t} dt = (-1)^n \frac{d^n e^{-j\omega t}}{dt^n}\bigg|_{t=0} = (j\omega)^n \tag{A.3-12}$$

so

$$\frac{d^n \delta(t)}{dt^n} \leftrightarrow (j\omega)^n \tag{A.3-13}$$

Similarly

$$\frac{1}{2\pi} \int_{-\infty}^{\infty} \frac{d^n \delta(\omega)}{d\omega^n} e^{j\omega t} d\omega = \frac{(-1)^n}{2\pi} \frac{d^n e^{j\omega t}}{d\omega^n}\bigg|_{\omega=0} = \frac{(-jt)^n}{2\pi} \tag{A.3-14}$$

so

$$\frac{(-jt)^n}{2\pi} \leftrightarrow \frac{d^n \delta(\omega)}{d\omega^n} \tag{A.3-15}$$

From (A.3-13) and (A.3-15) we have

$$\frac{d^n\delta(t)}{dt^n} = \frac{1}{2\pi} \int_{-\infty}^{\infty} (j\omega)^n e^{j\omega t} d\omega \tag{A.3-16}$$

$$\frac{d^n\delta(\omega)}{d\omega^n} = \frac{1}{2\pi} \int_{-\infty}^{\infty} (-jt)^n e^{-j\omega t} dt \tag{A.3-17}$$

which are useful representations of the derivatives of impulses.

Functional Argument

Let $g(t)$ be a real function of t, and let $t_1, t_2, \ldots, t_M$ be simple real roots of $g(t) = 0$. Then the symbolic representation

$$\delta[g(t)] = \sum_{m=1}^{M} \frac{\delta(t-t_m)}{|dg(t)/dt|_{t=t_m}|} \tag{A.3-18}$$

is valid (see Kanwal, 1983, pp. 52–53). If $g(t)$ has higher-order roots, no significance is attached to $\delta[g(t)]$.

Example A.3-1   Let $g(t) = t^2 - a^2$, so $g(t) = 0$ when $t = \pm a$ or $t_1 = a$ and $t_2 = -a$. Thus $dg(t)/dt|_{t=t_1} = 2a$ and $dg(t)/dt|_{t=t_2} = -2a$.
Finally

$$\delta(t^2 - a^2) = \frac{\delta(t-a)}{2|a|} + \frac{\delta(t+a)}{2|a|}$$

Shifting

$$\int_{-\infty}^{\infty} \delta(t-t_0)\phi(t)dt = \phi(t_0) \tag{A.3-19}$$

$$\int_{-\infty}^{\infty} \delta(t)\phi(t-t_0)dt = \phi(-t_0) \tag{A.3-20}$$

Scaling

$$\int_{-\infty}^{\infty} \delta(at-b)\phi(t)dt = \frac{1}{|a|}\phi\left(\frac{b}{a}\right) \tag{A.3-21}$$

Symmetry

$$\delta(-t) = \delta(t) \tag{A.3-22}$$

Area

$$\int_{-\infty}^{\infty} \delta(t)\,dt = 1 \tag{A.3-23}$$

Duration

$\delta(t)$ behaves as though it is zero for all $t \neq 0$. By analogy with an ordinary function, $\delta(t)$ appears as though its duration is zero.

Product of Function and Impulse

$$g(t)\,\delta(t) = g(0)\,\delta(t) \tag{A.3-24}$$

For the case where $g(t) = t$, an odd function,

$$t\delta(t) = 0 \tag{A.3-25}$$

Product of Function and Derivative of Impulse

$$g(t)\frac{d\delta(t)}{dt} = g(0)\frac{d\delta(t)}{dt} - \frac{dg(t)}{dt}\bigg|_{t=0} \delta(t) \tag{A.3-26}$$

If $g(t) = t$, an odd function,

$$t\,\frac{d\delta(t)}{dt} = -\delta(t) \tag{A.3-27}$$

Convolution of Two Impulses

$$\int_{-\infty}^{\infty} \delta(t)\delta(t_0 - t)\,dt = \delta(t_0) \tag{A.3-28}$$

PROBLEMS

A.1-1 Sketch the function $g(t) = 3\delta(t + 1) - 2\delta(t - 1) + 1.5 \text{ rect } (t/2) + 2\delta(t)$.

A.1-2 Sketch the function $g(t) = 2\delta(t) + 3 \text{ tri } (t/3) - 2\delta(t - 4)$.

★**A.1-3** If $g(t)$ is a good function and $G(\omega)$ is the Fourier transform of $g(t)$, show that $G(\omega)$ is a good function. [*Hint*: Differentiation $G(\omega)$ M times with respect to ω and then integrate with respect to t N times.]

A.1-4 Assume that the functions of the sequence

$$f_m(t) = e^{-t^2/m^2}$$

are good functions and show that the sequence is regular.

A.1-5 A generalized function $f(t)$ is defined by the sequence.

$$f_m(t) = \tfrac{1}{2} + \tfrac{1}{\pi} \tan^{-1}(mt)$$

Use definition 8 and this sequence to find a sequence to define the derivative $df(t)/dt$.

A.1-6 Work Problem A.1-5 except find the sequence to define $d^2f(t)/dt^2$.

A.1-7 From a rough sketch of $f_m(t)$ in Problem A.1-5, what are the generalized functions $f(t)$ and $df(t)/dt$? Correspondingly, what is the generalized function $d^2f(t)/dt^2$ of Problem A.1-6?

A.1-8 The third derivative of the sequence in Problem A.1-5 can be used to define the second derivative $d^2\delta(t)/dt^2$. Show that it can be put in the form

$$\frac{d^3f_m(t)}{dt^3} = \frac{-2m^3}{\pi} \frac{\cos[3\tan^{-1}(mt)]}{(1+m^2t^2)^{3/2}}$$

A.2-1 Use (A.2-1) and definition 8 to obtain sequences that define (a) $d\delta(t)/dt$, and (b) $d^2\delta(t)/dt^2$?

A.2-2 Show that the sequence (A.2-1) has area one for all m.

A.2-3 Use (A.2-4) and evaluate the following integrals:

(a) $\displaystyle\int_{-\infty}^{\infty} \delta(t)\cos\left(3t + \frac{\pi}{4}\right) dt$

(b) $\displaystyle\int_{-\infty}^{\infty} \delta(t)e^{-2t^2+1}\, dt$

(c) $\displaystyle\int_{-\infty}^{\infty} \delta(t)\frac{1}{1 + 6t^2 + 3t^4}\, dt$

A.2-4 Use (A.2-4) and evaluate the following integrals:

(a) $\displaystyle\int_{-\infty}^{\infty} \delta(t)\, e^{j\cos(6t)}\, dt$

(b) $\displaystyle\int_{-\infty}^{\infty} \delta(t)[-1 + 2t^2 + 4t^4]e^{-6t^2}\, dt$

(c) $\displaystyle\int_{-\infty}^{\infty} \delta(t)\tan^{-1}(2.2 + 3t^2)\, dt$

A.2-5 Use (A.2-7) and definition 8 to obtain a sequence to define $d\delta(t)/dt$.

A.2-6 Show that all three sequences of (A.2-8) through (A.2-10) have area one for any m.

A.2-7 Use (A.2-15) to evaluate the following integrals

(a) $\displaystyle\int_{-\infty}^{\infty} \frac{d\delta(t)}{dt} \cos\left(3t + \frac{\pi}{4}\right) dt$

(b) $\displaystyle\int_{-\infty}^{\infty} \frac{d\delta(t)}{dt} e^{-2t^2 + 1} dt$

(c) $\displaystyle\int_{-\infty}^{\infty} \frac{d^2\delta(t)}{dt^2} \frac{1}{1 + 6t^2 + 3t^4} dt$

A.2-8 Use (A.2-15) to evaluate the following integrals

(a) $\displaystyle\int_{-\infty}^{\infty} \frac{d^2\delta(t)}{dt^2} e^{j\cos(6t)} dt$

(b) $\displaystyle\int_{-\infty}^{\infty} \frac{d^3\delta(t)}{dt^3}[-1 + 2t^2 + 4t^4] dt$

(c) $\displaystyle\int_{-\infty}^{\infty} \frac{d\delta(t)}{dt} \tan^{-1}(2.2 + 3t^2) dt$

★**A.2-9** Replace $\phi(t)$ in (A.2-13) with $g(t)\phi(t)$, where $g(t)$ is a fairly good function, and show that

$$g(t)\frac{d\delta(t)}{dt} = g(0)\frac{d\delta(t)}{dt} - \frac{dg(t)}{dt}\bigg|_{t=0}\delta(t)$$

is true.

★**A.2-10** Replace $\phi(t)$ in (A.2-15) with $g(t)\phi(t)$, where $g(t)$ is a fairly good function, and show that

$$g(t)\frac{d^n\delta(t)}{dt^n} = \sum_{k=0}^{n} \frac{(-1)^k n!}{k!(n-k)!} \frac{d^k g(t)}{dt^k}\bigg|_{t=0} \frac{d^{n-k}\delta(t)}{dt^{n-k}}$$

is true. [*Hint*: Use the expansion formula for the derivative of a product:

$$\frac{d^n(g\phi)}{dt^n} = \sum_{k=0}^{n} \frac{n!}{k!(n-k)!} \frac{d^k g}{dt^k} \frac{d^{n-k}\phi}{dt^{n-k}}.]$$

★**A.2-11** Let $g(t) = t^m$ in the result of Problem A.2-10, and prove that

$$t^m\frac{d^n\delta(t)}{dt^n} = \begin{cases} 0, & n < m \\ (-1)^m m!\delta(t), & n = m \\ \dfrac{(-1)^m n!}{(n-m)!} \dfrac{d^{n-m}\delta(t)}{dt^{n-m}}, & n > m \end{cases}$$

A.3-1 Use (A.3-10) and (A.3-11) to show that

$$\delta(t) = \frac{1}{2\pi} \int_{-\infty}^{\infty} \cos{(\omega t)} \, d\omega = \frac{1}{\pi} \int_{0}^{\infty} \cos{(\omega t)} d\omega$$

A.3-2 Use (A.3-16) and (A.3-17) to show that

$$\frac{d\delta(t)}{dt} = \frac{-1}{2\pi} \int_{-\infty}^{\infty} \omega \sin{(\omega t)} \, d\omega = \frac{-1}{\pi} \int_{0}^{\infty} \omega \sin{(\omega t)} d\omega$$

A.3-3 If t_0 and ω_0 are real constants, find (a) the Fourier transform of $\delta(t - t_0)$ and (b) the inverse Fourier transform of $\delta(\omega - \omega_0)$.

A.3-4 Find the inverse Fourier transform of the sum of impulses $F(\omega) = \pi[\delta(\omega - \omega_0) + \delta(\omega + \omega_0)]$, where ω_0 is a real constant.

A.3-5 Work Problem A.3-4 except for the difference of impulses $F(\omega) = -j\pi[\delta(\omega - \omega_0) - \delta(\omega + \omega_0)]$.

A.3-6 Generalize Problems A.3-4 and A.3-5 by finding the inverse Fourier transform $f(t)$ of

$$F(\omega) = a_0\pi\delta(\omega) + \pi \sum_{n=1}^{\infty} a_n [\delta(\omega - n\omega_0) + \delta(\omega + n\omega_0)]$$

$$- j\pi \sum_{n=1}^{\infty} b_n [\delta(\omega - n\omega_0) - \delta(\omega + n\omega_0)]$$

where a_0, a_n, and b_n are real constants. Do you recognize $f(t)$?

A.3-7 Find the inverse Fourier transform $f(t)$ of the following sum of impulses when C_n are complex constants and ω_0 is a real constant:

$$F(\omega) = 2\pi \sum_{n=-\infty}^{\infty} C_n \delta(\omega - n\omega_0)$$

What is the form of $f(t)$?

A.3-8 Use (A.3-19) or (A.3-20) to solve the following integrals:

(a) $\displaystyle\int_{-\infty}^{\infty} \delta(t + 3) e^{-3\cos^2(6t)} \, dt$

(b) $\displaystyle\int_{-\infty}^{\infty} \delta(t - 1.5) \frac{\cos{(\pi t/2)}}{1 + 4t^2} \, dt$

(c) $\displaystyle\int_{-\infty}^{\infty} \delta(t) [1 + 3\cos^2{(t - 1)} + 4\cos^4{(t + 2)}] \, dt$

(d) $\displaystyle\int_{-\infty}^{\infty} \delta(t) \frac{e^{-2(t-2)^2}}{1 + 12(t - 2)^2} \, dt$

A.3-9 Use (A.3-18) to show that (A.3-21) is true.

A.3-10 Use (A.3-18) to show that

$$\delta[(t - a)(t - b)] = \frac{\delta(t - a)}{|a - b|} + \frac{\delta(t - b)}{|b - a|}$$

where a and b are real constants.

A.3-11 Extend Problem A.3-10 to show that

$$\delta[(t - a)(t - b)(t - c)] = \frac{\delta(t - a)}{|(a - b)(a - c)|}$$

$$+ \frac{\delta(t - b)}{|(b - a)(b - c)|} + \frac{\delta(t - c)}{|(c - a)(c - b)|}$$

where a, b, and c are real constants.

A.3-12 Use (A.3-18) to find right-side representations for

(a) $\delta(t^2 - \pi^2)$

(b) $\delta(t^3 - 7t)$

(c) $\delta(t^3 - 4t^2 + 3t)$

A.3-13 Use (A.3-18) to find right-side representations for

(a) $\delta[\cos(\pi t/2)]$

(b) $\delta[\sin(\pi t/2)]$

A.3-14 Symmetry of a generalized function $f(t)$ is called even, or odd, respectively, if

$$\int_{-\infty}^{\infty} f(t)\phi(t)dt = 0$$

for all odd, or even, good functions $\phi(t)$. Apply this definition to justify that $\delta(t)$ has even symmetry from (A.3-22).

A.3-15 If the "duration" of $\delta(t)$ is zero and its area is one from (A.3-23), give a plausible argument why $\delta(t)$ has an infinite "amplitude" at $t = 0$.

A.3-16 Prove that

$$g(t) \sum_{k=-\infty}^{\infty} \delta(t - kT) = \sum_{k=-\infty}^{\infty} g(kT)\delta(t - kT)$$

where T is a constant and $g(t)$ is a fairly good function.

APPENDIX B

REVIEW OF DETERMINISTIC SIGNAL THEORY

In this appendix we review several aspects of deterministic signal theory. Principally we summarize periodic and nonperiodic signals through their respective Fourier series and Fourier transforms. Network responses to these waveforms are included. The work forms a good refresher on signal theory which should give an adequate background prior to reading Chapter 6. Because this review is meant to stand alone, its notation is not designed to refer to any specific notation of the text, and the reader should be prepared to adjust the notation for whatever application is made of the subjects.

B.1 FOURIER SERIES

The Fourier series applies to periodic waveforms. That is, it applies to waveforms $f(t)$ where $f(t) = f(t + nT_0)$ for any integer value of n. T_0 is called the *fundamental period* of repetition. There are two main forms of Fourier series.

Real Fourier Series

The real form of the Fourier series for a real periodic waveform $f(t)$ having period T_0 is

$$f(t) = \frac{a_0}{2} + \sum_{n=1}^{\infty} a_n \cos(n\omega_0 t) + \sum_{n=1}^{\infty} b_n \sin(n\omega_0 t) \qquad \text{(B.1-1)}$$

where

$$\omega_0 = 2\pi/T_0 \qquad \text{(B.1-2)}$$

is the *fundamental angular frequency* of $f(t)$. The a_n and b_n are called *Fourier series coefficients*, and they are given by

$$a_n = \frac{2}{T_0} \int_{t_0 - (T_0/2)}^{t_0 + (T_0/2)} f(t) \cos(n\omega_0 t)\,dt, \qquad n = 0, 1, 2, \dots \qquad \text{(B.1-3)}$$

$$b_n = \frac{2}{T_0} \int_{t_0 - (T_0/2)}^{t_0 + (T_0/2)} f(t) \sin(n\omega_0 t)\,dt, \qquad n = 1, 2, 3, \dots \qquad \text{(B.1-4)}$$

The constant t_0 in (B.1-3) and (B.1-4) can be any real number. It is only required that the integrals be done over one period of $f(t)$.

Another form of the real Fourier series is

$$f(t) = \frac{a_0}{2} + \sum_{n=1}^{\infty} d_n \cos(n\omega_0 t + \phi_n) \qquad \text{(B.1-5)}$$

where

$$d_n = \sqrt{a_n^2 + b_n^2}, \qquad n = 1, 2, 3, \dots \qquad \text{(B.1-6)}$$

$$\phi_n = -\tan^{-1}\left(\frac{b_n}{a_n}\right), \qquad n = 1, 2, 3, \dots \qquad \text{(B.1-7)}$$

All periodic signals in the real world will have a Fourier series. Formally, however, a periodic signal is guaranteed to have a Fourier series representation (expansion) if it is bounded with period T_0, and has at most a finite number of maxima and minima and a finite number of discontinuities in any period. These conditions are often called the *Dirichlet conditions*. When satisfied, the Fourier series converges to $f(t)$ at all continuous points in time, and to the average of the right and left-hand limits of $f(t)$ at points of discontinuity.

The Fourier series reveals what frequencies are in a waveform (they can only be the discrete values $n\omega_0$) and the amplitudes (d_n) and phases (ϕ_n) of these frequencies.

Complex Fourier Series

By using the identities

$$\cos(x) = \frac{1}{2}(e^{jx} + e^{-jx}) \qquad \text{(B.1-8)}$$

$$\sin(x) = \frac{1}{2j}(e^{jx} - e^{-jx}) \qquad \text{(B.1-9)}$$

the real form of Fourier series can be converted to the form

$$f(t) = \sum_{n=-\infty}^{\infty} c_n e^{jn\omega_0 t} \qquad \text{(B.1-10)}$$

where

$$c_n = \frac{1}{T_0} \int_{t_0-(T_0/2)}^{t_0+(T_0/2)} f(t)\, e^{-jn\omega_0 t}\, dt, \qquad n = 0, \pm 1, \pm 2, \ldots \tag{B.1-11}$$

are also called Fourier series coefficients. However, the coefficients c_n are complex constants, while a_n and b_n are real. Coefficients are related by

$$c_0 = \frac{a_0}{2} \tag{B.1-12}$$

$$c_n = \frac{1}{2}(a_n - jb_n), \qquad n = 1, 2, 3, \ldots \tag{B.1-13}$$

$$c_{-n} = c_n^* = \frac{1}{2}(a_n + jb_n), \qquad n = 1, 2, 3, \ldots \tag{B.1-14}$$

where * represents complex conjugation.

The complex Fourier series of (B.1-10) is often preferred over the real forms of (B.1-1) and (B.1-5) because of its compactness. All carry the same information about $f(t)$.

Example B.1-1 We find the complex Fourier series for the periodic waveform of Fig. B.1-1a. The series is known if we find the coefficients. From (B.1-11) with $t_0 = 0$,

$$c_n = \frac{1}{T_0} \int_{-T_0/2}^{T_0/2} f(t)\, e^{-jn\omega_0 t}\, dt = \frac{1}{T_0} \int_{-\tau/2}^{\tau/2} A e^{-jn\omega_0 t}\, dt$$

$$= \frac{A}{T_0}\left[\frac{e^{-jn\omega_0 t}}{-jn\omega_0} \bigg|_{-\tau/2}^{\tau/2} \right] = \frac{A\tau}{T_0} \frac{\sin(n\omega_0 \tau/2)}{(n\omega_0 \tau/2)}$$

$$= A \frac{\sin(n\pi\tau/T_0)}{n\pi}, \qquad n = 0, \pm 1, \pm 2, \ldots$$

For this waveform c_n turn out real. A plot of the coefficients is shown in Fig. B.1-1b.

Fourier Series of Complex Waveforms

If $f(t)$ is complex and periodic, it can be written as

$$f(t) = r(t) + jx(t) \tag{B.1-15}$$

where $r(t)$ and $x(t)$ are real and periodic. The Fourier series of $f(t)$ becomes the sum of the Fourier series of the real part $r(t)$ plus j times the Fourier series of the imaginary part $x(t)$.

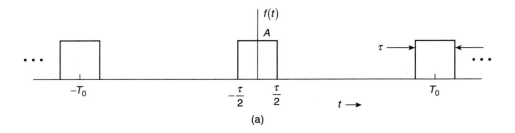

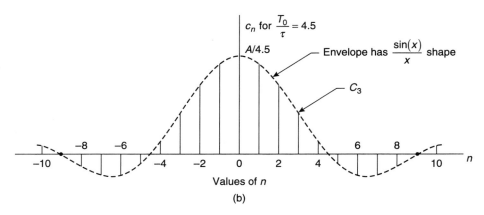

Figure B.1-1 (*a*) A signal of periodic rectangular pulses of period T_0, and (*b*) its Fourier coefficients c_n plotted versus n for $T_0/\tau = 4.5$.

B.2   PROPERTIES OF FOURIER SERIES

There are several useful properties of Fourier series that we next state without proof.

Even and Odd Functions

If a real signal $f(t)$ has even symmetry, which means that

$$f(-t) = f(t) \tag{B.2-1}$$

for all t, then the Fourier series of $f(t)$ will have only even terms (coefficients b_n are all zero).

If $f(t)$ has odd symmetry, which means that

$$f(-t) = -f(t) \tag{B.2-2}$$

for all t, then the Fourier series of $f(t)$ will have only odd terms (coefficients a_n are all zero).

Symmetry Decomposition

If $f(t)$ has arbitrary symmetry, it can be written as the sum of two component signals as

$$f(t) = f_e(t) + f_o(t) \tag{B.2-3}$$

where $f_e(t)$ has even symmetry and $f_o(t)$ has odd symmetry. These components are obtainable from $f(t)$ according to

$$f_e(t) = \frac{1}{2}[f(t) + f(-t)] \tag{B.2-4}$$

$$f_o(t) = \frac{1}{2}[f(t) - f(-t)] \tag{B.2-5}$$

Integration and Differentiation

Formally one can always integrate or differentiate a Fourier series term by term. Whether the resultant series represents (converges to) the integral or derivative of the given function depends on the rate of convergence of the original Fourier series. Generally, integration tends to produce more rapid convergence, whereas differentiation gives less rapid convergence.

If $f(t)$ satisfies the Dirichlet conditions, the integral of $f(t)$ will always equal the termwise integral of the Fourier series of $f(t)$. For differentiation, if $f(t)$ is everywhere continuous and its derivative satisfies the Dirichlet conditions, then, wherever it exists, the derivative of $f(t)$ may be found by termwise differentiation of the Fourier series of $f(t)$.

Average Power[1]

The average power P_{av} in a real periodic signal $f(t)$ is given by (Peebles, 1976)

$$P_{av} = \left(\frac{a_0}{2}\right)^2 + \sum_{n=1}^{\infty} \frac{a_n^2}{2} + \sum_{n=1}^{\infty} \frac{b_n^2}{2}$$

$$= \left(\frac{a_0}{2}\right)^2 + \sum_{n=1}^{\infty} \frac{d_n^2}{2} = \sum_{n=-\infty}^{\infty} |c_n|^2 \tag{B.2-6}$$

If $f(t)$ is complex, the average power is the sum of the average powers of the real and imaginary parts.

---

[1]Values given assume $f(t)$ is a voltage across, or a current through, a 1-Ω resistor, or $f(t)$ is normalized as discussed in Section 1.5.

B.3   FOURIER TRANSFORMS

Let $f(t)$ be a real or complex, periodic or nonperiodic function. Its *Fourier transform* [also called the *spectrum* of $f(t)$], denoted by $F(\omega)$, is given by

$$F(\omega) = \int_{-\infty}^{\infty} f(t) e^{-j\omega t} dt \tag{B.3-1}$$

The waveform $f(t)$ can be recovered from $F(\omega)$ by use of the *inverse Fourier transform*

$$f(t) = \frac{1}{2\pi} \int_{-\infty}^{\infty} F(\omega) e^{j\omega t} d\omega \tag{B.3-2}$$

Equations (B.3-1) and (B.3-2) are said to form a Fourier transform pair. Regardless of whether $f(t)$ is real or complex, $F(\omega)$ is a complex function of angular frequency ω, in general. If $f(t)$ is a voltage (or current), the function $F(\omega)$ is the voltage density (or current density) of the frequencies present in $f(t)$. The magnitude and phase of $F(\omega)$ are called the *amplitude spectrum* and *phase spectrum* of $f(t)$, respectively. It is often convenient to represent $f(t)$ and $F(\omega)$ by the notation

$$f(t) \leftrightarrow F(\omega) \tag{B.3-3}$$

where the double arrow implys the two functions are a Fourier pair.

There are no necessary conditions for $F(\omega)$ to exist for a given $f(t)$. Sufficient conditions that guarantee $F(\omega)$ will exist are (1) $f(t)$ satisfy the Dirichlet conditions on any finite time interval, and (2)

$$\int_{-\infty}^{\infty} |f(t)| \, dt < \infty \tag{B.3-4}$$

Many useful functions do not satisfy these sufficient conditions and do have transforms (e.g., the impulse).

Example B.3-1   We find $F(\omega)$ for the single rectangular pulse of Fig. B.3-1a. Here

$$f(t) = A \operatorname{rect}\left(\frac{t}{T}\right)$$

where the rectangular function is defined by

$$\operatorname{rect}(t) = \begin{cases} 1, & -\frac{1}{2} < t < \frac{1}{2} \\ 0, & \text{elsewhere} \end{cases}$$

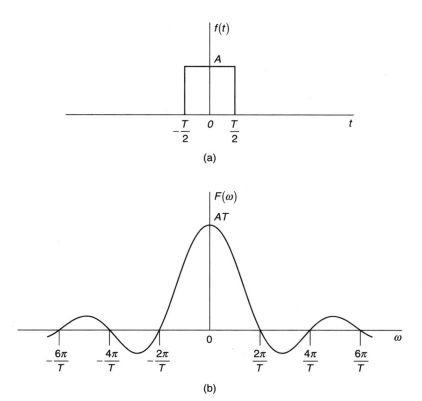

Figure B.3-1   A single rectangular pulse of duration T (a), and (b) its spectrum.

For $F(\omega)$,

$$F(\omega) = \int_{-\infty}^{\infty} f(t)\, e^{-j\omega t}\, dt = \int_{-\infty}^{\infty} A\, \text{rect}\left(\frac{t}{T}\right) e^{-j\omega t}\, dt$$

$$= A \int_{-T/2}^{T/2} e^{-j\omega t}\, dt = A T\, \text{Sa}\left(\frac{\omega T}{2}\right)$$

where the *sampling function* is defined by

$$\text{Sa}(x) = \frac{\sin(x)}{x}$$

$F(\omega)$ is shown in Fig. B.3-1b. Note that all frequencies are present in this waveform except at $\pm n2\pi/T$, $n = 1, 2, 3, \ldots$, where $F(\omega)$ passes through nulls.

B.4   PROPERTIES OF FOURIER TRANSFORMS

There are many properties of Fourier transforms that can save much labor in many transform problems. We list some of these without proofs in the following developments where $f_n(t)$ and $F_n(\omega)$ imply a number of signals and their corresponding Fourier transforms for $n = 1, 2, 3, \ldots, N$ and N may be infinite.

Linearity

$$f(t) = \sum_{n=1}^{N} \alpha_n f_n(t) \leftrightarrow \sum_{n=1}^{N} \alpha_n F_n(\omega) = F(\omega) \qquad \text{(B.4-1)}$$

where the α_n are real or complex constants

Time and Frequency Shifting

For ω_0 and t_0 both real constants,

$$f(t - t_0) \leftrightarrow F(\omega) e^{-j\omega t_0} \qquad \text{(B.4-2)}$$

$$f(t) e^{j\omega_0 t} \leftrightarrow F(\omega - \omega_0) \qquad \text{(B.4-3)}$$

Scaling

For α a real constant

$$f(\alpha t) \leftrightarrow \frac{1}{|\alpha|} F\left(\frac{\omega}{\alpha}\right) \qquad \text{(B.4-4)}$$

Duality

$$F(t) \leftrightarrow 2\pi f(-\omega) \qquad \text{(B.4-5)}$$

Differentiation

$$\frac{d^n f(t)}{dt^n} \leftrightarrow (j\omega)^n F(\omega) \qquad \text{(B.4-6)}$$

$$(-jt)^n f(t) \leftrightarrow \frac{d^n F(\omega)}{d\omega^n} \qquad \text{(B.4-7)}$$

Integration

$$\int_{-\infty}^{t} f(\xi) \, d\xi \leftrightarrow \pi F(0) \, \delta(\omega) + \frac{F(\omega)}{j\omega} \qquad \text{(B.4-8)}$$

$$\pi f(0) \, \delta(t) - \frac{f(t)}{jt} \leftrightarrow \int_{-\infty}^{\omega} F(\xi) \, d\xi \qquad \text{(B.4-9)}$$

Conjugation and Sign Change

$$f^*(t) \leftrightarrow F^*(-\omega) \qquad \text{(B.4-10)}$$

$$f^*(-t) \leftrightarrow F^*(\omega) \qquad \text{(B.4-11)}$$

$$f(-t) \leftrightarrow F(-\omega) \qquad \text{(B.4-12)}$$

Convolution

$$f(t) = \int_{-\infty}^{\infty} f_1(\xi) f_2(t - \xi) \, d\xi \leftrightarrow F_1(\omega) F_2(\omega) = F(\omega) \qquad \text{(B.4-13)}$$

$$f(t) = f_1(t) f_2(t) \leftrightarrow \frac{1}{2\pi} \int_{-\infty}^{\infty} F_1(\xi) F_2(\omega - \xi) \, d\xi = F(\omega) \qquad \text{(B.4-14)}$$

Correlation

$$f(t) = \int_{-\infty}^{\infty} f_1^*(\xi) f_2(\xi + t) \, d\xi \leftrightarrow F_1^*(\omega) F_2(\omega) = F(\omega) \qquad \text{(B.4-15)}$$

$$f(t) = f_1^*(t) f_2(t) \leftrightarrow \frac{1}{2\pi} \int_{-\infty}^{\infty} F_1^*(\xi) F_2(\xi + \omega) \, d\xi = F(\omega) \qquad \text{(B.4-16)}$$

Parseval's Theorem

$$\int_{-\infty}^{\infty} f_1^*(\xi) f_2(\xi) d\xi = \frac{1}{2\pi} \int_{-\infty}^{\infty} F_1^*(\omega) F_2(\omega) \, d\omega \qquad \text{(B.4-17)}$$

If $f_1(t) = f_2(t) = f(t)$,

$$\int_{-\infty}^{\infty} |f(t)|^2 dt = \frac{1}{2\pi} \int_{-\infty}^{\infty} |F(\omega)|^2 \, d\omega \qquad \text{(B.4-18)}$$

Area

$$F(0) = \int_{-\infty}^{\infty} f(t) \, dt \qquad \text{(B.4-19)}$$

$$f(0) = \frac{1}{2\pi} \int_{-\infty}^{\infty} F(\omega) \, d\omega \qquad \text{(B.4-20)}$$

B.5   SIGNAL RESPONSE OF NETWORKS

We define a real linear time-invariant network as one having a real impulse response, denoted by $h(t)$. If an arbitrary signal $f(t)$ (real, complex, random, nonrandom, periodic, or nonperiodic) is applied to the network's input, the response at its output, denoted by $g(t)$, is

$$g(t) = \int_{-\infty}^{\infty} f(\xi) h(t-\xi) d\xi = \int_{-\infty}^{\infty} f(t-\xi) h(\xi) d\xi \qquad \text{(B.5-1)}$$

These expressions are convolution integrals. If we next define $H(\omega)$ to be the Fourier transform of $h(t)$, called the *transfer function*, we can apply (B.4-13) to the convolutions to obtain the spectrum $G(\omega)$ of $g(t)$:

$$G(\omega) = F(\omega) H(\omega) \qquad \text{(B.5-2)}$$

When the input is specified in time by $f(t)$, the network is defined by $h(t)$, and (B.5-1) is used to find the output as a time function, the process is referred to as time-domain analysis. If the input is specified in the frequency domain by $F(\omega)$, then use of (B.5-2) to find the spectrum $G(\omega)$ of the output is called frequency-domain analysis.

Networks can also be defined using complex impulse responses. These methods are developed in Chapter 6 and have some special advantages to radar.

B.6 MULTIDIMENSIONAL FOURIER TRANSFORMS

Suppose that we are interested in defining the Fourier transform of a function of *two* "time" variables t_1 and t_2, as denoted by $f(t_1, t_2)$. We define the transform, denoted as $F(\omega_1, \omega_2)$, as the iteration of two transforms. On transforming with respect to t_1, first we obtain a function of t_2 and ω_1, the "frequency" variable of the first transform:

$$F(\omega_1, t_2) = \int_{-\infty}^{\infty} f(t_1, t_2) e^{-j\omega_1 t_1} dt_1 \qquad \text{(B.6-1)}$$

Next we form the second transform with respect to t_2 using a second "frequency" variable ω_2:

$$F(\omega_1, \omega_2) = \int_{-\infty}^{\infty} F(\omega_1, t_2) e^{-j\omega_2 t_2} dt_2$$

$$= \int_{-\infty}^{\infty} \int_{-\infty}^{\infty} f(t_1, t_2) e^{-j\omega_1 t_1 - j\omega_2 t_2} dt_1 dt_2 \qquad \text{(B.6-2)}$$

A similar inverse operation on $F(\omega_1, \omega_2)$ gives

$$f(t_1, t_2) = \frac{1}{(2\pi)^2} \int_{-\infty}^{\infty} \int_{-\infty}^{\infty} F(\omega_1, \omega_2) e^{j\omega_1 t_1 + j\omega_2 t_2} d\omega_1, d\omega_2 \qquad \text{(B.6-3)}$$

The two-dimensional Fourier transform is (B.6-2). The inverse two-dimensional Fourier transform is (B.6-3).

The extension of our procedures to an N-dimensional Fourier transform pair is straightforward:

$$F(\omega_1, \ldots, \omega_N) = \int_{-\infty}^{\infty} \cdots \int_{-\infty}^{\infty} f(t_1, \ldots, t_N) e^{-j\omega_1 t_1 - \cdots - j\omega_N t_N} dt_1 \cdots dt_N \tag{B.6-4}$$

$$f(t_1, \ldots, t_N) = \frac{1}{(2\pi)^N} \int_{-\infty}^{\infty} \cdots \int_{-\infty}^{\infty} F(\omega_1, \ldots, \omega_N) e^{j\omega_1 t_1 + \cdots + j\omega_N t_N} d\omega_1 \cdots d\omega_N \tag{B.6-5}$$

B.7 PROPERTIES OF TWO-DIMENSIONAL FOURIER TRANSFORMS

All the properties in Section B.4 for one-dimensional Fourier transforms can be extended to two dimensions. We list below only a few of the most important ones.

Linearity

For real or complex constants α_n and functions $f_n(t_1, t_2)$ having transforms $F_n(\omega_1, \omega_2)$ for $n = 1, 2, \ldots, N$, with N possibly infinite,

$$f(t_1, t_2) = \sum_{n=1}^{N} \alpha_n f_n(t_1, t_2) \leftrightarrow \sum_{n=1}^{N} \alpha_n F_n(\omega_1, \omega_2) = F(\omega_1, \omega_2) \tag{B.7-1}$$

Time and Frequency Shifting

For real constants t_{01}, t_{02}, ω_{01}, and ω_{02},

$$f(t_1 - t_{01}, t_2 - t_{02}) \leftrightarrow F(\omega_1, \omega_2) e^{-j\omega_1 t_{01} - j\omega_2 t_{02}} \tag{B.7-2}$$

$$f(t_1, t_2) e^{j\omega_{01} t_1 + j\omega_{02} t_2} \leftrightarrow F(\omega_1 - \omega_{01}, \omega_2 - \omega_{02}) \tag{B.7-3}$$

Differentiation

For n and $m = 0, 1, 2, \ldots$,

$$\frac{d^{n+m} f(t_1, t_2)}{dt_1^n dt_2^m} \leftrightarrow (j\omega_1)^n (j\omega_2)^m F(\omega_1, \omega_2) \tag{B.7-4}$$

$$(-jt_1)^n (-jt_2)^m f(t_1, t_2) \leftrightarrow \frac{d^{n+m} F(\omega_1, \omega_2)}{d\omega_1^n d\omega_2^m} \tag{B.7-5}$$

Convolution

$$f(t_1, t_2) = \int_{-\infty}^{\infty} \int_{-\infty}^{\infty} f_1(\xi_1, \xi_2) f_2(t_1 - \xi_1, t_2 - \xi_2) d\xi_1 d\xi_2$$

$$\leftrightarrow F_1(\omega_1, \omega_2) F_2(\omega_1, \omega_2) = F(\omega_1, \omega_2) \tag{B.7-6}$$

$$f(t_1, t_2) = f_1(t_1, t_2) f_2(t_1, t_2)$$

$$\leftrightarrow \frac{1}{(2\pi)^2} \int_{-\infty}^{\infty} \int_{-\infty}^{\infty} F_1(\xi_1, \xi_2) F_2(\omega_1 - \xi_1, \omega_2 - \xi_2) \, d\xi_1 d\xi_2$$

$$= F(\omega_1, \omega_2) \tag{B.7-7}$$

Correlation

$$f(t_1, t_2) = \int_{-\infty}^{\infty} \int_{-\infty}^{\infty} f_1^*(\xi_1, \xi_2) f_2(\xi_1 + t_1, \xi_2 + t_2) \, d\xi_1 d\xi_2$$

$$\leftrightarrow F_1^*(\omega_1, \omega_2) F_2(\omega_1, \omega_2) = F(\omega_1, \omega_2) \tag{B.7-8}$$

$$f(t_1, t_2) = f_1^*(t_1, t_2) f_2(t_1, t_2)$$

$$\leftrightarrow \frac{1}{(2\pi)^2} \int_{-\infty}^{\infty} \int_{-\infty}^{\infty} F_1^*(\xi_1, \xi_2) F_2(\xi_1 + \omega_1, \xi_2 + \omega_2) \, d\xi_1 d\xi_2$$

$$= F(\omega_1, \omega_2) \tag{B.7-9}$$

Parseval's Theorem

$$\int_{-\infty}^{\infty} \int_{-\infty}^{\infty} f_1^*(\xi_1, \xi_2) f_2(\xi_1, \xi_2) \, d\xi_1 d\xi_2$$

$$= \frac{1}{(2\pi)^2} \int_{-\infty}^{\infty} \int_{-\infty}^{\infty} F_1^*(\omega_1, \omega_2) F_2(\omega_1, \omega_2) \, d\omega_1 d\omega_2 \tag{B.7-10}$$

If $f_1(\xi_1, \xi_2) = f_2(\xi_1, \xi_2) = f(\xi_1, \xi_2)$

$$\int_{-\infty}^{\infty} \int_{-\infty}^{\infty} |f(\xi_1, \xi_2)|^2 \, d\xi_1 d\xi_2 = \frac{1}{(2\pi)^2} \int_{-\infty}^{\infty} \int_{-\infty}^{\infty} |F(\omega_1, \omega_2)|^2 d\omega_1 d\omega_2 \tag{B.7-11}$$

PROBLEMS

B.1-1 Find the real Fourier series of the waveform in Fig. PB.1-1, where in the central period

$$f(t) = \begin{cases} A \cos \left(\dfrac{\pi t}{2T} \right), & -T < t < T \\ 0, & \text{elsewhere} \end{cases}$$

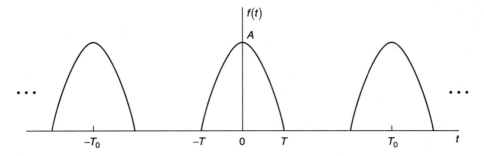

Figure PB.1-1   A periodic sequence of cosine-shaped pulses.

B.1-2 Find the real Fourier series of the waveform of Fig. PB.1-2 where in the central period

$$f(t) = \begin{cases} \dfrac{A}{T}t, & -T < t < T \\ 0, & \text{elsewhere} \end{cases}$$

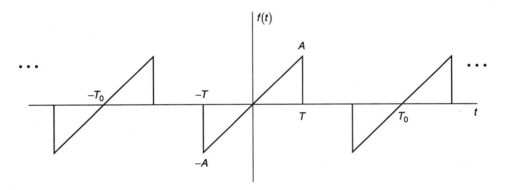

Figure PB.1-2   A periodic sequence of sawtooth-shaped pulses.

B.1-3 A signal is periodic and in the central period it is defined by

$$f(t) = \begin{cases} A \cos\left(\dfrac{2\pi}{t}\right), & -T < t < T \\ 0, & \text{elsewhere} \end{cases}$$

Does it satisfy the Dirichlet conditions? Discuss.

B.1-4 A periodic signal with period T_0 is defined in its central period by

$$f(t) = \begin{cases} \dfrac{A}{T}t, & 0 \le t < T \\ 0, & \text{elsewhere} \end{cases}$$

Find the complex Fourier series of $f(t)$.

B.1-5 The following functions define periodic signals in their central periods. In every case the given functions are nonzero only for $-T < t < T$, where $T < T_0/2$ with T_0 the period. Discuss what real Fourier series coefficients are nonzero and why.

(a) $\dfrac{5}{1 + 6t^2}$,   (b) $e^{-12|t|}$,   (c) $3\sin(15t)$ (d) $2\cos(2t)e^{-6t}$,

(e) $\dfrac{4t^3}{(1 + 7t^2)^2}$

B.2-1 A periodic signal $f(t)$ has no dc component. That is, $a_0 = 0$ or $c_0 = 0$ in its Fourier series. Write a general expression for the integral of $f(t)$ using (a) the real Fourier series and (b) the complex Fourier series.

B.2-2 A periodic signal is shown in Fig. PB.2-2. Find the even and odd components of the signal by using graphical methods.

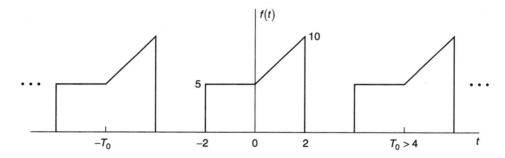

Figure PB.2-2   A periodic sequence of pulses.

B.2-3 A periodic signal $f(t)$ has a real Fourier series with coefficients

$$a_n = 0, \qquad n = 0, 1, 2, \dots,$$

$$b_n = \frac{2A}{n\pi}\left\{\frac{\sin(n2\pi T/T_0)}{(n2\pi T/T_0)} - \cos\left(n2\pi\frac{T}{T_0}\right)\right\}, \qquad n = 1, 2, \dots$$

The total average power in $f(t)$ is known to be $P_{av} = 2A^2 T/(3T_0)$. Find the percentage of total average power that is in the first (a) one, (b) two, and (c) three components of the Fourier series. Assume that $T_0 = 6T$.

B.2-4 A full-wave rectified cosine wave $f(t)$ with peak amplitude A is often filtered to generate dc. The rectified signal has a real Fourier series with only coefficients a_n given by

$$a_n = \frac{(-1)^n A}{\pi[0.25 - n^2]}, \qquad n = 0, 1, 2, \ldots$$

(a) Find the average power in the dc component of $f(t)$.
(b) Find the average power in the first and second non zero sinusoidal terms in the series for $f(t)$.
(c) If the total average power in $f(t)$ is $P_{av} = A^2/2$ what fraction of this total is in dc power after filtering all harmonic terms to zero?

B.3-1 Find the Fourier transform of the *triangular function* defined by

$$f(t) = A \operatorname{tri}\left(\frac{t}{T}\right) = \begin{cases} A\left[1 - \dfrac{|t|}{T}\right], & -T < t < T \\ 0, & \text{elsewhere} \end{cases}$$

where A and $T > 0$ are constants.

B.3-2 Find the Fourier transform of a single cosine-shaped pulse as defined by $f(t)$ in Problem B.1-1.

B.3-3 Find the Fourier transform of a pulse defined by

$$f(t) = A e^{-\alpha|t|}$$

where A and $\alpha > 0$ are constants.

B.4-1 Show that (B.4-2) is true.

B.4-2 Given a periodic sequence of impulses

$$f(t) = \sum_{n=-\infty}^{\infty} \delta(t - nT_0)$$

find the complex Fourier series for $f(t)$; then Fourier transform the series to obtain the following relationship:

$$\sum_{n=-\infty}^{\infty} \delta(t - nT_0) \leftrightarrow \frac{2\pi}{T_0} \sum_{n=-\infty}^{\infty} \delta(\omega - n\omega_0)$$

where $\omega_0 = 2\pi/T_0$. Your developments should also prove that

$$\sum_{n=-\infty}^{\infty} \delta(t - nT_0) = \frac{1}{T_0} \sum_{n=-\infty}^{\infty} e^{jn\omega_0 t}$$

★B.4-3 The waveform of Fig. PB.4-3 is approximated by three linear line segments. Find an approximate spectrum for $f(t)$ by using the segmented approximation, (B.4-6) with $n = 1$, and the transform pair of Example B.3-1.

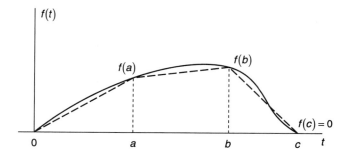

Figure PB.4-3   A pulse approximated by three linear segments.

B.4-4 Find the Fourier transform of

$$f(t) = [A + f_m(t)]\cos(\omega_0 t)$$

in terms of the spectrum $F_m(\omega)$ of $f_m(t)$, and A and ω_0 which are constants. [*Hint*: Use the fact that $A \leftrightarrow 2\pi A\delta(\omega)$.]

B.4-5 Integrate twice by parts to solve for $F(\omega)$ from (B.3-1) using the waveform

$$f(t) = \begin{cases} \dfrac{A}{T^2}t^2, & 0 \le t < T \\ 0, & \text{elsewhere} \end{cases}$$

where A and $T > 0$ are constants.

B.4-6 Use property (B.4-6) with $n = 2$ to find the Fourier transform of the signal of Fig. PB.4-6. [*Hint*: Use the known transform pair $\delta(t) \leftrightarrow 1$ with property (B.4-2).]

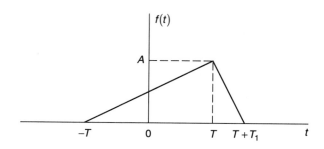

Figure PB.4-6 A triangular pulse.

B.4-7 Prove property (B.4-12).

B.4-8 Show that (B.4-17) is true.

B.5-1 A signal

$$f(t) = Au(t)e^{-Wt}$$

where A and $W > 0$ are constants, is applied to a linear network for which

$$H(\omega) = \frac{W_0}{W_0 + j\omega}$$

where $W_0 > 0$ is a constant. Use (B.5-1) to find the network's output signal $g(t)$.

B.6-1 A two-dimensional function is defined by

$$f(t_1, t_2) = \begin{cases} A, & -T_1 < t_1 < T_1, \quad -T_2 < t_2 < T_2 \\ 0, & \text{elsewhere} \end{cases}$$

where A, $T_1 > 0$, and $T_2 > 0$ are constants. Find the two-dimensional Fourier transform of $f(t_1, t_2)$.

B.7-1 Show that (B.7-6) is true.

APPENDIX C

REVIEW OF RANDOM SIGNAL THEORY

A waveform having properties that must be described in some probabilistic way can be called a *random signal*. The waveform may be either a desired signal, such as a computer waveform, or an undesired signal, such as noise. In either case the proper description of a random signal requires the theory of random processes. In this appendix we review the fundamental elements needed to understand random processes. Since our purpose is review, few proofs are given. For detail and proofs of the material reviewed, the reader is referred to the rich literature on random processes. A few references to begin further reading are Peebles (1993), Helstrom (1991), Shanmugan and Breipohl (1988), and Papoulis (1984).

Random processes first require an understanding of statistical averages and random variables, which in turn depend on basic topics in probability. We review all these subjects, starting with the basic elements of probability.

C.1 BASIC TOPICS AND PROBABILITY

Our intuition allows us to define a *random experiment* as some situation where the *outcome* of a *trial* of the experiment occurs in a random manner. An example experiment is to randomly choose a card from a deck of 52 cards. The choosing of a card is a trial, while the card chosen is the outcome.

Sample Spaces

The set of all possible outcomes in a given random experiment, denoted by S, is called the *sample space*. The sample space is called *discrete* if it has a finite number of outcomes (e.g., 52 in the example of choosing a card from a 52-card deck). It is *continuous* if a continuum of outcomes is possible (such as all the numbers from 0 to 1). Any one possible outcome of a sample space is called an *element* of the space.

A sample space can be any quantity. It can be a collection of chairs, numbers, people, or fish, to take some examples.

Events

An *event* is a subset of the sample space. It can be defined to relate to a *characteristic* of outcomes as opposed to the outcomes themselves. For example, an event K might be defined as "drawing a king" from an ordinary 52-card deck. Here four sample space elements are kings, so the event K is a set of these four elements, which is a subset of S. Since an event is a set, it can be discrete or continuous.

Two events that have no common elements are called *mutually exclusive* or *disjoint*.

Probability and Axioms

A nonnegative number called *probability* is assigned to each event defined on a sample space. Probability is therefore a function of the events defined. We use the notation $P(A)$ to denote the probability of event A.

Probabilities are chosen to satisfy three axioms; these are

Axiom 1

$$P(A) \geq 0 \tag{C.1-1}$$

Axiom 2

$$P(S) = 1 \tag{C.1-2}$$

Axiom 3

$$P\left(\bigcup_{n=1}^{N} A_n\right) = \sum_{n=1}^{N} P(A_n) \qquad \text{if } A_m \cap A_n = \emptyset \tag{C.1-3}$$

Axiom 1 simply states that probability is selected to be nonnegative. Because the sample space S contains all possible outcomes of an experiment, it *must occur* on any trial, so it is called the *certain event*. Axiom 2 simply states that the certain event has the largest probability, selected to be one. Thus for any event A, $0 \leq P(A) \leq 1$.

The notations $\cup$ and $\cap$ in axiom 3 refer to set *union* and *intersection*, respectively. The union is the set of elements contained in either or both of two events, while the intersection is the set of only those elements common to the two events. The symbol $\emptyset$ refers to a set with no elements; it is the *impossible event* and its probability is zero. The axiom states that the probability of the union of N disjoint events is the sum of their individual probabilities.

Joint and Conditional Probabilities

The probability that two events A and B occur together is called *joint probability*, denoted by $P(A \cap B)$; it corresponds to the set of sample space elements that are jointly present in the sets A and B. If $P(A \cup B)$ denotes the probability of the union of events A and B (elements of $A \cup B$ are those in A or B or both), then

$$P(A \cap B) = P(A) + P(B) - P(A \cup B) \tag{C.1-4}$$

$$P(A \cup B) = P(A) + P(B) - P(A \cap B) \le P(A) + P(B) \tag{C.1-5}$$

Conditional probability of an event A, given event B, is denoted by $P(A|B)$ and defined by

$$P(A|B) = \frac{P(A \cap B)}{P(B)} \tag{C.1-6}$$

where $P(B) > 0$ is assumed. Similarly, if $P(A) > 0$,

$$P(B|A) = \frac{P(B \cap A)}{P(A)} = \frac{P(A \cap B)}{P(A)} \tag{C.1-7}$$

so

$$P(A|B)P(B) = P(B|A)P(A) \tag{C.1-8}$$

which is called *Bayes's rule*.

Statistical Independence

Two events A and B are said to be *statistically independent* if

$$P(A|B) = P(A) \tag{C.1-9}$$

Also

$$P(B|A) = P(B) \tag{C.1-10}$$

Alternatively, they are statistically independent if

$$P(A \cap B) = P(A)P(B) \tag{C.1-11}$$

Three events A_1, A_2, and A_3 are statistically independent if *all* the following are true:

$$P(A_1 \cap A_2) = P(A_1)P(A_2) \tag{C.1-12a}$$

$$P(A_1 \cap A_3) = P(A_1)P(A_3) \tag{C.1-12b}$$

$$P(A_2 \cap A_3) = P(A_2)P(A_3) \tag{C.1-12c}$$

$$P(A_1 \cap A_2 \cap A_3) = P(A_1)P(A_2)P(A_3) \tag{C.1-12d}$$

Equations (C.1-12) indicate the events must be independent as all pairs and as a triple.

More generally, N events A_1, A_2, ..., A_N, are called statistically independent if independent as all pairs, as all triples, ..., and as an N-tuple. There is a total of $2^N - N - 1$ equations to be satisfied.

Example C.1-1   A box contains 100 thoroughly mixed resistors of identical shape. There are 57 100-Ω resistors, of which 20 have a 1/2-W power rating and 37 are 1 W. There are also 43 220-Ω units with 27 having 1/2-W rating and 16 are 1 W. If the experiment is to randomly select one resistor from the box and we define events $A = $ "select a 100 Ω resistor," and $B = $ "select a 1-W resistor," we find probabilities $P(A)$, $P(B)$, $P(A \mid B)$, $P(B \mid A)$, and $P(A \cap B)$.

Clearly, with 57 100-Ω resistors in the batch of 100, $P(A) = 57/100 = 0.57$. Similarly $P(B) = 53/100 = 0.53$. To find $P(A \mid B)$, we note that event B must occur (a condition), so we are limited to only 53 resistors that can be considered; of these only 37 satisfy event A, so $P(A \mid B) = 37/53$. Similarly $P(B \mid A) = 37/57$. Next only 37 resistors satisfy both events A and B, so $P(A \cap B) = 37/100$.

We note that $P(A \mid B) = 37/53 \neq P(A) = 57/100$, so events A and B are not statistically independent.

C.2   RANDOM VARIABLES, DISTRIBUTIONS, AND DENSITIES

Elements of a sample space may be anything whatsoever. For engineering purposes it is convenient to define a method whereby quantities related to a sample space are always numerical.

Random Variables

A *real random variable* is defined as a real function of the elements of a sample space S. Let X represent a random variable (capital letter) having values x. Then $X(s)$ is a function of the elements s of sample space S. The function $X(\cdot)$ essentially maps each element of S to some point on the real line x. The probability that X has some range of values is equal to the probability of the elements of S that map to the points in the range of values of X.

There can be more than one random variable defined on a sample space. A random variable is discrete or continuous if it takes on only discrete or continuous values, respectively. *Mixed random variables* are also possible; they have some discrete and some continuous values.

Distribution and Density Functions

The event $(X \leq x)$ is read to be the set of values that X can have that do not exceed x. The probability of this event is called the *cumulative probability distribution function* of X, denoted by $F_X(x)$; it is called the distribution function or just distribution for convenience. Thus

$$F_X(x) = P(X \leq x) \tag{C.2-1}$$

$F_X(x)$ for a discrete random variable is a stair-step function that is nondecreasing with increasing x. A continuous random variable has a continuously nondecreasing function $F_X(x)$.

Properties of $F_X(x)$ are

$$F_X(-\infty) = 0 \tag{C.2-2}$$

$$F_X(\infty) = 1 \tag{C.2-3}$$

$$0 \le F_X(x) \le 1 \tag{C.2-4}$$

$$F_X(x_1) \le F_X(x_2) \qquad \text{if } x_1 < x_2 \tag{C.2-5}$$

$$P(x_1 < X \le x_2) = F_X(x_2) - F_X(x_1) \tag{C.2-6}$$

$$F_X(x^+) = F_X(x) \tag{C.2-7}$$

The *probability density function*, denoted by $f_X(x)$, is defined as the derivative of the distribution

$$f_X(x) = \frac{dF_X(x)}{dx} \tag{C.2-8}$$

As short-form reference $f_X(x)$ is often just called the density function or just the density of X.

Properties of the density function are

$$0 \le f_X(x) \tag{C.2-9}$$

$$\int_{-\infty}^{\infty} f_X(x)dx = 1 \tag{C.2-10}$$

$$F_X(x) = \int_{-\infty}^{x} f_X(\xi)d\xi \tag{C.2-11}$$

$$P(x_1 < X \le x_2) = \int_{x_1}^{x_2} f_X(x)dx \tag{C.2-12}$$

Conditional Distribution and Density Functions

The *conditional distribution function* of a random variable X, denoted by $F_X(x\,|\,B)$, is defined by

$$F_X(x\,|\,B) = P(X \le x\,|\,B) = \frac{P(X \le x \cap B)}{P(B)} \tag{C.2-13}$$

for some conditioning event B. Properties of $F_X(x\,|\,B)$ are

$$F_X(-\infty\,|\,B) = 0 \tag{C.2-14}$$

$$F_X(\infty\,|\,B) = 1 \tag{C.2-15}$$

$$0 \le F_X(x\,|\,B) \le 1 \tag{C.2-16}$$

$$F_X(x_1\,|\,B) \le F_X(x_2\,|\,B) \qquad \text{if } x_1 < x_2 \tag{C.2-17}$$

$$P(x_1 < X \le x_2 \,|\, B) = F_X(x_2 \,|\, B) - F_X(x_1 \,|\, B) \tag{C.2-18}$$

$$F_X(x_1^+ \,|\, B) = F_X(x \,|\, B) \tag{C.2-19}$$

The *conditional density function*, denoted by $f_X(x \,|\, B)$, is defined by the derivative

$$f_X(x \,|\, B) = \frac{dF_X(x \,|\, B)}{dx} \tag{C.2-20}$$

Properties of $f_X(x \,|\, B)$ are

$$0 \le f_X(x \,|\, B) \tag{C.2-21}$$

$$\int_{-\infty}^{\infty} f_X(x \,|\, B) dx = 1 \tag{C.2-22}$$

$$F_X(x \,|\, B) = \int_{-\infty}^{x} f_X(\xi \,|\, B) d\xi \tag{C.2-23}$$

$$P(x_1 < X \le x_2 \,|\, B) = \int_{x_1}^{x_2} f_X(x \,|\, B) dx \tag{C.2-24}$$

Event B in (C.2-13) through (C.2-24) can be defined in many ways. It is often defined by values of a second random variable, such as Y. As examples, B might be defined by any of the events $B = (Y \le y)$, $B = (y_1 < Y \le y_2)$, or $B = (y_1 < Y)$, where y, y_1, and y_2 are specific values of Y.

Multiple Random Variables

For multiple random variables $X_1, X_2, \ldots, X_N$, we treat values of each as a component of an N-dimensional *vector random variable*. The *joint distribution function* and *joint density function* are defined by

$$F_{X_1, X_2, \ldots, X_N}(x_1, x_2, \ldots, x_N) = P(X_1 \le x_1, X_2 \le x_2, \ldots, X_N \le x_N) \tag{C.2-25}$$

$$f_{X_1, X_2, \ldots, X_N}(x_1, x_2, \ldots, x_N) = \frac{d^N F_{X_1, X_2, \ldots, X_N}(x_1, x_2, \ldots, x_N)}{dx_1 \, dx_2 \ldots dx_N} \tag{C.2-26}$$

For the important case of only two random variables X and Y the joint distribution function is

$$F_{X,Y}(x, y) = P(X \le x, Y \le y) \tag{C.2-27}$$

which has the properties

$$F_{X,Y}(-\infty, -\infty) = 0 \tag{C.2-28}$$

$$F_{X,Y}(-\infty, y) = 0 \tag{C.2-29}$$

$$F_{X,Y}(x, -\infty) = 0 \tag{C.2-30}$$

$$F_{X,Y}(\infty, \infty) = 1 \tag{C.2-31}$$

$$0 \le F_{X,Y}(x, y) \le 1 \tag{C.2-32}$$

$$F_{X,Y}(x, y) = \text{nondecreasing in both } x \text{ and } y \tag{C.2-33}$$

$$F_{X,Y}(x_2, y_2) + F_{X,Y}(x_1, y_1) - F_{X,Y}(x_1, y_2) - F_{X,Y}(x_2, y_1)$$
$$= P(x_1 < X \leq x_2, y_1 < Y \leq y_2) \tag{C.2-34}$$

$$F_{X,Y}(x, \infty) = F_X(x) \tag{C.2-35}$$

$$F_{X,Y}(\infty, y) = F_Y(y) \tag{C.2-36}$$

The two-variable (two-variate) density is

$$f_{X,Y}(x, y) = \frac{d^2 F_{X,Y}(x, y)}{dx \, dy} \tag{C.2-37}$$

which has the properties

$$0 \leq f_{X,Y}(x, y) \tag{C.2-38}$$

$$\int_{-\infty}^{\infty} \int_{-\infty}^{\infty} f_{X,Y}(x, y) \, dx \, dy = 1 \tag{C.2-39}$$

$$F_{X,Y}(x, y) = \int_{-\infty}^{y} \int_{-\infty}^{x} f_{X,Y}(\xi_1, \xi_2) \, d\xi_1 \, d\xi_2 \tag{C.2-40}$$

$$F_X(x) = \int_{-\infty}^{x} \int_{-\infty}^{\infty} f_{X,Y}(\xi_1, \xi_2) \, d\xi_2 \, d\xi_1 \tag{C.2-41}$$

$$F_Y(y) = \int_{-\infty}^{y} \int_{-\infty}^{\infty} f_{X,Y}(\xi_1, \xi_2) \, d\xi_1 \, d\xi_2 \tag{C.2-42}$$

$$P(x_1 < X \leq x_2, y_1 < Y \leq y_2) = \int_{y_1}^{y_2} \int_{x_1}^{x_2} f_{X,Y}(x, y) \, dx \, dy \tag{C.2-43}$$

$$f_X(x) = \int_{-\infty}^{\infty} f_{X,Y}(x, y) \, dy \tag{C.2-44}$$

$$f_Y(y) = \int_{-\infty}^{\infty} f_{X,Y}(x, y) \, dx \tag{C.2-45}$$

Joint conditional distribution and density functions for several random variables are direct extensions of (C.2-13) and (C.2-20).

Example C.2-1   The joint probability density function of two random variables X and Y is

$$f_{X,Y}(x, y) = \frac{1}{(b-a)(d-c)} \text{rect}\left(\frac{x}{b-a}\right) \text{rect}\left(\frac{y}{d-c}\right)$$

where $a, b > a, c,$ and $d > c$ are all positive constants. We find $F_{X,Y}(x, y)$ from (C.2-40).

$$F_{X,Y}(x, y) = \int_{-\infty}^{y} \int_{-\infty}^{x} f_{X,Y}(\xi_1, \xi_2) \, d\xi_1 \, d\xi_2$$

$$= \frac{1}{(d-c)} \int_{-\infty}^{y} \text{rect}\left(\frac{\xi_2}{d-c}\right) d\xi_2 \frac{1}{(b-a)} \int_{-\infty}^{x} \text{rect}\left(\frac{\xi_1}{b-a}\right) d\xi_1$$

$$= \begin{cases} 0, & x \le a \quad \text{and/or} \quad y \le c \\ \dfrac{(x-a)(y-c)}{(b-a)(d-c)}, & a < x \le b \quad \text{and} \quad c < y \le d \\ \dfrac{x-a}{b-a}, & a < x \le b \quad \text{and} \quad d < y \\ \dfrac{y-c}{d-c}, & b < x \quad \text{and} \quad c < y \le d \\ 1, & b < x \quad \text{and} \quad d < y \end{cases}$$

Statistical Independence

Two random variables X and Y are called statistically independent if

$$F_{X,Y}(x, y) = F_X(x) F_Y(y) \tag{C.2-46}$$

or

$$f_{X,Y}(x, y) = f_X(x) f_Y(y) \tag{C.2-47}$$

For N random variables $X_1, X_2, \ldots, X_N$, we define events

$$A_i = (X_i \le x_i), \qquad i = 1, 2, \ldots, N \tag{C.2-48}$$

and say the random variables are statistically independent if events A_i are statistically independent as described in Section C.1 [e.g., by (C.1-12) for $N = 3$].

C.3   STATISTICAL AVERAGES

More often than not we are interested in some characteristic of a random variable (often called a *statistic*) as opposed to the random variable itself. These characteristics are almost always some form of statistical average which is defined in this section.

Average of a Function of Random Variables

Let $g(X)$ be some real function of the real random variable X. The *statistical average* of $g(X)$, also variously called the *expected value*, *mean value*, or just *average value*, is defined by

$$E[g(X)] = \int_{-\infty}^{\infty} g(x) f_X(x) dx \tag{C.3-1}$$

Here $E[\cdot]$ is the statistical expectation operation. For a function $g(X_1, X_2, \ldots, X_N)$ of N random variables $X_1, X_2, \ldots, X_N$, the extension of (C.3-1) is

$$E[g(X_1, X_2, \ldots, X_N)]$$
$$= \int_{-\infty}^{\infty} \cdots \int_{-\infty}^{\infty} g(x_1, x_2, \ldots, x_N) f_{X_1, X_2, \ldots, X_N}(x_1, x_2, \ldots, x_N) dx_1 \, dx_2 \ldots dx_N \tag{C.3-2}$$

Moments

Of special interest are functions $g(\cdot)$ that lead to moments of random variables. For a single random variable X the function

$$g(X) = X^n \qquad\qquad\qquad (C.3\text{-}3)$$

gives moments about the origin, denoted by m_n, according to

$$m_n = E[X^n] = \int_{-\infty}^{\infty} x^n f_X(x)\,dx, \qquad n = 0, 1, 2, \ldots \qquad (C.3\text{-}4)$$

Moments m_1 and m_2 are called the *mean* and *power* of X, respectively. The function

$$g(X) = (X - m_1)^n \qquad\qquad\qquad (C.3\text{-}5)$$

defines *central moments* (about the mean m_1), denoted by μ_n, according to

$$\mu_n = E[(X - m_1)^n] = \int_{-\infty}^{\infty} (x - m_1)^n f_X(x)\,dx, \qquad n = 0, 1, 2, \ldots \qquad (C.3\text{-}6)$$

Moment μ_2 is called the *variance* of X and is usually given the special symbol σ_X^2, so

$$\sigma_X^2 = E[(X - m_1)^2] = \int_{-\infty}^{\infty} (x - m_1)^2 f_X(x)\,dx \qquad (C.3\text{-}7)$$

The positive square root of variance is called *standard deviation*.

For two random variables X and Y *joint moments* about the origin, denoted by m_{nk}, are

$$m_{nk} = E[g(X, Y)] = E[X^n Y^k] = \int_{-\infty}^{\infty} \int_{-\infty}^{\infty} x^n y^k f_{X,Y}(x, y)\,dx\,dy \qquad (C.3\text{-}8)$$

Clearly $m_{n0} = E[X^n]$ are the moments of X, while $m_{0k} = E[Y^k]$ are the moments of Y. The joint moment m_{11} is a special one given the special name *correlation*, denoted by R_{XY}:

$$R_{XY} = E[XY] = \int_{-\infty}^{\infty} \int_{-\infty}^{\infty} xy f_{X,Y}(x, y)\,dx\,dy \qquad (C.3\text{-}9)$$

If $R_{XY} = 0$, X and Y are said to be *orthogonal*. If $R_{XY} = m_{01} m_{10}$, X and Y are called *uncorrelated*.

Joint central moments of random variables X and Y are denoted by μ_{nk} and defined by

$$\mu_{nk} = E[(X - m_{10})^n (Y - m_{01})^k]$$
$$= \int_{-\infty}^{\infty} \int_{-\infty}^{\infty} (x - m_{10})^n (y - m_{01})^k f_{X,Y}(x, y)\,dx\,dy \qquad (C.3\text{-}10)$$

It is easy to show that $\mu_{20} = \sigma_X^2$ and $\mu_{02} = \sigma_Y^2$, the variances of X and Y, respectively. Central moment μ_{11} is very important and is given the special name *covariance* and notation C_{XY}. Thus

$$C_{XY} = E[(X - m_{10})(Y - m_{01})] = \int_{-\infty}^{\infty} \int_{-\infty}^{\infty} (x - m_{10})(y - m_{01}) f_{X,Y}(x, y) dx dy$$

$$(C.3\text{-}11)$$

If $C_{XY} = 0$, X and Y are uncorrelated. If $C_{XY} = -m_{10}m_{01}$, X and Y are orthogonal.

The *correlation coefficient* ρ_{XY} of X and Y is a normalized covariance defined by

$$\rho_{XY} = \frac{\mu_{11}}{\sqrt{\mu_{20}\mu_{02}}} = \frac{C_{XY}}{\sigma_X \sigma_Y} = E\left[\left(\frac{X - m_{10}}{\sigma_X}\right)\left(\frac{Y - m_{01}}{\sigma_Y}\right)\right]$$

$$(C.3\text{-}12)$$

Moments about the origin and central moments can be defined for any number of random variables. However, these are not given here, as most problems can be developed using only moments of two random variables (see Peebles, 1993, p. 138, for joint central moments).

Example C.3-1   Random variables X and Y have moments $E[X] = 1.5$, $\sigma_X^2 = 1.44$, $E[Y] = -0.8$, $\sigma_Y^2 = 4.41$, and $C_{XY} = -1.89$. We find $E[X^2]$, $E[Y^2]$, R_{XY}, and ρ_{XY}.

From (C.3-7), $\sigma_X^2 = E[X^2 - 2Xm_1 + m_1^2] = E[X^2] - m_1^2$ or $E[X^2] = \sigma_X^2 + (E[X])^2 = 1.44 + (1.5)^2 = 3.69$. Similarly, $E[Y^2] = \sigma_Y^2 + (E[Y])^2 = 4.41 + (-0.8)^2 = 5.05$.

From (C.3-11), $C_{XY} = E[XY - m_{10}Y - m_{01}X + m_{01}m_{10}] = R_{XY} - E(X)E(Y)$, so $R_{XY} = C_{XY} + E(X)E(Y) = -1.89 + (1.5)(-0.8) = -3.09$. From (C.3-12), $\rho_{XY} = -1.89/\sqrt{1.44(4.41)} = -0.75$.

Characteristic Functions

For a single random variable X, moments m_n can be found from

$$m_n = (-j)^n \frac{d^n \Phi_X(\omega)}{d\omega^n}\bigg|_{\omega=0}, \qquad n = 0, 1, 2, \ldots \qquad (C.3\text{-}13)$$

where $\Phi_X(\omega)$ is the *characteristic function* of X defined by

$$\Phi_X(\omega) = E[e^{j\omega X}] = \int_{-\infty}^{\infty} f_X(x) e^{j\omega x} dx \qquad (C.3\text{-}14)$$

For two random variables X and Y the moments m_{nk} are given by

$$m_{nk} = (-j)^{n+k} \frac{\partial^{n+k} \Phi_{X,Y}(\omega_1, \omega_2)}{\partial \omega_1^n \partial \omega_2^k}\bigg|_{\omega_1=0, \, \omega_2=0}, \qquad n \text{ and } k = 0, 1, \ldots \qquad (C.3\text{-}15)$$

where

$$\Phi_{X,Y}(\omega_1, \omega_2) = E[e^{j\omega_1 X + j\omega_2 Y}]$$

$$= \int_{-\infty}^{\infty} \int_{-\infty}^{\infty} f_{X,Y}(x, y) e^{j\omega_1 x + j\omega_2 y} dx dy \tag{C.3-16}$$

is called the *joint characteristic function* of X and Y.

In the above expressions ω_1 and ω_2 are parameters. The functions of (C.3-14) and (C.3-16) are Fourier transforms (with signs of ω, ω_1 and ω_2 changed). Because of the uniqueness of Fourier transforms the density functions $f_X(x)$ and $f_{X,Y}(x, y)$ can be found by inverse Fourier transformation of $\Phi_X(\omega)$ and $\Phi_{X,Y}(\omega_1, \omega_2)$ respectively (with signs of ω, ω_1 and ω_2 again reversed).

C.4 GAUSSIAN RANDOM VARIABLES

A random variable X is called *gaussian* if its density has the form

$$f_X(x) = \frac{e^{-(x-\bar{X})^2/2\sigma_X^2}}{\sqrt{2\pi\sigma_X^2}} \tag{C.4-1}$$

where we use the notation $\bar{X}$ to represent $E[X] = m_1$. For zero mean and unit variance we use the notation $f(x)$:

$$f(x) = \frac{1}{\sqrt{2\pi}} e^{-x^2/2} \tag{C.4-2}$$

The distribution functions cannot be found in simple closed form. That for the general case can be found from that for (C.4-2) which is

$$F(x) = \frac{1}{\sqrt{2\pi}} \int_{-\infty}^{x} e^{-\xi^2/2} d\xi \tag{C.4-3}$$

The general case is

$$F_X(x) = F\left(\frac{x - \bar{X}}{\sigma_X}\right) \tag{C.4-4}$$

The normalized function $F(x)$ is tabulated in Appendix E. The general case can be found using the table and (C.4-4).

Example C.4-1 A noise voltage is sampled. If the noise is gaussian with mean value $E[X] = \bar{X} = 1.8$ V and variance $\sigma_X^2 = 0.36$ V^2, find the probability that the sample value will exceed 1.08 V but not exceed 3.24 V.

Here we want the probability $P(1.08 < X \le 3.24) = F_X(3.24) - F_X(1.08)$. We use the normalized form (C.4-4) to find these two values. With $(x - \bar{X})/\sigma_x = (3.24 - 1.8)/\sqrt{0.36} = 2.4$ we have $F_X(3.24) = F(2.4)$. With $(x - \bar{X})/\sigma_x = (1.08 - 1.8)/\sqrt{0.36} = -1.2$

we have $F_X(1.08) = F(-1.2)$, so

$$P(1.08 < X \leq 3.24) = F(2.4) - F(-1.2)$$
$$= F(2.4) - 1 + F(1.2)$$

where (E.1-4) has been used. Finally we find $F(2.4) = 0.9918$ and $F(1.2) = 0.8849$ from Table E-1 and obtain

$$P(1.08 < X \leq 3.24) = 0.9918 - 1 + 0.8849 = 0.8767$$

N random variables $X_1, X_2, \ldots, X_N$, are said to be *jointly gaussian* if their joint density function has the form

$$f_{X_1,X_2,\ldots,X_N}(x_1, x_2, \ldots, x_N) = \frac{|[C_X]^{-1}|^{1/2}}{(2\pi)^{N/2}} e^{-[x-\bar{X}]^t [C_X]^{-1} [x-\bar{X}]/2} \qquad \text{(C.4-5)}$$

where $[\cdot]^t$ and $[\cdot]^{-1}$ denote the transpose and inverse, respectively, of a matrix $[\cdot]$ represented by heavy brackets. The determinant is denoted by $|[\cdot]|$. Matrices are defined as follows:

$$[x - \bar{X}] = \begin{bmatrix} x_1 - \bar{X}_1 \\ x_2 - \bar{X}_2 \\ \vdots \\ x_N - \bar{X}_N \end{bmatrix} \qquad \text{(C.4-6)}$$

$$[C_X] = \begin{bmatrix} C_{11} & C_{12} & \cdots & C_{1N} \\ C_{21} & C_{22} & \cdots & C_{2N} \\ \vdots & \vdots & & \vdots \\ C_{N1} & C_{N2} & \cdots & C_{NN} \end{bmatrix} \qquad \text{(C.4-7)}$$

where elements of the *covariance matrix* $[C_X]$ are

$$C_{ij} = E[(X_i - \bar{X}_i)(X_j - \bar{X}_j)] = \begin{cases} \sigma_{X_i}^2, & i = j \\ C_{X_i X_j}, & i \neq j \end{cases} \qquad \text{(C.4-8)}$$

and $\bar{X}_i$ are mean values

$$\bar{X}_i = E[X_i], \qquad i = 1, 2, \ldots, N \qquad \text{(C.4-9)}$$

It is to be noted that only first and second order moments[1] are needed to completely defined the joint density of any number of gaussian random variables; that is, we need only the means, variances, and covariances.

---

[1] Order is defined by n in (C.3-4) and (C.3-6), and the sum $n + k$ in (C.3-8) and (C.3-10).

Some important properties of gaussian random variables are (1) if uncorrelated, they are also statistically independent; (2) if N random variables are linearly transformed to N new random variables by a set of N linear nonsingular transformations, the new random variables are jointly gaussian; (3) N correlated random variables can be linearly transformed to N uncorrelated (and statistically independent) random variables, and vice versa.

C.5   RANDOM SIGNALS AND PROCESSES

Any waveform having a randomly varying amplitude with time, whether it is desired or undesired, will be called a random signal. The collection, or *ensemble*, of all possible random signals for a given problem is called a *random process*. Since there can be a wide variety of forms of random signals, there can be a wide variety of random processes. Processes are often classified according to their form.

For example, Fig. C.5-1 shows a typical random signal with continuous random amplitude fluctuations over continuous time. This signal might represent noise in a particular radar receiver. The random process, called a *continuous random process*, is the collection of all such signals as might be imagined for all such similar radar receivers.

There are other classifications that are important. For example, if the waveform of Fig. C.5-1 is quantized to have only a finite number of amplitudes (levels) between which the signal switches at random (any of a continuum of) times, the ensemble is called a *discrete random process*. For periodic sampling of the waveform of Fig. C.5-1, the ensemble is known as a *continuous* (in amplitude) *random sequence* (discrete time). As a last example, a periodically sampled version of the quantized version of the signal of Fig. C.5-1 leads to a *discrete random sequence*.

Random Process Concept

Since a random process is an ensemble or collection of all random signals for a given problem (or experiment), Fig. C.5-2 might be viewed as a random process. At any time t_1 the various amplitudes are values of a random variable X_1. At any other time, say t_2, the process is described by another random variable X_2. Any number of random variables X_i, $i = 1, 2, \ldots, N$, can be defined for a process where N can be infinite. The behavior of the process, denoted by $X(t)$, is determined by the joint probability density and distribution of the random variables $X(t_i) = X_i$ of interest.

Thus statistical properties (densities and distributions) and time behaviors of processes are of critical importance. Frequency (spectral) properties, which are appropriate transformations of temporal properties, are also of interest.

Correlation Functions

The most important time behaviors of a process can be determined from *correlation functions* that describe the statistical relationships between random variables defined at only two times. For a single random process $X(t)$ and times defined as $t_1 = t$ and $t_2 = t + \tau$, where τ is the offset $\tau = t_2 - t_1$ the two most important correlations are the

Figure C.5-1   Typical waveform of a continuous random process.

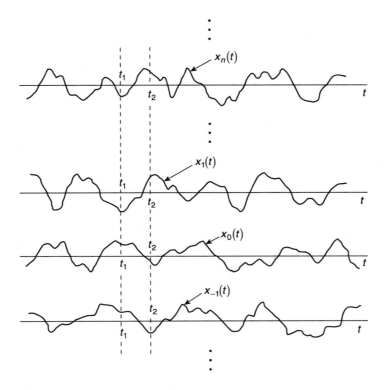

Figure C.5-2   An ensemble of random signals that constitutes a random process.

autocorrelation function, denoted by $R_{XX}(t, t + \tau)$, and the *autocovariance function*, denoted by $C_{XX}(t, t + \tau)$:[2]

$$R_{X,X}(t, t + \tau) = E[X(t)X(t + \tau)] \tag{C.5-1}$$

$$C_{XX}(t, t + \tau) = E[\{X(t) - E[X(t)]\}\{X(t + \tau) - E[X(t + \tau)]\}] \tag{C.5-2}$$

[2] We consider only real random processes.

When there are *two* random processes, denoted by $X(t)$ and $Y(t)$, the important correlations are the *cross-correlation function*, denoted by $R_{XY}(t, t + \tau)$, and the *cross-covariance function*, denoted by $C_{XY}(t, t + \tau)$, as defined by

$$R_{XY}(t, t + \tau) = E[X(t)Y(t + \tau)] \tag{C.5-3}$$

$$C_{XY}(t, t + \tau) = E[\{X(t) - E[X(t)]\}\{Y(t + \tau) - E[Y(t + \tau)]\}] \tag{C.5-4}$$

With two random processes a second cross-correlation function $R_{YX}(t, t + \tau)$ is defined by $E[Y(t)X(t + \tau)]$ similar to (C.5-3). Equation (C.5-4) can similarly be modified to define the second cross-variance function $C_{YX}(t, t + \tau)$.

Stationarity

Broadly stated, a random process is called *stationary* if its statistical properties do not change with time. More precisely, there are many orders of stationarity. We will not discuss these (see Peebles, 1993; Papoulis, 1984) but only define what is probably the most important practical from of stationarity. A random process $X(t)$ is said to be *wide-sense stationary* if

$$E[X(t)] = \bar{X} = \text{constant} \tag{C.5-5}$$

$$E[X(t)X(t + \tau)] = R_{XX}(\tau) \tag{C.5-6}$$

where $\bar{X}$ is a constant representing the mean value of $X(t)$ at *any* time t. Equation (C.5-6) indicates that the process autocorrelation function of (C.5-1) is dependent only on time separation τ and not on absolute time t. For a wide-sense stationary process the autocovariance function of (C.5-2) also reduces to a function of t only, and not absolute time t:

$$\begin{aligned} C_{XX}(\tau) &= R_{XX}(\tau) - \{E[X(t)]\}^2 \\ &= R_{XX}(\tau) - \bar{X}^2 \end{aligned} \tag{C.5-7}$$

Example C.5-1 A random process $X(t)$ is defined by $X(t) = A\cos[\omega_0 t + \Theta]$, where A and ω_0 are constants, and Θ is a random variable uniform (with constant probability density function $1/(2\pi)$ for $0 < \theta < 2\pi$) on $(0, 2\pi)$. This process is classified as *deterministic* because each sample function is known for all time once a value of Θ is determined for a given ensemble waveform (A and ω_0 are assumed known). We show that this process is wide-sense stationary.

First, its mean value for *any* time t is

$$E[X(t)] = E[A\cos(\omega_0 t + \Theta)] = A\int_0^{2\pi} \cos(\omega_0 t + \theta)\frac{1}{2\pi}d\theta = 0$$

which is the constant zero. The autocorrelation function for any time t and displacement τ is

$$R_{XX}(t, t + \tau) = E[X(t)X(t + \tau)]$$

$$= E[A^2 \cos(\omega_0 t + \Theta)\cos(\omega_0 t + \omega_0\tau + \Theta)]$$

$$= \frac{A^2}{2} E[\cos(\omega_0 \tau) + \cos(2\omega_0 t + \omega_0 \tau + 2\Theta)]$$

$$= \frac{A^2}{2} \left\{ \cos(\omega_0 \tau) + \int_0^{2\pi} \cos(2\omega_0 t + \omega_0 \tau + 2\theta) \frac{1}{2\pi} d\theta \right\}$$

$$= \frac{A^2}{2} \cos(\omega_0 \tau)$$

Since the mean value is constant and the autocorrelation function does not depend on t, $X(t)$ is wide-sense stationary.

Some useful properties of a wide-sense stationary random process $X(t)$ are

$$|R_{XX}(\tau)| \leq R_{XX}(0) \tag{C.5-8}$$

$$R_{XX}(-\tau) = R_{XX}(\tau) \tag{C.5-9}$$

$$R_{XX}(0) = E[X^2(t)] \tag{C.5-10}$$

For two random processes $X(t)$ and $Y(t)$, we say they are *jointly wide-sense stationary* if each is wide-sense stationary *and* their cross-correlation function of (C.5-3) depends on τ only and not absolute time t:

$$R_{XY}(\tau) = E[X(t)Y(t + \tau)] \tag{C.5-11}$$

Similarly,

$$C_{XY}(\tau) = R_{XY}(\tau) - E[X(t)]E[Y(t + \tau)]$$
$$= R_{XY}(\tau) - \bar{X}\bar{Y} \tag{C.5-12}$$

In general, the *variance* of a process $X(t)$, denoted by σ_X^2, is given by (C.5-2) with $\tau = 0$. For the wide-sense stationary case

$$\sigma_X^2 = R_{XX}(0) - \bar{X}^2 \tag{C.5-13}$$

from (C.5-7).

Two random processes $X(t)$ and $Y(t)$ are called *uncorrelated* if $C_{XY}(t, t + \tau) = 0$. They are called *orthogonal* if $R_{XY}(t, t + \tau) = 0$.

Some properties of the cross-correlation function of jointly wide-sense stationary processes $X(t)$ and $Y(t)$ are

$$R_{XY}(-\tau) = R_{YX}(\tau) \tag{C.5-14}$$

$$|R_{XY}(\tau)| \leq \sqrt{R_{XX}(0) R_{YY}(0)} \tag{C.5-15}$$

$$|R_{XY}(\tau)| \leq \frac{1}{2}[R_{XX}(0) + R_{YY}(0)] \tag{C.5-16}$$

The right side of (C.5-15) is a tighter upper bound on $|R_{XY}(\tau)|$ than the right side of (C.5-16) because the geometric mean of two positive numbers does not exceed the arithmetic mean of the same two numbers.

Time Averages

The time average of a quantity, denoted by the operation $A(\cdot)$ is

$$A(\cdot) = \lim_{T \to \infty} \frac{1}{2T} \int_{-T}^{T} (\cdot) dt \qquad \text{(C.5-17)}$$

Examples are the time average of a random process $X(t)$ and the time averages of autocorrelation and cross-correlation functions:[3]

$$A[X(t)] = \lim_{T \to \infty} \frac{1}{2T} \int_{-T}^{T} X(t) dt \qquad \text{(C.5-18)}$$

$$A[R_{XX}(t, t+\tau)] = \lim_{T \to \infty} \frac{1}{2T} \int_{-T}^{T} R_{XX}(t, t+\tau) dt \qquad \text{(C.5-19)}$$

$$A[R_{XY}(t, t+\tau)] = \lim_{T \to \infty} \frac{1}{2T} \int_{-T}^{T} R_{XY}(t, t+\tau) dt \qquad \text{(C.5-20)}$$

C.6 POWER SPECTRA

The spectral behaviors of random processes are related to their temporal behaviors through appropriate transforms.

Power Density Spectrum

The *power density spectrum* of a random process $X(t)$, which is often called a *power spectral density*, is defined as the Fourier transform of the time-averaged autocorrelation function. If it is denoted by $\mathscr{S}_{XX}(\omega)$, then

$$\mathscr{S}_{XX}(\omega) = \int_{-\infty}^{\infty} A[R_{XX}(t, t+\tau)] e^{-j\omega\tau} d\tau \qquad \text{(C.6-1)}$$

where ω is angular frequency. Since Fourier transforms are unique, knowing $\mathscr{S}_{XX}(\omega)$ allows recovery of $A[R_{XX}(t, t+\tau)]$

$$A[R_{XX}(t, t+\tau)] = \frac{1}{2\pi} \int_{-\infty}^{\infty} \mathscr{S}_{XX}(\omega) e^{j\omega\tau} d\omega \qquad \text{(C.6-2)}$$

For a wide-sense stationary process the autocorrelation function does not depend on t and

$$\mathscr{S}_{XX}(\omega) = \int_{-\infty}^{\infty} R_{XX}(\tau) e^{-j\omega\tau} d\tau \qquad \text{(C.6-3)}$$

---

[3] Time averages are useful in working with nonstationary processes and in making measurements on ensemble members (single random signals) of a random process.

$$R_{XX}(\tau) = \frac{1}{2\pi} \int_{-\infty}^{\infty} \mathscr{S}_{XX}(\omega) e^{j\omega\tau} d\omega \tag{C.6-4}$$

The unit of $\mathscr{S}_{XX}(\omega)$ is volts-squared per hertz if $X(t)$ represents voltage. However, it is customary to imagine $X(t)$ exists across a 1-ohm resistor or treat $X(t)$ as a normalized voltage so that $\mathscr{S}_{XX}(\omega)$ is said to have the unit of power per hertz. Because $d\omega/2\pi = df\,(\text{Hz})$, the total power in a random process derives from (C.6-4) when $\tau = 0$:

$$P_{XX} = R_{XX}(0) = \frac{1}{2\pi} \int_{-\infty}^{\infty} \mathscr{S}_{XX}(\omega) d\omega \tag{C.6-5}$$

Here P_{XX} denotes the average power in $X(t)$, and (C.6-5) is called the *power formula*.

The power density spectrum is a real, nonnegative function having even symmetry in the variable ω.

Example C.6-1 A random process $X(t)$ is wide-sense stationary and bandlimited to have a power density spectrum

$$\mathscr{S}_{XX}(\omega) = \frac{\mathscr{N}_0}{2} \left[\text{rect}\left(\frac{\omega - \omega_0}{W}\right) + \text{rect}\left(\frac{\omega + \omega_0}{W}\right) \right] \tag{1}$$

where $\mathscr{N}_0$, W, and $\omega_0 > W/2$ are all real positive constants. We find $R_{XX}(\tau)$ and the total power in this process.

From the known Fourier transform pair

$$\frac{\mathscr{N}_0 W}{4\pi} \text{Sa}\left(\frac{W}{2}\tau\right) \leftrightarrow \frac{\mathscr{N}_0}{2} \text{rect}\left(\frac{\omega}{W}\right)$$

we apply the frequency shifting property of Fourier transforms to obtain

$$\frac{\mathscr{N}_0 W}{4\pi} \text{Sa}\left(\frac{W}{2}\tau\right) e^{\pm j\omega_0 \tau} \leftrightarrow \frac{\mathscr{N}_0}{2} \text{rect}\left(\frac{\omega \mp \omega_0}{W}\right)$$

These results are applied to the inverse Fourier transform of (1) to get

$$R_{XX}(\tau) = \mathscr{F}^{-1}\{\mathscr{S}_{XX}(\omega)\} = \frac{\mathscr{N}_0 W}{2\pi} \text{Sa}\left(\frac{W}{2}\tau\right) \cos(\omega_0 \tau)$$

The total power in $X(t)$ is equal to $R_{XX}(0)$:

$$R_{XX}(0) = \frac{\mathscr{N}_0 W}{2\pi}$$

Cross-Power Density Spectrum

The *cross-power density spectrum* for two random processes $X(t)$ and $Y(t)$ is defined by the Fourier transform of the time-averaged cross-correlation function. On using the notation $\mathscr{S}_{XY}(\omega)$, we have

$$\mathscr{S}_{XY}(\omega) = \int_{-\infty}^{\infty} A[R_{XY}(t, t + \tau)]e^{-j\omega\tau} d\tau \tag{C.6-6}$$

The inverse Fourier transform also applies, so

$$A[R_{XY}(t, t + \tau)] = \frac{1}{2\pi} \int_{-\infty}^{\infty} \mathscr{S}_{XY}(\omega)e^{j\omega\tau} d\omega \tag{C.6-7}$$

For jointly wide-sense stationary processes these two results reduce to

$$\mathscr{S}_{XY}(\omega) = \int_{-\infty}^{\infty} R_{XY}(\tau)e^{-j\omega\tau} d\tau \tag{C.6-8}$$

$$R_{XY}(\tau) = \frac{1}{2\pi} \int_{-\infty}^{\infty} \mathscr{S}_{XY}(\omega)e^{j\omega\tau} d\omega \tag{C.6-9}$$

The cross-power, denoted by P_{XY}, due to the two processes is given by (C.6-9) with $\tau = 0$:

$$P_{XY} = \frac{1}{2\pi} \int_{-\infty}^{\infty} \mathscr{S}_{XY}(\omega) d\omega \tag{C.6-10}$$

This is just the power formula of (C.6-5) applied to the cross-power spectrum.

There are two cross-power spectra applicable to two random processes. The first is defined by (C.6-6); the second, denoted by $\mathscr{S}_{YX}(\omega)$, is also given by (C.6-6) if $A[R_{YX}(t, t + \tau)]$ replaces $A[R_{XY}(t, t + \tau)]$.

Some useful properties of the cross-power density spectra of jointly wide-sense stationary real processes $X(t)$ and $Y(t)$ are

$$\mathscr{S}_{XY}(\omega) = \mathscr{S}_{YX}(-\omega) = \mathscr{S}_{YX}^{*}(\omega) \tag{C.6-11}$$

$$\text{Re}[\mathscr{S}_{XY}(\omega)] \quad \text{and} \quad \text{Re}[\mathscr{S}_{YX}(\omega)] \quad \text{are even functions of } \omega \tag{C.6-12}$$

$$\text{Im}[\mathscr{S}_{XY}(\omega)] \quad \text{and} \quad \text{Im}[\mathscr{S}_{YX}(\omega)] \quad \text{are odd functions of } \omega \tag{C.6-13}$$

If $X(t)$ and $Y(t)$ are orthogonal, then

$$\mathscr{S}_{XY}(\omega) = 0 \quad \text{and} \quad \mathscr{S}_{YX}(\omega) = 0 \tag{C.6-14}$$

If $X(t)$ and $Y(t)$ are uncorrelated, then

$$\mathscr{S}_{XY}(\omega) = \mathscr{S}_{YX}(\omega) = 2\pi \bar{X}\bar{Y}\delta(\omega) \tag{C.6-15}$$

where $\bar{X}$ and $\bar{Y}$ are the respective mean values of $X(t)$ and $Y(t)$.

C.7 NETWORK RESPONSES TO RANDOM SIGNALS

When a random signal is applied to a network's input, it evokes a response (output) that is also a random signal. In this section we review means to define the response based on knowledge of the random input and the network. We will limit our review to only real linear time-invariant networks excited by random signals from wide-sense stationary random processes. For details on other cases, the reader should consult the literature (Peebles, 1993; Helstrom, 1991; Shanmugan and Breipohl, 1988; Papoulis, 1984).

Fundamental Result

The single, most fundamental, fact on which all characteristics of network responses can be calculated is the convolution theorem. The response $y(t)$ of any real linear time-invariant network, with impulse response $h(t)$, to an input $x(t)$ (deterministic or random) is given by the convolution of $h(t)$ and $x(t)$:

$$y(t) = \int_{-\infty}^{\infty} h(\xi)x(t - \xi)d\xi = \int_{-\infty}^{\infty} h(t - \xi)x(\xi)d\xi \tag{C.7-1}$$

For a random signal $x(t)$, (C.7-1) defines the network's random response signal. All possible waveforms $x(t)$ comprising an input process $X(t)$ are acted on as in (C.7-1) to define all possible waveforms $y(t)$ comprising the process $Y(t)$ that defines the response. Collectively we may view (C.7-1) as a transformation of process $X(t)$ to a new process $Y(t)$ according to

$$Y(t) = \int_{-\infty}^{\infty} h(\xi)X(t - \xi)d\xi = \int_{-\infty}^{\infty} h(t - \xi)X(\xi)d\xi \tag{C.7-2}$$

Temporal Descriptions of Network Response

The mean value of $Y(t)$ is[4]

$$E[Y(t)] = \bar{Y} = E\left[\int_{-\infty}^{\infty} h(\xi)X(t - \xi)d\xi \right] = \int_{-\infty}^{\infty} h(\xi)E[X(t - \xi)]d\xi$$

$$= \bar{X} \int_{-\infty}^{\infty} h(\xi)d\xi = \bar{X}H(0) \tag{C.7-3}$$

Here $H(\omega)$ is the *transfer function* of the network; it is the Fourier transform of $h(t)$. Since $E[X(t)]$ is constant for a wide-sense stationary process, we use an overbar to represent a constant value of expectation for the quantity under the bar.

---

[4] In all work we assume that the expected value and integral operations can be interchanged.

The autocorrelation function of the response can be shown to be

$$R_{YY}(\tau) = \int_{-\infty}^{\infty} \int_{-\infty}^{\infty} R_{XX}(\tau + \xi_1 - \xi_2)h(\xi_1)h(\xi_2)d\xi_1 d\xi_2 \qquad (C.7\text{-}4)$$

The mean-squared value of the response, which is the total power in $Y(t)$, derives from (C.7-4) with $\tau = 0$:

$$E[Y^2(t)] = R_{YY}(0) = \int_{-\infty}^{\infty} \int_{-\infty}^{\infty} R_{XX}(\xi_1 - \xi_2)h(\xi_1)h(\xi_2)d\xi_1 d\xi_2 \qquad (C.7\text{-}5)$$

Clearly (C.7-4) implies the output autocorrelation function depends on τ and not absolute time t, while (C.7-3) shows $E[Y(t)]$ is a constant. These two conditions indicate $Y(t)$ is wide-sense stationary when the input is wide-sense stationary.

Similarly the cross-correlation functions for input $X(t)$ and response $Y(t)$ are functions of τ only and not absolute time t:

$$R_{XY}(\tau) = \int_{-\infty}^{\infty} R_{XX}(\tau - \xi)h(\xi)d\xi \qquad (C.7\text{-}6)$$

$$R_{YX}(\tau) = \int_{-\infty}^{\infty} R_{XX}(\tau - \xi)h(-\xi)d\xi \qquad (C.7\text{-}7)$$

Thus, $Y(t)$ and $X(t)$ are *jointly* wide-sense stationary when $X(t)$ is wide-sense stationary.

Spectral Descriptions of Network Response

For our assumed real linear network the power density spectrum of the response is

$$\mathscr{S}_{YY}(\omega) = \mathscr{S}_{XX}(\omega)|H(\omega)|^2 \qquad (C.7\text{-}8)$$

Cross-power density spectra are

$$\mathscr{S}_{XY}(\omega) = \mathscr{S}_{XX}(\omega)H(\omega) \qquad (C.7\text{-}9)$$

$$\mathscr{S}_{YX}(\omega) = \mathscr{S}_{XX}(\omega)H(-\omega) = \mathscr{S}_{XX}(\omega)H^*(\omega) \qquad (C.7\text{-}10)$$

Example C.7-1 A random process $X(t)$ has a power density spectrum $\mathscr{S}_{XX}(\omega) = K_X[1 + (\omega/W)^2]^{-1}$, where K_X and W are positive constants. $X(t)$ is applied to a network for which $H(\omega) = K_H[1 + (\omega/W)^2]^{-1}$, where K_H is a positive constant. We find the total power P_{XX} in $X(t)$, the power density spectrum of the response $Y(t)$ and its total power.

From the power formula

$$P_{XX} = \frac{1}{2\pi} \int_{-\infty}^{\infty} \mathscr{S}_{XX}(\omega)d\omega = \frac{K_X}{2\pi} \int_{-\infty}^{\infty} \frac{d\omega}{1 + (\omega/W)^2} = \frac{K_X W}{2}$$

From (C.7-8),

$$\mathscr{S}_{YY}(\omega) = \mathscr{S}_{XX}(\omega)|H(\omega)|^2 = \frac{K_X K_H^2}{[1 + (\omega/W)^2]^3}$$

Again the power formula gives

$$P_{YY} = \frac{K_X K_H^2}{2\pi} \int_{-\infty}^{\infty} \frac{d\omega}{[1 + (\omega/W)^2]^3} = \frac{3K_X K_H^2 W}{16}$$

We observe that $P_{YY}/P_{XX} = 3K_H^2/8$; the output power can be larger or smaller than the input power depending on the network's "gain" constant K_H.

C.8   BANDPASS RANDOM PROCESSES

Let $N(t)$ be a random process with a power density spectrum having its most significant spectral content about a "carrier" frequency ω_0 (rad/s). It is therefore a bandpass process. $N(t)$ has two significant and important representations:

$$N(t) = A(t)\cos[\omega_0 t + \Theta(t)] \tag{C.8-1}$$

$$N(t) = X(t)\cos(\omega_0 t) - Y(t)\sin(\omega_0 t) \tag{C.8-2}$$

The first representation is in terms of low-pass (or baseband) "envelope" $A(t)$ and "phase" $\Theta(t)$ functions; it is useful in systems where envelope detectors are to be used. The second representation, sometimes called a *quadrature representation*, is important in systems with coherent, or synchronous, detectors. Here $X(t)$ and $Y(t)$ are low-pass processes called the *in-phase* and *quadrature-phase components* of $N(t)$ The various baseband components are related by

$$X(t) = A(t)\cos[\Theta(t)] \tag{C.8-3}$$

$$Y(t) = A(t)\sin[\Theta(t)] \tag{C.8-4}$$

$$A(t) = [X^2(t) + Y^2(t)]^{1/2} \tag{C.8-5}$$

$$\Theta(t) = \tan^{-1}\left[\frac{Y(t)}{X(t)}\right] \tag{C.8-6}$$

Analysis shows (see Peebles, 1976, 1993) that $X(t)$ and $Y(t)$ are jointly wide-sense stationary processes with zero means when $N(t)$ is a wide-sense stationary gaussian zero-mean bandpass process. Other very important properties of $X(t)$ and $Y(t)$ are

$$E[X^2(t)] = E[Y^2(t)] = E[N^2(t)] \tag{C.8-7}$$

$$R_{XX}(\tau) = \frac{1}{\pi} \int_0^{\infty} \mathscr{S}_{NN}(\omega)\cos[(\omega - \omega_0)\tau]d\omega \tag{C.8-8}$$

$$R_{YY}(\tau) = R_{XX}(\tau) \tag{C.8-9}$$

$$R_{XY}(\tau) = \frac{1}{\pi} \int_0^\infty \mathscr{S}_{NN}(\omega) \sin[(\omega - \omega_0)\tau]d\omega = - R_{XY}(-\tau) \quad \text{(C.8-10)}$$

$$R_{YX}(\tau) = - R_{XY}(\tau) \tag{C.8-11}$$

$$R_{XY}(0) = 0, \quad R_{YX}(0) = 0 \tag{C.8-12}$$

$$\mathscr{S}_{XX}(\omega) = L_p[\mathscr{S}_{NN}(\omega - \omega_0) + \mathscr{S}_{NN}(\omega + \omega_0)] \tag{C.8-13}$$

$$\mathscr{S}_{YY}(\omega) = \mathscr{S}_{XX}(\omega) \tag{C.8-14}$$

$$\mathscr{S}_{XY}(\omega) = jL_p[\mathscr{S}_{NN}(\omega - \omega_0) - \mathscr{S}_{NN}(\omega + \omega_0)] \tag{C.8-15}$$

$$\mathscr{S}_{YX}(\omega) = - \mathscr{S}_{XY}(\omega) \tag{C.8-16}$$

Here $L_p[\cdot]$ represents preserving only the low-pass, or baseband, part of the quantity within the brackets.

PROBLEMS

C.1-1 A single card is drawn from an ordinary deck of 52 cards. Find the probability it is (a) a queen, (b) a seven, (c) a face card, or (d) a numbered card.

C.1-2 You are dealt four cards from an ordinary deck of 52 cards. What is the probability that all four are aces?

C.1-3 Work Problem C.1-2 except let the four cards be *any* four face cards.

C.1-4 Work Problem C.1-2 except require the four cards to be all spades.

C.2-1 An *exponential* random variable X has a probability density function

$$f_X(x) = \frac{1}{b}u(x)e^{-x/b}$$

where $b > 0$ is a constant. (a) Find $F_X(x)$. (b) What is the probability that X will have values larger than b?

C.2-2 Work Problem C.2-1 for a *Rayleigh* random variable X for which

$$f_X(x) = \frac{2}{b}u(x)xe^{-x^2/b}$$

C.2-3 A uniformly distributed random variable X has a probability distribution function

$$F_X(x) = \begin{cases} 0, & x < a \\ \dfrac{x - a}{b - a}, & a \le x < b \\ 1, & b \le x \end{cases}$$

where a and $b > a$ are constants. Find and sketch $f_X(x)$.

C.2-4 X and Y are both discrete random variables. X has three possible values 1, 2, and 4, while Y has two possible values 3 and 5. Their joint probability density function is

$$f_{X,Y}(x, y) = 0.1\delta(x - 1)\delta(y - 3) + 0.2\delta(x - 2)\delta(y - 3)$$
$$+ 0.3\delta(x - 2)\delta(y - 5) + 0.4\delta(x - 4)\delta(y - 5)$$

Find: (a) The density of X alone, (b) the density of Y alone. (c) Are X and Y statistically independent?

C.2-5 For the random variables of Example C.2-1, find $P(1.2a < X \leq 0.8b, c < Y \leq 0.4d)$.

C.2-6 How many conditions must be satisfied by four random variables X_1, X_2, X_3, and X_4 if they are to be statistically independent? List these conditions.

C.3-1 Find the mean value and variance expressions for the exponential random variable of Problem C.2-1.

C.3-2 Work Problem C.3-1 except for the Rayleigh random variable of Problem C.2-2.

C.3-3 Find, by using (C.3-4), the general moment m_n for the uniform random variable of Problem C.2-3.

C.3-4 Suppose that a radar's time to failure (the time before a failure occurs requiring replacement) is a random variable X defined by

$$f_X(x) = \frac{x}{7000} u(x)e^{-x^2/14,000}$$

where the unit of X is weeks. (a) What is the probability the system will not last over 4 weeks? (b) What is the probability of at least a 2-year survival? (c) What minimum survival time occurs with 90% probability?

C.3-5 Two random variables X and Y have the joint density function

$$f_{X,Y}(x, y) = x \operatorname{rect}(x - 0.5)(y + 1.5)\operatorname{rect}(y - 0.5)$$

(a) Find the correlation R_{XY}. (b) Find the mean values m_{01} and m_{10}. (c) Find the covariance C_{XY}.

C.3-6 Extend Problem C.3-5 by finding all moments m_{nk}, n and $k = 0, 1, 2, \ldots$.

C.3-7 A random variable X is the weighted sum of N random variables X_i, $i = 1, 2, \ldots, N$, according to

$$X = \sum_{i=1}^{N} \alpha_i X_i$$

where the α_i are the weighting constants. (a) Find an expression for the mean value of X. (b) Find an expression for the variance of X. (c) To what does the variance expression reduce if the X_i are all uncorrelated?

C.3-8 A *Poisson* random variable X has the probability density

$$f_X(x) = e^{-b} \sum_{k=0}^{\infty} \frac{b^k}{k!} \delta(x - k)$$

where $b > 0$ is a constant. Find the characteristic function of X.

C.3-9 Find the characteristic function of the uniform random variable X for which

$$f_X(x) = \begin{cases} 0, & b < x < a \\ \dfrac{1}{b - a}, & a < x < b \end{cases}$$

where a and $b > a$ are constants.

C.3-10 Random variables X and Y have the joint characteristic function

$$\Phi_{X,Y}(\omega_1, \omega_2) = \frac{1}{[(1 - j2\omega_1)(1 - j2\omega_2)]^{N/2}}$$

Find (a) the correlation R_{XY}, (b) moments m_{20} and m_{02}, (c) the means of X and Y, and (d) the correlation coefficient ρ_{XY}.

C.4-1 A gaussian random variable has a mean of 2.6 and a standard deviation of 0.9. Find the probability $P(3.77 < X \le 1.43)$.

C.4-2 A gaussian random variable for which $\bar{X} = 0$ and $\sigma_X = 1.7$ V represents a good model for the noise at a point in a radar system. What threshold voltage is required if the probability of the noise voltage exceeding the threshold is to be no more than 0.0099?

C.4-3 Write (C.4-5) in expanded form for two random variables. Put the expression in terms of means $\bar{X}_1$ and $\bar{X}_2$, variances $\sigma_{X_1}^2$ and $\sigma_{X_2}^2$, and correlation coefficient $\rho_{X_1 X_2}$.

C.5-1 A random process is that of Example C.5-1 except that Θ is uniformly distributed on $(0, \pi)$ and not on $(0, 2\pi)$. Is the process wide-sense stationary?

C.5-2 A random process is defined by $X(t) = A \cos(\pi t)$, where A is a gaussian random variable with zero mean and variance σ_A^2. Random variables X_1 and X_2 are defined at times $t_1 = 0$ and t_2. Are there any conditions that can make X_1 and X_2 either uncorrelated or orthogonal? If so, state them.

C.5-3 A random process $Y(t)$ is defined by

$$Y(t) = X(t) \cos(\omega_0 t + \Theta)$$

where $X(t)$ is a wide-sense stationary random process and Θ is a random variable independent of $X(t)$ and uniformly distributed on $(0, 2\pi)$. (a) Find the mean value of $Y(t)$. (b) Find $R_{YY}(t, t + \tau)$. (c) Is $Y(t)$ wide-sense stationary?

C.5-4 Explain why the right sides of each of the following functions cannot represent two jointly wide-sense stationary zero-mean processes $X(t)$ and $Y(t)$ for which $\sigma_X^2 = 5$ and $\sigma_Y^2 = 12$.

(a) $R_{XX}(\tau) = 5 \dfrac{\sin(6\tau)}{\tau^2}$

(b) $R_{YY}(\tau) = 12e^{-|\tau - 3|^2}$

(c) $R_{XY}(\tau) = \dfrac{9}{1 + 2\tau^2}$

C.5-5 Two random processes are defined by

$$X(t) = A \cos(\omega_1 t + \Theta_1)$$
$$Y(t) = B \cos(\omega_2 t + \Theta_2)$$

where A, B, ω_1, and ω_2 are constants and Θ_1 and Θ_2 are statistically independent random variables, both uniformly distributed on $(0, 2\pi)$. (a) Are $X(t)$ and $Y(t)$ jointly wide-sense stationary? (*Hint*: Make use of results of Example C.5-1.) (b) If $\Theta_1 = \Theta_2$ are $X(t)$ and $Y(t)$ jointly wide-sense stationary? If not, are there any conditions on ω_1 and ω_2 that will make $X(t)$ and $Y(t)$ jointly wide-sense stationary?

C.6-1 Find the power density spectrum of the random process having the autocorrelation function

$$R_{XX}(\tau) = K \cos(\omega_0 \tau)$$

where K and ω_0 are positive constants.

C.6-2 Work Problem C.6-1 except for the function

$$R_{XX}(\tau) = K \operatorname{tri}\left(\frac{\tau}{T}\right)$$

where K and T are positive constants.

C.6-3 Work Problem C.6-1 except for the function

$$R_{XX}(\tau) = ae^{-b|\tau|} \cos(\omega_0 \tau)$$

where a, b, and ω_0 are all positive constants.

C.6-4 Explain why each of the following functions cannot be a valid power density spectrum:

(a) $\dfrac{24}{5 - 2\omega^2}$,   (b) $e^{-3\omega^2} \cos(12\omega)$,

(c) $\sin(7\omega)$,   (d) $\dfrac{e^{-j5\omega}}{(1 + j\omega)^2}$.

C.6-5 Find the autocorrelation function of a wide-sense stationary random process having the power density spectrum

$$\mathscr{S}_{XX}(\omega) = \frac{\mathscr{N}_0}{2} \lim_{W \to \infty} \left[\text{rect}\left(\frac{\omega}{W}\right) \right]$$

where $\mathscr{N}_0$ and W are positive constants.

C.6-6 The power density spectrum of a wide-sense stationary random process $X(t)$ is

$$\mathscr{S}_{XX}(\omega) = [a^2 - (b\omega)^2] \text{rect}\left(\frac{b\omega}{2a}\right)$$

where a and b are positive constants. Find the autocorrelation function of $X(t)$.

C.6-7 The cross-correlation function of jointly wide-sense stationary processes $X(t)$ and $Y(t)$ is

$$R_{XY}(\tau) = e^{-a|\tau| + jb\tau}$$

where a and b are positive constants. Find the cross-power spectrum.

C.6-8 Give at least one reason why each of the following functions cannot be a valid cross-power spectrum:

(a) $12\pi\delta(\omega - 10)$,   (b) $6\omega - j2\omega^2$,

(c) $\cos(\omega) + j\text{Sa}(2\omega)$,   (d) $e^{j4\omega + j(\pi/2)}$.

C.7-1 Explain how the last right-side form of (C.7-3) results.

C.7-2 A wide-sense stationary random process $X(t)$ has a mean value of 3.4 and a variance of 4.9; it is applied to the input of a network for which

$$h(t) = 8.4u(t)e^{-t/4}$$

(a) Find $\bar{Y}$, the mean of the network's output process. (b) Find the transfer function of the network.

C.7-3 White noise having a constant power density $\mathscr{N}_0/2$ at all frequencies $(-\infty < \omega < \infty)$ is applied to a network with impulse response

$$h(t) = Ku(t)e^{-Wt}$$

where K and W are positive constants. If the input noise's process is $X(t)$ and the network's response process is $Y(t)$, find (a) $R_{YY}(\tau)$, (b) $E[Y^2(t)]$, and (c) $R_{XY}(\tau)$.

C.7-4 For the network and noise of Problem C.7-3, find $\mathscr{S}_{YY}(\omega)$ and $\mathscr{S}_{XY}(\omega)$.

C.7-5 Work Problem C.7-3 except assume that

$$h(t) = Ku(t)te^{-Wt}$$

C.7-6 The input noise to a network is modeled as a random process $X(t)$ with auto-correlation function

$$R_{XX}(\tau) = Ae^{-W_0|\tau|}$$

where A and W_0 are positive constants. The network has the impulse response of Problem C.7-3. Find (a) $\mathscr{S}_{XX}(\omega)$, (b) $\mathscr{S}_{YY}(\omega)$, and (c) $\mathscr{S}_{XY}(\omega)$.

C.8-1 A bandpass random process $N(t)$ has the power density spectrum

$$\mathscr{S}_{NN}(\omega) = \begin{cases} \dfrac{\mathscr{N}_0}{2}, & -\omega_0 - W_2 < \omega < -\omega_0 + W_1 \\[2mm] \dfrac{\mathscr{N}_0}{2}, & \omega_0 - W_1 < \omega < \omega_0 + W_2 \\[2mm] 0, & \text{all other } \omega \end{cases}$$

where ω_0, W_1, and W_2 are positive constants and $W_1 + W_2 = W$, a constant. For the noise representation of (C.8-2), find and sketch (a) $\mathscr{S}_{XX}(\omega)$ and (b) $\mathscr{S}_{XY}(\omega)$.

APPENDIX D

USEFUL MATHEMATICAL FORMULAS

A number of mathematical formulas are listed that are useful in solving the problems at the ends of the chapters and in the interpretation of the text. Most formulas are very well-known and are readily available from many sources. Those that are more specialized are referenced by the source from which they came. These sources are coded as follows:

D: Dwight (1961), an excellent source of basic identities, derivatives, and integrals.

G: Gradshteyn and Ryzhik (1965), a good source of both basic and advanced formulas.

J: Jolley (1961), a good collection of series, both finite and infinite.

N: Abramowitz and Stegun (1964), an excellent source of mathematical formulas, including some approximations of functions, integral and series representations of functions, and tabulated data.

D.1 TRIGONOMETRIC IDENTITIES

$$\cos(x \pm y) = \cos(x)\cos(y) \mp \sin(x)\sin(y) \tag{D.1-1}$$

$$\sin(x \pm y) = \sin(x)\cos(y) \pm \cos(x)\sin(y) \tag{D.1-2}$$

$$\cos\left(x \pm \frac{\pi}{2}\right) = \mp \sin(x) \tag{D.1-3}$$

$$\sin\left(x \pm \frac{\pi}{2}\right) = \pm \cos(x) \tag{D.1-4}$$

$$\cos(2x) = \cos^2(x) - \sin^2(x) \tag{D.1-5}$$

$$\sin(2x) = 2 \sin(x)\cos(x) \tag{D.1-6}$$

$$2\cos(x) = e^{jx} + e^{-jx} \tag{D.1-7}$$

$$2j\sin(x) = e^{jx} - e^{-jx} \tag{D.1-8}$$

$$2\cos(x)\cos(y) = \cos(x - y) + \cos(x + y) \tag{D.1-9}$$

$$2\sin(x)\sin(y) = \cos(x - y) - \cos(x + y) \tag{D.1-10}$$

$$2\sin(x)\cos(y) = \sin(x - y) + \sin(x + y) \tag{D.1-11}$$

$$2\cos^2(x) = 1 + \cos(2x) \tag{D.1-12}$$

$$2\sin^2(x) = 1 - \cos(2x) \tag{D.1-13}$$

$$4\cos^3(x) = 3\cos(x) + \cos(3x) \tag{D.1-14}$$

$$4\sin^3(x) = 3\sin(x) - \sin(3x) \tag{D.1-15}$$

$$8\cos^4(x) = 3 + 4\cos(2x) + \cos(4x) \tag{D.1-16}$$

$$8\sin^4(x) = 3 - 4\cos(2x) + \cos(4x) \tag{D.1-17}$$

$$A\cos(x) - B\sin(x) = R\cos(x + \theta) \tag{D.1-18}$$

where

$$R = \sqrt{A^2 + B^2} \tag{D.1-19}$$

$$\theta = \tan^{-1}(B/A) \tag{D.1-20}$$

$$A = R\cos(\theta) \tag{D.1-21}$$

$$B = R\sin(\theta) \tag{D.1-22}$$

D.2   INDEFINITE INTEGRALS

Rational Algebraic Functions

$$\int (a + bx)^n \, dx = \frac{(a + bx)^{n+1}}{b(n + 1)}, \qquad 0 < n \tag{D.2-1}$$

$$\int \frac{dx}{a + bx} = \frac{1}{b} \ln|a + bx| \tag{D.2-2}$$

$$\int \frac{dx}{(a + bx)^n} = \frac{-1}{(n - 1)b(a + bx)^{n-1}}, \qquad 1 < n \tag{D.2-3}$$

$$\int \frac{dx}{c + bx + ax^2} = \frac{2}{\sqrt{4ac - b^2}} \tan^{-1}\left(\frac{2ax + b}{\sqrt{4ac - b^2}}\right), \quad b^2 < 4ac \tag{D.2-4}$$

$$= \frac{1}{\sqrt{b^2 - 4ac}} \ln\left|\frac{2ax + b - \sqrt{b^2 - 4ac}}{2ax + b + \sqrt{b^2 - 4ac}}\right|, \quad b^2 > 4ac$$

$$= \frac{-2}{2ax + b}, \quad b^2 = 4ac$$

$$\int \frac{xdx}{c + bx + ax^2} = \frac{1}{2a} \ln |ax^2 + bx + c| - \frac{b}{2a} \int \frac{dx}{c + bx + ax^2} \tag{D.2-5}$$

$$\int \frac{dx}{a^2 + b^2x^2} = \frac{1}{ab} \tan^{-1}\left(\frac{bx}{a}\right) \tag{D.2-6}$$

$$\int \frac{xdx}{a^2 + x^2} = \frac{1}{2} \ln(a^2 + x^2) \tag{D.2-7}$$

$$\int \frac{x^2dx}{a^2 + x^2} = x - a \tan^{-1}\left(\frac{x}{a}\right) \tag{D.2-8}$$

$$\int \frac{dx}{(a^2 + x^2)^2} = \frac{x}{2a^2(a^2 + x^2)} + \frac{1}{2a^3} \tan^{-1}\left(\frac{x}{a}\right) \tag{D.2-9}$$

$$\int \frac{xdx}{(a^2 + x^2)^2} = \frac{-1}{2(a^2 + x^2)} \tag{D.2-10}$$

$$\int \frac{x^2dx}{(a^2 + x^2)^2} = \frac{-x}{2(a^2 + x^2)} + \frac{1}{2a} \tan^{-1}\left(\frac{x}{a}\right) \tag{D.2-11}$$

$$\int \frac{dx}{(a^2 + x^2)^3} = \frac{x}{4a^2(a^2 + x^2)^2} + \frac{3x}{8a^4(a^2 + x^2)} + \frac{3}{8a^5} \tan^{-1}\left(\frac{x}{a}\right) \tag{D.2-12}$$

$$\int \frac{x^2dx}{(a^2 + x^2)^3} = \frac{-x}{4(a^2 + x^2)^2} + \frac{x}{8a^2(a^2 + x^2)} + \frac{1}{8a^3} \tan^{-1}\left(\frac{x}{a}\right) \tag{D.2-13}$$

$$\int \frac{x^4dx}{(a^2 + x^2)^3} = \frac{a^2x}{4(a^2 + x^2)^2} - \frac{5x}{8(a^2 + x^2)} + \frac{3}{8a} \tan^{-1}\left(\frac{x}{a}\right) \tag{D.2-14}$$

$$\int \frac{dx}{(a^2 + x^2)^4} = \frac{x}{6a^2(a^2 + x^2)^3} + \frac{5x}{24a^4(a^2 + x^2)^2} + \frac{5x}{16a^6(a^2 + x^2)}$$
$$+ \frac{5}{16a^7} \tan^{-1}\left(\frac{x}{a}\right) \tag{D.2-15}$$

$$\int \frac{x^2dx}{(a^2 + x^2)^4} = \frac{-x}{6(a^2 + x^2)^3} + \frac{x}{24a^2(a^2 + x^2)^2} + \frac{x}{16a^4(a^2 + x^2)}$$
$$+ \frac{1}{16a^5} \tan^{-1}\left(\frac{x}{a}\right) \tag{D.2-16}$$

$$\int \frac{x^4dx}{(a^2 + x^2)^4} = \frac{a^2x}{6(a^2 + x^2)^3} - \frac{7x}{24(a^2 + x^2)^2} + \frac{x}{16a^2(a^2 + x^2)}$$
$$+ \frac{1}{16a^3} \tan^{-1}\left(\frac{x}{a}\right) \tag{D.2-17}$$

$$\int \frac{dx}{a^4 + x^4} = \frac{1}{4a^3\sqrt{2}} \ln\left(\frac{x^2 + ax\sqrt{2} + a^2}{x^2 - ax\sqrt{2} + a^2}\right) + \frac{1}{2a^3\sqrt{2}} \tan^{-1}\left(\frac{ax\sqrt{2}}{a^2 - x^2}\right)$$

(D.2-18)

$$\int \frac{x^2 dx}{a^4 + x^4} = -\frac{1}{4a\sqrt{2}} \ln\left(\frac{x^2 + ax\sqrt{2} + a^2}{x^2 - ax\sqrt{2} + a^2}\right) + \frac{1}{2a\sqrt{2}} \tan^{-1}\left(\frac{ax\sqrt{2}}{a^2 - x^2}\right)$$

(D.2-19)

Trigonometric Functions

$$\int \cos(x)dx = \sin(x)$$

(D.2-20)

$$\int x\cos(x)dx = \cos(x) + x\sin(x)$$

(D.2-21)

$$\int x^2 \cos(x)dx = 2x\cos(x) + (x^2 - 2)\sin(x)$$

(D.2-22)

$$\int \sin(x)dx = -\cos(x)$$

(D.2-23)

$$\int x\sin(x)dx = \sin(x) - x\cos(x)$$

(D.2-24)

$$\int x^2 \sin(x)dx = 2x\sin(x) - (x^2 - 2)\cos(x)$$

(D.2-25)

$$\int \cos^2(x)dx = \tfrac{1}{4}[2x + \sin(2x)]$$

(D.2-26)

$$\int x\cos^2(x)dx = \tfrac{1}{8}[2x^2 + 2x\sin(2x) + \cos(2x)]$$

(D.2-27)

$$\int x^2 \cos^2(x)dx = \tfrac{1}{8}[\tfrac{4}{3}x^3 + (2x^2 - 1)\sin(2x) + 2x\cos(2x)]$$

(D.2-28)

$$\int \sin^2(x)dx = \tfrac{1}{4}[2x - \sin(2x)]$$

(D.2-29)

$$\int x\sin^2(x)dx = \tfrac{1}{8}[2x^2 - 2x\sin(2x) - \cos(2x)]$$

(D.2-30)

$$\int x^2 \sin^2(x)dx = \tfrac{1}{8}[\tfrac{4}{3}x^3 - (2x^2 - 1)\sin(2x) - 2x\cos(2x)]$$

(D.2-31)

Exponential Functions

$$\int e^{ax}dx = \frac{e^{ax}}{a} \tag{D.2-32}$$

$$\int xe^{ax}dx = e^{ax}\left[\frac{x}{a} - \frac{1}{a^2}\right] \tag{D.2-33}$$

$$\int x^2 e^{ax}dx = e^{ax}\left[\frac{x^2}{a} - \frac{2x}{a^2} + \frac{2}{a^3}\right] \tag{D.2-34}$$

$$\int x^3 e^{ax}dx = e^{ax}\left[\frac{x^3}{a} - \frac{3x^2}{a^2} + \frac{6x}{a^3} - \frac{6}{a^4}\right] \tag{D.2-35}$$

$$\int e^{ax}\sin(x)dx = \frac{e^{ax}}{a^2 + 1}[a\sin(x) - \cos(x)] \tag{D.2-36}$$

$$\int e^{ax}\cos(x)dx = \frac{e^{ax}}{a^2 + 1}[a\cos(x) + \sin(x)] \tag{D.2-37}$$

Bessel Functions

$$\int x^n J_{n-1}(x)\, dx = x^n J_n(x) \tag{D.2-38}$$

$$\int x^{-n} J_{n+1}(x)\, dx = -x^{-n} J_n(x) \tag{D.2-39}$$

$$\int x^n I_{n-1}(x)\, dx = x^n I_n(x) \tag{D.2-40}$$

$$\int x^{-n} I_{n+1}(x)\, dx = x^{-n} I_n(x) \tag{D.2-41}$$

D.3   DEFINITE INTEGRALS

$$\int_0^\infty \frac{\sin^p(x)}{x^m}\, dx = \frac{p}{m-1}\int_0^\infty \frac{\sin^{p-1}(x)}{x^{m-1}}\cos(x)dx, \qquad p > m - 1 > 0$$

$$= \frac{p(p-1)}{(m-1)(m-2)}\int_0^\infty \frac{\sin^{p-2}(x)}{x^{m-2}}\, dx$$

$$- \frac{p^2}{(m-1)(m-2)}\int_0^\infty \frac{\sin^p(x)}{x^{m-2}}\, dx, \qquad p > m - 1 > 1 \qquad \begin{array}{l}\text{(D.3-1)}\\ \text{(G, p. 447)}\end{array}$$

$$\int_0^\infty \left[\frac{\sin(x)}{x}\right]^n dx = \int_0^\infty \mathrm{Sa}^n(x)dx = \begin{cases} \pi/2, & n=1 \\ \pi/2, & n=2 \\ 3\pi/8, & n=3 \\ \pi/3, & n=4 \\ 115\pi/384, & n=5 \\ 11\pi/40, & n=6 \\ 52{,}983\pi/207{,}360, & n=7 \\ 4077\pi/17{,}010, & n=8 \end{cases}$$

(D.3-2)

$$\int_0^\infty \left[\frac{\sin(ax)}{x}\right]^2 \left[\frac{\sin(bx)}{x}\right]^2 dx = \begin{cases} \dfrac{a^2\pi}{6}(3b-a), & 0\le a\le b \\ \dfrac{b^2\pi}{6}(3a-b), & 0\le b\le a \end{cases}$$

(D.3-3)
(G, p. 451)

$$\int_0^\infty x^n e^{-ax}dx = \frac{n!}{a^{n+1}}, \qquad a>0,\ n=1,2,3,\dots$$

(D.3-4)
(D, p. 230)

$$\int_0^\infty x^{\nu-1}e^{-\gamma x-(\beta/x)}dx = 2(\beta/\gamma)^{\nu/2}K_\nu(2\sqrt{\gamma\beta}), \qquad \begin{cases} \mathrm{Re}(\gamma)>0 \\ \mathrm{Re}(\beta)>0 \end{cases}$$

(D.3-5)
(G, p. 340)

$$\int_{-\infty}^\infty e^{-a^2x^2+bx}\,dx = \frac{\sqrt{\pi}}{a}e^{b^2/(4a^2)}, \quad a>0$$

(D.3-6)

$$\int_0^\infty x^2 e^{-x^2}\,dx = \sqrt{\pi}/4$$

(D.3-7)

$$\int_0^{2\pi} e^{ja\cos(x-b)}\,dx = 2\pi J_0(a), \qquad \text{any real } b$$

(D.3-8)

$$\int_0^1 x^{\mu-1}\cos(ax)dx = \frac{1}{2\mu}\big[{}_1F_1(\mu;\mu+1;ja)$$

$$+ {}_1F_1(\mu;\mu+1;-ja)\big], \quad \begin{cases} a>0 \\ \mathrm{Re}(\mu)>0 \end{cases}$$

(D.3-9)
(G, p. 421)

$$\int_0^1 x^{\mu-1}\sin(ax)dx = \frac{-j}{2\mu}\big[{}_1F_1(\mu;\mu+1;ja)$$

$$- {}_1F_1(\mu;\mu+1;-ja)\big], \quad \begin{cases} a>0 \\ \mathrm{Re}(\mu)>-1 \end{cases}$$

(D.3-10)
(G, p. 420)

$$\int_0^1 x^{\nu+1}(1-x^2)^\mu J_\nu(bx)dx$$

$$= \frac{2^\mu \Gamma(\mu+1)}{b^{\mu+1}}J_{\nu+\mu+1}(b), \quad \begin{cases} b>0 \\ \mathrm{Re}(\nu)>-1 \\ \mathrm{Re}(\mu)>-1 \end{cases}$$

(D.3-11)
(G, p. 688)

D.4   INFINITE SERIES

$$\sum_{n=0}^{\infty} (ax)^n = \frac{1}{1-ax}, \qquad ax < 1 \tag{D.4-1}$$
(J, p. 2)

$$\sum_{n=0}^{\infty} (n+1)x^n = \frac{1}{(1-x)^2}, \qquad x < 1 \tag{D.4-2}$$
(J, p. 2)

$$e^x = 1 + x + \frac{x^2}{2!} + \frac{x^3}{3!} + \cdots = \sum_{n=0}^{\infty} \frac{x^n}{n!} \tag{D.4-3}$$

$$e^{-x^2} = \sum_{n=0}^{\infty} \frac{(-1)^n x^{2n}}{n!} \tag{D.4-4}$$

$$\ln(1+x) = \sum_{n=1}^{\infty} \frac{(-1)^{n-1} x^n}{n}, \qquad |x| < 1 \tag{D.4-5}$$

$$e^{j\beta \cos(x)} = \sum_{n=-\infty}^{\infty} (j)^n J_n(\beta) e^{jnx} \tag{D.4-6}$$
(G, p. 973)

$$\cos[\beta \cos(x)] = J_0(\beta) + 2\sum_{n=1}^{\infty} (-1)^n J_{2n}(\beta) \cos(2nx) \tag{D.4-7}$$
(N, p. 361)

$$\cos[\beta \sin(x)] = J_0(\beta) + 2\sum_{n=1}^{\infty} J_{2n}(\beta) \cos(2nx) \tag{D.4-8}$$
(N, p. 361)

$$\sin[\beta \cos(x)] = 2\sum_{n=0}^{\infty} (-1)^n J_{2n+1}(\beta) \cos[(2n+1)x] \tag{D.4-9}$$
(N, p. 361)

$$\sin[\beta \sin(x)] = 2\sum_{n=0}^{\infty} J_{2n+1}(\beta) \sin[(2n+1)x] \tag{D.4-10}$$
(N, p. 361)

$$\sum_{n=1}^{\infty} \frac{\cos(nx)}{n} = -\ln(2)\sin(x/2), \qquad 0 < x < 2\pi \tag{D.4-11}$$
(J, p. 96)

$$\sum_{n=1}^{\infty} \frac{\sin nx}{n} = \tfrac{1}{2}(\pi - x), \qquad 0 < x < 2\pi \tag{D.4-12}$$
(J, p. 96)

$$\sum_{n=1}^{\infty} \frac{\cos(nx)}{n^2} = \frac{\pi^2}{6} - \frac{\pi}{2}x + \tfrac{1}{4}x^2, \qquad 0 \le x \le 2\pi \tag{D.4-13}$$
(G, p. 39)

D.5   FINITE SERIES

$$\sum_{n=1}^{N} n = \frac{N(N+1)}{2} \tag{D.5-1}$$

$$\sum_{n=1}^{N} n^2 = \frac{N(N+1)(2N+1)}{6} \tag{D.5-2}$$

$$\sum_{n=1}^{N} n^3 = \frac{N^2(N+1)^2}{4} \tag{D.5-3}$$

$$\sum_{n=1}^{N} n^4 = \frac{N(N+1)(2N+1)(3N^2+3N-1)}{30} \tag{D.5-4}$$

$$\sum_{n=0}^{N} x^n = \frac{x^{N+1}-1}{x-1} \tag{D.5-5}$$

$$\sum_{n=0}^{N} \frac{N!}{n!(N-n)!} x^n y^{N-n} = (x+y)^N \tag{D.5-6}$$

$$\sum_{n=0}^{N} e^{j(\theta+n\phi)} = \frac{\sin[(N+1)\phi/2]}{\sin(\phi/2)} e^{j[\theta+(N\phi/2)]} \tag{D.5-7}$$

$$\sum_{n=0}^{N} \binom{N}{n} = \sum_{n=0}^{N} \frac{N!}{n!(N-n)!} = 2^N \tag{D.5-8}$$

APPENDIX E

GAUSSIAN, Q, AND ERROR FUNCTIONS

In this appendix we summarize the definitions of the gaussian distribution and density functions, the Q function, and the error and complementary error functions. We also show their interrelationships.

E.1 GAUSSIAN DENSITY AND DISTRIBUTION

The general gaussian probability density and cumulative probability distribution functions for a gaussian random variable X having a mean value $-\infty < a_X < \infty$ and variance $\sigma_X^2 > 0$ are

$$f_X(x) = \frac{1}{\sqrt{2\pi}\,\sigma_X}\,e^{-(x-a_X)^2/(2\sigma_X^2)} \tag{E.1-1}$$

$$F_X(x) = \int_{-\infty}^{x} f_X(\xi)d\xi = F\left(\frac{x - a_X}{\sigma_X}\right) \tag{E.1-2}$$

where

$$F(x) = \int_{-\infty}^{x} \frac{1}{\sqrt{2\pi}}\,e^{-\xi^2/2}\,d\xi \tag{E.1-3}$$

is the cumulative probability distribution function when $\sigma_X = 1$ and $a_X = 0$. $F(x)$ is sometimes called the "normalized" form of $F_X(x)$.

The function $F(x)$ is not known to exist in closed form; it is tabulated in Table E.1-1. For the general case when $\sigma_X \neq 1$ and $a_X \neq 0$, the value of $F_X(x)$

736

TABLE E.1-1   Values of $F(x)$ for $0 \leq x \leq 3.89$ in steps of 0.01

x	0.00	0.01	0.02	0.03	0.04	0.05	0.06	0.07	0.08	0.09
0.0	0.5000	0.5040	0.5080	0.5120	0.5160	0.5199	0.5239	0.5279	0.5319	0.5359
0.1	0.5398	0.5438	0.5478	0.5517	0.5557	0.5596	0.5636	0.5675	0.5714	0.5753
0.2	0.5793	0.5832	0.5871	0.5910	0.5948	0.5987	0.6026	0.6064	0.6103	0.6141
0.3	0.6179	0.6217	0.6255	0.6293	0.6331	0.6368	0.6406	0.6443	0.6480	0.6517
0.4	0.6554	0.6591	0.6628	0.6664	0.6700	0.6736	0.6772	0.6808	0.6844	0.6879
0.5	0.6915	0.6950	0.6985	0.7019	0.7054	0.7088	0.7123	0.7157	0.7190	0.7224
0.6	0.7257	0.7291	0.7324	0.7357	0.7389	0.7422	0.7454	0.7486	0.7517	0.7549
0.7	0.7580	0.7611	0.7642	0.7673	0.7704	0.7734	0.7764	0.7794	0.7823	0.7852
0.8	0.7881	0.7910	0.7939	0.7967	0.7995	0.8023	0.8051	0.8078	0.8106	0.8133
0.9	0.8159	0.8186	0.8212	0.8238	0.8264	0.8289	0.8315	0.8340	0.8365	0.8389
1.0	0.8413	0.8438	0.8461	0.8485	0.8508	0.8531	0.8554	0.8577	0.8599	0.8621
1.1	0.8643	0.8665	0.8686	0.8708	0.8729	0.8749	0.8770	0.8790	0.8810	0.8830
1.2	0.8849	0.8869	0.8888	0.8907	0.8925	0.8944	0.8962	0.8980	0.8997	0.9015
1.3	0.9032	0.9049	0.9066	0.9082	0.9099	0.9115	0.9131	0.9147	0.9162	0.9177
1.4	0.9192	0.9207	0.9222	0.9236	0.9251	0.9265	0.9279	0.9292	0.9306	0.9319
1.5	0.9332	0.9345	0.9357	0.9370	0.9382	0.9394	0.9406	0.9418	0.9429	0.9441
1.6	0.9452	0.9463	0.9474	0.9484	0.9495	0.9505	0.9515	0.9525	0.9535	0.9545
1.7	0.9554	0.9564	0.9573	0.9582	0.9591	0.9599	0.9608	0.9616	0.9625	0.9633
1.8	0.9641	0.9649	0.9656	0.9664	0.9671	0.9678	0.9686	0.9693	0.9699	0.9706
1.9	0.9713	0.9719	0.9726	0.9732	0.9738	0.9744	0.9750	0.9756	0.9761	0.9767
2.0	0.9773	0.9778	0.9783	0.9788	0.9793	0.9798	0.9803	0.9808	0.9812	0.9817
2.1	0.9821	0.9826	0.9830	0.9834	0.9838	0.9842	0.9846	0.9850	0.9854	0.9857
2.2	0.9861	0.9864	0.9868	0.9871	0.9875	0.9878	0.9881	0.9884	0.9887	0.9890
2.3	0.9893	0.9896	0.9898	0.9901	0.9904	0.9906	0.9909	0.9911	0.9913	0.9916
2.4	0.9918	0.9920	0.9922	0.9925	0.9927	0.9929	0.9931	0.9932	0.9934	0.9936
2.5	0.9938	0.9940	0.9941	0.9943	0.9945	0.9946	0.9948	0.9949	0.9951	0.9952
2.6	0.9953	0.9955	0.9956	0.9957	0.9959	0.9960	0.9961	0.9962	0.9963	0.9964
2.7	0.9965	0.9966	0.9967	0.9968	0.9969	0.9970	0.9971	0.9972	0.9973	0.9974
2.8	0.9974	0.9975	0.9976	0.9977	0.9977	0.9978	0.9979	0.9979	0.9980	0.9981
2.9	0.9981	0.9982	0.9982	0.9983	0.9984	0.9984	0.9985	0.9985	0.9986	0.9986
3.0	0.9987	0.9987	0.9987	0.9988	0.9988	0.9989	0.9989	0.9989	0.9990	0.9990
3.1	0.9990	0.9991	0.9991	0.9991	0.9992	0.9992	0.9992	0.9992	0.9993	0.9993
3.2	0.9993	0.9993	0.9994	0.9994	0.9994	0.9994	0.9994	0.9995	0.9995	0.9995
3.3	0.9995	0.9995	0.9996	0.9996	0.9996	0.9996	0.9996	0.9996	0.9996	0.9997
3.4	0.9997	0.9997	0.9997	0.9997	0.9997	0.9997	0.9997	0.9997	0.9998	0.9998
3.5	0.9998	0.9998	0.9998	0.9998	0.9998	0.9998	0.9998	0.9998	0.9998	0.9998
3.6	0.9998	0.9999	0.9999	0.9999	0.9999	0.9999	0.9999	0.9999	0.9999	0.9999
3.7	0.9999	0.9999	0.9999	0.9999	0.9999	0.9999	0.9999	0.9999	0.9999	0.9999
3.8	0.9999	0.9999	0.9999	0.9999	0.9999	0.9999	0.9999	1.0000	1.0000	1.0000

Source: Adapted from Peebles, P. Z., *Probability, Random Variables, and Random Signal Principles*, 3rd ed., 1993, McGraw-Hill, Inc., with permission.

can be found from (E.1-2) and the tabulated data. For negative values of x use

$$F(-x) = 1 - F(x) \qquad (\text{E.1-4})$$

Tabulated values of $F(x)$ exist in many sources. Abramowitz and Stegun (1964), as one example, gives $F(x)$ to 15 digits for many values of x.

E.2 Q FUNCTION

The Q function is closely related to the Gaussian function $F(x)$. Denoted by $Q(x)$, the Q function is

$$Q(x) = \frac{1}{\sqrt{2\pi}} \int_x^\infty e^{-\xi^2/2} \, d\xi = 1 - F(x) \qquad \text{(E.2-1)}$$

For negative values of x,

$$Q(-x) = 1 - Q(x) \qquad \text{(E.2-2)}$$

so

$$Q(-x) = F(x) \qquad \text{(E.2-3)}$$

As with $F(x)$, a closed form solution for $Q(x)$ is not known and approximations are often used. One excellent approximation is due to Börjesson and Sundberg (1979):

$$Q(x) \approx \left[\frac{1}{0.661x + 0.339\sqrt{x^2 + 5.51}} \right] \frac{1}{\sqrt{2\pi}} e^{-x^2/2}, \qquad x \geq 0 \qquad \text{(E.2-4)}$$

The approximation (E.2-4) is said to be accurate to a maximum absolute relative error of 0.27% for any $x \geq 0$. By using (E.2-4) with (E.2-1), an excellent approximation for $F(x)$ also is available.

E.3   ERROR AND COMPLEMENTARY ERROR FUNCTIONS

The error function, denoted by erf(x), is defined by

$$\text{erf}(x) = \frac{2}{\sqrt{\pi}} \int_0^x e^{-\xi^2} \, d\xi \qquad \text{(E.3-1)}$$

which can be written as

$$\text{erf}(x) = 1 - 2Q(\sqrt{2}x) = -1 + 2F(\sqrt{2}x) \qquad \text{(E.3-2)}$$

in terms of the functions $Q(\cdot)$ and $F(\cdot)$ from (E.2-1). For negative x use

$$\text{erf}(-x) = -\text{erf}(x) \qquad \text{(E.3-3)}$$

The complementary error function, denoted by erfc(x), is defined by

$$\text{erfc}(x) = 1 - \text{erf}(x) \qquad \text{(E.3-4)}$$

so

$$\text{erfc}(x) = 2Q(\sqrt{2}x) = 2[1 - F(\sqrt{2}x)] \tag{E.3-5}$$

Since (E.2-4) is an excellent approximation for $Q(x)$, it also serves as an excellent approximation for $\text{erfc}(x)$ from (E.3-5):

$$\text{erfc}(x) \approx \left[\frac{2}{0.661\sqrt{2}x + 0.339\sqrt{2x^2 + 5.51}}\right]\frac{1}{\sqrt{2\pi}}e^{-x^2}, \quad x \geq 0 \tag{E.3-6}$$

For negative values of x use

$$\text{erfc}(-x) = 1 - \text{erf}(-x)$$
$$= 1 + \text{erf}(x)$$
$$= 2 - \text{erfc}(x) \tag{E.3-7}$$

BIBLIOGRAPHY

The following abbreviations are used below:

ASTIA for Armed Service Technical Information Agency
IEE for Institute of Electrical Engineers
IEEE for The Institute of Electrical and Electronic Engineers
IRE for The Institute of Radio Engineers
NTIS for National Technical Information Service
RCA for Radio Corporation of America

Abramowitz, M., and I. A. Stegun (eds.): *Handbook of Mathematical Functions with Formulas, Graphs, and Mathematical Tables*, Applied Mathematics Series 55, National Bureau of Standards, Washington, D.C., 1964.

Al-Gwaiz, M. A.: *Theory of Distributions*, Marcel Dekker, Inc., New York, 1992.

Altshuler, E. A.: "A Simple Expression for Estimating Attenuation by Fog at Millimeter Wavelengths," *IEEE Transactions on Antennas and Propagation*, vol. AP-32, no. 7, July 1984, pp. 757–758.

Ament, W. S.: "Toward a Theory of Reflection by a Rough Surface," *Proceedings of the IRE*, vol. 41, no. 1, January 1953, pp. 142–146.

Amindavar, H., and J. A. Ritcey: "Padé Approximations for Detectability in K-Clutter and Noise," *IEEE Transactions on Aerospace and Electronic Systems*, vol. 30, no. 2, April 1994, pp. 425–434.

Anastassopolos, V, and G. A. Lampropoulos: "Optimal CFAR Detection in Weibull Clutter," *IEEE Transactions on Aerospace and Electronic Systems*, vol. 31, no. 1, January 1995, pp. 52–63.

Andreasen, M. G.: "Scattering from Bodies of Revolution," *IEEE Transactions on Antennas and Propagation*, vol. AP-13, no. 2, March 1965, pp. 303–310. See also correction in vol. AP-14, no. 5, September 1966, p. 659.

Asseo, S. J.: "Detection of Target Multiplicity Using Monopulse Quadrature Angle," *IEEE Transactions on Aerospace and Electronic Systems*, vol. AES-17, no. 2, March 1981, pp. 271–280.

Atlas, D., and E. Kessler, III: "A Model Atmosphere for Widespread Precipitation," *Aeronautical Engineering Review*, vol. 16, no. 2, February 1957, pp. 69–74, 82.

Balanis, C. A.: *Antenna Theory Analysis and Design*, Harper & Row, Publishers, New York, 1982.

Barankin, E. W.: "Locally Best Unbiased Estimates," *Annals of Mathematical Statistics*, vol. XX, no. 4, December 1949, pp. 477–501.

Barker, R. H.: "Group Synchronization of Binary Digital Systems." In Jackson, W. (ed.), *Communication Theory*, Academic Press, Inc., London, England, 1953, pp. 273–287.

Barton, D. K.: *Radar System Analysis*, Prentice-Hall, Inc., Englewood Cliffs, New Jersey, 1964.

Barton, D. K.: "Simple Procedures for Radar Detection Calculations," *IEEE Transactions on Aerospace and Electronic Systems*, vol. AES-5, no. 5, September 1969, pp. 837–846.

Barton, D. K. (ed.): *Radars, Volume 1, Monopulse Radar*, Artech House, Inc., Dedham, Massachusetts, 1977.

Barton, D. K.: *Modern Radar System Analysis*, Artech House, Inc., Norwood, Massachusetts, 1988.

Barton, D. K., C. E. Cook, and P. Hamilton (eds.): *Radar Evaluation Handbook*, Artech House, Inc., Norwood, Massachusetts, 1991.

Barton, D. K., and H. R. Ward: *Handbook of Radar Measurement*, Prentice-Hall, Inc., Englewood Cliffs, New Jersey, 1969.

Battan, L. J.: *Radar Observation of the Atmosphere*, University of Chicago Press, Chicago, Illinois, 1973.

Bayliss, E. T.: "Design of Monopulse Antenna Difference Patterns with Low Sidelobes," *Bell System Technical Journal*, vol. 47, no. 5, May–June 1968, pp. 623–650.

Bean, B. R.: "The Geographical and Height Distribution of the Gradient of Refractive Index," *Proceedings of the IRE*, vol. 41, no. 4, April 1953, pp. 549–550.

Bean, B. R., B. A. Cahoon, C. A. Samson, and G. D. Thayer: "A World Atlas of Atmospheric Radio Refractivity," U. S. Department of Commerce, *ESSA Monograph 1*, 1966.

Bechtel, M. E.: "Application of Geometric Diffraction Theory to Scattering From Cones and Disks," *Proceedings of the IEEE*, vol. 53, no. 8, August 1965, pp. 877–882.

Beck, J. V., and K. J. Arnold: *Parameter Estimation in Engineering and Science*, John Wiley & Sons, Inc., New York, 1977.

Beckman, P., and A. Spizzichino: *The Scattering of Electromagnetic Waves from Rough Surfaces*, Pergamon Press, Ltd., Oxford, England, 1963.

Bellegarda, J. R., and E. L. Titlebaum: "Time-Frequency Hop Codes Based Upon Extended Quadratic Congruences," *IEEE Transactions on Aerospace and Electronic Systems*, vol. 24, no. 6, November 1988, pp. 726–742.

Bellegarda, J. R., and E. L. Titlebaum: "Amendment to Time-Frequency Hop Codes Based Upon Extended Quadratic Congruences," *IEEE Transactions on Aerospace and Electronic Systems*, vol. 27, no. 1, January 1991, pp. 167–172.

Berkowitz, R. S. (ed.): *Modern Radar Analysis, Evaluation, and System Design*, John Wiley & Sons, Inc., New York, 1965.

Bhattacharyya, A.: "On Some Analogues of the Amount of Information and their Use in Statistical Estimation," *Sankhyā*, vol. 8, p. 1, November 1946, pp. 1–14. This paper is continued in vol. 8, pt. 3, October 1947, pp. 201–218.

Bhattacharyya, A. K., and D. L. Sengupta: *Radar Cross Section Analysis and Control*, Artech House, Inc., Norwood, Massachusetts, 1991.

Biernson, G.: *Optimal Radar Tracking Systems*, John Wiley & Sons, Inc., New York, 1990.

Bird, D.: "Target RCS Modelling", *Record of the 1994 IEEE National Radar Conference*, Atlanta, Georgia, March 29–31, 1994, pp. 74–79.

Bird, G. J. A.: *Radar Precision and Resolution*, John Wiley & Sons, Inc., New York, 1974.

Bird, J. S.: "Calculating Detection Probabilities for Adaptive Thresholds," *IEEE Transactions on Aerospace and Electronic Systems*, vol. AES-19, no. 4, July 1983, pp. 506–512.

Blake, L. V.: "Calculation of the Radar Cross Section of a Perfectly Conducting Sphere," NRL Memorandum Report 2419, April 1972, U. S. Naval Research Laboratory, Washington, D. C. Also NTIS document AD 704746.

Blake, L. V.: "Radar/Radio Tropospheric Absorption and Noise Temperature," NRL Report 7461, October 30, 1972, U. S. Naval Research Laboratory, Washington, D. C. Also NTIS document AD 753197.

Blake, L. V.: *Radar Range-Performance Analysis*, Lexington Books, Lexington, Massachusetts, 1980.

Blanchard, D. C.: "Raindrop Size-Distribution in Hawaiian Rains," *Journal of Meteorology*, vol. 10, no. 6, December 1953, pp. 457–473.

Blore, W. E.: "The Radar Cross Section of Ogives, Double-Backed Cones, Double-Rounded Cones, and Cone Spheres," *IEEE Transactions on Antennas and Propagation*, vol. AP-12, no. 5, September 1964, pp. 582–590.

Bochner, S.: "Stable Laws of Probability and Completely Monotone Functions," *Duke Mathematical Journal*, vol. 3, 1937, pp. 726–728.

Boehmer, A. M.: "Binary Pulse Compression Codes," *IEEE Transactions on Information Theory*, vol. IT-13, no. 2, April 1967, pp. 156–167.

Bogler, P. L.: "Detecting the Presence of Target Multiplicity," *IEEE Transactions on Aerospace and Electronic Systems*, vol. AES-22, no. 2, March 1986, pp. 197–203.

Bogush, Jr., A. J.: *Radar and the Atmosphere*, Artech House, Norwood, Massachusetts, 1989.

Bomer, L., and M. Antweiler: "Polyphase Barker Sequences," *Electronics Letters*, vol. 25, no. 23, November 9, 1989, pp. 1577–1579.

Börjesson, P. O., and C.-E. W. Sundberg: "Simple Approximations of the Error Function Q(x) for Communications Applications," *IEEE Transactions on Communications*, vol. COM-27, no. 3, March 1979, pp. 639–643.

Breit, G., and M. A. Tuve: "A Test of the Existence of the Conducting Layer," *Physical Review*, vol. 28, no. 3, 2d ser., September 1926, pp. 554–575.

Brennan, L. E., and I. S. Reed: "A Recursive Method of Computing the Q Function," *IEEE Transactions of Information Theory*, vol. IT-11, no. 2, April 1965, pp. 312–313.

Brunner, J. S.: "A Recursive Method of Calculating Binary Integration Detection Probabilities," *IEEE Transactions on Aerospace and Electronic Systems*, vol. 26, no. 6, November 1990, pp. 1034–1035.

Burdic, W. S.: *Radar Signal Analysis*, Prentice-Hall, Inc., Englewood Cliffs, New Jersey, 1968.

Burrows, C. R., and S. S. Attwood: *Radio Wave Propagation*, Academic Press, Inc., New York, 1949.

Byatt, D. W. G., and J. Wild: "Present and Future Radar Display Techniques," *International Conference on Radar – Present and Future*, October 23–25, 1973, IEE Conference Publication No. 105, pp. 201–206,

Cadzow, J. A.: *Foundations of Digital Signal Processing and Data Analysis*, Macmillan Publishing Company, New York, 1987.

Cantrell, B. H., and G. V. Trunk: "Angular Accuracy of a Scanning Radar Employing a Two-Pole Filter," *IEEE Transactions on Aerospace and Electronic Systems*, vol. AES-9, no. 5, September 1973, pp. 649–653 (see also Cantrell and Trunk, 1974).

Cantrell, B. H., and G. V. Trunk: "Correction to Angular Accuracy of a Scanning Radar Employing a Two-Pole Filter," *IEEE Transactions on Aerospace and Electronic Systems*, vol. AES-10, no. 6, November 1974, pp. 878–880.

Capon, J.: "Optimum Coincidence Procedures for Detecting Weak Signals in Noise," *1960 IRE International Convention Record, Part 4*, March 21–24, 1960, New York, pp. 154–166.

Carlson, A. B.: *Communication Systems, An Introduction to Signals and Noise in Electrical Communication*, McGraw-Hill Book Co., Inc., New York, 3d ed., 1986. See also 2d ed., 1975, and 1st ed., 1968.

CCIR: "Attenuation and Scattering by Precipitation and Other Atmospheric Particles," Report 721-2 in vol. V, *Propagation in Non-Ionized Media, Recommendation and Reports of the CCIR, 1986*, Geneva, International Telecommunications Union, 1986.

Chang, W., and K. Scarbrough: "Costas Arrays with Small Number of Cross-Coincidences," *IEEE Transactions on Aerospace and Electronic Systems*, vol. AES-25, no. 1, January 1989, pp. 109–112.

Chu, D. C.: "Polyphase Codes With Good Periodic Correlation Properties," *IEEE Transactions on Information Theory*, vol. IT-18, no. 4, July 1972, pp. 531–532.

Cohen, M. N., J. M. Baden, and P. E. Cohen: "Biphase Codes with Minimum Peak Sidelobes," *Proceedings of the 1989 IEEE National Radar Conference*, Dallas, Texas, March 29–30, 1989, pp. 62–66.

Cohen, M. N., M. R. Fox, and J. M. Baden: "Minimum Peak Sidelobe Pulse Compression Codes," *Record of the IEEE 1990 International Radar Conference*, Arlington, Virginia, May 7–10, 1990, pp. 633–638.

Collin, R. E.: *Antennas and Radio Wave Propagation*, McGraw-Hill Book Company, New York, 1985.

Conte, E., and M., Longo: "Characterization of Radar Clutter as a Spherically Invariant Random Process," *IEE Proceedings*, pt. F, vol. 134, no. 2, April 1987, pp. 191–197.

Conte, E., M. Lops, and G. Ricci: "Distribution-Free Radar Detection in Compound-Gaussian Clutter," (Record of) *Radar 92*, Brighton Conference Center, United Kingdom, October 12–13, 1992, pp. 98–101.

Cook, C. E., and M. Bernfeld: *Radar Signals: An Introduction to Theory and Application*, Academic Press, New York, 1967.

Cooley, J. W., and J. W. Tukey: "An Algorithm for the Machine Computation of Complex Fourier Series," *Mathematics of Computation*, vol. 19, April 1965, pp. 297–301.

Cooper, G. R., and R. D. Yates: "Design of Large Signal Sets with Good Aperiodic Correlation Properties," Purdue University Technical Report TR-EE 66–13, September 1966.

Costas, J. P.: "Medium Constraints on Sonar Design and Performance," General Electric Co., Class 1 Report R65EMH33, November 1965. A synopsis of this report appeared in *Eascon Convention Record*, 1975, pp. 68A–68L.

Costas, J. P.: "A Study of a Class of Detection Waveforms Having Nearly Ideal Range-Doppler Ambiguity Properties," *Proceedings of the IEEE*, vol. 72, no. 8, August 1984, pp. 996–1009.

Cramér, H.: *Mathematical Methods of Statistics*, Princeton University Press, Princeton, New Jersey, 1946, (10th printing, 1963; original edition printed in Upsala, Sweden, in 1945 by Almqvist & Wiksells).

Crane, R. K., and H-h. K. Burke: "The Evaluation of Models for Atmospheric Attenuation and Backscatter Characteristic Estimation at 95 GHz," ERT Document No. P-3606 by Environmental Research & Technology, Inc., for MIT Lincoln Laboratory, February 1978. Also NTIS document AD A088332.

Crispin, J. W., Jr. (ed.): *Methods of Radar Cross Section Analysis*, Academic Press, New York, 1968.

Crispin, J. W., Jr., and A. L. Maffett: "Radar Cross Section Estimation for Simple Shapes," *Proceedings of the IEEE*, vol. 53, no. 8, August 1965, pp. 833–848.

Crispin, J. W., Jr. and A. L. Maffett: "Radar Cross Section Estimation for Complex Shapes," *Proceedings of the IEEE*, vol. 53, no. 8, August 1965, pp. 972–982.

Cross, D. C.: "Low Jitter High Performance Electronic Range Tracker," *Record of the IEEE 1975 International Radar Conference*, Arlington, Virginia, April 21–23, 1975, pp. 408–411.

Currie, N. C. (ed.): *Radar Reflectivity Measurement: Techniques and Applications*, Artech House, Norwood, Massachusetts, 1989.

Dadi, M. I., and R. J. Marks, II: "Detector Relative Efficiencies in the Presence of Laplace Noise," *IEEE Transactions on Aerospace and Electronic Systems*, vol. AES-23, no. 4, July 1987, pp. 568–582.

Damonte, J. B., and D. J. Stoddard: "An Analysis of Conical Scan Antennas for Tracking," *IRE Convention Record, Part 1, Telemetry, Antennas, and Propagation*, IRE National Convention, March 19–22, 1956, pp. 39–47.

Davenport, W. B., Jr., and W. L. Root: *An Introduction to the Theory of Random Signals and Noise*, McGraw-Hill Book Co., Inc., New York, 1958.

David, P., and J. Voge: *Propagation of Waves*, Pergamon Press, New York, 1969.

Deutsch, R.: *Estimation Theory*, Prentice-Hall, Inc., Englewood Cliffs, New Jersey, 1965.

DiCaudo, V. J., and W. W. Martin: "Approximate Solution to Bistatic Radar Cross Section of Finite Length, Infinitely Conducting Cylinder," *IEEE Transactions on Antennas and Propagation*, vol. AP-14, no. 5, September 1966, pp. 668–669.

DiFranco, J. V., and W. L. Rubin: *Radar Detection*, Prentice-Hall, Inc., Englewood Cliffs, New Jersey, 1968.

Dillard, G. M.: "Generating Random Numbers Having Probability Distributions Occurring in Signal Detection Problems," *IEEE Transactions on Information Theory*, vol. IT-13, no. 4, October 1967, pp. 616–617.

Dirac, P. A. M.: *The Principles of Quantum Mechanics*, Oxford, 1930 (2d and 3d eds. in 1935 and 1947).

Dolph, C. L.: "A Current Distribution for Broadside Arrays Which Optimizes the Relationship Between Beamwidth and Side Lobe Level," *Proceedings of the IRE*, vol. 34, no. 6, June 1946, pp. 335–348.

Donaldson, Jr., R. J.: "The Measurement of Cloud Liquid Water Content by Radar," *Journal of Meteorology*, vol. 12, no. 3, June 1955, pp. 238–244.

Drumheller, D. M., and E. L. Titlebaum: "Cross-Correlation Properties of Algebraically Constructed Costas Arrays," *IEEE Transactions on Aerospace and Electronic Systems*, vol. 27, no. 1, January 1991, pp. 2–10.

Dwight, H. B.: *Tables of Integrals and Other Mathematical Data*, 4th ed., The Macmillan Company, New York, 1961.

Eaves, J. L., and E. K. Reedy (eds.): *Principles of Modern Radar*, Van Nostrand Reinhold, New York, 1987.

Edde, B.: *Radar Principles, Technology, Applications*, PTR Prentice-Hall, Englewood Cliffs, New Jersey, 1993.

Einersson, G.: "Address Assignment for a Time-Frequency-Coded Spread-Spectrum System," *Bell System Technical Journal*, vol. 59, no. 7, September 1980, pp. 1241–1255.

El Gamal, A. A., L. A. Hemachandra, I. Shperling, and V. K. Wei: "Using Simulated Annealing to Design Good Codes," *IEEE Transactions on Information Theory*, vol. IT-33, no. 1, January 1987, pp. 116–123.

Elliott, R. S.: "Beamwidth and Directivity of Large Scanning Arrays," part I. *Microwave Journal*, vol. VI, no. 12, December 1963, pp. 53–60.

Elliott, R. S.: "Beamwidth and Directivity of Large Scanning Arrays," part II. *Microwave Journal*, vol. VII, no. 1, January 1964, pp. 74–82.

Elliott, R. S.: "Design of Line Source Antennas for Difference Patterns with Sidelobes of Individually Arbitrary Heights," *IEEE Transactions on Antennas and Propagation*, vol. AP-24, no. 3, May 1976, pp. 310–316.

Elliott, R. S.: *Antenna Theory and Design*, Prentice-Hall, Inc., Englewood Cliffs, New Jersey, 1981.

Endresen, K., and R. Hedemark: "Coincidence Techniques for Radar Receivers Employing a Double-Threshold Method of Detection," *Proceedings of the IRE*, vol. 49, no. 10, October 1961, pp. 1561–1567 (see also comments of V. G. Hansen and the author's reply, vol. 50, no. 4, April 1962, p. 480).

Farison, J. B.: "On Calculating Moments for Some Common Probability Laws," *IEEE Transactions on Information Theory*, vol. IT-11, no. 4, October 1965, pp. 586–589.

Fehlner, L. F.: "Supplement to Marcum's and Swerling's Data on Target Detection by a Pulsed Radar," Technical Report TG-451A, Johns Hopkins University, Applied Physics Laboratory, September 1964.

Felhauer, T.: "New Class of Polyphase Pulse Compression Code With Unique Characteristics," *Electronics Letters*, vol. 28, no. 8, April 9, 1992, pp. 769–771.

Felhauer, T.: "Design and Analysis of New P(n,k) Polyphase Pulse Compression Codes," *IEEE Transactions on Aerospace and Electronic Systems*, vol. 30, no. 3, July 1994, pp. 865–874.

Fenster, W.: "The Application, Design, and Performance of Over-the-Horizon Radars," *International Conference RADAR-77*, October 25–28, 1977, IEE (London) Conference Publication No. 155, pp. 36–40.

Finn, H. M., and R. S. Johnson: "Adaptive Detection Mode with Threshold Control as a Function of Spatially Sampled Clutter-Level Estimates," *RCA Review*, vol. 29, no. 3, September 1968, pp. 414–464.

Fisher, R. A.: "On an Absolute Criterion for Fitting Frequency Curves," *Messenger of Mathematics*, vol. 41, May 1911–April 1912, 1912, pp. 155–160.

Fisher, R. A.: "On the Mathematical Foundations of Theoretical Statistics," *Philosophical Transactions of the Royal Society of London*, ser. A, vol. 222, May 1922, pp. 309–368.

Fisher, R. A.: "Theory of Statistical Estimation," *Proceedings of the Cambridge Philosophical Society*, vol. 22, 1925, p. 700.

Flock, W. L.: "Propagation Effects on Satellite Systems at Frequencies Below 10 GHz," NASA Reference Publication 1108(02), 1987.

Follin, J. W., Jr., F. C. Paddison, and A. L. Maffett: "Statistics of Radar Cross Section Scintillations," *Electromagnetics*, vol. 4, 1984, pp. 139–164.

Fowle, E. N.: "A General Method for Controlling the Time and Frequency Envelopes of FM Signals," MIT Lincoln Laboratory, Lexington, Massachusetts, Group Report 41G-0008, June 1961.

Fowle, E. N.: "The Design of FM Pulse-Compression Signals," *IEEE Transactions on Information Theory*, vol. IT-10, no. 1, January 1964, pp. 61–67.

Frank, R. L.: "Polyphase Codes with Good Nonperiodic Correlation Properties," *IEEE Transactions on Information Theory*, vol. IT-9, no. 1, January 1963, pp. 43–45.

Frank, R. L.: "Comments on Polyphase Codes With Good Correlation Properties," *IEEE Transactions on Information Theory*, vol. IT-19, no. 2, March 1973, p. 244.

Frank, R. L., and S. A. Zadoff: "Phase Shift Pulse Codes With Good Periodic Correlation Properties," *IRE Transactions on Information Theory*, vol. IT-8, no. 6, October 1962, pp. 381–382 (see also comments of R. C. Heimiller on p. 382).

Freedman A., and N. Levanon: "Any Two N × N Costas Signals Must Have at Least One Common Ambiguity Sidelobe if N > 3 – A Proof," *Proceedings of the IEEE*, vol. 73, no. 10, October 1985, pp. 1530–1531.

Freedman, A., and N. Levanon: "Staggered Costas Signals," *IEEE Transactions on Aerospace and Electronic Systems*, vol. AES-22, no. 6, November 1986, pp. 695–702.

Friedlander, F. G.: *Introduction to the Theory of Distributions*, Cambridge University Press, Cambridge, 1982.

Friedman, B.: *Principles and Techniques of Applied Mathematics*, John Wiley & Sons, Inc., New York, 1956.

Fritch, P. C. (ed.): "Special Issue on Radar Reflectivity," *Proceedings of the IEEE*, vol. 53, no. 8, August 1965.

Gabel, R. A., and R. A. Roberts: *Signals and Linear Systems*, John Wiley & Sons, Inc., New York, 1987.

Gerlach, K., and F. F. Kretschmer, Jr.: "Reciprocal Radar Waveforms," *IEEE Transactions on Aerospace and Electronic Systems*, vol. 27, no. 4, July 1991, pp. 646–654.

Getz, B., and N. Levanon: "Weight Effects on the Periodic Ambiguity Function," *IEEE Transactions on Aerospace and Electronic Systems*, vol. 31, no. 1, January 1995, pp. 182–193.

Golomb, S. W. (ed.): *Digital Communications With Space Applications*, Prentice-Hall, Inc., Englewood Cliffs, New Jersey, 1964.

Golomb, S. W.: *Shift Register Sequences*, Holden-Day, Oakland, California, 1967.

Golomb, S. W.: "Algebraic Constructions for Costas Arrays," *Journal of Combinatorial Theory*, ser. A, vol. 37, no. 1, July 1984, pp. 13–21.

Golomb, S. W., and H. Taylor: "Two-Dimensional Synchronization Patterns for Minimum Ambiguity," *IEEE Transactions on Information Theory*, vol. IT-28, no. 4, July 1982, pp. 600–604.

Golomb, S. W., and H. Taylor: "Constructions and Properties of Costas Arrays," *Proceedings of the IEEE*, vol. 72, no. 9, September 1984, pp. 1143–1163.

Gottesman, S. R., P. G. Grieve, and S. W. Golomb: "A Class of Pseudonoise-Like Pulse Compression Codes," *IEEE Transactions on Aerospace and Electronic Systems*, vol. 28, no. 2, April 1992, pp. 355–361.

Gradshteyn, I. S., and I. M. Ryzhik: *Tables of Integrals, Series, and Products*, Academic Press, New York, 4th ed. (prepared by Yu. V. Geronimus and M. Yu. Tseytlin, translated by A. Jeffrey), 1965.

Grasso, G., and P. F. Guarguaglini: "On the Optimum Coincidence Procedure in Presence of a Fluctuating Target," *IEEE Transactions on Information Theory*, vol. IT-13, no. 3, July 1967, pp. 522–524.

Green, Jr., B. A.: "Radar Detection Probability with Logarithmic Detectors," *IEEE Transactions on Information Theory*, vol. IT-4, no. 1, March 1958, pp. 50–52.

Griep, K. R., J. A. Ritcey, and J. J. Burlingame: "Poly-Phase Codes and Optimal Filters for Multiple User Ranging," *IEEE Transactions on Aerospace and Electronic Systems*, vol. 31, no. 2, April 1995, pp. 752–767.

Guida, M., M. Longo, and M. Lops: "Biparametric CFAR Procedures for Lognormal Clutter," *IEEE Transactions on Aerospace and Electronic Systems*, vol. 29, no. 3, July 1993, pp. 798–809.

Guinard, N. W., and J. C. Daley: "An Experimental Study of a Sea Clutter Model," *Proceedings of the IEEE*, vol. 58, no. 4, April 1970, pp. 543–550.

Gunn, K. L. S., and T. W. R. East: "The Microwave Properties of Precipitation Particles," *Quarterly Journal of the Royal Meteorological Society*, vol. 80, no. 346, October 1954, pp. 522–545 (see also corrections in vol. 83, no. 357, July 1957, p. 416).

Halperin, I.: *Introduction to the Theory of Distributions*, University of Toronto Press, Toronto, 1952.

Han, J., P. K. Varshney, and R. Srinivasan: "Distributed Binary Integration," *IEEE Transactions on Aerospace and Electronic Systems*, vol. 29, no. 1, January 1993, pp. 2–8.

Hannan, P. W.: "Microwave Antennas Derived From the Cassegrain Telescope," *IRE Transactions on Antennas and Propagation*, vol. AP-9, no. 2, March 1961, pp. 140–153.

Hannan, P. W.: "Optimum Feeds for All Three Modes of a Monopulse Antenna I: Theory," *IRE Transactions on Antennas and Propagation*, vol. AP-9, no. 5, September 1961, pp. 444–454.

Hannan, P. W.: "Optimum Feeds for All Three Modes of a Monopulse Antenna II: Practice," *IRE Transactions on Antennas and Propagation*, vol. AP-9, no. 5, September 1961, pp. 454–461.

Hannan, P. W., and P. A. Loth: "A Monopulse Antenna Having Independent Optimization of the Sum and Difference Modes," *IRE Convention Record, Part 1*, 1961, pp. 57–60. (This paper is also reprinted in Barton (ed.), 1977).

Hansen, R. C.: "Tables of Taylor Distributions for Circular Aperture Antennas," *IRE Transactions on Antennas and Propagation*, vol. AP-8, no. 1, January 1960, pp. 23–26.

Hansen, R. C. (ed.): *Microwave Scanning Antennas*, vol. I (*Apertures*), Academic Press, New York, 1964. See also Vols. II (*Array Theory and Practice*), 1966, and III (*Array Systems*), 1966.

Hansen, V. G.: "Simple Expressions for Determining Radar Detection Thresholds," *IEEE Transactions on Aerospace and Electronic Systems*, vol. AES-18, no. 4, July 1982, pp. 510–512. See also Urkowitz, H.: "Corrections to and Comments on 'Simple Expressions for Determining Radar Detections Thresholds,'" vol. AES-21, no. 4, July 1985, pp. 583–584.

Hardy, G. H., J. E. Littlewood, and G. Polya: *Inequalities*, Cambridge University Press, London, England, 1964.

Harrington, J. V.: "An Analysis of the Detection of Repeated Signals in noise by Binary Integration," *IRE Transactions on Information Theory*, vol. IT-1, no. 1, March 1955, pp. 1–9.

Harrington, R. F.: *Time-Harmonic Electromagnetic Fields*, McGraw-Hill Book Company, New York, 1968.

Heidbreder, G. R., and R. L. Mitchell: "Detection Probabilities for Log-Normally Distributed Signals," *IEEE Transactions on Aerospace and Electronic Systems*, vol. AES-3, no. 1, January 1967, pp. 5–13.

Heimiller, R. C.: "Phase Shift Pulse Codes with Good Periodic Correlation Properties," *IRE Transactions on Information Theory*, vol. IT-7, no. 6, October 1961, pp. 254–257.

Helstrom, C. W.: *Statistical Theory of Signal Detection*, 2d ed., Pergamon Press, New York, 1975.

Helstrom, C. W.: "Detectability of Signals in Laplace Noise," *IEEE Transactions on Aerospace and Electronic Systems*, vol. AES-25, no. 2, March 1989, pp. 190–196.

Helstrom, C. W.: "Performance of Receivers with Linear Detectors," *IEEE Transactions on Aerospace and Electronic Systems*, vol. 26, no. 2, March 1990, pp. 210–217.

Helstrom, C. W.: *Probability and Stochastic Processes for Engineers*, 2d ed., Macmillan Publishing Company, New York, 1991.

Helstrom, C. W.: "Detection Probabilities for Correlated Rayleigh Fading Signals," *IEEE Transactions on Aerospace and Electronic Systems*, vol. 28, no. 1, January 1992, pp. 259–267.

Helstrom, C. W.: *Elements of Signal Detection and Estimation*, PTR Prentice-Hall, Englewood Cliffs, New Jersey, 1995.

Helstrom, C. W.: "Approximate Inversion of Marcum's Q-Function," *IEEE Transactions on Aerospace and Electronic Systems*, vol. 34, no. 1, January 1998, pp. 317–319.

Helstrom, C. W., and J. A. Ritcey: "Evaluating Radar Detection Probabilities by Steepest Descent Integration," *IEEE Transactions on Aerospace and Electronic Systems*, vol. AES-20, no. 5, September 1984, pp. 624–633.

Hildebrand, F. B.: *Finite-Difference Equations and Simulations*, Prentice-Hall, Inc., Englewood Cliffs, New Jersey, 1968.

Hou, X.-Y., and N. Morinaga: "Detection Performance in K-Distributed and Correlated Rayleigh Clutters," *IEEE Transactions on Aerospace and Electronic Systems*, vol. 25, no. 5, September 1989, pp. 634–642.

Hou, X.-Y., N. Morinaga, and T. Namekawa: "Direct Evaluation of Radar Detection Probabilities," *IEEE Transactions on Aerospace and Electronic Systems*, vol. AES-23, no. 4, July 1987, pp. 418–424.

Hua, C. X., and J. Oksman: "A New Algorithm to Optimize Barker Code Sidelobe Suppression Filters," *IEEE Transactions on Aerospace and Electronic Systems*, vol. 26, no. 4, July 1990, pp. 673–677.

Ibragimov, I. A., and R. Z. Has'minskii: *Statistical Estimation Asymptotic Theory*, Springer-Verlag, New York, 1981.

IEEE: "IEEE Standard Radar Definitions," *IEEE Standard 686-82*, New York, 1982.

IEEE: "IEEE Standard Letter Designations for Radar-Frequency Bands," *Supplement to the Record of the IEEE 1985 International Conference on Radar*, Arlington, Virginia, May 6–9, 1985, pp. S-33–S-34.

Iglehart, S. C.: "Some Results on Digital Chirp," *IEEE Transactions on Aerospace and Electronic Systems*, vol. AES-14, no. 1, January 1978, pp. 118–127.

Ippolito, L. J.: "Propagation Effects Handbook for Satellite Systems Design," NASA Reference Publication 1082 (04), February 1989.

Jakeman, E., and P. N. Pusey: "A Model for Non-Rayleigh Sea Echo," *IEEE Transactions on Antennas and Propagation*, vol. AP-24, no. 6, November 1976, pp. 806–814.

Jakeman, E., and P. N. Pusey: "Significance of K Distributions in Scattering Experiments," *Physical Review Letters*, vol. 40, no. 9, February 27, 1978, pp. 546–550.

James, J. R.: "Engineering Approach to the Design of Tapered Dielectric-Rod and Horn Antennas," *Radio and Electronic Engineer*, vol. 42, no. 6, June 1972, pp. 251–259.

Jao, J. K.: "Amplitude Distribution of Composite Terrain Radar Clutter and the K-Distribution," *IEEE Transactions on Antennas and Propagation*, vol. AP-32, no. 10, October 1984, pp. 1049–1062.

Jolley, L. B. W.: *Summation of Series*, 2d ed., Dover Publications, Inc., New York, 1965.

Kamen, E. W.: *Introduction to Signals and Systems*, 2d ed., Macmillan Publishing Company, New York, 1990.

Kamen, E. W., and B. S. Heck: *Fundamentals of Signals and Systems Using MATLAB*, Prentice-Hall, Upper Saddle River, New Jersey, 1997.

Kanter, I.: "Exact Detection Probability for Partially Correlated Rayleigh Targets," *IEEE Transactions on Aerospace and Electronic Systems*, vol. AES-22, no. 2, March 1986, pp. 184–196.

Kanwal, R. P.: *Generalized Functions: Theory and Technique*, Academic Press, New York, 1983.

Keller, J. B.: "Backscattering from a Finite Cone," *IRE Transactions on Antennas and Propagation*, vol. AP-8, no. 2, March 1960, pp. 175–182.

Kelley, E. J., I. S. Reed, and W. L. Root: "The Detection of Radar Echoes in Noise I," *Journal of the Society for Industrial Applied Mathematics*, vol. 8, no. 2, June 1960, pp. 309–341.

Kelley, E. J., I. S. Reed, and W. L. Root: "The Detection of Radar Echoes in Noise II," *Journal of the Society for Industrial Applied Mathematics*, vol. 8, no. 3, September 1960, pp. 481–507.

Kerdock, A. M., R. Mayer, and D. Bass: "Longest Binary Pulse Compression Codes with Given Peak Sidelobe Levels," *Proceedings of the IEEE*, vol. 74, no. 2, February 1986, p. 366.

Kerker, M., M. P. Langleben, and K. L. S. Gunn: "Scattering of Microwaves by a Melting Spherical Ice Particle," *Journal of Meteorology*, vol. 8, no. 6, December 1951, p. 424.

Kerr, D. E. (ed.): *Propagation of Short Radio Waves*, Boston Technical Publishers, Inc., Lexington, Massachusetts, 1964. Originally published as vol. 13, MIT Radiation Laboratory series, by McGraw-Hill Book Company, New York, 1951.

Key, E. L., E. N. Fowle, and R. D. Haggarty: "A Method of Pulse Compression Employing Nonlinear Frequency Modulation," MIT Lincoln Laboratory, Lexington, Massachusetts, Technical Report No. 207, August 13, 1959.

Key, E. L., E. N. Fowle, and R. D. Haggarty: "A Method of Designing Signals of Large Time-Bandwidth Product," *IRE International Convention Record, Part 4*, 1961, pp. 146–155.

Kirkpatrick, G. M.: "Final Engineering Report on Angular Accuracy Improvement," General Electric report dated August 1, 1952. Reprinted in Barton, D. K. (ed.): *Radars Volume 1, Monopulse Radar*, Artech House, Inc., Dedham, Massachusetts, 1977.

Klauder, J. R., A. C. Price, S. Darlington, and W. J. Albersheim: "The Theory and Design of Chirp Radars," *Bell System Technical Journal*, vol. xxxix, no. 4, July 1960, pp. 745–820.

Knott, E. F., J. F. Shaeffer, and M. T. Tuley: *Radar Cross Section, Its Prediction, Measurement and Reduction*, Artech House, Inc., Dedham, Massachusetts, 1985.

Kohlenberg, A.: "Exact Interpolation of Band-Limited Functions," *Journal of Applied Physics*, December 1953, pp. 1432–1436.

Kraus, J. D.: *Antennas*, McGraw-Hill, Inc., New York, 1950 (2d ed., 1988).

Kretschmer, F. F., Jr., and K. Gerlach: "Low Sidelobe Radar Waveforms Derived From Orthogonal Matrices," *IEEE Transactions on Aerospace and Electronic Systems*, vol. 27, no. 1, January 1991, pp. 92–101.

Kretschmer, F. F., Jr., and B. L. Lewis: "Doppler Properties of Polyphase Coded Pulse Compression Waveforms," *IEEE Transactions on Aerospace and Electronic Systems*, vol. AES-19, no. 4, July 1983, pp. 521–531.

Lainiotis, D. G. (ed.): *Estimation Theory*, American Elsevier Publishing Company, Inc., New York, 1974.

Lathi, B. P.: *Communication Systems*, John Wiley & Sons, New York, 1968.

Leonov, A. I., and K. I. Fomichev: *Monopulse Radar*, Artech House, Inc., Norwood, Massachusetts, 1986.

Levanon, N.: *Radar Principles*, John Wiley & Sons, New York, 1988.

Levanon, N., and A. Freedman: "Ambiguity Function of Quadriphase Coded Radar Pulse," *IEEE Transactions on Aerospace and Electronic Systems*, vol. 25, no. 6, November 1989, pp. 848–853.

Lewis, B. L.: "Range-Time-Sidelobe Reduction Technique for FM-Derived Polyphase PC Codes," *IEEE Transactions on Aerospace and Electronic Systems*, vol. 29, no. 3, July 1993, pp. 834–840.

Lewis, B. L., and F. F. Kretschmer, Jr.: "A New Class of Polyphase Pulse Compression Codes and Techniques," *IEEE Transactions on Aerospace and Electronic Systems*, vol. AES-17, no. 3, May 1981, pp. 364–372 (with correction in vol. AES-17, no. 5, September 1981, p. 726).

Lewis, B. L., and F. F. Kretschmer, Jr.: "Linear Frequency Modulation Derived Polyphase Pulse Compression Codes," *IEEE Transactions on Aerospace and Electronic Systems*, vol. AES-18, no. 5, September 1982, pp. 637–641.

Lide, D. R. (ed.): *CRC Handbook of Chemistry and Physics*, 73rd ed., CRC Press, Inc., Boca Raton, Florida, 1992.

Lighthill, M. J.: *An Introduction to Fourier Analysis and Generalized Function*, Cambridge University Press, New York, 1959.

Linder, I. W., Jr., and P. Swerling: "Performance of the Double Threshold Radar Receiver in the Presence of Interference," RAND Corporation Research Memorandum RM-1719, May 28, 1956. Also ASTIA document AD-115366.

Lindner, J.: "Binary Sequences Up to Length 40 with Best Possible Autocorrelation Function," *Electronics Letters*, vol. 11, no. 21, October 1975, p. 507.

Ludeman, L. C.: *Fundamentals of Digital Signal Processing*, Harper & Row, Publishers, New York, 1986.

Lützen, J.: *The Prehistory of the Theory of Distributions*, Springer-Verlag, New York, 1982.

Mallinckrodt, A. J., and T. E. Sollenberger: "Optimum Pulse-Time Determination," *IRE Transactions on Information Theory*, vol. PGIT-3, March, 1954, pp. 151–159.

Manasse, R.: "Range and Velocity Accuracy from Radar Measurements," Group Report 312-26, MIT Lincoln Laboratory, Laxington, Massachusetts. Originally issued February 3, 1955; reissued July 16, 1959, and May 2, 1960.

Mao, Y-H.: "The Detection Performance of a Modified Moving Window Detector," *IEEE Transactions on Aerospace and Electronic Systems*, vol. AES-17, no. 3, May 1981, pp. 392–400.

Marcum, J. I.: "A Statistical Theory of Target Detection by Pulsed Radar," RAND Corporation Research Memorandum RM-754, December 1, 1947. This report is reprinted in a special issue of *IEEE Transactions on Information Theory*, vol. IT-6, no. 2, April 1960, pp. 59–144.

Marcum, J. I.: "A Statistical Theory of Target Detection by Pulsed Radar: Mathematical Appendix," RAND Corporation Research Memorandum RM-753, July 1, 1948. This report is reprinted in a special issue of *IEEE Transactions on Information Theory*, vol. IT-6, no. 2, April 1960, pp. 145–267.

Marcum, J. I.: "Table of Q Functions," RAND Corporation Research Memorandum RM-339, January 1, 1950.

Marić, S. V., I. Seskar, and E. L. Titlebaum: "On Cross-Ambiguity Properties of Welch-Costas Arrays," *IEEE Transactions on Aerospace and Electronic Systems*, (correspondence), vol. 30, no. 4, October 1994, pp. 1063–1071.

Marić, S. V., and E. L. Titlebaum: "Frequency Hop Multiple Access Codes Based Upon the Theory of Cubic Congruences," *IEEE Transactions on Aerospace and Electronic Systems* (correspondence), vol. 26, no. 6, November 1990, pp. 1035–1039.

Marić, S. V., and E. L. Titlebaum: "A Class of Frequency Hop Codes with Nearly Ideal Characteristics for Use in Multiple-Access Spread-Spectrum Communications and Radar and Sonar Systems," *IEEE Transactions on Communications*, vol. 40, no. 9, September 1992, pp. 1442–1447.

Marier, L. J., Jr.: "Correlated K-Distributed Clutter Generation for Radar Detection and Track," *IEEE Transactions on Aerospace and Electronic Systems*, vol. 31, no. 2, April 1995, pp. 568–580.

Marshall, J. S., and W. McK. Palmer: "The Distribution of Raindrops with Size," *Journal of Meteorology*, vol. 5, no. 4, August 1948, pp. 165–166.

Masuko, H., K. Okamoto, M. Shimada, and S. Niwa: "Measurement of Microwave Backscattering Signatures of the Ocean Surface Using X Band and K_a Band Airborne Scatterometers," *Journal of Geophysical Research*, vol. 91, no. C11, November 15, 1986, pp. 13,065–13,083.

McLoughlan, S. D., M. B. Thomas, and G. Watkins: "The Plasma Panel as a Potential Solution to the Bright Labelled Radar Display Problem," *International Conference on Displays for Man-Machine Systems*, April 4–7, 1977, IEE Conference Publication No. 150, pp. 14–16.

Meeks, M. L.: *Radar Propagation at Low Altitudes*, Artech House, Dedham, Massachusetts, 1982.

Melsa, J. L., and D. L. Cohn: *Decision and Estimation Theory*, McGraw-Hill Book Company, New York, 1978.

Meyer, D. P., and H. A. Mayer: *Radar Target Detection*, Academic Press, New York, 1973.

Miller, J. M.: "An Alternative Method for Analyzing the Double Threshold Detector," *IEEE Transactions on Aerospace and Electronic Systems*, vol. AES-21, no. 4, July 1985, pp. 508–513.

Milligan, T. A.: *Modern Antenna Design*, McGraw-Hill Book Company, New York, 1985.

Mitchell, R. L., and J. F. Walker: "Recursive Methods for Computing Detection Probabilities," *IEEE Transactions on Aerospace and Electronic Systems*, vol. AES-7, no. 4, July 1971, pp. 671–676.

Mood, A. M., and F. A. Graybill: *Introduction to the Theory of Statistics*, 2d ed., McGraw-Hill Book Co., Inc., New York, 1963.

Moore, R. K., K. A. Soofi, and S. M. Purduski: "A Radar Clutter Model: Average Scattering Coefficients of Land, Snow, and Ice," *IEEE Transactions on Aerospace and Electronic Systems*, vol. AES-16, no. 6, November 1980, pp. 783–799.

Morchin, W. C.: *Airborne Early Warning Radar*, Artech House, Boston, Massachusetts, 1990.

Mott, H.: *Antennas for Radar and Communications: A Polarimetric Approach*, John Wiley & Sons, New York, 1992.

Mumford, W. W., and E. H. Scheibe: *Noise Performance Factors in Communication Systems*, Horizon House Microwave, Inc., Dedham, Massachusetts, 1968.

Nathanson, F. E., J. P. Reilly, and M. N. Cohen: *Radar Design Principles Signal Processing and Environment*, 2d ed., McGraw-Hill, Inc., New York, 1991.

O'Brien, G. G.: "New Scanning Techniques in Conopulse Angle Tracking Radar," Ph.D. dissertation, University of Florida, 1995.

Oppenheim, A. V., and A. S. Willsky with S. H. Nawab: *Signals and Systems*, 2d ed., Prentice-Hall, Upper Saddle River, New Jersey, 1997.

Oppenheim, A. V., and R. W. Shaefer: *Discrete Time Signal Processing*, Prentice-Hall, Englewood Cliffs, New Jersey, 1989.

Pachares, J.: "A Table of Bias Levels Useful in Radar Detection Problems," *IRE Transactions on Information Theory*, vol. IT-4, no. 1, March 1958, pp. 38–45.

Papoulis, A.: *The Fourier Integral and its Applications*, McGraw-Hill Book Co., Inc., New York, 1962.

Papoulis, A.: *Probability, Random Variables, and Stochastic Processes*, 2d ed., McGraw-Hill Book Company, New York, 1984.

Peebles, P. Z., Jr.: "Signal Processors and Accuracy of Three-Beam Monopulse Tracking Radar," *IEEE Transactions on Aerospace and Electronic Systems*, vol. AES-5, no. 1, January 1969, pp. 52–57.

Peebles, P. Z., Jr.: "An Alternate Approach to the Prediction of Polynomial Signals in Noise from Discrete Data," *IEEE Transactions on Aerospace and Electronic Systems*, vol. AES-6, no. 4, July 1970, pp. 534–543.

Peebles, P. Z., Jr.: "Multipath Angle Error Reduction Using Multiple-Target Methods," *IEEE Transactions on Aerospace and Electronic Systems*, vol. AES-7, no. 6, November 1971, pp. 1123–1130.

Peebles, P. Z., Jr.: "The Generation of Correlated Log-Normal Clutter for Radar Simulations," *IEEE Transactions on Aerospace and Electronic Systems*, vol. AES-7, no. 6, November 1971, pp. 1215–1217.

Peebles, P. Z., Jr.: "Bounds on Radar Rain Clutter Cancellation for the M-P Distribution," *Proceedings of the 1975 IEEE Southeastcon*, Charlotte, North Carolina, April 6–9, 1975, vol. 2, pp. 5C-1-1–5C-1-5.

Peebles, P. Z., Jr.: "Radar Rain Clutter Cancellation Bounds Using Circular Polarization," *The Record of the IEEE 1975 International Radar Conference*, Arlington, Virginia, April 21–23, 1975, pp. 210–214.

Peebles, P. Z., Jr.: *Communication System Principles*, Addison-Wesley Publishing Company, Inc., Advanced Book Program, Reading, Massachusetts, 1976.

Peebles, P. Z., Jr.: "Monopulse Radar Angle Tracking Accuracy with Sum Channel Limiting," *IEEE Transactions on Aerospace and Electronic Systems*, vol. AES-21, no. 1, January 1985, pp. 137–143.

Peebles, P. Z., Jr.: *Digital Communication Systems*, Prentice-Hall, Inc., Englewood Cliffs, New Jersey, 1987.

Peebles, P. Z., Jr.: *Probability, Random Variables, and Random Signal Principles*, 3d ed., McGraw-Hill, Inc., New York, 1993.

Peebles, P. Z., Jr., and R. S. Berkowitz: "Multiple-Target Monopulse Radar Processing Techniques," *IEEE Transactions on Aerospace and Electronic Systems*, vol. AES-4, no. 6, November 1968, pp. 845–854.

Peebles, P. Z., Jr., and H. Sakamoto: "On Conopulse Radar Theoretical Angle Tracking Accuracy," *IEEE Transactions on Aerospace and Electronic Systems*, vol. AES-16, no. 3, May 1980, pp. 399–402.

Peebles, P. Z., Jr., and H. Sakamoto: "Conopulse Radar Angle Tracking Accuracy," *IEEE Transactions on Aerospace and Electronic Systems*, vol. AES-16, no. 6, November 1980, pp. 870–874.

Peebles, P. Z., Jr., and G. H. Stevens: "A Technique for the Generation of Highly Linear FM Pulse Radar Signals," *IEEE Transactions on Military Electronics*, vol. MIL-9, no. 1, January 1965, pp. 32–38.

Peebles, P. Z., Jr., and T. K. Wang: "Noise Angle Accuracy of Several Monopulse Architectures," *IEEE Transactions on Aerospace and Electronic Systems*, vol. AES-18, no. 4, November 1982, pp. 712–721. See also corrections in vol. AES-19, no. 2, March 1983, p. 330.

Peterson, W. W., and E. J. Weldon, Jr.: *Error Correcting Codes*, 2d ed., MIT Press, Cambridge, Massachusetts, 1972.

Popović, B. M.: "Generalized Chirp-Like Polyphase Sequences with Optimum Correlation Properties," *IEEE Transactions on Information Theory*, vol. 38, no. 4, July 1992, pp. 1406–1409.

Popović, B. M.: "Efficient Matched Filter for the Generalized Chirp-Like Polyphase Sequences," *IEEE Transactions on Aerospace and Electronic Systems*, vol. 30, no. 3, July 1994, pp. 769–777.

RCA: Final Report: Optimum Waveform Study for Coherent Pulse Doppler," Final report on contract NOnr 4649(00)(X), Office of Naval Research, Department of the Navy, Washington, D.C., February 28, 1965.

Raghavan, R. S.: "A Model for Spatially Correlated Radar Clutter," *IEEE Transactions on Aerospace and Electronic Systems*, vol. 27, no. 2, March 1991, pp. 268–275.

Ramo, S., J. R. Whinnery, and T. Van Duzer: *Fields and Waves in Communication Electronics*, 2d ed., John Wiley & Sons, New York, 1984.

Rao, C. R.: "Information and Accurracy Attainable in the Estimation of Statistical Parameters," *Bulletin of the Calcutta Mathematical Society*, vol. 37, 1945, pp. 81–91.

Rao, C. R.: "Minimum Variance and the Estimation of Several Parameters," *Proceedings of the Cambridge Philosophical Society*, vol. 43, pt. 2, April 1947, pp. 280–283.

Rao, C. R.: "Sufficient Statistics and Minimum Variance Estimates," *Proceedings of the Cambridge Philosophical Society*, vol. 45, no. 2, April 1949, pp. 213–218.

Rao, K. V., and V. U. Reddy: "Biphase Sequence Generation With Low Sidelobe Autocorrelation Function," *IEEE Transactions on Aerospace and Electronic Systems*, vol. AES-22, no. 2, March 1986, pp. 128–132.

Rhodes, D. R.: *Introduction to Monopulse*, McGraw-Hill Book Company, Inc., 1959.

Richard, V. W., J. E. Kammerer, and H. B. Wallace: "Rain Backscatter Measurements at Millimeter Wavelengths," *IEEE Transactions on Geoscience and Remote Sensing*, vol. 26, no. 3, May 1988, pp. 244–252.

Rifkin, R.: "Analysis of CFAR Performance in Weibull Clutter," *IEEE Transactions on Aerospace and Electronic Systems*, vol. 30, no. 2, April 1994, pp. 315–329.

Rihaczek, A. W.: *Principles of High-Resolution Radar*, McGraw-Hill Book Company, New York, 1969.

Rihaczek, A. W., and S. J. Hershkowitz: *Radar Resolution and Complex-Image Analysis*, Artech House, Norwood, Massachusetts, 1996.

Riley, J. R.: "Radar Cross Section of Insects," *Proceedings of the IEEE*, vol. 73, no. 2, February 1985, pp. 228–232.

Ross, R. A.: "Radar Cross Section of Rectangular Flat Plates as a Function of Aspect Angle," *IEEE Transactions on Antennas and Propagation*, vol. AP-14, no. 3, May 1966, pp. 329–335.

Rubin, W. L., and J. V. DiFranco: "Radar Detection," *Electro-Technology*, Science and Engineering Series 64, April 1964, pp. 61–90.

Ruck, G. T. (ed.): *Radar Cross Section Handbook*, vols. 1 and 2, Plenum Press, New York, 1970.

Ryde, J. W., and D. Ryde: "Attenuation of Centemetre and Millimetre Waves by Rain, Hail, Fogs and Clouds," GEC Report No. 8670, May 1945.

Sakamoto, H.: "Analysis of Conopulse Radar Systems," Ph.D. Dissertation, University of Tennessee, December 1975.

Sakamoto, H., and P. Z. Peebles, Jr.: "Conopulse Radar, A New Tracking Technique," *Proceedings of 1976 Southeastern Conference and Exhibit*, April 6–7, 1976, pp. 11C-1–11C-3.

Sakamoto, H., and P. Z. Peebles, Jr.: "Accuracy of Conopulse Tracking Radar," *Proceedings of 1976 Southeastern Conference and Exhibit*, April 6–7, 1976, pp. 11D-1–11D-2.

Sakamoto, H., and P. Z. Peebles, Jr.: "Conopulse Radar," *IEEE Transactions on Aerospace and Electronic Systems*, vol. AES-14, no. 1, January 1978, pp. 199–208. See also corrections in same journal, vol. AES-14, no. 4, July 1978, p. 673.

Sangston, K. J., and K. R. Gerlach: "Coherent Detection of Radar Targets in a Non-Gaussian Background," *IEEE Transactions on Aerospace and Electronic Systems*, vol. 30, no. 2, April 1994, pp. 330–340.

Sarwate, D. V.: "Bounds on Crosscorrelation and Autocorrelation of Sequences," *IEEE Transactions on Information Theory*, vol. IT-25, no. 6, November 1979, pp. 700–724.

Sarwate, D. V., and M. B. Pursley: "Crosscorrelation Properties of Pseudorandom and Related Sequences," *Proceedings of the IEEE*, vol. 68, no. 5, May 1980, pp. 593–619.

Saxton, J. A.: "The Influence of Atmospheric Conditions on Radar Performance," *Journal of the Institute of Navigation*, (London), vol. XI, no. 3, July 1958, pp. 290–303.

Saxton, J. A., and J. A. Lane: "Electrical Properties of Sea Water," *Wireless Engineer*, vol. 29, no. 349, October 1952, pp. 269–275.

Schleher, D. C.: "Radar Detection in Log-Normal Clutter," *The Record of the IEEE 1975 International Radar Conference*, Arlington, Virginia, April 21–23, 1975, pp. 262–267.

Schleher, D. C.: "Radar Detection in Weibull Clutter," *IEEE Transactions on Aerospace and Electronic Systems*, vol. AES-12, no. 6, November 1976, pp. 736–743.

Schleher, D. C.: *MTI and Pulsed Doppler Radar*, Artech House, Inc., Norwood, Massachusetts, 1991.

Schwartz, L.: "Théorie des distributions," *Actualités Scientifiques et Industrielles*, no. 1091 and 1122, Hermann & Cie, Paris, 1950–1951.

Schwartz, L.: *Théorie des distributions*, Hermann, Paris, 1966. This book (in French) is a reprint with correction, and two added chapters, of Schwartz's classic papers of 1950 and 1951 cited above.

Schwartz, M.: "A Coincidence Procedure for Signal Detection," *IRE Transactions on Information Theory*, vol. IT-2, no. 4, December 1956, pp. 135–139.

Sekine, M., and G. Lind: "Raindrop Shape Limitations on Clutter Cancellation Ratio Using Circular Polarization," *IEEE Transactions on Aerospace and Electronic Systems*, vol. AES-19, no. 4, July 1983, pp. 631–633.

Sekine, M., and Y. Mao: *Weibull Radar Clutter*, IEEE Radar, Sonar, Navigation and Avionics Series No. 3, IEEE, New York, 1990.

Sekine, M., T. Musha, Y. Tomita, and T. Irabu: "Suppression of Weibull-Distributed Clutters Using a Cell-Averaging LOG/CFAR Receiver," *IEEE Transactions on Aerospace and Electronic Systems*, vol. AES-14, no. 5, September 1978, pp. 823–826.

Senior, T. B. A.: "The Backscattering Cross Section of a Cone-Sphere," *IEEE Transactions on Antennas and Propagation*, vol. AP-13, no. 2, March 1965, pp. 271–277.

Shanmugan, K. S., and A. M. Breipohl: *Random Signals: Detection, Estimation and Data Analysis*, John Wiley & Sons, New York, 1988.

Shannon, H. H.: "Recent Refraction Data Corrects Radar Errors," *Electronics*, vol. 35, no. 49, December 7, 1962, pp. 52–56.

Shenoi, K.: *Digital Signal Processing in Telecommunications*, Prentice Hall PTR, Upper Saddle River, New Jersey, 1995.

Sherman, S. M.: *Monopulse Principles and Techniques*, Artech House, Dedham, Massachusetts, 1984.

Shnidman, D. A.: "Calculation of Probability of Detection for Log-Normal Target Fluctuations," *IEEE Transactions on Aerospace and Electronic Systems*, vol. 27, no. 1, January 1991, pp. 172–174.

Shnidman, D. A.: "Radar Detection Probabilities and Their Calculation," *IEEE Transactions on Aerospace and Electronic Systems*, vol. 31, no. 3, July 1995, pp. 928–950.

Siegel, K. M.: "Far Field Scattering from Bodies of Revolution," *Applied Scientific Research*, Section B, vol. 7, 1958, pp. 293–328.

Silver, S.: *Microwave Antenna Theory and Design*, McGraw-Hill Book Company, New York, 1949.

Sinclair, G.: "The Transmission and Reception of Elliptically Polarized Waves," *Proceedings of the IRE*, vol. 38, no. 2, February 1950, pp. 148–151.

Sites, M. J.: "Coded Frequency Shift Keyed Sequences with Applications to Low Data Rate Communication and Radar," Stanford Electronics Laboratory Report 3606-5, September 1969.

Skolnik, M. I.: "Theoretical Accuracy of Radar Measurements," *IRE Transactions on Aeronautical and Navigational Electronics*, vol. ANE-7, no. 4, December 1960, pp. 123–129.

Skolnik, M. I.: *Introduction to Radar Systems*, McGraw-Hill Book Company, Inc., New York, 1962.

Skolnik, M. I.: "An Empirical Formula for the Radar Cross Section of Ships at Grazing Incidence," *IEEE Transactions on Aerospace and Electronic Systems*, vol. AES-10, no. 2, March 1974, p. 292.

Skolnik, M. I.: *Introduction to Radar Systems*, 2d ed., McGraw-Hill Book Company, New York, 1980.

Skolnik, M. I. (ed.): *Radar Handbook*, 2d ed., McGraw-Hill Publishing Co., New York, 1990.

Slepian, D.: "Estimation of Signal Parameters in the Presence of Noise," *IRE Transactions on Information Theory*, vol. PGIT-3, March 1954, pp. 68–89.

Soliman, S. S., and M. D. Srinath: *Continuous and Discrete Signals and Systems*, Prentice Hall, Englewood Cliffs, New Jersey, 1990.

Sorenson, H. W.: *Parameter Estimation: Principles and Problems*, Marcel Dekker, Inc., New York, 1980.

Spellmire, R. J.: "Tables of Taylor Aperture Distributions," Hughes Aircraft Company, Culver City, California, Technical Memorandum No. 581, October 1958.

Staelin, D. H.: "Measurements and Interpretation of the Microwave Spectrum of the Terrestrial Atmosphere Near 1-cm Wavelength," *Journal of Geophysical Research*, vol. 71, 1966, pp. 2975–2981.

Stark, H., and J. W. Woods: *Probability, Random Processes, and Estimation Theory for Engineers*, Prentice-Hall, Englewood Cliffs, New Jersey, 1986.

Straiton, A. W., and C. W. Tolbert: "Anomolies in the Absorption of Radio Waves by Atmospheric Gases," *Proceedings of the IRE*, vol. 48, no. 5, May 1960, pp. 898–903.

Suehiro, N., and M. Hatori: "Modulatable Orthogonal Sequences and their Application to SSMA Systems," *IEEE Transactions on Information Theory*, vol. 34, no. 1, January 1988, pp. 93–100.

Swerling, P.: "The 'Double Threshold' Method of Detection," RAND Corporation Research Memorandum RM-1008, December 17, 1952. Also ASTIA document AD-210454.

Swerling, P.: "Probability of Detection for Fluctuating Targets," RAND Corporation Research Memorandum RM-1217, March 17, 1954. This report is reprinted in a special issue of *IEEE Transactions on Information Theory*, vol. IT-6, no. 2, April 1960, pp. 269–308.

Swerling, P.: "Maximum Angular Accuracy of a Pulsed Search Radar," *Proceedings of the IRE*, vol. 44, no. 9, September 1956, pp. 1146–1155.

Swerling, P.: "Parameter Estimation for Waveforms in Additive Gaussian Noise," *Journal of the Society for Industrial Applied Mathematics*, vol. 7, no. 2, June 1959, pp. 152–166.

Swerling, P.: "Parameter Estimation Accuracy Formulas," *IEEE Transactions on Information Theory*, vol. IT-10, no. 4, October 1964, pp. 302–314.

Swerling, P.: "More on Detection of Fluctuating Targets," *IEEE Transactions on Information Theory*, vol. IT-11, no. 3, July 1965, pp. 459–460.

Swerling, P.: "Radar Probability of Detection for Some Additional Fluctuating Target Cases," *IEEE Transactions on Aerospace and Electronic Systems*, vol. 33, no. 2, April 1997, pp. 698–709.

Taylor, F.: *Principles of Signals and Systems*, McGraw-Hill, Inc., New York, 1994.

Taylor, J. W. Jr., and H. J. Blinchikoff: "Quadriphase Code—A Radar Pulse Compression Signal With Unique Characteristics," *IEEE Transactions on Aerospace and Electronic Systems*, vol. 24, no. 2, March 1988, pp. 156–170.

Taylor, S. A., and J. L. MacArthur: "Digital Pulse Compression Radar Receiver," *Applied Physics Laboratory Technical Digest*, vol. 6, no. 4, March–April 1967, pp. 2–10.

Taylor, T. T.: "One Parameter Family of Line Sources Producing Modified $\mathrm{Sin}(\pi u)/\pi u$ Patterns," Hughes Aircraft Company, Culver City, California, Technical Memorandum 324, Contract AF 19(604)-262-F-14, September 4, 1953.

Taylor, T. T.: "Design of Line-Source Antennas for Narrow Beamwidth and Low Side Lobes," *IRE Transactions on Antennas and Propagation*, vol. AP-3, no. 1, January 1955, pp. 16–28.

Taylor, T. T.: "Design of Circular Apertures for Narrow Beamwidth and Low Sidelobes," *IRE Transactions on Antennas and Propagation*, vol. AP-8, no. 1, January 1960, pp. 17–22.

Temple, G: "Theories and Applications of Generalized Functions," *Journal of the London Mathematical Society*, vol. 28, pt. 2, no. 110, April 1953, pp. 134–148.

Temple, G.: "The Theory of Generalized Functions," *Proceedings of the Royal Society of London*, vol. 228, no. 1173, February 22, 1955, pp. 175–190.

Thomas, J. B.: *An Introduction to Statistical Communication Theory*, John Wiley & Sons, Inc., New York, 1969.

Titlebaum, E. L.: "Time-Frequency Hop Signals Part I: Coding Based Upon the Theory of Linear Congruences," *IEEE Transactions on Aerospace and Electronic Systems*, vol. AES-17, no. 4, July 1981, pp. 490–493.

Titlebaum, E. L., and L. H. Sibul: "Time-Frequency Hop Signals Part II: Coding Based Upon Quadratic Congruences," *IEEE Transactions on Aerospace and Electronic Systems*, vol. AES-17, no. 4, July 1981, pp. 494–500.

Titlebaum, E. L., S. V. Marić, and J. R. Bellegarda: "Ambiguity Properties of Quadratic Congruential Coding," *IEEE Transactions on Aerospace and Electronic Systems*, vol. 27, no. 1, January 1991, pp. 18–29.

Trizna, D. B.: "Statistics of Low Grazing Angle Radar Sea Scatter for Moderate and Fully Developed Ocean Waves," *IEEE Transactions on Antennas and Propagation*, vol. 39, no. 12, December 1991, pp. 1681–1690.

Trunk, G. V.: "Detection Results for Scanning Radars Employing Feedback Integration," *IEEE Transactions on Aerospace and Electronic Systems*, vol. AES-6, no. 4, July 1970, pp. 522–527.

Trunk, G. V.: "Comparison of Two Scanning Radar Detectors: The Moving Window and the Feedback Integrator," *IEEE Transactions on Aerospace and Electronic Systems*, vol. AES-7, no. 2, March 1971, pp. 395–398.

Trunk, G. V.: "Further Results on the Detection of Targets in Non-Gaussian Sea Clutter," *IEEE Transactions on Aerospace and Electronic Systems*, vol. AES-7, no. 3, May 1971, pp. 553–556.

Trunk, G. V.: "Radar Properties of Non-Rayleigh Sea Clutter," *IEEE Transactions on Aerospace and Electronic Systems*, vol. AES-8, no. 2, March 1972, pp. 196–204.

Trunk, G. V., and S. F. George: "Detection of Targets in Non-Gaussian Sea Clutter," *IEEE Transactions on Aerospace and Electronic Systems*, vol. AES-6, no. 5, September 1970, pp. 620–628.

Turyn, R.: "On Barker Codes of Even Length," *Proceedings of the IEEE* (correspondence), vol. 51, no. 9, September 1963, p. 1256.

Turyn, R., and J. Storer: "On Binary Sequences," *Proceedings of the American Mathematical Society*, vol. 12, no. 3, June 1961, pp. 394–399.

Ulaby, F. T.: "Vegetation Clutter Model," *IEEE Transactions on Antennas and Propagation*, vol. AP-28, no. 4, July 1980, pp. 538–545.

Ulaby, F. T., P. P. Batlivala, and M. C. Dobson: "Microwave Backscatter Dependence on Surface Roughness, Soil Moisture, and Soil Texture: Part I—Bare Soil," *IEEE Transactions on Geoscience Electronics*, vol. GE-16, no. 4, October 1978, pp. 286–295.

Ulaby, F. T., and M. C. Dobson: *Handbook of Radar Scattering Statistics for Terrain*, Artech House, Inc., Norwood, Massachusetts, 1989.

Urkowitz, H.: "Hansen's Method Applied to the Inversion of the Incomplete Gamma Functions, with Applications," *IEEE Transactions on Aerospace and Electronic Systems*, vol. AES-21, no. 5, September 1985, pp. 728–731.

Urkowitz, H., C. A. Hauer, and J. F. Koval: "Generalized Resolution in Radar Systems," *Proceedings of the IRE*, vol. 50, no. 10, October 1962, pp. 2093–2105.

Vakman, D. E.: *Sophisticated Signals and the Uncertainty Principle in Radar*, Springer-Verlag New York, Inc., New York, 1968.

Van Bladel, J.: *Electromagnetic Fields*, McGraw-Hill Book Company, New York, 1964.

Van Trees, H. L.: *Detection, Estimation, and Modulation Theory, Part I*, John Wiley & Sons, Inc., New York, 1968.

Walker, J. F.: "Performance Data for a Double-Threshold Detection Radar," *IEEE Transactions on Aerospace and Electronic Systems*, vol. AES-7, no. 1, January 1971, pp. 142–146 (see also comments of V. G. Hansen in same journal, vol. AES-7, no. 3, May 1971, p. 561).

Watts, S.: "Radar detection prediction in sea clutter using the compound K-distribution model," *IEE Proceedings*, vol. 132, pt. F, no. 7, December 1985, pp. 613–620.

Watts, S.: "Radar Detection Prediction in K-Distributed Sea Clutter and Thermal Noise," *IEEE Transactions on Aerospace and Electronic Systems*, vol. AES-23, no. 1, January 1987, pp. 40–45.

Watts, S., and D. C. Wicks: "Empirical Models for Detection Prediction in K-Distribution Radar Sea Clutter," *The Record of the IEEE 1990 International Radar Conference*, Arlington, Virginia, May 7–10, 1990, pp. 189–194.

Weickmann, H. K., and H. J. aufm Kampe: "Physical Properties of Cumulus Clouds," *Journal of Meteorology*, vol. 10, no. 3, 1953, pp. 204–211.

Weiner, M. A. : "Detection Probability for Partially Correlated Chi-Square Targets," *IEEE Transactions on Aerospace and Electronic Systems*, vol. 24, no. 4, July 1988, pp. 411–416.

Weiner, M. A.: "Binary Integration of Fluctuating Targets," *IEEE Transactions on Aerospace and Electronic Systems*, vol. 27, no. 1, January 1991, pp. 11–17.

Weinstock, W.: "Target Cross Section Models for Radar Systems Analysis," doctoral dissertation, University of Pennsylvania, Philadelphia, 1964.

Wilks, S. S.: *Mathematical Statistics*, John Wiley & Sons, Inc., New York, 1962,

Wilson, J. D.: "Probability of Detecting Aircraft Targets," *IEEE Transactions on Aerospace and Electronic Systems*, vol. AES-8, no. 6, November 1972, pp. 757–761.

Woodward, P. M.: *Probability and Information Theory, with Applications to Radar*, Pergamon Press, Inc., New York, 1953 (second printing, 1960).

Worley, R.: "Optimum Thresholds for Binary Integration," *IEEE Transactions on Information Theory*, vol. IT-14, no. 2, March 1968, pp. 349–353.

Wylie, C. R.: *Advanced Engineering Mathematics*, McGraw-Hill Book Co., Inc., New York, 1951.

Xu, X., and P. Huang: "A New RCS Statistical Model of Radar Targets," *IEEE Transactions on Aerospace and Electronic Systems*, vol. 33, no. 2, April 1997, pp. 710–714.

Ziemer, R. E., W. H. Tranter, and D. R. Fannin: *Signals and Systems: Continuous and Discrete*, 3d ed., Macmillan Publishing Company, New York, 1993.

INDEX